Sustainable Agriculture and New Biotechnologies

Advances in Agroecology

Series Editor: Clive A. Edwards

Agroecosystems in a Changing Climate, Paul C.D. Newton, R. Andrew Carran, Grant R. Edwards, and Pascal A. Niklaus

Agroecosystem Sustainability: Developing Practical Strategies, Stephen R. Gliessman

Agroforestry in Sustainable Agricultural Systems, Louise E. Buck, James P. Lassoie, and Erick C.M. Fernandes

Biodiversity in Agroecosystems, Wanda Williams Collins and Calvin O. Qualset

The Conversion to Sustainable Agriculture: Principles, Processes, and Practices, Stephen R. Gliessman and Martha Rosemeyer

Integrated Assessment of Health and Sustainability of Agroecosystems, Thomas Gitau, Margaret W. Gitau, and David Waltner-Toews

Interactions between Agroecosystems and Rural Communities, Cornelia Flora

Landscape Ecology in Agroecosystems Management, Lech Ryszkowski

Multi-Scale Integrated Analysis of Agroecosystems, Mario Giampietro

Soil Ecology in Sustainable Agricultural Systems, Lijbert Brussaard and Ronald Ferrera-Cerrato

Soil Organic Matter in Sustainable Agriculture, Fred Magdoff and Ray R. Weil

Soil Tillage in Agroecosystems, Adel El Titi

Structure and Function in Agroecosystem Design and Management, Masae Shiyomi and Hiroshi Koizumi

Sustainable Agriculture and New Biotechnologies, Noureddine Benkeblia

Sustainable Agroecosystem Management: Integrating Ecology, Economics and Society, Patrick J. Bohlen and Gar House

Tropical Agroecosystems, John H. Vandermeer

Sustainable Agriculture and New Biotechnologies

Edited by Noureddine Benkeblia

CRC Press
Taylor & Francis Group
Boca Raton London New York

CRC Press is an imprint of the
Taylor & Francis Group, an **informa** business

CRC Press
Taylor & Francis Group
6000 Broken Sound Parkway NW, Suite 300
Boca Raton, FL 33487-2742

Printed in the United States of America on acid-free paper
10 9 8 7 6 5 4 3 2 1

International Standard Book Number: 978-1-4398-2504-4 (Hardback)

Library of Congress Cataloging-in-Publication Data

Sustainable agriculture and new biotechnologies / editor, Noureddine Benkeblia.
p. cm. -- (Advances in agroecology)
Includes bibliographical references and index.
ISBN 978-1-4398-2504-4 (alk. paper)
1. Plant biotechnology. 2. Sustainable agriculture. I. Benkeblia, Noureddine. II. Title. III. Series: Advances in agroecology.

SB106.B56S87 2011
630--dc22 2010051450

Visit the Taylor & Francis Web site at
http://www.taylorandfrancis.com

and the CRC Press Web site at
http://www.crcpress.com

My Wife
Zahra and Mohamed

To whom I say "Sorry" prior to saying "Thanks" for being with me.

"Humility and Patience are the Keys of Knowledge, and Understanding leads to Passion and Passion leads to Perfection"

Contents

Foreword

I am writing the foreword to this book as the editor-in-chief of a series of books published under the general title "Advances in Agroecology," which focuses on the importance of agroecology in increasing crop production, maintaining soil health and fertility, and promoting a more sustainable agriculture on a global scale. These goals have to be achieved in the context of increasing human populations, and diminished and degraded natural resources including soils and water resulting from overcropping intensive chemical-based agriculture. The themes of the previous 15 volumes in the series have been diverse. They have addressed issues such as strategies for increasing agricultural sustainability; the role of agroforestry, biodiversity, and landscape ecology in improving crop and animal productivity; the structure and function of agroecosystem design and management; the role of rural communities in agriculture; multiscale integrated analyses of agroecosystems; integration of agroecosystem health and sustainability; the effects of a changing climate on agroecosystems; the importance of soil organic matter in agricultural sustainability; and the integration of ecology, economics, and society in sustainable agroecosystem management. These subjects are all very broad but relatively well-defined and readily applied aspects of agricultural sustainability.

This new volume in the series takes a much broader and innovative informational approach to addressing issues related to understanding and improving agricultural sustainability. It is based on a relatively new approach to issues associated with ecosystems and agricultural sustainability that is summarized by a new general term *omics*. The concept of *omics* has been adopted over the past 10–15 years as a broad general discipline of science that diverse groups of bioinformationists developed and used for analyzing interactions between biological information components in various *omes*.

These terms have been derived from the concept of a *biome*, which is one of the basic ideas of all ecologists. More recently, the use of the term *genome* by molecular scientists has become widely familiar in representing the complete genetic makeup of an organism. The concept of *omes* (which stems from the Greek word for all, whole, or complete) and the derived term *omics* is a holistic way of describing and analyzing complex biological systems. The discipline summarized as *omics* focuses on mapping informational components, seeking to identify interactions and relationships among these components, and studying networks of components to try to identify regulatory mechanisms. Finally, it seeks to integrate large bodies of information from the wide diversity of "omes" that have been described in an overall holistic pattern for different biological systems.

A broad range of "omes" have been defined, and some of these have much more relevance to sustainable agriculture than others, and hence are used more frequently in this book. Obviously, *genomes* and *genomics* have a major role to play in the development and breeding of new and improved crops that can withstand both biotic and abiotic stresses. *Metabolomics* has been described as a systematic study of the chemical fingerprints that specific cellular processes leave behind, that is, the study of small molecule metabolite profiles. The *metabolome* represents all of the metabolites in a biological cell, tissue, or organism that are the end products of cellular processes, which can help us to understand the molecular genetic basis of crop responses to stresses and improve the overall nutrient contents and quality of crops. The term *ionome* represents all of the properties of the genes, proteins, metabolites, nutrients, and elements within an organism and *Ionomics* provides a tool to study the functional state of an organism under different conditions, driven by genetic and developmental differences that are influenced by both biotic and abiotic factors. *Metagenomics* is another recently described "ome" that may include *population genomics*, *community genomics*, and *environmental genomics*. Traditional microbiology and the sequencing of microbial genomes depend up clonal cultures, whereas *metagenomics* facilitates the studies of organisms that are not readily cultured and can sequence genetic material from uncultured environmental samples.

Additionally, *metagenomics* may provide additional strategies for assessing the impact of pollutants on the environment and cleaning up contaminated environments.

Many of these approaches, as well as various other *omic*-based concepts and ideas, have been used in the discussions in various chapters in this book, which together address a broad range of very diverse subjects, including the use of plant genetic resources to produce more sustainable crops, sustainable herbivore production, rhizosphere metabolomics, microbial functionality and diversity, and nutritional genomics, improving wheat productivity, and assessing various effects of climatic changes on crop agriculture. The wide-ranging discussions in this book should provide an important catalyst to the future development of these concepts and additional practical tools that might play a valuable part in our efforts to improve agricultural sustainability and food productivity.

Clive A. Edwards
Professor of Entomology and Environmental Sciences
The Ohio State University

Preface

Nowadays, the contribution of new biotechnologies, such as *omics* to agriculture is spreading fast, and discoveries in genomics, transcriptomics, proteomics, and metabolomics are leading to a better understanding of how microorganisms, animals, and plants either function or respond to the environment. This understanding is one of the key factors in the present and future development of breeding programs, improvement of yield and resistances of crops, and management of pest diseases and other problems faced by agriculture. Thus, applications of these new technologies will definitely help to implement efficient and sustainable agricultural systems. However, achieving the benefits of omics technologies requires the involvement of both the scientific community and governments to assure appropriate investments in these emerging fields. Moreover, an efficient relationship among academic, industrial, social, and technical communities needs to be implemented and developed to move ahead and address the problems encountered by the human population such as hunger, food availability, pollution, and other related agricultural and environmental problems. Consequently, these new technologies are and will benefit the international community; however, this community needs to find the way to benefit from these technologies by developing and improving investments and funds availability, enough to allow this science to progress and satisfy what is expected. This book is the contribution of a wide range of scientists from different horizons, and its content attempts to give a comprehensive overview on the "omics" technologies and their potentials for a sustainable agriculture able to satisfy our needs and preserve our Earth. Scientists, researchers, and students will find recent and useful information to help them conduct their future research work, analyze their key findings, and identify areas of differing perspectives where there are gaps in knowledge.

The production of this book involved the efforts of many people, especially the authors who contributed chapters of outstanding clarity and who have shown patience during the assembling and editing of all the contributions.

Acknowledgments

The completion of this book needed much effort, and without the support of people around me, I could not have done it.

- My thanks to the University of the West Indies, and to its Office of Graduate Studies & Research (OGRS) for its support, with particular thanks to Professor Gordon Shirley, Professor Ronald Young, and Professor Ishenkumba Kahwa.
- My acknowledgment of the staff members of the Department of Life Sciences for their assistance, with particular thanks to Dr. Mona Webber (HOD), Dr. Eric Hyslop (former HOD), Professor Dale Webber, and Professor Ralph Robinson for their continuous encouragement.
- My gratitude to Professor Patrick Varoquaux (former INRA staff), who taught me life, humility, and perfection before science.
- My acknowledgment of SQPOV, INRA, Avignon (France), staff (retired and present), who were my second family.
- I will not forget Professor Norio Shiomi, Rakuno Gakuen University, Japan (retired) with whom I had the pleasure to work for many years, and whom I appreciate for his generosity.
- My thanks to all who have been of any support or assistance.

Contributors

Ganesh Kumar Agrawal
Research Laboratory for Biotechnology and Biochemistry (RLABB)
Kathmandu, Nepal

Shashi Bhushan Agrawal
Department of Botany
Banaras Hindu University
Varanasi, Uttar Pradesh, India

Sophie Alvarez
Proteomics and Mass Spectrometry Facility
Donald Danforth Plant Center
St. Louis, Missouri

Kandan Aravindaram
Agricultural Biotechnology Research Center
Academia Sinica
Taipei, Taiwan (Republic of China)

Atanas Atanassov
AgroBioInstitute
Sofia, Bulgaria

Ilian Badjakov
AgroBioInstitute
Sofia, Bulgaria

Felipe Bastida
Department of Soil and Water Conservation
Campus Universitario de Espinardo
Espinardo, Murcia, Spain

Urmila Basu
Department of Agricultural, Food and Nutritional Science
University of Alberta
Edmonton, Alberta, Canada

Noureddine Benkeblia
Department of Life Sciences
The University of the West Indies
Kingston, Jamaica

Laurence Bernard
Unité de Recherches sur les Herbivores
Saint-Genès, Champanelle, France

Hamid Boudra
Unité de Recherches sur les Herbivores
Saint-Genès, Champanelle, France

Isabelle Cassar-Malek
Unité de Recherches sur les Herbivores
Saint-Genès, Champanelle, France

Charles Chen
National Peanut Research Laboratory, USDA-ARS
Dawson, Georgia

Kyoungwon Cho
Environmental Biology Division
National Institute for Environmental Studies (NIES)
Tsukuba, Ibaraki, Japan

Ye Chu
Department of Horticulture
The University of Georgia
Tifton, Georgia

Dolores Corella
Department of Preventive Medicine and Public Health
University of Valencia
València, Spain

Agnès Cornu
Unité de Recherches sur les Herbivores
Saint-Genès, Champanelle, France

Allison C. Crawford
Department of Agriculture
Food Western Australia
South Perth, Western Australia, Australia

Angela Hendrickson Culler
Product Safety Center
Monsanto Company
St. Louis, Missouri

Denys Durand
Unité de Recherches sur les Herbivores
Saint-Genès, Champanelle, France

Anne Ferlay
Unité de Recherches sur les Herbivores
Saint-Genès, Champanelle, France

Michael G. Francki
Department of Agriculture
Food Western Australia
South Perth, Western Australia, Australia

Carlos García
Department of Soil and Water Conservation
Campus Universitario de Espinardo
Espinardo, Murcia, Spain

Kevin C. Glenn
Product Safety Center
Monsanto Company
St. Louis, Missouri

John Gray
Department of Biological Sciences
University of Toledo
Toledo, Ohio

Erich Grotewold
Plant Biotechnology Center and Department of Plant Cellular and Molecular Biology
The Ohio State University
Columbus, Ohio

Dominique Gruffat
Unité de Recherches sur les Herbivores
Saint-Genès, Champanelle, France

María Luisa Guillén
Department of Preventive Medicine and Public Health
University of Valencia
València, Spain

Baozhu Guo
Crop Protection and Management Research Unit, USDA-ARS
Tifton, Georgia

George G. Harrigan
Product Safety Center
Monsanto Company
St. Louis, Missouri

Teresa Hernández
Department of Soil and Water Conservation
Campus Universitario de Espinardo
Espinardo, Murcia, Spain

Leslie M. Hicks
Proteomics and Mass Spectrometry Facility
Donald Danforth Plant Center
St. Louis, Missouri

Jean-François Hocquette
Unité de Recherches sur les Herbivores
Saint-Genès, Champanelle, France

C. Corley Holbrook
Crop Genetics and Breeding Research Unit, USDA-ARS
Tifton, Georgia

Anna Janská
Crop Research Institute
and
Department of Experimental Plant Biology
Charles University in Prague
Prague, Czech Republic

Oliver Andrzej Hodgson Jones
School of Engineering and Computing Sciences
University of Durham
Durham, United Kingdom

Nam-Soo Jwa
Department of Molecular Biology
College of Life Sciences
Sejong University
Seoul, South Korea

Suman Khowala
Drug Development and Biotechnology
Indian Institute of Chemical Biology (CSIR)
Kolkata, India

Shoshi Kikuchi
National Institute of Agrobiological Sciences
Tsukuba, Ibaraki, Japan

Dea-Wook Kim
National Institute of Crop Science
Rural Development Administration (RDA)
Suwon, South Korea

Masakazu Komatsuzaki
Center for Field Science Research and Education
Ibaraki University
Inashiki, Ibaraki, Japan

Violeta Kondakova
AgroBioInstitute
Sofia, Bulgaria

Akihiro Kubo
Environmental Biology Division
National Institute for Environmental Studies (NIES)
Tsukuba, Ibaraki, Japan

Christine Leroux
Unité de Recherches sur les Herbivores
Saint-Genès, Champanelle, France

Petr Maršík
Laboratory of Plant Biotechnologies
Institute of Experimental Botany Academy of Sciences CR
Prague, Czech Republic

Thomas A. McKeon
Western Regional Research Center
USDA-ARS
Albany, California

José Luis Moreno
Department of Soil and Water Conservation
Campus Universitario de Espinardo
Espinardo, Murcia, Spain

Diego P. Morgavi
Unité de Recherches sur les Herbivores
Saint-Genès, Champanelle, France

Kalyan Chakravarthy Mynampati
Singapore Delft Water Alliance
National University of Singapore
Singapore

Muslima Nazir
School of Biosciences and Biotechnology
Baba Ghulam Shah Badshah University
Rajouri, Jammu & Kashmir, India

César Nicolás
Department of Soil and Water Conservation
Campus Universitario de Espinardo
Espinardo, Murcia, Spain

Klaus Oldach
Plant Genomics Centre
South Australia Research Development Institute
Urrbrae, South Australia, Australia

Jaroslava Ovesná
Crop Research Institute
Prague, Czech Republic

Peggy Ozias-Akins
Department of Horticulture
The University of Georgia
Tifton, Georgia

Swagata Pal
Drug Development and Biotechnology
Indian Institute of Chemical Biology (CSIR)
Kolkata, India

Brigitte Picard
Unité de Recherches sur les Herbivores
Saint-Genès, Champanelle, France

Randeep Rakwal
School of Medicine
Showa University
Shinagawa, Tokyo, Japan
and
Department of Biology
Toho University
Funabashi, Chiba, Japan

Sheela Reuben
Singapore Delft Water Alliance
National University of Singapore
Singapore

William P. Ridley
Product Safety Center
Monsanto Company
St. Louis, Missouri

Abhijit Sarkar
Department of Botany
Banaras Hindu University
Varanasi, Uttar Pradesh, India

Kouji Satoh
National Agricultural Research Center
Tsukuba, Ibaraki, Japan

Isabelle Savary-Auzeloux
Unité de Recherches sur les Herbivores
Saint-Genès, Champanelle, France

Junko Shibato
Environmental Biology Division
National Institute for Environmental Studies (NIES)
Tsukuba, Ibaraki, Japan

Mercedes Sotos-Prieto
Department of Preventive Medicine and Public Health
University of Valencia
València, Spain

H. Thomas Stalker
Department of Crop Science
North Carolina State University
Raleigh, North Carolina

Keita Suwabe
Graduate School of Bioresources
Mie University
Mie-Pref, Japan

Ayako Suzuki
Laboratory of Bioinformatics
Meiji University
Kawasaki, Japan

Sanjay Swarup
Department of Biological Sciences
National University of Singapore
Singapore

Faiz M. Syuaib
Department of Mechanical and Biosystem Engineering
Bogor Agricultural University
Bogor, Indonesia

Claudia Terlouw
Unité de Recherches sur les Herbivores
Saint-Genès, Champanelle, France

Traci Viinanen
Department of Ecology and Evolution
University of Chicago
Chicago, Illinois

Ning-Sun Yang
Agricultural Biotechnology Research Center
Academia Sinica
Taipei, Taiwan (Republic of China)

Kentaro Yano
Laboratory of Bioinformatics
Meiji University
Kawasaki, Japan

Sajad Majeed Zargar
School of Biosciences and Biotechnology
Baba Ghulam Shah Badshah University
Rajouri, Jammu & Kashmir, India

Sylva Zelenková
Department of Experimental Plant Biology
Charles University in Prague
Prague, Czech Republic

CHAPTER 1

The Use of Omics Databases for Plants

Ayako Suzuki, Keita Suwabe and Kentaro Yano

CONTENTS

1.1 INTRODUCTION

Owing to the rapid technological advances in recent years, the amount of large-scale omics data obtained from comprehensive genomics, transcriptomics, metabolomics and proteomics studies has been rapidly accumulating and is being stored in Web databases. The nucleotide sequence data (entries) in the International Nucleotide Sequence Databases (INSD) (Brunak et al., 2002) are maintained by members of the International Nucleotide Sequence Database Collaboration DDBJ (Kaminuma et al., 2011), EMBL (Cochrane et al., 2009) and GenBank (Benson et al., 2011), and the number of entries is steadily increasing. While the first version of DDBJ was released in July 1987, approximately 40% of the bases and entries in the latest version were released from 2007 onward. About 110 billion bases in more than 112 million DNA sequences are recorded in DDBJ (Release 80.0 as of December 2009). Such a rapid increase in genome and transcriptome sequence data is facilitated by innovations in the development of multicapillary sequencing and the next-generation sequencing (NGS) technologies. Further improvements in experimental methodology such as NGS will accelerate the increase in the volume of omics data.

Large-scale omics data can open a new avenue for understanding complex biological systems in organisms. Genome sequences of various organisms facilitate DNA sequence-based comparative genomics studies to understand the genomic relationship between related species, and such comparison helps to infer the evolutionary history of species. Together with large-scale gene-expression data obtained from microarray and DNA sequencing platforms, bioinformatics technology allows construction of gene-expression networks, which can provide information about coregulated genes. Such large-scale network analysis for genes or gene products has recently attracted attention as a high-throughput approach for predictions of genes with similar expression patterns (transcriptional modules), gene families, transcriptional regulators, conserved *cis*-element motifs and interactions between genes or proteins. Based on metabolomic data from mass spectrometry (MS) analysis, detailed maps of metabolic networks have also been developed in many model organisms. These large-scale omics data, as well as sequence data, have been stored and maintained in Web databases, and are accessible at any time.

This wealth of comprehensive online resources and Web databases for omics data allows effective and efficient extraction of essential data for specific queries. For example, it is possible to compare experimental data from any laboratory with data stored in the databases. It is also possible to design a new experimental strategy using online resources, without a pilot study. For the effective utilization of the accumulating omics data and databases, it is essential for all researchers to know which databases are beneficial for their requirements and how to search online databases that are relevant. In this chapter, we introduce the current status of Web databases and bioinformatics tools available for plant research, and discuss their availability in plant science. The Web sites mentioned in this chapter are summarized in Table 1.1.

1.2 INFORMATION ON WEB RESOURCES FOR DATABASES AND EXPERIMENTAL MATERIALS

The number of Web databases for molecular biology has proliferated since many projects for omics studies were launched. These projects, including genome sequencing projects, have led to construction and maintenance of Web-based databases to manage up-to-date information and provide their own large-scale data. The list is valuable for searching databases relevant to any research.

Table 1.1 Web Resources for Omics Data and Experimental Materials

	Sequence Databases	
DDBJ	Nucleotide Sequence Database	http://www.ddbj.nig.ac.jp/index-e.html
EMBL	Nucleotide Sequence Database	http://www.ebi.ac.uk/embl/
GenBank	Nucleotide Sequence Database	http://www.ncbi.nlm.nih.gov/Genbank/
INSDC	International Nucleotide Sequence Databases	http://www.insdc.org/
	Genome Projects	
BGI	Whole genome sequencing	http://www.genomics.cn/en/index.php
The Genomes On Line Database (GOLD)	Genome Projects	http://genomesonline.org/
JGI Genome Projects	Genome Projects	http://www.jgi.doe.gov/genome-projects/
NCBI Entrez Genome Project database	Genome Projects	http://www.ncbi.nlm.nih.gov/genomeprj
Phytozome	Genome annotations for model plants	http://www.phytozome.net/
	Structure, Function, and Expression of Genes and Proteins	
ArrayExpress	Gene expressions	http://www.ebi.ac.uk/microarray-as/aer/
BioCyc	Metabolic pathways	http://biocyc.org/
CIBEX	Gene expressions	http://cibex.nig.ac.jp/index.jsp
dbEST	ESTs	http://www.ncbi.nlm.nih.gov/dbEST/
euKaryotic Orthologous Groups (KOG)	Orthologous Groups	http://www.ncbi.nlm.nih.gov/COG/grace/shokog.cgi
Gene Ontology	GO	http://www.geneontology.org/
Gene Index	ESTs and unigenes	http://compbio.dfci.harvard.edu/tgi/
GEO	Gene expression	http://www.ncbi.nlm.nih.gov/geo/
IUBMB	EC number	http://www.chem.qmul.ac.uk/iubmb/
miRBase	microRNA	http://microrna.sanger.ac.uk/sequences/index.shtml
NASCArrays	Microarray data	http://affymetrix.arabidopsis.info
Plant Ontology	Plant Ontology	http://www.plantontology.org/
PRIME	Systems biology	http://prime.psc.riken.jp/
UniGene	ESTs and unigenes	http://www.ncbi.nlm.nih.gov/unigene

continued

Table 1.1 (continued) Web Resources for Omics Data and Experimental Materials

	Metabolic and Biological Pathways	
BioCyc	Pathway/Genome Databases	http://biocyc.org/
BioModels	Annotated Published Models	http://www.ebi.ac.uk/biomodels-main/
IntAct	Protein interactions	http://www.ebi.ac.uk/intact/main.xhtml
PlantCyc	Metabolic pathways	http://pmn.plantcyc.org/PLANT/
Reactome	Biological pathways	http://www.reactome.org
Rhea	Biochemical reactions	http://www.ebi.ac.uk/rhea/home.xhtml
UniPathway	Metabolic pathways	http://www.grenoble.prabi.fr/obiwarehouse/unipathway
	Links for Database, Tools and Resources	
BioConductor	Bioinformatics tools for R	http://www.bioconductor.org/
Bioinformatics Links Directory	Web-based software resources	http://bioinformatics.ca/links_directory/
CRAN	R packages	http://cran.r-project.org/
Molecular Biology Database Collection	Links of Web databases	http://www.oxfordjournals.org/nar/database/c/
SHIGEN	Experiment materials and databases	http://www.shigen.nig.ac.jp/
	Arabidopsis	
AraCyc	Metabolic pathways	http://www.arabidopsis.org/biocyc/index.jsp
AtGenExpress	Microarray data	http://www.arabidopsis.org/portals/expression/microarray/ATGenExpress.jsp
ATTED-II	Coexpression	http://www.atted.bio.titech.ac.jp/
CSB.DB—A Comprehensive Systems-Biology Database	Coexpression	http://csbdb.mpimp-golm.mpg.de/index.html
KATANA	Annotations	http://www.kazusa.or.jp/katana/
PRIMe	Metabolomics and transcriptomics	http://prime.psc.riken.jp/
RAFL	Full-length cDNAs	http://www.brc.riken.jp/lab/epd/catalog/cdnaclone.html
RARGE	RIKEN Arabidopsis resources	http://rarge.psc.riken.jp/index.html
SIGnAL	Full-length cDNAs, T-DNA lines	http://signal.salk.edu/
TAIR	Omics database	http://www.arabidopsis.org/
The 1001 Genomes Project	Whole-genome sequence in 1001 strains	http://www.1001genomes.org/
	Rice	
KOME	Full-length cDNAs	http://cdna01.dna.affrc.go.jp/cDNA/
RAP-DB	Genome annotations	http://rapdb.dna.affrc.go.jp/
MSU	Genome annotations	http://rice.plantbiology.msu.edu/
OryzaExpress	Annotations and coexpression	http://riceball.lab.nig.ac.jp/oryzaexpress/
GRAMENE	An integrated database of grass species	http://www.gramene.org/
RiceCyc	Metabolic pathways	http://www.gramene.org/pathway/ricecyc.html
OMAP	Physical maps of the genus *Oryza*	http://www.omap.org/
OryzaBASE	A comprehensive database for rice	http://www.shigen.nig.ac.jp/rice/oryzabase/top/top.jsp

Table 1.1 (continued) Web Resources for Omics Data and Experimental Materials

	Solanaceae	
KaFTom	Full-length cDNAs	http://www.pgb.kazusa.or.jp/kaftom/
LycoCyc	Tomato metabolic pathways	http://www.gramene.org/pathway/lycocyc.html
MiBASE	ESTs, unigenes, metabolic pathways	http://www.pgb.kazusa.or.jp/mibase/
PGSC	Potato Genome Sequencing Consortium	http://www.potatogenome.net/index.php
SGN	An integrated database of Solanaceae	http://www.sgn.cornell.edu/index.pl
SolCyc	Pathway tools for Solanaceae	http://solcyc.solgenomics.net//
Tomato SBM Database	Tomato genome sequences and annotations	http://www.kazusa.or.jp/tomato/
Tomato Functional Genomics Database	Expression, metabolite, and small RNA	http://ted.bti.cornell.edu/
	Legumes	
BRC-MTR	*Medicago truncatula* resources	http://www1.montpellier.inra.fr/BRC-MTR/accueil.php
CMAP	Comparative map viewers for Legumes	http://cmap.comparative-legumes.org/
The *Lotus japonicus* website	Resources of *Lotus japonicus*	http://www.lotusjaponicus.org/
miyakogusa.jp	*Lotus japonicus* genome browser	http://www.kazusa.or.jp/lotus/index.html
Lotus japonicus EST index	ESTs in *Lotus japonicus*	http://est.kazusa.or.jp/en/plant/lotus/EST/index.html
Medicago truncatula HAPMAP PROJECT	*Medicago truncatula* genome browser	http://www.medicagohapmap.org/index.php
NBRP Legume Base	Resources of *Lotus japonicus* and *Glycine max*	http://www.legumebase.brc.miyazaki-u.ac.jp/index.jsp
Soybase	Integrated databases of soybean	http://soybase.org/index.php
	Other Plants	
CAES Genomics	ESTs of citrus, prunus, apple, grape, walnut	http://cgf.ucdavis.edu/home/
Cassava full-length cDNA clone	Cassava full-length cDNAs	http://www.brc.riken.jp/lab/epd/Eng/catalog/manihot.shtml
Cassava Online Archive	Information resource of cassava	http://cassava.psc.riken.jp/index.pl
Cucumber Genome Database	Cucumber genome and annotations	http://cucumber.genomics.org.cn/page/cucumber/index.jsp
Cucurbit Genomics Database	ESTs, pathways, markers, and genetics maps	http://www.icugi.org/
HarvEST	EST database-viewing software	http://harvest.ucr.edu/
International Citrus Genome Consortium	Citrus genomic information	http://int-citrusgenomics.org/
IWGSC	Physical mapping and sequencing of wheat	http://www.wheatgenome.org/
NBRP/KOMUGI	Wheat ESTs, markers, strains, and maps	http://www.shigen.nig.ac.jp/wheat/komugi/top/top.jsp

continued

Table 1.1 (continued) Web Resources for Omics Data and Experimental Materials

Poplar full-length cDNA clone	Poplar full-length cDNAs	http://www.brc.riken.jp/lab/epd/Eng/catalog/poplar.shtml
RIKEN Cassava Full-length cDNA database	Cassava full-length cDNA	http://amber.gsc.riken.jp/cassava/
SABRE	An integrated database of plant resources	http://saber.epd.brc.riken.jp/sabre/SABRE0101.cgi
Sunflower CMap database	Sunflower genetic maps	http://www.sunflower.uga.edu/cmap/

The Molecular Biology Database Collection in the *Nucleic Acids Research* journal online lists 1230 databases, which were carefully selected by criteria of the journal in 2010 (Cochrane and Galperin, 2010). The collection is updated according to the journal's annual database issues. Users can browse the alphabetical list and search for databases by the categories shown on the Web site. Their collection contains databases for model plants and representative species, such as *Arabidopsis*, rice, *Brassica*, maize, soybean, tomato, potato, poplar and so on.

The SHared Information of GENetic resources (SHIGEN) Project maintains a Web server that provides information about experimental materials and databases. The experimental materials (i.e., live animal stocks, frozen embryos, plant seeds, cultured cells, DNA clones, etc.) stated in the databases are available upon request.

1.3 GENOME PROJECTS AND DATABASES

Genome sequencing projects and their sequence data facilitate various omics studies such as genomics, transcriptomics and metabolomics. The information obtained from genome sequencing projects is also available from the NCBI Entrez Genome Project database (Sayers et al., 2011) and the Genomes OnLine Database (GOLD; Liolios et al., 2008).

The NCBI Entrez Genome Project database is a searchable overview collection of completed and in-progress large-scale sequencing, assembly, annotation and mapping projects, and the database includes current status and information on 2211 projects. The projects are grouped into three categories according to the status: 'Complete', 'Draft assembly' and 'In progress'. The web page of statistics shows the categories of kingdom (e.g., Archaea, Bacteria and Plant) and group (e.g., green algae and land plants in the plant kingdom), a description of the organism, its genome size, current status of genome sequencing and the project ID. Each collection of data is organized into an organism-specific overview and all projects pertaining to the organism, such as genomic, EST, mRNA, protein and whole genome sequencing (WGS) information, are incorporated. Detailed information on each project can be browsed and retrieved by hyperlinks on the Web page. For example, the page for oilseed rape (*Brassica napus*) contains species information, genomic organization (e.g., AACC genome, $2n = 38$) and the genome sequencing status.

GOLD is a Web-based resource for comprehensive access to genome projects worldwide, and the current version contains information on 6604 projects. GOLD also includes information on metagenome sequencing projects on archaea, bacteria and eukaryotic organisms. Completed and ongoing projects are summarized by organisms on respective pages, called the GOLD CARD.

Public databases for genome sequencing projects are also available. The Joint Genome Institute (JGI) of the U.S. Department of Energy (DOE) was established in 1997 to integrate the expertise and resources of DNA sequencing, technology and informatics. The initial objective of the JGI was to complete genome sequencing of human chromosomes 5, 16 and 19. After completion of the sequencing of the human genome, their target mission has broadened to other species including nonhuman sequences. In plants, projects on a green alga (*Chlamydomonas reinhardtii*), a diatom (*Thalassiosira pseudonana*), the cottonwood tree (*Populus trichocarpa*), various plant pathogens and agriculturally important plants such as grape and soybean have been launched as JGI's genome sequencing projects. Due to the accumulating JGI sequencing data, it is essential to develop new analytical tools and data management systems in computational genomics. The Phytozome, developed by the JGI computational genomics program, is a tool for comparative genomics in green plants and provides genome and gene annotations for *Arabidopsis* (*Arabidopsis thaliana*), rice (*Oryza sativa*), maize (*Zea mays*), sorghum (*Sorghum bicolor*), *Brachypodium* (*Brachypodium distachyon*), soybean (*Glycine max*), *Medicago* (*Medicago truncatula*), poplar (*P. trichocarpa*), cassava (*Manihot esculenta*), grape (*Vitis vinifera*) and papaya (*Carica papaya*). It can allow visualization of comparative search results between species and also has links to functional and phenotypic information on the JGI and related Web sites.

GRAMENE is an integrated database for genetics, genomics and comparative genomics in grasses, including rice, maize, rye, sorghum, wheat and other close relatives (Liang et al., 2008). The data in GRAMENE include DNA/RNA sequences, functional annotations of genes and proteins, gene ontology (GO) (The Gene Ontology Consortium, 2008), genetic and physical maps and markers, quantitative trait loci (QTLs), pseudomolecule assembly, genetic diversity among germplasm and comparative genetics and genomics between rice and its wild relatives.

The Beijing Genomics Institute (BGI) has launched a variety of sequencing-based projects, such as WGS, genome/gene variation and metagenomics studies, in humans, animals, plants and bacteria. Recently, BGI has accomplished WGS on the giant panda using NGS technology, and their next target is genome sequencing of big cats. For plants, BGI has genome sequencing projects for cucumber and orchid. In the cucumber genome project, they are applying traditional Sanger sequencing and NGS technology (Huang et al., 2009).

1.4 GENE EXPRESSION AND COEXPRESSED GENE DATABASES

Gene-expression data are widely used to predict gene function and gene regulatory mechanisms. Expression data obtained from microarrays and sequencing have been collected in Web databases. When researchers publish their results from microarray experiments, as well as sequencing, the experimental data and associated protocols are required to be deposited into publicly accessible databases. For submitting microarray data, a description of the data set is prepared following MIAME (Minimum Information About a Microarray Experiment) standard format (Brazma et al., 2001) and is deposited into public databases, such as ArrayExpress at EBI (Parkinson et al., 2009), GEO at NCBI (Barrett et al., 2009) and CIBEX at DDBJ (Ikeo et al., 2003). These databases maintain and provide all kinds of microarray data. In addition, they also cover a wide range of high-throughput experiments by non-array-based technologies such as serial analysis of gene expression and MS peptide profiles. The expression profiles for all species can be queried by keywords and terms, for example, gene name, sample name and type of experiment. Results of a query can be obtained as a graph or table of gene-expression profiles.

Large-scale microarray data allow analysis and mining for coexpressed genes that show similar expression patterns. A list of coexpressed genes provides a whole image of the genes' network and has the potential to predict biological implications and characteristics of the network, such as gene functions, gene families and genetic regulatory mechanisms. Using large-scale microarray data in public databases, coexpressed gene network databases such as ATTED-II (*A. thaliana trans*-factor and *cis*-element prediction database), OryzaExpress and MiBASE have been constructed and maintained. ATTED-II is a specialized database for the prediction of *trans*-factors and *cis*-elements in *Arabidopsis* (Obayashi et al., 2009). Based on repositories of microarray data in AtGenExpress, ATTED-II predicts coexpressed gene networks and *cis*-elements in the region up to 200 bp upstream from the transcriptional starting point. OryzaExpress and MiBASE provide information on coexpressed genes in rice and tomato using expression data from GEO (see below for details).

1.5 GENE ONTOLOGY DATABASES

The GO project is a collaborative effort aiming to provide consistent descriptions of gene products in different databases (The Gene Ontology Consortium, 2008). It was launched in 1998 as to integrate databases for *Drosophila*, *Saccharomyces* and mouse data. To date, the GO consortium has grown to include many other animal, microbe and plant databases, and enables cross-linking between the ontologies and the products of the genes in the collaborative databases.

The GO collaborators developed three structured, controlled vocabularies (GO terms) that describe gene products in terms of three categories: molecular function, biological process and cellular component. These vocabularies function in a species-independent manner; thus, it is possible to link genes and gene products with the most relevant GO terms in each category. However, it is important to note that the GO is not a nomenclature for genes or gene products but a vocabulary for the description of the attributes of biological objectives and phenomena.

The association between a specific GO term and gene product is assigned an evidence code. The evidence code indicates how the association is supported. For example, the evidence code 'Inferred from Mutant Phenotype (IMP)' means that the properties of a gene product are inferred based on differences between wild-type and mutant alleles. GO Slim is a minimal version of GO that gives an overview of the ontology content. GO Slim is particularly useful for obtaining a summary of GO annotations.

1.6 EUKARYOTIC ORTHOLOGOUS GROUP DATABASE

Information on the phylogenetic classification of proteins encoded in complete genomes is provided by the Clusters of Orthologous Groups (COG) database (Tatusov et al., 2003). The Eukaryotic Orthologous Groups (KOG) database, the eukaryote-specific version of COG, contains data sets for *Arabidopsis* proteins and provides a full list of KOG classifications that are ordered by KOG group and letter. Orthologous or paralogous proteins are assigned KOG identifiers (IDs) and classified into one of four functional groups, cellular processes and signalling, information storage and processing, metabolism and poorly characterized. Within each classification group, orthologous or paralogous proteins are listed and can be analysed for either an 'individual gene model set' or 'multiple gene model sets' by selecting the model set(s) of interest to view. By clicking each protein name of interest, the sequence and sequence similarities with other proteins are shown. The list of gene model(s) also includes JGI's name and identification number for the predicted genes.

1.7 EXPRESSED SEQUENCE TAGS, UNIGENE SEQUENCES AND FULL-LENGTH cDNAs

Transcriptome contains expressed sequence tags (ESTs), a nonredundant sequence set (UniGene cluster) derived from these ESTs, and full-length cDNA sequences, and these kinds of sequence data are provided by INSD and other public databases. ESTs generated by 5′ and/or 3′ end-sequencing of cDNA libraries provide information about transcript sequences and expression patterns in tissues and organs at various developmental stages (Adams et al., 1991). The dbEST (Boguski et al., 1993) and UniGene (Wheeler et al., 2003) databases provide information about

ESTs from a variety of organisms, including plants. The UniGene database provides information about protein similarities, gene expression, cDNA clones and genomic locations, in addition to a list of accession numbers of ESTs. It is regularly updated weekly or monthly, thus providing more recent entry information (accession numbers) for each UniGene cluster. It covers a variety of plant species, such as *A. thaliana*, rice, maize, barley, wheat, oilseed, citrus, soybean and coffee. The tomato UniGene database MiBASE, for example, provides information about UniGene sequences obtained by assembling ESTs in tomato and a wild relative as well as information about SNPs, simple sequence repeats, GO terms, metabolic pathway names and gene-expression data (Yamamoto et al., 2005; Yano et al., 2006a,b).

The availability of nonredundant consensus sequences obtained from ESTs allows the use of computational approaches for protein domain searches, comparison of transcript sequences among different organisms and homology searches of nucleotide sequences. Although NCBI's UniGene does not provide consensus UniGene sequences, such nonredundant consensus sequence sets are available from the Gene Indices databases of the Dana-Farber Cancer Institute. Gene Indices provides information about a nonredundant consensus sequence set called a tentative consensus (TC), generated by assembling and clustering methods (Lee et al., 2005). In the databases, sequences of TCs are available together with their variants and functional and structural annotations. The detailed information in each TC includes homologous protein sequences, open reading frames (ORFs), GO terms, SNPs, alternative splicing sequences, cDNA libraries, Enzyme Commission numbers of the International Union of Biochemistry and Molecular Biology, Kyoto Encyclopedia of Genes and Genomes (KEGG) metabolic pathways (Kanehisa et al., 2010), unique 70-mer oligonucleotide sequences and orthologs in other organisms.

Full-length cDNA clones and sequences are fundamental resources in molecular biology for investigations of gene function, intron–exon structures and protein coding sequences. In several plant species, comprehensive sequencing of full-length cDNAs has been conducted. Information on full-length cDNA sequences is provided in organism-specific databases (see below).

1.8 METABOLIC PATHWAYS

Gas chromatography (GC) and MS are widely used to investigate metabolic pathways in many plant species. Information on metabolic pathways is stored and provided in the Web databases listed below. Understanding of the genes and enzymes involved in secondary metabolism should allow improvement of agronomically important traits in crops and production of industrial materials in plants.

1.8.1 KEGG

KEGG is an integrated database consisting of 16 main databases (Kanehisa et al., 2010). The databases are classified into three categories: systems information, genomic information and chemical information. KEGG PATHWAY, one of the KEGG databases, is a collection of manually drawn pathway maps of molecular interactions and reaction networks for metabolic processing, genetic information processing, environmental information processing, cellular processes, organismal systems and human diseases. Pathway maps for metabolism are divided into 12 subcategories: carbohydrate, energy, lipid, nucleotide, major amino acid, other amino acid, glycan, polyketide and nonribosomal peptide, cofactor/vitamin, secondary metabolites, xenobiotics and overviews.

All pathway maps can be viewed graphically, and genes and compounds are boxes and circles in the maps, respectively. A new graphical interface, KEGG Atlas, is also available for viewing the global metabolism map and the other pathway maps, with zoom and navigation capabilities.

1.8.2 BioCyc

BioCyc is a collection of pathway/genome databases (Karp et al., 2005). At present, the genomic data and metabolic pathways of 505 organisms are provided in BioCyc.

The databases in BioCyc are divided into three categories (tiers) according to the quality of data. Two databases, EcoCyc and MetaCyc, include intensively curated databases (Tier 1). EcoCyc is a metabolic pathway database for *Escherichia coli* K-12 (Keseler et al., 2009). MetaCyc is an integrated database for metabolic pathways and enzymes, elucidated from more than 1500 organisms and 19,000 publications (Caspi et al., 2010). Tier 2 contains computationally derived databases that are subjected to moderate reviewing. It currently consists of 23 databases. Tier 3 includes computationally derived databases without curation and currently consists of 482 databases.

MetaCyc and PlantCyc databases include plant metabolic pathway information. PlantCyc, released by the Plant Metabolic Network (PMN), provides manually curated or reviewed information about shared and unique metabolic pathways in over 250 plant species. PMN has released and maintained the AraCyc database for *Arabidopsis* (Zhang et al., 2005) and PoplarCyc for the model tree *P. trichocarpa*. Metabolic pathway databases for other organisms have also been constructed by PMN collaborators. To date, the CapCyc (pepper), CoffeaCyc (coffee), LycoCyc (tomato), MedicCyc (Medicago), NicotianaCyc (tobacco), PetuniaCyc (petunia), PotatoCyc (potato), RiceCyc (rice), SolaCyc (eggplant) and SorghumCyc (sorghum) databases have been released.

1.8.3 Other Pathway Databases

Other databases also provide information of metabolic pathways and chemical or protein reactions. A database UniPathway stores data of metabolic pathways curated by the UniProtKB/Swiss-Prot knowledgebase. The EMBL-EBI has pathway and network databases BioModels, IntAct, Rhea and Reactome. BioModels Database stores published mathematical models of biological interests (Li et al., 2010). Models in BioModels are annotated and linked to relevant data resources, such as publications, databases of compounds and pathways, controlled vocabularies and so on. IntAct provides protein interaction data which are derived from literature curation or direct user submissions (Aranda et al., 2010). Rhea is a manually annotated database of chemical reactions. Reactome, a curated knowledgebase of biological pathways, contains data for model organisms including *Arabidopsis* and rice (Croft et al., 2011).

1.9 ADVANCED TECHNOLOGY AND METHODS FOR LARGE-SCALE ANALYSES

1.9.1 High-Throughput Sequencing

DNA sequencing is a technique for determining the nucleotide sequence within DNA fragments. Sequences are based largely on results from sequencing methods developed by Frederick Sanger in 1977; some methodologies have been developed specifically for Sanger sequencing. Recently, a novel technology, MPSS technology or pyrosequencing, has been developed commercially. Several highly innovative high-throughput sequencing technologies now available are the Solexa Genome Analyzer IIx, Roche Genome Sequencer FLX and Applied Biosystems SOLiD 4 system platforms; others, using the so-called third-generation sequencing, will be released in the near future. The systems exceed Sanger sequencing in their ability to produce nucleotide sequence data of 400,000 to 2 billion nucleotides per operation. This means that rapid genome sequencing will enable us to conduct many kinds of research from the micro- to the macro-level, and it will also be possible to sequence genomes among varieties, subspecies and ecotypes. A good demonstration of this system's utility was the WGS of scientist James Watson (Wheeler et al., 2008). In plant research, this methodology will facilitate further studies of genome sequences and DNA polymorphisms among different genomes.

1.9.2 Tiling Array in *Arabidopsis*

A tiling array is a type of microarray chip in which short DNA fragments are designed to cover the whole genome, allowing the tiling array to be used to investigate gene expression, genome structure and protein binding on a genome-wide scale. The procedure is similar to traditional microarray technology, but it differs in a variety of objectives: unbiased gene-expression profiling, transcriptome mapping, chromatin immunoprecipitation (Chip)-chip, methyl-DNA immunoprecipitation (MeDIP)-chip and DNase Chip (e.g., Gregory et al., 2008; Stolc et al., 2005). Although this approach still has problems with operating cost and signal sensitivity, it will likely become one of the preferred tools for genome-wide investigation in the future.

1.9.3 Large-Scale Expression Analysis

Since high-throughput technologies have rapidly developed, the sizes of omics data sets have continued to increase. An increase in the size of a data set can make computer analyses too time-consuming. To solve this problem or to handle various large-scale omics data including microarray, MS and NGS data, many analysis methods have been proposed. Some analysis tools are available on the Web. The annual Web server issues of *Nucleic Acids Research* report on Web-based software resources for molecular biology research and provide links to them (Benson, 2010). The Bioinformatics Links Directory also includes links to the Web servers. The links are categorized by subject (e.g., literature, DNA, RNA, expression) and are searchable. From the Comprehensive R Archive Network (CRAN) and BioConductor Web sites, researchers can also search and get various tools for bioinformatics analyses.

With accumulation of expression data, various expression analysis methods have been developed and applied to find a novel gene or elucidate its biological functions. In microarray data analyses, hierarchical clustering methods have been widely used to cluster genes (probes) according to their expression profiles (Eisen et al., 1998). In clustering methods, viewers for dendrograms and expression maps help identify genes showing specific expression patterns related to phenotypes or genotypes. However, hierarchical clustering methods require large amounts of computer memory and calculation time due to the recent increases in the number of probes on a microarray platform and the amount of data obtained from microarray approaches. Some analysis tools used to perform analyses with a short calculation time, such as self-organizing maps and k-means, are also used to categorize genes with similar expression patterns. Recently, a new statistical approach combined with correspondence analysis was proposed as a high-throughput gene discovery method (Yano et al., 2006c). This method allows quick analysis of a large-scale microarray data set to identify up- or downregulated genes related to a trait of interest.

1.10 GENOME ANNOTATIONS AND COMPARATIVE GENOMICS FOR MODEL PLANTS

1.10.1 Arabidopsis

1.10.1.1 The 1001 Genomes Project

A. thaliana is a small flowering plant in the Brassicaceae and one of the important model species in plant biology. Its genome was the first to be completely sequenced in plants. However, based on the findings of a current genomics study that DNA polymorphisms exist also in protein-coding sequences, a reference genome from a single strain is not sufficient to understand the genome of *Arabidopsis*. Thus, the aim of the 1001 Genomes Project is to collect WGS information on 1001 strains of *Arabidopsis* to understand its genome thoroughly and to detect genetic variations in it (Weigel and Mott, 2009).

1.10.1.2 TAIR

The *Arabidopsis* Information Resource (TAIR) is a comprehensive database for *A. thaliana* (Swarbreck et al., 2008). Along with the completed and fully annotated *Arabidopsis* genome sequence, the database contains a broad range of information, such as genes, gene-expression data, clones, nucleotide sequences, proteins, DNA markers, mutants, seed stocks, members of the *Arabidopsis* research community, and research papers. The TAIR database contains various Web pages for efficient and intuitive querying and browsing of data sets, graphical and interactive map viewers for genes and the genome, and allows downloading of data stored in the database. TAIR also includes the AraCyc metabolic pathway database (Zhang et al., 2005). Although TAIR has stopped accepting deposition of new microarray data, it has links to more comprehensive microarray data set repositories at the ArrayExpress, GEO, NASCArrays and Stanford Microarray databases. All of these repositories still accept microarray data submissions. In addition, a multinational consortium has launched the AtGenExpress database for uncovering genome-wide gene-expression patterns in *A. thaliana* and this database is also available to obtain large-scale transcriptome data sets.

The TAIR Web site is updated every two weeks with new information from research publications and community data submissions. Gene structures are also updated a couple of times per year using computational and manual methods. TAIR and NCBI released the latest version of the *Arabidopsis* genome annotation (TAIR9).

The browser is simple and user-friendly. A user can retrieve information of interest after navigating only a few pages. By clicking a tab of interest such as 'search', 'tools' or 'stocks' on the topmost page, researchers can get basic information, with more detailed information accessible via hyperlinks to internal and external Web pages.

1.10.1.3 Full-Length cDNA Databases

Information on full-length *Arabidopsis* cDNA clones is available from databases. *Arabidopsis* full-length cDNA clones are accessible from the RIKEN *Arabidopsis* full-length (RAFL) database (Seki et al., 2002), and currently 251,382 full-length cDNA clones have been released. The RIKEN *Arabidopsis* Genome Encyclopedia (RARGE) database accumulates information on transposon mutants, gene expression, physical maps and alternative splicing, along with full-length cDNAs (Sakurai et al., 2005). Currently, 224,336 full-length cDNA clones, including 18,090 unique genes and 15,295 full reading sequences, are available in the database. In addition, the Salk, Stanford and PGEC Consortium (SSP) has determined over 22,000 protein coding ORFs (Yamada et al., 2003), and the SIGnAL database provides information on the *Arabidopsis* Gene/ORFeome and T-DNA collections.

1.10.1.4 PRIMe

The Platform for RIKEN Metabolomics (PRIMe) provides information on metabolomics and transcriptomics (Akiyama et al., 2008). Data on standard metabolites measured by multidimensional NMR spectroscopy, GC/MS, liquid chromatography (LC)/MS and capillary electrophoresis/MS are stored and released with MS data. PRIMe has hyperlinks to PubChem (NCBI) and KEGG.

AtMetExpress in PRIMe provides data on various phytochemicals detected by LC/MS (Matsuda et al., 2010).

1.10.2 Rice

1.10.2.1 RAP-DB

After sequencing of the rice genome was completed in 2004 (International Rice Genome Sequencing Project, 2005), the Rice Annotation Project (RAP) was launched to provide accurate and timely gene annotation of sequences in the rice genome. RAP collaborates closely with the International Rice Genome Sequencing Project (IRGSP). Genome annotations stored in the database called the RAP-DB include information about nucleotide and protein sequences, structures and functions of genes, gene families and DNA markers (Rice Annotation Project, 2008). RAP annotations, integrated with IRGSP genome build 5, are now available from RAP-DB. Due to a continuing effort to update RAP-DB, search functions (e.g., BLAST, BLAT, keyword searches and the GBrowse genome browser) are also effective. RAP helps to identify gene families and genes with biological functions that have not yet been elucidated. It is noteworthy that one function of the RAP, called the identifier (ID) converter, allows retrieval of IDs are assigned by RAP and another rice annotation project, the MSU Rice Genome Annotation Project described below. For example, the ID converter can determine that the sequence Os01g0100100 in the RAP-DB and sequences LOC_Os01g01010.1 and LOC_Os01g01010.2 in the MSU database are in fact assigned to the same gene.

1.10.2.2 The MSU Rice Genome Annotation Project

The MSU Rice Genome Annotation Project maintains a database that provides sequence and annotation data for the rice genome (Ouyang et al., 2007). In addition to standardized annotations for each gene model, functional annotations including expression data, gene ontologies and tagged lines have been updated. It should be noted that locus IDs (e.g., LOC_Os01g01010.1) assigned by the database differ from those in RAP-DB. Moreover, the methods of genome assembly and genome annotation in this database are distinct from those employed by RAP-DB. Thus, the genome sequences and corresponding transcripts in RAP-DB and MSU may not always have the same nucleotide sequences.

1.10.2.3 KOME

The Knowledge-based *Oryza* Molecular biological Encyclopedia (KOME) is a database for full-length cDNAs of japonica rice (The Rice Full Length cDNA Consortium, 2003). In this database, 170,000 full-length cDNA clones have been grouped into 28,000 independent groups, and all representative clones have been completely sequenced. The KOME database contains not only sequence information but also information on homologous sequences from the public databases, mapping information, alternative splicing, protein domains, transmembrane structure, cellular localization and gene function with corresponding GO term.

1.10.2.4 Oryzabase

Oryzabase is a comprehensive database for rice research established in 2000 by a committee of rice researchers in Japan (Kurata and Yamazaki, 2006). It includes a variety of resources: genetic lines and wild germplasm collected from all over the world, mutant lines classified based on tissue in which a mutated gene disrupts the phenotype, genes annotated by RAP-DB and MSU, literature references, linkage maps that are integrated into reference maps among various subspecies of *japonica* and between subspecies of *japonica* and *indica*, physical maps of SHIGEN and MSU databases, comparative maps between rice and other grass species, DNA and organelle databases and analytical tools and protocols. It also has links to related databases such as RAP-DB, KOME, MSU and GRAMENE. Thus, Oryzabase would be the optimum choice when a researcher needs a fundamental database for analysis in rice research.

1.10.2.5 GRAMENE

In addition to the above-mentioned feature of GRAMENE as an integrated database for genetics and genomics, it is noteworthy that substantial information about QTLs and GO terms in rice has been accumulated in the GRAMENE database. The RiceCyc rice metabolic pathway database is also accessible via GRAMENE. In RiceCyc, locus identifiers employed in the MSU Rice Genome Annotation Database are assigned.

Because GRAMENE allows easy access to other databases by hyperlinks, it is a good idea to start computational analysis at GRAMENE to obtain detailed data by general or in-depth searching. It also has the user-friendly feature of being available for downloading and local installation so that GRAMENE's data and tools can be customized for a researcher's requirements.

1.10.2.6 OMAP

The main objective of the Oryza Map Alignment Project (OMAP) is to develop a database for physical maps of the genomes of 11 wild species in the *Oryza* genus, namely *O. rufipogon*, *O. glaberrima*, *O. punctata*, *O. officinalis*, *O. minuta*, *O. australiensis*, *O. latifolia*, *O. schlechteri*, *O. ridleyi*, *O. brachyantha* and *O. granulata*. Currently, 830,000 fingerprints and 1,659,000 bacterial artificial chromosome (BAC) end sequences are annotated.

In addition to complete full-genome sequencing of a reference rice genome, this project offers the advantage of DNA fingerprint/BAC-end sequences of 11 wild relatives of rice for comparison. OMAP is a collaborative project by the Arizona Genomics Institute, Arizona Genomics Computational Lab, Cold Spring Harbor Lab and Purdue University (Wing et al., 2005). The database includes: (1) an alignment of maps between *japonica* and *indica* subspecies of *O. sativa*, (2) construction of high resolution physical maps of chromosomes 1, 3 and 10 across the 11 wild rice species and (3) development of convenient bioinformatics and educational tools for understanding the *Oryza* genome. These achievements provide an experimental system for understanding the evolutionary history of the *Oryza* genome. It will contribute to our understanding of evolutionary, physiological and biochemical aspects of the genus *Oryza*.

1.10.2.7 OryzaExpress

OryzaExpress provides functional annotations of genes, reaction names in metabolic pathways, locus identifiers (IDs) assigned from RAP and the MSU rice databases, gene-expression profiles and gene-expression networks.

OryzaExpress works as a bridge between corresponding databases. By searching keywords or IDs in OryzaExpress, users can find genes, reaction names of metabolic pathways in KEGG and RiceCyc, locus IDs in RAP and MSU, and probe names used in the Affymetrix Rice Genome Array and Agilent Rice Oligo Microarrays (22K). Although locus IDs and probe names for the same genes differ among the public databases and microarray platforms, OryzaExpress allows simultaneous retrieval of information from distinct public databases. In addition, gene-expression networks derived from microarray data of the Affymetrix Rice Genome Array are accessible with interactive viewers.

1.10.3 Solanaceae

The Solanaceae includes agriculturally important plants such as tomato (*Solanum lycopersicum*), potato (*Solanum tuberosum*), eggplant (*Solanum melongena*), tobacco (*Nicotiana tabacum*), pepper (*Capsicum annuum*) and petunia (*Petunia × hybrida*).

Tomato, a vegetable crop consumed worldwide, is a model plant of the Solanaceae. In 2004, a tomato genome sequencing program was launched by the internationally coordinated International Solanaceae Initiative (SOL) consortium (Mueller et al., 2005). The International Tomato Sequencing Project announced a preleased version of the tomato genome shotgun sequence in 2009. The SOL Genomics Network (SGN), which is part of SOL, provides information relevant to the tomato genome sequencing project such as linkage maps containing DNA markers, BAC clones anchored to these linkage maps (called seed BAC clones), and BAC sequences with genomic and functional annotations. Information on ESTs, UniGene clusters, genetic maps and DNA markers for tomato and other related species is also available. SGN includes SolCyc, which is a collection of Pathway Genome Databases for Solanaceae species.

1.10.3.1 Tomato SBM

The Tomato Selected BAC clone Mixture (SBM) Database provides information on nonredundant tomato genome sequences. The genome sequences were determined by shotgun libraries from two sets of BAC clones. A total of 4,248,000 reads obtained from sequencing was assembled into nonredundant sequences of 540,588,968 bp (100,783 contigs). About 4000 scaffolds including 15,119 contigs were constructed according to the positions of BAC and fosmid end sequences. In the Tomato SBM Database, users can obtain information on nucleotide and translated sequences, gene models and annotations from BLAST and InterPro.

1.10.3.2 TFGD

The Tomato Functional Genomics Database (TFGD) is an integrated database for gene expression, metabolites and small RNAs (Fei et al., 2006). In gene-expression database, digital expression data (RNA sequencing) of tomato ESTs and microarray data sets based on TOM1 cDNA array, TOM2 oligo array and Affymetrix Tomato Genome Array platforms are incorporated for an efficient exploration of the analysed gene expression data. The tomato metabolite database provides profile data sets of tomato metabolites related to fruit nutrition and flavour. It also contains metabolite profiles during fruit development and ripening of wild tomato and ripening deficient mutants. The tomato small RNA database accumulates all tomato small RNAs reported/deposited to date and currently contains ~15.4 million tomato small RNAs representing more than 5.3 million unique ones. TFGD incorporates all data sets into effective tools and queries in the database, for coexpression analysis of a given gene, identification of enriched GO terms and altered metabolite pathways.

1.10.3.3 MiBASE

The miniature tomato cultivar Micro-Tom has attracted attention as a model plant that can be cultivated even in ordinary laboratory spaces (Meissner et al., 1997; Shibata, 2005). The MiBASE database contains information on 125,883 ESTs from laboratory-grown Micro-Tom and the Kazusa Tomato Unigene (KTU) data set from Micro-Tom ESTs and other publicly available tomato ESTs from SGN. Data for KTU3 (KTU version 3) in MiBASE include gene functions, GO and GO Slim terms, metabolic pathways and DNA markers (Yano et al., 2006a). Gene-expression networks, based on Pearson's correlation coefficients and other statistical indices using the Affymetrix Tomato

Genome Array, are also available. This framework allows users to obtain information on probes that have similar expression profiles by an interactive graphical viewer.

1.10.3.4 KaFTom

Full-length cDNA libraries of Micro-Tom have been constructed from fruit at different developmental stages, pathogen-treated leaves and roots. To date, information about 13,227 draft full-length sequences (High-throughput cDNA: HTC), including gene functions, protein functional domains and GO terms are available in the KaFTom database (Aoki et al., 2007; Yano et al., 2007), which also includes structural annotations of sequences between HTC (cv. Micro-Tom) and tomato genome sequences available in SGN.

1.10.3.5 PGSC

As potato is a popular vegetable worldwide, an internationally collaborated group launched the Potato Genome Sequencing Consortium (PGSC) to completely sequence the potato genome. High-quality annotated genome sequences enhance the identification of allelic variants and contribute to unravelling important agronomical traits related to yield, quality, nutritional value and disease resistance of potato. The PGSC has two plant materials for sequencing projects, the diploid line RH89-039-16 of *S. tuberosum* and the doubled monoploid DM1-3 516R44 of *Solanum phureja*. On the PGSC Web site, researchers can obtain the sequence files and perform homology searches against BAC and BAC end sequences.

1.10.4 Legumes

Legume crops including soybean (*G. max*), alfalfa (*Medicago sativa*), clover (*Trifolium* species), beans, lentils, peas and *Arachis* show a high protein content. Legumes are utilized as food for human beings, forage for livestock and green manure. The ability of legumes to fix atmospheric nitrogen by a symbiotic relationship with rhizobia reduces quantities and costs of fertilizer. The mechanisms of nitrogen fixation and symbiosis in legumes have been deeply investigated.

Lotus japonicas ($n = 6$) (Sato et al., 2008) and *M. truncatula* ($n = 8$) (Cannon et al., 2006) are used as model legumes due to their short life cycles, small diploid genomes (about 500 Mb) and ability for self-fertilization. Genome annotation information is available from genome browsers provided by the genome sequencing projects. Over 105,000 ESTs and consensus sequences of *L. japonicas* are accessible from the *L. japonicus* EST index (Asamizu et al., 2004). Information on resources for *L. japonicus* and soybean is available from the National BioResource Project (NBRP) Legume Base (Yamazaki et al., 2010). Information on biological resources for *M. truncatula*, including inbred and TILLING lines, is also provided on the BRC-MTR Web site (Biological Resource Centre, INRA).

Soybean (*G. max*, genome size 1115 Mb) is an important crop for seed protein, oil content and crop rotation. Genome sequence and annotations for cultivar Williams 82 have been performed and are provided by the Phytozome database (Schmutz et al., 2010). Composite soybean genetic and physical maps with genomic sequence are also available from SoyBase. Besides maps, SoyBase stores data of the SoyCyc Soybean Metabolic Pathway Database and annotations from the Affymetrix GeneChip Soybean Genome Array, and corresponding to Plant Ontology and GRAMENE Trait Ontology. The CMap module provides a comparative map viewer for legumes including alfalfa, chickpea, common bean, cowpea, pea, peanut, soybean, *L. japonicas* and *M. truncatula*. The viewer contains information on ESTs, clones and DNA markers such as Amplified Fragment Length Polymorphism, Cleaved Amplified Polymorphic Sequences (CAPS), derived Cleaved Amplified Polymorphic Sequences (dCAPS), Finger-Printing Contig (FPC).

1.10.5 Brassica

The genus *Brassica* in the Brassicaceae is widely cultivated in the world and includes a variety of important agronomical crops such as oilseed rape, broccoli, cabbage, Chinese cabbage, black mustard and other leafy vegetables. The genetic relationships of *Brassica* have been well studied and are referred to as the U's triangle (U, 1935). The genomes of three diploid species, *B. rapa*, *B. nigra* and *B. oleracea*, have been designated A, B and C, respectively, and those of the amphidiploids *B. juncea*, *B. napus* and *B. carinata*, as AB, AC and BC, respectively. It is noteworthy that cultivated *Brassica* species represent the group of crops most closely related to *A. thaliana*, the model plant for dicots. Segmental chromosomal collinearity and physically and functionally conserved gene(s), known as orthologs and paralogs, have been identified between *Brassica* and *Arabidopsis*. Such investigation may lead to a better understanding of plant genetics, including molecular biological and physiological aspects, which have emerged via evolution. Thus, the remarkable diversity of *Brassica* is an excellent example for understanding how genetic and morphological variations have developed during evolution of plant genomes. The international research community has launched the Multinational *Brassica* Genome Project (MBGP), and their current objective is to sequence the genome of *B. rapa*. The *Brassica rapa* Genome Sequencing Project, organized by MBGP, will announce the whole genome sequence of *B. rapa* in the near future. In addition, the MBGP has an integrated database Web site that provides a variety of resource information such as DNA sequences, DNA markers, genetic maps, proteomics, metabolomics, transcriptomes, QTLs and plant materials. All information is available on request.

1.10.6 Cucurbitaceae

1.10.6.1 *The Cucumber Genome Initiative*

Cucurbitaceae is a plant family that includes pumpkin, melon and watermelon. Cucumber, *Cucumis sativus*, is an important vegetable of the Cucurbitaceae, as well as a model plant for cucurbit-specific research such as sex determination. The Cucumber Genome Initiative was launched to obtain the complete sequence of the cucumber genome. Its objective is to develop genetic and molecular bases for understanding important agronomical traits and a comprehensive toolbox for molecular breeding. In the genome sequencing project, a common line in modern cucumber breeding, Chinese long inbred line 9930, was selected. The Cucumber Genome Database has also been developed for cucumber genome data and related information, as mentioned below.

1.10.6.2 *CuGenDB and Polish Consortium of Cucumber Genome Sequencing*

The Cucurbit Genomics Database (CuGenDB) was developed by the International Cucurbit Genomics Initiative (ICuGI), supported by BARD grants. The database provides information on ESTs, pathways, DNA markers and genetic maps for melon, cucumber and watermelon. Data on pumpkin small RNAs and a melon cDNA array are also available in the database.

The Polish Consortium of Cucumber Genome Sequencing was launched to explore the cucumber genome and to improve the research environment for cucumber.

1.10.7 Other Plants

Maize (*Z. mays*) is an important crop for bioethanol as well as food and feed. The Maize Genome Sequencing Project, funded by the NSF, USDA and DOE, started in 2005. Using a minimum tiling path of BAC clones annotated by physical and genetic maps, the project sequenced the genome of maize cultivar B73 (Schnable et al., 2009). Sequence information on B73 is available in the Maize Genetics and Genomics Database (MaizeGDB) (Schnable et al., 2009). MaizeGDB provides information on the draft sequence of the 2.3-Gb genome, transcriptomes and genome annotations. In addition, it provides a variety of biological information including nucleotide sequences, gene products, genetic maps, DNA markers, phenotypes, metabolic pathways and microarray data. In MaizeGDB, the MaizeGDB Genome Browser (GBrowse-based) can integrate and visualize data on genomic regions, genetic markers and nucleotide sequences based on B73 sequence assembly. MaizeSequence.org also provides information on sequences and annotations resulting from the Maize Genome Sequencing Project. Recently, the genome sequence of an ancient maize landrace, Palomero Toluqueño, has been released. Genome sequencing data of Palomero Toluqueño are temporarily available from the Web site, which is accessible from MaizeGDB.

Grape, *V. vinifera* ($2n = 38$,500 Mb genome), is consumed as fresh fruit, juice and wine, and is a model plant for fruit tree genetics. In grape, two genome sequence projects have been launched, the Pinot Noir Grapevine Genome Initiative and International Grape Genome Program (IGGP). The former is an international consortium for sequencing of the *V. vinifera* cv. Pinot Noir genome, coordinated by the IASMA Research Center. The latter is a multinational collaborative research programme for grape genomics and molecular biology. IGGP consists of international members of the major wine regions of the world and focuses on the accumulation of biological resources on grape research, such as genetic markers, a BAC library, genetic and physical maps, ESTs, transcriptomes, functional analysis, bioinformatics and genome sequencing. This knowledge is submitted to public databases at NCBI and will contribute to improve agronomical traits such as quality for fruit and wine grapes, disease resistance, abiotic stress resistance and physiological features.

Poplar (*P. trichocarpa*) is a black cottonwood tree that has a genome estimated to be approximately 480 Mb in 19 chromosomes. A draft poplar genome sequence and its initial analysis have been reported (Tuskan et al., 2006). Information on the poplar genome and additional data including a genome browser, functional annotations such as GO terms, protein domains and KOG, are available from the Genome Portal and Phytozome in JGI.

The Rosaceae is a large family including fruit-producing crops and ornamental plants such as apple, peach, almond, cherry, raspberry, strawberry and rose. The Genome Database for Rosaceae (GDR) provides integrated genetics and genomic data of the Rosaceae (Jung et al., 2008). The Comparative Map Viewer for Rosaceae (CMap) allows users to view and compare various kinds of maps including genetic maps, physical maps and sequence-based maps of Rosaceae species. GDR also provides data on ESTs, contig assembly and additional annotations that include gene functions, DNA markers, ORFs, GO terms and anchored positions to the peach physical map.

S. bicolor (730 Mb with $2n = 20$) is a drought-tolerant crop and a model plant for comparative genomic studies with rice. A genome project on *S. bicolor* BT × 623 has been started as a part of the DOE-JGI Community Sequencing Program (CSP). Genome information on a genome browser and relevant data including nucleotide and peptide sequences, gene families, KOG, protein domains and publicly available annotations are provided in Phytozome. In GRAMENE, information on genes, transcripts and proteins is also provided via the genome browser.

Common wheat (*Triticum aestivum*) is one of the most important cereal crops. It has an allohexaploid genome composed of three subgenomes, AA, BB and DD. KOMUGI is a wheat genetic resource database managed by NBRP in Japan. NBRP/KOMUGI provides resources and information on wheat strains, genes, DNA markers, comparative maps and microarray data. The International Wheat Genome Sequencing Consortium (IWGSC) has been established for complete sequencing of the wheat genome and for understanding the structure and function of the genome. The IWGSC project consists of two main categories, sequencing and physical mapping (Paux et al., 2008).

Cassava (*M. esculenta*) is a woody shrub with edible roots containing starch. Information on full-length cDNA of cassava is available in the RIKEN Cassava Full-length cDNA database (Sakurai et al.,

2007). The Cassava Online Archive in RIKEN provides mRNA sequences and ESTs from NCBI, and additional annotations including gene function, protein functional domains and GO terms.

Citrus is an important fruit tree crop encompassing oranges, lemons, limes and grapefruit. The International Citrus Genome Consortium (ICGC) integrates citrus genomic information. The College of Agricultural and Environmental Sciences' Genomics Facility at the University of California, Davis, provides analysis data on citrus ESTs.

Sunflower (*Helianthus annuus*) is an important Asteraceae food and oil crop. Other commercially important plants in the Asteraceae include the food crops *Lactuca sativa* (lettuce) and *Tanacetum cinerariifolium* (chrysanthemum), the flower of which is the main source of the insecticide pyrethrum. The sunflower genome (with a size of about 3 Gb) has been sequenced by a collaboration among Genome Canada, Genome British Columbia, the US DOE and USDA, and France's INRA. The number of ESTs generated from sunflower is about 133,000 at present, and the genome is being sequenced with NGS technology. Genetic maps of sunflower are provided by the Compositae Genome Project (CGP).

1.11 GOAL TO GLOBAL UNDERSTANDING OF BIOLOGICAL EVENTS

With the completion of full-genome nucleotide sequencing of model plants, there is now a great need for systems that analyse these nucleotide sequence data in a public database. In addition, with improvements in technology for molecular biology, many kinds of data such as cDNA sequences, EST sequences, DNA markers, and microarray data have been produced all over the world. These developments mean that an integrated assembly system covering every piece of data is essential in coming studies of plant biology. Since any finding contributes to progress and may do so at a tipping point, any such finding can shift the equilibrium and accelerate the rate at which valuable data accumulate for a particular subdiscipline. Although the research environment for plant biology is improving, there still seems to be a large gap between single experimental studies and bioinformatics. To bridge such gaps and to fully exploit the available data, all researchers in every specialized field need to contribute in order for plant biology to advance. This is a challenge of omics in plant biology.

REFERENCES

Adams, M. D., J. M. Kelley, J. D. Gocayne et al. 1991. Complementary DNA sequencing: Expressed sequence tags and human genome project. *Science* 252:1651–1656.

Akiyama, K., E. Chikayama, H. Yuasa et al. 2008. PRIMe: A web site that assembles tools for metabolomics and transcriptomics. *In Silico Biol* 8:339–345.

Aoki, K., K. Yano, N. Sakurai et al. 2007. Expression-based approach toward functional characterization of tomato genes that have no or weak similarity to *Arabidopsis* genes. *Acta Hortic* 745:457–464.

Aranda, B., P. Achuthan, Y. Alam-Faruque et al. 2010. The IntAct molecular interaction database in 2010. *Nucleic Acids Res* 38:D525–D531.

Asamizu, E., Y. Nakamura, S. Sato and S. Tabata. 2004. Characteristics of the *Lotus japonicus* gene repertoire deduced from large-scale expressed sequence tag (EST) analysis. *Plant Mol Biol* 54:405–414.

Barrett, T., D. B. Troup, S. E. Wilhite et al. 2009. NCBI GEO: Archive for high-throughput functional genomic data. *Nucleic Acids Res* 37:D885–D890.

Benson, D. A., I. Karsch-Mizrachi, D. J. Lipman, J. Ostell and E. W. Sayers. 2011. GenBank. *Nucleic Acids Res* 39:D32–D37.

Benson, G. 2010. Editorial. *Nucleic Acids Research* annual Web Server Issue in 2010. *Nucleic Acids Res* 38:W1–W2.

Boguski, M. S., T. M. J. Lowe and C. M. Tolstoshev. 1993. dbEST—Database for "expressed sequence tags". *Nat Genet* 4:332–333.

Brazma, A., P. Hingamp, J. Quackenbush et al. 2001. Minimum information about a microarray experiment (MIAME)—Toward standards for microarray data. *Nat Genet* 29:365–371.

Brunak, S., A. Danchin, M. Hattori et al. 2002. Nucleotide sequence database policies. *Science* 298:1333.

Cannon, S. B., L. Sterck, S. Rombauts et al. 2006. Legume evolution viewed through the *Medicago truncatula* and *Lotus japonicus* genomes. *Proc Natl Acad Sci USA* 103:14959–14964.

Caspi, R., T. Altman, J. M. Dale et al. 2010. The MetaCyc database of metabolic pathways and enzymes and the BioCyc collection of pathway/genome databases. *Nucleic Acids Res* 38:D473–D479.

Cochrane, G. R. and M. Y. Galperin. 2010. The 2010 Nucleic Acids Research Database Issue and online Database Collection: A community of data resources. *Nucleic Acids Res* 38:D1–D4.

Cochrane, G., R. Akhtar, J. Bonfield et al. 2009. Petabyte-scale innovations at the European Nucleotide Archive. *Nucleic Acids Res* 37:D19–D25.

Croft, D., G. O'Kelly, G. Wu et al. 2011. Reactome: A database of reactions, pathways and biological processes. *Nucleic Acids Res* 39:D691–D697.

Eisen, M. B., P. T. Spellman, P. O. Brown and D. Botstein. 1998. Cluster analysis and display of genome-wide expression patterns. *Proc Natl Acad Sci USA* 95:14863–14868.

Fei, Z., X. Tang, R. Alba and J. Giovannoni. 2006. Tomato Expression Database (TED): A suite of data presentation and analysis tools. *Nucleic Acids Res* 34:D766–D770.

Gregory, B. D., J. Yazaki and J. R. Ecker. 2008. Utilizing tiling microarrays for whole-genome analysis in plants. *Plant J* 53:636–644.

Huang, S., R. Li, Z. Zhang et al. 2009. The genome of the cucumber, *Cucumis sativus* L. *Nat Genet* 41:1275–1281.

Ikeo, K., J. Ishi-I, T. Tamura, T. Gojobori and Y. Tateno. 2003. CIBEX: Center for information Biology gene EXpression database. *C R Biol* 326:1079–1082.

International Rice Genome Sequencing Project. 2005. The map-based sequence of the rice genome. *Nature* 436:793–800.

Jung, S., M. Staton, T. Lee et al. 2008. GDR (Genome Database for Rosaceae): Integrated web-database for Rosaceae genomics and genetics data. *Nucleic Acids Res* 36:D1034–D1040.

Kaminuma, E., T. Kosuge, Y. Kodama et al. 2011. DDBJ progress report. *Nucleic Acids Res* 39:D22–D27.

Kanehisa, M., S. Goto, M. Furumichi, M. Tanabe and M. Hirakawa. 2010. KEGG for representation and analysis of molecular networks involving diseases and drugs. *Nucleic Acids Res* 38:D355–D360.

Karp, P. D., C. A. Ouzounis, C. Moore-Kochlacs et al. 2005. Expansion of the BioCyc collection of pathway/genome databases to 160 genomes. *Nucleic Acids Res* 33:6083–6089.

Keseler, I. M., C. Bonavides-Martínez, J. Collado-Vides et al. 2009. EcoCyc: A comprehensive view of *Escherichia coli* biology. *Nucleic Acids Res* 37:D464–D470.

Kurata, N. and Y. Yamazaki. 2006. Oryzabase. An integrated biological and genome information database for rice. *Plant Physiol* 140:12–17.

Lee, Y., J. Tsai, S. Sunkara et al. 2005. The TIGR gene indices: Clustering and assembling EST and known genes and integration with eukaryotic genomes. *Nucleic Acids Res* 33:D71–D74.

Li, C., M. Courtot, N. Le Novère and C. Laibe. 2010. BioModels.net Web Services, a free and integrated toolkit for computational modelling software. *Brief Bioinform* 11:270–277.

Liang, C., P. Jaiswal, C. Hebbard et al. 2008. Gramene: A growing plant comparative genomics resource. *Nucleic Acids Res* 36:D947–D953.

Liolios, K., K. Mavromatis, N. Tavernarakis and N. C. Kyrpides. 2008. The Genomes On Line Database (GOLD) in 2007: Status of genomic and metagenomic projects and their associated metadata. *Nucleic Acids Res* 36:D475–D479.

Matsuda, F., M. Y. Hirai, E. Sasaki et al. 2010. AtMetExpress development: A phytochemical atlas of *Arabidopsis* development. *Plant Physiol* 152:566–578.

Meissner, R., Y. Jacobson, S. Melamed et al. 1997. A new model system for tomato genetics. *Plant J* 12:1465–1472.

Mueller, L. A., S. D. Tanksley, J. J. Giovannoni et al. 2005. The tomato sequencing project, the first cornerstone of the international Solanaceae project (SOL). *Comp Funct Genomics* 6:153–158.

Obayashi, T., S. Hayashi, M. Saeki, H. Ohta and K. Kinoshita. 2009. ATTED-II provides coexpressed gene networks for Arabidopsis. *Nucleic Acids Res* 37:D987–D991.

Ouyang, S., W. Zhu, J. Hamilton et al. 2007. The TIGR Rice Genome Annotation Resource: Improvements and new features. *Nucleic Acids Res* 35:D883–D887.

Parkinson, H., M. Kapushesky, N. Kolesnikov et al. 2009. ArrayExpress update—From an archive of functional genomics experiments to the atlas of gene expression. *Nucleic Acids Res* 37:D868–D872.

Paux, E., P. Sourdille, J. Salse et al. 2008. A physical map of the 1-gigabase bread wheat chromosome 3B. *Science* 322:101–104.

Rice Annotation Project. 2008. The Rice Annotation Project Database (RAP-DB): 2008 update. *Nucleic Acids Res* 36:D1028–D1033.

Sakurai, T., G. Plata, F. Rodríguez-Zapata et al. 2007. Sequencing analysis of 20,000 full-length cDNA clones from cassava reveals lineage specific expansions in gene families related to stress response. *BMC Plant Biol* 7:66.

Sakurai, T., M. Satou, K. Akiyama et al. 2005. RARGE: A large-scale database of RIKEN *Arabidopsis* resources ranging from transcriptome to phenome. *Nucleic Acids Res* 33:D647–D650.

Sato, S., Y. Nakamura, T. Kaneko et al. 2008. Genome structure of the legume, *Lotus japonicus. DNA Res* 15:227–239.

Sayers, E. W., T. Barrett, D. A. Benson et al. 2011. Database resources of the National Center for Biotechnology Information. *Nucleic Acids Res* 39:D38–D51.

Schmutz, J., S. B. Cannon, J. Schlueter et al. 2010. Genome sequence of the palaeopolyploid soybean. *Nature* 463:178–183.

Schnable, P. S., D. Ware, R. S. Fulton et al. 2009. The B73 maize genome: Complexity, diversity, and dynamics. *Science* 326:1112–1115.

Seki, M., M. Narusaka, A. Kamiya et al. 2002. Functional annotation of a full-length *Arabidopsis* cDNA collection. *Science* 296:141–145.

Shibata, D. 2005. Genome sequencing and functional genomics approaches in tomato. *J Gen Plant Pathol* 71:1–7.

Stolc, V., M. P. Samanta, W. Tongprasit et al. 2005. Identification of transcribed sequences in *Arabidopsis thaliana* by using high-resolution genome tiling arrays. *Proc Natl Acad Sci USA* 102:4453–4458.

Swarbreck, D., C. Wilks, P. Lamesch et al. 2008. The Arabidopsis Information Resource (TAIR): Gene structure and function annotation. *Nucleic Acids Res* 36:D1009–D1014.

Tatusov, R. L., N. D. Fedorova, J. D. Jackson et al. 2003. The COG database: An updated version includes eukaryotes. *BMC Bioinformatics* 4:41.

The Gene Ontology Consortium. 2008. The Gene Ontology project in 2008. *Nucleic Acids Res* 36:D440–D444.

The Rice Full-Length cDNA Consortium. 2003. Collection, mapping, and annotation of over 28,000 cDNA clones from *japonica* rice. *Science* 301:376–379.

Tuskan, G. A., S. DiFazio, S. Jansson et al. 2006. The genome of black cottonwood, *Populus trichocarpa* (Torr. & Gray). *Science* 313:1596–1604.

U N. 1935. Genome-analysis in *Brassica* with special reference to the experimental formation of *B. napus* and peculiar mode of fertilization. *Jpn J Bot* 7:389–452.

Weigel, D. and R. Mott. 2009. The 1001 genomes project for *Arabidopsis thaliana. Genome Biol* 10.107.

Wheeler, D. A., M. Srinivasan, M. Egholm et al. 2008. The complete genome of an individual by massively parallel DNA sequencing. *Nature* 452:872–876.

Wheeler, D. L., D. M. Church, S. Federhen et al. 2003. Database resources of the National Center for Biotechnology. *Nucleic Acids Res* 31:28–33.

Wing, R. A., J. S. S. Ammiraju, M. Luo et al. 2005. The *Oryza* map alignment project: The golden path to unlocking the genetic potential of wild rice species. *Plant Mol Biol* 59:53–62.

Yamada, K., J. Lim, J. M. Dale et al. 2003. Empirical analysis of transcriptional activity in the *Arabidopsis* genome. *Science* 302:842–846.

Yamamoto, N., T. Tsugane, M. Watanabe et al. 2005. Expressed sequence tags from the laboratory-grown miniature tomato (*Lycopersicon esculentum*) cultivar Micro-Tom and mining for single nucleotide polymorphisms and insertions/deletions in tomato cultivars. *Gene* 356:127–134.

Yamazaki, Y., R. Akashi, Y. Banno et al. 2010. NBRP databases: Databases of biological resources in Japan. *Nucleic Acids Res* 38:D26–D32.

Yano, K., K. Aoki and D. Shibata. 2007. Genomic databases for Tomato. *Plant Biotechnol* 24:17–25.

Yano, K., K. Imai, A. Shimizu and T. Hanashita. 2006c. A new method for gene discovery in large-scale microarray data. *Nucleic Acids Res* 34:1532–1539.

Yano, K., M. Watanabe, N. Yamamoto et al. 2006a. MiBASE: A database of a miniature tomato cultivar Micro-Tom. *Plant Biotechnol* 23:195–198.

Yano, K., T. Tsugane, M. Watanabe et al. 2006b. Non-biased distribution of tomato genes with no counterparts in *Arabidopsis thaliana* in expression patterns during fruit maturation. *Plant Biotechnol* 23:199–202.
Zhang, P., H. Foerster, C. P. Tissier et al. 2005. MetaCyc and AraCyc. Metabolic pathway databases for plant research. *Plant Physiol* 138:27–37.

CHAPTER 2

High-Throughput Approaches for Characterization and Efficient Use of Plant Genetic Resources

Jaroslava Ovesná, Anna Janská, Sylva Zelenková and Petr Maršík

CONTENTS

2.1 INTRODUCTION

The genetic resources of crop species represent a reservoir of genes conferring resistance to viral, fungal and other diseases (Kuraparthy et al., 2009; Mew et al., 2004; Poland et al., 2009; Sip et al., 2004) and tolerance to various abiotic stresses (Ashraf and Akram, 2009), as well as encoding other economically valuable traits (Ovesna et al., 2006). A number of non-crop plant species are also of value as sources of pharmaceutically active compounds (Li and Vederas, 2009) and other fine biochemicals used in industry (Carpita and McCann, 2008; Spiertz and Ewert, 2009). Germplasm accessions typically need to be characterized both in the field (Anothai et al., 2009; Janovska et al., 2007) and in the laboratory (Engindeniz et al., 2010; Hornickova et al., 2009; Nevo and Chen, 2010; Wei et al., 2010). The assessment of genetic diversity relies both on phenotype and genotype, the latter generally being accomplished using one or several DNA-based assays (Laurentin, 2009; Ovesna et al., 2001; Roussel et al., 2005; Rout and Mohapatra, 2006; Shegwu et al., 2003). A combination of genotypic and phenotypic assessments can be used to uncover an association between a genetic marker(s) and a specific trait, based on genetic linkage (Araus et al., 2008; Duran et al., 2009; Todorovska et al., 2009). Where markers can be shown to be linked to a useful trait, they can then be used as a selection tool to transfer the trait between breeding lines (Landjeva et al., 2007; Ordon et al., 2009; Varshney and Dubey, 2009).

2.2 GENOMIC APPROACHES TO MEASURING GENETIC DIVERSITY

DNA sequence polymorphism represents the basis of molecular diversity. The principle of DNA/DNA hybridization exploited by Southern (1975) and the amplification of DNA fragments achieved via the polymerase chain reaction (PCR) (Mullis et al., 1986) together underpin most of the methods of DNA analysis. DNAs are extracted either from individual plants or from bulked samples, and processed in various ways depending on the sort of analysis being undertaken (Dziechciarkova et al., 2004; Kumar et al., 2009; Park et al., 2009; Zhao et al., 2006). A number of derived analytical methods, some more cost effective than others, have been used to generate large amounts of genotypic data to be used for the quantification of the genetic diversity present in populations, germplasm collections and so on. It is DNA sequence itself, however, which represents the gold standard for the determination of genotype, and since the initial demonstrations of the feasibility of determining DNA sequence (Maxam and Gilbert, 1977; Sanger et al., 1977), a number of highly efficient platforms have been developed. A large (and growing) collection of DNA sequence obtained from various organisms is curated in various databases (NCBI, http://www.ncbi.nlm.nih.gov/genbank/, EMBL http://www.ebi.ac.uk/embl/, KOMUGI, http://www.shigen.nig.ac.jp/wheat/komugi/top/top.jsp, Grain Genes http://wheat.pw.usda.gov/GG2/index.shtml, etc.). Allelic variants can include major changes in sequence, but most frequently the changes are small, even as little as 1 bp. Such polymorphisms, dubbed 'single-nucleotide polymorphisms' (SNPs), require specific detection technologies. The simplest of these detects the presence/absence of a restriction enzyme recognition site within a PCR amplicon, which is recognized by electrophoresis through an agarose gel; more polymorphisms employ fluorescently labelled primers or dideoxyterminators (ddNTPs) in combination with capillary electrophoresis. The so-called pyrosequencing method is based on the *de novo* synthesis of a short single-stranded DNA fragment, in which the identity of each added base is detected by a chemiluminescent signal—this method has enjoyed significant popularity. Its relative simplicity has encouraged its use to define allelic variation in various germplasm collections (Novaes et al., 2008; Pacey-Miller and Henry, 2003; Polakova et al., 2003).

Comparisons of the sequences of orthologues allow inferences regarding phylogeny, while within-species comparisons of homologues can be used to hypothesize adaptation to specific environments. Increasingly, DNA sequence is being used alongside marker genotype to assess genetic diversity (Soltis et al., 2009). Full genome sequence is still only available for a limited number of organisms, as its acquisition, except for small genome organisms such as viruses and bacteria, remains a major investment. At present, some 1000 prokaryotic whole genome sequences are in the public domain (http://www.ncbi.nlm.nih.gov/genomes/lproks.cgi). The first plant genome sequence to be completed was that of the model species *Arabidopsis thaliana* (Arabidopsis Genome Initiative, 2000), but in the subsequent years this has been followed by a number of crop species, such as rice (*Oryza sativa* L.), maize (*Zea mays* L.) and grape (*Vitis vinifera* L.) (Choi et al., 2004; International Rice Genome Sequencing Project, 2005; Schmnutz et al., 2010; Troggio et al., 2007). A number of other genomes of crop species have been/are in the process of being sequenced (www.jgi.doe.gov/genome-projects/pages/projects.jsf?taxonomy = Whole + Genome&kingdom = Plant).

Until rather recently, sequencing technology has relied on single reactions, each producing a read of up to 1000 bp (Batley and Edwards, 2009). However, current sequencing platforms exploit parallel reactions, in which many thousands of sequences can be analysed in a single pass; this high-throughput approach has substantially reduced the per base pair cost of sequencing (Church, 2006; Margulies et al., 2005; Schuster, 2008) and has made feasible the acquisition of large tracts of DNA sequence (Sandberg et al., 2010). The barley genome is large in comparison with those of the model species like *A. thaliana* or rice, but is being fully sequenced using current sequencing technology (Cronn et al., 2008; Imelfort and Edwards, 2009; Steuernagel et al., 2009). Therefore, the way has now been opened for both the large-scale discovery of SNPs by re-sequencing (McKernan et al., 2009), and the direct assessment of intra- and inter-species diversity using gene sequences, rather than markers. The identification and tracking of genetic variation has become efficient and precise enough to follow thousands of variants within large populations (Varshney et al., 2009).

SNPs are abundant and uniformly distributed throughout the genome, and have become the markers of choice in human and animal plant diagnostics (Gupta et al., 2008). While SNP detection technology is now extremely efficient, SNP discovery, especially rare alleles, remains a bottleneck. Alternatives to re-sequencing have emerged from efforts to detect novel alleles following mutagenesis, particularly the technique based on heterodimer formation referred to as TILLING (Gilchrist and Haughn, 2005; McCallum et al., 2000). Along with the development of advanced analytical tools, there has also been progress in the context of targeted mutagenesis; in particular, a large panel of *A. thaliana* plants mutagenized by the introduction of T-DNA has been established. This resource simplifies the identification of the gene mutation responsible for a given observable phenotype (Martin et al., 2009; Sreelakshmi et al., 2010).

Microarrays can also be exploited for the characterization of genetic diversity, since they can readily be designed to detect SNPs (e.g., Das et al. (2008), who used Affymetrix soybean genome array to detect single-feature polymorphisms). Higher level of data hunting is represented by next-generation sequencing (NGS) techniques that are bringing new and more precise information also on genes and their variants that have not been described by previously used approaches (de Carvalho and da Silva, 2010; Gilad et al., 2009; Zhou et al., 2010). Enormous number of rather scrappy data generated by NGS nowadays places scientists before a complicated task of their processing and building up a comprehensive picture on natural variability.

Current genomic technology has become increasingly efficient at recognizing variation in DNA sequence, and thus has a prominent role to play in the characterization of plant genetic resources. For species where the full genome sequence has not yet been acquired, a number of marker technologies can be applied for this purpose. One of the most recently developed one is 'Diversity Array Technology' (DArT), which combines the principles of amplified fragment-length polymorphism (AFLP), hybridization and DNA arrays (Wenzl et al., 2004). In barley, DArT markers have been linked genetically with other well-established markers, such as microsatellites and RFLPs (restriction fragment-length polymorphisms) (Wenzl et al., 2006), and attempts have already been made to correlate phenotypic variation with various DArT markers (Akbari et al., 2006).

2.3 TRANSCRIPTOMICS

Global patterns of gene expression (or the 'transcriptome') can be determined from the RNA content of the cell. This aspect of genomics is referred to as transcriptomics. While the nucleus of each cell carries the full complement of the organism's genes, not all cells show an identical pattern of gene expression; rather these tend to be specific to cell type, developmental stage, and are affected by the external environment. The expression of many genes is regulated by transcription factors and various gene-silencing mechanisms (miRNA and siRNA). Whole transcriptome analysis platforms were initially developed in order to understand the predominant mechanisms and processes underlying growth and development, responses to biotic and abiotic stresses and plant ecology. The principle of these platforms is to capture the identity and assess the quantity present of every gene transcribed under set environments and/or at a set developmental stage or particular tissue/organ of the plant. The principal technology exploited for this purpose is the microarray. Because the probe sequences mounted on a microarray (typically cDNAs, PCR amplicons or synthetic oligonucleotides) are intended to hybridize with specific genes, some prior knowledge of genome sequence is necessary. The identity and type of probe sequence, along with the means by which the chip is manufactured, vary somewhat from platform to platform. Two prominent products are the Illumina Bead array system (http://www.illumina.com) and the Affymetrix photolithography-based system (http://www.affymetrix.com). Microarray analysis can give a substantial first insight into the transcriptome, because it reflects the expression of thousands of genes in a single pass. However, microarray data tend to be relative (i.e., they provide a comparison between two or more samples) rather than quantitative in nature. Their specific advantage is that they identify small subsets of genes for which a more detailed picture of gene expression is subsequently obtained using real-time PCR.

A rather different approach to transcriptomic analysis is based on the parallel sequencing of populations of cDNA. The serial analysis of gene expression (SAGE) platform, developed in 1995 (Velculescu et al., 1995), was designed to identify the components of a cDNA population from short stretches of sequence. These when concatenated allow the acquisition of many tagged sequences from a single sequencing reaction. In an updated version ('SuperSAGE'), the tagged sequences are longer, thus reducing the risk of mis-identification of the target genes. The principle behind the quantification of expression achieved by SAGE is a count of the frequency with which the same tag reoccurs. Thus, unlike the microarray platform, unknown transcripts can be detected, and an absolute abundance of each transcript is generated (http://www.sagenet.org). Whole transcriptome shotgun sequencing (also referred to as RNA-Seq) works on a similar principle, but exploits the potential of parallel sequencing. Some have speculated that these platforms spell the beginning of the end for the microarray (Shendure, 2008), but their cost remains uncompetitive as yet, and so microarrays continue to be widely used for the study of many aspects of plant development (e.g., Aprile et al., 2009; Endo et al., 2009; Fujino and Matsuda, 2010; Jain et al., 2010; Sarkar et al., 2009; Seo et al., 2009; Wang et al., 2010) and the responses to biotic (e.g., Desmond et al., 2008) and abiotic stresses at the cell, organ or whole-plant level (see Table 2.1).

An improved understanding of both the acclimation process to stress and of the mechanisms used for recovery from stress, is a key research area in the field of breeding for improved abiotic stress tolerance. The lack of a proper understanding of their regulation continues to hamper efforts to tackle

Table 2.1 Examples of Papers Focused on Abiotic Stress of Cereals Using Microarrays

Cereal Species	Stress Conditions	References
Barley (*Hordeum vulgare*)	Drought, salt	Ozturk et al. (2002)
	Cold, dehydration, salinity, high light, copper and their combinations	Atienza et al. (2004)
	Osmotic stress	Ueda et al. (2004)
	Cold acclimation	Svensson et al. (2006)
	Salt stress	Ueda et al. (2006)
	Dehydration shock and drought stress	Talame et al. (2007)
	Drought	Guo et al. (2009)
Wheat (*Triticum aestivum*)	Salt	Mott and Wang (2007)
	Drought	Xue et al. (2008)
	Drought	Mohammadi et al. (2008)
	Salt	Kawaura et al. (2008)
	Aluminium	Houde et al. (2008)
	Drought	Aprile et al. (2009)
	Cold	Kocsy et al. (2010)
Rice (*Oryza sativa*)	High temperature	Sarkar et al. (2009)
	Cold, salt, dehydration	Sharma et al. (2009)
	High temperature	Endo et al. (2009)
	Cadmium	Ogawa et al. (2009)
	Oxidative stress	Liu et al. (2010)
	Low phosphorus	Li et al. (2010)
	Drought	Ning et al. (2010)
	Oxidative stress	Frei et al. (2010)

this major breeding goal. Abiotic stress tolerance is typically quantitatively inherited, and interacts with other traits and pathways. Some components of wheat's responses to drought, salinity and freezing stress are similar to one another (Langridge et al., 2006), while other responses, such as those involved in stomatal closure in the face of heat and drought stress, can be antagonistic (Rizhsky et al., 2002). Much information has been acquired regarding stress tolerance mechanisms and the genes underlying them in *A. thaliana*, but the transferability of these data to crop species, especially the monocotyledons, has proven to be unreliable. Wheat and barley are far more drought and cold tolerant than *A. thaliana*; so it would seem more logical to concentrate on mechanisms and genes present in these species themselves. While whole genome sequences are not yet to hand for these cereals, those of both rice and *Brachypodium distachyon* are, and these species are closely enough related to wheat and barley to allow many inferences regarding gene content (Moore et al., 1995).

Microarray analyses have been used to generate a substantial amount of genetic data regarding the acclimative response of barley (Close et al., 2004), and other cereals (Table 2.1). However, still many questions surrounding cereal abiotic stress tolerance remain to be resolved. Cold and drought acclimation induces a number of enzymes involved in osmolyte metabolism, especially those responsible for the accumulation of soluble sugars (sucrose, raffinose, stachyose, trehalose), sugar alcohols (sorbitol, ribitol, inositol), the amino acid proline and betaines (Gusta et al., 2004; Hare et al., 1998; Iba, 2002; Wang et al., 2003). Because many abiotic and certain biotic stresses involve oxidative stress, the expression of the anti-oxidant enzymes superoxide dismutase, glutathione peroxidase, glutathione reductase, ascorbate peroxidase and catalase, as well as that of the non-enzymatic anti-oxidants tripeptidthiol, glutathione, ascorbic acid and α-tocoferol, is promoted (Chen and Li, 2002). Both cold and drought stress also compromise protein and membrane stability; so acclimation to both stresses is associated with the synthesis of chaperons and various heat-shock proteins, as well as dehydrins and particular cold-regulated or drought-responsive proteins. All these proteins participate in the stabilization of the plasma membrane, membrane or cytosolic proteins, and are active in the scavenging of reactive oxygen species (Guo et al., 2009; Iba, 2002). Changes also occur at the level of lipid metabolism. At low temperatures, unsaturated fatty acids of phospholipids are preferentially synthesized, while there is evidence that the activity of phospholipase D (involved in lipid catabolism) is suppressed (Rajashekar, 2000). Some plant species avoid tissue damage caused by extracellular ice formation by synthesizing antifreeze proteins, which bind to the ice crystals and change their shape or reduce their size (Griffith et al., 1992). These proteins are highly similar to the so-called plant pathogen-related (PR) proteins (Moffatt et al., 2006).

Transcriptional microarrays can also be applied to detect the activity of transcription factors and other signalling sequences involved in the regulation of the acclimation response. The C-repeat-binding factor/dehydration-responsive element-binding protein (CBF/DREB) transcription factor is part of the cold-responsive signalling pathway in plants, and has been extensively explored in *A. thaliana*. The same signal transduction pathway is also active in both crop species and in woody plants (e.g., Kayal et al., 2006; Navarro et al., 2009). In *A. thaliana*, the CBF/DREB1 transcription factor is associated with the low-temperature response, whereas DREB2 is involved in osmotic stress signalling (Nakashima and Yamaguchi-Shinozaki, 2006). The freezing tolerance of potato was successfully improved by the over-expression of *AtCBF* genes driven by a stress-inducible promoter (Pino et al., 2007). The over-expression of *AtCBF1* alone also enhanced freezing tolerance (Pino et al., 2008). Similarly, freezing tolerance in maize was improved by the transgenic over-expression of *AtCBF3* (Al-Abed et al., 2007). Tobacco plants over-expressing the rice orthologue *OsCBF1* showed not only an improved freezing and oxidative stress tolerance, but also up-regulated the expression of *PR* genes, suggesting that this transcription factor plays a role in both the abiotic and biotic stress response (Gutha and Reddy, 2008). Both Fowler and Thomashow (2002) and Svensson et al. (2006), however, were able to show that the majority of cold-regulated genes are *CBF*-independent; rather they are chloroplast and light dependent. Similarly, Crosatti et al. (1999) suggested that the chloroplast was the primary sensor of cold/light stress in barley. The excitation of photosystem II provides a stimulus for the expression of cold-regulated genes (Gray et al., 1997; Ndong et al., 2001). In effect, therefore, full cold acclimation appears to require both *CBF* activity, and the contribution of certain chloroplast factors (Svensson et al., 2006). Few examples have yet been provided in the literature of the use of SAGE to characterize the abiotic stress response of plants (but see Lee and Lee, 2003), although it has been more commonly applied to determine the molecular response to biotic stress (Fregene et al., 2004; Thomas et al., 2002; Uehara et al., 2007). The RNA-Seq platform is as yet restricted to animal and human research.

2.4 PROTEOMICS

Plant proteomics is a relatively new research field. It seeks to analyse the full set of proteins present in a specific organ, tissue, cell or organelle. High-throughput proteomics has become possible through technical advances in protein separation, mass spectrometry, genome sequencing/annotation and protein search algorithms (Thelen, 2007). The rapidly growing influence of proteomics, both in basic and applied plant research, demonstrates the centrality of this research field in current plant science. The application of proteomics has led to the discovery of a number of important proteins in genetic resources accessions, and has facilitated attempts to deduce their influence over the determination of economic yield and tolerance to environmental stress (Mochida and Shinozaki, 2010).

The two basic technologies required for proteomic analysis are gel-based electrophoresis and mass-spectrometry-based peptide identification. Both have become increasingly automated and so less time consuming. Further innovations, notably high-resolution two-dimensional electrophoresis, multidimensional protein identification technology (MudPit), fluorescent difference gel electrophoresis (DIGE), isotope-coded affinity tag (ICAT) and isobaric tag for relative and absolute quantification (iTRAQ), have recently been established. These promise to increase the opportunities to apply a proteomic approach to diverse areas within plant and plant breeding research (Agrawal et al., 2006; Lee and Cooper, 2006; Renaut et al., 2006; Roberts, 2002; Rose et al., 2004; Salekdeh and Komatsu, 2007; Shulaev and Olivier, 2006).

Large-scale, high-throughput approaches are becoming an essential part of the study of complex biological systems. In addition to two-dimensional electrophoresis and mass spectrometry, antibodies, yeast two-hybrid technology and protein microarrays have been deployed to acquire relevant data. The latter, in particular, are emerging as a valuable tool for the profiling and functional characterization of cDNA products (Glockner and Angenendt, 2003; Salekdeh and Komatsu, 2007). Protein arrays have also been combined with immobilized antibodies. The microarray format is of especial interest because of its capacity for parallel data analysis, miniaturization and automatization. Thus, protein arrays have the potential for high-throughput identification of novel protein markers. A pioneering plant protein microarray consisted of a set of 95 immobilized *A. thaliana* proteins (Kersten et al., 2002), followed by Kramer et al.'s (2004) demonstration of a microarray-based assay for barley kinase CK2α. A better-established means of characterizing protein function is the yeast two-hybrid system, which is designed to identify protein–protein interactions, and to allow for the straightforward isolation of the genes encoding the interacting partners. A high-throughput version of this technology has been applied to a set of >1000 *A. thaliana* cDNAs, representing a range of signal transduction pathways. The hope is that such large-scale two-hybrid screens will lead to the development of protein interaction maps in crop species, and thereby facilitate novel insights into the metabolic, regulatory and signal transduction pathways involved in the expression of agronomically significant traits. As more complete genome sequences are acquired, there is an expectation that high-throughput technologies of this kind will help to clarify the mechanisms underlying many key plant processes (Kersten et al., 2002).

The proteome of *A. thaliana* remains the best understood of all plant species, and special attention has been focused on the gene products of its various plant cell organelles and compartments.

Proteomic approaches have also been applied to its pollen and seed development. Bearenfaller et al. (2008) created the first *A. thaliana* organ proteome map based on a population of 13,000 distinct proteins, and used it to examine correlations between transcriptomic and proteomic data, and to identify proteins specifically expressed in particular organs. Since the proteome is the sum total of all proteins present in a cell/tissue/organ/plant at a particular time, the expectation is that it will be closely associated with the plant's phenotype.

Outside of *A. thaliana*, a growing volume of proteomic data has been collected from the major crop species—rice, wheat, barley, maize, soybean and oilseed rape (Baginsky and Gruissem, 2007; Bourguignon and Jaquinod, 2008; Hajduch et al., 2007; Hirano, 2007; Hurkman et al., 2008; Komatsu, 2006; Nagaraj et al., 2008). The number of protein spots resolved by two-dimensional electrophoresis depends greatly on the choice of plant tissue/organ and species, but typically ranges from a few hundred and to ~2000. While rare proteins may be present at a concentration of as little as 10 molecules per cell, as many as 10^7 copies of some abundant ones may be present (Lee and Cooper, 2006). Significant progress has been made towards the identification of these proteins in rice (Moreno-Risueno et al., 2010; Salekdeh and Komatsu, 2007). Although the estimated number of genes present in rice lies in the range of 35,000–60,000, up to 500,000 different proteins may be produced by the plant (Lee and Cooper, 2006). Many of these variants reflect various post-translational modifications (PTMs), of which some 300 types have been observed, including phosphorylation, glycosylation, sulphation, acetylation, myristoylation, palmitoylation, methylation, prenylation and ubiquitination. Phosphorylation is the commonest PTM of proteins involved in regulatory and signalling activity. PTM can affect both the activity and the binding of a protein, and thereby alter its biological role. Unravelling the complexity of PTM and linking it to physiological phenomena remain major challenges for the study of signalling and gene regulation. High-throughput proteomics represents a link between genomics, genetics and physiology (Zivy and de Vienne, 2000), because of its capacity to define the pleiotropic effects of single gene mutants. Changes in a proteomic profile can be used to indicate the involvement of a specific protein(s) in a given metabolic pathway. The benefit of a proteomic approach is that it not only enables the identification of the structural genes involved in the determination of a given trait, but also the loci controlling their expression.

The suitability of proteomic data sets to clarify the complexities of plant metabolism and its regulation means that they can be expected to contribute significantly to our understanding of the physiology of the plant response to stress, and through this, leads to identifying stress-tolerant germplasm. The hope is that proteomic experiments will increasingly be able to identify key proteins, which can then be used as markers in abiotic stress breeding programmes.

A major responsibility of gene banks is the characterization of the accessions they curate. A deal of such characterization has been dedicated to cataloguing the seed storage protein profiles of both wild and cultivated cereals (particularly rice, barley, maize, rye and wheat) (Agrawal and Rakwal, 2006; Consoli et al., 2002; Hirano, 2007). A more proteomics-based approach has been employed to determine the identity of hundreds of proteins synthesized during seed filling in oilseed rape, soybean and castor (Catusse et al., 2008). Similarly, global proteomic assays have been applied to barley in a search for ways to predict favourable malting quality and to understand the molecular basis for this trait (Perrocheau et al., 2005). A similar approach has been taken in wheat, where the research question revolves around the baking quality of the flour, and one outcome of these experiments has been an insight into many of the biochemical processes taking place during the development of the wheat grain (Skylas et al., 2005).

Genetic manipulation can generate a variety of changes at the protein level, and some of these, particularly in rice and wheat, have been successfully detected and characterized via a proteomic approach. As a result, proteomic technology is viewed by some regulatory bodies as providing an objective means of defining the differences between genetically modified and non-modified food (Horváth-Szanicz et al., 2005). Proteomic analysis has also been used to describe the individual proteinaceous components of secondary metabolite biosynthesis, because of the complexity of the enzymes involved (Jacobs et al., 2000).

Large-scale proteomic methods have been used as a means of assessing genetic similarity between gene bank accessions as well as in similar experiments comparing the proteomes of mutants with that of the wild type (Agrawal and Rakwal, 2006). A contribution of proteomics to crop breeding could begin with the identification of a set of stress-responsive proteins from comparisons made between stressed and non-stressed plants. Following this, the critical question to address is whether inter-accession variation for the expression of any of these proteins can be correlated with superior performance when challenged by stress. This correlation can later be verified by taking a genetic mapping approach. In this way, proteomics has contributed to the important area of marker-assisted selection, especially in the field of abiotic stress tolerance breeding.

Forecast changes in the global climate and the continuing increase in the human population raise major issues of the sustainability of agricultural production. Proteomic analysis of the plant stress response may enable the identification of multiple proteins associated with various types of biotic and/or abiotic stress. Proteomic descriptions of the plant response to a series of abiotic stresses (high temperature, salinity, drought, excess light, ozone, heavy metal contamination, herbicide and soil mineral depletion) have been produced, and similarly to a large range of biotic stresses, including aerial and soil microbial pathogens, insects, nematodes and beneficial microorganisms such as arbuscular mycorrhizal fungi and nitrogen-fixing bacteria (Amme et al., 2006; Hashimoto and Komatsu, 2007; Kim and Kang, 2008; Vincent and Zivy, 2007). The detailed proteomes of several model species are in the public domain (Qureshi et al., 2007), giving many leads for the major protein types over-expressed as a result of a particular treatment, and informing as to the overall impact of such treatments on cellular metabolism and protein localization. A number of proteins have emerged as potential cellular markers of stress. These include actin depolymerizing factor, various oxidative stress defence enzymes (e.g., superoxide dismutase, ascorbate peroxidase, dehydroascorbate reductase, peroxiredoxin), small heat-shock proteins, nucleoside diphosphate kinase, RuBisCO activase, the RuBisCO subunits, PR proteins, phytochelatins and metallothioneins. The diversity of these proteins underlines the complexity of the plant response to environmental challenge, and suggests strategies for the identification of markers for stress tolerance of utility to plant breeders.

Thus, in summary, high-throughput proteomic approaches enable the acquisition of a detailed picture of the cultivated plant proteome, and represent a new generation of biotechnological tools of relevance to plant breeding. In some respects proteomics offers an alternative, or certainly a complementary technology to genotypic analysis for the monitoring of biodiversity and screening of genetic resources. The technology links the transcriptome and the metabolome in the context of research into the response of the plant to abiotic and biotic stresses. Integration of plant proteomics with systems biology can be expected to enhance the understanding of the signalling and metabolic networks underlying plant growth, development and interaction with the environment, with the hope that it may ultimately become possible to engineer plant processes important for crop yield, nutrition and defense against pathogens. The large-scale data sets generated by proteomics research represent an important building block for deciphering gene function and describing complex biological systems. In a number of ways, therefore, proteomics promises to contribute to the development of tomorrow's productive and sustainable agriculture.

2.5 METABOLOMICS

The term metabolomics describes the analysis of the metabolic content of a cell, both in terms of identifying and quantifying the individual molecules present (Sumner et al., 2003). The metabolome is sensitive to the external environment and changes over development, and this represents a dynamic system under the influence of both exogenous and endogenous stimuli (Harrigan et al., 2007). This fluidity raises the possibility that a description of the metabolome could be correlated with aspects of plant phenotype, and that its manipulation might have potential as a breeding tool (Fernie and Schauer, 2008; Larkin and Harrigan, 2007). It has even been suggested that the metabolome reflects the biology of a plant more accurately than either the transcriptome or the proteome (Fiehn, 2002). The metabolomics concept is sometimes targeted to specific types of compounds, rather than applied *in senso strictu* to the entire complement of metabolic products (Goodcare et al., 2004). The former focus on either a specific class of metabolites (e.g., lipids, sugars or organic acids (see Chen et al., 2003; Harrigan et al., 2007), and have spawned a number of sub-disciplines, such as lipidomics (Welti et al., 2007) and glycomics (Feizi et al., 2003). Particular groups of crop metabolites have been identified for the purpose of metabolic engineering, breeding and the classification of cultivars (Bovy et al., 2007; Fernie and Schauer, 2008). In contrast, global analyses of metabolic content are predicated on no particular *a priori* hypothesis, and generate profiles depending on the analytical method chosen by the experimenter (Harrigan et al., 2007). A major advantage of a non-targeted method such as fingerprinting is simplicity of sample preparation, and possibly a reduction in the time needed for analysis; this then allows for the measurement of larger numbers of biological and technical samples, thereby improving the statistical validity of the conclusions (Enot and Draper, 2007). An option favoured by Goodcare et al. (2004) is to initially profile the material in a non-targeted way, and then to subsequently perform more specific and quantitative measurements of selected metabolites using more exacting analytical methods (Catchpole et al., 2005).

The measurement of primary metabolites involves the use of almost a full set of available analytical methods (Sumner et al., 2003). Mass spectrometry is the commonest analytical platform in metabolomic studies, and is frequently combined with a separation method(s) such as chromatography and/or capillary electrophoresis (Harada and Fukusaki, 2009; Schauer et al., 2005). In particular, the combination of mass spectrometry with gas chromatography (GC/MS) allows for a relatively simple interpretation of spectral data within a low-maintenance technical environment (Jonsson et al., 2005). However, this platform requires that all analytes are volatile, lipophilic and that their molecular weight does not exceed 600 Da. Some of these restrictions can be overcome by derivatization (silylation, methylation or methoxymination) (Hope et al., 2005), although this pre-treatment necessarily reduces the precision of the assay. A better solution is offered by the combination of mass spectrometry with liquid chromatography (LC/MS), which can cope with polar non-volatile molecules of higher molecular weight (Villas-Boas et al., 2005). Although LC/MS is a more technically demanding technology, it has become the staple platform for metabolomic applications (t'Kindt et al., 2008). An alternative, widely used analytical platform is nuclear magnetic resonance (NMR), which offers a simple protocol for sample preparation, and a good level of identification for well-characterized metabolites (Choi et al., 2004; Fumagalli et al., 2009). Its major disadvantage is the high purchase and operating cost of the equipment. An NMR-based approach was used for the detection of biomarkers induced by xenobiotic agents (Bailey et al., 2000) in maize. Other analytical methods include ultraviolet absorption spectrometry, infrared spectrometry, evaporative light scattering detection and fluorescent or pulsed amperometric detection are generally used as supplementary techniques to NMR and LC/MS (or GC/MS). Nevertheless, they could often be full-value equivalent to NMR and mass spectrometry (Chen et al., 2003). The post-acquisition processing of metabolomic data is a critical part of the analysis. This can involve data standardization, transformation and correction, as well as data set reduction, before any attempt can be made to identify and quantify the individual components of the metabolome (van den Berg et al., 2006). Outputs are generally interpreted in conjunction with transcriptomic and proteomic data, and can be applied for association mapping (Fernie and Schauer, 2008; Zhao et al., 2005).

2.6 CONCLUSIONS

Plants generally serve as the reservoir of food, feed, industrial fibres or phytochemicals for mankind. Diversity of plant species and their ability to adopt upon environmental stimuli have to be therefore investigated to fully understand principles of evolution and impact of selection. Nowadays, many analytical tools are available to study diversity on different level ranging from genome up to metabolome to answer the questions.

ACKNOWLEDGEMENTS

The work was supported by the Czech Republic National Agency for Agricultural Research (project no. QH 81287) and by the Czech Ministry of Agriculture (project no. Mze0002700604).

REFERENCES

Agrawal, G. K. and R. Rakwal. 2006. Rice proteomics: A cornerstone for cereal food crop proteomes. *Mass Spectrom Rev* 25:1–53.

Agrawal, G. K., N.–S. Jwa, Y. Iwahashi, M. Yonekura and R. Rakwal. 2006. Rejuvenating rice proteomics: Fact, challenges, and vision. *Proteomics* 6:5549–5576.

Akbari, M., P. Wenzl, V. Caig et al. 2006. Diversity arrays technology (DArT) for high-throughput profiling of the hexaploid wheat genome. *Theor Appl Genet* 113:1409–1420.

Al-Abed, D., P. Madasamy, R. Talla, S. Goldman and S. Rudrabhatla. 2007. Genetic engineering of maize with the *Arabidopsis* DREB1A/CBF3 gene using split-seed explants. *Crop Sci* 47:2390–2402.

Amme, S., A. Matros, B. Schlesier and H. P. Mock. 2006. Proteome analysis of cold stress response in *Arabidopsis thaliana* using DIGE-technology. *J Exp Bot* 57:1537–1546.

Anothai, J., A. Patanothai, K. Pannangpetch, S. Jogloy, K. J. Boote and G. Hoogenboom. 2009. Multi-environment evaluation of peanut lines by model simulation with the cultivar coefficients derived from a reduced set of observed field data. *Field Crop Res* 110:111–122.

Aprile, A., A. M. Mastrangelo, A. M. De Leonardis et al. 2009. Transcriptional profiling in response to terminal drought stress reveals differential responses along the wheat genome. *BMC Genomics* 10:279.

Arabidopsis Genome Initiative. 2000. Analysis of the genome sequence of the flowering plant *Arabidopsis thaliana*. *Nature* 408:796–815.

Araus, J. L., G. A. Slafer, C. Royo and M. D. Serret. 2008. Breeding for yield potential and stress adaptation in cereals. *Crit Rev Plant Sci* 27:377–412.

Ashraf, M. and N. A. Akram. 2009. Improving salinity tolerance of plants through conventional breeding and genetic engineering: An analytical comparison. *Biotechnol Adv* 27:744–752.

Atienza, S. G., P. Faccioli, G. Perrotta et al. 2004. Large scale analysis of transcripts abundance in barley subjected to several single and combined abiotic stress conditions. *Plant Sci* 167:1359–1365.

Baerenfaller, K., J. Grossmann, M. A. Grobel et al. 2008. Genome-scale proteomics reveals *Arabidopsis thaliana* gene models and proteome dynamics. *Science* 320:938–941.

Baginsky, S. and W. Gruissem. 2007. Current status of *Arabidopsis thaliana* proteomics. In *Plant Proteomics*, eds. J. Šamaj and J. J. Thelen, pp. 105–120. Berlin: Springer-Verlag.

Bailey, N. J. C., P. D. Stanley, S. T. Hadfield, J. C. Lindon and J. K. Nicholson. 2000. Mass spectrometrically detected directly coupled high performance liquid chromatography/nuclear magnetic resonance spectroscopy/mass spectrometry for the identification of xenobiotic metabolites in maize plants. *Rapid Commun Mass Spectrosc* 14:679–684.

Batley, J. and D. Edwards. 2009. Genome sequence data: Management, storage, and visualization. *Biotechniques* 46:333–336.

Bourguignon, J. and M. Jaquinod. 2008. An overview of the *Arabidopsis* proteome. In *Plant Proteomics. Technologies, Strategies and Application*, eds. G. K. Agarwal and R. Rakwal, pp. 143–164. Hoboken, NJ: John Wiley & Sons Inc.

Bovy, A., E. Schijlen and R. D. Hall. 2007. Metabolic engineering of flavonoids in tomato (*Solanum lycopersicum*): The potential for metabolomics. *Metabolomics* 3:399–412.

Carpita, N. C. and M. C. McCann. 2008. Maize and sorghum: Genetic resources for bioenergy grasses. *Trends Plant Sci* 13:415–420.

Catchpole, G. S., M. Beckmann, D. P. Enot et al. 2005. Hierarchical metabolomics demonstrates substantial compositional similarity between genetically modified and conventional potato crops. *Proc Natl Acad Sci USA* 102:14458–14462.

Catusse, J., L. Rajjou, C. Job and D. Job. 2008. Proteome of seed development and germination. In *Plant Proteomics. Technologies, Strategies and Application*, eds. G. K. Agarwal and R. Rakwal, pp. 191–206. Hoboken, NJ: John Wiley & Sons, Inc.

Chen, F., A. L. Duran, J. W. Blount, L. W. Sumner and R. A. Dixon. 2003. Profiling phenolic metabolites in transgenic alfalfa modified in lignin biosynthesis. *Phytochemistry* 64:1013–1021.

Chen, W. P. and P. H. Li. 2002. Attenuation of reactive oxygen production during chilling in ABA-treated maize cultured cells. In *Plant Cold Hardiness*, eds. C. Li and E. T. Palva, pp. 223–233. Dordrecht: Kluwer Academic Publishers.

Choi, H. K., D. Kim, T. Uhm et al. 2004. A sequence-based genetic map of *Medicago truncatula* and comparison of marker colinearity with *M. sativa*. *Genetics* 166:1463–1502.

Choi, Y. H., E. C. Tapias, H. K. Kim et al. 2004. Metabolic discrimination of *Catharanthus roseus* leaves infected by phytoplasma using 1H-NMR spectroscopy and multivariate data analysis. *Plant Physiol* 135:2398–2410.

Church, G. M. 2006. Genomes for all. *Sci Am* 294:46–54.

Close, T. J., S. I. Wanamaker, R. A. Caldo et al. 2004. A new resource for cereal genomics: 22K Barley GeneChip comes of age. *Plant Physiol* 134:960–968.

Consoli, L., A. Lefevre, M. Zivy, D. de Vienne and C. Damerval. 2002. QTL analysis of proteome and transcriptome variations for dissecting the genetic architecture of complex traits in maize. *Plant Mol Biol* 48:575–581.

Cronn, R., A. Liston, M. Parks, D. S. Gernandt, R. Shen and T. Mockler 2008. Multiplex sequencing of plant chloroplast genomes using Solexa sequencing-by-synthesis technology. *Nucleic Acids Res* 36:e122.

Crosatti, C., P. P. De Laureto, R. Bassi and L. Cattivelli. 1999. The interaction between cold and light controls the expression of the cold-regulated barley gene *cor14b* and the accumulation of the corresponding protein. *Plant Physiol* 119:671–680.

Das, S., P. R. Bhat, C. Sudhakar et al. 2008. Detection and validation of single feature polymorphisms in cowpea (*Vigna unguiculata* L. Walp) using a soybean genome array. *BMC Genomics* 9:107.

De Carvalho, M. C. D. G. and D. C. G. da Silva. 2010 Next generation DNA sequencing and its applications in plant genomics. *Cienc Rural* 40:735–744.

Desmond, O. J., J. M. Manners, P. M. Schenk, D. J. Maclean and K. Kazan. 2008. Gene expression analysis of the wheat response to infection by *Fusarium pseudograminearum*. *Physiol Mol Plant P* 73:40–47.

Duran, C., N. Appleby, D. Edwards and J. Batley. 2009. Molecular genetic markers: Discovery, applications, data storage and visualisation. *Curr Bioinform* 4:16–27.

Dziechciarkova, M., A. Lebeda, I. Dolezalova and D. Astley. 2004. Characterization of *Lactuca* spp. germplasm by protein and molecular markers—A review. *Plant Soil Environ* 50:47–58.

Endo, M., T. Tsuchiya, K. Hamada et al. 2009. High temperatures cause male sterility in rice plants with transcriptional alterations during pollen development. *Plant Cell Physiol* 50:1911–1922.

Engindeniz, S., I. Yilmaz, E. Durmusoglu, B. Yagmur, R. Z. Eltez, B. Demirtas, D. Engindeniz and A. H. Tatarhan. 2010. Comparative input analysis of greenhouse vegetables. *Ekoloji* 19:122–130.

Enot, D. and J. Draper. 2007. Statistical measures for validating plant genotype similarity assessments following multivariate analysis of metabolome fingerprint data. *Metabolomics* 3:349–355.

Feizi, T., F. Fazio, W. C. Chai and C. H. Wong. 2003. Carbohydrate microarrays—A new set of technologies at the frontiers of glycomics. *Curr Opin Struc Biol* 13:637–645.

Fernie, A. R. and N. Schauer. 2008. Metabolomics-assisted breeding: A viable option for crop improvement? *Trends Genet* 25:39–48.

Fiehn, O. 2002. Metabolomics—The link between genotype and phenotype. *Plant Mol Biol* 48:155–171.

Fowler, S. and M. F. Thomashow. 2002. Arabidopsis transcriptome profiling indicates that multiple regulatory pathways are activated during cold acclimation in addition to the CBF cold response pathway. *Plant Cell* 14:1675–1690.

Fregene, M., H. Matsumura, A. Akano, A. Dixon and R. Terauchi. 2004. Serial analysis of gene expression (SAGE) of host–plant resistance to the cassava mosaic disease (CMD). *Plant Mol Biol* 56:563–571.

Frei, M., J. P. Tanaka, C. P. Chen and M. Wissuwa. 2010. Mechanisms of ozone tolerance in rice: Characterization of two QTLs affecting leaf bronzing by gene expression profiling and biochemical analyses. *J Exp Bot* 61:1405–1417.

Fujino, K. and Y. Matsuda 2010. Genome-wide analysis of genes targeted by qLTG3–1 controlling low-temperature germinability in rice. *Plant Mol Biol* 72:137–152.

Fumagalli, E., E. Baldoni, P. Abbruscato et al. 2009. NMR techniques coupled with multivariate statistical analysis: Tools to analyse *Oryza sativa* metabolic content under stress conditions. *J Agron Crop Sci* 195:77–88.

Gilad, Y., J. K. Pritchard and K. Thornton, K. 2009. Characterizing natural variation using next-generation sequencing technologies. *Trends Genet* 25:463–471.

Gilchrist, E. J. and G. W. Haughn. 2005. TILLING without a plough: A new method with applications for reverse genetics. *Curr Opin Plant Biol* 8:211–215.

Glockner, J. and P. Angenendt. 2003. Protein and antibody microarray technology. *J Chromatogr B* 797:229–240.

Goodcare, R., S. Vaidyanathan, W. B. Dunn, G. G. Harrigan and D. B. Kell. 2004. Metabolomics by numbers: Acquiring and understanding global metabolite data. *Trends Biotechnol* 5:245–252.

Gray, G. R., L.-P. Chauvin, F. Sarhan and N. P. A. Huner. 1997. Cold acclimation and freezing tolerance: A complex interaction of light and temperature. *Plant Physiol* 114:467–474.

Griffith, M., C. Lumb, P. Ala, D. S. C. Yang, W.-C. Hon and B. A. Moffatt. 1992. Antifreeze protein produced endogenously in winter rye leaves. *Plant Physiol* 100:593–596.

Guo, P., M. Baum, S. Grando et al. 2009. Differentially expressed genes between drought-tolerant and drought-sensitive barley genotypes in response to drought stress during the reproductive stage. *J Exp Bot* 60:3531–3544.

Gupta, P. K., S. Rustgi and R. R. Mir. 2008. Array-based high-throughput DNA markers for crop improvement. *Heredity* 101:5–18.

Gusta, L. V., M. Wisniewski, N. T. Nesbitt and M. L. Gusta. 2004. The effect of water, sugars, and proteins on the pattern of ice nucleation and propagation in acclimated and nonacclimated canola leaves. *Plant Physiol* 135:1642–1653.

Gutha, L. R. and A. R. Reddy. 2008. Rice DREB1B promoter shows distinct stress-specific responses, and the overexpression of cDNA in tobacco confers improved abiotic and biotic stress tolerance. *Plant Mol Biol* 68:533–555.

Hajduch, M., G. K. Agarwal and J. J. Thelen. 2007. Proteomics of seed development in oilseed crops. In *Plant Proteomics*, eds. J. Šamaj and J. J. Thelen, pp. 137–154. Berlin: Springer-Verlag.

Harada, K. and E. Fukusaki 2009. Profiling of primary metabolite by means of capillary electrophoresis–mass spectrometry and its application for plant science. *Plant Biotechnol* 26:47–52.

Hare, P. D., W. A. Cress and J. Van Staden. 1998. Dissecting the roles of osmolyte accumulation during stress. *Plant, Cell Environ* 21:535–553.

Harrigan, G. G., S. Martino-Catt and K. C. Glenn. 2007. Metabolomics, metabolic diversity and genetic variation in crops. *Metabolomics* 3:259–272.

Hashimoto, M. and S. Komatsu. 2007. Proteomic analysis of rice seedling during cold stress. *Proteomics* 7:1293–1302.

Haynes, P. A. and T. H. Roberts. 2007. Subcellular shotgun proteomics in plants: Looking beyond the usual suspects. *Proteomics* 7:2963–2975.

Hirano, H. 2007. Cereal proteomics. In *Plant Proteomics*, eds. J. Šamaj and J. J. Thelen, pp. 87–104. Berlin: Springer-Verlag.

Hope, J. L., B. J. Prazen, E. J. Nilsson, M. E. Lidstrom and R. E. Synovec. 2005. Comprehensive two-dimensional gas chromatography with time-of-flight mass spectrometry detection: Analysis of amino acid and organic acid trimethylsilyl derivatives, with application to the analysis of metabolites in rye grass samples. *Talanta* 65:380–388.

Hornickova, J., J. Velisek, J. Ovesna and H. Stavelikova. 2009. Distribution of *S*-alk(en)yl-L-cysteine sulfoxides in garlic (*Allium sativum* L.). *Czech J Food Sci* 27:S232–S235.

Horváth-Szanics, E., Z. Szabo, T. Janáky, J. Pauk and G. Hajás. 2006. Proteomics as an emergent tool for identification of stress-induced proteins in control and genetically modified wheat lines. *Chromatographia* 63:S143–S147.

Houde, M. and A. O. Diallo. 2008. Identification of genes and pathways associated with aluminum stress and tolerance using transcriptome profiling of wheat near-isogenic lines. *BMC Genomics* 9:400.

Hurkman, W. J., W. H. Vensel, F. M. DuPont, S. B. Altenbach and B. B. Buchana 2008. Endosperm and amyloplast proteomes of wheat grain. In *Plant Proteomics. Technologies, Strategies and Application*, eds. G. K. Agarwal and R. Rakwal, pp. 207–222. Hoboken, NJ: John Wiley & Sons, Inc.

Iba, K. 2002. Acclimative response to temperature stress in higher plants: Approaches of gene engineering for temperature tolerance. *Annu Rev Plant Biol* 53:225–245.

Imelfort, M. and D. Edwards. 2009. De novo sequencing of plant genomes using second-generation technologies. *Brief Bioinform* 10:609–618.

International Rice Genome Sequencing Project. 2005. The map-based sequence of the rice genome. *Nature* 436:793–800.

Jacobs, D. I., R. van der Heijden and R. Verpoorte. 2000. Proteomics in plant biotechnology and secondary metabolism research. *Phytochem Analysis* 11:277–287.

Jain, M., C. Ghanashyam and A. Bhattacharjee. 2010. Comprehensive expression analysis suggests overlapping and specific roles of rice glutathione *S*-transferase genes during development and stress responses. *BMC Genomics* 11:73.

Janovska, D., Z. Stehno and P. Cepkova. 2007. Evaluation of common buckwheat genetic resources in Czech Gene Bank. In *Advances in Buckwheat Research, Proceedings of the 10th International Symposium on Buckwheat*, pp. 31–40. Yangling, China, August 14–18.

Jonsson, P., A. I. Johansson, J. Gullberg et al. 2005. High-throughput data analysis for detecting and identifying differences between samples in GC/MS-based metabolomic analyses. *Anal Chem* 77:5635–5642.

Kawaura, K., K. Mochida and Y. Ogihara. 2008. Genome-wide analysis for identification of salt-responsive genes in common wheat. *Funct Integr Genomic* 8:277–286.

Kayal, W. E., M. Navarro, G. Marque, G. Keller, C. Marque and C. Teulieres. 2006. Expression profile of CBF-like transcriptional factor genes from *Eucalyptus* in response to cold. *J Exp Bot* 57:2455–2469.

Kersten, B., L. Burkle, E. J. Kuhn et al. 2002. Large-scale proteomics. *Plant Mol Biol* 48:133–141.

Kim, S. T. and K. Y. Kang. 2008. Proteomics in plant defence response. In *Plant Proteomics. Technologies, Strategies and application*, eds. G. K. Agarwal and R. Rakwal, pp. 587–601. Hoboken, NJ: John Wiley & Sons, Inc.

Kocsy, G., B. Athmer, D. Perovic et al. 2010. Regulation of gene expression by chromosome 5A during cold hardening in wheat. *Mol Genet Genomics* 283:351–363.

Komatsu, S. 2006. Cereal proteomics. In *Annual Plant Review, Vol. 28, Plant Proteomics*, ed. C. Finnie, pp. 129–149. Oxford: Blackwell Publishing Ltd.

Kramer, A., T. Feilner, A. Possling et al. 2004. Identification of barley CK2α targets by using the protein microarray technology. *Phytochemistry* 65:1777–1784.

Kumar, P., V. K. Gupta, A. K. Misra, D. R. Modi and B. K. Pandey. 2009. Potential of molecular markers in plant biotechnology. *Plant Omics J* 2:141–162.

Kuraparthy, V., S. Sood and B. S. Gill. 2009. Molecular genetic description of the cryptic wheat-*Aegilops geniculata* introgression carrying rust resistance genes Lr57 and Yr40 using wheat ESTs and synteny with rice. *Genome* 52:1025–1036.

Landjeva, S., V. Korzun and A. Borner. 2007. Molecular markers: Actual and potential contributions to wheat genome characterization and breeding. *Euphytica* 156:271–296.

Langridge, P., N. Paltridge and G. Fincher. 2006. Functional genomics of abiotic stress tolerance in cereals. *Brief Funct Genom Proteom* 4:343–354.

Larkin, P. and G. G. Harrigan. 2007. Opportunities and surprises in crops modified by transgenic technology: Metabolic engineering of benzylisoquinoline alkaloid, gossypol and lysine biosynthetic pathways. *Metabolomics* 3:371–382.

Laurentin, H. 2009. Data analysis for molecular characterization of plant genetic resources. *Genet Resour Crop Evol* 56:277–292.

Lee, J. and B. Cooper. 2006. Alternative workflows for plant proteomic analysis. *Mol Syst Biol* 2:621–626.

Lee, J.-Y. and D.-H. Lee. 2003. Use of serial analysis of gene expression technology to reveal changes in gene expression in *Arabidopsis* pollen undergoing cold stress. *Plant Physiol* 132:517–529.

Li, J. W. H. and J. C. Vederas. 2009. Drug discovery and natural products: End of an era or an endless frontier? *Science* 325:161–165.

Li, L., C. Liu and X. Lian. 2010. Gene expression profiles in rice roots under low phosphorus stress. *Plant Mol Biol* 72:423–432.

Liu, F., W. Xu, Q. Wei et al. 2010. Gene expression profiles deciphering rice phenotypic variation between Nipponbare (*japonica*) and 93–11 (*indica*) during oxidative stress. *Plos One* 5:e8632.

Margulies, M., M. Egholm, W. E. Altman et al. 2005. Genome sequencing in microfabricated high-density picolitre reactors. *Nature* 437:376–380.

Martin, B., M. Ramiro, J. M. Martinez-Zapater and C. Alonso-Blanco. 2009. A high-density collection of EMS-induced mutations for TILLING in Landsberg erecta genetic background of *Arabidopsis*. *BMC Plant Biol* 9:147.

Maxam, A. M. and W. Gilbert 1977. A new method for sequencing DNA. *Proc Natl Acad Sci USA* 74:560–564.

McCallum, C. M., L. Comai, E. A. Greene and S. Henikoff. 2000. Targeting induced local lesions in genomes (TILLING) for plant functional genomics. *Plant Physiol* 123:439–442.

McKernan K. J., H. E. Peckham, G. Costa et al. 2009. Sequence and structural variation in a human genome uncovered by short-read, massively parallel ligation sequencing using two base encoding. *Genome Res* 19:1527–1541.

Mew, T. W., H. Leung, S. Savary, C. M. V. Cruz and J. E. Leach. 2004. Looking ahead in rice disease research and management. *Crit Rev Plant Sci* 23:103–127.

Mochida, K. and K. Shinozaki. 2010. Genomics and bioinformatics resources for crop improvement. *Plant Cell Physiol* 51:497–523.

Moffatt, B., V. Ewart and A. Eastman. 2006. Cold comfort: Plant antifreeze proteins. *Physiol Plantarum* 126:5–16.

Mohammadi, M., H. N. V. Kav and M. K. Deyholos. 2008. Transcript expression profile of water-limited roots of hexaploid wheat (*Triticum aestivum* "Opata"). *Genome* 51:357–367.

Moore, G., K. M. Devos, Z. Wang and M. D. Gale. 1995. Grasses, line up and form a circle. *Curr Biol* 5:737–739.

Moreno-Risueno, M., W. Busch and P. N. Benfey. 2010. Omics meet networks—Using system approaches to infer regulatory networks in plants. *Curr Opin Plant Biol* 13:126–131.

Mott, I. W. and R. R. C. Wang. 2007. Comparative transcriptome analysis of salt-tolerant wheat germplasm lines using wheat genome arrays. *Plant Sci* 173:327–339.

Mullis, K., F. Faloona, S. Scharf, R. Saiki, G. Horn and H. Erlich. 1986. Specific enzymatic amplification of DNA *in vitro*—The polymerase chain reaction. *Cold Spring Harb Sym* 51:263–273.

Nagaraj, S., Z. Lei, B. Watson and L. W. Sumner. 2008. Proteomics of legume plants. In *Plant Proteomics. Technologies, Strategies and Application*, eds. G. K. Agarwal and R. Rakwal, pp. 179–190. Hoboken, NJ: John Wiley & Sons, Inc.

Nakashima, K. and K. Yamaguchi-Shinozaki. 2006. Regulons involved in osmotic stress-responsive and cold stress-responsive gene expression in plants. *Physiol Plantarum* 126:62–71.

Navarro, M., G. Marque, C. Ayax et al. 2009. Complementary regulation of four *Eucalyptus* CBF genes under various cold conditions. *J Exp Bot* 60:2713–2724.

Ndong, C., J. Danyluk, N. P. A. Huner and F. Sarhan. 2001. Survey of gene expression in winter rye during changes in growth temperature, irradiance or excitation pressure. *Plant Mol Biol* 45:691–703.

Nevo, E. and G. X. Chen. 2010. Drought and salt tolerances in wild relatives for wheat and barley improvement. *Plant Cell Environ* 33:670–685.

Ning, J., X. Li, L. M. Hicks and L. Xiong. 2010. A Raf-like MAPKKK gene *DSM1* mediates drought resistance through reactive oxygen species scavenging in rice. *Plant Physiol* 152:876–890.

Novaes, E., D. R. Drost, W. G. Farmerie et al. 2008. High-throughput gene and SNP discovery in *Eucalyptus grandis*, an uncharacterized genome. *BMC Genomics* 9:312.

Ogawa, I., H. Nakanishi, S. Mori and N. K. Nishizawa. 2009. Time course analysis of gene regulation under cadmium stress in rice. *Plant Soil* 325:97–108.

Ordon, F., A. Habekuss, U. Kastirr, F. Rabenstein and T. Kühne. 2009. Virus resistance in cereals: Sources of resistance, genetics and breeding. *J Phytopathol* 157:535–545.

Ovesna, J., L. Leisova and L. Kucera. 2001. Characterisation of barley (*Hordeum vulgare* L.) varieties and breeding lines using RAPD and QTL associated PCR markers. *Rost Vyroba* 47:141–148.

Ovesna, J., K. Polakova, L. Kucera,K. Vaculova and J. Milotova. 2006. Evaluation of Czech spring malting barleys with respect to the beta-amylase allele incidence. *Plant Breed* 125:236–242.

Ozturk, Z. N., V. Talamè, M. Deyholos et al. 2002. Monitoring large-scale changes in transcript abundance in drought- and salt-stressed barley. *Plant Mol Biol* 48:551–573.

Pacey-Miller, T. and R. Henry. 2003. Single-nucleotide polymorphism detection in plants using a single-stranded pyrosequencing protocol with a universal biotinylated primer. *Anal Biochem* 317:165–170.

Park, Y. J., J. K. Lee and N. S. Kim. 2009. Simple sequence repeat polymorphisms (SSRPs) for evaluation of molecular diversity and germplasm classification of minor crops. *Molecules* 14:4546–4569.

Perrocheau, L., H. Rogniaux, P. Boivin and D. Marion. 2005. Probing heat-stable water-soluble proteins from barley to malt and beer. *Proteomics* 5:2849–2858.

Pino, M.-T., J. S. Skinner, Z. Jeknic´ et al. 2008. Ectopic AtCBF1 over-expression enhances freezing tolerance and induces cold acclimation-associated physiological modifications in potato. *Plant Cell Environ* 31:393–406.

Pino, M. T., J. S. Skinner, E. J. Park et al. 2007. Use of a stress inducible promoter to drive Ectopic AtCBF expression improves potato freezing tolerance while minimizing negative effects on tuber yield. *Plant Biotechnol J* 5:591–604.

Polakova, K., D. Laurie, K. Vaculova and J. Ovesna. 2003. Characterization of beta-amylase alleles in 79 barley varieties with pyrosequencing. *Plant Mol Biol Rep* 21:439–447.

Poland, J. A., P. J. Balint-Kurti, R. J. Wisser, R. C. Pratt and R. J. Nelson. 2009. Shades of gray: The world of quantitative disease resistance. *Trends Plant Sci* 14:21–29.

Qureshi, M. I., S. Qadir and L. Zolla. 2007. Proteomics-based dissection of stress-responsive pathways in plants. *J Plant Physiol* 164:1239–1260.

Rajashekar, C. B. 2000. Cold response and freezing tolerance in plants. In *Plant–Environment Interactions*, 2nd edition, ed. Wilkinson, R. E, pp. 321–341. New York, NY: Marcel Dekker, Inc.

Renaut, J., J.-F. Hausman and M. E. Wisniewski. 2006. Proteomic and low-temperature studies: Bridging the gap between gene expression and metabolism. *Physiol Plantarum* 126:97–109.

Rizhsky, L., H. Liang and R. Mittler. 2002. The combined effect of drought stress and heat shock on gene expression in tobacco. *Plant Physiol* 130:1143–1151.

Roberts, J. K. 2002. Proteomics and a future generation of plant molecular biologists. *Plant Mol Biol* 48:143–54.

Rose, J. K. C., S. Bashir, J. J. Giovannoni, M. M. Jahn and R. S. Saravanan. 2004. Tackling the plant proteome: Practical approaches, hurdles and experimental tools. *Plant J* 39:715–733.

Roussel, V., L. Leisova, F. Exbrayat, Z. Stehno and F. Balfourier. 2005. SSR allelic diversity changes in 480 European bread wheat varieties released from 1840 to 2000. *Theor Appl Genet* 111:162–170.

Rout, G. R. and A. Mohapatra. 2006. Use of molecular markers in ornamental plants: A critical reappraisal. *Eur J Hortic Sci* 71:53–68.

Salekdeh, G. H. and S. Komatsu. 2007. Crop proteomics: Aim at sustainable agriculture of tomorrow. *Proteomics* 7:2976–2996.

Sandberg, J., M. Neiman, A. Ahmadian and J. Lundeberg. 2010. Gene-specific FACS sorting method for target selection in high-throughput amplicon sequencing. *BMC Genomics* 11:140.

Sanger, F., S. Nicklen and A. R. Coulson. 1977. DNA sequencing with chain-terminating inhibitors. *Proc Natl Acad Sci USA* 74:5463–5467.

Sarkar, N. K., Y. K. Kim and A. Grover. 2009. Rice *sHSP* genes: Genomic organization and expression profiling under stress and development. *BMC Genomics* 10:393.

Schauer, N., D. Steinhauser, S. Strelkov et al. 2005. GC–MS libraries for the rapid identification of metabolites in complex biological samples. *FEBS Lett* 579:1332–1337.

Schmutz, J., S. B. Cannon, J. Schlueter et al. 2010. Genome sequence of the palaeopolyploid soybean. *Nature* 463:178–183.

Schuster, S. C. 2008. Next-generation sequencing transforms today's biology. *Nat Methods* 5:16–18.

Seo, E., H. Lee, J. Jeon et al. 2009. Crosstalk between cold response and flowering in *Arabidopsis* is mediated through the flowering-time gene SOC1 and its upstream negative regulator FLC. *Plant Cell* 21:3185–197.

Sharma, R., R. K. M. Singh, G. Malik et al. 2009. Rice cytosine DNA methyltransferases—Gene expression profiling during reproductive development and abiotic stress. *FEBS J* 276:6301–6311.

Shendure, J. 2008. The beginning of the end for microarrays? *Nat Methods* 5:585–587.

Shengwu, H., J. Ovesna and L. Kucera. 2003. Evaluation of genetic diversity of *Brassica napus* germplasm from China and Europe assessed by RAPD markers. *Plant Soil Environ* 49:106–113.

Shulaev, V. and D. J. Oliver. 2006. Metabolic and proteomic markers for oxidative stress. New tools for reactive oxygen species research. *Plant Physiol* 141:367–372.

Sip, V., J. Chrpova, J. Vacke and J. Ovesna. 2004. Possibility of exploiting the Yd2 resistance to BYDV in spring barley breeding. *Plant Breeding* 123:24–29.

Skylas, D., D. Van Dyk. and C. W. Wrigley. 2005. Proteomics of wheat grain. *J Cereal Sci* 41:165–179.

Soltis, D. E., M. J. Moore, G. Burleigh and P. S. Soltis. 2009. Molecular markers and concepts of plant evolutionary relationships: Progress, promise, and future prospects. *Crit Rev Plant Sci* 28:1–15.

Southern, E. M. 1975. Detection of specific sequences among DNA fragments separated by gel-electrophoresis. *J Mol Biol* 98:503–517.

Spiertz, J. H. J. and F. Ewert. 2009. Crop production and resource use to meet the growing demand for food, feed and fuel: Opportunities and constraints. *NJAS-Wagen J Life Sc* 56:281–300.

Sreelakshmi, Y., S. Gupta, R. Bodanapu et al. 2010. NEATTILL: A simplified procedure for nucleic acid extraction from arrayed tissue for TILLING and other high-throughput reverse genetic applications. *Plant Methods* 6:3.

Steuernagel, B., S. Taudien, H. Gundlach et al. 2009. *De novo* 454 sequencing of barcoded BAC pools for comprehensive gene survey and genome analysis in the complex genome of barley. *BMC Genomics* 10:547.

Sumner, L. W., P. Mendes and R. A. Dixon. 2003. Plant metabolomics: Large-scale phytochemistry in the functional genomics era. *Phytochemistry* 62:817–836.

Svensson, J. T., C. Crosatti, C. Campoli et al. 2006. Transcriptome analysis of cold acclimation in barley *Albina* and *Xantha* mutants. *Plant Physiol* 141:257–270.

Talamè, V., N. Z. Ozturk, H. J. Bohnert and R. Tuberosa. 2007. Barley transcript profiles under dehydration shock and drought stress treatments: A comparative analysis. *J Exp Bot* 58:229–240.

Thelen, J. J. 2007. Introduction to proteomics: A brief historical perspective on contemporary approaches. In *Plant Proteomics*, eds. J. Šamaj and J. J. Thelen, pp. 1–13. Berlin: Springer-Verlag.

Thomas, S. W., M. A. Glaring, S. W. Rasmussen, J. T. Kinane and R. P. Oliver. 2002. Transcript profiling in the barley mildew pathogen *Blumeria graminis* by serial analysis of gene expression (SAGE). *Mol Plant Microbe In* 15:847–886.

t'Kindt, R., L. De Veylder, M. Storme, D. Deforce, J. Van Bocxlaer. 2008. LC–MS metabolic profiling of *Arabidopsis thaliana* plant leaves and cell cultures: Optimization of pre-LC–MS procedure parameters. *J Chromatogr B* 871:37–43.

Todorovska, E., N. Christov, S. Slavov, P. Christova and D. Vassilev. 2009. Biotic stress resistance in wheat—Breeding and genomic selection implications. *Biotechnol Biotec Eq* 23:1417–1426.

Troggio, M., G. Malacarne, G. Coppola et al. 2007. A dense single-nucleotide polymorphism-based genetic linkage map of grapevine (*Vitis vinifera* L.) anchoring pinot noir bacterial artificial chromosome contigs. *Genetics* 176:2637–2650.

Ueda, A., A. Kathiresan, J. Bennett and T. Takabe. 2006. Comparative transcriptome analyses of barley and rice under salt stress. *Theor Appl Genet* 112:1286–1294.

Ueda, A., A. Kathiresan, M. Inada et al. 2004. Osmotic stress in barley regulates expression of a different set of genes than salt stress. *J Exp Bot* 55:2213–2218.

Uehara, T., S. Sugiyama and C. Masuta. 2007. Comparative serial analysis of gene expression of transcript profiles of tomato roots infected with cyst nematode. *Plant Mol Biol* 63:185–194.

van den Berg, R. A., H. C. J. Hoefsloot, J. A. Westerhuis, A. K. Smilde and M. J. van der Werf. 2006. Centering, scaling, and transformations: Improving the biological information content of metabolomics data. *BMC Genomics* 7:142.

Varshney, R. K. and A. Dubey. 2009. Novel genomic tools and modern genetic and breeding approaches for crop improvement. *J Plant Biochem Biot* 18:127–138.

Varshney, R. K., S. N. Nayak, G. D. May and S. A. Jackson. 2009. Next-generation sequencing technologies and their implications for crop genetics and breeding. *Trends Biotechnol* 27:522–530.

Velculescu, V. E., L. Zhang, B. Vogelstein and K. W. Kinzler. 1995. Serial analysis of gene expression. *Science* 270:484–487.

Villas-Boas, S. G., S. Mas, M. Akesson, J. Smedsgaard and J. Nielsen. 2005. Mass spectrometry in metabolome analysis. *Mass Spectrom Rev* 24:613–646.

Vincent, D. and M. Zivy. 2007. Plant proteome response to abiotic stress. In *Plant Proteomics*, eds. J. Šamaj and J. J. Thelen, pp. 346–360. Berlin: Springer-Verlag.

Wang, K., H. Peng, E. Lin et al. 2010. Identification of genes related to the development of bamboo rhizome bud. *J Exp Bot* 61:551–561.

Wang, W., B. Vinocur and A. Altman. 2003. Plant responses to drought, salinity and extreme temperatures: Towards genetic engineering for stress tolerance. *Planta* 218:1–14.

Wei, H., L. Sun, Z. Tai, S. Gao, W. Xu and W. Chen. 2010. A simple and sensitive HPLC method for the simultaneous determination of eight bioactive components and fingerprint analysis of *Schisandra sphenanthera*. *Anal Chim Acta* 662:97–104.

Welti, R., M. R. Roth, Y. Deng, J. Shah and C. Wang. 2007. Lipidomics: ESI-MS/MS-based profiling to determine the function of genes involved in metabolism of complex lipids. In *Concepts in Plant Metabolomics*, eds. B. J. Nikolau and E. S. Wurtele, pp. 87–92. Dodrecht: Springer-Verlag.

Wenzl, P., J. Carling, D. Kudrna et al. 2004. Diversity Arrays Technology (DArT) for whole-genome profiling of barley. *Proc Natl Acad Sci USA* 101:9915–9920.

Wenzl, P., H. B. Li, J. Carling et al. 2006. A high-density consensus map of barley linking DArT markers to SSR, RFLP and STS loci and agricultural traits. *BMC Genomics* 7:206.

Xue, G.-P., C. L. McIntyre, D. Glassop and R. Shorter. 2008. Use of expression analysis to dissect alterations in carbohydrate metabolism in wheat leaves during drought stress. *Plant Mol Biol* 67:197–214.

Zhao, J., L. C. Davis and R. Verpoorte. 2005. Elicitor signal transduction leading to production of plant secondary metabolites. *Biotechnol Adv* 23:283–333.

Zhao, R., Z. Cheng, W. F. Lu and B. R. Lu. 2006. Estimating genetic diversity and sampling strategy for a wild soybean (*Glycine soja*) population based on different molecular markers. *Chinese Sci Bull* 51:1219–1227.

Zhou, X.G., L. F. Ren, Y. T. Li, M. Zhang, Y. D. Yu and J. Yu. 2010. The next-generation sequencing technology: A technology review and future perspective. *Sci China Life Sci* 53:44–57.

Zivy, M. and D. de Vienne. 2000. Proteomics: A link between genomics, genetics and physiology. *Plant Mol Biol* 44:575–580.

CHAPTER 3

Breeding for Sustainability

Utilizing High-Throughput Genomics to Design Plants for a New Green Revolution

Traci Viinanen

CONTENTS

3.1 INTRODUCTION

Natural variation is a term that is not often uttered in agricultural research. There is, however, an abundant use of the phrase—the genetic variation for trait X is significant, so we expect a sufficient variation for Y% increase in yield. These results are not unimportant; on the contrary, studies that investigate the genetic basis of traits are key to a better understanding of plant biology. The issue lies with the implications of these studies in that only by continuing to improve the existing annual varieties of major crops will we be able to feed the world's growing population. There are certainly limits to what can be achieved in our current agricultural system when so much of the research is biased toward short-term improvements or solutions that will maximize industrial profit but do little to help small-scale farmers or increase sustainability. There is a general support for a more sustainable agricultural system, but we firmly believe that this will not be achieved by solely focusing on research efforts to improve existing annual varieties of corn, soybeans, wheat, and rice. Long-term solutions, such as perennial crops, are becoming viable alternatives with the assistance of new and improved biotechnologies.

Currently, over 40% of the earth's ice-free land surface is intensely managed by humans, much of that producing grain to feed animals (Foley et al., 2005). The increase in land-use change toward grain production, along with major breakthroughs in crop breeding, has stabilized food supplies that have saved millions of people from starvation and allowed for some nations, for example, China, to move from recipients of food aid to food exporters. The "miracle seed" and commercial fertilizer of the Green Revolution more than doubled grain yields per hectare (Evenson and Gollin, 2003), and although modern agriculture has been successful in increasing food production, the resulting high-yielding grain varieties have come at some expense to the environment. The costs of soil, water, and air (especially N_2O) degradation have been externalized, such that natural resources are consumed, reducing the total amount of prime farmland (Wang et al., 2009). Furthermore, agriculture on sloping upland soils and rain-fed tropics are estimated to lose about 10 million hectares to erosion annually (Dixon and Gibbon, 2001; Pimental, 2006). Thus, as global demand for grain increases, the supply of prime farmland is decreasing due to overexploitation. The development of new crops optimized for alternative sustainable farming practices is urgently needed.

With the advent of genome-wide molecular markers, which can be generated for any species with second-generation sequencing technologies, genetic loci controlling agronomically important traits can be identified. Combining phenotypic characterization with genomic data will allow for direct selection on the genetic basis for yield traits through marker-assisted selection, linking the Genome Revolution to a new Green Revolution. New biotechnologies will aid breeders in utilizing natural variation in important traits through a better understanding of genetic trade-offs and physiological restrictions, informing the limitations and possibilities of designing a plant for the twenty-first century and creating a more sustainable cropping system.

3.2 PRACTICING SUSTAINABLE AGRICULTURE

3.2.1 The Concept of Sustainable Agriculture

The concept of sustainable agriculture is complex, spanning the environmental, economic, and social sciences. Three goals consistently rise from the literature as crucial to a sustainable agricultural system: high environmental health, economic profitability, and economic and social equity (Feenstra et al., 2008). Although these goals are generally agreed upon, to what extent each should be emphasized is highly debated. There are many trade-offs in agriculture. Since the mid-twentieth century, high-yielding varieties have been rewarded; because of this reward system, agricultural industries are motivated by the potential economic profit that could be realized by the next high-yielding variety, and not by long-term sustainability. This race for profit and narrow scope of research has resulted in an unfair dichotomy of feeding this generation or ensuring the future of food production. Norman Borlaug gave us the "miracle" seed that provided a quick fix for the rising food demand and bought us time to find a technical fix for soil erosion and soil and water degradation. While conservation strategies have made great strides, especially with regard to soil conservation, dangerous trade-offs remain.

The fact remains that agriculture was built on cheap oil (Pollan, 2009). Before the industrialization of our food system, it required two calories of energy (from fossil fuels) to produce one calorie of food. That ratio is now 10:1. As Michael Pollan (2009) has so delicately described it, we are eating oil and spewing greenhouse gas. How we produce our food has implications far beyond agriculture. We will not be able to make significant progress on health care costs, climate change, and energy independence without addressing our food system. "The way we're feeding ourselves is at the heart of these issues" (Pollan, 2009). Of the $2 trillion spent on health care, 1.5 trillion is spent on preventable diseases. We cannot have a healthy population without a healthy diet, and we cannot have a healthy diet without a healthy agriculture. According to Foley et al. (2005), a sustainable agricultural system should provide most of the ecosystem services of natural ecosystems while also providing a grain harvest. This concept is already used by the forestry and fisheries industries to ensure that resources are not used at an unsustainable rate and that the base product, or population, will be there for future generations. The same standards should be applied to our agricultural systems. If wheat, for example, were a perennial, as are most of the dominant plants in natural ecosystems, we would have hope of achieving grain production and soil/water conservation simultaneously.

3.2.2 Issues and Potential Solutions

In the 1970s, farmers in the Bolivian highlands were surveyed to learn of their priorities for improving crop performance (Wall, 1999). They listed fertilizers, high-yielding varieties, and weed control as priorities for boosting yields. Twenty years later, farmer surveys in the same region identified soil erosion and moisture stress as the two primary limits to yield of wheat on their farms. There are certain limitations to annual crops that cannot be addressed by simply practicing conservation agriculture (Rogoff and Rawlins, 1987). Annual cropping practices generally leave the soil fallow (and consequently unstable) during a portion of the year, up to 6 months in some regions [see Figure 3.1 for a comparison of annual wheat and perennial intermediate wheatgrass (IWG)]. Fallow soil is sensitive to erosion by wind, water, and especially tillage.

Soil erosion not only depletes soil's fertility, but also reduces soil's organic matter content and available rooting depth. Soil degradation has direct impacts on yield not only through loss of nutrient fertility, but also by reducing soil's capacity to absorb and store water between rain events, increasing a crop's vulnerability to drought (Richards, 2000). In contrast, perennial cropping systems, for example, land areas under forage production or pasture, maintain a protective organic mulch covering the soil. Annual species, such as wheat and rice, have gone through tight genetic bottlenecks and were not selected for their ability to condition soil to prevent erosion or hold water (see Figure 3.2 for a comparison of the root systems of annual wheat and perennial IWG). Addressing issues of water, soil, and air qualities, combined with the ominous effects of climate disruption, require alternative agricultural solutions. The "business-as-usual" path of industrial agriculture is not sustainable; it exploits our natural resources and places our future food production at risk (Cox et al., 2006; DeHaan et al., 2005; Glover et al., 2010; Wang et al., 2009). Exasperating this issue is the fact that much of the farmlands of the Mid-western United States have been laid with drainage tiles, resulting in non-point-source pollution to our water ways. While there is not likely to be a simple solution for the surmounting environmental issues and ominous threat of climate disruption, perennial crops could play a significant role. However, a better understanding of seed production limitations and potential trade-offs is needed in order to achieve a high-yielding perennial.

Figure 3.1 Comparison of fall ground cover between annual wheat and perennial IWG. (From Glover, J. D. et al. 2010. *Science* 328:1638–1639. With permission from The Land Institute.)

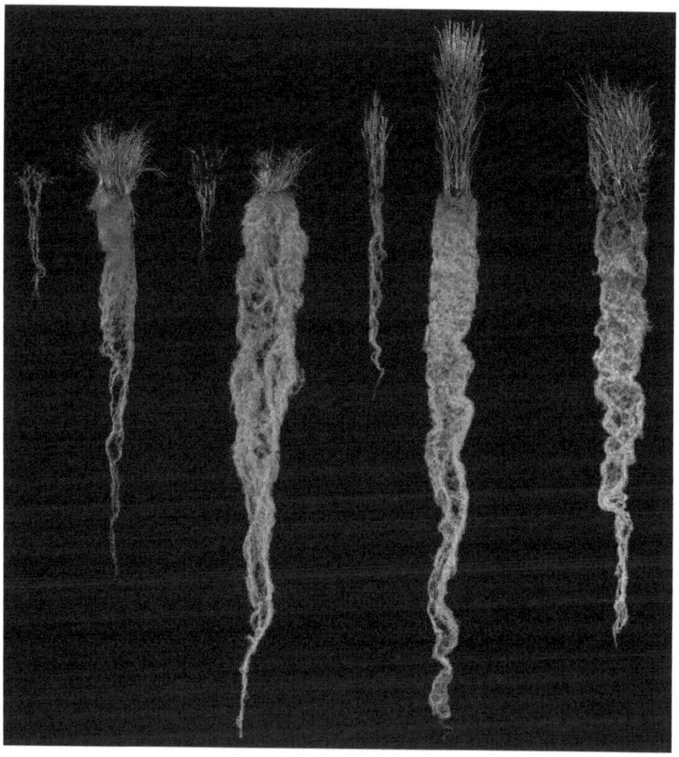

Figure 3.2 From left: fall, winter, spring, and summer comparisons of annual wheat (left in each panel) and perennial IWG (right in each panel). The roots of intermediate wheatgrass extend to 3 m below the surface. (From Glover, J. D. et al. 2010. *Science* 328:1638–1639. With permission from The Land Institute.)

Fortunately, new (and increasingly affordable) biotechnologies exist that will provide more data than anyone in even the previous generation could have imaged. Computational power and statistical methods are also improving and will be essential to the analysis of these data. Solutions to complex issues will soon be limited by the questions asked by the researcher.

3.3 NATURAL VARIATION

3.3.1 What We Have Learned

Natural variation, as defined by Alonso-Blanco et al. (2009, p. 1877), is "the within-species phenotypic variation caused by spontaneously arising mutations that have been maintained in nature by any evolutionary process including, among others, artificial and natural selection." Understanding natural variation is a major objective for any research, from studying linkage and pleiotropy among genes to determining effects of biodiversity on restoration efforts. Much of the early efforts to understand and identify the molecular basis of the ecological and evolutionary processes that maintain natural variation have come from studies of the annual weed *Arabidopsis thaliana*. This species quickly became a model for studying natural variation and adaptation because of its small genome, rapid life cycles, and global distribution. *Arabidopsis* research will likely continue to lead the quest for the foundation of natural genetic diversity and describe with ever greater accuracy the networks in which genes cooperate to produce the phenotype. Advances in -omics technologies have not only aided in the discovery pace of genes in model species such as *A. thaliana*, but also provided a template to pursue research in nonmodel plants. Some of the major achievements in plant development and physiology highlighted (see Alonso-Blanco et al., 2009 for a full review), and how these results, in combination with the latest -omics technologies, have paved the way for research in nonmodel species.

Alonso-Blanco et al. (2009) review what is known regarding the molecular mechanisms underlying important developmental processes such as flowering time, plant architecture and morphology, and seed dormancy and germination. The natural variation of these life-history traits has been well studied, leading to a better, although far from complete, understanding of how plants adapt to a range of natural and agricultural environments. Flowering time is a complex trait that has received much attention because of its role in local adaptation. Over 25 causal genes and their nucleotide polymorphisms of major effect have been identified in species ranging from *A. thaliana* to major crops, short lived perennials, and long-lived woody trees (Alonso-Blanco et al., 2009). Over 20 loci of large effect have been identified that are associated with plant architecture; many of these loci underlie domestication traits and genetic studies have revealed that loss-of-function of transcription factor genes are often involved (Alonso-Blanco et al., 2009), including two rice shattering loci. Gain-of-function alleles encoding transcription factors have also been identified, including *tb1* in maize that influences branching patterns and reduced height genes (Rht) in wheat that confer dwarfing. Several quantitative trait loci (QTLs) have also been identified that affect root growth and architecture, as well as biomass allocation and mineral accumulation, and will be discussed in Section 3.5.

Seed dormancy is a mechanism that regulates the timing of seed germination, causing a temporary failure of an intact viable seed to complete germination under favorable conditions (Bewley, 1997). Seed dormancy and germination vary greatly in nature, and are especially important agronomical traits, as preharvest sprouting, problems with uniform germination, and some seed processing properties (e.g., malting in barley) are largely determined by seed dormancy characteristics. Seed dormancy and germination are controlled by endogenous hormonal (abscisic acid and gibberellin) and metabolic signals and greatly influenced by environmental factors, that is, temperature, light, and moisture (Finch-Savage and Leubner-Metzger, 2006; Holdsworth et al., 2008). The QTL delay of germination (DOG)1 in *A. thaliana* is the first locus identified to be involved in seed dormancy variation. Since its discovery, genetic analyses in crop species have uncovered seed dormancy and germination QTLs respective to each species, aiding the plant breeding process.

Kliebenstein (2010) emphasizes the need for "more intricate, genomic, and broad phenotypic analyses of natural variation to address the fundamentals of quantitative genetics" in his update on the potential role of systems biology in uncovering the generation and maintenance of natural genetic variation. Genomics approaches are not only resulting in faster identification of causal genes, but also giving researchers an opportunity to build a systems view of the genetic architecture of natural variation and identify contributing evolutionary events (Kliebenstein, 2010). Integrating the knowledge of such life-history traits as flowering time, plant architecture and morphology, and seed dormancy and germination allows for the development of plant breeding strategies.

3.3.2 Beyond One Trait

Genetic markers represent the genetic differences between individuals, or even species. A molecular marker is a type of genetic marker and has a particular location in a genome that can be associated with a trait of interest. Molecular markers are used to detect sequence variation in sites underlying phenotypic traits of interest (such as flower color, height, or seed size) and are particularly useful for selecting targeted alleles (Collard and Mackill, 2008). Phenotypic traits that are controlled by many factors are called quantitative traits and molecular markers can be used to divide these traits into their individual components (QTL).

The use of molecular markers in plant breeding is known as marker-assisted selection, and its implementation has resulted in accelerating the plant selection process. Marker-assisted selection is especially advantageous when used to screen populations at the seedling stage, because plants with desirable gene combinations can be identified early and those with undesirable traits can be eliminated (Collard et al., 2008). A single-nucleotide polymorphism (SNP) is a type of molecular marker and is the difference in a nucleotide at a particular location between individuals. SNP markers are currently the marker of choice in the scientific and breeding communities because they occur in practically unlimited numbers (Angaji, 2009; Ganal et al., 2009; Nordborg and Weigel, 2008); however, new algorithms are being developed to simultaneously identify and utilize several types of sequence variants. Genome-wide association mapping is used to detect associations between quantitative traits of interest and the DNA sequence variants present in an individual's genome by determining the genotype at tens or even hundreds of thousands of SNPs (Ganal et al., 2009; Nordborg and Weigel, 2008).

It is important to point out that advances in sequence technology will enable researchers to have a more interdisciplinary approach to their studies than ever before. Plant breeders will be able to take advantage of the decreasing costs of genotyping large numbers of individuals, identifying genetic markers for agronomically important traits with increasing ease. At the same time, they will be able to contribute to important questions regarding the genetic architecture of the pursued traits, such as: How many genes control a given quantitative trait? What is pleiotropy? Where is the heritability? How is conditional genetic variation generated? (Kliebenstein, 2010). Hybrid sequencing (utilizing two or more sequencing technologies) has become a popular genomics approach to gather sequence data for the organism of choice, generally using a long-read platform to generate a reference genome, than a short-read platform for deeper sequencing.

Roche's 454 technology is the current leader in read length, generating over 1 million reads with a modal length of 500 bp and an average length of 400 bp (http://454.com/products-solutions/system-features.asp, 2010). Both Illumina and Applied Biosystems (ABI) specialize in generating a greater abundance of shorter reads. Illumina's paired-end method is gaining popularity because of the resulting high-quality reads and its ability to indels, inversions, and other rearrangements (Illumina Sequencing, 2009).

These sequencing platforms are becoming an ever greater economically viable option for both marker discovery and genotyping, as multiplexing methods become more established. Currently,

Illumina offers a kit for pooling 12 samples per lane. Floragenex's management team collaborates with researchers to provide advice and services for sequencing approaches and data analysis, and currently (May, 2010) uses the barcoding technology to pool up to 30 samples per Illumina lane. Ed Buckler's laboratory at Cornell and the Borevitz Laboratory at the University of Chicago have been developing a barcoding approach similar to that of Illumina, and are now pooling 48 samples per lane. Everyone working on the barcoding approach, including 454 and Illumina, is promising 96 samples (potentially more) per lane in the next year or so, which will make genotyping more cost effective than ever. We, along with Floragenex and other labs, are using a restriction-site-associated DNA (RAD) marker approach to reduce the complexity of the large-grass genomes with which we are working. RAD markers cut at the same sites throughout the genome every time, increasing the coverage of the sites that are sequenced. Pacific Biosciences' DNA sequencing technology is representative of third-generation sequencing and is based on single-molecule real-time sequencing, and promises long reads (1000 + bp) with high accuracy and the potential to pool many samples. Their first version of the sequencing technology was released in 2010, although is not cheap, it is advertised as low cost per base.

Many of the examples of QTL mapping and gene discovery in Section 3.3 were the result of extensive linkage mapping (genetic mapping based on the idea that the farther the two linked genes, the more likely a recombination event will occur between them). High-resolution linkage mapping is both time and labor intensive, and usually interrogates just one or two phenotypes. A typical linkage mapping analysis has low mapping resolution and low statistical power in identifying modest effect loci. With the increasing quantity of sequencing data, new approaches have been developed and continue to be improved in order to be able to interrogate tens to hundreds of thousands of polymorphisms. Genome-wide association mapping has become the analytical method of choice for the incredible amounts of sequence data becoming available. This approach is best suited for studies of natural populations with a large geographic distribution, as the high mapping resolution comes from the assumption that recombination will have removed association between a QTL and a marker that is not tightly linked to it (Jannink and Walsh, 2002). The only disadvantages of genome-wide association studies are that they require a high number of SNPs, usually over 100,000, and are very sensitive to genetic heterogeneity.

There have been several alternative approaches developed on an "as needed" basis, for example, a nested association mapping approach for maize mapping (Yu et al., 2008). Experimental crosses were used to map genes so that there would not be a confounding population structure due to controlled crosses and would only need a moderate number of markers. A known genotype was crossed with 25 genetically diverse lines to create 5000 recombinant inbred lines (RILs), sequenced the founder lines, and genotyped the RILs, resulting in a haplotype map for each of the RILs. Admixture mapping is another approach built on the same basic concept as nested association mapping, in that it removes the confounding factors of population structure. It utilizes a population termed "advanced intercross line," which is derived from two inbred lines, and although it results in a lower mapping resolution than association mapping, it has higher statistical power and requires fewer SNPs (Jannink and Walsh, 2005).

3.4 DOMESTICATION AND THE FUTURE OF SELECTION

3.4.1 A Brief History

The first selection pressures to progenitors of some of our major crops dates to approximately 10,000 years before present (reviewed by Cox, 2009). Archaeological evidence shows that our Neolithic ancestors collected seed from not only annual plants, but also perennials. However, it is thought that they began selecting on annual plants because of the plants' ability to establish quickly, as well as the convenience of an annual sowing/harvest to provide a steady supply of grains. Although intentional selection for seed size was applied to some of the wild progenitors of modern annual crops, including wheat and barley, most of the selection for agronomically important traits was likely unintentional. Irrespective of intention, important alleles for traits such as seed size, nonshattering, and free-threshing in annuals increased in frequency with each year's sowing, and the unchanged, wild perennials were eventually left behind.

Domesticated species have received much attention in studies of adaptation, as well as studies seeking to identify the genetic mechanisms underlying phenotypic change in agronomically important traits. Darwin saw the usefulness in studying domestication, and used it as a way to understand the evolutionary processes underlying adaptation, that is, natural selection (Darwin, 1868). Opponents of Darwin's analogy argued that domesticated species arose from artificial selection, from conscious decisions made from intelligent designers (Ross-Ibarra et al., 2007). There were certainly traits that humans saw as desirous; however, Darwin concluded that most traits regarded as "domestication traits" were the result of unconscious selection, and no different from that of natural selection (Darwin, 1868).

The four major cereal crops (rice, maize, wheat, and barley) experienced parallel phenotypic shifts through domestication, including reduced seed dispersal, reduced branching or tillering, decreased seed dormancy, synchronized seed maturation, an increase in grain size, and larger inflorescences (Ross-Ibarra et al., 2007). Since the development of genomic resources for important agricultural species such as rice, maize, barley, and tomato, there have also been advances in the understanding of the genetic basis of these traits involved in domestication. Comparative genetics studies among the major crops in the grass family has revealed major conservation of gene content and order (Gale and Devos, 1998), with several of the genes being responsible for major phenotypic shifts (e.g., *Q* and Rht in wheat, *FW2.2* in tomato, and *tb1* in maize). Natural variation has provided all of the pieces (or genes) to create a variety of new crops; an evaluation of several candidate perennial species (see Section 3.7) has revealed rare alleles that will play a crucial role in the development of perennial crops. Omics technologies will be imperative to the rapid development of lines with combinations of the most desirable traits, which can then be bred for regional/local conditions.

3.4.2 Trade-Offs and Limitations

There has been a call for a "Second Green Revolution," highlighting the need for varieties of crops that will produce high yields in low-fertility soils, instead of high-fertility soils. The research focus is mostly on selection of belowground traits associated with the root system, especially nutrient-use and water-use efficiencies. While root system biology is an incredibly underdeveloped area of study and deserves more attention, research is largely biased toward annual species with limited plasticity for root growth. Again, I am not discounting the importance of research that will improve the sustainability and/or yield of our major annual crops; these will not be replaced in the immediate future and we need short-term fixes for environmental change and degradation. However, our dependence on these major crops cannot continue, as the concept of sustainability is not about short-term fixes; we need long-term solutions to the environmental issues surrounding agriculture.

One solution may be the implementation of perennials in agriculture, which have many benefits, including large root systems that prevent erosion, reduce lodging, and increase access to water and nutrients; reduced fertilization requirements; less leaching of nutrients; and multiyear harvests from the same crop. Nonetheless, perennials have limitations, especially with regard to seed production. There are numerous basic research endeavors to increase seed yield that are useful for both perennials and annuals; understanding the genetic trade-offs and physiological limitations is crucial to overcome low seed production. This is the question guiding our own research, and an area in which genomics could shed new light; this section concludes with a review of trade-offs and limitations and how this information can be utilized by plant breeders.

The central concepts of life-history evolution are that natural selection maximizes fitness and that trait combinations are constrained by trade-offs. This basically means how often to reproduce and how many offspring. There are several evolved strategies to survival and reproduction, but in the plant kingdom, there are primarily annuals and perennials. Annual species tend to have a semelparous life history, in which there have been selection pressures to grow, reproduce, and die. Perennial species tend to have an iteroparous life-history strategy, in which selection pressures have been for longevity, with bouts of reproduction along the way. Artificial selection pressures can only go so far to improve certain traits (like rooting depth) in annuals, as they have a limited growing season and are programmed to allocate all their energy to reproduction. However, a perennial's strategy is to be conservative, investing much of its energy in growth, especially to its root system to ensure survival to the next growing season. Understanding what maintains plant strategies is of interest to evolutionary ecologists and agronomists, as we are certain to gain insight into not only the patterns of terrestrial plant diversity, but also the possible evolutionary trajectories when natural variation is utilized with foresight.

A plant strategy, according to Craine (2009, p. 6), is "a set of interlinked adaptations that arose as a consequence of natural selection and that promote growth and successful reproduction in a given environment." Plant strategies are sets of correlations that describe relationships among functional traits, which are "morpho-physio-phenological traits which impact fitness indirectly via their effects on growth, reproduction, and survival" (Violle et al., 2007, p. 882). Natural selection does not act on individual traits, but on the individual, or modules within an individual. Therefore, while there are excellent examples of adaptive radiation of individual traits (e.g., the beaks of Darwin's finches), "the pattern left behind by natural selection is one of coordinated changes in multiple traits ... encoded in the genomes of ... species" (Craine, 2009, p. 5). This pattern often involves trade-offs, which, like the finches' beaks analogy, can be misleading, as they are maintained by a combination of phenotypic plasticity and genetic variation among individuals (although the variation can be the result of just one or the other). The relative importance of the two sources of variation can be assessed through pedigree experiments by determining the genetic basis of the trade-off and any genetic variation in phenotypic plasticity (Roff and Fairbairn, 2007).

Roff and Fairbairn (2007) suggest four mechanisms that could maintain the genetic variation underlying a trade-off: mutation–selection balance, antagonistic pleiotropy, correlational selection, and spatiotemporal heterogeneity. Mutation–selection balance likely plays an important part in maintaining the phenotypic and genetic covariation in trade-offs, as much of the standing genetic variance in life-history traits in natural populations is due to mutational input (Houle, 1991; Lynch et al., 1998). Mutations that increase the fitness contribution of both traits in a trade-off are predicted to fix quickly in the population, while mutations that are deleterious to the components of fitness tend to be purged from the population (Estes et al., 2005). However, individuals lacking the ability to purge these mutations (due to sterility mutations, competition, etc.) will accumulate a genetic load, originally defined as the total amounts of potentially deleterious mutations in the genome of an organism (Muller, 1950). Most plants contain many mutated genes in their genomes that are potentially harmful or even deleterious, but the individuals may not show reduced fitness because of their polyploid nature and the redundancy in gene copy number and function. Genetic load (in plants) tends to be associated with outcrossing species because many deleterious mutations are masked in the heterozygous state (Wiens et al., 1987). The number of deleterious mutations and the genetic load in an organism's genome are difficult to measure directly because of the complexity of the genome. However, the amount of genetic load in a species' genome can be indirectly measured by its radiosensitivity, which is defined as the observable damage in fitness-related traits caused by mutagenic treatments (Xu et al., 2006).

Mutations with antagonistically pleiotropic effects (increasing fitness due to a single mutation's effect on one trait while decreasing it due to the same mutation's effect on another trait) would remain in the population for a longer period of time, generating variation in the trade-off (Roff and

Fairbairn, 2007). Although antagonistic pleiotropy is common, its role in maintaining genetic variance is debated; the negative effects of pleiotropic mutations and the restrictive conditions required for it to maintain genetic variation in trade-offs suggest that antagonistic pleiotropy is insufficient to maintain genetic variation in trade-offs. The concept of antagonistic pleiotropy is similar to that of correlational selection in that both predict fitness to be maximized at certain combinations of traits. However, pleiotropy involves gene action, while correlational selection refers to a fitness surface (Roff and Fairbairn, 2007). Correlational selection can generate many combinations of trait values that have equal fitnesses (not just a single fitness peak), which may help to maintain the variation in trade-offs (Roff and Fairbairn, 2007). While correlational selection can be statistically difficult to identify (Roff and Fairbairn, 2007), it has been shown to occur, for example, the trade-off between water-use efficiency and leaf size in *Cakile edentula* (Dudley, 1996) and antagonistic pleiotropy is likely part of the mechanism (Roff and Fairbairn, 2007). Additionally, spatiotemporal heterogeneity, differences in the environment over space and time, has often been shown to be a factor in maintaining variation in numerous studies, but especially those that investigate changes in disturbance and resource availability over time.

It is likely a combination of these factors, as well as neutrality, that maintain the genetic variation in the trade-off between sexual and asexual production, which is evidenced by the variation found in empirical studies of this trade-off (Abrahamson, 1975; Liu et al., 2009; Reekie, 1991; Thompson and Eckert, 2004). Observations of the relationship between sexual versus vegetative reproduction suggest that seed production is less effective and costlier than ramet production, and that seed production will only be favored in high-density situations and unstable conditions, when there are no favorable conditions for ramet dispersal (Gardner and Mangel, 1999; Liu et al., 2009; Olejniczak, 2001, 2003). Most of the assumptions underlying this trade-off are based on the theoretical expectation of an evolutionary stable strategy, in which there is an optimum balance in allocation to sexual and asexual reproduction. If this is the case, there is a very low probability of the two extremes of the trade-off occurring in the same place, and an even smaller chance of ever-observing recombinant individuals in natural conditions. Studies examining the effect of nutrient levels on trade-offs between sexual reproduction and vegetative reproduction have found high levels of plasticity in the relationship (Cheplick, 1995; Liu et al., 2009; Venable, 1992). Trade-offs tend to be more obvious when available resources are limited, but less evident when the available resource level is high (Ashmun et al., 1985; Lambers and Poorter, 1992). However, Reekie (1991) found no evidence of this trade-off across sites with a range of nutrient availability. The inconsistency of results could be indicative of several factors, for example, the experimental conditions may not have been correct for observing the trade-off, or perhaps the trade-off is maintained by more than resource availability, and is potentially a by-product of seed production limitation.

Plants generally produce more ovules than seeds, with an average ovule survivorship of ~50% for outcrossing perennials and 85% for inbreeding species (Wiens, 1984). Pollen limitation, resource limitation, and genetic quality (compatibility between genotypes; genetic load) are proximate (ecological) mechanisms, while bet-hedging for stochastic pollination is the leading ultimate (evolutionary) cause for low seed-to-ovule ratios. Pollen limitation and resource limitation, that is, an insufficient quantity of resources for all fertilized ovules to mature into seeds, are often considered the primary drivers of low seed-to-ovule ratios. However, several studies provide evidence for genetic control of the ovule survivorship (Allphin et al., 2002; Marshall and Ludlam, 1989; Wiens, 1984). Wiens' (1984) data show that the majority of abortion involved partially developed ovules, that is, initial embryo development had occurred, suggesting that pollen and resource limitation were not responsible for ovule abortion. Additional evidence of a broader pattern of seed limitation is based on consistent seed-to-ovule ratios: among interpopulations (including intercontinental populations) within and across years; among species compared between natural and greenhouse conditions; and between perennial and annual species found in the same habitat (Wiens, 1984). This evidence is supported by a study of perennial ryegrass, *Lolium perenne*, which did not find nutrient-level effects on seed-to-ovule ratios (Marshall and Ludlam, 1989). The authors concluded that the low seed-to-ovule ratio pattern is likely due to the species' outcrossing breeding system, and reflects the effects of genetic load and developmental selection, rather than a reliance of resource supply or pollination failure. The assumed trade-off between sexual and asexual reproduction and specifically the mechanism underlying seed production limitation deserves further investigation. We are currently examining the relationship between seed and rhizome production in wild accessions and breeding lines of IWG, *Thinopyrum intermedium*, as well as experimentally manipulating lines by crossing in rare, major alleles important for increasing seed production.

3.5 INTELLIGENT DESIGN

3.5.1 Seed Yield

Yield potential (YP) is a complex trait, but can be broken into its components: light intercepted (LI), radiation-use efficiency (RUE), and harvest index (HI):

$$YP = LI \times RUE \times HI.$$

The product of LI and RUE is biomass, and the partitioning of that biomass to yield is the HI (Reynolds et al., 2009). While high YP has been regarded as one of the most important standards for a successful crop, it is no longer the most attractive characteristic for many regions of the world that either cannot afford the fertilizers or lack the necessary environmental conditions to achieve the upper YP. Annual crops, for example, millet, in sub-Saharan Africa experience infrequent, heavy rainfall events and lose 70–85% of available water from rainfall due to high rates of loss below root zones, runoff, and evaporation (Wallace, 2000). Water loss has been recorded to be as high as 95% for fields in West Africa (Rockstrom, 1997). This inability to efficiently capture available water partly explains the 1 Mg/ha yields of annual grains typical of such regions (Wallace, 2000). In other regions of the world, the use of annual rice on sloping landscapes results in soil erosion and increased deforestation. Therefore, while YP is an important consideration for planting a certain crop, if it cannot be achieved, or if the external costs associated with its production outweigh the benefits, it is no longer profitable. While the per-year YP of perennials may never reach that of annuals, this may not be necessary in order for perennials to be viewed as a more attractive alternative. Several perennial species evaluated by research institutions possess substantial variation for important agronomic traits, many of which are rare and some of which may be difficult to select for due to their quantitative behavior. Markers for these traits are necessary for the success of perennial breeding programs, and the new wave of sequencing technology provides the means to achieve perennial crops. An approach that can tease apart genetic trade-offs and physiological limitations will inform the optimum strategy for increasing seed yield, and will involve selection for traits that will maximize photosynthetic rate/capacity, sink capacity, and seed set.

3.5.2 Photosynthetic Rate/Capacity

More than 90% of the dry biomass of a plant is composed of compounds of carbon, hydrogen, and oxygen derived from photosynthesis. The challenge for researchers will be to better understand the mechanisms that regulate the partitioning of these compounds among the different parts of the plant. The partitioning of carbon is, in ecological terms, referred to as "source, path, sink" dynamics, transporting carbon from the leaf (source), through the phloem in the stem (path), and eventually to the fruit, roots, flowers, grain, and so on (sinks). Throughout the growing season, the various sinks compete with each other for the carbon fixed by the leaves. The harvestable portion of a plant mostly relies on the transfer of carbohydrates from more photosynthetically active parts of the plant, namely the chloroplasts in the cells of leaves. Whole plant photosynthesis is a good indicator of whole plant productivity; the photosynthetic rate is closely related to the photosynthetic rate per unit leaf area and the production of new leaf area, but also depends on rate of transfer between the source and the sink.

Much of the research to attain further increases in annual crops' yields is directed toward improving RUE via increasing leaf/cellular level photosynthesis. Several major crop species, including rice, soybeans, sunflower, and wheat, are being used to investigate the genetic basis of traits involved in photosynthetic rate and capacity. Stomatal conductance (Kawamitsu and Agata, 1987) and leaf N content (Cook and Evans, 1983) largely determine leaf photosynthetic capacity, which is a major determinant of YP for crops. The chlorophyll content is also positively correlated with photosynthetic rate and yield (Zhang et al., 2009). Zhang et al. (2009) identified 11 additive and six pairs of epistatic QTLs for yield and chlorophyll content in a wheat-mapping population. Photosynthesis could also be improved by utilizing natural variation in Rubisco's catalytic rate, increasing carbon fixation efficiency in C_3 plants (Reynolds et al., 2009). The increasing accessibility of new biotechnologies will undoubtedly contribute to the understanding of the genetic basis of photosynthetic processes, and will allow for the development of marker-assisted selection breeding programs for photosynthetic capacity in any species.

There are also possibilities to improve RUE through modifications of inflorescence photosynthesis, leaf canopy architecture, and source–sink dynamics (Reynolds et al., 2009). While the total assimilate production will increase with improved RUE, the optimal partitioning will depend on several factors. There is evidence in several species for underutilization of photosynthetic capacity during grain filling, coupled with unnecessary seed abortion; a better understanding of the underlying physiological and genetic mechanisms will aid selection strategies by plant breeders. The ability to adjust photosynthate partitioning would be a major accomplishment not only for the improvement of existing crops, but also for the development of new crops, especially perennial grains. Perennials already have the advantage of a longer growing season; after the first year of growth, the root system is established and regrowth can begin early in spring. If a longer growing season can be coupled with increased assimilate to seed production, the yield of perennial grains could be higher than anticipated.

3.5.3 Root System

The root system is arguably the most important, but least understood, part of a plant. A healthy, functional, and efficient root system is directly related to plant growth and yield. The uptake of water and nutrients is determined by the combination of the uptake rate per unit root length and the root total length and distribution. Rooting depth variation plays a significant role in ecosystem fluxes of carbon (C) and water, and consequently has an effect on climate (Lee et al., 2005). It is well established that the climate disruption observed over the past century and predicted for the next century is mostly due to the increase in anthropogenic green house gas concentrations, especially (CO_2) (IPCC, 2007). Mitigating CO_2 emissions has been one argument for the implementation of perennial crops, as their production could result C neutral, or C negative food and fuel production (Glover et al., 2007). Deep rooting benefits plants living in dry or infertile conditions, but are not necessarily most efficient in areas that are P-limited. In these areas, a far-spreading root system is desirable, as the majority of P is located in upper layers of soil and decreases with depth (Lynch, 1995).

Root system architecture refers to the spatial configuration of the root system, that is, the explicit geometric deployment of root axes (Lynch, 1995). Although root system architecture is a highly plastic trait, several molecular factors controlling morphology have been identified.

Although genes and major QTL controlling root traits have been identified in arabidopsis, rice, maize, and other model species, relatively little is known about how root traits influence yield. Hammer et al. (2009) examined whether changes in root system architecture and water capture could help explain the observed increases in maize yield, associated with an interaction with plant density. They found that these changes had a direct effect on biomass accumulation and historical yield trends.

Plants exhibit a range of variation for nutrient and water uptake efficiency that is mostly dependent on root architectural traits, which can be used for phenotypic screening or tagged with molecular markers for marker-assisted selection (Lynch, 1995). For example, the transcriptional regulator RSL4 promotes postmitotic elongation of root-hair cells in *A. thaliana* by controlling the expression of genes that encode proteins affecting cell extension (Yi et al., 2010). This growth is controlled by endogenous developmental and exogenous environmental signals, including auxin and low phosphate availability. Selection for an increased proportion of finer root types, including root hair formation and elongation, may reduce the metabolic cost of soil exploration and therefore increase nutrient uptake efficiency. Several studies have also made important discoveries regarding the control of root spreading and depth and the associated patterns of resource acquisition, for example, revealing genes associated with increased phosphorus use-efficiency, which coincides with a more shallow rooting depth and greater spread of the root system. Liu et al. (2008) found a major QTL for root average axial root length at low-nitrogen conditions that is closely linked with other published QTLs for grain yield and nitrogen uptake, suggesting that root length plays an important role in N uptake and grain yield. Although not the focus of this chapter, the rest of the rhizosphere must not be ignored (see Chapters 11 and 12). Plant roots release a wide range of compounds that attract organisms, often forming mutualistic associations in the rhizosphere. The most important of the known mutualisms are between plants and mycorrhizae or rhizobacteria (Badri et al., 2009). Mycorrhizal associations are found in almost all land plants, and involve nutrient exchange. While little is known about the signaling networks between mycorrhizal fungi and plants, even less is known about the interaction between mycorrhizae and microbial communities. New and powerful biotechnology will continue to provide information about the complex chemical and biological interactions that occur in the rhizosphere that will surely benefit productivity and improve the sustainability of agricultural systems.

Perennials have a definite advantage over annuals in that they produce a much larger root system and possess a vast amount of variation for root architecture traits that can be selected for based on the nutrient- and water-use efficiencies needs of different environments. We are aware of three primary approaches to the development of more sustainable crops. The first approach seems to be the most popular; whenever there is a problem that needs to be overcome in an annual crop, researchers tend to search for important disease- or drought-resistant genes in perennial relatives to introgress into the annual. This is the mirror image of the second approach, which is to take the perennial with all the desirable traits for sustainability, and select for agronomically important traits associated with seed yield. The third approach utilizes crosses and backcrosses between annuals and perennials, creating interspecific hybrids, a few of which may be high-yielding perennials. There are benefits and caveats to all of these approaches, which will be discussed in the case studies in Section 3.7, but in order to maximize the efficiency of any of these methods, molecular markers are needed.

3.6 THE LAND INSTITUTE

The Land Institute is a nonprofit research and education organization that began answering the call for a Second Green Revolution before there was such a call. Founded on the principles of natural systems agriculture, research is focused on the development of perennial breeding programs. The inspiration for such breeding programs came from functional observations of the prairie, the majority of which was plowed for modern agriculture. Prairies produce high amounts of biomass without fertilizer or irrigation inputs and are resilient to many perturbations, such as disease incidents, drought, and increased CO_2 levels. These ecosystems accumulate "ecological capital," building the soil fertility, preventing erosion, and storing carbon.

The Land Institute is in the early stages of domesticating several native North American grassland species. Each species has been treated differently, but a pattern has emerged. Seeds from a candidate species are collected from several native populations and grown in an agricultural field at agricultural plant and row spacing to determine whether the species can take advantage of reduced competition and fertile soils. Plants from different populations are allowed to intermate to create a breeding population. From this population, individuals with the highest yield are selected, intermated, and then the cycle of selection is repeated.

In the case of perennial sunflowers (e.g., *Helianthus maximiliani*, *H. rigidus*), evidence for large-effect genes controlling seed dispersal have not been identified. This contrasts with the majority of domestic crops where a few recessive genes confer shatter resistance but is consistent with the genetically complex shatter resistance found in annual sunflower (Wills and Burk, 2007). To select for shattering as a quantitative trait, plants with partial shatter resistance are intermated; unfortunately, the actual rate of shattering is highly dependent on plant maturity and weather conditions and is easily confused with seed predation by birds or insects. Getting high-quality phenotypic data on thousands of plants is difficult, expensive, and will have to be done for many generations to get maximum shatter resistance. Mapping or sequencing *Helianthus* genes involved in seed dispersal, combined with high-throughput genotyping would accelerate progress on this and other domestication-related traits.

Illinois bundleflower (*Desmanthus illinoensis*) is a native legume with relatively large seeds. The Land Institute researchers collected ecotypes with different seed dispersal mechanisms and have recovered segregants in generations following hybridization between them with virtually no seed dispersal—an agronomically desirable trait. This illustrates the need to be able to make controlled crosses, but emasculation is very difficult in this genus and a large proportion of "crossed" seed is always self-pollinated. Genotyping the parents and seedlings would allow the F1 hybrids to be immediately identified. Currently, hundreds of seedlings must be grown to guarantee a few true F1s based on phenotype. Additionally, some phenotypes cannot be confidently scored the first year, requiring the maintenance of a large population for two years just to find hybrids.

Researchers at the Land Institute have paved the way and evaluated the genetic diversity present in the species, but creating a new crop in a time-efficient way will not happen without developing incorporating marker-assisted selection. All of these species, as well as the following case studies, have the potential to make a significant impact on the way we think about plant breeding and food production.

3.7 CASE STUDIES

3.7.1 Intermediate Wheatgrass

IWG, *T. intermedium*, is a perennial grass (family Poaceae, subfamily Triticeae) with a native distribution from Portugal east to Kyrgyzstan and southern Iran and Greece north to Poland and Belarus. IWG was introduced into the United States from the Maikop region of Russia in 1932 to increase the forage quantity and quality of pastures and hay mixtures (Hendrickson et al., 2005). IWG was bred in the 1980s and 1990s for use as a perennial grain by Rodale and the USDA plant materials center at Big Flats (Wagoner, 1995). After more than a decade of evaluation, researchers at Rodale chose 14 individuals with which to continue breeding for increased yield. Progeny from these 14 individuals, as well as another breeding population of 12 of those 14 individuals, were used as the starting population of 1000 genotypes of IWG at The Land Institute. The five best individuals and the 50 best individuals (including the five best) were intermated in isolation and 4266 progeny were planted in another plot as the Cycle 2 population. The 50 best individuals, six best short/high seed-yielding combination individuals, and four best "naked-seed" individuals were intermated in isolation and 4800 progeny were planted as the Cycle 3 population. The Cycle 3 population will be evaluated in Spring/Summer 2010 (L. R. DeHaan, personal communication).

These established breeding populations are ideal for investigating seed limitation, as well as the potential trade-offs associated with increasing seed production, as several traits that are rare in natural populations and appear to be controlled by major-effect alleles have been identified and increased in frequency through artificial selection. An inverse relationship between seed production and rhizome production is expected (Abrahamson, 1975; Reekie and Bazzaz, 1992); however, increased allocation to vegetative reproduction is not likely the causal factor limiting seed production (Allphin et al., 2002; Liu et al., 2009; Reekie, 1991; Reynolds et al., 2009). Resource availability has been shown to limit seed production in other species, and specific conditions have provided evidence for the trade-off (Liu et al., 2009); however, not all studies find evidence for a trade-off between sexual and asexual reproduction (Reekie, 1991; Thompson and Eckert, 2004). I propose that seed yield is more limited by sterility mutations and/or a lack of plasticity in seed size than its cost of production, as assumed in the trade-off between seed and rhizome production.

If a plant has a gene that confers sterility, no amount of water, nitrogen, sunlight, or reduced rhizomes would increase seed production. However, if this plant is grown in ideal conditions that increases its photosynthetic capacity, it could allocate those additional resources to increase its biomass, perhaps by growing taller stems, longer roots, or longer rhizomes, but the plant would not be able to utilize those resources for seed production. Similarly, if an individual has a gene that confers dwarfing, its expression could constrain the plasticity for stem elongation/vegetative growth. A plant that would respond to high resource availability by increasing height would not be able to utilize the resources for this purpose, which could "force" more photosynthate into other sinks. However, if the sinks also lack plasticity, the plant is sink limited and would simply decrease photosynthetic effort. Major-effect alleles/genes have been identified in breeding population of IWG that constrain tillering, height, leaf size, head size, and fertility. Would a plant with alleles that constrain tillering, height, leaf size, and head size, but increase seed fertility respond to high resource availability by allocating the additional photosynthate to seed production? Can a perennial be "programmed" to allocate more resources to sexual reproduction? This depends on the potential pleiotropic effects and epistatic interactions of combining these alleles into a single individual. The fact remains that alleles that confer increased seed set are rare in natural populations, which could be due to trade-offs; these potential trade-offs will be investigated by crossing breed ing lines with higher seed set with wild lines, utilizing both field and greenhouse experiments.

A series of crosses were made at the Land Institute in June 2010 as part of an experiment to investigate the source–sink dynamics of IWG and reveal costs (physiological limitations) associated with increased plasticity for seed production. All of the mapping populations will be constructed the same way: two parental types; F1 progeny will be backcrossed to parents, and F2 recombinants will be evaluated in Spring/Summer 2011. The first mapping population will be established by crossing individuals with high seed set with "wild" individuals (high seed production × high rhizome production). F2 recombinants should reveal trade-offs associated with high seed yield, and may provide evidence for independent inheritance of rhizome and seed production. The second population will be established by crossing "dwarf" individuals with high seed set with "wild" individuals. We predict that some F2 recombinants will reveal the consequences of constrained growth, potentially that resources are "forced" to other sinks, including seed production. The third population will be established by crossing ompact "head" individuals with high fertility with "wild" individuals. F2 recombinants that have higher seed yield will reveal any changes in allocation patterns, potentially providing evidence for an increased ability to "pull" resources to aid seed development. The parents and progeny of all of these crosses will also be used for genetic studies, including investigating the role of linkage and/or pleiotropy in resource allocation patterns, and specifically, rhizome and seed production.

Three experiments are planned to investigate genetic load as a mechanism limiting seed production. The first experiment utilizes gamma-ray irradiation to investigate tolerance differences between highly rhizomatous individuals and highly fertile individuals. Gamma-ray irradiation is the commonest way to generate mutations in plants (Xu et al., 2006); it tends to cause random deletions of large genomic segments and serious phenotypic consequences in treated plants, which are linearly related to the treatment dosage (Nuclear Institute for Agriculture and Biology 2003; Soriano, 1961). The more sensitive to gamma-ray irradiation a plant is, the larger the genetic load it has in its genome; thus, this technique provides an indirect measure of the relative size of the genetic load. Tall trees end up with a huge mutational load due to the number of mitotic divisions between sexual cycles. Highly rhizomatous plants likely experience the same phenomena, which is supported by the evidence of almost no self-pollinating rhizomatous grasses (Vallejo-Marin and O'Brien, 2007) and no self-pollinating trees (Scofield and Schultz, 2006). However, caespitose selfing grasses are common. It is possible that nearly caespitose populations of IWG have fewer sterility alleles than the highly rhizomatous populations. To test this hypothesis, 100 seeds will be collected from each rhizomatous and severely rhizomatous wild plants; 100 seeds will also be collected from each elite breeding lines and the Rodale population (early breeding material) to test for purging of genetic load. Seeds will be irradiated with 300 Gy of gamma-rays at an exposure rate of 1.4 Gy min^{-1} from a ^{60}Co source (Xu et al., 2006). Fifty seeds from each population type will also be collected and grown as a control. All 600 seeds will be planted in the U of C greenhouse in mid-September 2010, and seedling height will be measured one week after germination. The individuals will be evaluated for seed set and height in August 2011; QTLs affecting seedling height and seed fertility, as well as resistance to gamma-ray irradiation, should be identified.

The effects of genetic load will be investigated by crossing half-siblings and measuring inbreeding depression in the progeny, and comparing the progeny to those derived from crosses between unrelated individuals. All of the parental lines will be chosen based on the similarity of morphological characteristics and fertility. The third experiment will utilize individuals with the ability to self-pollinate. These individuals will undergo multiple rounds of selfing, in an attempt to purge their genetic load, and then will be crossed with each other. Progeny should exhibit recovered fitness. If the genetic load (sterility mutations) can be purged, selfing could be used to increase seed yield, potentially without incurring trade-offs.

Greenhouse experiments will be used to investigate the effects of density on the expression of optimal (high seed production) and nonoptimal (high rhizome production and/or high sterility) traits. There will be several scenarios with different combinations of genotypes and resource levels: highly rhizomatous individuals and highly fertile individuals with high, moderate, and low levels of nutrients. When highly rhizomatous individuals are planted at high density, reproductive allocation will be favored. High-density situations bring out competitive strategies for survival, as local conditions for growth are unfavorable due to resource depletion, and tend to result in increased allocation to reproduction. If seed production and rhizome production are plastic traits, the individuals are likely to favor seed production, as there will not be as much room for favorable ramet establishment. However, individuals characterized with high levels of sterility (i.e., low seed production) will not increase allocation to seed production when planted at high density. If sterility alleles are present, no amount of additional resources or increased planting density will result in increased seed production.

3.7.2 Perennial Rice

Perennial rice is just one cereal crop that could allow subsistence farmers to indefinitely produce food on one plot of land, instead of continually clearing plots in the forest when the land is no longer suitable for crop production. Clearing a new plot is currently necessary for the highlands of southern China and across southeast Asia, as the annual upland rice is not able to hold the soil together on slopes, allowing rainfall to wash soil and/or nutrients away and resulting in severe erosion (Valentin et al., 2008). Each plot may only last for a year or two, forcing the farmers to abandon the plot and clear a new plot. The plots are often replanted after a few years of laying fallow; population growth and agricultural intensification is decreasing this fallow period (de Rouw et al., 2005), increasing the risk for soil degradation.

Perennial upland rice breeding was initiated at the International Rice Research Institute in the 1990s to address soil erosion (International Rice Research Institute, 1988). Annual rice, *Oryza sativa*, was crossed with two distantly related perennial species, *O. rufipogon*, and *O. longistaminata*, and the resulting progeny were backcrossed to the perennial parents to create recombinant populations. The crosses between *O. sativa* and *O. longistaminata* were most successful, and the derived F2 population was used to map two dominant complimentary genes, rhz2 and rhz3, controlling rhizomatousness. Efforts to fine map these genes continue; however, rhz2 and rhz3 have been found to correspond with two QTLs associated with rhizomatousness in the genus *Sorghum*, suggesting that the evolution of the annual habit occurred independently. While additional genes likely contribute to perenniality and rhizomatousness, these two are required in rice.

The Yunnan Academy of Agricultural Sciences took over the breeding program in 2001, and has been working to develop markers for marker-assisted selection. Sterility has been a major factor in preventing rapid advances in the breeding program, as there are as many as 35 sterility genes in *Oryza*. Finding the individuals with the rare combination of rhizome production and high fertility requires screening of large populations. Markers for rhz2 and rhz3 are currently being used to screen large F2 populations for rhizome production; however, fine-mapping of these loci would increase the efficiency of marker-assisted selection, and potentially lead to the cloning of rhizome genes, which could be utilized in recombinant gene techniques (Tao et al., 2001).

3.7.3 Sunflowers (Compositae)

Major efforts are underway in the Rieseberg Lab (University of British Columbia) to sequence the genome of the cultivated sunflower, *H. annuus*, the first genome sequence for the Compositae (Asteraceae), the largest family of flowering plants, whose ~24,000 named species comprise roughly 10% of all flowering plants. Many of these species are economically important crops, medicinal plants, or weeds, and this diversity is only beginning to be tapped by modern growers and breeders. The most economically important of these is the sunflower, which is also the most widely studied and genetically well-developed representative of the family. In addition to its importance as an oil seed crop, the sunflower has tremendous potential for cellulosic biomass production, both as a primary source and as a residue of oilseed production. Thus, this large-scale sequencing effort will be accompanied by the development of genomic resources and knowledge needed for manipulating cellulosic biomass traits and other key agronomic traits in hybrid sunflower breeding programs. Additionally, improved genetic resources would be useful for those working on other valuable Compositae crops, medicinal varieties, and horticultural varieties, as well as in control efforts for the many noxious Compositae weeds. The objectives associated with these efforts are to extend the genetic map for sunflower; construct a physical map of the sunflower genome; generate a reference sequence for sunflower and the Compositae; determine the genetic basis of agriculturally important traits; and develop a xylem expressed sequence tag (EST) database and determine the genetic basis of cellulosic biomass traits. Because sunflower has a large genome (3500 Mbp) with abundant repetitive sequences (mainly long terminal repeat (LTR) transposons), a shotgun sequencing approach is unlikely to reliably link, order, and orient sequenced contigs. Therefore, the Rieseberg lab took a hybrid approach, which incorporates both the whole genome shotgun (WGS) sequencing to ~40× with the Illumina GA platform and sequencing of BAC pools to ~20× with Roche's GS FLX platform. The sequencing of bacterial artificial chromosome (BAC) pools mitigates the assembly problems associated with WGS, while concomitantly reducing the number of libraries required for the traditional BAC sequencing approach. Likewise, the blending of Illumina and GS FLX reads is effective because the two sequencing platforms bring different strengths: Illumina brings great depth, but cannot bridge regions of low complexity, whereas GS FLX reads can span longer repeats, but their higher cost makes deep coverage expensive. This strategy will provide sequence equivalent in accuracy to high coverage Sanger sequence, but at a small fraction of the cost. Finally, gaps in the assembly will be filled by targeted BAC sequencing in a final, finishing stage. This approach thus takes advantage of next-generation sequencing platforms, using them to avoid costly and time-consuming pitfalls with traditional genome sequencing approaches.

The availability of a full genome sequence will have immediate benefits, allowing several high-value projects to move forward at a rapid pace. First, an association-mapping project will characterize genetic and morphological variation among domesticated sunflower lines and the relationship between genetic differences and important crop traits. Several hundred inbred cultivated lines will be grown and phenotyped in two different field environments, measuring a suite of important morphological traits throughout the growing season as well as seed traits after final harvest. Each line will be genotyped at 96 candidate genes for important traits, enabling the detection of statistical associations between alleles at each gene and traits of interest. The discovery of sequenced-based markers for favorable alleles will facilitate molecular breeding programs underway in the private sector.

A second major focus is the development of sunflower as a new biofuel source with unique advantages as an annual woody plant. The biofuel development will exploit wood-producing ecotypes of two extremely drought tolerant, wood-forming, desert-dwelling wild species: silverleaf sunflower (*H. argophyllus*) and Algodones dune sunflower (*H. tephrodes*). Preliminary characterization of the wood-producing ecotypes of silverleaf sunflower indicates that they are, for all practical purposes, miniature annual trees. The woody ecotypes grow to 4 m in height and 10 cm in diameter in a single year. Silverleaf sunflower differs from switchgrass and other herbaceous annual plants in a fundamentally important way—the sunflower wood is similar in quality to quaking aspen and poplar, much denser than forage- and silage-type cellulosic biomass feedstocks. Denser materials store more energy/volume and are less costly to transport and store. Moreover, as a drought-tolerant annual dicot, sunflower occupies a unique ecological and production niche, and is capable of growing in marginal land not suitable for most crop or timber species. To develop sunflowers as a viable biofuel source, a better understanding of the genetic basis of wood production in these species is needed. Comparisons of EST sequences from woody and nonwoody lines will provide candidate genes to examine. A complementary approach will involve QTL mapping for woody traits and chemical phenotypes in BC1′s (brain cytoplasmic 1) and RILs derived from crosses between woody and nonwood lines. QTL-NILs for wood formation will be advanced from these populations and will provide a basis for the development of high-biomass woody cultivars.

In summary, a fully sequenced sunflower genome will be a fundamentally important resource, enabling major advances for sunflower and the entire Compositae and providing the data necessary for functional and comparative genomic analyses related to agricultural, biological, and environmental research. Immediate applications of information from the project include the development of second-generation expression and genotyping arrays for molecular breeding of sunflower, whereas in the longer term the project will lead to the development of woody, high-biomass sunflower cultivars. These woody cultivars could be used for food and fuel, providing great benefits for subsistence farmers in developing countries.

3.7.4 Other Perennials

There are several research groups that are working on various perennial species, but they all share a primary goal: to increase the sustainability (both environmental and economic) of local, regional, and global agricultural production. For example, researchers at Texas A&M are working to better understand the genetic basis of perennialism and tillering with the intent of perennializing sorghum and maize (S. Murray, personal communication). Perennial ryegrass and switchgrass are quickly becoming model species as genetic/genomic resources are developed. With the advent of new sequencing technologies, the perennialization of the major crops and/or the domestication of new perennial crops are/is within reach.

3.8 ECOSYSTEM SERVICES

3.8.1 Economic Valuation

The environmental impact of current agricultural systems calls for the creation of ecosystem health indicators for use in economic valuation models. Intensive cropland provides food, but delivers little other ecosystem services. Perennial grains, while not identical to a cropland with fully restored ecosystem services, would not only provide food, but also regulate water flow and quality, hold soil and restore soil fertility, sequester carbon, and have a reduced need for fossil fuel inputs. However, as with any resource, the economic potential needs to be demonstrated in order to effectively inform policy. Bottom-up collaborations with economists can be easily formed to properly understand the jointness of ecology and economics. Data sets collected over the period of a study can inform and improve parameterization of economic models. In order to develop indicators of ecosystem health, we need to have a better understanding of how traits associated with factors of sustainability, such as carbon sequestration and nutrient-use efficiency, can be measured, selected for, and maximized. Currently, non-point-source pollution and soil degradation are not considered in the equation of an annual cropping system's profitability. Selection for traits mentioned throughout this chapter, balanced with the existing perennial characteristics, should allow for the reduction of nutrient inputs (fertilizer), which will reduce monetary costs of production, as well as the cleanup costs associated with non-point-source pollution. When fertilizer application is reduced, there will likely be a reduction yield. However, if costs saved are economically and environmentally significant, that loss of yield could end up being more profitable once direct and indirect costs are factored into the equation. We predict that when ecosystem services are taken into account in a cost–benefit model for sustainable yield, perennial grains will be shown to be economically competitive with their annual counterparts.

3.8.2 Implications for Policy

The first half of this section is inspired from a conglomeration of Michael Pollan's speeches and writings, as well as ideas from Wendell Berry and Wes Jackson, all of whom speak and write very eloquently and passionately about sustainable agriculture. Pollan has focused mostly on food reform, emphasizing the need to wean the agricultural system off fossil fuel and rely once again on sunshine. He draws inspiration from Wendell Berry and others, in stating that annual monocultures have been rewarded for too long. Consequently, our food system is not a result of the free market, but of incentives programs. The government, not the farmers, decides what and how much should be grown, resulting in cheap food and a country in which 67% of its adult population is overweight (that number is expected to be 75% by 2015). Cheap food has hidden costs, including increased costs in health care; as mentioned in Section 3.2.1, \$1.5 of the \$2 trillion spent on health care went toward the treatment of preventable diseases (75% of health care costs are going to preventable diseases and almost 75% of American adults are overweight; coincidence?). The majority of the American population does not know how to eat as they should or grow their own food; most of us have been taught since early childhood how to be fast-food consumers. Most kids (and many adults) are completely ignorant as to the source of their food. Pollan suggests simple solutions to at least partially combat this lack of knowledge; make lunch an academic subject (kids should learn how to grow and prepare their own food) and put a second label on foods (e.g., 60 calories in a head of lettuce, 160 calories were used to make and bring it to the store). The process of bringing the food to the table needs to become more transparent so that the average person can make better-informed decisions about the food that they buy (Pollan, 2009).

Agricultural subsidies mostly go to those who need subsidization the least, that is, the biggest farms, which are most profitable because they have economies of scale. Almost 75% of subsidy money is distributed to the top 10% of recipients (Riedl, 2002; US General Accounting Office, 2006). Much of the federal subsidies are actually being used to buy out small farms, thus consolidating the agricultural industry, and making it even harder for young farmers to get established. Over 90% of money goes to staple crops such as corn, wheat, soybeans, and rice, discouraging diversity. Subsidies are not a bad idea; research has shown that small farms receive more payments in proportion to the value of their crops than big farms. The Common Agricultural Policy in Europe has provisions that encourage local varieties and pays out subsidies based upon total area and not production. Subsidies could be paid for fulfilling the public interest; farmers could be rewarded for diversification, that is, the number of different crops grown as well as the carbon sequestered. Carbon could be returned to the soil through sustainable agriculture and perennials could play a major role. Subsidies are also needed to get young farmers the land and education necessary to start farming. Animals also need to be brought back to the farm; in reference to the fertility problem on the farm and the fertility surplus on the feedlots, Wendell Berry conveys that we have taken a brilliant solution and neatly divided it into two problems.

The public health problem used to be hunger; the Green Revolution took care of that in the United States; however, another transition is needed (as evidenced by the obesity epidemic). There are many other issues with the way our food is produced; however, there are common-sense solutions. We need a resilient agricultural system; with new biotechnology, crops can be designed for the twenty-first century, built on principles of sustainability and farmer independence, not cheap oil and industrial profit.

3.9 CONCLUDING REMARKS

The focus of agricultural research in past decades has been on the provision of resources to crops and the development of high-yielding varieties to make use of the inputs (such as fertilizer and water). This strategy was, and remains, largely effective on prime (rich, nonsloping) farmland, although soil in the United States continues to be degraded at 10 times the sustainable rate (Pimental, 2006). As agricultural inputs become more expensive, species that are better able to utilize resources will become imperative to the profitability of farmers and survival of nations. Perennial species have a longer growing season than annual species and consequently produce much more aboveground biomass and maintain the root system throughout the year. This excess biomass production could be an answer to critics of the biofuels boom, who are rightfully concerned with the prospect of growing fuel instead of food. A solution to this problem could be perennial grains; the heads could be harvested for food and the remaining biomass could be harvested for fuel or left for forage.

While seed yield is an important consideration for a crop, the rest of the plant should not be ignored; we need to design the entire plant, not just the aboveground organs (Waines and Ehdaie, 2007). The Second Revolution calls for a focus on traits within the root system, which are important for nutrient- and water-use efficiencies. These traits, as well as other desirable traits, are often found in wild relatives and introgressed into the annual crop. Alternatively, plant breeders have selected several wild perennial species for direct domestication because of the presence of genetic variation for so many desirable traits, such as disease resistance, drought-resistance, cold-tolerance, nutrient-use efficiency, fertility, seed yield, and so on. While these perennials may never outyield major annual crops when compared in optimum conditions, it is proposed that perennials are a better option for marginal land.

The full cost of agricultural production is not currently considered; millions of dollars are poured into cleanup efforts of our waterways and coasts. Nitrogen loads have polluted our rivers and caused "dead zones" in once productive areas. Perennials will require much less fertilization and their root systems should prevent nutrient leaching. Some perennials, such as switchgrass and IWG, are ready to be planted on marginal lands. The first lines of a perennial grain could be ready for commercial use within the next 10 years, but only with the incorporation of marker-assisted selection and increased funding to long-term solutions to the current issues surrounding mainstream agriculture. Breeders at the Land Institute have doubled seed size and increased yield in IWG by ~60% during the past 5 years with the use of traditional breeding methods (L. R. DeHaan, personal communication). New biotechnology has given us the capability of turning any species into a model organism. As evidenced by the sequencing of the maize genome, research will not have to be limited to simple organisms with small genomes. We now have the technology to cost effectively identify the genetic basis of desired traits, gaining a better understanding of the limitations and the possibilities of breeding for sustainability.

REFERENCES

Abrahamson, W. G. 1975. Reproductive strategies in dewberries. *Ecology* 56:721–726.

Allphin, L., D. Wiens, and K. T. Harper. 2002. The relative effects of resources and genetics on reproductive success in the rare Kachina Daisy, *Erigeron kachinensis* (Asteraceae). *Int J Plant Sci* 163:599–612.

Alonso-Blanco, C., M. G. M. Aarts, L. Bentsink et al. 2009. What has natural variation taught us about plant development, physiology, and adaptation? *Plant Cell* 21:1877–1896.

Angaji, S. A. 2009. Single nucleotide polymorphism genotyping and its application on mapping and marker-assisted plant breeding. *Afr J Biotechnol* 8:908–914.

Ashmun, J. W., R. L. Brown, and L. F. Pitelka. 1985. Biomass allocation in *Aster acuminatus*: Variation within and among populations over 5 years. *Can J Bot* 63:2035–2043.

Badri, D. V., T. L. Weir, D. van der Lelie, and J. M. Vivanco. 2009. Rhizosphere chemical dialogues: Plant–microbe interactions. *Curr Op Biotech* 20:642–650.

Bewley, J. D. 1997. Seed germination and dormancy. *Plant Cell* 9:1055–1066.

Cheplick, G. P. 1995. Life history trade-offs in *Aphibromus scabrivalvis* (Poaceae): Allocation to clonal growth, storage, and cleistogamous reproduction. *Am J Bot* 82:621–629.

Collard, B. C. Y. and D. J. Mackill. 2008. Marker-assisted selection: An approach for precision plant breeding in the twenty-first century. *Philos T R Soc B* 363:557–572.

Cook, M. G. and L. T. Evans. 1983. Nutrient responses of seedlings of wild and cultivated *Oryza* species. *Field Crop Res* 6:205–218.

Cox, T. S. 2009. Domestication and the first plant breeders. In *Plant Breeding and Farmer Participation*, eds. S. Ceccarelli, E. P. Guimarães, and E. Weltizien. Rome: Food and Agriculture Organization of the United Nations.

Cox, T. S., J. D. Glover, D. L. Van Tassel, C. M. Cox, and L. R. DeHaan. 2006. Prospects for developing perennial-grain crops. *Bioscience* 56:649–659.

Craine, J. M. 2009. *Resource Strategies of Wild Plants*. Princeton, NJ: Princeton University Press.

Darwin, C. 1868. *The Variation of Animals and Plants under Domestication*, Vol. II. London: John Murray, Albemarle Street.

De Rouw, A., B. Soulileuth, K. Phanthavong, and B. Dupin. 2005. The adaptation of upland rice cropping to ever-shorter fallow periods and its limit. In *Poverty Reduction and Shifting Cultivation Stabilisation in the Uplands of Lao PDR: Technologies, Approaches and Methods for Improving Upland Livelihoods*, eds. B. Bouahom, A. Glendinning, S. Nilsson, and M. Victor., pp. 139–148. Vientiane: National Agriculture and Forestry Research Institute.

DeHaan, L. R., D. L. Van Tassel, and T. S. Cox. 2005. Perennial grain crops: A synthesis of ecology and plant breeding. *Renew Agr Food Sys* 20:5–14.
Dixon, J., A. Gulliver, and D. Gibbon (eds.) 2001. *Farming Systems and Poverty*. Food and Agriculture Organization, Rome.
Dudley, S. A. 1996. Differing selection on plant physiological traits in response to environmental water availability: A test of adaptive hypotheses. *Evolution* 50:92–102.
Estes, A., B. C. Ajie, M. Lynch, and P. C. Phillips. 2005. Spontaneous mutational correlations for life-history, morphological, and behavioral characters in *Caenorhabditis elegans*. *Genetics* 170:5–53.
Evenson, R. E. and D. Gollin. 2003. Assessing the impact of the Green Revolution, 1960 to 2000. *Science* 300:758–762.
Feenstra, G., C. Ingels, and D. Campbell. 2008. What is sustainable agriculture? UC Sustainable Agriculture Research and Education Program. http://www.sarep.ucdavis.edu. Accessed July 3, 2010.
Finch-Savage, W. E. and G. Leubner-Metzger. 2006. Seed dormancy and the control of germination. *New Phytol* 171:501–523.
Foley, J. A., R. DeFries, G. P. Asner et al. 2005. Global consequences of land use. *Science* 309:570–574.
Gale, M. D. and K. M. Devos. 1998. Plant comparative genetics after 10 years. *Science* 282:656–659.
Ganal, M. W., T. Altmann, and M. S. Roder. 2009. SNP identification in crop plants. *Curr Opin Plant Biol* 12:211–217.
Gardner, S. N. and M. Mangel. 1999. Modeling investments in seeds, clonal offspring, and translocation in a clonal plant. *Ecology* 80:1202–1220.
Glover, J. D., C. M. Cox, and J. P. Reganold. 2007. Future farming: A return to roots? *Sci Am* 297:82–89.
Glover, J. D., J. P. Reganold, L. W. Bell et al. 2010. Increased food and ecosystem security via perennial grains. *Science* 328:638–639.
Hammer, G. L., Z. Dong, G. McLean et al. 2009. Can changes in canopy and/or root system architecture explain historical maize yield trends in the U.S. corn belt? *Crop Sci* 49:299–312.
Hendrickson, J. R., J. D. Berdahl, M. A. Liebig, and J. F. Karn. 2005. Tiller persistence of eight intermediate wheatgrass entries grazed at three morphological stages. *Agron J* 97:1390–1395.
Holdsworth, M. J., L. Bentsink, and W. J. J. Soppe. 2008. Molecular networks regulating Arabidopsis seed maturation, after-ripening, dormancy, and germination. *New Phytol* 179:33–54.
Houle, D. 1991. Genetic covariance of fitness correlates: What genetic correlations are made of and why it matters. *Evolution* 45:30–48.
http://454.com/products-solutions/system-features.asp. Accessed July 15, 2010.
Illumina Sequencing. 2009. Genomic sequencing. http://www.illumina.com/applications/sequencing/rna.ilmn. Accessed July 15, 2010.
International Rice Research Institute. 1988. Toward 2000 and Beyond. IRRI, Manila.
IPCC. 2007. Climate change 2007: The physical science basis (summary for policy makers). Intergovernmental Panel on Climate Change, Geneva.
Jannink, J. L. and B. Walsh. 2002. Association mapping in plant populations. In *Quantitative Genetics, Genomics and Plant Breeding*, ed. M. S. Kang, pp. 59–68. New York, NY: CAB International.
Kawamitsu, Y. and W. Agata. 1987. Varietal differences in photosynthetic rate, transpiration rate and leaf conductance for leaves of rice plants. *Jpn J Crop Sci* 53:563–565–570 [in Japanese with English summary].
Kliebenstein, D. J. 2010. Systems biology uncovers the foundation of natural genetic diversity. *Plant Physiol* 152:480–486.
Lambers, H. and H. Poorter. 1992. Inherent variation in growth rate between higher plants: A research for physiological causes and ecological consequences. *Adv Ecol Res* 23:187–261.
Lee, J. E., R. S. Oliveira, T. E. Dawson, and I. Fung. 2005. Root functioning modifies seasonal climate. *Proc Natl Acad Sci USA* 102:17576–17581.
Liu, F., J. M. Chen, and W. F. Wang. 2009. Trade-offs between sexual and asexual reproduction in a monoecious species *Sagittaria pygmaea* (Alismataceae): The effect of different nutrient levels. *Plant Syst Evol* 277:61–65.
Liu, J., J. Li, F. Chen, F. Zhang, T. Ren, Z. Zhuang, and G. Mi. 2008. Mapping QTLs for root traits under different nitrate levels at the seedling stage in maize (*Zea mays* L.). *Plant Soil* 305:1–2.
Lynch, J. P. 1995. Root architecture and plant productivity. *Plant Physiol* 109:7–13.
Lynch, M., L. Latta, J. Hicks, and M. Giorgianni. 1998. Mutation, selection, and the maintenance of life-history variation in a natural population. *Evolution* 52:727–733.

Marshall, C. and D. Ludlam. 1989. The pattern of abortion of developing seeds in *Lolium perenne* L. *Ann Bot-London* 63:19–27.
Muller, H. J. 1950. Our load of mutations. *Am J Hum Genet* 2:11–76.
Nordborg, M. and D. Weigel. 2008. Next-generation genetics in plants. *Nature* 456:720–723.
Nuclear Institute for Agriculture and Biology (NIAB). 2003. Radiosensitivity studies in Basmati rice. *Pak J Bot* 35:197–207.
Olenjniczak, P. 2001. Evolutionarily stable allocation to vegetative and sexual reproduction in plants. *Oikos* 95:156–160.
Olenjniczak, P. 2003. Optimal allocation to vegetative and sexual reproduction in plants: The effect of ramet density. *Evol Ecol* 17:265–275.
Pimental, D. 2006. Soil erosion: A food and environmental threat. *Environ Dev Sustain* 8:119–137.
Pollan, M. May 2009. Deep agriculture. (Speech).
Reekie, E. G. 1991. Cost of seed versus rhizome production in *Agropyron repens*. *Can J Bot* 69:2678–2683.
Reekie, E. G. and F. A. Bazzaz. 1992. Cost of reproduction as reduced growth in genotypes of two congeneric species with contrasting life histories. *Oecologia* 90:21–26.
Reynolds, M., M. J. Foulkes, G. A. Slafer et al. 2009. Raising yield potential in wheat. *J Exp Bot* 60:7:1899–1918.
Richards, R. A. 2000. Selectable traits to increase crop photosynthesis and yield of grain crops. *J Exp Bot* 51:447–458.
Riedl, B. M. 2002. Still at the Federal trough: Farm subsidies for the rich and famous shattered records in 2002. Heritage Foundation. http://www.heritage.org. Accessed July 15, 2010.
Rockstrom, J. 1997. *On-farm agrohydrological analysis of the Sahelian yield crisis: Rainfall partitioning, soil nutrients and water use efficiency of pearl millet*. PhD Thesis, University of Stockholm.
Roff, D. A. and D. J. Fairbairn. 2007. The evolution of trade-offs: Where are we? *J Evol Biol* 20:3347.
Rogoff, M. H. and S. L. Rawlins. 1987. Food security: A technological alternative. *Bioscience* 37:800–808.
Ross-Ibarra, J., P. L. Morrell, and B. S. Gaut. 2007. Plant domestication, a unique opportunity to identify the genetic basis of adaptation. *Proc Natl Acad Sci USA* 104:8641–8648.
Scofield, D. G. and S. T. Schultz. 2006. Mitosis, stature and evolution of plant mating systems: Low-mitotic and high-mitotic plants. *Proc Roy Soc B-Biol Sci* 273:275–282.
Soriano, J. D. 1961. Mutagenic effects of gamma radiation on rice. *Botanical Gazette* 123:57–63.
Tao, D., F. Hu, E. Sacks, K. McNally et al. 2001. Several lines with Rhizomatous were obtained from interspecific BC2F1 progenies of RD23/*O. longistaminata* backcrossed to RD23. Division Seminars of PBGB, International Rice Research Institute, Los Banos, Philippines.
Thompson, F. L. and C. G. Eckert. 2004. Trade-offs between sexual and clonal reproduction in an aquatic plant: Experimental manipulations vs. phenotypic correlations. *J Evol Biol* 17:581–592.
US General Accounting Office. 2006. Farm programs: Information on recipients of Federal Payments. 2001–06. pp. http://www.gao.gov/new.items/d01606.pdf. Accessed July 15, 2010.
Valentin, C., F. Agusb, R. Alambanc et al. 2008. Runoff and sediment losses from 27 upland catchments in Southeast Asia: Impact of rapid land use changes and conservation practices. *Agr Ecosyst Environ* 128:25–38.
Vallejo-Marin, M. and H. E. O'Brien. 2007. Correlated evolution of self-incompatibility and clonal reproduction in *Solanum* (Solanaceae). *New Phytol* 173:415–421.
Venable, D. L. 1992. Size-number trade-off and the variation of seed size with plant resource status. *Am Nat* 140:287–304.
Violle, C., M. L. Navas, D. Vile, E. Kazakou, C. Fortunel, I. Hummel, and E. Garnier. 2007. Let the concept of trait be functional! *Oikos* 116:882–892.
Wagoner, P. 1995. Wild Triga Intermediate Wheatgrass. Rodale Institute Research Center, Kutztown, PA. http://www.hort.purdue.edu/newcrop/cropfactsheets/triga.html. Accessed July 15, 2010.
Waines, J. G. and B. Ehdaie. 2007. Domestication and crop physiology: Roots of Green-Rev Wheat. *Ann Bot-London* 100:991–998.
Wall, P. C. 1999. Experiences with crop residue cover and direct seeding in the Bolivian Highlands. *Mt Res Dev* 19:313–317.
Wallace, J. S. 2000. Increasing agricultural water use efficiency to meet future food production. *Agr Ecosyst Environ* 82:105–119.
Wang, X. B., W. N. Liu, and W. L. Wu. 2009. A holistic approach to the development of sustainable agriculture: Application of the ecosystem health model. *Int J Sust Dev World* 16:339–345.

Wiens, D., C. L. Calvin, C. A. Wilson, C. I. Davern, D. Frank, and S. R. Seavey. 1987. Reproductive success, spontaneous embryo abortion, and genetic load in flowering plants. *Oecologia* 71:1–09.

Wiens, D. 1984. Ovule survivorship, brood size, life history, breeding systems and reproductive success in plants. *Oecologia* 64:47–53.

Wills, D. M. and J. M. Burke. 2007. QTL analysis of the early domestication of sunflower. *Genetics* 176:2589–2599.

Xu, J. L., J. M. Wang, Y. Q. Sun et al. 2006. Heavy genetic load associated with the subspecific differentiation of japonica rice (*Oryza sativa* ssp. *japonica* L.). *J Exp Bot* 57:2815–2824.

Yi, K., B. Menand, E. Bell, and L. Dolan. 2010. A basic helix–loop–helix transcription factor controls cell growth and size in root hairs. *Nat Genet* 42, 264–267.

Yu, J., J. B. Holland, M. D. McMullen, and E. S. Buckler. 2008. Genetic design and statistical power of nested association mapping in maize. *Genetics* 178, 539–551.

Zhang, K., Y. Zhang, G. Chen, and J. Tian. 2009. Genetic analysis of grain yield and leaf chlorophyll content in common wheat. *Cereal Res Commun* 37:499–511.

CHAPTER 4

Transcription Factors, Gene Regulatory Networks, and Agronomic Traits

John Gray and Erich Grotewold

CONTENTS

4.1 INTRODUCTION

In this chapter, our goal is to describe the components and tools that already exist, and which are required for the future study and application of transcription factors (TFs) by agricultural biotechnologists. The components include the TF genes that are being isolated along with the promoters and the *cis*-regulatory elements (CRE) that they regulate. Their study is being facilitated by a growing cadre of tools that have been enhanced by recent developments in genomics technologies. This chapter will focus primarily on monocots because of the relative importance of cereals in agriculture, but will discuss dicot species where important insights and tools are being developed. The ultimate goal of these studies is to manipulate traits guided by the detailed knowledge of the underlying complexity in gene regulation. As we hope to make clear, there are many examples of TFs that have been manipulated for thousands of years as part of the domestication of cereal crops. The hope is that a deeper knowledge of TFs and their global regulatory roles will permit the further improvement of cereals by marker-assisted selection or transgene technologies, and hasten the time frame in which these improvements can be attained.

4.1.1 QTLs and TFs

The manipulation of agronomic traits has been greatly facilitated by the development of molecular marker linkage maps and their use in identifying quantitative trait loci (QTLs). Although knowledge of the gene(s) underlying particular QTLs is not essential for their use in marker-assisted selection, this information can help improve the choice of germplasm for a breeding objective. The advent of complete or near-complete plant genomes has increased the pace at which the genes underlying QTLs can be identified. This research is being facilitated by resources such as *Gramene QTL* (http://www.gramene.org/qtl/index.html) that hosts a comprehensive collection of 8646 annotated QTLs for rice and 1747 for corn (Liang et al., 2008). It should perhaps be of little surprise that many of the genes underlying QTLs for important agronomic traits have been shown to encode TFs. A list of some TFs underlying QTLs that have been isolated in the past decade is provided in Table 4.1.

Table 4.1 List of TFs Underlying Known QTLs in Crop Plants

Agronomic Trait	Species	Transcription Factor Family	Reference
Flowering time	*Brassica rapa*	Flowering Locus C (MADS)	Yuan et al. (2009)
Anthocyanin content	*Rubus* sp. (Red Raspberry)	bHLH, NAM/CUC2-like, and bZIP	Kassim et al. (2009)
Anthocyanin content	*Vitis vinifera*	MYB	Salmaso et al. (2008)
Grain protein content	*Hordeum vulgare*	NAC	Distelfeld et al. (2008)
Tillering	*Leymus cinereus*	GRAS family member	Kaur et al. (2008)
Leaf development	*Populus trichocarpa*	YABBY and HB family members	Street et al. (2008)
Submergence and bacterial blight resistance	*Oryza sativa*	EREBP	Kottapalli et al. (2007)
Cold tolerance	*Hordeum vulgare* and *Triticum aestivum*	CBF	Miller et al. (2006); Francia et al. (2007)
Seed shattering	*Oryza sativa*	BEL-type homeodomain	Konishi et al. (2006)
Disease resistance	*Oryza sativa*	JAMYB	Ramalingam et al. (2003)
Lignin content	Eucalyptus	EgMYB2	Goicoechea et al. (2005)
Corn earworm resistance (maysin content)	*Zea mays*	ZmMYB3 (P1)	Byrne et al. (1998); Grotewold et al. (1998)

The complexities that accompany TFs that underlie QTLs can be exemplified by the study of the corn earworm resistance in maize (*P1* as QTL for maysin production). The *P1* gene encodes an R2R3-MYB regulator [ZmMYB3 according to (Gray et al., 2009)] that controls the accumulation of 3-deoxy flavonoids in maize floral organs (Grotewold et al., 1991, 1994) resulting in the red coloration characteristic of Indian Corn in the pericarp, through the formation of the phlobaphene pigments. The expression of *P1* in maize Black Mexican Sweet cultured cells resulted in the accumulation of the phlobaphene precursors, the flavan 4-ols apiferol and luteoferol, as well as in the formation of *C*-glycosylflavones (Grotewold et al., 1998) that resembled in structure insecticidal compounds, such as maysin, present in the silks of maize and responsible for providing major resistance against the corn earworm, *Helicoverpa zea* (Snook et al., 1994; Wiseman et al., 1993). Simultaneously, a QTL analysis was performed for maysin accumulation and corn earworm resistance, and *P1* was identified as a major loci contributing to the this important agronomic trait (Byrne et al., 1996; Lee et al., 1998; Dias et al., 2003; McMullen et al., 1998). It is likely that many other TFs will be shown to underlie important agronomic traits such as those listed in Table 4.1, and thus it is important to develop and maintain resources that facilitate QTL study (see Section 4.6).

4.1.2 TFs and the Domestication of Crops

In the period between 10,000 and 4000 years ago wild plants (and animals) were domesticated as humans transitioned from a hunting and gathering lifestyle to a settled agricultural lifestyle. In recent years, scientists have begun to identify the genes that were actively selected for during this period of domestication. Although many developmental pathways were targeted in this selection process, several of the genes involved turned out to encode TFs. As crop breeders seek to introduce genes from wild relatives (e.g., disease-resistance genes), they also need to keep the alleles of genes of traits that were selected for during domestication. Therefore, knowledge of the underlying genes is helpful. In this section we review briefly those TFs that have been linked to domestication of major crops (Table 4.2) with an emphasis on the cereals.

Table 4.2 TFs Implicated in Crop Domestication

Crop	Gene	Trait	TF Family/Type of Change	Reference
Maize	*Teosinte branched 1 (Tb1)*	Repression of axillary meristem growth	TCP family. Regulatory change	Doebley et al. (1997)
Maize	*Teosinte glume architecture1 (Tga1)*	Cell lignification, silica deposition in cells, three-dimensional organ growth, and organ size	SBP family that regulate MADS box TFs—amino acid change	Wang et al. (2005)
Rice	*QTL for shattering 1 (qSH1)*	BEL2 homeodomain protein	Homeodomain	Konishi et al. (2006)
Rice	*SHA1*	Seed shattering	Trihelix	Lin et al. (2007)
Rice	*Shattering 4 (Sh4)*	Abscission layer and formation, seed shattering	Myb3	Li et al. (2006)
Wheat	*Q*	Several, but especially flowering	AP2 family, increase in gene expression	Simons et al. (2006)
Maize	*C1*	Kernel color	MYB	Cone et al. (1986)
Maize	*R1*	Kernel color	bHLH	Ludwig et al. (1989)
Tomato	*Rin*	Fruit ripening	MADS	Vrebalov et al. (2002)
Rice	*Hd1*	Flowering time	Zinc finger	Yano et al. (2000)
Wheat	*Rht*	Plant height	SH2	Peng et al. (1999)
Wheat	*Vrn1 and Vrn2*	Vernalization	MADS and ZCCT	Yan et al. (2004, 2003)
Cauliflower	*boCal*	Inflorescence structure	MADS	Purugganan et al. (2000)
Maize	*Barren1 (Ba1)*	Inflorescence structure	bHLH	Skirpan et al. (2008)
Maize	*Ra1 (Ramosa1)*	Inflorescence structure	MYB	McSteen (2006)

4.1.2.1 Domestication of Maize

The origins of domesticated maize (*Zea mays*) have been genetically traced to southern Mexico by analyzing populations of its closest modern relative, teosinte. A few major mutations are thought to have given rise to the species now known as maize (Doebley et al., 2006). Teosinte plants are branched and the *Teosinte branched1* (*tb1*) gene of maize was identified as a major QTL controlling the difference in apical dominance between the two species (Doebley et al., 1997). *Tb1* belongs to the TCP family of TFs that regulate cell cycle genes. Comparison of this gene in the two species indicates that it is changes in gene expression that result in a higher level of this TF in maize (Doebley et al., 1997). The lack of any fixed amino acid differences between maize and teosinte in the TB1 protein supports this hypothesis. In contrast, a single amino acid change in the *TGA1* (*TEOSINTE GLUME ARCHITECTURE 1*) gene appears to be responsible for the loss of the hardened, protective casing that envelops the kernel in teosinte (Wang et al., 2005). This interpretation is supported by the lack of discernable differences in *tga1* gene expression between maize and teosinte. The *tga1* gene encodes a member of the SBP-box TF family (Cardon et al., 1999).

4.1.2.2 Domestication of Rice

Two key events in the domestication of rice were the loss of a prostrate growth habit and the loss of seed shattering. The genes underlying these two traits have been identified and they encode TFs that exhibit reduced or loss-of-function alleles in domesticated rice varieties (Konishi et al., 2006; Lin et al., 2007; Tan et al., 2008). Wild varieties of rice carry the semidominant gene *Prostrate Growth 1* (*Prog1*) that encodes a C2H2 type zinc-finger domain that is common in many TFs. Identical mutations are present in the *prog1* coding region in 182 varieties of cultivated rice, indicating that this event became fixed during domestication (Tan et al., 2008). Similarly, mutations in other TFs underlie the loss of seed shattering. A single SNP in the promoter and 12 kb upstream of the *QTL of seed shattering in chromosome 1 (qsh1)* gene, that encodes a BEL1-type homeodomain protein, leads to the absence of an abscission layer in *japonica* rice (Konishi et al., 2006). Another rice locus, *Shattering 1* (*SHA1*) encodes a member of the trihelix family of plant-specific TFs. A single amino acid substitution (K79N) in the trihelix domain results in the loss of seed shattering by affecting cell separation in the abscission zone (Lin et al., 2007). The *shattering4* (*sh4*) locus corresponds to another major QTL controlling whether the seed adheres to the plant or not (Li et al., 2006). This locus encodes a gene with homology to R2R3-MYB TFs. Interestingly, the amino acid substitution in the R3 domain that is present in the domesticated allele weakens, but does not eliminate, shattering, and this may have been selected for in order to permit easier threshing after harvest (Doebley et al., 2006). Another TF that has contributed to the domestication of rice is the *Sub1* locus that confers tolerance to complete submergence. Duplication and divergence within this locus, which encodes an ethylene response factor (ERF), has been shown to have continued after domestication (Fukao et al., 2009).

4.1.3 Examples of TFs Linked to Other Agronomic Traits

Agricultural biotechnologists increasingly aim to breed crop species to be better adapted to local climate conditions. Thus, plant responses to biotic and abiotic stresses need to be understood. It may be that the manipulation of single TFs regulating such responses is more amenable than the altering of multiple downstream genes. Here we provide some examples of the TFs underlying traits that are already being incorporated into breeding programs.

4.1.3.1 Flowering Time

A recent compilation of the flowering time gene network in *Arabidopsis* included 52 different genes including several TFs (Flowers et al., 2009). By performing association mapping of 51 of these loci in 275 *Arabidopsis thaliana* accessions, it was estimated that 4–14% of known flowering-time genes harbor common alleles that contribute to natural variation in this life-history trait. In keeping with this finding, a large nested association mapping (NAM) study in maize found no evidence for any single large-effect QTLs for flowering time. Instead, evidence was found for numerous small-effect QTLs shared among families (Buckler et al., 2009). A question that arises is how many of these loci are conserved between dicots and monocots. Although some of these genes are conserved, many are not. For example, FLOWERING LOCUS C (FLC) encodes a MADS-domain TF that functions as a repressor of flowering involved in the vernalization pathway by repressing Flowering locus T (FT) to delay flowering until plants experience winter. Because *Arabidopsis* is a member of the Brassicaceae, it serves as a good reference model for *Brassica* crops such as canola (*Brassica rapa*). Analysis of this locus in 121 *B. rapa* accessions revealed that a naturally occurring splicing mutation in the *BrFLC1* gene contributes greatly to flowering-time variation in this species and also in *Capsella* (Yuan et al., 2009). However, many varieties of wheat and barley require vernalization to flower, but unlike dicots, FLC-like genes have not been identified in cereals. Instead, VERNALIZATION2 (VRN2) inhibits long-day induction of FT-like1 (FT1) prior to winter. In rice, other TF genes, including *Early heading date* (*Ehd1*), *Oryza sativa* MADS51 (*OsMADS51*), and *INDETERMINATE1* (*OsID1*) upregulate the FT1 homolog *Hd3a* in short days, but homologs of these genes are not present in *Arabidopsis*. It appears that different TF genes regulate FT orthologs to elicit seasonal flowering responses in *Arabidopsis* and the cereals (Greenup et al., 2009). Thus, more studies are required in cereals using model plants such as *Brachypodium distachyon* (Vogel et al., 2010).

4.1.3.2 Cold Tolerance

A variety of TFs have been linked to cold stress in plants. These include members of the AP2/EREBP, bZIP, NAC, MYB, MYC, and WRKY families (Bhatnagar-Mathur et al., 2008; Umezawa et al., 2006). The TFs that have been studied the most in relation to cold tolerance and dehydration are the C-repeat Binding Factor genes (CBF) that are also known as the Dehydration-Responsive Element Binding Proteins (Century et al., 2008). Overexpression of these TFs in a variety of different crops has indeed resulted in increased tolerance to cold and drought (Umezawa et al., 2006). However, it has been observed in several species that this also results in a reduction in growth rate. In a few instances, overexpression of CBF under the control of a stress inducible gene can lead to increased stress tolerance without growth retardation (Oh et al., 2005; Pino et al., 2008). Another approach is to overexpress individual select genes downstream of CBF as a means for avoiding or reducing excessive plant stress responses in transgenic plants (Dai et al., 2007; Kim et al., 2009; Ma et al., 2009). Although the CBF and related TF genes have been studied for over a decade, it remains to be seen how their use will lead to effective commercial products (see Section 4.6). In addition, the conservation of CBF pathways cannot be presumed for different species. In an attempt to engineer cold-tolerant papaya (*Carica papaya* L.) trees (whose genome does not appear to harbor CBF genes), CBF genes were introduced. Although *CBF* was expressed, the presence of *CBF*-responsive genes could not be detected (Dhekney et al., 2007).

4.1.3.3 Plant Architecture

The three-dimensional structure of the aerial portion of a plant is influenced by traits such as branching (tillering) pattern, plant height, leaf, and reproductive organ arrangement. The modification of plant height to produce shorter but sturdier plants was one of the main successes of the Green Revolution and scientists continue to try to modify plant architecture. Here we provide a few classical as well as new examples of TFs that have been linked to plant architecture.

The domestication of maize from teosinte involved the reduction of lateral branching and this was brought about by increased expression of the *ZmTb1* locus that encodes a TCP family TF (see above). It may be desirable to reduce tillering in other grass species, but the mechanisms by which TB1 regulates tillering are not understood. When a *Tb1* homolog (*OsTb1*) was overexpressed in rice, it was found that tillering was also greatly reduced. Using a proteomics approach, it was found that a rice serine proteinase inhibitor, OsSerpin, accumulated to much greater levels in high-tillering but not in low-tillering rice, suggesting that it is one of the downstream genes, which regulates tillering (Yeu et al., 2007). TFs with a demonstrated role in rice tillering include MONOCULM1 (MOC1), which controls the initiation and outgrowth of rice tiller buds. MOC1 is a member of the plant-specific GRAS family of TFs. MOC1 appears to play a role in a very early step of axillary meristem initiation, but target(s) of this TF are not yet known (Yang and Hwa, 2008). The *OsTil1* gene (*Oryza sativa Tillering 1*) was isolated by activation tagging and shown to encode a NAC TF (OsNAC2). Overexpression of this TF indicates that it does not promote tiller bud initiation but it promotes outgrowth of existing tiller buds (Mao et al., 2007).

When axillary branching occurs during the reproductive phase of grasses, it affects the architecture of infloresence development (McSteen, 2009). A number of TFs have been identified that influence this trait. One of these in maize is a bHLH family member named *BARREN STALK 1* (*BA1*) which is phosphorylated by the Ser/Thr protein kinase *BARREN INFLORESCENCE 2* (*BIF2*) (Gallavotti et al., 2004; McSteen, 2009). By sampling nucleotide diversity in the *barren stalk1* region, it was shown that two haplotypes entered the maize gene pool from its wild progenitor, teosinte. Yet, only one haplotype was incorporated into modern inbred lines, suggesting that *ba1* like *tb1* was also selected for during maize domestication (Gallavotti et al., 2004).

Leaf rolling is considered another important agronomic trait with moderate leaf rolling increasing photosynthesis and hence raising grain yield (Shi et al., 2007). Class III homeodomain leucine zipper (HD-ZIPIII) family members REVOLUTA (REV), PHABULOSA (PHB), and PHAVOLUTA (PHV) are involved in specifying the adaxial side of the leaf whereas the KANADI (KAN) and YABBY (YAB) gene families are responsible for abaxialization. The roles of this class of homeodomain proteins are conserved between dicots and monocots, as shown by the study of the *Rld1* locus in maize, which is an ortholog of the *Revoluta* gene. A HD-ZIPIII family member (OsHB3) was found to be expressed in the shoot apical meristem in response to auxin and appears to play a role in leaf initiation (Itoh et al., 2008). In one study, the *Shallot-like1* (*Sll1*) gene from rice, that encodes a SHAQKYF class R2R3-MYB family TF belonging to the KANADI family was studied. *Sll1* deficiency resulted in defective programmed cell death of abaxial mesophyll cells, whereas overexpression stimulated phloem development on the abaxial side of the leaf and suppressed bulliform cell and sclerenchyma development on the adaxial side (Zhang et al., 2009).

4.1.3.4 Metabolite Production

There may be as many as 200,000 different metabolites produced in the plant kingdom. These metabolites have been exploited by mankind for many purposes including as foods and pharmaceuticals (Iwase et al., 2009). The metabolic pathways that produce these metabolites involve multiple steps and are often branched and expressed in a temporal, and tissue, specific manner. Overexpressing the genes encoding rate-limiting enzymes has been successful as an approach to overproducing certain metabolites; however, it is often difficult to identify single rate-limiting steps and the overproduction of carotenoids in rice required overexpression of three different genes (Dixon, 2005; Falco et al., 1995; Mahmoud and Croteau, 2001; Yanagisawa et al., 2004; Ye et al., 2000). Interest has grown in utilizing TFS to manipulate the overall flux through metabolic pathways (Broun, 2004; Grotewold, 2008; Iwase et al., 2009; Tian et al., 2008). Pioneering work on the COLORLESS1 (C1) and Red (R) MYB and bHLH TFs in corn demonstrated that these two TFs control flavonoid gene expression and anthocyanin accumulation in plants (Grotewold et al., 1998). In a good example of the application of this knowledge, overexpression of the maize anthocyanin regulators *Leaf color* (*Lc*) and *C1* in tomato resulted in an increase of health-beneficial flavonols (Bovy et al., 2002; Butelli et al., 2008; Ubi, 2007) throughout the fruit. In another application, the LAP1 (Legume Anthocyanin Production 1) MYB factor was overexpressed in *Medicago* and induced massive accumulation of anthocyanin pigments comprising multiple glycosidic conjugates of cyanidins (Peel et al., 2009). Between 70 and 260 downstream genes were activated by this TF, including many involved directly in anthocyanin biosynthesis. Similarly, overexpression of a plant-specific *Dof1* TF induced the upregulation of many genes encoding enzymes for carbon skeleton production, causing a marked increase in amino acid contents, and a reduction of the glucose level in transgenic *Arabidopsis* (Yanagisawa et al., 2004). Thus, a single TF could coordinate carbon and nitrogen metabolism and the transgenic plants exhibited improved growth under low-nitrogen conditions, which is an important agronomic trait.

These examples serve to demonstrate the potential of TFs for metabolic engineering in plants but significant challenges remain. One of the challenges is to cope with the complexity of metabolic flux in plants (Allen et al., 2009; Kruger and Ratcliffe, 2009; Libourel and Shachar-Hill, 2008; Morandini and Salamini, 2003). As we have noted above, overexpression of a single TF affects many downstream genes whose functions may not be known. The term "silent metabolism" has been coined to describe occult metabolic capacities already present or induced in plants (Lewinsohn and Gijzen, 2009). For example, the absence of lycopene in "Golden Rice" shows that the pathway proceeds beyond the transgenic endpoint and thus that an endogenous pathway was also acting—in fact real-time polymerase chain reaction (PCR) revealed that most other carotenoid enzymes were already present in rice and that the wild ancestor of rice had a pigmented endosperm (Schaub et al., 2005). Thus, the "predictability" of altering metabolism by manipulating TFs will depend not only on its position in the overall hierarchy of gene regulation but also on the conservation of target genes across species. It may be necessary to use a TF to drive flux through a pathway and alter expression of an enzyme to divert pathway intermediates to the desired final product (Grotewold, 2008).

TFs can also act as repressors of natural product accumulation. Silencing of the MYB4 gene in Arabidopsis resulted in elevated sinapate esters in the leaves and an increased tolerance of UV-B irradiation (Jin et al., 2000). There is considerable interest in reducing the lignin content of biofuel plants (Weng et al., 2008) and maize MYB factors have been identified that down-regulate lignin biosynthesis (Goicoechea et al., 2005; Newman et al., 2004; Patzlaff et al., 2003; Tamagnone et al., 1998). When two related MYBs from maize are overexpressed in *Arabidopsis*, a significant reduction in lignin content is observed (Caparros-Ruiz et al., 2007; Fornale et al., 2006; Sonbol et al., 2009). It remains to be seen if these MYB TFs also downregulate other pathways and if there are side effects to their overexpression on overall plant fitness. In general, a better understanding of regulatory networks is needed for forward manipulation of metabolic pathways and developmental pathways. Some new techniques such as global identification of TF targets by ChIP-Seq (see below) should aid in revealing the complexities of these networks and thus lead to greater predictability in using TFs for metabolic flux manipulation (Grotewold, 2008).

4.2 FROM GENOME SEQUENCES TO TF COLLECTIONS

Within the monocots, the grasses include the most important economic species, and have therefore received most attention in initial genome sequencing projects. By 2010, several cereal genomes were completed or near completed including rice, sorghum, *Brachypodium*, and maize. There is also interest in studying nongrass monocot species, including palms, banana, onion, asparagus, agave, yucca, irises, orchids, and Acorus, the basal-most monocot lineage in the angiosperm phylogeny. As outlined above, TFs underlie many important agronomic traits and so there is a great need to establish collections of TFs that are well annotated and permit rapid application of knowledge among related species. In this section we outline some of the progress in establishing databases for plant TFs and how complete public TFome collections may be established.

4.2.1 General Characteristics

With complete genomes on hand, the first challenge is to use bioinformatics to identify all TFs and annotate them in cross-relational databases. TFs are classified into families, based on the presence of conserved DNA-recognition domains (Pabo and Sauer, 1992). Different authors utilize slightly different classifications, and thus it is difficult to compare from one study to another the exact number of families.

The Pfam database (www.pfam.org) is a useful starting point for the identification of genes containing one or more DNA-binding motifs or other protein domains. This comprehensive database provides curated alignments of families of related proteins and profile hidden Markov models (HMMs) built from the seed alignments. This database also provides an automatically generated full alignment, which contains all detectable protein sequences belonging to a given family, as defined by profile HMM searches of primary sequence databases (Finn et al., 2010). However, not all plant TF families are represented in the Pfam database, and in the case of the most recent PlnTFDB release, new profile HMMs were generated for the TF families NOZZLE and VARL

(Perez-Rodriguez et al., 2010; Riano-Pachon et al., 2007). Having identified domains, then families are usually defined based on a single or sometimes a combination of domains. For example, in the third release of the PlnTFDB (http://plntfdb.bio.uni-potsdam.de/v3.0/), 77 of 84 families exhibited a single domain. The AGRIS database (http://arabidopsis.med.ohio-state.edu/), which is a comprehensive catalog of TFs in *A. thaliana* first identified TFs using a combination of BLAST and motif searches based on the available literature on known TFs, or on motifs conserved among TFs from a family (Davuluri et al., 2003; Palaniswamy et al., 2006). The GRASSIUS database (http://grassius.org/), which currently catalogs TFs from four grass species, adopts the same family organization that was utilized for *Arabidopsis* TFs in AGRIS (Yilmaz et al., 2009). According to this, plant TFs can be classified into 50–60 discrete families (Grotewold and Gray, 2009). The relative number of members in each family is different between monocots and dicots, as specific TF families have undergone more recent amplifications than others, as, for example, found for R2R3-MYB TFs in the grasses (Dias et al., 2003). However, it is unlikely that TF families will be identified that are restricted to either monocots or dicots (Shiu et al., 2005). In this section we will describe some of the major TF families present in grasses as well as a brief description of some of the major plant and eukaryote TF databases.

4.2.2 Major TF Families in Grasses

In this chapter we emphasize the TF repertoires that are present in cereal species. Although there are about 60 TF families present in these species, the top 10 largest TF families comprise a large percentage of the overall TF present in a given species (Table 4.3). In maize, these 10 families comprise more than half of the TF repertoire and reflect the expansion of these families in plant species, although the higher numbers in maize are also a consequence of recent genome duplication. A brief review of the top five largest families is provided here.

4.2.2.1 bHLH Family

The basic helix–loop–helix (bHLH) family of proteins is found in both plants and animals and although they evolved before the plant/animal split they appear to function in plant-specific or animal-specific processes. In animals, bHLH proteins are involved in the regulation of essential developmental processes. The few bHLH proteins that have been characterized in plants participate in diverse functions that include anthocyanin biosynthesis, phytochrome signaling, globulin

Table 4.3 Top 10 Largest TF Families in Grass Genomes[a]

	Species		
TF Family	**Maize**	**Rice**	**Sorghum**
bHLH	330	143	148
AP2-EREBP	330	161	161
HB	268	97	71
MYB	247	127	122
bZIP	238	89	89
NAC	216	145	112
C2H2	204	102	105
WRKY	202	111	94
MYB-related	190	85	85
MADS	142	73	70
Total	2367	1133	1057

[a] Estimated from PlnTFDB 3.0 April 2010.

expression, fruit dehiscence, carpel development, and epidermal cell differentiation. These TFs are characterized by a highly evolutionary conserved bHLH domain that often, but not always, mediates DNA-binding through specific dimerization.

4.2.2.2 AP2-EREBP Family

The AP2 (APETALA2)/EREBP (Ethylene Responsive Element Binding Protein) TF family includes many developmentally and physiologically important TFs (Aharoni et al., 2004; Kizis and Pages, 2002; Ohto et al., 2005; Zhu et al., 2003). These proteins are divided into two subfamilies: AP2 genes with two AP2 domains and EREBP genes with a single AP2/ERF (Ethylene Responsive element binding Factor) domain. The expression of AP2-like genes is regulated by the microRNA miR172, and the target site of miR172 is significantly conserved in gymnosperm AP2 homologs, suggesting that regulatory mechanisms of these TFs using microRNA have been conserved over the 300 million years (Shigyo et al., 2006).

4.2.2.3 Homeodomain (HB) Family

The homeodomain (homeobox) is considered an ancient DNA-binding domain that is present in all multicellular species and was fundamental for their evolution and diversification (Kappen, 2000). The homeodomain-leucine zipper (HD-Zip) genes, which are characterized by the presence of both a homeodomain and a leucine zipper (LZip) dimerization motif, are unique to the plant kingdom (Ariel et al., 2007). They can be classified into four subfamilies, based on DNA-binding specificities, gene structures, additional common motifs, and physiological functions. Roles for HD-Zip proteins include organ and vascular development, meristem maintenance, signal transduction via hormones, and responses to environmental conditions (Elhiti and Stasolla, 2009).

4.2.2.4 MYB Family

The MYB factors represent a heterogeneous group of proteins that is ubiquitous in eukaryotes, most notably in plants, and which contain 1–4+ MYB repeats. They are usually classified according to the number of repeats that form the MYB domains. Most MYB proteins from vertebrates consist of three imperfect repeats (R1, R2, and R3), and 3R-MYB proteins are also found in the plants (Braun and Grotewold, 1999), where they form a small gene family involved in cell cycle progression (Ito, 2005; Ito et al., 2001). However, the large majority of plant MYB proteins correspond to the R2R3-MYB family, characterized by the presence of two MYB repeats, R2 and R3. The R2R3-MYB family is large, with ~130 members in *Arabidopsis* (Stracke et al., 2001) and 247 in maize (Table 4.3). The amplification of the R2R3-MYB family occurred 450–200 million years ago (MYA), likely after plants began to colonize land (Rabinowicz et al., 1999). In the grasses, specific subgroups of R2R3-MYB genes appear to be still undergoing amplification (Rabinowicz et al., 1999), and this is has been linked to the diversification of plant metabolic pathways (Grotewold, 2005).

4.2.2.5 bZIP Family

Members of the bZIP superfamily, which are present exclusively in eukaryotes, bind to DNA homo- or hetero-dimers and recognize related, yet distinct, palindromic sequences. The bZIPs DNA-binding domain (DBD) consists of a positively charged segment, the basic region, linked to a sequence of heptad repeats of Leu residues, the Zip, that mediates dimerization (Amoutzias et al., 2007). Outside of the bZIP domain, these TFs exhibit great diversity in protein structure which aids in defining subgroups of this family (Jakoby et al., 2002). In plants, they have been shown to regulate diverse plant-specific phenomena, including seed maturation and germination, floral induction and development, and photomorphogenesis, and are also involved in stress and hormone signaling (Nijhawan et al., 2008).

4.2.3 TF Databases: Monocot and Dicot

4.2.3.1 AGRIS

The Arabidopsis Gene Regulatory Information Server (http://arabidopsis.med.ohio-state.edu/)—AGRIS is composed of three databases that integrate information on Arabidopsis TFs (AtTFDB), promoters and CREs (AtcisDB) and experimentally validated interactions between TFs and promoters (RegNet) (Davuluri et al., 2003; Palaniswamy et al., 2006). Currently (February 2010), AGRIS contains information on ~1770 TFs with direct links to clones and resources available to them at the Arabidopsis Biological Resource Center (ABRC, http://abrc.osu.edu/), ~25,516 promoters and 10,653 TF-promoter interactions that have been experimentally identified by various methods. In addition, AGRIS now contains information on all the words (sequence motifs) of length 5–15 that represent the Arabidopsis genome (Lichtenberg et al., 2009), adding another powerful resource to determine if a particular sequence in a genome segment is overrepresented with regard to other genomic regions.

4.2.3.2 GRASSIUS

The Grass Regulatory Information Server (http://www.grassius.org/)—GRASSIUS is a knowledgebase Web resource that integrates information on TFs and gene promoters across the grasses (Yilmaz et al., 2009). GRASSIUS currently consists of two separate, yet linked, databases. GrassTFDB holds information on TFs from maize, sorghum, sugarcane, and rice. *Brachypodium* is likely to be soon added, thanks to the recently completed genome sequence (Vogel et al., 2010). TFs are classified into families, and phylogenetic relationships are beginning to uncover orthologous relationships among the participating species. GRASSIUS also provides a centralized clearinghouse for TF synonyms in the grasses, benefiting from clear rules for the naming of grasses TFs (Gray et al., 2009). GrassTFDB is linked to the grasses TFome collection, which provides clones in recombination-based vectors corresponding to full-length open-reading frames (ORFs) for a growing number of grass TFs. GrassPROMDB contains promoter and CRE information for those grass species and genes for which enough data are available. The integration of GrassTFDB and GrassPROMDB is being accomplished through GrassRegNet, representing the architecture of grasses regulatory networks.

4.2.3.3 PlnTFDB

The Plant Transcription Factor Database (http://plntfdb.bio.uni-potsdam.de/v3.0/)—The Plant Transcription Factor Database maintained since 2006 at the University of Potsdam and the Max-Planck Institute of Molecular Plant Physiology provides a catalogue of TFs and other transcription regulator families, encompassing 62 and 22 families, respectively (Riano-Pachon et al., 2007). This resource (latest release July 2009) includes more than 28,000 regulatory proteins identified in 19 plant species ranging from unicellular red algae to angiosperms. Regulatory proteins in different species are related through orthology relationships, providing effective means for cross-species navigation. For each regulatory protein information including hits to expressed sequence tags (ESTs), domain architecture, homolog PDB entries, and cross-references to external resources is provided (Perez-Rodriguez et al., 2010). PlnTFDB is employed in various genome annotation projects (Correa et al., 2008; Velasco et al., 2007, *Galdieria sulphuraria*, *Emiliana huxleyi*,

Selaginella moellendorfii, unpublished) and phylogenetic studies of regulatory proteins (Correa et al., 2008), and it has served as a starting point for the development of gene expression profiling platforms in different species (Caldana et al., 2007; Richardt et al., 2010), and *Chlamydomonas reinhardtii* quantitative polymerase chain reaction platform.

4.2.3.4 PlantTFDB

Plant Transcription Factor Databases (http://planttfdb.cbi.pku.edu.cn/)—The main focus of this database is on *Arabidopsis*, rice, poplar, moss (*Physcomitrella patens*), and algae (*Chlamydomonas*) and EST sequences from 39 plant species including crops (maize, barley, wheat, etc.), fruits (apple, orange, grape, etc.), trees (pine, spruce, etc.), and other economically important plants (cotton, potato, soybean, etc.) (Guo et al., 2008).

4.2.3.5 SoyDB

A knowledge database of soybean TFs (http://casp.rnet.missouri.edu/soydb/)—Analysis of the soybean genome leads to the prediction of nearly 6000 TFs. This database contains protein sequences, predicted tertiary structures, putative DNA-binding sites, domains, homologous templates in the Protein Data Bank (PDB), protein family classifications (64 families), multiple sequence alignments, consensus protein sequence motifs, a web logo of each family, and web links to the soybean TF database PlantTFDB, known EST sequences, and other general protein databases including Swiss-Prot, Gene Ontology, KEGG, EMBL, TAIR, InterPro, SMART, PROSITE, NCBI, and Pfam (Wang et al., 2010).

4.2.3.6 LEGUMETFDB

An integrative database of *Glycine max*, *Lotus japonicus*, and *Medicago truncatula* TFs (http://legumetfdb.psc.riken.jp/)—This is a newer database that is an extension of SoybeanTFDB (http://soybeantfdb.psc.riken.jp/) aimed at integrating knowledge on the legume TFs, and providing public resource for comparative genomics of TFs of legumes, nonlegume plants and other organisms. This database integrates unique information for each TF gene and family, including sequence features, gene promoters, domain alignments, gene ontology (GO) assignment, and sequence comparison data derived from comparative analysis with TFs found within legumes, in *Arabidopsis*, rice and poplar as well as with proteins in NCBI nr and UniProt (Mochida et al., 2010).

4.2.3.7 DBD

Transcription Factor Prediction Database 2.0 (www.transcriptionfactor.org)—DBD is a database of predicted TFs in completely sequenced genomes. All the predicted TFs contain assignments to sequence specific DNA-binding domain families. The predictions are based on domain assignments from the SUPERFAMILY and Pfam HMM libraries. Benchmarks of the TF predictions show that they are accurate and have wide coverage on a genomic scale. The DBD consists of predicted TF repertoires for 927 completely sequenced genomes (Wilson et al., 2008).

4.2.3.8 TRANSFAC® 7.0 Public 2005

www.generegulation.com—This database contains data on TFs, their experimentally proven binding sites, and regulated genes. Its broad compilation of binding sites allows the derivation of positional weight matrices. An older public version of the database is available for free and a more comprehensive database (>1,000,000 binding sites) on a subscription basis. This database includes information on Arabidopsis, rice, and soybean (Matys et al., 2006; Wingender, 2008). This website also hosts *PathoDB® 2.0 Public 2005*, which is a database on pathologically relevant mutated forms of TFs and their binding sites. It comprises numerous cases of defective TFs or mutated TF binding sites, which are known to cause pathological defects.

4.2.4 TFome Collections

Concomitant with the need for databases, there is a need for researchers to have unrestricted access to physical full-length clones of TFs and other genes. For example, in yeast two-hybrid screens, previously researchers had to rely on libraries that were incomplete but now, in theory, libraries can be created in which every gene and thus every possible interacting protein is present in the screen (Rual et al., 2004). Thus, complete or near complete ORFeome collections can serve as the platform for functional genomics and systems approaches to biological questions. For example, in order to screen for TFs that bind a particular promoter using yeast one-hybrid approach it would be very beneficial to have a complete TF ORFeome collection.

ORFeome collections were first established for bacteria and yeast where genomic DNA could be used since most genes do not contain introns. The establishment of eukaryotic ORFeomes is more laborious and long or rare transcripts are often very difficult to obtain and so the collections remain incomplete. The *Caenorhabditis elegans* was one of the first eukaryotic ORFeomes to be initiated, and currently offers more than 12,000 complete ORF clones (http://worfdb.dfci.harvard.edu/). Later the human ORFeome collaborative (www.orfeomecollaboration.org) was established and provided a model for the establishment of other ORFeome collections. The human ORFeome collaborative was established in 2005 and a total of nearly 16,000 ORFs are now (2010) available through a searchable database (http://horfdb.dfci.harvard.edu/). This effort aims at providing at least one full-length clone for currently defined human genes and in a format that allows for easy transfer of the ORF sequences into virtually any type of expression vector. The Gateway™ system was adopted to meet this goal as the site-specific recombination system enables efficient generation of expression constructs with an extremely low risk of mutation, thus reducing the need for additional sequence analysis (Hartley et al., 2000). The collaborative first aims at providing ORFs without a stop codon so as to permit C-terminal fusions but eventually hopes to provide clones with stop codons also. The construction of this collection is a collaborative effort with contributions from at least nine institutions and a few authorized commercial distributors in the United States, Europe, and Asia (Lamesch et al., 2007; Lennon et al., 1996). Whereas entire ORFeome collections are being established for some model organisms, the effort involved may be too expensive for others, but an entire TF ORFeome may be feasible. In the case of maize, a small-scale TF ORFeome has been started with a modest goal of including at least one member from each TF family (www.grassius.org). Such collections lend themselves to inclusion in undergraduate laboratories where students learning cloning techniques can contribute to incremental completion of such ORFeomes (Yilmaz et al., 2009).

4.2.5 Synthetic TFs Zinc Fingers for Gene Regulation

It has been found that manipulation of a sole TF is sometimes insufficient for the alteration of an amount of a metabolite of interest (van der Fits and Memelink, 2000). In some instances combining the manipulation of a gene encoding an enzyme with that of a TF is needed to achieve the required alteration of a metabolic pathway (Grotewold, 2008; Iwase et al., 2009; van der Fits and Memelink, 2000; Xie et al., 2006). When TFs have a repressive function, it may be desirable to reduce expression of that TF in order to increase flux through a pathway. When there are redundant TFs then multiple knockdowns may be required to achieve a reduction in target gene expression (Gonzalez et al., 2008). Artificial TFs are being developed that hold promise in addressing the challenges of these situations (Sera, 2009). One method termed CRES-T (Chimeric Repressor

Gene Silencing Technology) enables activator TFs to be converted into dominant negative regulators by fusion with a short repressor motif (Hiratsu et al., 2003). This approach has the added advantage of regulating all targets genes even when there is functional redundancy of the positive regulator (Matsui et al., 2004).

Another approach is the creation of artificial TFs that can be used to target the regulation of many endogenous genes. The approach that currently holds most promise is the use of Cys_2His_2-type zinc-finger proteins that contain one of the most common DNA binding motifs in eukaryotes (Papworth et al., 2006). The ability to custom-design zinc-finger proteins with sufficient sequence specificity was elegantly demonstrated by their use in targeted gene replacement in plants including tobacco and maize (Shukla et al., 2009; Townsend et al., 2009; Weinthal et al., 2010). The Zinc Finger Consortium (ZFC) (www.zincfingers.org) has been established to promote the development of this technology and make it publicly available (Ahern, 2009; Maeder et al., 2008; Mandell and Barbas, 2006). The ZFC has also established a Zinc Finger Database ZiFDB (http://bindr.gdcb.iastate.edu:8080/ZiFDB/) that organizes information on more than 700 individual zinc-finger modules and engineered zinc-finger arrays (ZFAs) (Fu et al., 2009). Another database ZifBASE provides a collection of various natural and engineered zinc-finger proteins (Jayakanthan et al., 2009). Currently (April 2010), information is stored in ZifBASE on 89 and 50 natural zinc-finger proteins respectively. With these advances and resources, it is anticipated that this technology will begin to mature in the coming years and find a place in crop improvement.

4.2.6 Use of TFs in Transgenic Crops: Potential versus Practice

A review of recent US patents related to the use of plant TFs to modify plant traits reveals that the many companies and universities continue to seek protection for their use. Indeed Mendel Biotechnology (www.mendelbio.com) has been awarded several patents governing the use of a large set of plant TFs for crop improvement (Century et al., 2008). Proof of concept has been provided for the ability of several TFs to improve plant tolerance to plant stresses such as drought and cold. However a review of actual applications for APHIS approval of transgenic plants (http://www.aphis.usda.gov/brs/not_reg.html) from 1992 to date indicates not a single one in which expression of a TF is modified. Most of the approved and pending applications involve overexpression of enzymes or proteins associated with herbicide tolerance, insect resistance, virus resistance, oil content, and male sterility. It may be that the next wave of transgenic crops will begin to incorporate TFs as means of improving agronomic traits.

4.3 PROMOTERS: INDISPENSABLE BUT ELUSIVE

4.3.1 Finding Promoters

Traditionally, promoters, the DNA sequences responsible for controlling the expression of a gene, were positioned upstream of the transcription start site (TSS), their length requiring to be experimentally determined (Hehl et al., 2004). Thus, a first step toward experimentally or computationally predicting the 3′ end of a promoter is to identify the TSS. This involves mapping the start of the mRNA, or obtaining full-length cDNAs, for example, using the cap trapper method (Carninci et al., 1997, 1996), which has been efficiently applied to plants (Seki et al., 2009). Today, based primarily on genome-wide studies carried out in animals, for example, as part of the Encyclopedia of DNA Elements (ENCODE) project (Birney et al., 2007), it is clear that gene control regions are present not just in the 5′ regulatory region, but also in 5′-UTRs, introns, 3′-UTRs, and 3′ genic regions. Indeed the frequent role of first introns in enhancing plant gene expression has been known for several years (Rose, 2004; Rose and Beliakoff, 2000; Rose et al., 2008). Thus, first introns are routinely included when making constructs for expression in maize (Grotewold et al., 1994; Hernandez et al., 2004).

Experimentally establishing the regulatory region of a gene responsible for recapitulating the normal gene expression is very time consuming and experimentally uncertain. Transient expression experiments provide a convenient tool to assay rapidly the activity of a regulatory region when fused to a reporter such as luciferase, GUS or GFP. However, in most transient expression assays, such as particle bombardment or electroporation, the introduced DNA is not integrated in the genome and is present in a large number of copies, thus often obscuring important regulatory mechanisms sensitive to copy number or chromosomal integration. Stable integration resulting from the transformation and selection of transgenic plants harboring the promoter–reporter construct are limited to those cases in which plant transformation is possible. Even then, because of the inability to perform in plants gene replacement or targeted integration, many independent transformed lines need to be analyzed to overcome what is normally known as "positional effect" variations in gene expression (De Buck and Depicker, 2004; Mueller and Wassenegger, 2004).

4.3.2 Many Promoters but Few Used

In theory, there are as many different promoters as there are genes, but the fact that after two decades of genetic engineering there are only a handful of constitutive promoters available for use in plants (Table 4.4) underscores the point that suitable promoters are hard to find. Even fewer are employed with any widespread frequency especially in commercial applications and in some cases several different promoters are required within the one construct. In addition many of these genes have not been tested over the entire spectrum of plant development.

The availability of complete genomes combined with large microarray databases can help facilitate the discovery of candidate genes that exhibit a desired expression profile. This approach was recently employed to find a promoter that is cell type specific. By screening microarray data derived from *Arabidopsis* guard cells, several candidate genes were identified and then the promoters tested by examining GUS and yellow Cameleon YC3.60 reporters *in vivo*. The best promoter (At1g22690) proved to also work in a stomata-specific manner in tobacco plants (Yang et al., 2008). In a more recent study, five candidate constitutive expressed genes (*APX*, *SCP1*, *PGD1*, *R1G1B*, and *EIF5*) were identified by analysis of rice microarray data. The activity of the corresponding constitutive gene

Table 4.4 List of Genes with Promoters Used for Constitutive Expression in Transgenic Plants

Gene Name	Source	Activity	Reference
CaMV 35S	CaMV	Constitutive	Odell et al. (1985), Omirulleh et al. (1993)
Ocs/Mas chimera	Agrobacterium	Strongest constitutive	Ni et al. (1995)
Actin 1/2	Rice	Constitutive	McElroy et al. (1991), Zhong et al. (1996)
Ubiquitin 1	Rice and maize	Constitutive	Katiyar-Agarwal et al. (2002), Lu et al. (2008), Cornejo et al. (1993)
Actin 2/8	Arabidopsis	Constitutive	An et al. (1996)
Ubiquitin	Arabidopsis	Constitutive	Callis et al. (1990)
tCUP	Tobacco	Constitutive (Cryptic)	Malik et al. (2002)
Acetolactate synthase	Arabidopsis	Constitutive	Ahmad et al. (2009)
ibAGP1 ADP-glucose pyrophosphorylase	Sweet Potato	Constitutive	Kwak et al. (2007)
IF4a Initiation Factor 4a	Tobacco	Constitutive	Mandel et al. (1995)
Bch1 beta-carotene hydroxylase 1 gene	Arabidopsis	Constitutive	Liang et al. (2009)

promoters was then quantitatively analyzed in transgenic rice plants using a GFP reporter construct. It was found that three of these (*APX*, *PGD1*, and *R1G1B*) correspond to novel gene promoters that are highly active at all stages of plant growth (Park et al., 2010). This approach can be applied to find suitable promoters in other plant genomes and it is expected that analysis of the homologs in plants with incomplete genomes will also provide a larger collection of suitable promoters.

4.3.2.1 Promoter Collections

In theory, once a genome is complete one can PCR amplify the relevant region of any particular gene for use in promoter studies. In the absence of complete genomes, it is also possible to use PCR-based methods to amplify the upstream or downstream regions from any starting point within a gene of interest. However, research progress would be enhanced by the availability of physical clones (as described for TFome collections above) preferentially in Gateway® vectors to facilitate ease of over-expression of target genes. Currently, there is no public or commercial resource for cloned promoters from plant species. A commercial library for one-third of the (6000) predicted promoters from *C. elegans* is available in a Gateway® vector (www.openbiosystems.com) and another for over 17,000 human promoters (www.switchgeargenomics.com). Given the value of such collections, it is likely that they will be established by private companies before public resources are developed.

4.3.2.2 Tools and Databases for Promoter Analysis

Several tools and databases have been established to facilitate analysis of eukaryotic and more specifically plant promoters. One of the tools that has emerged to help define transcriptional start sites (TSSs) is "*in silico* primer extension" (Schmid et al., 2004). This tool makes use of available 5′ EST sequences to help define TSS for a given gene. These estimations are provided in addition to the annotation of experimental evidence derived from literature sources in the Eukaryotic Promoter Database (EDB) (http://www.epd.isb-sib.ch/index.html) (Schmid et al., 2004). There are a few plant specific databases that include information on plant promoters. These include the RIKEN Arabidopsis Genome Encyclopedia (RARGE) (http://rarge.gsc.riken.jp/), which has used the information from over 14,000 nonredundant cDNAs to predict and annotate transcription units (Sakurai et al., 2005). The AGRIS database has been described above and includes the AtcisDB containing more than 25,000 promoter sequences and descriptions of CRE (Davuluri et al., 2003; Palaniswamy et al., 2006). Likewise the GrassPromDB within GRASSIUS provides information on promoters and *cis* elements for monocot species (Yilmaz et al., 2009). The PlantProm DB (http://linux1.softberry.com) provides annotated information on 150 and 430 monocot and dicot promoters with a further 7723 mapped promoters based on cDNA/EST 5′ sequence evidence (Shahmuradov et al., 2003). The OSIRIS database (http://www.bioinformatics2.wsu.edu/cgi-bin/Osiris/cgi/home.pl) is specific for rice and contains promoter sequences, predicted TF binding sites, gene ontology annotation, and microarray expression data for 24,209 genes in the rice genome (Morris et al., 2008), and statistical tools permit the user to visualize TF binding sites in multiple promoters; analyze the statistical significance of enriched TF binding sites; query for genes containing similar promoter regulatory logic or gene function and visualize the microarray-expression patterns of queried or selected gene sets.

A number of other databases function to catalog and define the CRE within plant promoters. The AthaMap database (www.athamap.de/) generates a map for *cis*-regulatory sequences for the whole *A. thaliana* genome. This database was initially developed by matrix-based detection of putative TF-binding sites (TFBS), mostly determined from random binding site selection experiments. Experimentally verified TFBS have also been included for 48 different *Arabidopsis* TFs. Using this resource it is possible to search for TFBS in a genomic segment and find enriched (overrepresented) sequences, for example, for drought responsive elements in cold-induced genes (Buelow et al., 2006; Galuschka et al., 2007). PPDB (http://www.ppdb.gene.nagoya-u.ac.jp) is a plant promoter database that provides promoter annotation of both *Arabidopsis* and rice (Yamamoto and Obokata, 2008). This database contains information on promoter structures, TSSs that have been identified from full-length cDNA clones, and also a large amount of TSS tag data. In PPDB, the promoter structures are determined by sets of promoter elements identified by a position-sensitive extraction method called local distribution of short sequences (LDSS) (Yamamoto et al., 2006, 2007). By using this database, the core promoter structure, the presence of regulatory elements and the distribution of TSS clusters can be identified (Yamamoto et al., 2009). Each regulatory sequence is hyperlinked to literary information, a PLACE entry served by a plant *cis*-element database, and a list of promoters containing the regulatory sequence. The Plant *Cis*-acting Regulatory DNA Elements (PLACE) database (http://www.dna.affrc.go.jp/PLACE/) provides information on motifs found in plant promoters, but since 2007 updates have been discontinued (Higo et al., 1999). PlantCARE (http://bioinformatics.psb.ugent.be/webtools/plantcare/html/) is a database of plant *cis*-acting regulatory elements, enhancers, and repressors (Lescot et al., 2002). Regulatory elements are represented by positional matrices, consensus sequences, and individual sites on particular promoter sequences. Data about the transcription sites are extracted mainly from the literature, supplemented with *in silico* predicted data. Apart from a general description for specific TF sites, levels of confidence for the experimental evidence, functional information and the position on the promoter are also provided. In order to find the conserved regions among promoters across several species, the VISTA alignment tool (http://www-gsd.lbl.gov/vista/) is useful for aligning noncoding regions (Guo and Moose, 2003). Lastly, the Transcription Regulatory Regions Database (TRRD) (http://www.bionet.nsc.ru/trrd/) contains experimentally confirmed data on (1) TFBSs; (2) regulatory units (promoter regions, enhancers, and silencers); and (3) locus control regions, mainly for human genes but also for other eukaryotic genes (Kolchanov et al., 2002, 2006).

4.3.3 Synthetic Promoters

An area that has been slow to take off but which is gaining some momentum is to generate synthetic promoters that deliver expression to particular cell types/tissues (Klein-Marcuschamer et al., 2010; Rushton et al., 2002). Synthetic promoters can significantly benefit from information available on the frequency and distribution of particular motifs (words) in plant genomes (Lichtenberg et al., 2009; Molina and Grotewold, 2005). More recently, a screen of a library containing 52,429 unique synthetic 100 bp oligomers yielded promoters that were as potent at activating transcription as the WT Cauliflower Mosaic Virus immediate early enhancer (Schlabach et al., 2010). This study suggests that the 10-mer synthetic enhancer space is sufficiently rich to allow the creation of synthetic promoters of all strengths in most, if not all, cell types.

4.4 ESTABLISHING GENE REGULATORY NETWORKS

4.4.1 Tools for Establishing Gene Regulatory Networks

A key step in linking TFs to the genes that they regulate is to investigate which regulatory sequences TFs are bound *in vivo*, in particular tissues and at a given time. Recruitment of a TF to a particular promoter in no way guarantees that changes in gene expression will be detected. Indeed, from several recent plant genome-wide comparisons of TF locations and alterations of gene expression, the picture that emerges is that less than 20% of the genes targeted by a particular TF show a detectable change of gene expression (Kaufmann et al., 2009; Lee et al., 2007; Morohashi et al., 2009; Oh et al., 2009; Thibaud-Nissen et al., 2006; Zheng et al., 2009). Whether this reflects that the targets gene are expressed at times not captured in the experiment, or that TFs can "rest" on promoter sequences without consequences for gene expression, it is not known. Despite these limitations, establishing the genomic sites to which TFs bind has become an important aspect of studying gene regulatory networks. Several tools have therefore been developed to achieve this goal.

4.4.1.1 Chromatin Immunoprecipitation (ChIP)-Based Techniques

In ChIP, intact tissues or cells are treated with a cross-linking agent such as formaldehyde that will covalently link *in vivo* proteins (TFs, histones or other proteins that interact with DNA) with the DNA. The chromatin is then extracted, sheared, for example, by sonication or using enzymes, and the covalently linked protein–DNA complex is recovered by immunoprecipitation (IP) using antibodies specific to the protein, or to a tag incorporated into it. After ChIP, the protein–DNA complex is reverse crosslinked and the DNA is purified (Morohashi et al., 2009; Wells and Farnham, 2002). Establishing which DNA fragments are present in the ChIPed material can be accomplished by normal PCR, interrogating for the presence of a particular promoter fragment. Alternatively, the ChIPed DNA can be amplified and used to hybridize a microarray composed of oligonucleotides either corresponding to promoters or covering the entire genome (tiling array). This method to identify all the sequences to which a TF binds has been termed ChIP-chip (Beyer et al., 2006; Buck and Lieb, 2004; O'Geen et al., 2006). Alternatively, specific linkers can be ligated to the ChIPed material and subjected to high-throughput sequencing, using any of the second-generation sequencing platforms available (e.g., SOLiD, SOLEXA, or 454). This technique, known as ChIP-Seq (Johnson et al., 2007; Mardis, 2007), requires a robust computational platform for the analysis of the short reads, but in contrast to ChIP-Seq, is not dependent on prior knowledge of the complete genome sequence. Both ChIP-chip and ChIP-Seq have been used in plants, primarily Arabidopsis, to investigate the genome-wide location of various TFs (Kaufmann et al., 2009; Lee et al., 2007; Morohashi et al., 2009; Oh et al., 2009; Thibaud-Nissen et al., 2006; Zheng et al., 2009). A recent modification of ChIP (known as serial ChIP, re-ChIP, or double ChIP) allows one to investigate whether two or more DNA-binding proteins are localized at the same time on the same DNA molecule, or whether the binding of one excludes a second from binding. This technique was recently implemented in plants (Xie and Grotewold, 2008), and its implementation, coupled with the hybridization of microarrays or high-throughput sequencing, is likely to provide novel and important insight on combinatorial mechanisms underlying plant regulation of gene expression.

4.4.1.2 Using Fusions of TFs with the Glucocorticoid Receptor

An alternative, or perhaps even more powerful as a complement to ChIP methods, is the use of TF fusions to the hormone-binding domain of the glucocorticoid receptor (GR) receptor (TF-GR fusions), followed by the identification of the mRNAs induced/repressed in the presence of the GR ligand (dexamethasone, DEX), in the presence of an inhibitor of translation (e.g., cycloheximide, CHX). This method has been used in several occasions in plants (Morohashi and Grotewold, 2009; Morohashi et al., 2007; Sablowski and Meyerowitz, 1998; Shin et al., 2002; Spelt et al., 2002; Wang et al., 2006; Wellmer et al., 2006; Zhao et al., 2008). Significant limitations include the need to generate transgenic plants for the TF-GR construct in which the TF is shown to recapitulate the function of the endogenous one, and problems associated with genome-wide alterations in mRNA accumulation induced by CHX.

4.4.1.3 Yeast One-Hybrid Experiments

Gene-centered approaches in which one or a collection of promoter or gene regulatory sequences are used as "baits" to fish out TFs that can bind to them have gained significant momentum since the identification of 238+ interactions in *C. elegans* using a yeast one-hybrid approach (Deplancke et al., 2006; Walhout et al., 2000). These studies resulted in the generation of vectors and yeast strains (Deplancke et al., 2004) amenable for applying similar tools to the study of gene interaction networks in plants.

4.4.1.4 Coexpression Analyses

Another strategy that is significantly contributing to predicting hierarchical relationships between genes is coexpression analysis. In a nutshell, if a TF regulates the expression or two or more genes, then it is very likely that these two genes, and likely the TF as well, will be expressed more often under the same conditions or tissues than two randomly chosen genes. Such coexpression analyses have not only been used to identify the TFs regulating a particular plant metabolic pathway, but also to identify additional pathway genes when a few are known (Hirai et al., 2007; Saito et al., 2008; Yonekura-Sakakibara et al., 2007). Databases such as Genevestigator (https://www.genevestigator.com) (Zimmermann et al., 2004, 2005) and ATTED-II (http://atted.jp/) (Hirai et al., 2007) provide information and tools to explore the coexpression of *Arabidopsis* genes.

4.4.2 Gene Regulatory Networks

Ultimately, predicting the consequences on gene expression of perturbing a particular regulatory step in an organism requires a clear understanding of the architecture and dynamics of the underlying gene regulatory network (GRN) that describe all the interactions between TFs, and of TFs with non-TF genes, and how they evolve over space–time. Studies in *Saccaromyces cerevisiae* and *Escherichia coli* (Yu and Gerstein, 2006) have suggested a pyramid-shaped structure with discrete hierarchical levels. Master TFs are situated near the top of the pyramid and they directly control few centrally located TFs. At the bottom of the pyramid are the regulatory factors that control structural proteins and enzymes responsible for carrying out most of the cellular processes.

GRNs can be represented statically by using graphs, such that the nodes represent the proteins, and the edges represent the interactions between TFs and their targets. By using directed graphs, directionality (hierarchy) to these interactions can be added, therefore making it clear which factor controls which other. The structure and topology of the derived graphs has been the subject of intense study in the past few years (Barabasi, 2009; Barabasi and Oltvai, 2004). In plants, the study of GRN is only beginning, and most of the data available are from *Arabidopsis*. If all the available information on this reference organism is put together, static networks like the one shown in Figure 4.1 can be obtained.

4.5 THE COMPLICATING ISSUES OF HETEROSIS AND EPIGENETICS

Although it is not our intention to provide here a complete description of epigenetics, issues like heterosis and the role of small RNAs in modulating TF gene activity are central to plant breeding. Heterosis, or hybrid vigor, is the improved performance of the first filial generation (F1) of the cross of two inbred lines. Heterosis is the premise behind the success of hybrid seeds for many crops. The genetic and molecular bases for heterosis are not fully understood. Single- and multiple-locus models combined with the multiplicative effect of particular interactions (dominance and overdominance hypotheses) have been proposed (Charlesworth and Willis, 2009; Hochholdinger and Hoecker, 2007). A complicating issue is the recent suggestion, based on results obtained from a panel of crosses between B73 and several other maize inbred lines that the genetic basis of heterosis might be dependent on the particular trait being examined (Buckler et al., 2009). In addition, gene-expression analyses suggested that transcriptional variation between parental inbred lines might be more important in hybrid vigor than higher levels of additive or nonadditive gene expression (Stupar et al., 2008).

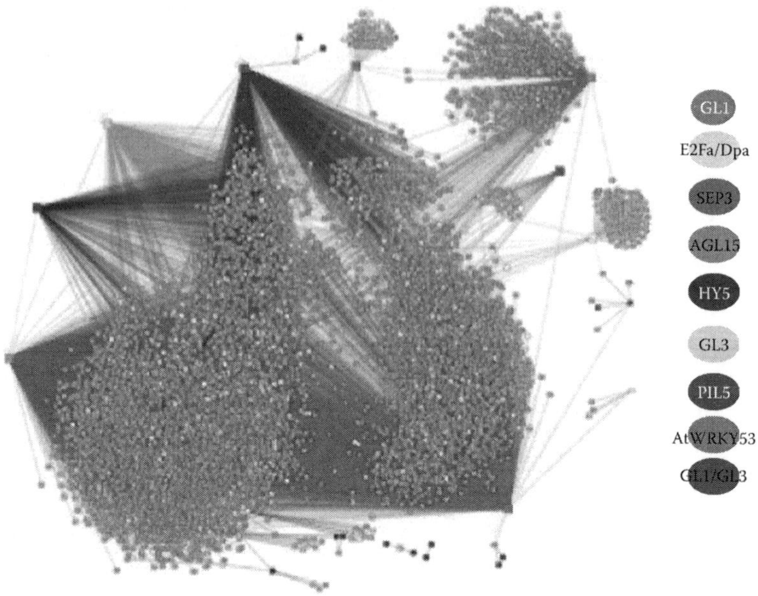

Figure 4.1 Graphical representation of the experimentally determined Arabidopsis gene regulatory network, as obtained from the RegNet database of AGRIS using the ReIN application (Palaniswamy et al., 2006). The current network represents ~10,000 interactions by nine different TFs (shaded gray dots) with their targets (white dots). The TFs in this figure are as follows: GL1, GLABRA1: E2Fa/Dpa, E2Fa, and DPa heterodimer; SEP3, SEPELLATA3; AGL15, AGAMOUS-LIKE15: HY5, ELONGATED HYPOCOTYL5; GL3, GLABRA3; PIL5, PHYTOCHROME INTERACTING FACTOR 3-LIKE 5; AtWRKY53; and GL1/GL3 GLABRA1 and 3 heterodimer.

Two main classes of regulatory small RNAs (smRNAs, 20–24 bp long in plants) are the microRNAs (miRNAs) and several classes of endogenous small interfering (siRNAs). smRNAs have emerged over the past 10 years as key players in control of gene expression. Plant miRNAs direct the cleavage, and sometimes the translational repression, of mRNA target transcripts by a number of mechanisms that involve evolutionarily conserved ARGONAUTE (AGO) proteins (Mallory and Bouche, 2008). In Arabidopsis, many TF mRNAs are targets of smRNAs. Some of these smRNAs are conserved in other plants, suggesting perhaps similar regulatory mechanisms (Rajagopalan et al., 2006). A complete list of miRNA and tasiRNAs (*trans*-acting small interfering RNAs) targets for *Arabidopsis* is available at the Arabidopsis Small RNA Project (ASRP, http://asrp.cgrb.oregonstate.edu/db/) database. The Cereal Small RNAs Database (CSRDB, http://sundarlab.ucdavis.edu/smrnas/) has also an increasing number of smRNAs for grasses, particularly maize and rice. We anticipate that mixed GRNs in which the interaction of TFs with their target genes will be complemented by information on the targets of small RNAs and how they fine modulate gene expression, will significantly expand our current view of how gene expression is controlled.

4.6 FUTURE PERSPECTIVES

A significant challenge in further harnessing the power of TFs for agronomical purposes is how little continues to be known with regard to the functions and targets of TFs in important crops. Thus, identifying the GRNs in which TFs participate is a very important priority. This knowledge can then be used to directly control agronomic traits, either using gene knock-out or overexpression strategies, which in most cases will involve the generation of transgenic plants. However, given the complex nature of GRNs it is not necessary to share the same genes to generate natural variations.

Alternatively, knowledge on the function of TFs can be utilized by breeders for marker-assisted breeding. To this end, there is a need for extensive databases housing TF information and increased efforts to link TF location with QTLs for important agronomic traits. Once the genes underlying such QTLs are defined, there is strong potential for this knowledge to be applied across multiple plant species.

REFERENCES

Aharoni, A., S. Dixit, R. Jetter et al. 2004. The SHINE clade of AP2 domain transcription factors activates wax biosynthesis, alters cuticle properties, and confers drought tolerance when overexpressed in Arabidopsis. *Plant Cell* 16:2463–2480.

Ahern, K. 2009. Nailing fingers. *Biotechniques* 46:499.

Ahmad, A., I. Kaji, Y. Murakami et al. 2009. Transformation of *Arabidopsis* with plant-derived DNA sequences necessary for selecting transformants and driving an objective gene. *Biosci Biotechnol Biochem* 73:936–938.

Allen, D. K., I. G. L. Libourel, and Y. Shachar-Hill. 2009. Metabolic flux analysis in plants: Coping with complexity. *Plant Cell Environ* 32:1241–1257.

Amoutzias, G. D., A. S. Veron, J. Weiner et al. 2007. One billion years of bZIP transcription factor evolution: Conservation and change in dimerization and DNA-binding site specificity. *Mol Biol Evol* 24:827–835.

An, Y.-Q., J. M. McDowell, S. Huang et al. 1996. Strong, constitutive expression of the *Arabidopsis* ACT2/ACT8 actin subclass in vegetative tissues. *Plant J* 10:107–121.

Ariel, F. D., P. A. Manavella, C. A. Dezar, and R. L. Chan. 2007. The true story of the HD-Zip family. *Trends Plant Sci* 12:419–426.

Barabasi, A. L. 2009. Scale-free networks: A decade and beyond. *Science* 325:412–413.

Barabasi, A. L. and Z. N. Oltvai. 2004. Network biology: Understanding the cell's functional organization. *Nat Rev Genet* 5:101–113.

Beyer, A., C. Workman, J. Hollunder et al. 2006. Integrated assessment and prediction of transcription factor binding. *PLoS Comput Biol* 2:e70.

Bhatnagar-Mathur, P., V. Vadez, and K. K. Sharma. 2008. Transgenic approaches for abiotic stress tolerance in plants: Retrospect and prospects. *Plant Cell Reports* 27:411–424.

Birney, E., J. A. Stamatoyannopoulos, A. Dutta et al. 2007. Identification and analysis of functional elements in 1% of the human genome by the ENCODE pilot project. *Nature* 447:799–816.

Bovy, A., R. de Vos, M. Kemper et al. 2002. High-flavonol tomatoes resulting from the heterologous expression of the maize transcription factor genes LC and C1. *Plant Cell* 14:2509–2526.

Braun, E. L. and E. Grotewold. 1999. Newly discovered plant c-myb-like genes rewrite the evolution of the plant myb gene family. *Plant Physiol* 121:21–24.

Broun, P. 2004. Transcription factors as tools for metabolic engineering in plants. *Curr Opin Plant Biol* 7:202–209.

Buck, M. J. and J. D. Lieb. 2004. ChIP-chip: Considerations for the design, analysis, and application of genome-wide chromatin immunoprecipitation experiments. *Genomics* 83:349–360.

Buckler, E. S., J. B. Holland, P. J. Bradbury et al. 2009. The genetic architecture of maize flowering time. *Science* 325:714–718.

Buelow, L., N. O. Steffens, C. Galuschka, M. Schindler, and R. Hehl. 2006. AthaMap: From *in silico* data to real transcription factor binding sites. *In Silico Biol* 6:243–252.

Butelli, E., L. Titta, M. Giorgio et al. 2008. Enrichment of tomato fruit with health-promoting anthocyanins by expression of select transcription factors. *Nat Biotechnol* 26:1301–1308.

Byrne, P. F., M. D. McMullen, M. E. Snook et al. 1996. Quantitative trait loci and metabolic pathways: Genetic control of the concentration of maysin, a corn earworm resistance factor, in maize silks. *Proc Natl Acad Sci USA* 93:8820–8825.

Byrne, P. F., M. D. McMullen, B. R. Wiseman et al. 1998. Maize silk maysin concentration and corn earworm antibiosis: QTLs and genetic mechanisms. *Crop Science* 38:461–471.

Caldana, C., W. R. Scheible, B. Mueller-Roeber, and S. Ruzicic. 2007. A quantitative RT-PCR platform for high-throughput expression profiling of 2500 rice transcription factors. *BMC Plant Methods* 3:7.

Callis, J., J. A. Raasch, and R. D. Vierstra. 1990. Ubiquitin extension proteins of *Arabidopsis thaliana* structure localization and expression of their promoters in transgenic tobacco. *J Biol Chem* 265:12486–12493.

Caparros-Ruiz, D., M. Capellades, S. Fornale, P. Puigdomenech, and J. Rigau. 2007. Downregulation of structural lignin genes to improve digestibility and bioethanol production in maize. *J Biotechnol* 131:S29–S30.

Cardon, G., S. Hohmann, J. Klein et al. 1999. Molecular characterisation of the *Arabidopsis* SBP-box genes. *Gene* 237:91–104.

Carninci, P., A. Westover, Y. Nishiyama et al. 1997. High efficiency selection of full-length cDNA by improved biotinylated cap trapper. *DNA Res* 4:61–66.

Carninci, P., C. Kvam, A. Kitamura et al. 1996. High-efficiency full-length cDNA cloning by biotinylated CAP trapper. *Genomics* 37:327–336.

Century, K., T. L. Reuber, and O. J. Ratcliffe. 2008. Regulating the regulators: The future prospects for transcription-factor-based agricultural biotechnology products. *Plant Physiol* 147:20–29.

Charlesworth, D. and J. H. Willis. 2009. The genetics of inbreeding depression. *Nat Rev Genet* 10:783–796.

Cone, K. C., F. A. Burr, and B. Burr. 1986. Molecular analysis of the maize anthocyanin regulatory locus *c1*. *Proc Natl Acad Sci USA* 83:9631–9635.

Cornejo, M.-J., D. Luth, K. M. Blankenship, O. D. Anderson, and A. E. Blechl. 1993. Activity of a maize ubiquitin promoter in transgenic rice. *Plant Mol Biol* 23:567–581.

Correa, L. G., D. M. Riano-Pachon, C. G. Schrago et al. 2008. The role of bZIP transcription factors in green plant evolution: Adaptive features emerging from four founder genes. *PLoS One* 3:e2944.

Dai, X., Y. Xu, Q. Ma et al. 2007. Overexpression of an R1R2R3 MYB gene, OsMYB3R-2, increases tolerance to freezing, drought, and salt stress in transgenic *Arabidopsis*. *Plant Physiol* 143:1739–1751.

Davuluri, R. V., H. Sun, S. K. Palaniswamy et al. 2003. AGRIS: Arabidopsis gene regulatory information server, an information resource of *Arabidopsis cis*-regulatory elements and transcription factors. *BMC Bioinform* 4:25.

De Buck, S. and A. Depicker. 2004. Gene expression and level of expression. In *Handbook of Plant Biotechnology*, eds. P. Christou, and H. Klee, pp. 331–351. Hoboken, NJ: John Wiley & Sons Inc.

Deplancke, B., A. Mukhopadhyay, W. Ao et al. 2006. A gene-centered *C. elegans* protein–DNA interaction network. *Cell* 125:1193–1205.

Deplancke, B., D. Dupuy, M. Vidal, and A. J. Walhout. 2004. A gateway-compatible yeast one-hybrid system. *Genome Res* 14:2093–2101.

Dhekney, S. A., R. E. Litz, D. A. M. Amador, and A. K. Yadav. 2007. Potential for introducing cold tolerance into papaya by transformation with C-repeat binding factor (CBF) genes. *In Vitro Cell Dev Biol Plant* 43:195–202.

Dias, A. P., E. L. Braun, M. D. McMullen, and E. Grotewold. 2003. Recently duplicated maize *R2R3 Myb* genes provide evidence for distinct mechanisms of evolutionary divergence after duplication. *Plant Physiol* 131:610–620.

Distelfeld, A., A. Korol, J. Dubcovsky et al. 2008. Colinearity between the barley grain protein content (GPC) QTL on chromosome arm 6HS and the wheat Gpc–B1 region. *Molecular Breeding* 22:25–38.

Dixon, R. A. 2005. Engineering of plant natural product pathways. *Curr Opin Plant Biol* 8:329–336.

Doebley, J. F., B. S. Gaut, and B. D. Smith. 2006. The molecular genetics of crop domestication. *Cell* 127:1309–1321.

Doebley, J., A. Stec, and L. Hubbard. 1997. The evolution of apical dominance in maize. *Nature* 386:485–488.

Elhiti, M. and C. Stasolla. 2009. Structure and function of homodomain-leucine zipper (HD-Zip) proteins. *Plant Signal Behav* 4:86–88.

Falco, S. C., T. Guida, M. Locke et al. 1995. Transgenic canola and soybean seeds with increased lysine. *Bio-Technology (New York)* 13:577–582.

Finn, R. D., J. Mistry, J. Tate et al. 2010. The Pfam protein families database. *Nucleic Acids Res* Database Issue 38:D211–222.

Flowers, J. M., Y. Hanzawa, M. C. Hall, R. C. Moore, and M. D. Purugganan. 2009. Population genomics of the *Arabidopsis thaliana* flowering time gene network. *Mol Biol Evol* 26:2475–2486.

Fornale, S., F.-M. Sonbol, T. Maes et al. 2006. Down-regulation of the maize and *Arabidopsis thaliana* caffeic acid *O*-methyl-transferase genes by two new maize R2R3-MYB transcription factors. *Plant Mol Biol* 62:809–823.

Francia, E., D. Barabaschi, A. Tondelli et al. 2007. Fine mapping of a *HvCBF* gene cluster at the frost resistance locus *Fr-H2* in barley. *Theoretical and Applied Genetics* 115:1083–1091.

Fu, F., J. D. Sander, M. Maeder et al. 2009. Zinc Finger Database (ZiFDB): A repository for information on C2H2 zinc fingers and engineered zinc-finger arrays. *Nucleic Acids Res* 37:D279–D283.

Fukao, T., T. Harris, and J. Bailey-Serres. 2009. Evolutionary analysis of the *Sub1* gene cluster that confers submergence tolerance to domesticated rice. *Ann Bot* 103:143–150.

Gallavotti, A., Q. Zhao, J. Kyozuka et al. 2004. The role of *barren stalk1* in the architecture of maize. *Nature* 432:630–635.

Galuschka, C., M. Schindler, L. Buelow, and R. Hehl. 2007. AthaMap web tools for the analysis and identification of co-regulated genes. *Nucleic Acids Res* 35:D857–D862.

Goicoechea, M., E. Lacombe, S. Legay et al. 2005. EgMYB2, a new transcriptional activator from *Eucalyptus* xylem, regulates secondary cell wall formation and lignin biosynthesis. *Plant J* 43:553–567.

Gonzalez, A., M. Zhao, J. M. Leavitt, and A. M. Lloyd. 2008. Regulation of the anthocyanin biosynthetic pathway by the TTG1/bHLH/Myb transcriptional complex in *Arabidopsis* seedlings. *Plant J* 53:814–827.

Gray, J., M. Bevan, T. Brutnell et al. 2009. A recommendation for naming transcription factor proteins in the grasses. *Plant Physiol* 149:4–6.

Greenup, A., W. J. Peacock, E. S. Dennis, and B. Trevaskis. 2009. The molecular biology of seasonal flowering-responses in *Arabidopsis* and the cereals. *Ann Bot* 103:1165–1172.

Grotewold, E. 2005. Plant metabolic diversity: A regulatory perspective. *Trends in Plant Science* 10:57–62.

Grotewold, E. 2008. Transcription factors for predictive plant metabolic engineering: Are we there yet? *Curr Opin Biotechnol* 19:138–144.

Grotewold, E. and J. Gray. 2009. Maize transcription factors. In *The Maize Handbook*, eds. S. Hake and J. Bennetzen, pp. 693–713. New York, NY: Springer.

Grotewold, E., B. J. Drummond, B. Bowen, and T. Peterson. 1994. The myb-homologous P gene controls phlobaphene pigmentation in maize floral organs by directly activating a flavonoid biosynthetic gene subset. *Cell* 76:543–553.

Grotewold, E., M. Chamberlin, M. Snook et al. 1998. Engineering secondary metabolism in maize cells by ectopic expression of transcription factor. *Plant Cell* 10:721–740.

Grotewold, E., P. Athma, and T. Peterson. 1991. Alternatively spliced products of the maize P gene encode proteins with homology to the DNA-binding domain of myb-like transcription factors. *Proc Natl Acad Sci USA* 88:4587–4591.

Guo, A.-Y., X. Chen, G. Gao et al. 2008. PlantTFDB: A comprehensive plant transcription factor database. *Nucleic Acids Res* 36:D966–D969.

Guo, H. and S. P. Moose. 2003. Conserved noncoding sequences among cultivated cereal genomes identify candidate regulatory sequence elements and patterns of promoter evolution. *Plant Cell* 15:1143–1158.

Hartley, J. L., G. F. Temple, and M. A. Brasch. 2000. DNA cloning using *in vitro* site-specific recombination. *Genome Res* 10:1788–1795.

Hehl, R., N. O. Steffens, and E. Wingender. 2004. Isolation and analysis of gene regulatory sequences. In *Handbook of Plant Biotechnology*, eds. P. Christou, and H. Klee, pp. 81–102. Hoboken, NJ: John Wiley & Sons Inc.

Hernandez, J., G. Heine, N. G. Irani et al. 2004. Different mechanisms participate in the R-dependent activity of the R2R3 MYB transcription factor C1. *J Biol Chem* 279:48205–48213.

Higo, K., Y. Ugawa, M. Iwamoto, and T. Korenaga. 1999. Plant *cis*-acting regulatory DNA elements (PLACE) database: 1999. *Nucleic Acids Res* 27:297–300.

Hirai, M. Y., K. Sugiyama, Y. Sawada et al. 2007. Omics-based identification of *Arabidopsis* Myb transcription factors regulating aliphatic glucosinolate biosynthesis. *Proc Natl Acad Sci USA* 104:6478–6483.

Hiratsu, K., K. Matsui, T. Koyama, and M. Ohme-Takagi. 2003. Dominant repression of target genes by chimeric repressors that include the EAR motif, a repression domain, in *Arabidopsis*. *Plant J* 34:733–739.

Hochholdinger, F. and N. Hoecker. 2007. Towards the molecular basis of heterosis. *Trends Plant Sci* 12:427–432.

Initiative, I. B., J. P. Vogel, D. F. Garvin et al. 2010. Genome sequencing and analysis of the model grass *Brachypodium distachyon*. *Nature* 463:763–768.

Ito, M. 2005. Conservation and diversification of three-repeat Myb transcription factors in plants. *Journal of Plant Research* 118:61–69.

Ito, M., S. Araki, S. Matsunaga et al. 2001. G2/M-phase-specific transcription during the plant cell cycle is mediated by c-Myb-like transcription factors. *Plant Cell* 13:1891–1905.

Itoh, J.-I., K.-I. Hibara, Y. Sato, and Y. Nagato. 2008. Developmental role and auxin responsiveness of class III homeodomain leucine zipper gene family members in rice. *Plant Physiol* 147:1960–1975.

Iwase, A., K. Matsui, and M. Ohme-Takagi. 2009. Manipulation of plant metabolic pathways by transcription factors. *Plant Biotechnology* 26:29–38.

Jakoby, M., B. Weisshaar, W. Droge-Laser et al. 2002. bZIP transcription factors in *Arabidopsis. Trends Plant Sci* 7:106–111.

Jayakanthan, M., J. Muthukumaran, S. Chandrasekar et al. 2009. ZifBASE: A database of zinc finger proteins and associated resources. *BMC Genomics* 10:421.

Jin, H., E. Cominelli, P. Bailey et al. 2000. Transcriptional repression by AtMYB4 controls production of UV-protecting sunscreens in *Arabidopsis. EMBO Journal* 19:6150–6161.

Johnson, D. S., A. Mortazavi, R. M. Myers, and B. Wold. 2007. Genome-wide mapping of *in vivo* protein–DNA interactions. *Science* 316:1497–1502.

Kappen, C. 2000. The homeodomain: An ancient evolutionary motif in animals and plants. *Comput Chem* 24:95–103.

Kassim, A., J. Poette, A. Paterson et al. 2009. Environmental and seasonal influences on red raspberry anthocyanin antioxidant contents and identification of quantitative traits loci (QTL). *Molecular Nutrition & Food Research* 53:625–634.

Katiyar-Agarwal, S., A. Kapoor, and A. Grover. 2002. Binary cloning vectors for efficient genetic transformation of rice. *Curr Sci* 82:873–876.

Kaufmann, K., J. M. Muino, R. Jauregui et al. 2009. Target genes of the MADS transcription factor SEPALLATA3: Integration of developmental and hormonal pathways in the *Arabidopsis* flower. *PLoS Biol* 7:e1000090.

Kaur, P., S. R. Larson, B. S. Bushman et al. 2008. Genes controlling plant growth habit in *Leymus* (Triticeae): *maize barren stalk1* (*ba1*), rice *lax panicle*, and wheat *tiller inhibition* (*tin3*) genes as possible candidates. *Functional & Integrative Genomics* 8:375–386.

Kim, S.-J., S.-C. Lee, S. K. Hong et al. 2009. Ectopic expression of a cold-responsive OsAsr1 cDNA gives enhanced cold tolerance in transgenic rice plants. *Mol Cells* 27:449–458.

Kizis, D. and M. Pages. 2002. Maize DRE-binding proteins DBF1 and DBF2 are involved in *rab17* regulation through the drought-responsive element in an ABA-dependent pathway. *Plant J* 30:679–689.

Klein-Marcuschamer, D., V. G. Yadav, A. Ghaderi, and G. N. Stephanopoulos. 2009. *De novo* metabolic engineering and the promise of synthetic DNA. *Adv Biochem Eng Biotechnol* 120:101–131.

Kolchanov, N. A., E. V. Ignatieva, E. A. Ananko et al. 2002. Transcription Regulatory Regions Database (TRRD): Its status in 2002. *Nucleic Acids Res* 30:312–317.

Kolchanov, N., E. Ignatieva, O. Podkolodnaya et al. 2006. Transcription Regulatory Regions Database (TRRD): A source of experimentally confirmed data on transcription regulatory regions of eukaryotic genes. In *Bioinformatics of Genome Regulation and Structure II*, eds. N. Kolchanov, R. Hofestaedt, and L. Milanesi. pp. 43–53. New York, NY: Springer Sciences + Business Media Inc.

Konishi, S., T. Izawa, S. Y. Lin et al. 2006. An SNP caused loss of seed shattering during rice domestication. *Science* 312:1392–1396.

Kottapalli, K. R., K. Satoh, R. Rakwal et al. 2007. Combining *in silico* mapping and arraying: An approach to identifying common candidate genes for submergence tolerance and resistance to bacterial leaf blight in ice. *Molecules and Cells* 24:394–408.

Kruger, N. J. and R. G. Ratcliffe. 2009. Insights into plant metabolic networks from steady-state metabolic flux analysis. *Biochimie (Paris)* 91:697–702.

Kwak, M. S., M.-J. Oh, S. W. Lee et al. 2007. A strong constitutive gene expression system derived from ibAGP1 promoter and its transit peptide. *Plant Cell Reports* 26:1253–1262.

Lamesch, P., N. Li, S. Milstein et al. 2007. h0RFeome v3.1: A resource of human open reading frames representing over 10,000 human genes. *Genomics* 89:307–315.

Lee, E. A., P. F. Byrne, M. D. McMullen et al. 1998. Genetic mechanisms underlying apimaysin and maysin synthesis and corn earworm antibiosis in maize (*Zea mays* L.). *Genetics* 149:1997–2006.

Lee, J., K. He, V. Stolc et al. 2007. Analysis of transcription factor HY5 genomic binding sites revealed its hierarchical role in light regulation of development. *Plant Cell* 19:731–749.

Lennon, G., C. Aufrray, M. Polymeropoulos, and M. B. Soares. 1996. The I.M.A.G.E. consortium: An integrated molecular analysis of genomes and their expression. *Genomics* 33:151–152.

Lescot, M., P. Dehais, G. Thijs et al. 2002. PlantCARE, a database of plant *cis*-acting regulatory elements and a portal to tools for *in silico* analysis of promoter sequences. *Nucleic Acids Res* 30:325–327.

Lewinsohn, E. and M. Gijzen. 2009. Phytochemical diversity: The sounds of silent metabolism. *Plant Science* 176:161–169.

Li, C., A. Zhou, and T. Sang. 2006. Rice domestication by reducing shattering. *Science* 311:1936–1939.

Liang, C., P. Jaiswal, C. Hebbard et al. 2008. Gramene: A growing plant comparative genomics resource. *Nucleic Acids Res* 36:D947–D953.

Liang, Y. S., H.-J. Bae, S.-H. Kang et al. 2009. The *Arabidopsis* beta-carotene hydroxylase gene promoter for a strong constitutive expression of transgene. *Plant Biotechnol Rep* 3:325–331.

Libourel, I. G. L. and Y. Shachar-Hill. 2008. Metabolic flux analysis in plants: From intelligent design to rational engineering. *Ann Rev Plant Biol* 59:625–650.

Lichtenberg, J., A. Yilmaz, J. D. Welch et al. 2009. The word landscape of the non-coding segments of the *Arabidopsis thaliana* genome. *BMC Genomics* 10:463.

Lin, Z., M. E. Griffith, X. Li et al. 2007. Origin of seed shattering in rice (*Oryza sativa* L.). *Planta* 226:11–20.

Lu, J., E. Sivamani, X. Li, and R. Qu. 2008. Activity of the 5′ regulatory regions of the rice polyubiquitin rubi3 gene in transgenic rice plants as analyzed by both GUS and GFP reporter genes. *Plant Cell Rep* 27:1587–1600.

Ludwig, S. R., L. F. Habera, S. L. Dellaporta, and S. R. Wessler. 1989. *Lc* a member of the maize *R* gene family responsible for tissue-specific anthocyanin production encodes a protein similar to transcriptional activators and contains the myc-homology region. *Proc Natl Acad Sci USA* 86:7092–7096.

Ma, Q., X. Dai, Y. Xu et al. 2009. Enhanced tolerance to chilling stress in OsMYB3R-2 transgenic rice is mediated by alteration in cell cycle and ectopic expression of stress genes. *Plant Physiol* 150:244–256.

Maeder, M. L., S. Thibodeau-Beganny, A. Osiak et al. 2008. Rapid "Open-Source" engineering of customized zinc-finger nucleases for highly efficient gene modification. *Mol Cell* 31:294–301.

Mahmoud, S. S. and R. B. Croteau. 2001. Metabolic engineering of essential oil yield and composition in mint by altering expression of deoxyxylulose phosphate reductoisomerase and menthofuran synthase. *Proc Natl Acad Sci USA* 98:8915–8920.

Malik, K., K. Wu, X. Q. Li et al. 2002. A constitutive gene expression system derived from the tCUP cryptic promoter elements. *Theoret Appl Genet* 105:505–514.

Mallory, A. C., and N. Bouche. 2008. MicroRNA-directed regulation: To cleave or not to cleave. *Trends Plant Sci* 13:359–367.

Mandel, T., A. J. Fleming, R. Kraehenbuehl, and C. Kuhlemeier. 1995. Definition of constitutive gene expression in plants: The translation initiation factor 4A gene as a model. *Plant Mol Biol* 29:995–1004.

Mandell, J. G. and C. F. Barbas, III. 2006. Zinc finger tools: Custom DNA-binding domains for transcription factors and nucleases. *Nucleic Acids Res* 34:W516–W523.

Mao, C., W. Ding, Y. Wu et al. 2007. Overexpression of a NAC-domain protein promotes shoot branching in rice. *New Phytologist* 176:288–298.

Mardis, E. R. 2007. ChIP-seq: Welcome to the new frontier. *Nat Methods* 4:613–614.

Matsui, K., H. Tanaka, and M. Ohme-Takagi. 2004. Suppression of the biosynthesis of proanthocyanidin in *Arabidopsis* by a chimeric PAP1 repressor. *Plant Biotechnol J* 2:487–493.

Matys, V., O. V. Kel-Margoulis, E. Fricke et al. 2006. TRANSFAC (R) and its module TRANSCompel (R): Transcriptional gene regulation in eukaryotes. *Nucleic Acids Res* 34:D108–D110.

McElroy, D., A. D. Blowers, B. Jenes, and R. Wu. 1991. Construction of expression vectors based on the rice *Actin 1* 5″ region for use in monocot transformation. *Mol Gen Genet* 231:150–160.

McMullen, M. D., P. F. Byrne, M. E. Snook et al. 1998. Quantitative trait loci and metabolic pathways. *Proc Natl Acad Sci USA* 95:1996–2000.

McSteen, P. 2006. Branching out: The ramosa pathway and the evolution of grass inflorescence morphology. *Plant Cell* 18:518–522.

McSteen, P. 2009. Hormonal regulation of branching in grasses. *Plant Physiol* 149:46–55.

Miller, A. K., G. Galiba, and J. Dubcovsky. 2006. A cluster of 11 CBF transcription factors is located at the frost tolerance locus Fr-A(m)2 in Triticum monococcum. *Molecular Genetics and Genomics* 275:193–203.

Mochida, K., T. Yoshida, T. Sakurai et al. 2010. LegumeTFDB: An integrative database of *Glycine max*, *Lotus japonicus* and *Medicago truncatula* transcription factors. *Bioinformatics* 26:290–291.

Molina, C. and E. Grotewold. 2005. Genome wide analysis of *Arabidopsis* core promoters. *BMC Genomics* 6:25.

Morandini, P. and F. Salamini. 2003. Plant biotechnology and breeding: Allied for years to come. *Trends Plant Sci* 8:70–75.

Morohashi, K. and E. Grotewold. 2009. A systems approach reveals regulatory circuitry for *Arabidopsis* trichome initiation by the GL3 and GL1 selectors. *PLoS Genet* 5:e1000396.

Morohashi, K., M. Zhao, M. Yang et al. 2007. Participation of the *Arabidopsis* bHLH factor GL3 in trichome initiation regulatory events. *Plant Physiol* 145:736–746.

Morohashi, K., Z. Xie, and E. Grotewold. 2009. Gene-specific and genome-wide ChIP approaches to study plant transcriptional networks. *Methods Mol Biol* 553:3–12.

Morris, R. T., T. R. O'Connor, and J. J. Wyrick. 2008. Osiris: An integrated promoter database for *Oryza sativa* L. *Bioinformatics* 24:2915–2917.

Mueller, A. and M. Wassenegger. 2004. Control and silencing of transgene expression. In *Handbook of Plant Biotechnology*, eds. P. Christou, and H. Klee, pp. 291–330. Hoboken, NJ: John Wiley & Sons Inc.

Newman, L. J., D. E. Perazza, L. Juda, and M. M. Campbell. 2004. Involvement of the R2R3-MYB, AtMYB61, in the ectopic lignification and dark-photomorphogenic components of the det3 mutant phenotype. *Plant J* 37:239–250.

Ni, M., D. Cui, J. Einstein et al. 1995. Strength and tissue specificity of chimeric promoters derived from the octopine and mannopine synthase genes. *Plant J* 7:661–676.

Nijhawan, A., M. Jain, A. K. Tyagi, and J. P. Khurana. 2008. Genomic survey and gene expression analysis of the basic leucine zipper transcription factor family in rice. *Plant Physiol* 146:333–350.

Odell, J. T., F. Nagy, and N. H. Chua. 1985. Identification of DNA sequences required for activity of the Cauliflower Mosaic Virus 35s promoter. *Nature* 313:810–812.

Oh, E., H. Kang, S. Yamaguchi et al. 2009. Genome-wide analysis of genes targeted by PHYTOCHROME INTERACTING FACTOR 3-LIKE5 during seed germination in Arabidopsis. *Plant Cell* 21:403–419.

Oh, S.-J., S. I. Song, Y. S. Kim et al. 2005. *Arabidopsis* CBF3/DREB1A and ABF3 in transgenic rice increased tolerance to abiotic stress without stunting growth. *Plant Physiol* 138:341–351.

Ohto, M.-a., R. L. Fischer, R. B. Goldberg, K. Nakamura, and J. J. Harada. 2005. Control of seed mass by APETALA2. *Proc Natl Acad Sci USA* 102:3123–3128.

Omirulleh, S., M. Abraham, M. V. G. Stefanov et al. 1993. Activity of a chimeric promoter with the doubled CaMV 35S enhancer element in protoplast-derived cells and transgenic plants in maize. *Plant Mol Biol* 21:415–428.

O'Geen, H., C. M. Nicolet, K. Blahnik, R. Green, and P. J. Farnham. 2006. Comparison of sample preparation methods for ChIP-chip assays. *Biotechniques* 41:577–580.

Pabo, C. O. and R. T. Sauer. 1992. Transcription factors: Structural families and principles of DNA recognition. *Annu Rev Biochem* 61:1053–1095.

Palaniswamy, K., S. James, H. Sun et al. 2006. AGRIS and AtRegNet: A platform to link *cis*-regulatory elements and transcription factors into regulatory networks. *Plant Phyisiol* 140:818–829.

Papworth, M., P. Kolasinska, and M. Minczuk. 2006. Designer zinc-finger proteins and their applications. *Gene* 366:27–38.

Park, S. H., N. Yi, Y. S. Kim et al. 2010. Analysis of five novel putative constitutive gene promoters in transgenic rice plants. *J Exp Bot* 2:2.

Patzlaff, A., S. McInnis, A. Courtenay et al. 2003. Characterisation of a pine MYB that regulates lignification. *Plant J* 36:743–754.

Peel, G. J., Y. Pang, L. V. Modolo, and R. A. Dixon. 2009. The LAP1 MYB transcription factor orchestrates anthocyanidin biosynthesis and glycosylation in *Medicago. Plant J* 59:136–149.

Peng, J., D. E. Richards, N. M. Hartley et al. 1999. "Green revolution" genes encode mutant gibberellin response modulators. *Nature (London)* 400:256–261.

Perez-Rodriguez, P., D. M. Riano-Pachon, L. G. Correa et al. 2010. PlnTFDB: Updated content and new features of the plant transcription factor database. *Nucleic Acids Res* 38:D822–D827.

Pino, M.-T., J. S. Skinner, Z. Jeknic et al. 2008. Ectopic AtCBF1 over-expression enhances freezing tolerance and induces cold acclimation-associated physiological modifications in potato. *Plant Cell Environ* 31:393–406.

Purugganan, M. D., A. L. Boyles, and J. I. Suddith. 2000. Variation and selection at the CAULIFLOWER floral homeotic gene accompanying the evolution of domesticated *Brassica oleracea. Genetics* 155:855–862.

Rabinowicz, P. D., E. L. Braun, A. D. Wolfe, B. Bowen, and E. Grotewold. 1999. Maize R2R3 Myb genes: Sequence analysis reveals amplification in the higher plants. *Genetics* 153:427–444.

Rajagopalan, R., H. Vaucheret, J. Trejo, and D. P. Bartel. 2006. A diverse and evolutionarily fluid set of microRNAs in *Arabidopsis thaliana. Genes Dev* 20:3407–425.

Ramalingam, J., C. M. V. Cruz, K. Kukreja et al. 2003. Candidate defense genes from rice, barley, and maize and their association with qualitative and quantitative resistance in rice. *Mol Plant Microbe Interact* 16:14–24.

Riano-Pachon, D. M., S. Ruzicic, I. Dreyer, and B. Mueller-Roeber. 2007. PlnTFDB: An integrative plant transcription factor database. *BMC Bioinform* 8: Article No.: 42.

Richardt, S., G. Timmerhaus, D. Lang et al. 2010. Microarray analysis of the moss *Physcomitrella patens* reveals evolutionarily conserved transcriptional regulation of salt stress and abscisic acid signalling. *Plant Mol Biol* 72:27–45.

Rose, A. B. 2004. The effect of intron location on intron-mediated enhancement of gene expression in *Arabidopsis*. *Plant J* 40:744–751.

Rose, A. B. and J. A. Beliakoff. 2000. Intron-mediated enhancement of gene expression independent of unique intron sequences and splicing. *Plant Physiol* 122:535–542.

Rose, A. B., T. Elfersi, G. Parra, and I. Korf. 2008. Promoter-proximal introns in *Arabidopsis thaliana* are enriched in dispersed signals that elevate gene expression. *Plant Cell* 20:543–551.

Rual, J.-F., D. E. Hill, and M. Vidal. 2004. ORFeome projects: Gateway between genomics and omics. *Curr Opin Chem Biol* 8:20–25.

Rushton, P. J., A. Reinstadler, V. Lipka, B. Lippok, and I. E. Somssich. 2002. Synthetic plant promoters containing defined regulatory elements provide novel insights into pathogen- and wound-induced signaling. *Plant Cell* 14:749–762.

Sablowski, R. W. M. and E. M. Meyerowitz. 1998. A Homolog of *NO APICAL MERISTEM* is an immediate target of the floral homeotic genes *APETALA3/PISTILLATA*. *Cell* 92:93–103.

Saito, K., M. Y. Hirai, and K. Yonekura-Sakakibara. 2008. Decoding genes with coexpression networks and metabolomics—"Majority report by precogs". *Trends Plant Sci* 13:36–43.

Sakurai, T., M. Satou, K. Akiyama et al. 2005. RARGE: A large-scale database of RIKEN Arabidopsis resources ranging from transcriptome to phenome. *Nucleic Acids Res* 33:D647–D650.

Salmaso, M., G. Malacarne, M. Troggio et al. 2008. A grapevine (*Vitis vinifera* L.) genetic map integrating the position of 139 expressed genes. *Theoretical and Applied Genetics* 116:1129–1143.

Schaub, P., S. Al-Babili, R. Drake, and P. Beyer. 2005. Why is Golden Rice golden (yellow) instead of red? *Plant Physiol* 138:441–450.

Schlabach, M. R., J. K. Hu, M. Li, and S. J. Elledge. 2010. Synthetic design of strong promoters. *Proc Natl Acad Sci USA* 107:2538–2543.

Schmid, C. D., V. Praz, M. Delorenzi, R. Perier, and P. Bucher. 2004. The Eukaryotic Promoter Database EPD: The impact of *in silico* primer extension. *Nucleic Acids Res* 32:D82–D85.

Seki, M., A. Kamiya, P. Carninci, Y. Hayashizaki, and K. Shinozaki. 2009. Generation of full-length cDNA libraries: Focus on plants. *Methods Mol Biol* 533:49–68.

Sera, T. 2009. Zinc finger based artificial transcription factors and their applications. *Advanced Drug Delivery Reviews* 61:513–526.

Shahmuradov, I. A., A. J. Gammerman, J. M. Hancock, P. M. Bramley, and V. V. Solovyev. 2003. PlantProm: A database of plant promoter sequences. *Nucleic Acids Res* 31:114–117.

Shi, Z., J. Wang, X. Wan et al. 2007. Over-expression of rice OsAGO7 gene induces upward curling of the leaf blade that enhanced erect-leaf. *Planta* 226:99–108.

Shigyo, M., M. Hasebe, and M. Ito. 2006. Molecular evolution of the AP2 subfamily. *Gene (Amsterdam)* 366:256–265.

Shin, B., G. Choi, H. Yi et al. 2002. AtMYB21, a gene encoding a flower-specific transcription factor, is regulated by COP1. *Plant J* 30:23–32.

Shiu, S. H., M. C. Shih, and W. H. Li. 2005. Transcription factor families have much higher expansion rates in plants than in animals. *Plant Physiol* 139:18–26.

Shukla, V. K., Y. Doyon, J. C. Miller et al. 2009. Precise genome modification in the crop species *Zea mays* using zinc-finger nucleases. *Nature* 459:437–441.

Simons, K. J., J. P. Fellers, H. N. Trick et al. 2006. Molecular characterization of the major wheat domestication gene Q. *Genetics* 172:547–555.

Skirpan, A., X. Wu, and P. McSteen. 2008. Genetic and physical interaction suggest that BARREN STALK1 is a target of BARREN INFLORESCENCE2 in maize inflorescence development. *Plant J* 55: 787–797.

Snook, M. E., N. W. Widstrom, B. R. Wiseman et al. 1994. New flavone *c*-glycosides from corn (*Zea mays* L.) for the control of the corn earworm (*Helicoverpa zea*). In *Bioregulators for Crop Protection and Pest Control*, eds. P. A. Hedin, pp. 122–135. Washington, DC: American Chemical Society Symposium Series.

Sonbol, F.-M., S. Fornale, M. Capellades et al. 2009. The maize ZmMYB42 represses the phenylpropanoid pathway and affects the cell wall structure, composition and degradability in *Arabidopsis thaliana. Plant Mol Biol* 70:283–296.

Spelt, C., F. Quattrocchio, J. Mol, and R. Koes. 2002. ANTHOCYANIN1 of petunia controls pigment synthesis, vacuolar pH, and seed coat development by genetically distinct mechanisms. *Plant Cell* 14:2121–2135.

Stracke, R., M. Werber, and B. Weisshaar. 2001. The R2R3-MYB gene family in Arabidopsis thaliana. *Curr Opin Plant Biol* 4:447–456.

Street, N. R., A. Sjodin, M. Bylesjo et al. 2008. A cross-species transcriptomics approach to identify genes involved in leaf development. *BMC Genomics* 9:Article No.: 589.

Stupar, R. M., J. M. Gardiner, A. G. Oldre et al. 2008. Gene expression analyses in maize inbreds and hybrids with varying levels of heterosis. *BMC Plant Biol* 8:33.

Tamagnone, L., A. Merida, A. Parr et al. 1998. The AmMYB308 and AmMYB330 transcription factors from Antirrhinum regulate phenylpropanoid and lignin biosynthesis in transgenic tobacco. *Plant Cell* 10:135–154.

Tan, L., X. Li, F. Liu et al. 2008. Control of a key transition from prostrate to erect growth in rice domestication. *Nat Genet* 40:1360–1364.

Thibaud-Nissen, F., H. Wu, T. Richmond et al. 2006. Development of *Arabidopsis* whole-genome microarrays and their application to the discovery of binding sites for the TGA2 transcription factor in salicylic acid-treated plants. *Plant J* 47:152–162.

Tian, L., Y. Pang, and R. A. Dixon. 2008. Biosynthesis and genetic engineering of proanthocyanidins and (iso) flavonoids. *Phytochem Rev* 7:445–465.

Townsend, J. A., D. A. Wright, R. J. Winfrey et al. 2009. High-frequency modification of plant genes using engineered zinc-finger nucleases. *Nature* 459:442–445.

Ubi, B. E. 2007. Molecular mechanisms underlying anthocyanin biosynthesis: A useful tool for the metabolic engineering of the flavonoid pathway genes for novel products. *J Food Agric Environ* 5:83–87.

Umezawa, T., M. Fujita, Y. Fujita, K. Yamaguchi-Shinozaki, and K. Shinozaki. 2006. Engineering drought tolerance in plants: Discovering and tailoring genes to unlock the future. *Curr Opin Biotechnol* 17:113–122.

Van der Fits, L. and J. Memelink. 2000. ORCA3, a jasmonate-responsive transcriptional regulator of plant primary and secondary metabolism. *Science* 289:295–297.

Velasco, R., A. Zharkikh, M. Troggio et al. 2007. A high quality draft consensus sequence of the genome of a heterozygous grapevine variety. *PLoS One* 2:e1326.

Vogel, J. P., D. F. Garvin, T. C. Mockler et al., 2010. Genome sequencing and analysis of the model grass *Brachypodium distachyon. Nature* 463:763–768.

Vrebalov, J., D. Ruezinsky, V. Padmanabhan et al. 2002. A MADS-box gene necessary for fruit ripening at the tomato ripening-inhibitor (rin) locus. *Science (Washington D C)* 296:343–346.

Walhout, A. J. M., R. Sordella, X. Lu et al. 2000. Protein interaction mapping in *C. elegans* using proteins involved in vulval development. *Science* 287:116–122.

Wang, D., N. Amornsiripanitch, and X. Dong. 2006. A genomic approach to identify regulatory nodes in the transcriptional network of systemic acquired resistance in plants. *PLoS Pathog* 2:e123.

Wang, H., T. Nussbaum-Wagler, B. Li et al. 2005. The origin of the naked grains of maize. *Nature* 436:714–719.

Wang, Z., M. Libault, T. Joshi et al. 2010. SoyDB: A knowledge database of soybean transcription factors. *BMC Plant Biology* 10: Article No.:14.

Weinthal, D., A. Tovkach, V. Zeevi, and T. Tzfira. 2010. Genome editing in plant cells by zinc finger nucleases. *Trends Plant Sci* 26:26.

Wellmer, F., M. Alves-Ferreira, A. Dubois, J. L. Riechmann, and E. M. Meyerowitz. 2006. Genome-wide analysis of gene expression during early *Arabidopsis* flower development. *PLoS Genet* 2:e117.

Wells, J., and P. J. Farnham. 2002. Characterizing transcription factor binding sites using formaldehyde cross-linking and immunoprecipitation. *Methods* 26:48–56.

Weng, J.-K., X. Li, N. D. Bonawitz, and C. Chapple. 2008. Emerging strategies of lignin engineering and degradation for cellulosic biofuel production. *Curr Opin Biotechnol* 19:166–172.

Wilson, D., V. Charoensawan, S. K. Kummerfeld, and S. A. Teichmann. 2008. DBD taxonomically broad transcription factor predictions: New content and functionality. *Nucleic Acids Res* Database Issue: D88–D92.

Wingender, E. 2008. The TRANSFAC project as an example of framework technology that supports the analysis of genomic regulation. *Brief Bioinform* 9:326–332.

Wiseman, B. R., M. E. Snook, and D. J. Isenhour. 1993. Maysin content and growth of corn earworm larvae (Lepidoptera: Noctuidae) on silks from first and second ears of corn. *J Econ Entomol* 86:939–944.

Xie, D.-Y., S. B. Sharma, E. Wright, Z.-Y. Wang, and R. A. Dixon. 2006. Metabolic engineering of proanthocyanidins through co-expression of anthocyanidin reductase and the PAP1 MYB transcription factor. *Plant J* 45:895–907.

Xie, Z., and E. Grotewold. 2008. Serial ChIP as a tool to investigate the co-localization or exclusion of proteins on plant genes. *Plant Methods* 4:25.

Yamamoto, Y. Y. and J. Obokata. 2008. ppdb: A plant promoter database. *Nucleic Acids Res* 36:D977–D981.

Yamamoto, Y. Y., H. Ichida, M. Matsui et al. 2007. Identification of plant promoter constituents by analysis of local distribution of short sequences. *BMC Genomics* 8: Article No.:67.

Yamamoto, Y. Y., H. Ichida, T. Abe et al. 2006. LDSS—A novel *in silico* approach for extraction of promoter constituents from plant and mammalian genomes. *Genes Genet Syst* 81:449.

Yamamoto, Y. Y., T. Yoshitsugu, T. Sakurai et al. 2009. Heterogeneity of *Arabidopsis* core promoters revealed by high-density TSS analysis. *Plant J* 60:350–362.

Yan, L., A. Loukoianov, A. Blechl et al. 2004. The wheat VRN2 gene is a flowering repressor down-regulated by vernalization. *Science (Washington DC)* 303:1640–1644.

Yan, L., A. Loukoianov, G. Tranquilli et al. 2003. Positional cloning of the wheat vernalization gene VRN1. *Proc Natl Acad Sci USA* 100:6263–6268.

Yanagisawa, S., A. Akiyama, H. Kisaka, H. Uchimiya, and T. Miwa. 2004. Metabolic engineering with Dof1 transcription factor in plants: Improved nitrogen assimilation and growth under low-nitrogen conditions. *Proc Natl Acad Sci USA* 101:7833–7838.

Yang, X. C. and C. M. Hwa. 2008. Genetic modification of plant architecture and variety improvement in rice. *Heredity* 101:396–404.

Yang, Y., A. Costa, N. Leonhardt, R. S. Siegel, and J. I. Schroeder. 2008. Isolation of a strong *Arabidopsis* guard cell promoter and its potential as a research tool. *Plant Methods* 4:6.

Yano, M., Y. Katayose, M. Ashikari et al. 2000. Hd1, a major photoperiod sensitivity quantitative trait locus in rice, is closely related to the *Arabidopsis* flowering time gene CONSTANS. *Plant Cell* 12:2473–2483.

Ye, X., S. Al-Babili, A. Kloti et al. 2000. Engineering the provitamin A (beta-carotene) biosynthetic pathway into (carotenoid-free) rice endosperm. *Science* 287:303–305.

Yeu, S. Y., B. S. Park, W. G. Sang et al. 2007. The serine proteinase inhibitor OsSerpin is a potent tillering regulator in rice. *Journal of Plant Biology* 50:600–604.

Yilmaz, A., M. Y. Nishiyama, Jr., B. G. Fuentes et al. 2009. GRASSIUS: A platform for comparative regulatory genomics across the grasses. *Plant Physiol* 149:171–180.

Yonekura-Sakakibara, K., T. Tohge, R. Niida, and K. Saito. 2007. Identification of a flavonol 7-O rhamnosyltransferase gene determining flavonoid pattern in Arabidopsis by transcriptome coexpression analysis and reverse genetics. *J Biol Chem* 282:14932–14941.

Yu, H. and M. Gerstein. 2006. Genomic analysis of the hierarchical structure of regulatory networks. *Proc Natl Acad Sci USA* 103:14724–14731.

Yuan, Y.-X., J. Wu, R.-F. Sun et al. 2009. A naturally occurring splicing site mutation in the *Brassica rapa* FLC1 gene is associated with variation in flowering time. *J Exp Bot* 60:1299–1308.

Zhang, G.-H., Q. Xu, X.-D. Zhu, Q. Qian, and H.-W. Xue. 2009. SHALLOT-LIKE1 is a KANADI transcription factor that modulates rice leaf rolling by regulating leaf abaxial cell development. *Plant Cell* 21:719–735.

Zhao, M., K. Morohashi, G. Hatlestad, E. Grotewold, and A. Lloyd. 2008. The TTG1–bHLH–MYB complex controls trichome cell fate and patterning through direct targeting of regulatory loci. *Development* 135:1991–1999.

Zheng, Y., N. Ren, H. Wang, A. J. Stromberg, and S. E. Perry. 2009. Global identification of targets of the *Arabidopsis* MADS domain protein AGAMOUS-Like15. *Plant Cell* 21:2563–2577.

Zhong, H., S. Zhang, D. Warkentin et al. 1996. Analysis of the functional activity of the 1.4-kb 5′-region of the rice actin 1 gene in stable transgenic plants of maize (*Zea mays* L.). *Plant Science (Shannon)* 116:73–84.

Zhu, Q.-H., M. S. Hoque, E. S. Dennis, and N. M. Upadhyaya. 2003. Ds tagging of branched floretless 1 (BFL1) that mediates the transition from spikelet to floret meristem in rice (Oryza sativa L). *BMC Plant Biology* 3: Article No.:6.

Zimmermann, P., L. Hennig, and W. Gruissem. 2005. Gene-expression analysis and network discovery using Genevestigator. *Trends Plant Sci* 10:407–409.

Zimmermann, P., M. Hirsch-Hoffmann, L. Hennig, and W. Gruissem. 2004. GENEVESTIGATOR. *Arabidopsis* microarray database and analysis toolbox. *Plant Physiol* 136:2621–2632.

CHAPTER 5

Contribution of 'Omics' Approaches to Sustainable Herbivore Production

Jean-François Hocquette, Hamid Boudra, Isabelle Cassar-Malek, Christine Leroux, Brigitte Picard, Isabelle Savary-Auzeloux, Laurence Bernard, Agnès Cornu, Denys Durand, Anne Ferlay, Dominique Gruffat, Diego P. Morgavi and Claudia Terlouw

CONTENTS

5.1 INTRODUCTION

A major goal of animal science research today is to contribute towards the development of sustainable livestock systems in agreement with social expectations, or in other words reconciling socio-economic viability, environmental friendliness, product quality and animal well-being. To achieve this goal, researchers have various levels of approach at their disposal (territory, farms, herd, animal) requiring the use of a variety of disciplines. Among these, biology provides the means for analysing mechanisms involved in the animals' physiological functions by taking advantage of modern tools, notably those provided by genomics.

The molecular mechanisms underlying biological processes are governed by the expression of many genes. Methods to study these complex processes have experienced tremendous technological advances over the past few years. Whereas in the past molecular biology targeted a few genes chosen for their biological function, today genomics, proteomics and metabolomics allow simultaneous study of a large number of genes, proteins or metabolites, irrespective of their biological function. The Clermont–Theix INRA Centre recently placed at the disposal of its researchers a platform dedicated to 'omic' approaches by pooling all the means of the various laboratories. These new technologies will help improve our understanding of the biology of herbivores in a context of sustainable production.

So the aim of this chapter is to give details on some of the other possible applications for 'omics' techniques with respect to sustainable breeding of herbivores, apart from those in genetics which have been the subject of other studies reported elsewhere (Bidanel et al., 2008). We shall look successively at the contribution of 'omics' approaches to various subjects: improvement of animal performance and metabolic efficiency, control of nitrogen discharge by herbivores, the impact of the presence of mycotoxins in foodstuffs on the health of animals and man, the sensorial and nutritional qualities of dairy and meat products, traceability of production systems and also stress at the time of slaughter.

5.2 PRINCIPLES OF 'OMICS' APPROACHES

Physiological processes are governed by a certain number of genes acting together rather than by only one or a few genes. Over the past decade, genomics techniques have made it possible to study simultaneously thousands of genes, proteins or metabolites by the so-called high-throughput approaches. By studying all the transcripts, proteins or metabolites in tissues, scientists aim to detect potentially interesting genes that may, for example, be biomarkers for product quality and/or the animals' physiological state. Detailed reviews describing the limits and advantages of genomics have been published (e.g., see Hocquette, 2005; Hocquette et al., 2009b; Mullen et al., 2006). One of the strategies used is identification of the genes or proteins that are expressed differentially between extreme animals without any prior knowledge of the biological processes involved. Another consists of studying the links between their levels of expression and the phenotypic traits of the animal (Bernard et al., 2007, 2009; Kwasiborski et al., 2009; Morzel et al., 2008). The results expected are a better understanding of the biochemical mechanisms and identification of genes enabling animals with the desirable characteristics to be detected. Unlike the 'candidate gene' approach, the 'genomic' approach allows data to be collected without any prior hypothesis concerning the biological processes involved, and it thus produces new working hypotheses. Among the 'omics' approaches (Hocquette, 2007), the study of the transcriptome (all the transcripts for a given tissue) using DNA chips and of the proteome (all of the proteins) after two-dimensional electrophoresis (2DE) is the most widespread. By analogy, the more recently developed field of metabolomics studies the metabolome which covers all the metabolites present in a biological sample (cells, tissues, plasma and urine). Metabolomic techniques can also be put to good use in association with other approaches to study or predict all the metabolic flows (Nielsen, 2007) in the cellular context or even at the whole organism level (exchanges between tissues and organs). Finally, overall knowledge of genes, proteins and metabolites allows the interactome to be studied (Figure 5.1), that is, the interactions between macromolecules, notably between proteins which contribute to the complexity of living beings and the determinism of biological functions. Generally speaking, the 'omics' sciences lie at the interface between several disciplines and skills such as biology, physiology, chemistry, mass spectrometry, statistics and bioinformatics.

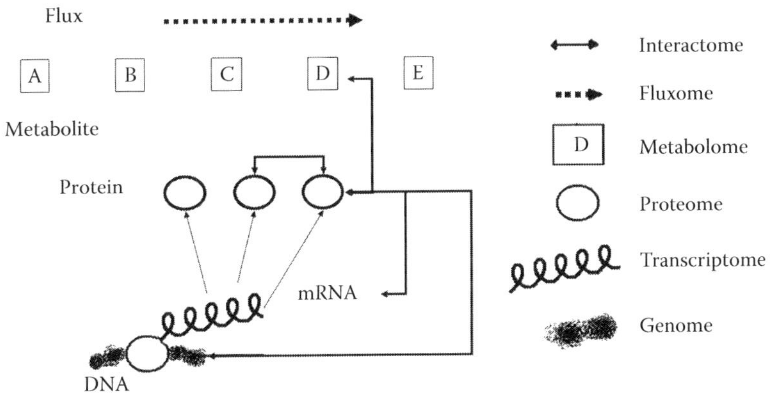

Figure 5.1 The various steps in the expression of genes and the associated various 'ome' concepts. Genome: all the genes; transcriptome: all the transcripts; proteome: all the proteins expressed; metabolome: all the metabolites present in a biological fluid; fluxome: all the metabolic fluxes in the cellular context; interactome: all the interactions between the macromolecules in a cell. (From Hocquette, J.-F. et al. 2010. *INRA Prod Anim* 22:385–396. With permission.)

5.2.1 Study of Animal Transcripts

Initially focussed on a few genes chosen according to their function in the physiological process being studied, quantification of gene transcripts (or mRNA) became more general, thanks to transcriptomic analysis tools that allow large-scale study of the transcriptome (Figure 5.1), that is, all the mRNA present in a given cellular type or tissue at a given moment and in a very specific biological condition. Several technologies allow these transcriptomic analyses to be carried out today. Among them, DNA chips are based on the principle of molecular hybridisation. This technology was born at the beginning of the 1990s but started to be applied in herbivores later. Indeed, before the development of bovine cDNA microarrays, some groups used available human microarrays in cross-species hybridisation studies (Sudre et al., 2003) before the development of bovine-specific chips (for review, see Hocquette et al., 2007b).

The completion of the bovine genome (The Bovine Genome Sequencing Analysis Consortium, 2009) and the assembly of the ovine genome (International Sheep Genome Consortium, http://www.sheephapmap.org/) are spurring the development of denser, more complete DNA microarrays. In addition, direct identification and quantification of transcripts using new-generation sequencing technologies as an alternative to microarray-based methods is expected to bring new insights in transcriptome studies (Morozova et al., 2009).

5.2.2 Study of Animal Proteins

Proteomics is the large-scale study of proteins that involves several steps: protein extraction from complex biological samples, protein separation most often by electrophoresis in 2DE, analysis of 2DE gels after staining using specific image analysis tools in order to compare protein abundance, identification of proteins of interest (e.g., with differential abundance) with mass spectrometry (MS) with or without trypsin digestion. Using tandem mass spectrometry (MS/MS) and Edman sequencing, digested peptides can be easily sequenced. Proteomic techniques are rapidly developing and lead to new large-scale technologies, as illustrated by protein chips (Dhamoon et al., 2007).

The study of proteins is extremely important because a gene may result in the synthesis of a transcript which is not translated as a protein. Moreover, a protein can exist in different forms, termed 'isoforms', which may be the products of various genes or all synthesised from one and the same gene. For example using 2DE, Scheler et al. (1999) detected 59 isoforms of the 'heat-shock' protein HSP27 in cardiac muscle, many of which corresponded to phosphorylated forms of the protein. The various protein spots may also be the result of alternative splicing of mRNA. An example of this are the T troponin isoforms (TnT), which are encoded by three different genes characteristic of the cardiac slow and fast types, and for which 17 isoforms (six slow and 11 fast) were found in bovine *semitendinosus* (ST) *muscle* using 2DE. MS/MS analysis showed that the 11 fast TnT isoforms result from exclusive alternative splicing of exons 16 and 17 (Bouley et al., 2005).

5.2.3 Study of Animal Metabolites

As for transcriptomics and proteomics, the metabolomics approach is a global one that provides integrative information concerning all the metabolites involved in the functioning of cells or organs (Figure 5.1). There are two common ways to analyse the metabolome: the metabolic fingerprint and targeted analysis. The fingerprint of the metabolome or fingerprinting is a global approach that gives a snapshot of all the metabolites. Its main objective is to search for and identify new markers by comparing fingerprints obtained from a matrix under different situations such as diverse physiological conditions, changes in feeding or pathological situations. The second approach, also called 'target and profiling method' focuses on the analysis of known metabolites, for example, lipidogram.

A metabolomic study is commonly divided into four steps: analysis of samples, data analysis (alignment, filtering and exporting of data), statistical analysis and identification of potential markers and their biological interpretation. One major improvement is necessary for data analysis, where the crucial first step is the extraction of peak variables (width, shape, intensity, resolution, etc.). The second critical hurdle is the need to develop the existing databases in order to facilitate the identification of markers. This issue is especially important for herbivores since metabolomics is not well developed for these species compared to the situation in plants and humans.

5.3 PHYSIOLOGICAL PERFORMANCE AND METABOLIC EFFICIENCY

5.3.1 Muscle Development

Muscle development in beef animals is the subject of many studies which aimed at improving carcass yield. Particular attention is paid to muscle growth during foetal life which determines the subsequent potential for muscle growth. Myogenesis occurs during several phases which take place at separate times. The first corresponds to proliferation of the precursor cells (myoblasts) in the premuscular masses, under the control of growth factors. The myoblasts then withdraw from the cell cycle, and align and fuse to give rise to differentiated multinuclear cells called myotubes. These cells go through many more biochemical changes and gradually acquire their adult contractile and metabolic characteristics. This process is complicated by the existence of several generations of cells, probably three in large animals, which appear at around 60, 90 and 110 days of foetal life (postconception, p.c.) in bovines (for review, see Picard et al., 2002). Proliferation takes place prior to 180 days p.c. and contractile and metabolic differentiation of the fibres usually occurs after this stage and right through the final trimester of gestation. Analysis of the transcriptome (Lehnert et al., 2007; Sudre et al., 2003) and proteome (Chaze et al., 2008a,b) in bovines has provided the means for describing certain gene-expression profiles and proteins regulated during the development of muscle tissue. Thus, Sudre et al. (2003) showed that a few genes appear to be strongly regulated at around 180 days p.c., 260 days p.c. or between foetal life and 15 months old. This study confirmed in particular the physiological importance of the stage at 180 days p.c. as an ontogenetic transition. In addition, the importance of the last 3 months of foetal life for differentiation of bovine muscle, previously

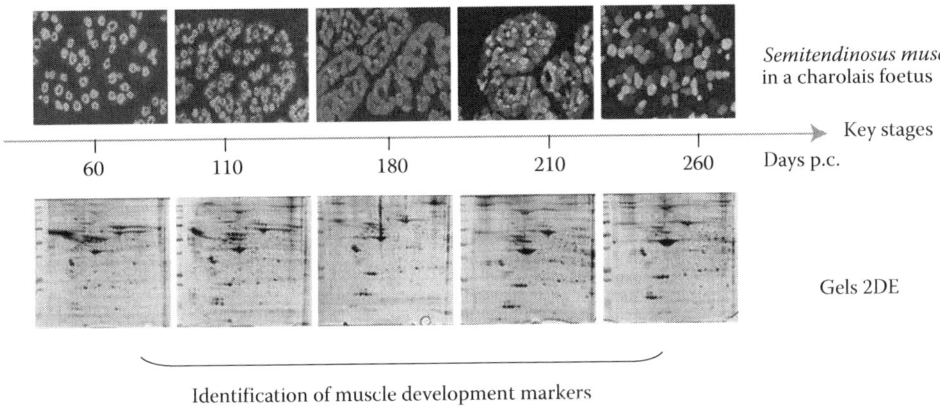

Figure 5.2 (See color insert.) Strategy for the identification of muscle development in bovines at key stages in myogenesis by analysis of the changes in muscle proteome using two-dimensional electrophoresis (2DE) at different stages in days post-conception (p.c.).

demonstrated by biochemical approaches (Picard et al., 2003), was also confirmed (Sudre et al., 2003). Finally, transcriptomic analysis has allowed various genes to be identified (Leu5, Trip15, Siat8, etc.), which are regulated during myogenesis but whose role in this developmental process remains to be clarified. Comparison of the muscle transcriptome in animals carrying mutations leading to a loss of myostatin function has also underlined the importance for muscle hypertrophy of genes involved in the constitution of muscle connective tissue and muscle energy and lipid metabolism (Cassar-Malek et al., 2007), together with the apoptosis processes (Chelh et al., 2009).

The proteome of ST muscle of bovines has been monitored by 2DE and MS at key stages in myogenesis: 60 days p.c. corresponding to the proliferation and establishment of the first generation of cells, 110 days p.c. characterised mainly by the establishment of second- and third-generation cells, the stage at 180 days p.c in which the total number of fibres is fixed, and two stages (210 and 260 days p.c.) characterised by contractile and metabolic differentiation of the muscle fibres (Picard et al., 2002) (Figure 5.2). A total of 248 proteins have been identified by MS and analysed using bioinformatic tools. Analysis by hierarchical classification of the first three stages in gestation has revealed quite a large number of expression clusters related mostly with cell proliferation and death (Chaze et al., 2008a,b), suggesting that the balance between cell proliferation and apoptosis is of prime importance in the control of the total number of muscle fibres during the first two-thirds of foetal life. In addition, statistical analysis has revealed new markers connected with regulation of the total number of fibres (WARS and DJ1 protein) and other markers which appear to be specific to proliferation of both myoblast generations (CLIC4 for the primary myoblasts, HnRNPK for the second generation of myoblasts). The final third of foetal life is above all marked by a considerable number of changes in the contractile and metabolic protein isoforms (myosin light chains, slow or fast T troponins, and also alpha and beta enolases) (Chaze et al., 2009). So a very rich map of ST muscle proteins during foetal life has been drawn up to accompany the map produced for adult ST muscle (Bouley et al., 2004a; Chaze et al., 2006). These data concerning modifications in the bovine muscle proteome during foetal life represent a reference for myogenesis studies in other types of bovines, and also for other species in comparative biology studies.

5.3.2 Regulation of Gene Expression by Nutrients

In the context of sustainability of livestock systems, the control of animal performances is a major economic issue. It is of particular importance to predict the effects of nutritional changes at the whole animal level (in terms of metabolism of various tissues and organs). This approach often focused on specific metabolic pathways or rate-limiting metabolic enzymes. For many years, scientists studied individually the expression of individual genes of interest subjected to various nutritional regulations. The development of functional genomic tools allows global analyses of the metabolism and physiology by analysing thousands of genes, proteins or metabolites simultaneously without knowledge *a priori* of the underlying biological processes. Genomics also allows, for example, another view of the molecular links between nutrition and physiology and in particular the interactions between genes and nutrients. This has led to the concept of 'nutrigenomics' (Chadwick, 2004). Nutrigenomics, more generally devoted to the interaction between nutrition and health for humans, can also be extrapolated to animal sciences, in particular for the improvement of sustainable husbandry practices for herbivores. In addition, the increasing availability of bovine DNA chips (for reviews, see Cassar-Malek et al., 2008; Hocquette, 2005) as well as the recent sequencing of the bovine genome can also improve knowledge of the genome of other species such as the sheep or the goat (Everts-van der Wind et al., 2005), now making it possible to analyse modifications of gene expression in answer to different animal production factors (nutritional, variations of growth curve, etc.). The integration of genomic and physiology (measurements of tissues and body metabolism) in livestock, in particular herbivores, should improve our knowledge concerning the regulation of gene expression by physiological and nutritional factors. This should lead to an optimisation of the herbivore production systems by novel strategies of nutrition and management in addition to an improvement in animal health and the sensorial and nutritional quality of the products (milk and meat), contributing in consequence to the sustainability of livestock systems.

Studies are undertaken in several laboratories in the world with the goal of understanding how the genetic and environmental factors can control gene expression in livestock. One such study consisted in identifying the genes implicated in the control of global metabolism by comparing gene expression in various bovine breeds presenting different capacities to use nutrients (Schwerin et al., 2006). In another study, gene-expression profiles were characterised in the liver in response to different *prepartum* nutrition plans in milking cows (Loor et al., 2006). This study showed that an *ad libitum* feeding favoured the expression of genes implicated in lipid synthesis, thus predisposing the cows to hepatic steatosis. On the other hand, energy restriction induced an increase in the expression of the genes involved in fatty acid (FA) oxidation, neoglucogenesis and cholesterol synthesis. Thus, overfeeding cows before parturition, as is the usual practice, would have negative consequences on the health of the liver, in particular via transcriptional changes leading to an increase in the susceptibility to oxidising stress and DNA alterations. These data raise the question of reviewing the nutrition of milking cows before their parturition.

During the lactating period, intense metabolic activity takes place in the mammary gland, to allow synthesis and secretion of milk components. This involves the expression of a large number of genes whose regulation is still poorly understood. Nutrition has a considerable effect on the milk lipid composition which has in turn a considerable effect on the nutritional, technological and sensorial qualities of milk products (Chilliard and Ferlay, 2004). Thus, nutritional studies were first focused on the regulation of expression of a few genes encoding the key lipogenic enzymes in the mammary gland. A few *in vivo* studies were performed with different lipid supplements (in terms of amounts and nature) in interaction with the type and proportion of forage and concentrate (level of starch) in the diet. In goats, these studies have shown that the mammary lipogenic genes exhibited specific responses to nutritional regulation sometimes less pronounced than the response of the corresponding milk FAs (reviewed by Bernard et al., 2008). Thus, it was suggested that for the lipoprotein lipase the substrate availability was a more limiting factor than the potential enzyme activity (Bernard et al., 2005a,b, 2008). Conversely, the observed variation of expression of the gene encoding acetyl-CoA carboxylase was related to the milk secretion of medium chain FAs. The stearoyl-CoA desaturase gene expression varies only little (Bernard et al., 2005a, 2008), probably due to its implication in the maintenance of milk fat fluidity. These results in goats contrast with those obtained in dairy cows where a milk fat depression together with a strong down-regulation of lipogenic gene expression was observed, thus underlining the interest of nutritional studies comparing these two species.

More recently, the availability of new tools for transcriptomic analysis in ruminants made it possible to understand the nutritional regulation of the milk components biosynthesis in greater detail. As a result, in the context of the European Lipgene and ANR-Genomilkfat programmes, global analysis of the mammary transcriptome was carried out using a microarray containing 8379 signatures genes. For the first nutrigenomic studies reported in ruminants, a comparison of the mammary gene-expression profiles between extreme nutritional statuses (*ad libitum* feeding versus food deprivation for 48 h) was performed in lactating goats. Ollier et al. (2007) showed that food deprivation modifies the expression of 161 genes in the mammary gland. The gene classification based on their function revealed that among the down-regulated genes some are involved in milk component biosynthesis related with milk performances (milk production, protein and fat yield). In addition to the expected genes, a class of genes involved in lipid metabolism also contains genes whose role in the mammary gland is not yet well understood. They thus constitute new candidate genes. In the same way, a class of genes whose sequence or function are still unknown represents potentially interesting targets. Moreover, a modulation of expression of genes implied in differentiation, cellular proliferation or programmed cell death was also highlighted (Figure 5.3). These latter genes could be associated with an orientation of the mammary gland towards an early process of involution. The nutrigenomic studies to compare the effects of different feeding plans (various forages supplemented or not, various ratios forages/concentrates supplemented or not, etc.) on the mammary transcriptome are still in progress.

In addition to nutrition, genetics also largely impacts biosynthesis and secretion of milk lipids. Thus, the impact of the polymorphism of gene encoding alphaS1 casein on the mammary transcriptome has been studied, with the aim of evaluating its interaction with the nutritional factor. This polymorphism is associated with modifications of milk composition. Among 41 genes differentially expressed according to the genotype, five genes involved in lipid metabolism were

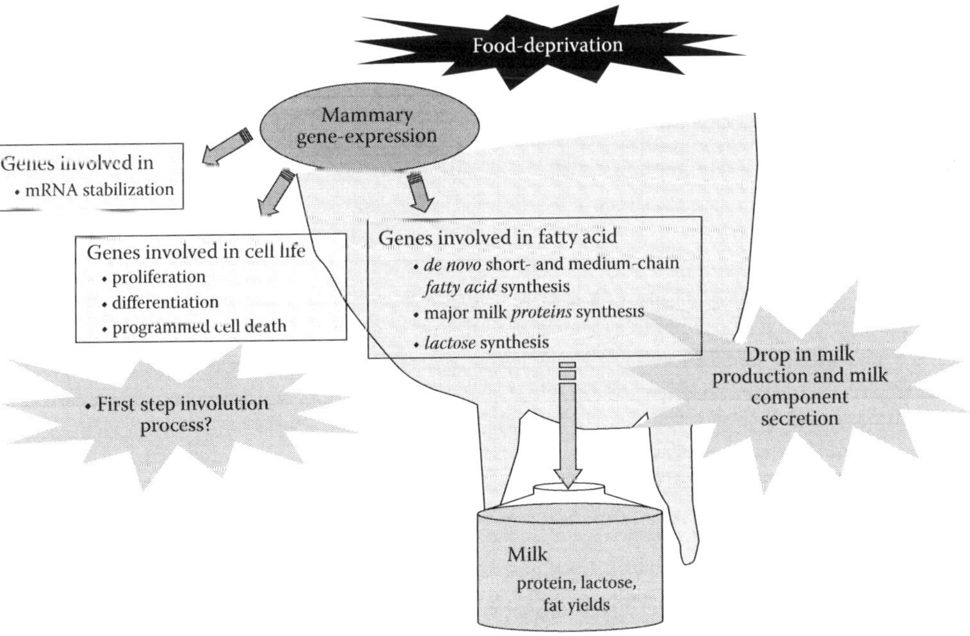

Figure 5.3 Food deprivation alters, in the mammary gland, the expression of genes involved in different mechanisms such as the milk component biosynthesis and programmed cell death.

downexpressed in the 'defective' animals (carrying two 'defective' alleles) compared to the 'reference' animals (carrying two 'reference' alleles). The expression of genes involved in other mechanisms such as cell–cell interactions or nucleic acid metabolism was also pointed out (Ollier et al., 2008). The effects of the interaction between this genetic polymorphism and the nutritional factor are currently under evaluation.

Finally, genomics has allowed the influence on muscle characteristics of two management approaches representative of intensive and extensive production systems to be analysed. Clearly, muscles of animals at pasture were more oxidative, which could be explained by an increased mobility at pasture and to a lower extent by the effect of the grass (versus maize silage) based diet (Jurie et al., 2006). Gene-expression profiling in two muscle types was performed on both production groups using a multi-tissue bovine cDNA repertoire. Variance analysis showed an effect of muscle type and of production system on gene expression. Among the genes differentially expressed, Selenoprotein W was found to be underexpressed in animals at pasture and could be proposed as a putative gene marker of the grass-based system (Cassar-Malek et al., 2009). A complementary analysis by Northern blot indicated that the differential expression of this gene was due to the diet and not to mobility at pasture.

5.3.3 Interactions between Tissues and Organs

Management of production efficiency and the improvement of meat and milk quality is a crucial issue in ruminants. The efficiency of conversion of diet proteins into milk or meat proteins is low in lactating or growing ruminants (20–30%, reviewed by Lobley, 2003). These low conversion rates are due to digestive processes (rumination) and also metabolic processes at the splanchnic level (rumen, digestive tract, liver). Indeed, these tissues represent 50–60% of oxygen consumption and about 50% of AA utilisation at the whole body level in ruminants. Hence, via the digestive processes, the splanchnic tissues supply nutrients to the muscle and the udder but are also 'competitors' with these tissues for the utilisation of nutrients. Consequently, it is necessary to measure the metabolic activity of the splanchnic area and the metabolic outcome of certain nutrients within splanchnic tissues in various nutritional and physiological situations. The aim of these studies was to assess the metabolic processes involved in the low conversion efficiency of nutrients into products and to improve the adequacy between nutrient requirements and feed supply.

Among the most recent techniques used to study the utilisation of nutrients at the whole body, tissues and organ levels, stable isotopes are frequently used. The target nutrients (amino acids, glucose, long/short chain FAs, for instance) are labelled on the carbon (^{13}C), hydrogen (^{2}H) or nitrogen (^{15}N) atoms to be 'tracked' within the body (and to estimate their utilisation within tissues and organs) or within a metabolic pathway (nutrients metabolised into metabolites) (Ortigues et al., 2003). The apparatus used to detect the isotopic enrichment in labelled molecules in organic fluids or in various tissues and organs is MS (Wolfe, 1992). These machines have greatly improved over recent years with the development of liquid chromatography (LC)/MS technologies (used in metabolomic studies) (Hollywood et al., 2006) which are complementary with less recent gas chromatography/SM technologies (Wolfe, 1992). These new LC–MS technologies allow new fields of investigation in the tracking of molecules within metabolic pathways (sulphur amino acids, hepatic gluconeogenesis, Krebs cycle and so on which are metabolic knots more difficult to assess with gas chromatography technology).

5.4 LIMITATION OF NITROGEN WASTE DISCHARGE INTO THE ENVIRONMENT

In order to increase milk or meat production rates or improve profitability, farmers have increased nitrogen supplies in ruminants' feed. However, the nitrogen cycle in livestock production systems

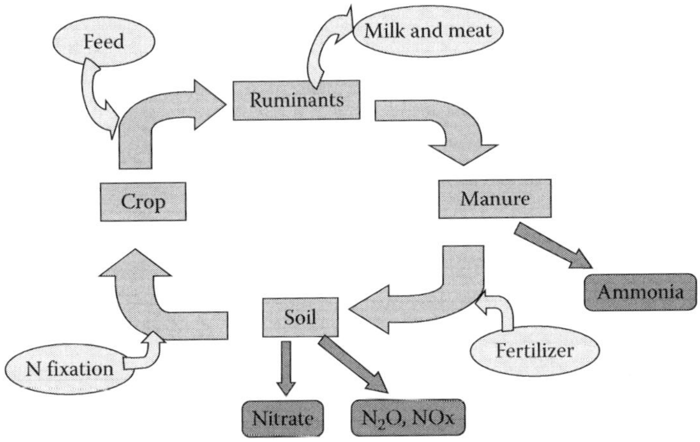

Figure 5.4 Nitrogen cycle on farms producing milk and meat from ruminants (European project RedNex). The compounds highlighted in dark grey correspond to polluting compounds.

generates numerous polluting products (Figure 5.4). The increase in nitrogen waste discharge into the environment impacts on surface and groundwater (pollution with nitrates) and on nitrogen emission in the atmosphere (nitrogen oxides: N_2O and NOx), leading to eutrophication and acidification of soils (Food and Agriculture Organization, 2006).

The same report from Food and Agriculture Organization (2006) estimates, for instance, that 65% of nitrogen hemioxide emissions (essentially due to manure) is due to the nitrogen cycle in livestock production (Figure 5.4).

The reduction of nitrogen waste requires the development of new breeding systems (for lactating cows), and particularly the optimisation of rumen function and a better assessment of the outcome of the ingested nitrogen (N) in secreted milk, urea and faeces. Thus, better knowledge of the efficiency of conversion of feed nitrogen into milk or meat proteins, using labelled molecules and MS apparatus is essential (see Section 5.2.3) to study nutrients metabolism, optimise their utilisation and limit the waste. Recent data obtained on growing lambs show that when the animals' energy requirements are met, INRA recommendation in terms of digestible proteins in the intestine (Jarrige et al., 1995) overestimate the nitrogen requirements of the animals, leading to increased nitrogen waste (as urea). Indeed, the amino acids supplied in excess in the food of the animals are catabolised in the liver and the resulting nitrogen excreted into urea. Conversely, the energy requirements (estimated by the INRA feeding system) should not be decreased in order to maintain the animals' net protein accretion (Kraft et al., 2009). These concepts concerning the interaction between nitrogen and energy supplied in ruminants' feed will be reassessed in dairy cows in a European Programme (RedNex: Reduction of Nitrogen excretion by ruminants). Associated with the utilisation of labelled molecules and SM, a metabolomic approach will also be used to highlight some new biomarkers linked with nitrogen utilisation.

5.5 METABOLOMICS TO HELP MYCOTOXICOSIS DIAGNOSIS

In spite of the progress made in preventive approaches such as the improvement in agricultural practices, hazardous concentrations of mycotoxins are still found in animal feeds (Garon et al., 2006; Scudamore and Livesey, 1998). Many of the known mycotoxins are toxic to animals affecting production and causing economic losses to farmers. Acute intoxication by mycotoxins has become rare due to improved harvesting and storage techniques. However, repeated ingestion of low doses of mycotoxins alone or in a mixture are still frequent in the field. Chronic exposure to mycotoxins is difficult to diagnose as it does not induce specific clinical signs but may alter some metabolic functions in animals. At low doses, some mycotoxins have shown different effects on cell function that can go from simple changes in the metabolism, such as a decrease in protein synthesis, up to apoptosis. The immune system is also the target of most mycotoxins. In addition, we do not know the economic consequences of this type of exposure to low doses in terms of productivity losses and the negative effect on the animal's health. Due to these difficulties, the majority of animal mycotoxicoses remain undiagnosed.

Normally, the aetiological diagnosis of mycotoxicosis is based on the analysis of mycotoxins and/or their metabolites in feeds and, more rarely, in biological matrices. This approach is not effective for two reasons. First, the decision to analyse the suspected feeds is usually performed 2–3 weeks after the initial observation of the production or health problem in the herd and when commonest infectious or metabolic diseases have been eliminated. Second, after this lapse of time the contaminated feeds may have been consumed and the majority or all of the ingested mycotoxins may also have been excreted from the animal body.

Recently, we tested the metabolomic tool to search for urine markers that could improve the diagnosis of mycotoxicosis. We used ochratoxin A (OTA), which is a mycotoxin commonly found in foods and feeds (Pittet, 1998; Van Egmond, 2004). It is nephrotoxic, hepatotoxic, teratogenic and carcinogenic in animals, and it was recently classified by the International Agency of Research on Cancer as a class 2B, possible human carcinogen. When ingested by ruminants, OTA is largely metabolised by rumen microorganisms, in particular protozoa, into a less toxic metabolite, ochratoxin α (Ozpinar et al., 1999).

In a first experiment, six lactating dairy ewes were randomly divided into two lots that received 30 μg OTA/kg bw/day for 4 weeks. Urine was collected from animals placed in individual cages fitted with a mesh to separate faeces. Four sampling periods were used during the experiment: before exposure to OTA U0 (day –3 to 0); and at different time points during the period of exposure: U1 (day 1–4); U2 (day 15–18) and U3 (day 29–31). The urine metabolic fingerprints were established by a combined LC–MS/nuclear magnetic resonance (NMR) system (Metabolic Profiler®, Bruker Biospin, Germany). Principal component analysis (PCA) of NMR data uncovered differences in the metabolic profiles of animals before and following OTA exposure (Figure 5.5).

In a second experiment (Figure 5.6), we investigated the effect of OTA at a same dose and over the same period of exposure. It was carried out with four couples of twin lambs ($n = 8$ males) that were divided into two groups (one twin per group). In the first period, the animals were kept fauna free and in the second period they were allowed to become contaminated with rumen protozoa (faunated). The use of twin lambs decreases the genetic variability that could be important in metabolomic studies. As with the previous study, we obtained many potential markers. The PCA of data obtained from LC/MS-QTOF analysis showed a clear separation between the fauna-free and faunated animals (grey points) and after exposure for both groups of animals (fauna-free and faunated animals).

Results obtained from both studies suggest that a metabolomic approach can detect early subtle metabolic changes in animals due to mycotoxin exposure (Boudra et al., 2008).

5.6 STRATEGIES FOR IMPROVEMENT OF THE QUALITY OF DAIRY AND MEAT PRODUCTS

5.6.1 Mechanisms of Bioconversion and Stability of Long-Chain FAs

Dairy products provide about 40% of the overall fat consumed by humans. When consumed in excessive amounts, they are considered as metabolic syndrome risk factors because of their high

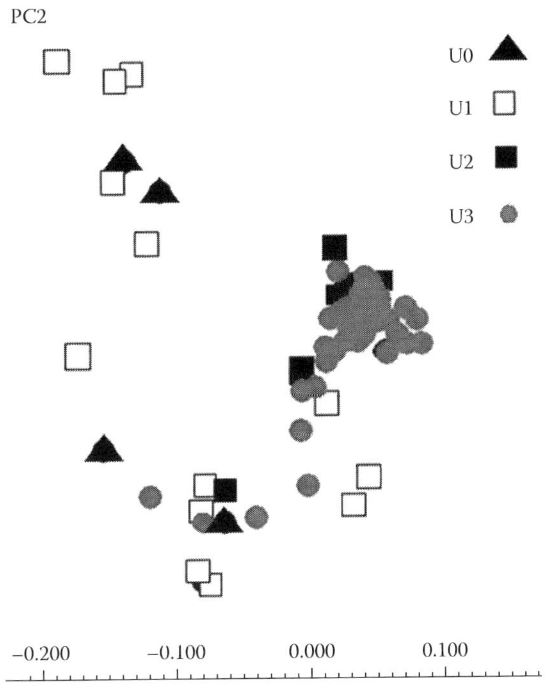

Figure 5.5 Principal component analysis (PCA) overview of NMR data obtained during four different periods of experiment: U0 (day −3 to 0; no toxin); U1 (day 1–4); U2 (day 15–18); and U3 (day 29–31) after exposition to ochratoxin A (OTA).

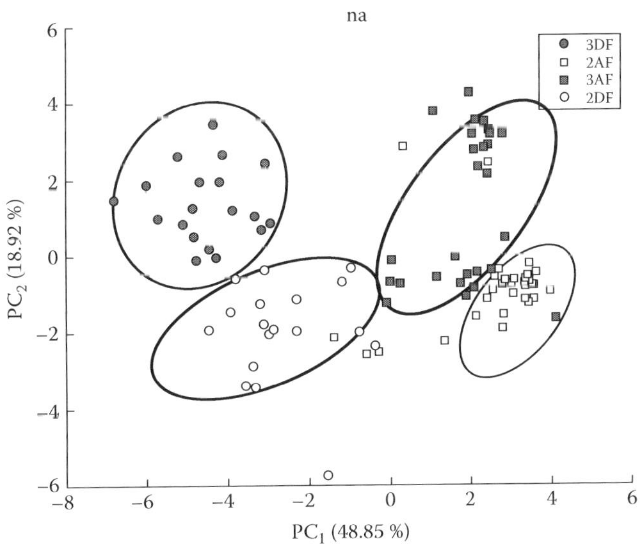

Figure 5.6 Eight lambs in two groups: faunated (AF) versus fauna-free (DF). The picture is the plot of a PCA with LC–MS data after analyses of samples before exposure (2AF versus 2DF) or after exposure everyday with ochratoxin A (OTA) during 4 weeks (3AF versus 3DF).

saturated FAs content (69%) and of some *trans* unsaturated FAs. In contrast, dairy products contain omega 3 FA, which have positive effects on human health or minor FA (conjugated linoleic acid (CLA) or branched-chain FA) which present putative anti-cancer activity or anti-atherogenic effects (reviewed by Shingfield et al., 2008). The dietary polyunsaturated FAs (linoleic and linolenic acids) are extensively metabolised and biohydrogenated by the rumen bacteria, resulting in the production of *cis* and *trans* isomers of 18:1, 18:2 and 18:3 and stearic acid (Bauman and Griinari, 2003). The different constituents of the diet (nature of forage, forage/concentrate ratio, starch content of the diet) and oilseed supplementation are the main dietary factors able to modify the rumen metabolism of PUFA, and consequently to influence the FAs profile leaving the rumen (Chilliard et al., 2007). Grass-based diets (grazed or conserved), when compared with diets rich in concentrates and/or maize silage, decrease the milk-saturated FA content and increase that of *cis*9–*trans*11-CLA and 18:3n-3 (Ferlay et al., 2006, 2008). Likewise, supplementation with oilseeds (rich in 18:2n-6 or 18:3n-3) strongly decreases the milk content of saturated FA and increases that of 18:0, *cis*9-18:1 and *cis*9–*trans*11-CLA, but also the concentration of *trans* isomers of 18:1 and 18:2 (Chilliard et al., 2007).

In milk, rumenic acid (RA) (C18:2 *cis*-9, *trans*-11), which is the predominant isomer of CLA, is derived from ruminal metabolism of 18:2n – 6 and further absorption and uptake by the mammary gland, and mainly from endogenous Δ9-desaturation of *trans*-11–18:1 (vaccenic acid, VA) in the mammary gland (Mosley et al., 2006; Shingfield et al., 2007). In goats, despite the close and linear relationship between the quantities of milk VA and RA (Chilliard et al., 2003), no data were available on the endogenous synthesis of RA from VA. Thus, the metabolism of VA was studied in this species by using an *in vivo* chemical tracer approach to measure the conversion of VA into RA in goats fed two different dietary treatments. To achieve this goal, a single dose of ^{13}C-VA was delivered intravenously to lactating goats receiving two diets with lipid supplements in interaction with the level of starch. The kinetics of enrichment of milk ^{13}C-VA and ^{13}C-RA and of milk VA and RA secretions were determined to calculate the *in vivo* conversion of VA into RA. This study showed using ^{13}C-labelled VA that in goats, 32% of VA was Δ9-desaturated into RA and that milk RA originating from VA represents 63 – 73% of total milk RA (Mouriot et al., 2009). These data are close to the estimations of the contribution of endogenous synthesis of RA in cows (64–91%) obtained using a similar tracer methodology (Mosley et al., 2006) or involving a quantification of the duodenal or abomasal flow and milk secretion of VA and RA (for review, see Glasser et al., 2008).

Beef is also considered to be a food with an excessive fat content and notably a high proportion of saturated FA and a low level of PUFA. Several nutritional studies were developed to improve this composition, especially by increasing the beef content in long-chain n-3 PUFA (LC n-3 PUFA) and in particular in DHA known for its beneficial biological properties for humans (Simopoulos, 1999). However, the results of these studies suggest that the endogenous synthesis of LC n-3 PUFA from 18:3 n-3 is very limited in ruminants. It is therefore considered necessary to (1) better understand the precise mechanisms regulating the expression of the different proteins involved in the conversion of 18:3 n-3 into DHA in the tissues of ruminants by a transcriptomic approach using a dedicated DNA chip manufactured under the GENOTEND programme (Hocquette et al., 2009a) and (2) determine nutritional strategies to adopt to stimulate the expression of these proteins and thus the synthesis and the deposition of LC n-3 PUFA in muscles of ruminants.

It is now well recognised that the enrichment of meat with LC n-3 PUFA can cause a deterioration of its sensory and nutritional qualities by peroxidation phenomena (Durand et al., 2005), and these processes are strongly amplified during ageing and packaging (Gobert et al., 2008). So for efficient protection against these processes, it is essential to better understand the mechanisms involved in the different stages of peroxidation including (1) to obtain discriminating markers of the orientation of peroxidation processes by an approach using MS (Gobert et al., 2009) to quantify specific products of peroxidation of the two major families of PUFA: 4-hydroxy-2-hexenal and F4-isoprostanes for ω3 and 4-hydroxy-2-nonenal and F2-isoprostanes for ω6, (2) to develop transcriptomic approaches using dedicated chips focused on identified targets of oxidative stress and lipid metabolism and (3) to develop a proteomics approach for the analysis of proteins modified by the lipoperoxidation processes (aldehyde adduct formation and carbonylation).

5.6.2 Characterising the Qualities of Animal Products through Micronutrient Analysis

Volatile compounds were initially studied for their participation in product aroma, owing to their odour-active properties. As many of them are terminal metabolites originating from protein, lipid and carbohydrate catabolic pathways, several authors demonstrated that animal feeding influenced the composition of the volatile fraction in milk (Urbach, 1990) and meat (reviewed by Vasta and Priolo, 2006) products. For example, 2,3-octanedione in sheep adipose tissue was proposed as a tracer for grazing (Priolo et al., 2004). Young et al. (1997) hypothesised that this compound originates from the action of lipoxygenase on linolenic acid, both the enzyme and its substrate being more abundant in green leafy tissue than in preserved forages. Among the volatile compounds, terpenes are a particular case: these plant secondary metabolites, directly transferred from the forage to the milk or the meat, are also valuable candidates as animal feed tracers and carry information about the geographical origin of the forage consumed by the animal.

Beside the energy and nitrogen required for animal growth and performances, forage plants provide the animal with a pool of secondary metabolites such as phenolic compounds, terpenes, carotenoid pigments and vitamins, referred to as micronutrients. Most of them possess specific properties favouring product conservation or human health, such as antioxidant phenolic compounds, antimicrobial terpenes and so on, and many of them possess aromatic properties. For these reasons, micronutrients are strong contributors to the diverse aspects of product quality: sensorial characteristics, safety and preservation, benefits to human health and also 'typicity' for local productions. Todays' research aims at understanding the relationship between animal feed and the micronutrient profiles in animal products, in order to evaluate the different rearing practices from the product quality point of view.

Several studies showed that terpene diversity and amount in milk, cheese and meat were considerably increased when the animals were fed on pasture. The same is true for carotenoid pigments, with the advantage that their global amount can be estimated by a simple and rapid spectrophotometric method, owing to the yellow colour they give to the product. These compounds are considered good tracers for pasture feeding (reviewed by Prache et al., 2007). Phenolic compound profiles in milk obtained from groups of cows offered different diets comprised compounds that were ubiquitous and others that were specifically found with a given diet (Besle et al., 2005). Unlike terpenes, for which forage profiles are fairly accurately reflected in the milk, phenolic compound profiles are strongly modified. These compounds undergo bioconversions under the action of rumen microorganisms before being absorbed at gut level or under the action of liver enzymes before being eliminated. Understanding the relationship between feed and product requires micronutrients to be studied not only in the forage and the product, but also in the digestive tract, blood, urine and faeces. Given the very high number of plant secondary metabolites encountered in both the terpenoid and the phenolic family, these studies will benefit greatly from high flow rate analysis techniques.

5.6.3 Meat Tenderness Predictors

Genomic techniques have also been put to use to look for beef quality markers or predictors (for review, see Hocquette et al., 2007b) (Figure 5.7).

One important result is the identification of a negative relation between the expression of the *DNAJA1* gene, and meat sensorial tenderness after 14 days of ageing, not only in young bulls but also Charolais steers (Bernard et al., 2007). This gene encodes a chaperone protein of the HSP40 family. The expression level of other stress proteins (notably *HSPB1* encoding HSP27) is positively corre-

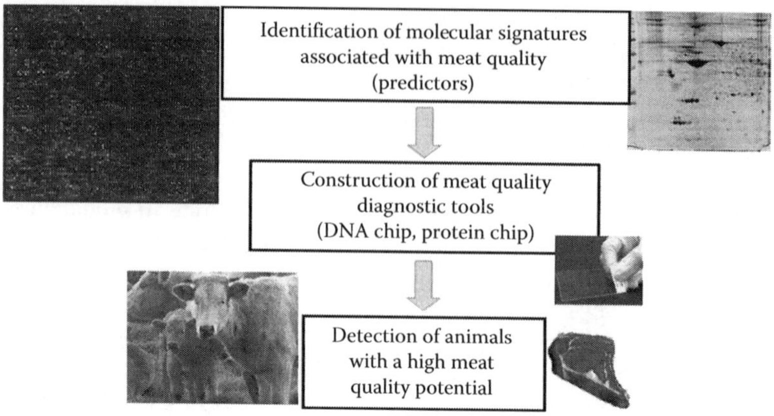

Figure 5.7 **(See color insert.)** Use of genomic techniques to identify beef sensorial quality markers. It is necessary to integrate the results obtained by proteomics and transcriptomics in order to identify a combination of quality predictors. This should lead in the short term to the development of diagnostic tools allowing the identification of live animals whose muscle molecular phenotype is liable to give good-quality meat.

lated with the shear force (and so the toughness) whether at mRNA or protein level. These proteins have an anti-apoptotic activity and thus may slow down the cellular death process and consequently meat ageing, with a positive effect on the muscle tenderising process after slaughter. During *post-mortem* ageing, it has also been shown that levels of HSP27 in fresh muscle and its subsequent fragmentation explain up to 91% of the variability in sensorial tenderness (Morzel et al., 2008).

Moreover, analysis of the proteome of the ST and *longissimus* muscles of bovines classified according to meat tenderness shows that the superior tenderness groups are characterised globally by overexpression of the proteins related with slow twitch speed and an oxidative metabolism (Bouley et al., 2004b; Hocquette et al., 2007a). In agreement with these results, Morzel et al. (2008) showed in ST muscle of young Blond d'Aquitaine bulls that abundance of succinate dehydrogenase (an oxidative metabolism enzyme) appears to be the best enzymatic predictor of initial and overall tenderness of meat, explaining, respectively, 65.6% and 57.8% of their variability.

However, a study carried out on several beef (Charolais and Limousin) and hardy (Salers) breeds showed that potential markers for tenderness appear to differ between beef and hardy breeds. Indeed, differences in expression of proteins such as parvalbumin, myosin light chain (MLC2) and acyl-coA-binding protein, which are all connected with energy or calcium metabolism, are found between tenderness groups in the two beef breeds, but not in the Salers breed. In the latter breed the tenderness groups differ with respect to other proteins such as phosphoglucomutase and lactate dehydrogenase B, which are underexpressed in the higher tenderness group (Bouley et al., 2004b). These results suggest that potential meat tenderness indicators appear to differ between breeds, as also suggested by other studies in pigs (Kwasiborski et al., 2008). Complementary studies currently under way will provide the means for checking these initial data.

Proteomic analysis is also used to track protein changes during meat ageing or even after cooking. In particular, using MS, Bauchart et al. (2006) revealed seven peptides of interest corresponding to five proteins in beef *pectoralis profundus* muscle during storage and after cooking. Three of these proteins are known targets of *post-mortem* proteolysis: TroponinT, Nebuline and Cypher protein. The two others correspond to connective tissue proteins, that is, procollagen proteins type I and IV. More generally, the proteolysis markers detected during the first 48 h *post-mortem* correspond mainly to structural proteins such as actin, myosin and troponin T, and metabolic enzymes such as myokinase, pyruvate kinase and glycogen phosphorylase (for review, see Bendixen, 2005). These protein fragments appear to be correlated with meat tenderness.

These studies will be continued thanks to complementary approaches currently under way. First, a DNA chip is being developed within the framework of the GENOTEND project (Hocquette et al., 2009a) with all the genes known to be involved in beef sensorial quality. These genes were identified, thanks to biochemical, proteomic, transcriptomic or genetic approaches developed in France and elsewhere. In addition, the dot-blot technique currently being developed has the potential to provide the means for analysing the expression of a large number of proteins associated with meat sensorial quality in a large number of samples (Guillemin et al., 2009).

5.7 PRE-SLAUGHTER STRESS

It is well known that stress reactions at slaughter change *ante-* and *post-mortem* muscle physiology. Depending on the context, pre-slaughter stress may cause acceleration in post-mortem muscle pH decline or a reduction in the amplitude of this decline (Tarrant et al., 1992; Warriss and Brown, 1985). Rate and amplitude of pH decline may influence many meat quality characteristics, including colour, water-holding capacity, taste and tenderness (Beltran et al., 1997; Hwang and Thompson, 2001; Renerre, 1990). Recently, it was shown that pre-slaughter stress may also enhance oxidative processes, including lipoperoxidation (Durand et al., 2005; Durand et al., unpublished results). Consequently, the nutritional value of meat may be altered (Biasi et al., 2008). In the following, we will discuss the effects of stress on energy metabolism and lipoperoxidation in the *post-mortem* muscle, and the interest of omics to increase our understanding of the underlying mechanisms.

As indicated above, pre-slaughter stress may influence meat tenderness (Beltran et al., 1997; Hwang et al., 2001) and tenderness was repeatedly found to be correlated with HSP content of the early *post-mortem* muscle (Bernard et al., 2007; Kim et al., 2008; Morzel et al., 2008). Many reports describe that stress may increase the expression of various HSPs. For example, HSP70 and HSP27 levels increased rapidly following exercise in all fractions of muscle cell extracts, including cytosolic, cytoskeletal, membrane and nuclei fractions supporting a general up-regulation of the HSPs (Folkesson et al., 2008; Paulsen et al., 2009). In addition, exercise induced small HSPs (HSP27 and αB-crystallin) to translocate and accumulate in areas of myofibrillar disruption and on sarcomeric structures (Paulsen et al., 2009). The HSPs are believed to protect myofibrillar structures from exercise-induced damage (Paulsen et al., 2007, 2009).

The mechanisms underlying the relationship between the HSP content of the early *post-mortem* muscle and subsequent tenderness are so far unknown. Several mechanisms could be involved. First, tenderisation is believed to rely, in part, on apoptotic processes (Herrara-Mendez et al., 2006; Laville et al., 2009) and various HSPs including HSP27 and HSP70 are known to have anti-apoptotic action (Concannon et al., 2003; Creagh et al., 2000). Increased HSP levels could thus slow down the tenderisation process. Second, tenderisation relies also on proteolytic processes (Koohmaraie, 1992; Ouali, 1992; Taylor et al., 1995a,b). HSP27 is known to associate with myofibrils and to protect desmin, a myofibrillar protein, from proteolysis (Blunt et al., 2007; Sugiyama et al., 2000) and could thus hamper the tenderisation process.

In addition, HSPs may not have a direct effect on the tenderisation process, but simply be an indicator of the physiological or biochemical status of the *post-mortem* muscle. First, HSP contents differ according to muscle fibre type and may therefore be indicative of muscle fibre composition. Thus, in the unstressed and stressed muscles, the HSP72 content is higher in muscles with a higher proportion of type I (slow twitch) muscle fibres (Hernando and Manso, 1997; Locke et al., 1991). Various reports indicate that tenderness depends in part on relative proportions of different fibre types (Hocquette et al., 1998; Morzel et al., 2008). Second, as indicated above, stress-induced changes in energy metabolism influence the tenderisation process and HSP levels could simply reflect pre-slaughter stress status (Kwasiborski et al., 2009). However, this relationship appears to be complex. Although increased HSP levels are in principle indicative of increased stress levels, various studies found that in the *post-mortem* muscle, at lower pH, extractability of small HSPs declines, particularly HSP27, which is known to associate with myofibrillar proteins (Kwasiborski et al., 2008; Laville et al., 2009; Pulford et al., 2008). It is believed that at lower pH or at increased myofibrillar contraction, these HSPs are less easily dissociated from myofibrils during the extraction procedure (Kwasiborski et al., 2008; Laville et al., 2009; Pulford et al., 2008). This phenomenon may explain certain opposite relationships reported. For example, young bulls producing tender meat had increased abundance of HSP27 in their muscle extract (Morzel et al., 2008) which could thus indicate that they were less, rather than more stressed. This is in contrast with various other reports (Bernard et al., 2007; Kim et al., 2008) and could thus be related to the effect of animal stress or physiological status of the muscle on extractability.

Other studies (Terlouw et al., unpublished results) found that the effect of pre-slaughter stress depends on the characteristics of the animal. In this experiment, indoor (slatted floor, 2.1 m^2/pig) and outdoor (142 m^2/pig) reared pigs were slaughtered either immediately after 20 min of transport, or after transport to the abattoir the evening before slaughter. Indoor and outdoor pigs slaughtered immediately after transport had a faster pH decline than their counterparts. This reflects a faster *post-mortem* energy metabolism, explained by increased energy demand just before slaughter, during transport (Terlouw et al., 2009). Analysis of the early *post-mortem* muscle proteome found that in indoor pigs, the faster pH decline was associated with increased abundance of muscular pyruvate kinase isoform 1 (PKM1) and glycerol-3-phosphate dehydrogenase (cG3PDH), both enzymes of the glycolytic pathway, and fatty-binding protein isoform 3, involved in the transport of long-chain FAs. These results are coherent with a faster energy metabolism of the *post-mortem* muscle resulting in a faster pH decline. Outdoor pigs slaughtered immediately after transport, compared to their counterparts transported the evening before slaughter, had a lower abundance of HSP27 which is in agreement with the above suggestion that at faster energy metabolism, HSP27 extractability is reduced. However, in contrast to indoor pigs, outdoor pigs slaughtered immediately after transport had decreased, rather than increased abundance of cG2PDH and PKM1. This indicates that other biochemical mechanisms, which remain to be elucidated, underlie the faster muscle metabolism in these pigs.

Several situations can lead to increased oxidative stress in animals including slaughtering. Indeed, the transport of animals at slaughter as well as the handling and mixing with unfamiliar conspecifics can cause emotional and physical stresses well known to significantly increase the production of radical reactive to oxygen species (ROS) which are particularly unstable and harmful for health. Many studies have investigated the relationship between physical activity and free radical production. Thus, studies conducted in humans describe many situations in which exercise induces an increase of ROS, especially in athletes (Bloomer and Goldfarb, 2004). Similarly, in horse, racing exercise causes an overproduction of ROS (De Moffarts et al., 2007). In sheep, physical stress associated with emotional stress causes an increase of plasma peroxide products (Durand et al., unpublished results). Finally, in cattle, a very recent study shows that a moderate preslaughter stress (doubled cortisol) induces an increase in lipoperoxidation process in muscle (from + 50 to + 100%, $P = 0.08$) (Gobert et al., 2009).

On the other hand, it is now accepted that ROS and other radical species may act on the expression of protective enzymes against lipoperoxidation such as SOD, catalase, glutathione synthase and transferase and NADPH: quinone oxidoreductase (Nguyen et al., 2000). In mammals, two classes of transcription factors, nuclear factor kappa B and activator protein 1, are involved in the response to oxidative stress (Scandalios, 2005) (Figure 5.8). Moreover, the induction of specific antioxidant genes involved in xenobiotic metabolism is mediated by 'elements sensitive to antioxidants' ('antioxidant responsive element') usually located in the promoter region of these genes (Scandalios, 2005). Finally, the highlighting of the regulation of specific genes by the process of peroxidation suggests that ROS act as messengers for cellular regulation and translation of these genes (Allen and Tresini, 2000).

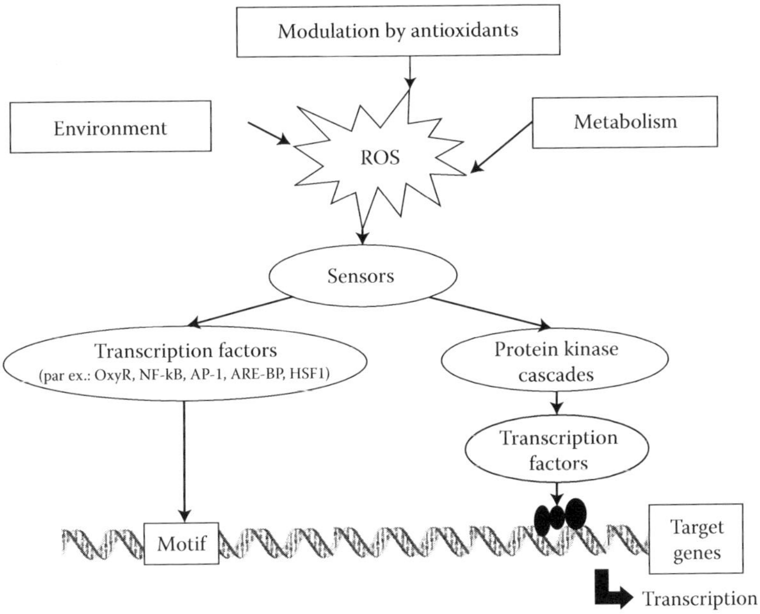

Figure 5.8 Major molecular mechanisms activated in response to oxidative stress. ROS: radical species reactive to oxygen. (From Scandalios, J. G. 2005. *Braz J Med Biol Res* 38:995–1014. With permission.)

All these data confirm the interest of deepening the understanding of the mechanisms involved in the development of oxidative stress due to poor conditions for slaughter by using transcriptomic and proteomic approaches to precisely identify the proteins regulated by the peroxidation process. This work will be conducted in close collaboration with the studies described above on the organoleptic quality of meat.

5.8 CONCLUDING REMARKS

Today, omic techniques are used in research on all aspects of biosciences. Many research institutes meet the increased demand for proteomic research by the development of shared-access specialised laboratories or platforms, which offer both technological equipment and expertise in their use and data exploitation. The use of omic approaches significantly increases our research capacity, particularly in terms of throughput. In addition, the process of research is fundamentally altered. While research was originally hypothesis driven, in omics experiments data can be collected without an existing hypothesis, followed by generation and testing of biological hypotheses. These new conditions are promising regarding the prospects for discovering mechanisms underlying disease, and also of obtaining new fundamental knowledge in different disciplines including metabolism, microbiology and nutrition, among others, with benefits not only for humans but also livestock production sciences.

REFERENCES

Allen, R. G. and M. Tresini. 2000. Oxidative stress and gene regulation. *Free Radic Bio Med* 28:463–499.

Bauchart, C., D. Rémond, C. Chambon et al. 2006. Small peptides (< 5 kDa) found in ready-to-eat beef meat. *Meat Sci* 74:658–666.

Bauman D. E. and J. M. Griinari. 2003. Nutritional regulation of milk fat synthesis. *Annu Rev Nutr* 23:203–227.

Beltran, J. A., I. Jaime, P. Santolaria, C. Sanudo, P. Alberti and P. Roncales. 1997. Effect of stress-induced high post-mortem pH on protease activity and tenderness of beef. *Meat Sci* 45:201–207.

Bendixen, E. 2005. The use of proteomics in meat science. *Meat Sci* 71:138–149.

Bernard, C., I. Cassar-Malek, M. Le Cunff, H. Dubroeucq, G. Renand and J.-F. Hocquette. 2007. New indicators of beef sensory quality revealed by expression of specific genes. *J Agric Food Chem* 55:5229–5237.

Bernard, C., I. Cassar-Malek, G. Renand and J.-F. Hocquette. 2009. Changes in muscle gene expression related to metabolism according to growth potential in young bulls. *Meat Sci* 82:205–212.

Bernard, L., C. Leroux and Y. Chilliard. 2008. Expression and nutritional regulation of lipogenic genes in the ruminant lactating mammary gland. *Adv Exp Med Biol* 606:67–108.

Bernard, L., C. Leroux, M. Bonnet, J. Rouel, P. Martin and Y. Chilliard. 2005a. Expression and nutritional regulation of lipogenic genes in mammary gland and adipose tissues of lactating goats. *J Dairy Res* 72:250–255.

Bernard, L., J. Rouel, C. Leroux et al. 2005b. Mammary lipid metabolism and milk fatty acid secretion in alpine goats fed vegetable lipids. *J Dairy Sci* 88:1478–1489.

Besle, J. M., J. L. Lamaison, B. Dujol et al. 2005. Flavonoids and other phenolics in milk as a putative tool for traceability of dairy production systems. In *Indicators of Milk and Beef Quality*, eds. J.-F. Hocquette and S. Gigli, EAAP Publication 112, pp. 345–350. Wageningen: Wageningen Academic Publishers.

Biasi, F., C. Mascia and G. Poli. 2008. The contribution of animal fat oxidation products to colon carcinogenesis, through modulation of TGF-beta1 signaling. *Carcinogenesis* 29:890–894.

Bidanel, J. P., D. Boichard and C. Chevalet. 2008. From genetics to genomics. *INRA Prod Anim* 21:15–31.

Bloomer, R. J. and A. H. Goldfarb. 2004. Anaerobic exercise and oxidative stress: A review. *Can J Appl Physiol* 29:245–263.

Blunt, B. C., A. T. Creek, D. C. Henderson and P. A. Hofmann. 2007. H2O2 activation of HSP25/27 protects desmin from calpain proteolysis in rat ventricular myocytes. *Am J Physiol Heart Circ Physiol* 293:H1518–H1525.

Boudra, H., M. Godejohann and D. P. Morgavi. 2008. Uncovering metabolic changes using a combined LC–MS/NMR approach: Ochratoxin's effect in sheep. 3èmes *Journées Scientifiques du Réseau Français de Métabolomique et Fluxomique 7 et 8 Février*, Bordeaux, France.

Bouley, J., C. Chambon and B. Picard. 2004a. Mapping of bovine skeletal muscle proteins using two-dimensional gel electrophoresis and mass spectrometry. *Proteomics* 4:1811–1824.

Bouley, J., B. Meunier, J. Culioli and B. Picard. 2004b. Analyse protéomique du muscle de Bovin appliquée à la recherche de marqueurs de la tendreté de la viande. *Renc Rech Ruminants* 11:87–89.

Bouley, J., B. Meunier, C. Chambon, S. De Smet, J.-F. Hocquette and B. Picard. 2005. Proteomic analysis of bovine skeletal muscle hypertrophy. *Proteomics* 5:490–500.

Cassar-Malek, I., C. Jurie, C. Bernard, I. Barnola, D. Micol and J.-F. Hocquette. 2009. Pasture-feeding of charolais steers influences skeletal muscle metabolism and gene expression. *J Physiol Pharmacol* 60: Suppl 2:83–90.

Cassar-Malek, I., F. Passelaigue, C. Bernard, J. Léger and J.-F. Hocquette. 2007. Target genes of myostatin loss-of-function in muscles of late bovine fetuses. *BMC Genom* 8:63.

Cassar-Malek, I., B. Picard, C. Bernard and J.-F. Hocquette. 2008. Application of gene expression studies in livestock production systems: A European perspective. *Aust J Exp Agr* 48:701–710.

Chadwick, R. 2004. Nutrigenomics, individualism and public health. *Proc Nutr Soc* 63:161–166.

Chaze, T., J. Bouley, C. Chambon, C. Barboiron and B. Picard 2006. Mapping of alkaline proteins in bovine skeletal muscle. *Proteomics* 6:2571–2575.

Chaze, T., B. Meunier, C. Chambon, C. Jurie and B. Picard. 2008a. *In vivo* proteome dynamics during early bovine myogenesis. *Proteomics* 8:4236–4248.

Chaze, T., B. Meunier, C. Chambon, C. Jurie and B. Picard. 2009. Proteome dynamics during contractile and metabolic differentiation of bovine foetal muscle. *Animal* 3:980–1000.

Chaze, T., B. Meunier and B. Picard. 2008b. Description of *in vivo* bovine myogenesis using proteomic approach gives new insights for muscle development. *Arch Tierzucht* 51:51–57.

Chelh, I., B. Meunier, B. Picard et al. 2009. Molecular profiles of Quadriceps muscle in myostatin-null mice reveal P13K and apoptotic pathways as myostatin targets. *BMC Genomics* 10:196.

Chilliard, Y. and A. Ferlay. 2004. Dietary lipids and forages interactions on cow and goat milk fatty acid composition and sensory properties. *Reprod Nutr Dev* 44:467–492.

Chilliard, Y., A. Ferlay, J. Rouel and G. Lamberet. 2003. A review of nutritional and physiological factors affecting goat milk lipid synthesis and lipolysis. *J Dairy Sci* 86:1751–1770.

Chilliard, Y., F. Glasser, A. Ferlay, L. Bernard, J. Rouel and M. Doreau. 2007. Diet, rumen biohydrogenation and nutritional quality of cow and goat milk fat. *Eur J Lipid Sci Tech* 109:828–855.

Concannon, C. G., A. M. Gorman and A. Samali. 2003. On the role of Hsp27 in regulating apoptosis. *Apoptosis* 8:61–70.

Creagh, E. M., R. J. Carmody and T.G. Cotter. 2000. Heat shock protein 70 inhibits caspase-dependent and -independent apoptosis in Jurkat T cells. *Exp Cell Res* 257:58–66.

De Moffarts, B., K. Portier, N. Kirschvink et al. 2007. Effects of exercise and oral supplementation enriched in (n-3) fatty acids on blood oxidant markers and erythrocyte membrane fluidity in horses. *Vet J* 174:113–121.

Dhamoon, A. S., E. C. Kohn and N. S. Azad. 2007. The ongoing evolution of proteomics in malignancy. *Drug Discov Today* 12:700–708.

Durand, D., V. Scislowski, Y. Chilliard, D. Gruffat and D. Bauchart. 2005. High fat rations and lipid peroxidation in ruminants; consequences on animal health and quality of products. In *Indicators of milk and beef quality*, eds. J.-F. Hocquette and S. Gigli, EAAP Publication 112, pp. 137–150. Wageningen: Wageningen Academic Publishers.

Everts-van der Wind, A., D. M. Larkin, C. A. Green et al. 2005. A high-resolution whole-genome cattle–human comparative map reveals details of mammalian chromosome evolution. *Proc Natl Acad Sci USA* 102:18526–18531.

Ferlay, A., C. Agabriel, C. Sibra, C. Journal, B. Martin and Y. Chilliard. 2008. Tanker milk variability in fatty acids according to farm feeding and husbandry practices in a French semi-mountain area. *Dairy Sci Technol* 88:193–215.

Ferlay, A., B. Martin, P. Pradel, J. B. Coulon and Y. Chilliard. 2006. Influence of grass-based diets on milk fatty acid composition and milk lipolytic system in tarentaise and Montbeliarde cow breeds. *J Dairy Sci* 89:4026–4041.

Folkesson, M., A. L. Mackey, L. Holm et al. 2008. Immunohistochemical changes in the expression of Hsp27 in exercised human *Vastus Lateralis* muscle. *Acta Physiol* 194:215–222.

Food and Agriculture Organization. 2006. World agriculture towards 2015/2030: Summary report. Rome.

Garon, D., E. Richard, L. Sage, V. Bouchart, D. Pottier and P. Lebailly. 2006. Mycoflora and multimycotoxin detection in corn silage: Experimental study. *J Agric Food Chem* 54:3479–3484.

Glasser, F., A. Ferlay, M. Doreau, P. Schmidely, D. Sauvant and Y. Chilliard. 2008. Long-chain fatty acid metabolism in dairy cows: A meta-analysis of milk fatty acid yield in relation to duodenal flows and *de novo* synthesis. *J Dairy Sci* 91:2771–2785.

Gobert, M., D. Bauchart, E. Parafita, R. Jailler and D. Durand. 2008. Dietary vitamin E associated with plant polyphenols efficiently protect from lipoperoxidation in processed meats in the finishing bovine given an n-3 PUFA-rich diet. *Proceedings of the 54th International Congress of Meat Science and Technology*, Cape Town, South Africa. 3A4, 10–15 August, 2008.

Gobert, M., C. Bourguet, C. Terlouw et al. 2009. Pre-slaughter stress and lipoperoxidation: Protective effect of vitamin E and plant extracts rich in polyphenols given to finishing cattle. In *Ruminant Physiology*, eds. Y. Chilliard, F. Glasser, Y. Faulconnier, F. Bocquier, I. Veissier and M. Doreau, p. 814. Wageningen: Wageningen Academic Publishers.

Guillemin, N., B. Meunier, C. Jurie, I. Cassar-Malek, J.-F. Hocquette and B. Picard. 2009. Validation of a dot-blot quantitative technique for large scale analyses of beef tenderness biomarkers. *J Physiol. Pharmacol* 60:91–97.

Hernando, R. and R. Manso. 1997. Muscle fibre stress in response to exercise: Synthesis, accumulation and isoform transitions of 70-kDa heat-shock proteins. *Eur J Biochem* 243:460–467.

Herrera-Mendez, C. H., S. Becila, A. Boudjellal and A. Ouali. 2006. Meat ageing: Reconsideration of the current concept. *Trends Food Sci Tech* 17:394–405.

Hocquette, J.-F. 2005. Where are we in genomics? *J Physiol Pharmacol* 56:37–70.

Hocquette, J.-F. 2007. Introduction to the session "Omics in metabolism and nutrition studies": Lexic of the "Omics". In *Energy and Protein Metabolism and Nutrition*, 9–13 September, 2007, Vichy, France, ed. I. Ortigues, EAAP Publication 124, pp. 255–258. Wageningen: Wageningen Academic Publishers.

Hocquette, J.-F., C. Bernard, I. Cassar-Malek et al. 2007a. Mise en évidence de marqueurs de tendreté de la viande bovine par des approches de génomique fonctionnelle (projet MUGENE). *Renc Rech Ruminants* 14:117–120.

Hocquette, J.-F., C. Bernard-Capel, M. L. Vuillaume, B. Jesson, H. Levéziel, I. Cassar-Malek. 2009a. The GENOTEND chip: A tool to analyse gene expression in muscles of beef cattle. *BIT's 2nd Annual Congress and Expo of Molecular Diagnostics (CEMD-2009)*, 19–21 November, 2009, Beijing, China.

Hocquette, J.-F., H. Boudra, I. Cassar-Malek et al. 2010. Perspectives offered by "omics" approaches to sustainable herbivore production. *INRA Prod Anim* 22:385–396.

Hocquette, J.-F., I. Cassar-Malek, A. Scalbert and F. Guillou. 2009b. Contribution of genomics to the understanding of physiological functions. *J Physiol Pharmacol* 60:5–16.

Hocquette, J.-F., S. Lehnert, W. Barendse, I. Cassar-Malek and B. Picard. 2007b. Recent advances in cattle functional genomics and their application to beef quality. *Animal* 1:159–173.

Hocquette J.-F., I. Ortigues-Marty, D. W. Pethick, P. Herpin and X. Fernandez. 1998. Nutritional and hormonal regulation of energy metabolism in skeletal muscles of meat-producing animals. *Livest Prod Sci* 56:115–143.

Hollywood, K., D. R. Brison and R. Goodacre. 2006. Metabolomics: Current technologies and future trends. *Proteomics* 6:4716–4723.

Hwang, I. H. and J. M. Thompson. 2001. The interaction between pH and temperature decline early postmortem on the calpain system and objective tenderness in electrically stimulated beef *longissimus dorsi* muscle. *Meat Sci* 58:167–174.

Jarrige, R., Y. Ruckebush, C. Demarquilly, M. H. Farce and M. Journet. 1995. *Nutrition des ruminants domestiques. Ingestion et digestion*. Paris: INRA Editions.

Jurie, C., I. Ortigues-Marty, B. Picard, D. Micol and J.-F. Hocquette. 2006. The separate effects of the nature of diet and grazing mobility on metabolic potential of muscles from Charolais steers. *Livest Sci* 104:182–192.

Kim, N. K., S. Cho, S. H. Lee et al. 2008. Proteins in longissimus muscle of Korean native cattle and their relationship to meat quality. *Meat Sci* 80:1068–1073.

Koohmaraie, M. 1992. The role of Ca^{2+}-dependent proteases (calpains) in post mortem proteolysis and meat tenderness. *Biochimie*73:239–245.

Kraft, G., D. Gruffat, D. Dardevet, D. Rémond, I. Ortigues-Marty and I. Savary-Auzeloux. 2009. Nitrogen and energy imbalanced diets differentially affected hepatic protein synthesis and gluconeogenesis in growing lambs. *J Anim Sci* 87:1747–1758.

Kwasiborski, A., D. Rocha and C. Terlouw. 2009. Gene expression in large white or Duroc-sired female and castrated male pigs and relationships with pork quality. *Anim Gen* 40:852–862.

Kwasiborski, A., T. Sayd, C. Chambon, V. Santé-Lhoutellier, D. Rocha and C. Terlouw 2008. Pig Longissimus lumborum proteome: Part II: Relationships between protein content and meat quality. *Meat Sci* 80:982–996.

Laville, E., T. Sayd, M. Morzel et al. 2009. Proteome changes during meat ageing in tough and tender beef suggest the importance of apoptosis and chaperone proteins for beef ageing and tenderisation. *J Agric Food Chem* 57:10755–10764.

Lehnert, S. A., A. Reverter, K. A. Byrne et al. 2007. Gene expression studies of developing bovine longissimus muscle from two different beef cattle breeds. *BMC Dev Biol* 7:95.

Lobley, G. E. 2003. Protein turnover—What does that mean for animal production? *Can J Anim Sci* 83:327–340.

Locke, M., E. G. Noble and B. G. Atkinson. 1991. Inducible isoform of HSP70 is constitutively expressed in a muscle fiber type specific pattern. *Am J Physiol* 261:C774–C779.

Loor, J. J., H. M. Dann, N. A. Janovick Guretzky et al. 2006. Plane of nutrition prepartum alters hepatic gene expression and function in dairy cows as assessed by longitudinal transcript and metabolic profiling. *Physiol Genomics* 27:29–41.

Morozova, O., M. Hirst and M. A. Marra, 2009. Applications of new sequencing technologies for transcriptome analysis. *Annu Rev Genomics Hum Genet* 10:135–151.

Morzel, M., C. Terlouw, C. Chambon, D. Micol and B. Picard. 2008. Muscle proteome and meat eating qualities of Longissimus thoracis of "Blonde d'Aquitaine" young bulls: A central role of HSP27 isoforms. *Meat Sci* 78:297–304.

Mosley, E. E., B. Shafii Dagger, P. J. Moate and M. A. McGuire. 2006. *cis*-9, *trans*-11 conjugated linoleic acid is synthesized directly from vaccenic acid in lactating dairy cattle. *J Nutr* 136:570–575.

Mouriot, J., L. Bernard, P. Capitan et al. 2009. Quantitative estimation of the endogenous synthesis of rumenic acid in goats fed lipid supplements. In *Ruminant Physiology*, eds. Chilliard, Y., F. Glasser, Y. Faulconnier, F. Bocquier, I. Veissier and M. Doreau, p. 448, Wageningen: Wageningen Academic Publishers.

Mullen, A. M., P. C. Stapleton, D. R.M. Corcoran Hamill and A. White. 2006. Understanding meat quality through the application of genomic and proteomic approaches. *Meat Sci* 74:3–16.

Nguyen, A. N., A. Lee, W. Place and K. Shiozaki. 2000. Multistep phosphorelay proteins transmit oxidative stress signals to the fission yeast stress-activated protein kinase. *Biol Cell* 14:1169–1181.

Nielsen, J. 2007. Metabolomics in functional genomics and system biology. In *Metabolome Analysis: An Introduction*, eds. S. G. Villas-Bôas, J. Nielsen, J. Smedsgaard and M. A. E. Hansen, 3–14. New Jersey, NJ: Wiley Interscience.

Ollier, S., S. Chauvet, P. Martin, Y. Chilliard and C. Leroux. 2008. Goat's αS1-casein polymorphism affects gene expression profile of lactating mammary gland. *Animal* 2:566–573.

Ollier, S., C. Robert-Granié, L. Bernard, Y. Chilliard and C. Leroux. 2007. Mammary transcriptome analysis of food-deprived lactating goats highlights genes involved in milk secretion and programmed cell death. *J Nutr* 137:560–567.

Ortigues, I., C. Obled, D. Dardevet and I. Savary-Auzeloux. 2003. Role of the liver in the regulation of energy and protein status. In *Progress in Research on Energy and Protein Metabolism*, eds. W. B. Souffrant and C. C. Metges, EAAP Publication 109, pp. 13–18. Wageningen: Wageningen Academic Publishers.

Ouali, A. 1992. Proteolytic and physicochemical mechanisms involved in meat texture development. *Congrès OECD Workshop*, Nottigham, 74, pp. 251–265.

Ozpinar, H., G. Augonyte and W. Drochner. 1999. Inactivation of ochratoxin in ruminal fluid with variation of pH-value and fermentation parameters in an *in vitro* system. *Environ Toxicol Pharmacol* 7:1–9.

Paulsen, G., F. Lauritzen, M. L. Bayer et al. 2009. Subcellular movement and expression of HSP27, {alpha} B-crystallin, and HSP70 after two bouts of eccentric exercise in humans. *J Appl Physiol* 107:570–582.

Paulsen, G., K. Vissing, J. M. Kalhovde et al. 2007. Maximal eccentric exercise induces a rapid accumulation of small heat shock proteins on myofibrils and a delayed HSP70 response in humans. *Am J Physiol Regul Integr Comp Physiol* 293:R844–R853.

Picard, B., C. Jurie, I. Cassar-Malek and J.-F. Hocquette. 2003. Typologie et ontogenèse des fibres musculaires chez le bovin. *INRA Prod Anim* 16:125–131.

Picard, B., L. Lefaucheur, C. Berri and M. J. Duclos. 2002. Muscle fibre ontogenesis in farm animal species. *Reprod Nutr Dev* 42:1–17.

Pittet, A. 1998. Natural occurrence of mycotoxins in food and feed—An update review. *Rev Met Vet* Toulouse 149:479–492.

Prache, S., B. Martin, P. Nozière et al. 2007. Authentification de l'alimentation des ruminants à partir de leurs produits et tissus. *INRA Prod Anim* 20:295–308.

Priolo, A., A. Cornu, S. Prache et al. 2004. Fat volatile tracers of grass feeding in sheep. *Meat Sci* 66:475–481.

Pulford, D. J., S. F. Vazquez, D. F. Frost et al. 2008. The intracellular distribution of small heat shock proteins in post-mortem beef is determined by ultimate pH. *Meat Sci* 79:623–630.

Renerre, M. 1990. Review: Factors involved in the discoloration of beef meat. *Int J Food Sci Tech* 25:613–630.

Scandalios, J. G. 2005. Oxidative stress: Molecular perception and transduction of signals triggering antioxidant gene defences. *Braz J Med Biol Res* 38:995–1014.

Scheler, C., X. P. Li, J. Salnikow, M. J. Dunn and P. R. Jungblut. 1999. Comparison of two-dimensional electrophoresis patterns of heat shock protein Hsp27 species in normal and cardiomyopathic hearts. *Electrophoresis* 20:3623–3628.

Schwerin, M., C. Kuehn, S. Wimmers, C. Walz and T. Goldammer. 2006. Trait-associated expressed hepatic and intestine genes in cattle of different metabolic type-putative functional candidates for nutrient utilization. *J Anim Breed Genet* 123:307–314.

Scudamore, K. A. and C. Livesey. 1998. Occurrence and significance of mycotoxins in forage crops and silage: A review. *J Sci Food Agric* 77:1–17.

Shingfield, K. J., S. Ahvenjarvi, V. Toivonen, A. Vanhatalo and P. Huhtanen. 2007. Transfer of absorbed *cis*-9, *trans*-11 conjugated linoleic acid into milk is biologically more efficient than endogenous synthesis from absorbed vaccenic acid in lactating cows. *J Nutr* 137:1154–1160.

Shingfield, K. J., Y. Chilliard, V. Toivonen. P. Kairenius and D. I. Givens. 2008. Trans fatty acids and bioactive lipids in ruminant milk. *Adv Exp Med Biol* 606:3–65.

Simopoulos, A. P. 1999. Essential fatty acids in health and chronic disease. *Am J Clin. Nutr* 70:560S–5609S.

Sudre, K., C. Leroux, G. Pietu et al. 2003. Transcriptome analysis of two bovine muscles during ontogenesis. *J Biochem* 133:745–756.

Sugiyama, Y., A. Suzuki, M. Kishikawa et al. 2000. Muscle develops a specific form of small heat shock protein complex composed of MKBP/HSPB2 and HSPB3 during myogenic differentiation. *J Biol Chem* 275:1095–1104.

Tarrant, P. V., F. J. Kenny, D. Harrington and M. Murphy. 1992. Long distance transportation of steers to slaughter: Effect of stocking density on physiology, behaviour and carcass quality. *Livest Prod Sci* 30:223–238.

Taylor, R. G., M. Briand, N. Robert, Y. Briand and A. Ouali. 1995a. Proteolytic activity of proteasome on myofibrillar structures. *Mol Biol Rep* 21:71–73.

Taylor, R. G., V. F. Thompson, M. Koohmaraie and D. E. Goll. 1995b. Is Z-Disk degradation responsible for postmortem tenderzation. *J Anim Sci* 73:1351–1367.

Terlouw, C., A. Berne and T. Astruc. 2009. Effect of rearing and slaughter conditions on behaviour, physiology and meat quality of large white and Duroc-sired pigs. *Livest Sci* 122:199–213.

The Bovine Genome Sequencing Analysis Consortium, 2009. The genome sequence of taurine cattle: A window to ruminant biology and evolution. *Science* 324:522–528.

Urbach, G. 1990. Effect of feed on flavour in dairy foods. *J Dairy Sci* 73:3639–3650.

Van Egmond, H. P. 2004. Natural toxins: Risks, regulations and the analytical situation in Europe. *Anal Bioanal Chem* 378:1152–1160.

Vasta, V. and A. Priolo. 2006. Ruminant fat volatiles as affected by diet. A review. *Meat Sci* 73:218–228.

Warriss, P. D. and S. N. Brown. 1985. The physiological responses to fighting in pigs and the consequences for meat quality. *J Sci Food Agr* 36:87–92.

Wolfe, R. R. 1992. *Radioactive and Stable Isotope Tracers in Biomedicine*. New York, NY: Wiley and Sons.

Young, O. A., J. L. Berdagué, C. Viallon, S. Rousset-Akrim and M. Theriez. 1997. Fat-borne volatiles and sheepmeat odour. *Meat Sci* 45:183–200.

CHAPTER 6

Mining Omic Technologies and Their Application to Sustainable Agriculture and Food Production Systems

Noureddine Benkeblia

CONTENTS

6.1 INTRODUCTION

The sustainability of agricultural production is one of the major contemporary concerns for humans. Because agricultural production is the unique source of food for the increasing world population, it is a great challenge for man to ensure food production for this population using the finite resources. Sustainability is dependent to a large degree on environment quality, and management and maintenance of soils are considered central factors of the sustainable agricultural and food production systems. The incommensurate development in molecular biology and genetic engineering, as well as the new emerging technologies, for example, metabolomics, metagenomics, nutrigenomics and more recently ionomics, are translated to agricultural research and productive technologies. This translation would be of great help for human to examine how available and limited natural resources and environment would let him reach the goal of the global sustainable agricultural and food production systems. Thus, the principles that regulate the new agricultural and food production systems including soil management and environment preservation become fundamental rules to the philosophy of sustainability. While natural resources are limited and environmental degradation is progressing dangerously, it is necessary to find an issue to the two major concerns which are population growth and good life. On the other hand, the transformation from industrial to sustained agriculture could be considered as a success, because this concept of sustainability will make and ensure the 'balance' between the needs and the developments.

From the past century, the concerns of the scientific community involved in agricultural science focused on quantitative relations between resource availability and land, water and other natural resources and so sustain the continuous growth of humanity. However, in the late 1960s, the natural resource scarcity and the limited capacity of the global environment to assimilate the multiple forms of pollution generated by human activities and growth, an intense conflict emerged between the rising demand and the rapid growth (Ruttan, 1992). From not so far and since the mid-1980s, a new concept has been emerged from these two conflicting situations, and it centres on the implications for environmental quality, food production and human health that are occurring. During this period, effects of intensive agricultural production on resource degradation—through erosion, salinization and depletion of groundwater—and the quality of soil and water—though runoff and leaching of plant nutrients and pesticides—have been of intense concern. Thus, term *sustainable agriculture* began to enter the vocabulary of the people responsible for agricultural research resource allocation (Dorfman, 1991; Ruttan, 1992). The present period is known by an important 'turn' for agriculture and the trend is to move from an industrial to a sustainable agriculture which focuses on the ability to sustain the growth in the demands and this is considered to be the most remarkable transition in the history of agriculture (Horrigan et al., 2002; Ruttan, 1994).

As we shall look towards the future of the human life which is undoubtedly dependent on the environmental life and qualities, it is necessary to create new opportunities for advancing agricultural research to reverse the urgency of the concerns cited above. Advances in molecular biology and genetic engineering, as well as other new emerging technologies such as metabolomics, metagenomics, nutrigenomics and ionomics, are occurring rapidly. These promising advances are nowadays translated to agricultural research and productive technologies, and this translation could be a key answer to respond these concerns and examine how resources and environment may impose constraints for global sustainable agricultural systems.

6.2 SUSTAINABLE AGRICULTURE

From the first enumeration by Rodale (1988) of the reasons for the interest in agricultural sustainability, many definitions have been suggested for this concept. These suggestions were analysed by Okigbo (1991, p. 66) who defined 'agricultural sustainability as one that maintains an acceptable and increasing level of productivity that satisfies prevailing needs and is continuously adapted to meet the future needs for increasing the carrying capacity of the resource base and other worthwhile human needs', and accordingly when resources, inputs and technologies are within the abilities of the farmers to own, hire and manage with increasing efficiency. Thus, desirable levels of production in perpetuity with minimal or no adverse effects resources, environment quality and life will be the achievements of this concept.

6.3 FOOD PRODUCTION IN SUSTAINABLE AGRICULTURAL SYSTEM

Foods are the most basic objects of production and offer a valuable entry point in exploring the challenge of achieving sustainable production patterns (Figure 6.1). This challenge is essentially the task of establishing sustainable development as the world's method of providing 'a higher quality of life for everyone', for both current and future generations. Sustainable food production shall improve the quality of life while ensuring social, environmental and economic sustainability, and this concept should be considered as the goal of sustainable agriculture. The greatest challenge is to protect and sustainably manage the natural resource base on which food and fiber productions depend, while feeding and housing a population that is still growing. Industrial agriculture and conventional food industrial systems tend to produce cheap and abundant food; however, the unsustainable environmental and social costs of these systems are becoming increasingly apparent and the impacts of industrial agriculture are highlighted by the following points (OECD, 2001; UN, 1999).

6.3.1 Soil Degradation

The area of cropland per capita has been steadily declining, from 0.43 hectares (ha) in 1961 to about 0.26 ha in 1996, and to about 0.24 in 2004 (Figure 6.1) (UN, 1999; UNDP, 1998). Since 1945, around 2 billion hectares became uncultivated because of poor agricultural practices (WRI, 1998), while about 5–6 million hectares are lost each year due to soil degradation. Thus, it becomes urgent to put under strict control and/or ban the 'agricultural practices' that damage soils such as the use of chemical and synthetic pesticides and easily soluble mineral fertilizers.

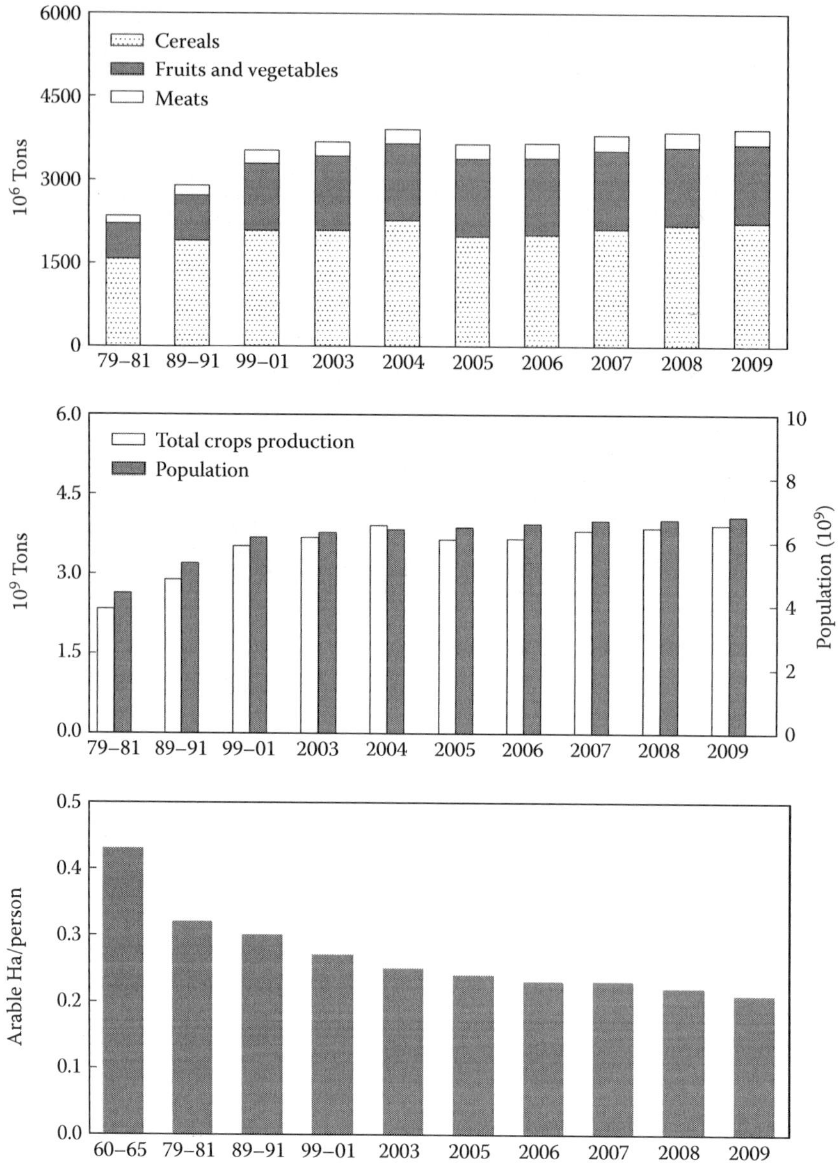

Figure 6.1 Crops production (a), total crop production and population evolution (b) and arable land available (c) worldwide. (Adapted from FAO. Food and Agriculture Organization. 2000. *Biotechnology for Sustainable Agriculture*. Report of the Commission on Sustainable Development. Food and Agriculture Organization, Rome, Italy.)

6.3.2 Cropland and Yield Losses

While the future jump in demand for food should motivate protection of arable land, this is generally not the case. Prime farmland continues to be lost at a serious rate due to urban development. Although losing almost million hectares of cropland, many countries continue towards a goal of building new cities. On the other hand, the dramatic rise in grain yields during the 1960s and 1970s tended to outweigh the loss of arable land. However, since the 1990s, grain yields have slowed to such a degree that they no longer compensate for the steady elimination of grainland.

6.3.3 Water Pollution and Overpumping

The increasing use of inorganic fertilizers, pesticides and other chemicals is resulting in some areas in the contamination of drinking water with nitrates and damages to aquatic ecosystems from eutrophication. Moreover, in many irrigation-dependent countries, water tables are falling because of overpumping.

6.3.4 Overfishing

Overfishing has resulted in reduced productivity of fisheries, with the marine fish harvest now stagnant. UNDP claims that fish stocks are declining, with about one-fourth currently depleted or in danger of depletion and another 44% being fished at their ecological limit.

6.4 ORGANIC FOOD PRODUCTION AND AGRICULTURAL BIOTECHNOLOGY

Industrial agriculture is associated with high chemical inputs and higher yields would mean more food. However, high-yield output from these methods was behind the degradation of natural fertility of soils, creation of pesticide-resistant pests, in addition to the damage caused to environment and health. The other major chemical input associated with industrial agriculture, for example, nitrogen fertilizers, increased in use more than fivefold during the past 40 years. One of the main pollution concerns with fertilizers is contamination of surface and groundwater with nitrates, threatening both health and environmental quality, and presently nitrate is now one of the most common chemical contaminants in drinking water. Thus, organic farming avoids such chemical dependency, and could 'dramatically reduces external inputs'. As a result, organic agriculture improves the fertility and structure of soils, using crop rotation, recycling of crop residues, and applying organic manures and mulches to encourage the development of soil microorganisms. Organic farming methods also protect surface and ground water and help preserve biodiversity, making this agroecological agriculture the most suited for meeting the challenge of creating food security while conserving the environment and natural resource base of agriculture. The goal of this sustainable farming is 'seeking how to make the best use of nature's goods and services whilst not damaging the environment' and this goal could be reached by (1) integrating natural processes such as nutrient cycling, nitrogen fixation and soil regeneration into food production processes and (2) minimizing the use of non-renewable inputs (pesticides and fertilizers) that damage the environment or harm the health. Unfortunately, only a small per cent of agricultural producers are organic farmers, for example, only 1.3% of all farms are organic within the European Union, while it tends to be more common in Sweden: 12%, Austria: 9% and Finland: 4% (EC, 1999). The major task and challenge is to find the strategy by which sustainable agriculture can successfully make the transition to become the standard operating framework for producing food and achieving sustainable food security. Thus, it is obvious that the transition from industrial agricultural system to a sustainable global agricultural system is going to need a lot more.

Biotechnology is another controversial development of industrial agriculture by integrating genetically modified (GM) food into the global food system. The Food and Agriculture Organization (FAO, 2000) claimed that the application of agricultural biotechnology techniques can increase food security. In their recommendations, FAO stresses 'the moral imperative' for making modern agricultural biotechnologies, such as GM (transgenic) crops readily and economically available to developing countries. However, genetically modified organisms are still under intense debate. Rather than an opportunity to enhance food security, consumers will have to pay the corporate owners of this new technology for permission to eat.

6.5 ROUTE OF OMIC TECHNOLOGIES TO GLOBAL AND AGRICULTURAL SUSTAINABILITY

Products obtainable through modern biotechnology are of crucial importance within the context of sustainable development (Figure 6.2). One of the greatest ultimate impacts will be in agricultural genomics, especially for breeding programs in crop and crop genetic engineering, and to overcome biotic and abiotic stresses. It is imperative that sustainable agriculture becomes a 'key target' in the 'gene revolution' because the cost of 'damage reparation' of these stresses may be higher than the cost of developing new concepts.

6.6 METABOLOMICS

6.6.1 Metabolomics for Biotic and Abiotic Stresses Assessment

The agronomic fall of sustainable agriculture can be biotic (e.g., resistance to diseases, parasites, insects and weeds) or abiotic (e.g., better adaptation to heat, drought, salinity, acidity, heavy metals, waterlogging and nutrient (especially nitrogen and phosphorus) availability) (Jahangir et al., 2009; Shao et al., 2007, 2008). Biotic and abiotic stresses cause significant losses in crops and significantly affect their productivity. The emerging metabolomics technology is promising strategy for

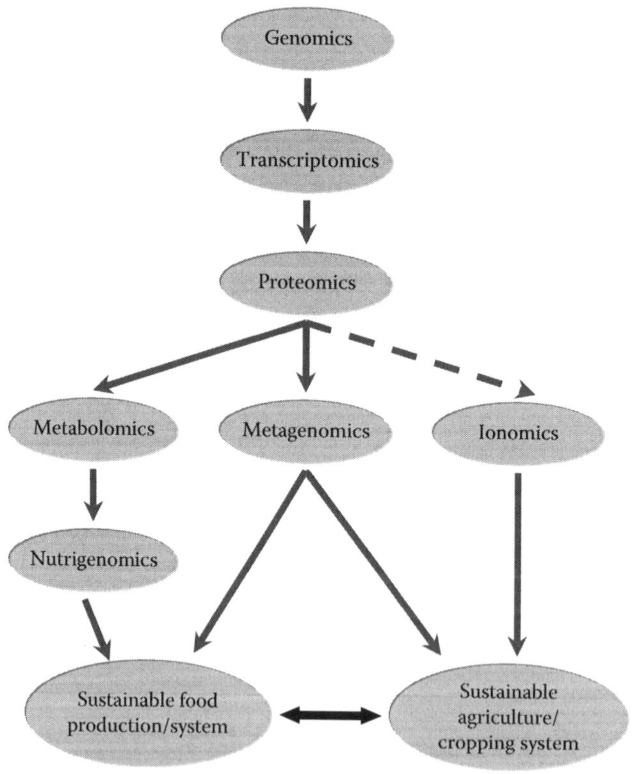

Figure 6.2 Schematic insight into 'omic' technologies and their serviceability in sustainable agricultural and food production systems.

understanding the molecular genetic basis of stress response and resistance, and then constraints affecting crops. Understanding the mechanisms that regulate the expression of stress-related genes is a fundamental issue in plant biology and will be necessary for the genetic improvement of crops mainly plant foods (Dita et al., 2006).

The major biotic stress affecting cultivated crops is fungal diseases, insects, nematodes, viruses, bacteria and parasitic weeds. The severity of fungal diseases and their effects on crops vary through years and regions. However, some of them affect large areas and cause considerable losses in quantity and quality (Nene and Reddy, 1987; Rubiales et al., 2002; Warkentin et al., 1996). Several soil-borne diseases are also common in crops and are major constraints in the production of crops. Most of these attack the seedling stage of the crops and are referred to as damping-off diseases and can result in up to 80% of plant death (Denman et al., 1995; Kolkman and Kelly, 2003; Navas-Cortés et al., 2000; Wang et al., 2003a). Moreover, viruses also cause losses for most crops, and some viruses have been considered the most important yield-limiting viruses (Coyne et al., 2003), while insects, another important biotic stress faced by crops, cause important damages both through direct feeding or by providing infection sites for pathogens (Garza et al., 1996; Romero-Andreas et al., 1986; Yoshida et al., 1997).

From the biological point of view, abiotic stresses, although a broad term, include multiple stresses such as heat, cold, excessive light, drought, waterlogging, wounding, exposure, salinity, heavy metals, nutrient availability and so on. It was estimated that only 90% of arable lands where crops grown-on experience one or more environmental stresses, and some of these stresses such as drought, extreme temperature and high salinity dramatically limit crops productivity (Dita et al., 2006). Water deficit constitutes the major abiotic factor affecting crops (Sharma and Lavanya, 2002), and in many non-tolerant drought legumes, this stress is particularly important because of aflatoxin formation after harvest (Arrus et al., 2005; Mahmoud and Abdalla, 1994). On the other hand, waterlogging causes severe yield losses (Dennis et al., 2000), and limits O_2 diffusion of the soil and subsequently nitrification is replaced as the most important N-transforming process, by denitrification and/or nitrate ammonification (Laanbroek, 1990). Furthermore, under waterlogging stress, plant potassium, sodium, iron and manganese uptake are limited and crops become more susceptible to diseases (McDonald and Dean, 1996). Salinity of soil has also an impact on nitrogen uptake, which causes yield reduction (van Hoorn et al., 2001). Deficiencies and toxicities of micronutrients are also considered as important factors affecting the growth of crops (Dwivedi et al., 1992; Hirai et al., 2004; Poulain and Almohammad, 1995). For example, deficiency or toxicity is more critical for roots nodulation than for the direct growth of soybean (Rahman et al., 1999).

To face the problems generated by these stresses, metabolomics are one of new strategies developed, and a number of targeted metabolome analyses were carried out to assess the involvement of subsets of metabolites in various stresses (Urano et al., 2010). For the stresses mentioned, metabolomics could be considered a powerful tool that has the potential to contribute to sustainable agriculture. The main research work on plant stress responses focused on phenolics accumulation observed in response to pathogens infection (Baldridge et al., 1998; BorejszaWysocki et al., 1997; Lozovaya et al., 2004; Saunders and O'Neill, 2004; Shimada et al., 2000). For example, phytoalexin accumulation was observed after copper or mercury stress (Mithofer et al., 2002) and other metabolite classes have also been described as potential defense or signal molecules such as terpenoids (He and Dixon, 2000; Wu and VanEtten, 2004). However, the exact roles of these molecules remain still not understood, but they are thought to act as attractants for natural predators of herbivorous insects and as systemic signal defense. Currently, large-scale analyses by using metabolomics technology are providing large data sets that would help identify potential candidate biomolecules for the increase of intrinsic resistance and/or tolerance levels in crops.

6.6.2 Metabolomics for Food Quality Attributes, Microbiology and Nutrition

The food quality and nutritional aspects of sustainable agriculture include content and quality of macronutrients such as starch, protein, oil, as well as nutritional or micronutrients (Mazur et al., 1999) and metabolomics could be considered as one of the foundations of sustainable agriculture and environment care (Campbell et al., 2003; Fernie and Schauer, 2009; Mazur et al., 1999; Somerville and Somerville, 1999; Tada et al., 1996). Metabolomics, combined with other omic technologies, were used to enhance the contents of essential and non-essential macronutrients and micronutrients, such as vitamins (e.g., A, C, E, folate), minerals (e.g., iron and zinc) and proteins (Beyer et al., 2002; Graham et al., 2001; Oresic, 2009; Potrykus, 2001; Wu et al., 2003; Zarate, 2010), while vitamin pathways have been designed for the synthesis of many other 'nonessential' compounds and macronutrients (DellaPena, 1999; DellaPenna, 2001). Microbiology, both food and general, is also profiting from metabolomics to study microbial metabolism which is of special interest concerning human and animal health (Grivet et al., 2003). Microbial metabolome also includes applications in basic studies of foodborne pathogens and foods metabolism resulting from its microbial ecology such as cheeses ripening (Rager et al., 2000).

More recently, because foods or ingredients derived from GM crops were perceived disparagingly by consumers because of concerns about unintended effects on human health (Frewer et al., 2004), risk assessment of potential adverse effects on humans and the environment became a necessity since these changes are connected to changes in metabolite levels in the plants. Thus, metabolomics are offering a good tool for comparing conventional cultivars to those GM (Risher and Oksman-Caldentey, 2006). For example, Catchpole et al. (2005) reported that the application of metabolomics methodology has been shown to be useful for the investigation of compositional similarity in GM potatoes.

6.6.3 Metabolomics and Environmental Concerns

Pollution with metals and xenobiotics is a global environmental problem that has resulted from mining, industrial, agricultural and other activities. The most important environmental concern today is the behaviour and the effects of chemical and organic pollutants, mainly pesticides and heavy metals, and their potential degradation by microorganisms. Currently, these pollutants and/or microbial degradation of xenobiotics are assessed using metabolomics. Moreover, metabolomics encompasses the study of the metabolic response of biological systems to these toxic contaminants (Grivet et al., 2003; Lommen et al., 2007).

Using a 'rhizosphere metabolomics' approach, root-associated microbes were used to enhance the depletion of some persistent organic pollutants such as PCBs (polychlorinated biphenyls). This depletion was effective by the production of secondary metabolites, such as phenylpropanoids which constitute 84% of the secondary metabolites exuded from *Arabidopsis* roots. Phenylpropanoid-utilizing microbes are more competitive and able to grow at least 100-fold better than their auxotrophic mutants on roots of plants that are able to synthesize or overproduce phenylpropanoids, such as flavonoids. This strategy is also likely to be applicable to improving the competitive abilities of biocontrol and biofertilization strains (Kuiper et al., 2001; Mohn et al., 1997; Narasimhan et al., 2003). Metabolomic approaches were also used to study the reductive cleavage of demethon-*S*-methyl by *C. glutamicum* (Girbal et al., 2000) and the biodegradation of fluorinated aromatic compounds (Boersma et al., 2001).

Microelements in excess, including essential or nonessential elements, and heavy or trace elements that are putatively non-essential, are potentially toxic or highly toxic to all classes of living organisms, respectively. Soils and water may be contaminated with these elements as a result of mining or industrial and domestic activities, use of phosphorus containing fertilizers, land applications of sewage sludge and atmospheric deposition (Cheng, 2003; di Toppi and Gabbrielli, 1999;

Meharg, 2004; Nriagu and Pacyna, 1988). As an example, Roessner et al. (2006) studied Boron which is an essential micronutrient that affects plant growth at either deficient or toxic concentrations in soil, and they used a metabolomics approach to compare metabolite profiles in root and leaf tissues. Bailey et al. (2003) also used metabolomics analysis to study the consequences of cadmium (Cd) exposure. On the other hand, a metabolomics comparison of drinking water sample from a farm to distillated water showed the presence of contaminants specific to this drinking water which unexpectedly contained an abnormal level of chrysene obviously not eliminated during clean-up (Lommen et al., 2007).

6.7 IONOMICS

6.7.1 What Is the Concept of 'Ionomics'?

A living biological system is regulated by its genome via the action of the four biochemical keys: (1) transcriptome, (2) proteome, (3) metabolome and (4) ionome which represent the sum of all the expressed genes, proteins, metabolites and elements within an organism. Aside genomics, transcriptomics and proteomics, metabolomics is currently under way (Fiehn et al., 2000), while the study of the ionome is still in its first steps because neither genes nor gene networks involved in its regulation are known (Hirschi, 2003; Lahner et al., 2003; Rea, 2003). The challenge to understand the ionome and its interaction with other cellular systems and environment would be of real asset to the understanding of the organic and inorganic metabolisms of plants (Baxter, 2009; Salt, 2004).

The term 'ionome' was first described by Lahner et al. (2003), and this description concerns all the metals, metalloids and non-metals present in an organism, while and extension to 'metallome' was made to include biologically significant nonmetals such as nitrogen, phosphorus, sulphur, selenium, chlorine and iodine (Outten and O'Halloran, 2001; Szpunar, 2004; Williams, 2001). However, the elements to be measured in the ionome will be determined by their biological importance or environmental relevance, in conjunction with their amenability to quantitation.

6.7.2 Plant Ionome

During the past decades, plant nutrition has undergone considerable progress (Marschner, 1995), and the new developed technologies, such as inductively coupled plasma mass spectrometry (ICP-MS), have accelerated this progress although much remains not discovered yet. One of the key advances enabling organisms to survive was the evolution of ion metabolism and many ion transporters, which have been the primary focus of most work involved in characterizing the ionome in plants, have been characterized (Mäser et al., 2001). The plant ionome involves simultaneously the measurement of the quantitative and the qualitative composition of plants and the potential changes in response to biotic, abiotic, environmental and genetic modifications. As for other omics technologies, ionomics also requires the application of high-throughput analysis technologies integrating bioinformatic and genetic. One of the most important features of ionomics is to offer the possibility to have information on the functional state of plants under different conditions, driven by the intrinsic and extrinsic factors. This technology offers the advantage to either study simultaneously the functional genetic analysis which is behind the ionome control, and beyond the genetic networks controlling both developmental and physiological processes affecting the ionome indirectly (Karley and White, 2009; Salt, 2008). For example, a comprehensive analysis of the *Arabidopsis* genome revealed the existence of *ca.* 1000 ion transporters although major parts have not yet been characterized. Ii was also reported that only 5% of the genes in the *Arabidopsis* genome are involved in regulating the ionome (Lahner et al., 2003). In addition, ionomics demonstrated its capability to provide a rapid way to identify genes that control the accumulation of elements in plants when combined with new genotyping technologies, and this combination therefore would help to better understand the regulation of these elements (Baxter, 2009). Moreover, multiple membrane proteins may be needed for ion uptake from the soil to adapt to varying extracellular conditions and nutrient availability, and typical examples exist on the regulation of some ions in plants, such as phosphorus, potassium, sodium, calcium and zinc (Curie and Briat, 2003; Rausch and Bucher, 2002; Sanders et al., 2002; Véry and Sentenac, 2003; Zhu, 2003).

Furthermore, plant mineral nutrition is closely related to the development of analytical chemistry, and more recent developments in analytical techniques have boosted incommensurately the development of omics biology, highlighting the response of the genes to changes in nutrient availability and others (Maathuis et al., 2003; Negishi et al., 2002; Thimm et al., 2001; Wang et al., 2003; Wintz et al., 2003). It is obvious from these studies that plants react specifically to the availability or deficiency of each suggesting that the existence of regulatory pathways. Thus, a high-throughput ion-profiling strategy based on the ICP-MS technology for genomic scale profiling of nutrients and trace elements was developed by Lahner et al. (2003).

However, the advent of high-throughput phenotyping technologies has generated a deluge difficult to manage. To unravel this problem, a system, Purdue Ionomics Information Management System (PiiMS, www.purdue.edu/dp/ionomics) has been developed by Baxter et al. (2007). This system provides integrated workflow control, data storage and analysis, and currently contains data on shoot concentrations of P, Ca, K, Mg, Cu, Fe, Zn, Mn, Co, Ni, B, Se, Mo, Na, As and Cd in over 60,000 shoot tissue samples of Arabidopsis (*Arabidopsis thaliana*).

6.7.3 Heavy Metals and Ionomics Approaches in Phytoremediation

Plants, as other organisms, live in their specific habitats and that constitute their environment. However, industrial and mining activities have negative impacts on the ecological balance and particularly on flora and fauna, and hence, people and their environments are being affected, strongly. Nowadays, environmental awareness and studies concerning environmental issues draw considerable attention in the world. Thus, and from a sustainable development point of view, both the continuity of industrial and mining activities and the minimization of possible adverse effects of these activities are required and constitute the goal of global sustainability.

The environment and more particularly soils are being increasingly polluted with heavy metals and a number of other potentially toxic elements, mainly in urban agglomerations. Heavy metals (or trace metals) in soils, as well as in the environment, have been shown to be very useful indicators of environmental pollution worldwide and have been the subject of great attention because of their peculiar pollutant characteristics. During the past decades, the natural input of heavy metals to soils due to pedogenesis has been exceeded by the anthropogenic input, even on a regional scale (Bacon et al., 1992).

The heavy metals accumulate in the soils from polluted air via rain, near roads via splash water, or by irrigation with wastewater and fertilization with waste compost. Special fertilizers (phosphates with Cd) or pesticides (fungicides with Cu) could also enrich heavy metals in soils over a period of time, and excessive heavy-metal contents are detrimental to soil organisms, crops and groundwater quality (Blume and Brümmer, 1991). On the other hand, soil concentrations of heavy metals are the result of both natural and anthropogenic activities, and higher concentrations of Cd, Cu, Pb and Zn were noted in the topsoil of woodlands than in the topsoil of arable land (Bloemen et al., 1995), while the greatest accumulation of heavy metals in forest ecosystems occurs in the litter layer (Martin et al., 1982). Therefore, large proportions of heavy-metal particles in dust deposits on leaves are conveyed to the soil by precipitation and shedding of the leaves (Bloemen et al., 1995). Because diversity and activity of microbial communities are important indices of soils quality, many reports have also shown that short-term or long-term exposure to toxic metals results in the reduction of microbial diversity and activities in soils (Lasat, 2002; McGrath et al., 2001; Wang et al., 2007).

Environmental pollution with metals and xenobiotics is a global problem, and the development of phytoremediation for soils is of significant interest. The term phytoremediation has been coined for the concept that plants could be used for low-cost environmental clean-up and this has attracted considerable attention in the past decade (Krämer, 2005; Lahner et al., 2003; McGrath and Zhao, 2003). Currently, phytoremediation is not widely available for pollution problems. However, further genomics of genes for phytoremediation is getting benefits from the recent development of segregating populations for a genetic analysis of naturally selected metal hyperaccumulation in plants, and from comprehensive ionomics data of *Arabidopsis* mutants (Krämer, 2005; Lahner et al., 2003).

Even though stopping all kinds of pollution should be the primary objective, this method was not applied previously. The depollution of water and soils is very costly, while no feasible technologies are yet available for many pollutants. From the biological point of view, plants have the potential to acquire and concentrate nutrients as well as numerous metabolic activities powered by photosynthesis (Krämer, 2005).

It is likely that effective but paradoxically difficult phytoremediation strategy is mineral elements extraction from soil or 'phytoextraction' which needs tolerant plants having ability to concentrate soil contaminants in their tissues. Metal-accumulating plants are naturally capable of accumulating trace elements, such as arsenic (As), cadmium (Cd), nickel (Ni), selenium (Se), aluminium (Al) or zinc (Zn), in their tissues, and the level of accumulation in the leaf dry biomass is up to 100-fold higher than that in the soil (McGrath and Zhao, 2003), and these plants do not show any toxicity symptoms (Baker and Brooks, 1989). For example, the brake fern *Pteris vittata* accumulated up to 7500 mg g^{-1} As on a contaminated site without visible toxicity symptoms (Ma et al., 2001). Moreover, encouraging trials in field have been conducted (Salido et al., 2003; Tripathi et al., 2007), and currently a commercial fern cultivar is available for As phytoextraction (http://www.edenspace.com/index.html).

On the other hand, another strategy of phytoremediation, namely chelator-assisted phytoremediation, is also available. This method consists in the application of chelators, for example, EDTA capable of solubilizing some poorly available metals (e.g., lead) in the soil, and then metal complexes are passively accumulated in plant shoots (Salt et al., 1998).

Thus, ionomics screens have been initiated for multi-element profiling in *A. thaliana* mutant populations to identify (1) mutants with altered elemental composition of rosette leaves and (2) genes with a potential role in metal hyper-accumulation (Lahner et al., 2003; Salt, 2004).

6.8 METAGENOMICS

The discoveries over the last three decades have demonstrated that microbiology is a central scientific discipline having wide applications in agriculture, as well as in medicine, bioremediation, biotechnology, engineering and other fields. Clearly, the roles of microbes in nature are so diverse that the process of mining their genetic variation for new applications will continue long into the future (Maloy and Schaechter, 2006; Rodríguez-Valera, 2004). Moreover, the rapid rate of microbial evolution ensures that there will be no permanent solution to agricultural, medical or environmental problems caused by microbes. These problems will demand a long and continual series of creative new approaches evolving along with the microbes. Consequently, a deep understanding of basic microbial physiology, genetics and ecology are needed to fulfill these opportunities (Beardsley, 2006; Fraser-Liggett, 2005; Handelsman, 2004, 2006; Xu, 2006).

6.8.1 What Is Metagenomics?

The term 'metagenomics' has been coined to describe an analysis of similar but not identical items, and means globally *analysis of analyses*, such as a meta-analysis, while community genomics, environmental genomics and population genomics are synonyms for this approach (Glass, 1976; Handelsman et al., 1998). Afterwards, cloning DNA directly from environmental samples was proposed by Pace (1995), and the next advance was the construction of a metagenomic library using DNA of organisms enriched on dried grasses (Healy et al., 1995). More recently, the term 'metagenomics' has been simplified and refers to the study of the collective genomes in an environmental community, such as soil or marine water sample, which contains substantially more genetic information than is available in the cultured subset. Studies of metagenomes typically involve cloning fragments of DNA isolated directly from microbes in natural environments, followed by sequencing and functional analysis of the cloned fragments (Ferrer et al., 2005; Liu and Zhu, 2005; Ward, 2006; Xu, 2006). Although the phylogenetic analysis of environmental microbial diversity was an early form of metagenomics and most of the techniques used for metagenomics studies already existed and are used routinely in molecular biology research, their application in analysing unknown environmental DNA samples have opened a floodgate of exciting research findings giving rise to this new field (Deutschbauer et al., 2006; Field et al., 2006; Green and Keller, 2006; Schloss and Handelsman, 2003; Ward and Fraser, 2005). Furthermore, the continuous development of omics technologies and the rate of discovery is high, metagenomics is becoming synonymous with *megagenomics* referring to the study of genomics on a massive scale, because the communities are large and complex, necessitating large libraries to achieve coverage (Chen and Murell, 2010; Handelsman, 2005). More recently, metagenomic was applied to study viruses suggesting novel patterns of evolution and changing the existing ideas of the composition of the virus world by revealing novel groups of viruses and virus-like agents (Kristensen et al., 2010).

6.8.2 Metagenomics and Soil Science

Soil habitat is the most challenging of environmental factors, with respect to the microbial community size and the diversity of species present. Based on the reassociation kinetics of DNA isolated from various soil samples, the estimated number of distinct prokaryotic genomes is ranging from 2000 to 18,000 genomes per gram of soil, and it is likely that this range might be far from the real number because genomes of rare and/or unrecovered species might have been excluded from these analyses (Torsvik et al., 1998; Torsvik et al., 2002).

Moreover, the spatial heterogeneity, the multitude of the phases of the nature (gases, water and solid), and the complexity of chemical and biological properties of soils could contribute to the microbial diversity present in soil samples because soil also would be considered as a 'living microworld' with minerals, shapes and chemical characteristics, with its *biota* and organic compounds (Cavicchioli et al., 2006; Handelsman, 2004; Liu and Zhu, 2005).

On the other hand, the metabolism and the survival of soil microorganisms are under the strong influence of water and nutrients availability, and surfaces of soils undergo dramatic rhythmic and cyclic changes in both water content, ranging from water saturation to extreme aridity, and mineral content, and a fraction of its microbial community does not survive during these excess or scarce periods (Foster, 1988; Hassink et al., 1993). Consequently, microbial soil communities fluctuate; however, the alteration of microbial populations resulting from changes in water content and other environmental factors such as pH, availability of oxygen, or temperature have neither been studied extensively nor intensively (Daniel, 2005). Although this emerging field is still in its first steps, two recent studies concerning direct applications of metagenomics in soil and environmental fields were reported. The first study concerns the analysis of two lab-scale-enhanced biological phosphorus removal sludges dominated by the uncultured bacterium *Candidatus accumulibacter* phosphates, and this study illustrated that metagenomics enables detailed insights into even well-studied biological systems (Martín et al., 2006). The second study, by extending metagenomics-based approaches to very complex and methodologically recalcitrant soil environments, has found that ammonia-oxidizing archaea are more abundant in many soils than bacteria (Leininger et al., 2006). However, further improvement of omic technologies and bioinformatics for analysing the enormous amount of data would make soil metagenome progressing better boosted by synergetic actions of each field (Figure 6.3).

6.8.3 Shift from Metagenomics to Industry

Because different industries have different motivations to probe the enormous resource that is uncultivated microbial diversity, metagenomics have the potential to substantially impact industrial production. In this perspective, different industries are interested in exploiting the resource of uncultivated microorganisms that has been identified through metagenomics (Lorenz and Eck, 2005).

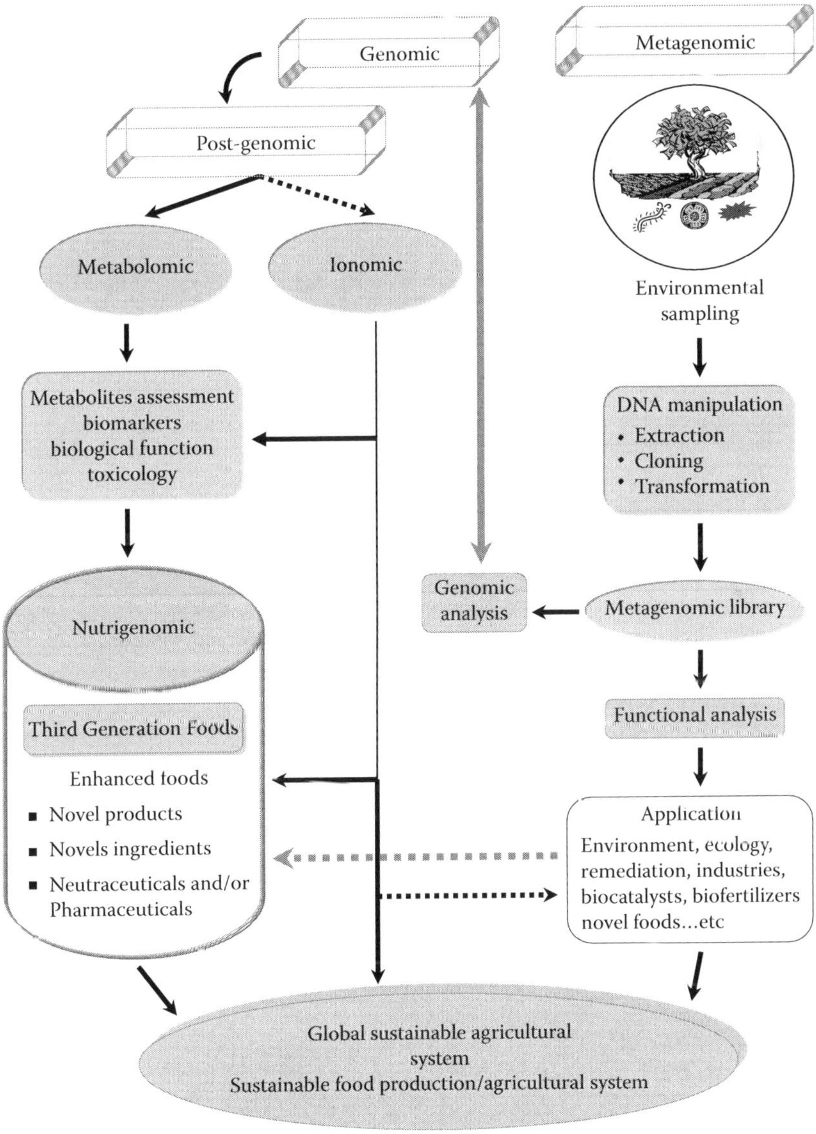

Figure 6.3 Inter- and intrarelation of metabolomic and metagenomic technologies, and their involvement in the global agroecological sustainable system.

Thus, the movement towards implementing sustainable technologies and processes is gaining momentum and the notion of white biotechnology by the development of the 'ideal catalyst' is arousing further interest (Ferrer et al., 2005).

Enzymes are remarkable and natural catalysts using numerous and complex molecules as substrates. They are exquisitely selective and catalyse reactions with unparalleled chiral and positional selectivities, often difficult to achieve by chemical synthesis. They also possess a real potential for sustainable industrial chemistry, avoiding the utilization of polluting catalysers. Biocatalysts therefore offer 'green' solutions for the production of a variety of chiral chemicals that can be used as basic building blocks for the synthesis of pharmaceuticals, agrochemicals, pesticides and insecticides (Davis and Boyer, 2001; Koeller and Wong, 2001).

Typically, the agro-food, cosmetic and detergent industries utilize few and limited number of enzyme reactions and substrates. On the contrary, the chemical and pharmaceutical industries need thousands of diverse molecules differing chemically and structurally, and production of each of these molecules requires specific enzymatic solutions. Consequently, biotechnological applications of microbial resources are for the modern organic chemist (Schoemaker et al., 2003). With the number of industrialized biotransformations, it is estimated that biocatalysis could provide a synthetic solution over classical chemistry (Straathof et al., 2002); however, the unavailability of an adequate biocatalyst would be a limiting factor for any biotransformation process (Schmid et al., 2002).

6.9 OMICS AND SOIL SCIENCE

6.9.1 Omics-Pipe from Soil Fertility to Agricultural Sustainability

Soil fertility management is a continuous process that begins well before the crops are established, while the pre-establishment phase targets to adjusted soils conditions for providing optimum fertility when the crops are established. Thus, it is obvious that efficient soil fertility management is the key to sustainable agriculture. Soil fertility is primarily concerned with the essential plant nutrients including their levels in soils, their availability to crops and their behaviour and becoming in soils, as well as the factors affecting these characters. However, the concept of soil fertility for plants is simple but difficult to explain. Since we do not know miracle fertilizers, only plain and simple chemical nutrients that are absolutely essential for plant growth are available commercially. In moist sand loams, plant nutrients are generally more accessible than in silt or clay soils although clay soils contain higher-fertility resources. In very wet or very dry soils, nutrients become restricted or unavailable for obvious physical reasons such as a lack of root absorption by most crop species (Cakmak, 2002; Lorenagan, 1999; Tisdale et al., 1985).

Whereas hydrogen (H), oxygen (O) and carbon (C) are copiously available in water and carbon dioxide, the macronutrients nitrogen (N), phosphorus (P) and sulphur (S) are less present. The elements H, O, C, silicon (Si), aluminium (Al) and iron (Fe) are abundant contributing to soil hard matter, water and carbon dioxide, while the remaining ones are more important to the chemistry of plants and animals (Deevey, 1970).

Plant nutrients are divided into macronutrients and micronutrients; however, they are indistinguishable whether they are from organic or chemical sources since all must be water soluble in order to enter plant root systems. Based on the criteria of essentiality, 16 elements are considered as essential for the growth of higher plants, These are nine macronutrients: nitrogen (N), phosphate (P), phosphorus (K), sulphur (S), calcium (Ca), magnesium (Mg), oxygen (O), hydrogen (H) and carbon (C), and seven micronutrients, needed in small quantities but are essential as macronutrients to normal plant growth: boron (B), chloride (Cl), copper (Cu), iron (Fe), manganese (Mn), molybdenum (Mo) and zinc (Zn) (Epstein, 1972; Tisdale et al., 1985). Aside from the fact that O, H and C are special cases and are taken from the air and soil water, the other remaining 13 elements must be supplied by the soil, and hence they are classified based on their uptake by plants into (1) primary nutrients (N, P and K), (2) secondary nutrients (CA, Mg and S) and (3) micronutrients (B, Cl, Cu, Fe, Mn, Mo and Zn) (Broyer, 1959; Leeper, 1952; Prasad and Power, 1997). Nevertheless, Section 6.9.2 will be limited to a limited number of these elements taking into account their agricultural importance.

6.9.2 Macronutrients

The term 'macronutrient' came into popular usage since the general acceptance of the term 'micronutrient'. This term suggests one of the nine elements needed by plants in larger quantity, namely nitrogen, phosphorus, sulphur, calcium, magnesium, potassium, carbon, hydrogen and oxygen. However, only the first six that can be supplied conveniently as fertilizers are ordinarily considered among the macronutrients. Thus, it is not common that chemical elements needed for maximal plant growth are supplied adequately by soils, and enormous tonnages of calcium, magnesium, potassium, nitrogen, phosphorus and sulphur are mined, processed and transported in the interest of sustained agricultural production (Broyer and Stout, 1959).

6.9.2.1 Nitrogen

The availability of nitrogen is considered as the major limiting factor in agricultural productivity, resulting in the heavy use of chemical nitrogenizer, and acquisition and assimilation of nitrogen are necessary for plant growth and development (Newbould, 1989). Moreover, production of high-quality and protein-rich foods is extremely dependent on the availability of nitrogen (Vance, 1997). Despite that, major crops, particularly cereals, respond favourably to high-nitrogen fertilizers. Concomitantly, there are volatilization of nitrogen (greenhouse gases) into atmosphere, depletion of nonrenewable resources, an imbalance in the global nitrogen cycle and leaching of NO^{-3} into groundwater, while bacteria also convert more of this leached nitrogen into gas (NH_3, N_2 and N_2O). Nitric oxides (such as N_2O, laughing gas; NO, etc.) are very powerful greenhouse gases, about 280 times more potent than carbon dioxide (Kinzing and Socolow, 1994).

Hence, one of the driving forces behind agricultural sustainability is effective management of nitrogen in the environment, and one of the most judicious methods to manage nitrogen is biological nitrogen fixation (Bohlool et al., 1992; Vance and Graham, 1994). This biologically fixed nitrogen, in addition to be bound in soil organic matter and consequently less susceptible to chemical transformation or physical factors that lead to volatilization and leaching, reduces the need to nitrogen fertilizers and increase soil tilth (Vance and Graham, 1994; Vance, 1997). Many diverse associations contribute to symbiotic N_2 fixation (Sprent, 1984); however, in agricultural settings, 80% of the biologically fixed nitrogen is by the soil bacteria namely *Rhizobium*-, *Bradyrhizobium*-, *Sinorhazobium*- and *Azorhizobium*-legume symbiosis (Vance, 1996), and this symbiotically fixed nitrogen may become available to crops through several ways (Peoples et al., 1995). Thus, economic and social benefits of N_2-fixing crops must be redefined in sustainable agriculture, and the biotechnological and omics approaches would be of great help to increase yield and maintain high N_2 fixation rates in crops targeted for use in sustainable agriculture.

6.9.2.2 Phosphorus

Phosphorus is considered the second important mineral for crops; however, in contrast to nitrogen, P is present as mineral deposit and its fixation by most soils requires the addition of a large excess of P-fertilizers to meet the requirements of crops (Loneragan, 1997). Nevertheless, P presents the positive point to remain in soil due to its immobilization by reactions ions in soils such as Ca, Fe and Al, and the soluble phosphorus fertilizers once applied revert to less soluble or insoluble forms and only 15–20% of the applied quantity would become available to crops (Prasad and Power, 1997).

In soils, the total phosphorus content is generally less than nitrogen or potassium, and is present under both organic and inorganic forms that are continuously transformed. Moreover, crops, particularly cereals and forages, have a high demand for phosphorus, and because it is relatively immobile in the soil it must be thoroughly mixed with the soil to bring the rooting zone into the optimum level for crops (Brady, 1990). Phosphorus is also involved in eutrophication of water through its runoff from soils into rivers and lakes. Consequently, to repeated applications of P-containing fertilizers together with manure, phosphorus concentrations in topsoil increases, leading to saturation of soil-binding sites with phosphorus and thus facilitating its movement into surface waters (Cakmak, 2002). Holford (1997) reported that phosphorus leaching could occur when the sorption capacity of soils was saturated by 17–38%. On the other hand, a major source of phosphorus in surface water showing eutrophication is the application of compost, manure and sewage (Eghball and Gilley, 1999; Reynolds and Davies, 2001), whereas phosphorus from manure application moves much deeper in soil than the phosphorus from fertilizers (Eghball et al., 1996). Organic compounds released during the decomposition of manures increase the availability of P from soil or fertilizers (Iyamuremye and Dick, 1996), and this positive effect of manure was ascribed to the mobilization of native soil phosphorus and improved physicochemical properties of the soil by manure use (Reddy et al., 1999).

One of the objectives of omics approaches could be an integrated research between omics and plant nutrition which would help to develop new plant genotypes with greater efficiency for phosphorus acquisition from soils and will greatly contribute to reduce the massive addition of P-fertilizers. For example, Baxter et al. (2008) defined physiological responses of phosphorus homeostasis by evaluating the shoot ionome in plants grown under different P-nutritional conditions, and demonstrated that multivariable ionomic signatures of physiological states associated with mineral nutrient homeostasis do exist.

6.9.2.3 *Potassium*

Potassium is present in much large amount in soils and has the particularity to be in inorganic form and fairly well distributed throughout the profile (Prasad and Power, 1997). This macroelement is present in soils in four different forms, namely primary mineral, fixed (nonexchangeable), exchangeable and in solution (Mulder, 1950; Tisdale, 1985; Sparks, 1987). Although potassium fixation by its conversion from exchangeable to nonexchangeable form was once considered a negative property of soils at the beginning; this view is no longer considered and this fixation is seen positively because it reduces K losses by leaching during the wet period and luxury consumption by plants (Brady, 1990; Tisdale, 1985). Potassium (K) is important for photosynthesis and in the formation of amino acids and protein from ammonium ions, while in most soils potassium adsorbed by clay or humus colloids is of much more direct importance for plant nutrition than that derived directly from soil minerals (Mulder, 1950).

Although a little is known on potassium compared to the extensive literature existing on nitrogen and phosphorus, genotypic difference in crop species with respect to potassium nutrition are known (Glass and Perley, 1980). As for phosphorus, omics approaches could be used to select genotypes that have better potassium acquisition from soils.

6.9.2.4 *Sulphur*

Sulphur is the least abundant of the macroelements found in plants. It is present at ~0.1% of dry matter compared to ~1.5% for nitrogen. Sulphur is not required in large quantities by growing plants, but is nevertheless an important nutrient, and has a similar cycle to that of nitrogen. Unlike nitrogen, sulphur has numerous biological functions; however, it is generally not a structural component of biomolecules, but nearly always directly involved in the catalytic functions of the molecules of which it is a component (Leustek et al., 2000; Schmidt and Jäger, 1992). Sulphur in its reduced form plays an important role in plants, being involved in the biosynthesis of primary and secondary metabolites and in the synthesis of coenzymes (Schmidt and Jäger, 1992).

Plants occupy a prominent position in the biological sulphur cycle, and have the ability to assimilate inorganic sulphate, which is first reduced to sulphide and then incorporated into cysteine (Leustek et al., 2000). Although the level of free cysteine in plants is very low, it is considered as the central intermediate from which most sulphur compounds are synthesized such as methionine, protein and glutathione synthesis (Giovanelli et al., 1980; Hell, 1997; Leustek et al., 2000). During the past decade, advances in understanding plant sulphur metabolism have been made and this knowledge has been significantly influenced by the isolation of genes for each of the metabolic steps and these works have been issued (see the special issue of *J Exp Bot* 55(404), 2004). Although significant progress has been made towards elucidating the structure, organization and regulation of glutathione *S*-transferases (GSTs), the *in vivo* catalytic function of most GSTs is still unknown. Thus, there are still clear opportunities for applying omics approaches to learn more about how glutathione and glutathione *S*-conjugates are transported and degraded in plants, while important questions remain concerning the transport of most sulphur-compounds (Hell, 1997; Leustek et al., 2000). Recently, postgenomic investigations have been carried out to study the metabolism of sulphur (Hirai and Saito, 2004), and the metabolome of leaf and root samples was analysed (Hirai et al., 2004). Because the pattern of the gene expression is regulated by a metabolite accumulation pattern and vice versa, metabolomics approaches could also be considered indispensable for the understanding of the whole mechanism of sulphur (Hirai and Saito, 2004).

6.9.2.5 Calcium

It has long been established that calcium is necessary for the growth of higher plants that need large quantities of this element (Tisdale et al., 1985). As an essential plant nutrient and a divalent cation (Ca^{2+}), it is required for structural roles in the cell wall and membranes, as a counter-cation for inorganic and organic anions in the vacuole, and as an intracellular messenger in the cytosol (Marschner, 1995; Martinoia et al., 2007). Calcium deficiency is rare in nature, but may occur in soils at low base saturation and/or high levels of acidic deposition, while its excess restricts plant communities on calcareous soils (McLaughlin and Wimmer, 1999; White and Broadley, 2003). By contrast, several costly Ca-deficiency disorders occur in horticulture (Shear, 1975).

Besides, it has long been known that calcium is an essential element for plants, and it is also established that plant species differ in both the amounts of calcium they require and their tolerance of calcium in the rhizosphere (Hepler, 2005; Martinoia et al., 2007; McLaughlin and Wimmer, 1999; Plieth, 2005; White and Broadley, 2003). These differences between plant species not only influence the natural flora of calcareous soils but also have consequences for breeding plants and crops (McLaughlin and Wimmer, 1999; White and Broadley, 2003).

More recently, research on calcium in plants has been focused at the cellular level, and the application of omics technologies will provide further insights into the biology of calcium in plants, and the responses to changes in calcium might include the differential expression of enzymes in the biochemical pathways containing compounds or ions that chelate calcium, and these compounds could be detected by a comprehensive metabolomics and ionomics profiling. Such approaches could ultimately lead to the development of crops that accumulate less of contaminants or deliver more calcium to the food chain.

6.9.2.6 Magnesium

Magnesium, which is fairly mobile in plants, is the most abundant divalent cation in a living cell and has few common characteristics to calcium such as (1) both are taken up by plants as cations and (2) they are basic or basic forming elements (Tisdale et al., 1985).

In higher plants, magnesium plays an even more prominent role because it is an essential component and the core cation in the structure of chlorophyll molecules. It also serves as a structural component in ribosome and plays consequently an important role in protein biosynthesis. Despite these critical cellular functions, magnesium uptake, transport and homeostasis in eukaryotes are poorly understood at both the physiological and molecular level (Li et al., 2001; Prasad and Power 1997), and little is reported on this cation (Li et al., 2001). However, there is some evidence that magnesium transport may be important, because its deficiency is induced in Aluminium-treated plants (Tan et al., 1991), and application of higher levels of magnesium can reduce aluminium toxicity (Tan et al., 1991; Matsumoto, 2000), while magnesium uptake by roots is inhibited by aluminium (Rengel and Robinson, 1989).

Thus, the omics approaches would be real issues to answer the numerous questions on magnesium in soils and plants. Probably genomics, associated with metabolomics and ionomics, would help to understand the physiological and biochemical pathways of this cation in soils and plants. These technologies could also help to determine plants that can be genetically engineered to improve their tolerance to aluminium.

6.9.3 Micronutrients

Micronutrient deficiencies in plants are becoming increasingly important globally, because intensive cultivation of high-yielding cultivars with heavy applications of N, P and K fertilizers lead to the occurrence of micronutrient deficiencies (Cakmak, 2002).

6.9.3.1 *Boron*

Boron is an essential micronutrient, highly mobile in soil and is taken up by plant as other anions; however, its role in plant growth development and metabolism remains still not clear (Eaton, 1940; Sommer and Lipman, 1926; Tisdale et al., 1985). Boron was first thought to be essentially immobile and fixed in the apoplast; nevertheless, recent evidence has shown that in some species boron is present as soluble complexes of sorbitol- or mannitol-esters, which are phloem mobile (Hu et al., 1997). The most functional roles ascribed to boron are related to its capacity to form diester bounds between molecules, such as oligosaccharides and complex sugars. Because up to 90% of cellular boron is present in the cell wall fraction, it plays a major role in maintaining the cell wall structure and membrane function, as well as supporting metabolic activities (Power and Woods, 1997).

6.9.3.2 *Chlorine*

Chlorine (Cl) is an essential micronutrient for higher plants, and a minimal requirement for crops growth of 1 g kg^{-1} dry weight has been suggested (Engvild, 1986; Marschner, 1995). This quantity is generally supplied by rainfall, and Cl-deficient plants are rarely observed in agriculture or nature. Nevertheless, high tissues Cl-concentrations can be toxic to plants, and may restrict the agriculture of the saline regions. Moreover, a part of the present interest in studying chlorine uptake and accumulation in plants is related to the needs to develop more salt-tolerant varieties for use in saline environments (Xu et al., 2000). Chlorine occurs predominantly in soils as Cl^- and its movement is determined by water fluxes, particularly the relationship between precipitation and evapotranspiration (Tisdale et al., 1985). In plant tissues, chlorine is an active solute in vacuoles and is involved in both turgor and osmoregulation. It also acts as a counter anion, Cl^- fluxes are implicated in membrane stabilization (White and Boadley, 2001), and its involvement in O_2 photosynthetic activity and growth was also reported (Terry, 1977). Chloride toxicity depends on the plants that can be Cl^- sensitive or Cl^- tolerant (Greenway and Munns, 1980), and many important cereal, fruit and vegetable crops are susceptible to Cl^- toxicity during cultivation and this is considered as a major constraint to horticultural production on irrigated or saline soils (Maas and Hoffman, 1977; Xu et al., 2000). Thus, there are real opportunities for omics technologies to study how plants can limit Cl^- accumulation and the results could lead to generate varieties that withstand Cl^- toxicity.

6.9.3.3 Copper

Copper (Cu) is an essential micronutrient for plants because of its numerous roles, and its stimulating effect on plants was noted early in the use of copper salts as fungicides (Sommer, 1931). In plants, copper plays a vital role in both photosynthetic and respiratory electron transport, and functions as a cofactor for a variety of enzymes such as super-oxide dismutase, cytochrome-*c* oxidase and plastocyanin (Clemens, 2001); however, like other redox-active metals in excess, copper is harmful to most plants (Fernandes and Henriques, 1991; Ke et al., 2007). On the other hand, plants regulate their intracellular Cu-levels by regulating its uptake or by minimizing free Cu-concentrations within the cell through metallochaperones (the soluble copper-binding proteins) (Grotz and Guerinot, 2006; O'Halloran and Culotta, 2000; Rosenzweig, 2002).

Because the development of copper-tolerant populations in crops is also an objective for local adaptation and microevolution of these crops, specific mechanisms might be developed in the resistant populations via natural selection in response to excess or deficiency in nutrients through omics approaches. More specifically, metabolomics and ionomics could help to make crops, as well as higher plants, to evolve in such unbalanced environments with acquisition of either exclusion or accumulation and sequestration basic strategies of copper tolerance.

6.9.3.4 Iron

Iron (Fe), although is the fourth most abundant element in the earth's crust, it is the third most limiting nutrient for plant growth primarily due to the low solubility of iron in aerobic environments, and over one-third of the world's soils are considered iron deficient (Yi et al., 1994). This deficiency often occurs in calcareous soils where chemical availability of iron to plant roots is extremely low since it was estimated that iron deficiency occurs in about 30% of the cultivated soils on the world, and results in large decreases in crop production and quality (Chen and Barak, 1982; Vose, 1982). Iron is an essential microelement for plants, as well as for all forms of life, and its limitation has a profound impact on the productivity of photosynthetic organisms. Iron takes part in photosynthesis, respiration, DNA synthesis and hormone structure and action. However, it exists primarily in its insoluble form and is therefore not freely available to plants (Martin et al., 1994); however, in order to deal with the limiting amounts of iron, plants have evolved to several strategies to obtain it from the soil (Curie and Briat, 2003; Hell and Stephan, 2003; Schmidt, 2003).

Complex strategies involving chelators, production of reductive agents, reductase activities, proton-mediated processes, specialized storage proteins, and others are currently acting together to mobilize iron from the environment into the plant and within the plant because of its fundamental role (Briat and Lobreaux, 1997; Curie and Briat, 2003; Hell and Stephan, 2003; Schmidt, 2003).

Although some recent studies have focused on the mechanism of iron transportation in plants and organelles (DiDonato et al., 2004; Ladouceur et al., 2006; Thomin et al. 2003), these important aspects of plant iron mobilization are not well understood yet, and several unsolved and central questions concerning sensing, trafficking, homeostasis, and delivery of iron in plants are currently a matter of intense debate (Cakmak, 2002; Ghandilyan et al., 2006; Graziano and Lamattina, 2005; Grotz and Guerinot, 2006). Moreover, recent studies on iron nutrition in plants reported evidence from iron homeostasis by proposing a new scenario involving the formation of nitric oxide and iron–nitrosyl complexes as part of the dynamic network that governs plant iron homeostasis (Stamler et al., 1992). Furthermore, it is likely that most of the iron in a cell is chelated but how proteins access this pool of iron remains unknown (Grotz and Guerinot, 2006). Thus, proteomics conjugated to metabolomics and ionomicsomics are considered as real alternative approaches not only to track these proteins, but also to exploit the genetic and phenotypic variations of crops and their metabolism of iron. Recently, Baxter et al. (2008) evaluated the shoot ionome in plants grown under different Fe-nutritional conditions, and have established a multivariable ionomic signature for the Fe-response status of *Arabidopsis*. This signature has been validated against known Fe-response proteins and allows the high-throughput detection of the Fe status of plants. Their study established that multivariable ionomic signatures of physiological states associated with mineral nutrient homeostasis do exist in *Arabidopsis* and are in principle robust enough to detect specific physiological responses to environmental or genetic perturbations.

6.9.3.5 *Manganese*

Manganese (Mn), although present in soils, is usually found under a complex unavailable form in plants (Marschner, 1995). Manganese is recognized as an essential microelement for the growth of plants and participates in a number of cellular activities including stabilization of structural proteins, ultrastructure of chloroplasts, and photosynthesis (Popelkova et al., 2003; Simpson and Robinson, 1984; Weiland et al., 1975).

The interpretation of plant responses to differing manganese concentrations is difficult because of (1) its close relationship with iron (Fe) (Izaguirre-Mayoral and Sinclair, 2005; Korshunova et al., 1999) and nitrogen (Tong et al., 1997), (2) differences among cultivars in the expression of visual manganese deficiency symptoms in the field (Graham et al., 1994; Pallotta et al., 2000) and (3) rhizosphere acidification has an important role in mobilizing soil manganese (Petrie and Jackson, 1984a; Petrie and Jackson, 1984b). Moreover, manganese deficiency is difficult to overcome by the addition of manganese-supplemented fertilizers because the added manganese is quickly converted to an unavailable form (Reuter et al., 1973). Superior internal utilization of manganese, a faster specific rate of Mn-absorption and better root geometry are unlikely to be major contributors to its efficiency (Graham, 1988). The possible mechanisms of manganese efficiency are either its better internal compartmentalization and remobilization (Huang et al., 1993) or roots excretion of Mn-efficient genotypes of high amounts of mobilizing substances of insoluble manganese (Rengel et al., 1996).

Taking into account these particularities of manganese, omics technologies could be carried out to examine root surface and soil interactions of plants grown from different soil types. A possible direct role of other compounds in the determination of manganese availability as well as in the expression of differential manganese efficiency among crop genotypes would be also examined using proteomics and metabolomics.

6.9.3.6 *Molybdenum*

Molybdenum (Mo) is a trace element found in soils and is required for the growth of green plants. Molybdenum is a transition element that can exist in several oxidation states ranging from zero to VI, where VI is the commonest form found in most agricultural soils. Similar to most metals required for plant growth, molybdenum has been utilized by specific plant enzymes to participate in reduction and oxidative reactions (Arnon and Stout, 1939; Kaiser et al., 2005). This element itself is not biologically active but is rather predominantly found to be an integral part of an organic pterin complex called the molybdenum co-factor (Moco) which binds to molybdenum-requiring enzymes (molybdoenzymes) found in most biological systems including plants (Mandel and Hänsch, 2002; Williams and Frausto da Silva, 2002). The availability of molybdenum for plants growth strongly depends on the soil pH, concentration of adsorbing oxides (e.g., Fe oxides), extent of water drainage and organic compounds found in the soil colloids (Reddy et al., 1997).

Molybdenum deficiencies are primarily associated with poor nitrogen, particularly when nitrate is the predominant nitrogen form available for plant growth, and the inability to synthesize

Moco will reduce the activity of the critical nitrogen-reducing and assimilatory enzymes (Jones et al., 1976) causing a decrease in plants growth and yield (Chatterjee et al., 1985; Unkles et al., 2004).

Compared to other elements, molybdenum has not been extensively studied and much more research work is needed to ascertain more clearly the processes involved in how plants gain access to molybdenum and how the element may be used in the future to expand growing areas where soil molybdate profiles limit plant growth. Thus, it is a significant scope in exploring how to optimize molybdenum fertilization in crops using omics technologies mainly ionomics, and this scope could be more pertinent where nitrate is the predominant available N-source. There is also a large lack in the understanding of how molybdate enters plant cells and is redistributed between tissues of the plants and metabolomics could help to understand this process.

6.9.3.7 *Zinc*

Zinc (Zn) deficiency has been found to be the most widespread micronutrient deficiency worldwide, and nearly 50% of the soils cultivated for cereal have low levels available Zn (Sillanpää, 1990; Sillanpää and Vlek, 1985). The essential roles of zinc for the growth of higher green plants have been first demonstrated by Sommer (1928). Zinc is an essential co-factor required for the structure and function of numerous proteins and is involved with enzyme systems that regulate various metabolic activities (Grotz and Guerinot, 2006), and it does not need to be reduced before transport and the chelation strategy is used in order to obtain zinc from soils (Welch, 1995) using zinc–phytosiderophores (Zn–PS) complexes (von Wirén et al., 1996). The molecular mechanisms for zinc efficiency remain still unclear; however, it was suggested that it arises from an increased secretion of PSs by zinc-efficient plants (Cakmak et al., 1996a,b; Hacisalihoglu and Kochian, 2003).

Despite significant progress in the molecular and genetic understanding of zinc nutrition in plants (Ghandilyan et al., 2006; Grotz and Guerinot, 2006), there is a need to improve zinc contents in crop plants by means of new approaches. To achieve this goal, knowledge about the physiological, biochemical and molecular mechanisms of zinc uptake and homeostasis, as well as other minerals, using omics technologies will be very helpful. The metabolome and ionome mapping of the existing variation will allow exploiting the available natural variation in crop plants.

6.10 OMICS AND RHIZOSPHERE SUSTAINABILITY: THE MICROWORLD IN THE MACROWORLD

From its first introduction by Hiltner (1904), the term 'rhizosphere' was defined as the zone of soil immediately adjacent to legume roots that supports high levels of bacterial activity. However, despite the scientific ability to characterize a number of environmental variables, until recently scientific community had almost no basis for doing so for plant and soil chemistry (Reich, 2005; Sen, 2005).

Recently, the term 'rhizosphere' was broadened out to include both the volume of soil influenced by the root and the root tissues colonized by microorganisms (Pinton et al., 2001). In the rhizosphere, the microorganisms react also to many metabolites released by plant roots, and their products also interact with plant roots in a variety of positive, negative and neutral ways. These interactions influence plants growth and development, change nutrient dynamics and alter plants' susceptibility to disease and abiotic stresses (Morgan and Whipps, 2001; Morgan et al., 2005; Smith, 2002).

From the purely physical point of view, the rhizosphere can be divided into (1) the endorhizosphere (the root tissue including the endodermis and cortical layers), (2) the rhizoplane (the root surface with the epidermis and mucilaginous polysaccharide layer) and (3) the ectorhizosphere (the soil immediately adjacent to the root). Moreover, plants that are colonized by mycorrhizal fungi also have a zone termed 'mycorrhizosphere', and these mycorrhizal fungi can extend out from the plant root for a significant region, while in well-colonized soils by plants, the plant roots may affect all the soil present in a particular area (Lindermann, 1988; Lynch, 1987). Conversely, the microorganisms in the rhizosphere can also influence plants in different ways, such as their growth, nutrition, development, susceptibility to disease, resistance to heavy metals, and the degradation of xenobiotics (Barea et al., 2005). Furthermore, the perpetual functioning of the rhizosphere ecosystem is essential for soil sustainability and productivity, and the understanding of the processes occurring in this ecosystem would help to improve the management of agricultural practices. Furthermore, understanding soil's tolerance to changes that may arise from processes, such as environmental changes, could enable better decision making for the sustainability concept (van Elsas et al., 1997).

Consequently, these interactions have considerable potential for biotechnological exploitation (Schloss and Handelsman, 2003); however, traditional microbial population studies, based on identification, quantification and the measurement of processes that occur in the rhizosphere, are often difficult or tedious (Barea et al., 2005). In parallel, the collection of relevant samples or the simulation of natural conditions in the laboratory can be problematical, but molecular techniques allow expecting improvements in understanding the rhizosphere microbial communities (Barea et al., 2005; Johnson et al., 2005).

By considering the omic technologies as a whole, the concept of ion homeostasis networks and root-associated microbes arise, and the establishment of large numbers of metabolically active roots and populations of beneficial soil microbes are critical for the success of agricultural practices, soils sciences, environmental remediation and so on (Metting, 1992; Narasimhan, 2003; O'Connell et al., 1996). The understanding of the soil chemistry and diversity and functional characteristics of microbial communities by their omics profiling may have great consequences for nutrients cycling processes, characteristics of the rhizosphere and productivity of natural plant communities.

6.11 CONCLUDING REMARKS

During the past three decades, there has been tremendous progress in the field of plant and microbial genomics, transcriptomics, proteomics, and metabolomics and more recently in ionomics and metagenomics. However, work to date represents undoubtedly just the tip of the iceberg, given the estimated enormous number of species and genomes on Earth. Despite the early struggle to understand and dissect the different layers of complexity, comparative omics approaches are well suited to tackle many new and exciting questions.

Moreover, fundamental questions are nowadays being explored regarding the functional capabilities of plants and microbial communities that directly affect the environment including man. Sustainable effects may depend on lasting changes in soil, environment and microbial composition, their metabolome, ionome and/or metagenome, and, more specifically, on alterations in the metabolic profile in response to the natural and stress conditions.

Because omics technologies have become integrative sciences that simultaneously adopt the tools of diverse scientific disciplines and impact diverse scientific disciplines, strategies to improve these new approaches will further accelerate the speed of discovery of the amazing diversity of the microlife. Moreover, with the accumulation of more data, it is likely that the next decade will be filled with new insights into the strange and often unpredictable plants and microbial worlds. Systems-based approaches that integrate DNA sequence with data from transcriptome, proteome and metabolome studies will begin to reveal the intricate workings of a living cell. Although understanding issues in systems biology is not realistic yet, it is philosophical to consider that identifying and conceptualizing the systems central to specific need is an integral issue to the success of the science. To understand how cells, organisms and communities act and react in a hierarchy of dynamic processes may provide important insights into the problem and answer some of our multitude questions. Furthermore, the discoveries of the 'microworld' were somewhat less compared with those of the 'macroworld', and this lack need to be considered because the two are undissociable and each depends on the other.

Finally, man has to bear in mind that he has not the ability to create a new world or bring neither additional lands nor water. So he is condemned to preserve what he has because his surviving depends undoubtedly on the survival of his global environment. To be simplistic, it is necessary to meet the needs of the present without compromising the needs of either the next or the far future.

REFERENCES

Arnon, D. I. and P. R. Stout. 1939. Molybdenum as an essential element for higher plants. *Plant Physiol* 14:595–602.

Arrus, K., G. Blank, D. Abramson, R. Clear and R. A. Holley. 2005. Aflatoxin production by *Aspergillus flavus* in Brazil nuts. *J Stored Prod Res* 41:513–527.

Bacon, J. R., M. L. Berrow and C. A. Shand. 1992. Isotopic composition as an indicator of origin of lead accumulations in surface soils. *Int J Environ Anal Chem* 46:71–76.

Bailey, N. J. C., M. Oven, E. Holmes, J. K. Nicholson and M. H. Zenk. 2003. Metabolomic analysis of the consequences of cadmium exposure in *Silene cucubalus* cell culture via 1H NMR spectroscopy and chemometrics. *Phytochemistry* 62:851–858.

Baker, A. J. M. and R. R. Brooks. 1989. Terrestrial higher plants which hyperaccumulate metallic elements—A review of their distribution, ecology and phytochemistry. *Biorecovery* 1:81–126.

Baldridge, G. D., N. R. O'Neill and D. A. Samac. 1998. Alfalfa (*Medicago sativa* L.) resistance to the root-lesion nematode, *Pratylenchus penetrans*: Defense–response gene mRNA and isoflavonoid phytoalexin levels in roots. *Plant Mol Biol* 38:999–1010.

Barea, J. M., M. J. Pozo, R. Azcón and C. Azcón-Aguilar. 2005. Microbial cooperation in the rhizosphere. *J Exp Bot* 56:1761–1778.

Baxter, I., M. Ouzzani, S. Orcun, B. Kennedy, S. S. Jandhyala and D. E. Salt. 2007. Purdue ionomics information management system. An integrated functional genomics platform. *Plant Physiol* 143:600–611. Baxter, I. R. 2009. Ionomics: Studying the social network of mineral nutrients. *Curr Opin Plant Biol* 12:381–386.

Baxter, I. R., O. Vitek, B. Lahner et al. 2008. The leaf ionome as a multivariable system to detect a plant's physiological status. *Proc Natl Acad Sci USA* 105:12081–12086.

Beardsley, T. M. 2006. Metagenomics reveals microbial diversity. *BioScience* 56:192–196.

Beyer, J. A., A. Salim, Y. Xudong et al. 2002. "Golden rice": Introducing the β–carotene biosynthetic pathway into rice endosperm by genetic engineering to defeat vitamin A-deficiency. *J Nutr* 132:S506–S510.

Bloemen, M. L., B. Markert and H. Lieth. 1995. The distribution of Cd, Cu, Pb and Zn in topsoils of Osnabrück in relation to land use. *Sci Total Environ* 166:137–148.

Blume, H. P. and G. Brümmer. 1991. Prediction of heavy metal behavior in soil by means of simple field tests. *Ecotox Environ Saf* 22:164–174.

Boersma, M. G., I. P. Solyanikova, W. L. H. van Berkel, J. Vervoort, L. A. Golovleva and I. M C. M. Rietjens. 2001. ^{19}F NMR metabolomics for elucidation of microbial degradation of microbial pathways of fluorophenols. *J Ind Microbiol Biotechnol* 26:22–24.

Bohlool, B. B., J. K. Ladha, D. P. Garrity and T. George. 1992. Biological nitrogen fixation for sustainable agriculture: A perspective. *Plant Soil* 141:1–11.

BorejszaWysocki, W., E. Borejsza Wysocka and G. Hrazdina. 1997. Pisatin metabolism in pea (*Pisum sativum* L.) cell suspension cultures. *Plant Cell Rep* 16:304–309.

Brady, N. C. 1990. *The Nature and Properties of Soils*, 10th edition. New York, NY: John Wiley & Sons.

Briat, J. F. and S. Lobreaux. 1997. Iron transport and storage in plants. *Trends Plant Sci* 2:187–193.

Broyer, T. C. and P. R. Stout. 1959. The macronutrient elements. *Annu Rev Plant Physiol* 10, 277–300.

Cakmak, I. 2002. Plant nutrition research: Priorities to meet human needs for food in sustainable ways. *Plant Soil* 247:3–24.

Cakmak, I., N. Sari, H. Marschner, M. Kalayci, A. Yilmaz and H. J. Braun. 1996a. Phytosiderophore release in bread and durum wheat genotypes differing in zinc efficiency. *Plant Soil* 180:183–189.

Cakmak, I., L. Öztürk, S. Karanlik, H. Marschner and H. Ekiz. 1996b. Zinc-efficient wild grasses enhance release of phytosiderophores under zinc deficiency. *J Plant Nutr* 19:551–563.

Campbell, M. M., A. M. Brunner, M. Jones and S. H. Strauss. 2003. Forestry's fertile crescent: The application of biotechnology to forest trees. *Plant Biotechnol J* 1:141–154.

Catchpole, G. S., M. Beckmann, D. P. Enot et al. 2005. Hierarchical metabolomics demonstrates substantial compositional similarity between genetically modified and conventional potato crops. *Proc Natl Acad Sci USA* 102:14458–14462.

Cavicchioloi, R., M. Z. DeMaere and T. Thomas. 2006. Metagenomic studies reveal the critical and wide-ranging ecological importance of uncultivated *Archaea*: The role of ammonia oxidizers. *BioEssays* 29:11–14.

Chatterjee, C., N. Nautiyal and S. C. Agarwala. 1985. Metabolic changes in mustard plants associated with molybdenum deficiency. *New Phytol* 100:511–518.

Chen, Y. and P. Barak. 1982. Iron nutrition of plants in calcareous soils. *Adv Agronomy* 35:217–240.

Chen, Y. and J. Y. Murell. 2010. When metagenomics meets stable-isotope probing: Progress and perspectives. *Trends Microbiol* 18:157–163.

Cheng, S. 2003. Heavy metal pollution in China: Origin, pattern and control. *Environ Sci Pollut Res Int* 10:192–198.

Clemens, S. 2001. Molecular mechanisms of plant metal tolerance and homeostasis. *Planta* 212:475–486.

Coyne, D. P., J. R. Steadman, G. Godoy-Lutz et al. 2003. Contribution of the bean/cowpea CRSP to management of bean diseases. *Field Crop Res* 82:155–168.

Curie, C. and J. F. Briat. 2003. Iron transport and signaling in plants. *Annu Rev Plant Biol* 54:183–206.

Daniel, R. 2005. The metagenomics of soil. *Nat Rev Microbiol* 3:470–478.

Davis, B. G. and V. Boyer. 2001. Biocatalysis and enzymes in organic synthesis. *Nat Prod Rep* 18:618–640.

Deevey, E. S. 1970. Mineral cycles. *Sci Am* 223:149–158.

DellaPenna, D. 1999. Nutritional genomics: Manipulating plant micronutrients to improve human health. *Science* 285:375–379.

DellaPena, D. 2001. Plant metabolic engineering. *Plant Physiol* 125:160–163.

Denman, S., P. S. Knoxdavis, F. J. Calitz and S. C. Lamprecht. 1995. Pathogenicity of *Pythium irregulare*, *Pythium sylvaticum* and *Pythium ultimum* var. *ultimum* to Lucerne (*Medicago sativa*). *Aust Plant Pathol* 24:137–143.

Dennis, E. S., R. Dolferus, M. Ellis et al. 2000. Molecular strategies for improving waterlogging tolerance in plant. *J Exp Biol* 51:89–97.

Deutschbauer, A. M., D. Chivian and A. P. Arkin. 2006. Genomics for environmental microbiology. *Curr Opin Biotechnol* 17:229–235.

DiDonato, R. J. Jr., L. A. Roberts, T. Sanderson, R. B. Sisley and E. L. Walker. 2004. Arabidopsis Yellow Stripe-Like2 (YSL2): A metal-regulated gene encoding a plasma membrane transporter of nicotianamine–metal complexes. *Plant J* 39:403–414.

Dita, M., N. Rispail, E. Prats, D. Rubiales and K. B. Singh. 2006. Biotechnology approaches to overcome biotic and abiotic stress constraints in legume. *Euphytica* 147:1–24.

di Toppi, L. S. and R. Gabbrielli. 1999. Response to cadmium in higher plants. *Environ Exp Bot* 41:105–130.

Dorfman, R. 1991. Protecting the global environment: An immodest proposal. *World Dev* 19:103–110.

Dwivedi, B. S., M. Ram, B. P. Singh, M. Das and R. N. Prasad. 1992. Effect of liming on boron nutrition of pea (*Pisum sativum* L.) and corn (*Zea mays* L.) grown in sequence in an acid alfisol. *Fertil Res* 31:257–262.

Eaton, S. V. 1940. Effects of boron deficiency and excess on plants. *Plant Physiol* 15:95–107.

EC (European Commission). 1999. *Agriculture, Environment, Rural Development: Facts and Figures*. Report No. 113, European Commission, Brussels, Belgium. http://ec.europa.eu/geninfo. Accessed 20 March, 2010.

Eghball, B. and J. E. Gilley. 1999. Phosphorus and nitrogen in runoff following beef cattle manure or compost application. *J Environ Qual* 28:1201–1210.

Eghball, B., G. D. Binford and D. D. Baltensperger. 1996. Phosphorus movement and adsorption in a soil receiving long-term manure and fertilizer application. *J Environ Qual* 25:1339–1343.

Engvild, K. C. 1986. Chlorine containing natural products in higher plants. *Phytochemistry* 15:781–791.

Epstein, F. 1972. *Mineral Nutrition of Plants: Principles and Perspectives*. New York, NY: John Wiley & Sons.

FAO. Food and Agriculture Organization. 2000. *Biotechnology for Sustainable Agriculture*. Report of the Commission on Sustainable Development. Food and Agriculture Organization, Rome, Italy.

Fernandes, J. C. and F. S. Henriques. 1991. Biochemical, physiological, and structural effects of excess copper in plants. *Bot Rev* 57:246–273.

Fernie, A. and N. Schauer. 2009. Metabolomics-assisted breeding: A viable option for crop improvement? *Trends Genet* 25:38–48.

Ferrer, M., F. Martínez-Abarca and P. N. Golyshin. 2005. Mining genomes and 'metagenomes' for novel catalysts. *Curr Opin Biotechnol* 16:588–593.

Fiehn, O., J. Kopka, P. Dömann, T. Altmann, R. N. Trethewey and L. Willmitzer. 2000. Metabolite profiling for plant functional genomics. *Nat Biotechnol* 18:1157–1161.

Field, D., G. Wilson and C. van der Gast. 2006. How do we compare hundreds of bacterial genomes? *Curr Opin Microbiol* 9:499–504.

Frewer, L., J. L. Lassen, B. Kettlitz, J. Scholderer, V. Beekman and K. G. Berdal. 2004. Social aspects of genetically modified foods. *Food Chem Toxicol* 42:1181–1193.

Foster, R. C. 1988. Microenvironments of soil microorganisms. *Biol Fertil Soils* 6:189–203.

Fraser-Liggett, C. M. 2005. Insights on biology and evolution from microbial genome sequencing. *Genome Res* 15:1603–1610.

Garza, R., C. Cardona and S. P. Singh. 1996. Inheritance of resistance to bean-pod werevil (*Apion godmani* Wagner) in common beans from Mexico. *Theor Appl Genet* 92:357–362.

Ghandilyan, A., D. Vreugdenhil and M. G. M. Aarts. 2006. Progress in the genetic understanding of plant iron and zinc nutrition. *Physiol Plant* 126:407–417.

Giovanelli, J., S. H. Mudd and A. H. Datko. 1980. Sulfur amino acids in plants. In *The Biochemistry of Plants*, Vol. 5, ed. B. J. Miflin, pp. 454–500. New York, NY: Academic Press.

Girbal, L., J. L. Rols and L. D. Lindley. 2000. Growth rate influences reductive biodegradation of the organophosphorus pesticide demethon by *Corynebacterium glutamicum*. *Biodegradation* 11:371–376.

Glass, G. V. 1976. Primary, secondary, and meta-analysis of research. *Educ Res* 5:3–8.

Glass, A. D. M. and J. E. Perley. 1980 Varietal differences in potassium uptake by barley. *Plant Physiol* 65:160–164.

Graham, R. D. 1988. Genotypic differences in tolerance to manganese deficiency. In *Manganese in Soils and Plants*, eds. R. D. Graham, R. J. Hannam and N. C. Uren, pp. 261–276. Dordrecht: Kluwer Academic Publishers.

Graham, M. J., C. D. Nickell and R. G. Hoeft. 1994. Effect of manganese deficiency on seed yield of soybean cultivars. *J Plant Nutr*17:1333–1340.

Graham, R. D., R. M. Welch and H. E. Bouis. 2001. Addressing micronutrient malnutrition through enhancing the nutritional quality of staple foods: Perspectives and knowledge gaps. *Adv Agron* 70:77–142.

Graziano, M. and L. Lamattina. 2005. Nitric oxide and iron in plants: An emerging and converging story. *Trends Plant Sci* 10:4–8

Green, B. D. and M. Keller. 2006. Capturing the uncultivated majority. *Curr Opin Biotechnol* 17:236–240.

Greenway, H. and R. Munns. 1980. Mechanisms of salt tolerance in non halophytes. *Annu Rev Plant Physiol* 31:149–190.

Grivet, J. P., A. M. Delort and J. C. Portais. 2003. NMR and microbiology: From physiology to metabolomics. *Biochimie* 85:823–840.

Grotz, N. and M. L. Guerinot. 2006. Molecular aspects of Cu, Fe and Zn homeostasis in plants. *Biochim Biophys Acta* 1763:595–608.

Hacisalihoglu, G. and V. L. Kochian. 2003. How do some plants tolerate low levels of soil zinc? Mechanisms of zinc efficiency in crop plants. *New Phytol* 159:341–350.

Handelsman, J. 2004. Metagenomics: Application of genomics to uncultured microorganisms. *Microbiol Mol Biol Rev* 68:669–685.

Handelsman, J. 2005. Metagenomics or megagenomics? *Nat Rev/Microbiol* 3:457–458.

Handelsman, J. M. R., S. F. Rondon, B. J. Clardy and R. M. Goodman. 1998. Molecular biological access to the chemistry of unknown soil microbes: A new frontier for natural products. *Chem Biol* 5:R245–R249.

Hassink, J., L. A. Bouwman, K. B. Zwart and L Brussaard. 1993. Relationship between habitable pore space, soil biota, and mineralization rates in grassland soils. *Soil Biol Biochem* 25:47–55.

He, X. Z. and R. A. Dixon. 2000. Genetic manipulation of isoflavone 7-*O*-methyltransferase enhances biosynthesis of 4′-O-methylated isoflavonoid phytoalexins and disease resistance in alfalfa. *Plant Cell* 12:1689–1702.

Healy, F. G., R. M. Ray, H. C. Aldrich, A. C. Wilkie, L. O. Ingram and K. T. Shanmugam. 1995. Direct isolation of functional genes encoding cellulases from the microbial consortia in a thermophilic, anaerobic digester maintained on lignocellulose. *Appl Microbiol Biotechnol* 43:667–674.

Hell, R. 1997. Molecular physiology of plant sulfur metabolism. *Planta* 202:138–148.

Hell, R. and U. W. Stephan. 2003. Iron uptake, trafficking and homeostasis in plants. *Planta* 216:541–551.

Hepler, P. K. 2005. Calcium: A Central regulator of plant growth and development. *Plant Cell* 17:2142–2155.

Hirai, M. Y., M. Yano, D. B. Goodenowe et al. 2004. Integration of transcriptomics and metabolomics for understanding of global responses to nutritional stresses in *Arabidopsis thaliana. Proc Natl Acad Sci USA* 101:10205–10210.

Hirai, M. Y. and K. Saito. 2004. Post-genomics approaches for the elucidation of plant adaptive mechanisms to sulphur deficiency. *J Exp Bot* 55:1871–1879.

Hirschi, K. D. 2003. Striking while the ionome id hot: Making the most of plant genomic advances. *Trends Biotechnol* 21:520–521.

Holford, I. C. R. 1997. Soil phosphorus, its measurements and its uptake by plants. *Aust J Soil Res* 35:227–239.

Horrigan, L., R. S. Lawrence and P. Walker. 2002. How sustainable agriculture can address the environmental and human health harms of industrial agriculture. *Environ Health Perspect* 110:445–456.

Hu, H., S. G. Penn, C. B. Lebrilla and P. H. Brown. 1997. Isolation and characterization of soluble boron complexes in higher plants: The mechanism of phloem mobility of boron. *Plant Physiol* 113:649–655.

Huang, C., M. J. Webb and R. D. Graham. 1993. Effect of pH on Mn absorption by barley genotypes in a chelate-buffered nutrient solution. In *Plant Nutrition—From Genetic Engineering to Field Practice*, ed. N. J. Barrow, pp. 653–656. Dordrecht: Kluwer Academic Publishers.

Iyamuremye, E. and R. P. Dick. 1996. Organic amendments and phosphorus sorption by soils. *Adv Agron* 56:139–185.

Izaguirre-Mayoral, M. L. and T. R. Sinclair. 2005. Soybean genotypic difference in growth, nutrient accumulation and ultrastructure in response to manganese and iron supply in solution culture. *Ann Bot* 96:149–158.

Jahangir, M., I. B. Abdel-Farid, K. Kim, Y. H. Choi and R. Verpoorte. 2009. Healthy and unhealthy plants: The effect of stress on the metabolism of Brassicaceae. *Environ Exp Bot* 67:23–33.

Johnson, D., M. Ijdo, D. R. Genney, I. C. Anderson and I. J. Alexander. 2005. How do plants regulate the function, community structure and diversity of mycorrhizal fungi? *J Exp Bot* 56:1751–1760.

Jones, R. W., A. J. Abbott, E. J. Hewitt, D. M. James and G. R. Best. 1976. Nitrate reductase activity and growth in Paul's scarlet rose suspension cultures in relation to nitrogen source and molybdenum. *Planta* 133:27–34.

Kaiser, B. N., K. L. Gridley, J. N. Brady, T. Phillips and S. D. Tyerman. 2005. The role of molybdenum in agricultural plant production. *Ann Bot* 96:745–754.

Karley, A. J. and P. J. White. 2009. Moving cationic minerals to edible tissues: Potassium, magnesium, calcium. *Curr Opin Plant Biol* 12:291–298.

Ke, W., Z. T. Xiong, C. S. Chen and J. Chen. 2007. Effects of copper and mineral nutrition on growth, copper accumulation and mineral element uptake in two *Rumex japonicus* populations from a copper mine and an uncontaminated field sites. *Environ Exp Bot* 59:59–67.

Kinzing, A. P. and R. H. Socolow. 1994. Human impacts on the nitrogen cycle. *Phys Today* 47:24–35.

Koeller, K. M. and C. H. Wong. 2001. Enzymes for chemical synthesis. *Nature* 409, 232–240.

Kolkman J. M. and J. D. Kelly. 2003. QTL, conferring resistance and avoidance to white mold in common beans. *Crop Sci* 43:539–548.

Korshunova, Y. O., D. Eide, W. G. Clark, M. L. Guerinot and H. B. Pakrasi. 1999. The IRT1 protein from *Arabidopsis thaliana* is a metal trans-porter with a broad substrate range. *Plant Mol Biol* 40:37–44.

Krämer, U, 2005. Phytoremediation: Novel approaches to cleaning up polluted soils. *Curr Opin Biotechnol* 16:133–141.

Kristensen, D. M., A. R. Mushegian, V. V. Dolja and E. V. Koonin. 2010. New dimensions of the virus world discovered through metagenomics. *Trends Microbiol* 18:11–19.

Kuiper, I., G. V Bloemberg and B. J. J. Lugtenberg. 2001. Selection of a plant–bacterium pair as a novel tool for rhizostimulation of polycyclic aromatic hydrocarbon-degrading bacteria. *Mol Plant–Microbe Interact* 14:1197–1205.

Laanbroek, H. 1990. Bacterial cycling of minerals that affect plant-growth in waterlogged soils: A review. *Aquat Bot* 38:109–125.

Ladouceur, A., S. Tozawa, S. Alam, S. Kamei and S. Kawai. 2006. Effect of low phosphorus and iron-deficient conditions on phytosiderophore release and mineral nutrition in barley. *Soil Sci Plant Nutr* 52:203–210.

Lahner, B, J. Gong, M. Mahmoudian et al. 2003. Genomic scale profiling of nutrient and trace elements in *Arabidopsis thaliana. Nat Biotechnol* 21:1215–1221.

Lasat, M. M. 2002. Phytoextraction of toxic metals: A review of biological mechanisms. *J Environ Qual* 31:109–120.

Leeper, G. W. 1952. Factors affecting availability of inorganic nutrients in soils with special reference to micronutrient metals. *Annu Rev Plant Physiol* 3:1–16.

Leininger, S., T. Urich, M. Schloter et al. 2006. Archaea predominate among ammonia-oxidizing prokaryotes in soils. *Nature* 442:806–809.

Leustek, T., M. N. Martin, J. A. Bick and J. P. Davies. 2000. Pathways and regulation of sulfur metabolism revealed through molecular and genetic. *Annu Rev Plant Physiol Plant Mol Biol* 51:141–165.

Li, L., A. F. Tutone, R. S. M. Drummond, R. C. Gardner and S. Luan. 2001. A novel family of magnesium transport genes in Arabidopsis. *Plant Cell* 13:2761–2775.

Linderman, R. G. 1988. Mycorrhizal interactions with the rhizosphere microflora—The mycorrhizosphere effect. *Phytopathology* 78:366–371.

Liu, T. W. and L. Zhu. 2005. Environmental microbiology-on-a-chip and its future impacts. *Trends Biotechnol* 23:174–179.

Lommen, A., G. van der Weg, M. C. van Engelen, G. Bor, L. A. P. Hoogenboom and M. W. F. Nielen. 2007. An untargeted metabolomics approach to contaminant analysis: Pinpointing potential unknown compounds. *Anal Chim Acta* 584:43–49.

Loneragan, J. F. 1997. Plant nutrition in the 20th and perspectives for the 21st century. *Plant Soil* 196:163–174.

Lorenz, P. and J. Eck. 2005. Metagenomics and industrial applications. *Nat Rev Microbiol* 3:510–516.

Lozovaya, V. V., A. V. Lygin, S. Li, G. L. Hartman and J. M. Widhohn. 2004. Biochemical response of soybean roots to *Fusarium solani* f. sp glycines infection. *Crop Sci* 44:819–826.

Lynch, J. M. 1987. *The Rhizosphere*. Chichester: Wiley Inter-science.

Ma, L. Q., K. M. Komar, C. Tu, W. Zhang, Y. Cai and E. D. Kennelley. 2001. A fern that hyperaccumulates arsenic. *Nature* 409:579.

Maathuis, F. J., V. Filatov, P. Herzyk et al. 2003. Transcriptome analysis of root transporters reveals participation of multiple gene families in the response to cation stress. *Plant J* 35:675–692.

Mahmoud, A. L. E. and M. H. Abdalla. 1994. Natural occurrence of mycotoxin in broad bean (*Vicia faba* L.) seeds and their effect *Rhizobium*–Legume symbiosis. *Soil Biol Biochem* 26:1081–1085.

Maloy, S. and M. Schaechter. 2006. The era of microbiology: A golden phoenix. *Int Microbiol* 9:1–7.

Mandel, R R, and R. Hänsch. 2002. Molybdoenzymes and molybdenum cofactor in plants. *J Exp Bot* 53:1689–1698.

Marschner, H. 1995. *Mineral Nutrition of Higher Plants*, 2nd edition. London. Academic Press.

Martin, M. H., E. M. Duncan and P. J. Coughtrey 1982. The distribution of heavy metals in contaminated woodland ecosystem. *Environ Pollut* 3:147–157.

Martin, J. H., K. H. Coale, K. S. Johnson et al. 1994. Testing the iron hypothesis in ecosystems of the equatorial Pacific Ocean. *Nature* 371:123–129.

Martín, H. G., N. Ivanova, V. Kunin et al. 2006. Metagenomic analysis of two enhanced biological phosphorus removal (EBPR) sludge communities. *Nat Biotechnol* 24:1263–1269.

Martinoia, E., M. Maeshima and H. E. Neuhaus. 2007. Vacuolar transporters and their essential role in plant metabolism. *J Exp Bot* 58:83–102.

Mäser, P., S. Thomine, J. I. Schroeder et al. 2001. Phylogenetic relationships within cation-transporter families of *Arabidopsis thaliana. Plant Physiol* 126:1646–1667.

Maas, E. V. and G. J. Hoffman. 1977. Crop salt tolerance—Current assessment. *J Irrig Drain E-ASCE* 103:115–134.

Mathur, A., A. K. Mathur, R. S. Sangwan, A. Gangwar and G. C. Uniyal. 2003. Differential morphogenetic responses, ginsenoside metabolism and RAPD patterns of three *Panax* species. *Genet Resour Crop Evol* 50:245–252.

Matsumoto, H. 2000. Cell biology of aluminum toxicity and tolerance in higher plants. *Int Rev Cytol* 200:1–46.

Mazur, B., E. Krebbers and S. Tingey. 1999. Gene discovery and product development for grain quality traits. *Science* 285:372–375.

McDonald, G. K. and G. Dean. 1996. Effect of waterlogging on the severity of disease caused by *Mycosphaerella pinodes* in peas (*Pisum sativum* L.). *Aust J Exl Agric* 36:219–222.

McGrath, S. P. F. J. Zhao and E. Lombi. 2001. Plant and rhizosphere processes involved in phytoremediation of metal-contaminated soils. *Plant Soil* 232:207–214.

McGrath, S. P. and F. J. Zhao. 2003. Phytoextraction of metals and metalloids from contaminated soils. *Curr Opin Biotechnol* 14:277–282.

McLaughlin, S. B. and R. Wimmer. 1999. Calcium physiology and terrestrial ecosystem processes. *New Phytol* 142:373–387.

Meharg, A. A. 2004. Arsenic in rice—Understanding a new disaster for South-East Asia. *Trends Plant Sci* 9:415–417.

Metting, F. B. Jr. 1992. Structure and physiological ecology of soil microbial communities. In *Soil Microbial Ecology*, ed. F. B. Jr. Metting, pp. 3–25. New York, NY: Marcel Dekker Inc.

Mithofer, A., B. Muller, G. Wanner and L. A. Eichacker. 2002. Identification of defence-related cell wall proteins in *Phytophthora sojae*-infected soybean roots by ESI-MS/MS. *Mol Plant Pathol* 3:163–166.

Mohn, W. W., K. Westerberg, W. R. Cullen and K. J. Reimer. 1997. Aerobic biodegradation of biphenyl and polychlorinated biphenyls by Arctic soil micro-organisms. *Appl Environ Microbiol* 63:3378–3384.

Morgan, J. A. W. and J. M. Whipps. 2001. Methodological approaches to the study of rhizosphere carbon flow and microbial population dynamics. In *The Rhizosphere: Biochemistry and Organic Substances at the Soil–Plant Interface*, eds. R. Pinton, Z. Varanini and P. Nannipieri, pp. 373–410. New York, NY: Marcel Dekker Inc.

Morgan, J. A. W., G. D. Bending and P. J. White. 2005. Biological costs and benefits to plant–microbe interactions in the rhizosphere. *J Exp Bot* 56:1729–1739.

Mulder, E. G. 1950. Mineral nutrition of plants. *Annu Rev Plant Physiol* 1:1–24.

Narasimhan, K., C. Basheer, V. B. Bajic and S. Swarup. 2003. Enhancement of plant–microbe interactions using a rhizosphere metabolomics-driven approach and its application in the removal of polychlorinated biphenyls. *Plant Physiol* 132:146–153.

Navas-Cortés, J. A., B. Hau and R. M. Jimenez-Diaz. 2000. Yield loss in chickpea in relation to development of *Fusarium* wilt epidermics. *Phytopathology* 90:1269–1278.

Negishi, T., H. Nakanishi, J. Yazaki et al. 2002. cDNA microarray analysis of gene expression during Fe-deficiency stress in barley suggests that polar transport of vesicles is implicated in phytosiderophore secretion in Fe-deficient barley roots. *Plant J* 30:83–94.

Nene, Y. L. and M. V. Reddy. 1987. Chickpea diseases and their control. In *The Chickpea*, eds. M. C. Saxena and K. B. Singh, pp. 233–270. Oxon: CAB International.

Newbould, P. 1989. The use of nitrogen fertiliser in agriculture. Where do we go practically and ecologically? *Plant Soil* 115:297–311.

Nriagu, J. O. and J. M. Pacyna. 1988. Quantitative assessment of worldwide contamination of air water and soils by trace metals. *Nature* 333:134–139.

OECD—Organization for Economic Cooperation and Development-2001. *The Application of Biotechnology to Industrial Sustainability*. Paris: OECD.

O'Connell, K. P., R. M. Goodman and J. Handelsman. 1996. Engineering the rhizosphere: Expressing a bias. *Trends Biotechnol* 14:83–88.

O'Halloran, T. V. and V. C. Culotta. 2000. Metallochaperones, an intracellular shuttle service for metal ions. *J Biol Chem* 275:25057–25060.

Okigbo, B. N. 1991. Development of sustainable agricultural production systems in Africa. International Institute of Tropical Agriculture, Ibadan.

Oresic, M. 2009. Metabolomics, a novel tool for studies of nutrition, metabolism and lipid dysfunction. *Nutr Met Cardiovas Dis* 19:816–824.

Outten, C. E. and T. V. O'Halloran. 2001. Femtomolar sensitivity of metalloregulatory proteins controlling zinc homeostasis. *Science* 292:2488–2492.

Pace, N. R. 1995. Opening the door onto the natural microbial world: Molecular microbial ecology. *Harvey Lect* 91:59–78.

Pallotta, M. A., R. D. Graham, P. Langridge, D. H. B. Sparrow and S. J. Barker. 2000. RFLP mapping of manganese efficiency in barley. *Theor Appl Genet* 101:1100–1108.

Peoples, M. B., D. E. Herridge and J. K. Ladha. 1995. Biological nitrogen fixation: An efficient source of nitrogen for sustainable agricultural production? *Plant Soil* 174:3–28.

Petrie, S. E. and T. L. Jackson. 1984a. Effect of fertilization on soil solution pH and manganese concentration. *Soil Sci Soc Am J* 48:315–318.

Petrie, S. E. and T. L. Jackson. 1984b. Effects of nitrogen fertilization on manganese concentration and yield of barley and oats. *Soil Sci Soc Am J* 48:319–322.

Pinton, R., Z. Varanini and P. Nannipieri. 2001. *The Rhizosphere: Biochemistry and Organic Substances at the Soil–Plant Interface*. New York, NY: Marcel Dekker.

Plieth, C. 2005. Calcium: Just another regulator in the machinery of life? *Ann Bot* 96:1–8.

Popelkova, H., A. Wyman and C. Yocum. 2003. Amino acid sequences and solution structures of manganese stabilizing protein that affect reconstitution of photosynthesis II activity. *Photosynth Res* 77:21–34.

Potrykus, I. 2001. Golden Rice and beyond. *Plant Physiol* 125:1157–1161.

Poulain, D. and H. Almohammad. 1995. Effects of boron deficiency and toxicity on faba bean (*Vicia faba* L.). *Eur J Agron* 4:127–134.

Power, P. P. and W. G. Woods. 1997. The chemistry of boron and its specification in plants. *Plant Soil* 193:1–13.

Prasad, R. and J. F. Power. 1997. *Soil Fertility Management for Sustainable Agriculture*. Boca Raton, FL: CRC Press.

Rager, M. H., M. R. B. Binet, G. Ionescu and O. M. M. Bouvet. 2000. ^{31}P and ^{13}C NMR studies of mannose metabolism in *Plesiomonas shigelloides*. *Eur J Biochem* 267:5136–5141.

Rahman, M. H. H., Y. Arima, K. Watanabe and H. Sekimoto. 1999. Ad-equate range of boron nutrition is more restricted for root nodule development than for plant growth in young soybean plant. *Soil Sci Plant Nutr* 45:287–296.

Rausch, C. and M. Bucher. 2002. Molecular mechanisms of phosphate transport in plants. *Planta* 216:23–37.

Rea, P. A. 2003. Ion genomics. *Nat Biotechnol* 21:1149–1151.

Reddy, K. J., L. C. Munn and L. Wang. 1997. Chemistry and mineralogy of molybdenum in soils. In *Molybdenum in Agriculture*, ed. U. C. Gupta, pp. 4–22. Cambridge: Cambridge University Press.

Reddy, D. D., A. S. Rao, K. S. Reddy and P. N. Takkar. 1999. Yield sustainability and phosphorus utilisation in soybean–wheat system on vertisols in response to integrated use of manure and fertilizer phosphorus. *Field Crops Res* 62:181–190.

Reich, P. A. 2005. Global biogeography of plant chemistry: Filling in the blanks. *New Phytol* 168:263–266.

Rengel, Z. and D. L. Robinson. 1989. Competitive aluminum ion inhibition of net magnesium ion uptake by intact *Lolium multiflorum* roots. *Plant Physiol* 91:1407–1413.

Rengel, Z., R. Guterridge, P. Hirsch and D. Hornby. 1996. Plant genotype, micronutrient fertilization and take-all infection influence bacterial populations in the rhizosphere of wheat. *Plant Soil* 183:269–277.

Reuter, D. J., T. G. Heard and A. M. Alston. 1973. Correction of manganese deficiency in barley crops on calcareous soils. I. Manganous-sulphate applied at sowing and as foliar sprays. *Aust J Exp Agric Anim Husb* 13:434–439.

Reynolds, C. S. and P. S. Davies. 2001. Sources and bioavailability of phosphorus fractions in freshwaters: A British perspective. *Biol Rev* 76:27–64.

Risher, H. and K. M. Oksman-Caldentcy. 2006. Unintended effects in genetically modified crops: Revealed by metabolomics? *Trends Biotechnol* 24:102–104.

Rodale, R. 1988. Agricultural systems: The importance of sustainability. *Natl Forum* 68:2–6.

Rodríguez-Valera, F. 2004. Environmental genomics, the big picture? *FEMS Microbiol Lett* 231:153–158.

Roessner, U., J. H. Patterson, M. G. Forbes, G. B. Fincher, P. Langridge and A. Bacic. 2006. An investigation of boron toxicity in barley using metabolomics. *Plant Physiol* 142:1087–1101.

Romero-Andreas, J., B. S. Yandell and F. A. L. Bliss. 1986. Inheritance of a novel seed protein of *Phaseolus vulgaris* L. and its effect on seed composition. *Theor Appl Genet* 72:123–128.

Rosenzweig, A. C. 2002. Metallochaperones: Bind and deliver. *Chem Biol* 9:673–677.

Rubiales, D., A. A. Emeran and J. C. Sillero. 2002. Rusts on legumes in Europe and North Africa. *Grain Legumes* 37:8–9.

Ruttan, V. W. 1992. *Sustainable Agriculture and the Environment. Perspectives on Growth and Constraints*. Boulder, CO: Westview Press.

Salido, A. L., K. L. Hasty, J. M. Lim and D. J. Butcher. 2003. Phytoremediation of arsenic and lead in contaminated soil using Chinese brake ferns (*Pteris vittata*) and Indian mustard (*Brassica juncea*). *Int J Phytoremed* 5:89–103.

Salt, D. E. 2004. Update on plant ionomics. *Plant Physiol* 136:2451–2456.

Salt, D. E. 2008. Ionomics and the study of the plant ionome. *Annu Rev Plant Biol* 59:709–733.

Salt, D. E., R. D. Smith and I. Raskin. 1998. Phytoremediation. *Annu Rev Plant Physiol Plant Mol Biol* 49:643–668.

Sanders, D., J. Pelloux, C. Brownless and J. F. Harper. 2002. Calcium at the crossroads of signaling. *Plant Cell* 14:S410–S417.

Saunders, J. and N. O'Neill. 2004. The characterization of defense responses to fungal infection in alfalfa. *BioControl* 49:715–728.

Schloss, P. D. and J. Handelsman. 2003. Biotechnological prospects from metagenomics. *Curr Opin Biotechnol* 14:303–310.

Schmid, A., F. Hollmann, J. B. Park and B. Buhler. 2002. The use of enzymes in the chemical industry in Europe. *Curr Opin Biotechnol* 13:359–366.

Schmidt, W. 2003. Iron solutions: Acquisition strategies and signaling pathways in plants. *Trends Plant Sci* 8:188–193.

Schmidt, A. and K. Jäger. 1992. Open question about sulfur metabolism in plants. *Annu Rev Plant Physiol Plant Mol Biol* 43:325–349.

Schoemaker, H. E., D. Mink and M. G. Wubbolts. 2003. Dispelling the myths—Biocatalysis in industrial synthesis. *Science* 299:1694–1697.

Sen, R. 2005. Towards a multifunctional rhizosphere concept: Back to the future? *New Phytol* 168:266–268.

Sharma, K. K. and M. Lavanya. 2002. Recent developments in transgenics for abiotic stress in legumes of the semi-arid tropics. In *Genetic Engineering of Crop Plants for Abiotic Stress,* ed. M. Ivanaga, pp. 61–73. Working Report No. 23. Tsukuba: JIRCAS.

Shao, H. B., Q. J. Guo, L. Y. Chu, X. N. Zhao, Z. L. Su, Y. C. Hu and J. f. Cheng. 2007. Understanding molecular mechanism of higher plant plasticity under abiotic stress. *Colloid Surf B* 54:35–54.

Shao, H. B., L. Y. Chu, C. Abdul Jaleel and C. X. Zhao. 2008. Water-deficit stress-induced anatomical changes in higher plants. *C R Biol* 331:215–225.

Shear, C. B. 1975. Calcium-related disorders of fruits and vegetables. *Hort Sci* 10:361–365.

Shimada, N., T. Akashi, T. Aoki and S. Ayabe. 2000. Induction of isoflavonoid pathway in the model legume *Lotus japonicus*: Molecular characterization of enzymes involved in phytoalexin biosynthesis. *Plant Sci* 160:37–47.

Sillanpää, M. 1990. *Micronutrients Assessment at the Country Level: An International Study*. FAO Soils Bulletin No. 63, Food and Agriculture Organisation, Rome.

Sillanpää, M. and P. L. G. Vlek. 1985. Micronutrients and the agroecology of tropical and Mediterranean regions. *Fertil Res* 7:151–167.

Simpson, D. J. and S. P. Robinson. 1984. Freeze-fracture ultrastructure of thylakoid membranes in chloroplasts from manganese-deficient plants. *Plant Physiol* 74:735–741.

Smith, S. E. 2002. Soil microbes and plants—Raising interest, mutual gains. *New Phytol* 156:142–144.

Somerville, C. and S. Somerville. 1999. Plant functional genomics. *Science* 285:380–383.

Sommer, A. L. 1928. Further evidence of the essential nature of zinc for the growth of higher green plants. *Plant Physiol* 3:217–221.

Sommer, A. L. 1931. Copper as an essential for plant growth. *Plant Physiol* 6:339–345.

Sommer, A. L. and C. B. Lipman. 1926. Evidence on the indispensable nature of zinc and boron for higher green plants. *Plant Physiol* 1:231–249.

Sparks, D. L. 1987. Potassium dynamics in soils. *Adv Soils Sci* 6:1–63.

Sprent, J. I. 1984. Evolution of nitrogen fixing symbioses. In *Advances in Plant Physiology*, ed. M. B. Wilkins, pp. 249–276. London: Pitman.

Stamler, J. S., D. J. Singel and J. Loscalzo. 1992. Biochemistry of nitric oxide and its redox-activated forms. *Science* 258:1898–1902.

Straathof, A. J., S. Panke and A. Schmid. 2002. The production of fine chemicals by biotransformations. *Curr Opin Biotechnol* 13:548–556.

Szpunar, J. 2004. Metallomics: A new frontier in analytical chemistry. *Anal Bioanal Chem* 378:54–56.

Tada, Y., M. Nakase, T. Adachi, R. Nakamura et al. 1996. Reduction of 14–16kDa allergenic proteins in transgenic rice plants by antisense gene. *FEBS Lett* 391:341–345.

Tan, K., W. G. Keltjens and G. R. Findenegg. 1991. Role of magnesium in combination with liming in alleviating acid–soil stress with the aluminum-sensitive sorghum genotype CV323. *Plant Soil* 136:65–72.

Terry, N. 1977. Photosynthesis, growth, and the role of chloride. *Plant Physiol* 60:69–75.

Thimm, O., B. Essigmann, S. Kloska, T. Altmann and T. J. Buckhout. 2001. Response of Arabidopsis to iron deficiency stress as revealed by micro-array analysis. *Plant Physiol* 127:1030–1043.

Thomine, S., F. Lelièvre, E. Debarbieux, J. I. Schroeder and H. Barbier-Brygoo. 2003. AtNRAMP3, a multi-specific vacuolar metal transporter involved in plant responses to iron deficiency. *Plant J* 34:685–695.

Tisdale, S. L., W. L. Nelson and J. D. Beaton. 1985. *Soil Fertility and Fertilizers*, Vol. 4. New York, NY: MacMillan.

Tong, Y., Z. Rengel and R. D. Graham. 1997. Interactions between nitrogen and manganese nutrition of barley genotypes differing in manganese efficiency. *Ann Bot* 79:53–58.

Torsvik, V., F. L. Daae, R. A. Sandaa and L. Øvreås. 1998. Novel techniques for analysing microbial diversity in natural and perturbed environments. *J Biotechnol* 64:53–62.

Torsvik, V., F. L. Daae, R. A. Sandaa and L. Øvreås. 2002. Microbial diversity and function in soil: From genes to ecosystems. *Curr Opin Microbiol* 5:240–245.

Tripathi, R. D., S. Srivastava, S. Mishra et al. 2007. Arsenic hazards: Strategies for tolerance and remediation by plants. *Trends Biotechnol* 25:158–165.

UN—United Nations. 1999. *Guidelines for Consumer Protection with Proposed New Elements on Sustainable Consumption*. Agenda 21. New York, NY: UN.

UNDP—United Nations Program for Development (1998) *Human Development Report*. Report No. 4, New York, NY: UNPD.

Unkles, S. E., R. Wang, Y. Wang, A. D. M. Glass, N. M. Crawford and J. R. Kinghorn. 2004. Nitrate reductase activity is required for nitrate uptake into fungal but not plant cells. *J Biol Chem* 279:28182–28196.

Urano, U., Y. Kurihara, M. Seki and K. Shinozaki. 2010. Omics' analyses of regulatory networks in plant abiotic stress responses. *Curr Opin Plant Biol* 13:132–138.

Vance, C. P. 1996. Root–bacteria interactions: Symbiotic N_2 fixation. In *Plant Roots: The Hidden Half*, eds. Y. Waisel, A. Eshel and U. Kafkafi, pp. 723–756. New York, NY: Marcel Dekker.

Vance, C. P. 1997. Enhanced agricultural sustainability through biological nitrogen fixation. In *Biological Fixation of Nitrogen for Ecology and Sustainable Agriculture*, eds. A. Legocki, H. Bothe and A. Pühler, pp. 179–186. Berlin: Springer-Verlag.

Vance, C. P. and P. H. Graham. 1994. Nitrogen fixation in agriculture: Application and erspectives. In *Nitrogen Fixation: Fundamentals and Applications*, eds. A. I. Tikhonovich, N. A. Provorov, V. I. Romanov and W. E. Newton, pp. 77–86. Dordrecht: Kluwer Academic.

Van Elsas, D., E. M. H. Wellington and J. T. Trevors. 1997. *Modern Soil Microbiology*. New York, NY: Marcel Dekker.

Van Hoorn, J. W., N. Katerji, A. Hamdy and M. Mastrorilli. 2001. Effect of salinity on yield and nitrogen uptake of four grain legumes and on biological nitrogen contribution from the soil. *Agr Water Manage* 51:87–98.

Véry, A. A. and H. Sentenac. 2003. Molecular mechanisms and regulation of K^+ transport in higher plants. *Annu Rev Plant Biol* 54:575–603.

Von Wirén, N., H. Marschner and V. Römheld. 1996. Roots of iron-efficient maize also absorb phytosiderophore-chelated zinc. *Plant Physiol* 111:1119–1125.

Vose, P. B. 1982. Iron nutrition in plants: A world overview. *J Plant Nutr* 5:238–249.

Wang, H., S. F. Hwang, K. F. Chang, G. D. Turnbull and R. J. Howard. 2003a. Suppression of important pea diseases by bacterial antagonists. *Biocontrol* 48:447–460.

Wang, R., M. Okamoto, X. Xing and N. M. Crawford. 2003b. Microarray analysis of the nitrate response in Arabidopsis roots and shoots reveals over 1000 rapidly responding genes and new linkages to glucose, trehalose-6-phosphate, iron, and sulfate metabolism. *Plant Physiol* 132:556–567.

Wang, Y. P., J. Y. Shi, H. Wang, Q. Lin, X. C. Chen and Y. X. Chen. 2007. The influence of soil heavy metals pollution on soil microbial biomass, enzyme activity, and community composition near a copper smelter. *Ecotox Environ Safe* 67:75–81.

Ward, N. 2006. New directions and interactions in metagenomics research. *FEMS Microbiol Ecol* 55:331–338.

Ward, N. and C. M. Fraser. 2005. How genomics has affected the concept of microbiology. *Curr Opin Microbiol* 8:564–571.

Warkentin, T. D., K. Rashid and A. G. Xue. 1996. Fungicidal control of ascochyta in field. *Can J Plant Sci* 76:67–71.

Weiland, R. T., R. D. Noble and R. E. Crang. 1975. Photosynthetic and chloroplast ultrastructural consequences of manganese deficiency in soybean. *Am J Bot* 62:501–508.

Welch, R. M. 1995. Micronutrient nutrition of plants. *Crit Rev Plant Sci* 14:49–82.

White, P. J. and M. R. Broadley. 2001. Chloride in soils and its uptake and movement within the plant: A review. *Ann Bot* 88:967–988.

Williams, R. J. P. 2001. Chemical selection of elements by cells. *Coord Chem Rev* 216–217:583–595.

Williams, R. J. P. and J. J. R. Frausto da Silva. 2002. The involvement of molybdenum in life. *Biochem Biophys Res Co* 292:293–299.

Wintz, H., T. Fox, Y. Y. Wu et al. 2003. Expression profiles of *Arabidopsis thaliana* in mineral deficiencies reveal novel transporters involved in metal homeostasis. *J Biol Chem* 278:47644–47653.

WRI—World Resources Institute (1998) *World Resources: A Guide to the Global Environment*. Report. Washington, DC: WRI.

Wu, Q. D. and H. D. VanEtten. 2004. Introduction of plant and fungal genes into pea (*Pisum sativum* L.) hairy roots reduces their ability to produce pisatin and affects their response to a fungal pathogen. *Mol Plant–Microbe Int* 17:798–804.

Wu, X. R., Z. H. Chen and W. R. Folk. 2003. Enrichment of cereal protein lysine content by altered tRNA[sup lys] coding during protein synthesis. *Plant Biotechnol* 1:187–194.

Xu, J. P. 2006. Microbial ecology in the age of genomics and metagenomics: Concepts, tools, and recent advances. *Mol Ecol* 15:1713–1731.

Xu, G., H. Magen, J. Tarchitzky and U. Kafkafi. 2000. Advances in chloride nutrition. *Adv Agron* 68:96–150.

Yi, Y., J. Saleeba and M. L. Guerinot. 1994. Iron uptake in *Arabidopsis thaliana*. In *Biochemistry of Metal Micronutrients in the Rhizosphere*, eds. J. Manthey, D. Luster and D. E. Crowley, pp. 295–307. Boca Raton, FL: CRC Press.

Yoshida, M., S. E. Cowgill and J. A. Wightman. 1997. Roles of oxalic and malic acids in chickpea trichome exudates in host-plant resistant to *Helicoverpa armigera*. *J Chem Ecol* 23:1195–1210.

Zarate, R. 2010. Plant secondary metabolism engineering: Methods, strategies, advances, and omics. *Comp Nat Prod* 3:629–668.

Zhu, J. K. 2003. Regulation of ion homeostasis under salt stress. *Curr Opin Plant Biol* 6:441–445.

CHAPTER 7

Identification of Molecular Processes Underlying Abiotic Stress Plants Adaptation Using 'Omics' Technologies

Urmila Basu

CONTENTS

7.1 INTRODUCTION

Abiotic stresses such as high salinity, drought, high or low temperature and metal toxicity can result in extensive losses to agricultural production (Bray et al., 2000). Exposure to different abiotic stresses can cause impaired cellular function by disrupting cellular osmotic potential and inhibition of cell division can lead to the production of reactive oxygen species (ROS) resulting in subsequent damage to proteins or lipids (Bray et al., 2000; Hasegawa et al., 2000; Mittler, 2002, 2006; Zhu, 2001). A remarkable feature of plant adaptation to abiotic stresses is the activation of multiple responses involving complex networks that are interconnected at several levels. One of the biggest challenges to modern sustainable agriculture development is to understand and regulate the relationship between the plants and the corresponding environmental stresses, which can be the main constraints for crop productivity (Leung, 2008; Yi and Bo, 2008). Extensive genetic studies have clearly indicated different degrees of variation for abiotic stress tolerance and it is difficult to study complex interconnected pathways involved in multiple responses to abiotic stresses, using traditional approaches. Current 'omics' technologies are now widely seen as promising tools for dissecting abiotic stress responses in plants.

7.1.1 Genotype/Environment Interaction

Both the genetic complexity of the organism and environment play an integrated role in defining the phenotype showing resistance to abiotic stress. By combining gene sequencing, gene cloning and plant transformation with biochemical and genome databases, different studies have identified genes, linked to adaptation to abiotic environment, including herbicide tolerance, disease resistance and product quality (Collins et al., 2008). However, there is a lack of the common abiotic stress resistance mechanism; for example, drought stress response has been found to be greatly influenced by genotype and environment variation (Blum, 2004; Bressan et al., 2009; Witcombe et al., 2008). For these genotype–environment systems, thousands of genes interact in complex ways to generate crop responses to the environment. These responses could be both over short timescales (e.g., cellular response to environment shocks such as cold and drought) and long timescales (e.g., morphological growth responses leading to crop development and morphology).

7.1.2 Model Systems and High-Throughput Technologies

In recent years, various high-throughput functional genomics approaches including expressed sequence tags (ESTs) profiling, microarray expression profiling, targeted mutagenesis and marker-assisted selection have been used to identify the complex network of genes or gene combinations involved in abiotic stress responses (Bohnert et al., 2006; Chen et al., 2002; Cushman and Bohnert, 2000; Sreenivasulu et al., 2007; Umezawa et al., 2006). The information gained from these studies has helped in the identification of interactions between interconnecting signalling pathways and promoters. The physiological basis of the abiotic resistance mechanisms has been pursued in the model system; *Arabidopsis thaliana*, the most popular plant species is used for molecular plant research, with the aim to elucidate the biochemical pathways involved in stress perception, signal transduction and adaptive responses (Liu et al., 2005; Seki et al., 2003). The quantitative differences in the expression of some genes have also been observed in resurrection model plants such as *Craterostigma plantagineum*, when compared with non-tolerant plants in drought tolerance studies (Deng et al., 2002).

In recent years, *Thellungiella halophila* (salt cress, Figure 7.1), a close relative of *Arabidopsis*, has emerged as an extremophile model plant for the molecular elucidation of abiotic stress tolerance (Amtmann et al., 2005; Amtmann, 2009; Vinocur and Altman, 2005). With twice the genome size of *Arabidopsis*, *Thellungiella* shares the similar morphology and sequence identity with *Arabidopsis*. The Shandong ecotype of *Thellungiella* has been collected from the high-salinity coastal areas of the Shandong province in northeast China (Inan et al., 2004) and the Yukon ecotype is native to Yukon Territories, Canada, a subarctic and semiarid region (Wong et al., 2005, 2006). Both these ecotypes have been shown to possess tolerance to salinity, drought and/or cold stresses (Griffith et al., 2007; Inan et al., 2004; Kant et al., 2006) and tolerate low nitrogen conditions (Kant et al., 2008a,b). Comparative transcript profiling of *Thellungiella* using *Arabidopsis* oligo and cDNA arrays revealed the expression of fewer genes in *Thellungiella* than in *Arabidopsis* (Amtmann et al., 2005). Several studies in *Thellungiella* (Amtmann, 2009 and references within) have reported up-regulation of the genes, by salt, ABA, H_2O_2 (*Thhsc70*, encoding a heat-shock protein), heat shock

Figure 7.1 *T. halophila*, an extremophile model plant for abiotic stress tolerance studies. With twice the genome of *A. thaliana*, it shares morphological and sequence identity with Arabidopsis. This figure shows the salinity tolerance of *T. halophila* and *A. thaliana* plants grown for 7 days in potting medium saturated with 0 mM, 200 mM and 600 mM NaCl. (From *Curr. Opin. Biotech.*, 16, Vinocur, B. and A. Altman, Recent advances in engineering plant tolerance to abiotic stress: Achievements and limitations, 123–132, Copyright (2005), with permission from Elsevier.)

(*ThCYP1*, encoding a cyclophilin), ABA, NaCl, PEG (*CBL9*, encoding a calcineurin-B-like protein) and by salt and drought (*ThZF1*, encoding a Cys-2/His-2-type transcription factor (TF)). All these studies have indicated that *Thellungiella* operates more specific regulation under stress as compared to *Arabidopsis*.

Recently, binary bacterial artificial chromosome (BIBAC) library was constructed for *Thellungiella halophila* and BIBAC clones with inserts over 50 kb were transformed into *Arabidopsis* for salt-tolerant plant screening. One transgenic line was found to be more salt tolerant than wild-type plants, indicating that the BIBAC library approach could potentially identify more important salt-related genes in *T. halophila* by functional complement and comparative genomics in the future (Wang et al., 2009a). The ongoing sequencing of the *Thellungiella* genome, generation and sequencing of cDNA libraries, collection of promoter regions of stress-induced genes, availability of RNAi lines for further testing and creation of T-DNA insertion mutants will provide *Thellungiella* as a promising system for abiotic stress resistance studies.

7.2 FUNCTIONAL GENOMICS TO UNDERSTAND THE GENE REGULATORY NETWORKS INVOLVED IN ABIOTIC STRESS-TOLERANCE MECHANISMS

Plant responses to different abiotic stresses include the regulatory changes to thousands of genes and activation of multiple genes and pathways (Chen et al., 2002; Chen and Zhu, 2004; Kant et al., 2008a,b; Kreps et al., 2002). Most of these gene functions are involved in stress tolerance through the regulation of gene expression and signal transduction in stress responses (Bartels and Sunkar, 2005; Shinozaki et al., 2003; Shinozaki and Shinozaki, 2006). The complex gene regulation in response to different abiotic stresses in plants could be modulated at different levels in plants. The interactions could occur at the level of transcriptional or post-transcriptional levels; however, the epigenetic regulation could also modulate these transcriptional regulatory networks (Chen et al., 2002; Urnov, 2003).

The group of proteins functioning in stress tolerance include chaperones, late embryogenesis abundant (LEA) proteins, osmotin, antifreeze proteins, mRNA-binding proteins (RBPs), transporters, detoxification enzymes, enzymes for fatty acid metabolism and inhibitors (Shinozaki and

Shinozaki, 2006). Another set of proteins involved in the regulation of signal transduction pathway and gene expression in stress response include various TFs (Seki et al., 2003). These TFs could comprise complex gene networks regulating responses to different stresses, for example, the dehydration-responsive element-binding protein (DREB)/C-repeat-binding factor (*CBF*) family of TFs binding the *DRE/CRT cis*-acting element in the promoters of stress-responsive genes (Yamaguchi-Shinozaki and Shinozaki, 2006). Several stress-related TF family members such as DREB, ERF, WRKY, MYB and bZIP were identified in *Arabidopsis* in response to cold, drought and salinity (Seki et al., 2002). Furthermore, these studies also indicated that there is a cross-talk between salt and drought stress signalling processes.

The stress-inducible expression of *DREB/CBF* genes led to increased expression of stress-responsive genes providing tolerance to drought, freezing and salt stress (Maruyama et al., 2004; Sakuma et al., 2006). The role of the members of the *Arabidopsis* group C/S1 basic leucine zipper (bZIP) TF network has been investigated in response to abiotic stress in different studies (Kilian et al., 2007; Weltmeier et al., 2009). The four group C and five group S1 members of C/S1 network form specific heterodimers and are, therefore, considered to cooperate functionally. AtGenExpress transcriptome data set revealed that the application of different abiotic stresses led to an induction of *S1 AtbZIP* genes in one part of the plant, whereas no effect or an inverse effect was observed in the other (Kilian et al., 2007). The transcription of group S1 genes strongly responded to various abiotic stresses, such as salt (*AtbZIP1*) or cold (*AtbZIP44*); however, the expression of group *C bZIPs* was not affected (Weltmeier et al., 2009).

Recently, a novel functional genomics-driven approach was used to identify genes, comprising upstream components of the stress signalling pathways that may regulate *Arabidopsis* responses to multiple abiotic stresses (Kant et al., 2008a,b). In this study, data from published *Arabidopsis* abiotic stress microarray analyses were combined with the data from microarray analysis of heat stress-responsive gene expression, to identify a set of 289 multiple stress (*MST*) genes. A subset of MST genes designated multiple stress Regulatory (*MSTR*) genes encoding regulatory proteins such as TFs, kinases and phosphatases were identified (Kant et al., 2008a,b). Functional genomics is now widely seen as a promising tool for dissecting abiotic stress responses to identify networks of stress perception, signal transduction and defensive responses.

7.2.1 Plant Engineering for Analysis of the Stress-Tolerance Mechanism

Plant engineering strategies for abiotic stress tolerance have been focused on the expression of genes that encode proteins conferring tolerance. These include different enzymatic components leading to the synthesis of functional and structural metabolites, including osmoprotectants or antioxidants or genes involved in signalling and regulatory pathways (Bhatnagar-Mathur et al., 2008; Sreenivasulu et al., 2007; Valliyodan and Nguyen, 2006). Different osmoprotectants that accumulate during osmotic adjustment during exposure to abiotic stresses include organic compounds such as amino acids (e.g., proline) and other amines (e.g., glycinebetaine and polyamines) as well as a variety of sugars and sugar alcohols such as mannitol, trehalose and galactinol (Vincour and Altman, 2005). Several studies have used the transgenic approach to alter the levels of osmoprotectants for providing stress tolerance, for example, glycine-betaine (Holmstrom et al., 2000; Ishitani et al., 1997; Sakamoto et al., 2000), proline (Yamada et al., 2005) and sugar alcohols (Cortina and Culiáñez, 2005; Garg et al., 2002). The stress-resistant plants, to salt, dehydration or freezing, have also been developed by engineering LEA proteins, involved in sequestering ions during cellular dehydration (Cheng et al., 2002; Goyal et al., 2005).

Another way of achieving tolerance to multiple stress conditions is to overexpress TF genes that subsequently control multiple genes from various pathways. Several molecular and genomic studies have revealed that many genes respond to multiple stresses such as salinity, dehydration and freezing at the transcriptional level, modulated by key TFs, which are induced under abiotic stress treatments

(Chen and Zhu, 2004; Chinnusamy et al., 2005). Several TFs may act cooperatively in the plant response to stress or may activate cascades of genes providing tolerance to multiple stresses (Vincour and Altman, 2005). Over 1600 TFs, representing ~6% of the total number of genes, have been reported in *Arabidopsis*. Based on the structure of their DNA-binding domains, several large TF families have been identified including dehydration-responsive element-binding protein 1 (*DREB*), -*CBF*s and ABA-responsive element (*ABRE*)-binding protein, ABRE-binding factor (*ABF*), *AP2/ERF*, *bZIP*, *MYB*, Zinc-finger and *WRKY*, which have been implicated in the regulation of stress responses (Gong et al., 2004; Nakashima et al., 2009).

In *Arabidopsis*, three genes encoding *DREB1/CBF*, *DREB1B/CBF1*, *DREB1A/CBF3* and *DREB1C*/CBF2 are placed in tandem on chromosome 4, and are probably the major TFs involved in cold-induced gene expression. Besides, there are two *DREB2* and *DREB2A* proteins that are involved in high-salinity- and drought-induced gene expression (Shinozaki and Shinozaki, 2006). Overexpression of *CBF1/DREB1B* in *Arabidopsis* (Jaglo-Ottosen et al., 1998); *Brassica* (Jaglo et al., 2001) and *CBF3/DREB1a* in rice (Oh et al., 2005) provided increased resistance to freezing, drought and salinity stress. In subsequent studies, the overexpression of *DREB1a* has been shown to improve the drought and low-temperature stress tolerance in tobacco (Kasuga et al., 2004), wheat (Pellegrineschi et al., 2004) and potato (Behnam et al., 2006).

Another bZIP TF family member, ABA-responsive element-binding factor (ABF) showed distinct roles in sugar metabolism, ABA and stress responses (Uno et al., 2000). Introduction of these TFs in the ABA signalling pathway could lead to improved plant stress tolerance. Enhanced drought tolerance has been observed after constitutive expression of *ABF3* or *ABF4* with altered the expression of ABA/stress-responsive genes, such as *rd29B*, *rab18*, *ABI1* and *ABI2* in *Arabidopsis* (Kagaya et al., 2002; Kang et al., 2002) and rice (Oh et al., 2005). In cereals, *lip19* genes encoding bZIP-type TFs play a regulatory role in gene expression during the cold acclimation process, and transgenic tobacco overexpressing a wheat *lip19* homologue, *Wlip19*, showed a significant increase in abiotic stress tolerance, especially to freezing tolerance (Kobayashi et al., 2008). In another recent study, transgenic rice overexpressing *OsbZIP23*, a member of the bZIP TF family, showed a significant improvement in tolerance to drought and high-salinity stresses and sensitivity to ABA (Xiang et al., 2008). Very few studies have been carried out on the function of other TFs, modulated by other phytohormones such as ethylene, jasmonic acid and salicylic acid to confer abiotic stress resistance. Ethylene-responsive element-binding proteins (*EREBP*) TF from the AP2-type family can mediate abiotic stress response in plants and overexpression of EREBP TF led to increased resistance to salt in tobacco (Park et al., 2001) and tomato (Wang et al., 2004a).

Stress perception and signalling pathways are interconnected pathways in abiotic stress response (Sreenivasulu et al., 2007) and the manipulation of signalling factors can modulate the downstream events resulting in increased tolerance to different stresses (Umezawa et al., 2006). Overexpression of an osmotic-stress-activated protein kinase, *SRK2C*, resulted in increased drought tolerance in *A. thaliana* (Umezawa et al., 2004). Increased expression of a truncated tobacco mitogen-activated protein kinase kinase kinase (MAPKKK), *NPK1*, resulted in cold, heat, salinity and drought tolerance in transgenic plants (Shou et al., 2004). Inositol phosphates (IPs) and their turnover products have been known to play important roles in stress signalling in eukaryotic cells. In recent studies, transgenic tobacco plants constitutively expressing *Arabidopsis* inositol polyphosphate 6-/3-kinase (*AtIpk2*β) have been found to exhibit improved tolerance to diverse abiotic stresses, when compared to wild-type plants (Yang et al., 2008).

In summary, TFs can be powerful targets for genetic engineering of stress resistance because TFs play an important role in the regulation of cascades of genes, in response to abiotic stress. More systematic study is needed to obtain better understanding of abiotic stress signalling pathways. Furthermore, there are several challenges that need to be addressed including the use of an effective expression system, selection of the suitable promoters for developing abiotic stress-resistant transgenic plants and screening of stress-resistant plants under field conditions.

7.2.2 Comparative Transcriptome Profiling: From EST Libraries and Microarrays to Next-Generation Sequencing

Different high-throughput functional genomics approaches have been employed to identify the stress-inducible genes involved in the abiotic stress-tolerance mechanism. A major approach is to compare ESTs from cDNA libraries prepared from different tissues exposed to various stress conditions and developmental stages. The abundance of a particular gene can provide information on the relative expression levels of stress-responsive genes and possibly gene families under that particular stress. This approach has been used in several species such as oat (Bräutigam et al., 2005), barley (Close et al., 2004) and halophyte *Thellungiella halophila* (Wang et al., 2004b). To identify genes involved in abiotic stress responses under natural conditions, normalized and subtracted cDNA libraries were prepared from control, stress-exposed and desert-grown *Populus euphratica* salt-tolerant trees. About 14,000 EST sequences were obtained with a good representation of genes putatively involved in resistance and tolerance to salt and other abiotic stresses (Brosché et al., 2005). Houde et al. (2006) generated 73,521 quality-filtered ESTs from 11 cDNA libraries, constructed from wheat plants, exposed to various abiotic stresses and at different developmental stages. The gene annotation of these ESTs helped in the identification of several pathways associated with abiotic stress. Recently, a total of 41,516 ESTs were generated from nine cDNA libraries of tall fescue (*Festuca arundinacea* Schreb). These libraries represented different plant organs, developmental stages and abiotic stress conditions and ESTs represented gene homologues of heat-shock and oxidative stress proteins, and various TF protein families (Mian et al., 2008).

Gene expression profiling of transcript levels by DNA microarrays provides a novel and promising approach towards the identification and functional analysis of genes underlying multiple-stress tolerance (Clarke and Zhu, 2006). Since the development of this technology in the *Arabidopsis* model system, microarrays have been used extensively to compare gene expression measurements under a range of abiotic stress conditions in different experimental systems (Cheong et al., 2002; Kreps et al., 2002; Liu et al., 2005; Seki et al., 2002; Swindell, 2006). Seki et al. (2002) identified 351 differentially expressed genes showing more than fivefold increase using 7000 *Arabidopsis* cDNAs array and 22 of these genes responded under all three stress conditions. In contrast, 2409 of 8100 *Arabidopsis* cDNAs were induced more than twofold under salt, osmotic or cold stress and the number of differentially expressed genes varied with the exposure time to stress indicating that stress resistance may shift from general stress to specific stress (Kreps et al., 2002).

In studies on whole-genome transcriptional profiles from *A. thaliana* under nine abiotic stress conditions (cold, osmotic stress, salt, drought, genotoxic stress, ultraviolet light, oxidative stress, wounding and high temperature) and at six different time points of stress exposure, most genes were identified from early to middle (1–6 h) time points of stress exposure. Further analysis of these genes indicated that cell rescue/defense/virulence, energy and metabolism functional classes were over-represented, providing novel insight into the functional basis of multiple stress tolerance in *Arabidopsis* (Swindell, 2006). In another intensive cDNA microarray study for abiotic stress responses in chickpea, 476 differentially expressed transcripts were detected in response to all stresses, genotypes, tissue-types and time points tested (Mantri et al., 2007). Microarray analysis of 16,000 genes from *Medicago truncatula* revealed 824 genes (including 84 TFs) to be differentially expressed in root apex exposed to salt stress (Gruber et al., 2009). Robinson and Parkin (2008) have employed serial analysis of gene expression (SAGE); a sequence-based high-throughput technology, to compare the transcriptome changes of *A. thaliana* over a time course of low-temperature stress. Clustering of the differentially regulated genes facilitated the identification of novel loci associated with freezing tolerance.

Large-scale comparative gene expression analysis in different species can help in the identification of set of genes or regulatory networks, regulated in a similar manner, in response to different abiotic stresses. Besides, stress-induced epigenetic regulation through changes in histone variants, histone N-tail modifications and DNA methylation are known to regulate stress-responsive genes (Chinnusamy and Zhu, 2009). The next-generation DNA sequencing technology (Kahvejian et al., 2008; Shendure and Ji, 2008) with enhanced sequencing throughput will further allow one to uncover the huge diversity of novel genes that are currently inaccessible. High-throughput sequencing would also help to unravel the role of microRNAs and epigenetic regulators which will further lead to improved integration of biological information, for a complete picture of many traits at an individual level. However, to verify the roles of these TFs in regulatory networks of abiotic stress response, further studies are necessary including yeast two-hybrid data, or testing the putative function of the genes by forward genetics or the loss of function. Recent study characterized the function of *WRKY25* and *WRKY33* as regulators of abiotic stress responses, by analysing both potential upstream and downstream targets of these genes, and their loss-of-function and overexpression phenotypes. These TFs, *WRKY25* and *WRKY33*, appear to be independent of SOS signalling, and partly dependent on ABA signalling (Jiang and Deyholos, 2009). Comparative epigenomics and transcritpomics studies will be necessary for increased understanding of stress adaptation in plants.

7.2.3 Protein Profiling

Large-scale high-throughput transcriptome analyses need to be integrated with proteome studies to obtain a comprehensive understanding of the abiotic stress response and to develop strategies to improve stress resistance. Usually the amounts of proteins fail to show correlation to the levels of expression of the respective mRNAs because many proteins undergo post-transcriptional modifications, such as removal of signal peptides, phosphorylation and glycosylation. This is especially true for low copy number mRNAs that are potentially very important for regulation (Gygi et al., 1999; Rose et al., 2004).

In recent years, improved proteomic technologies including separation and detection of proteins using mass spectrometry-based protein identification have impacted the study of plant responses to abiotic stress. In Chicory (*Cichorium intybus*) root material before the first freezing period, 881 proteins were identified in total, 619 of these proteins were classified into functional categories and seven proteins including antioxidant proteins were shown to be involved in cold acclimation (Degand et al., 2009). Proteomic analysis of the moss *Physcomitrella patens* which can tolerate several abiotic stresses, including salinity, cold and desiccation has shown that majority of the desiccation-responsive proteins are involved in metabolism, cytoskeleton, defense and signalling (Wang et al., 2009a,b).

Several proteomics studies have been carried out in response to salt, a major abiotic stress for plants. Proteomic profiles have been produced for NaCl-treated roots of pea (*Pisum sativum* L) (Kav et al., 2004). Proteomics analysis of rice roots exposed to salt stress-identified proteins involved in the regulation of carbohydrate, nitrogen and energy metabolism, ROS scavenging, mRNA and protein processing and cytoskeleton stability (Yan et al., 2005). Comparative proteomics analysis in potato shoots revealed differential expression of 47 proteins under NaCl treatment. Among the differentially expressed proteins, photosynthesis- and protein-synthesis-related proteins were down-regulated, whereas osmotin-like proteins, *TSI-1* protein, heat-shock proteins, protein inhibitors, calreticulin and five novel proteins were markedly up-regulated (Aghaei et al., 2008). In soybean, late embryogenesis-abundant protein, beta-conglycinin, elicitor peptide precursor and basic/helix-loop-helix protein were up-regulated, while protease inhibitor, lectin and stem 31-kDa glycoprotein precursor were down-regulated in response to salt stress (Aghaei et al., 2009).

In *Arabidopsis*, >1000 protein spots were reproducibly detected on each gel, with 112 protein spots decreased and 103 increased in response to NaCl treatment. The proteins identified included many stress-responsive proteins related to processes including scavenging for ROS; signal transduction; translation, cell wall biosynthesis, protein translation, processing and degradation; and metabolism of energy, amino acids and hormones (Jiang et al., 2007). Comparative proteomics analysis in the mangrove plant *Bruguiera gymnorhiza* grown under conditions of salt revealed the identity of fructose-1,6-bisphosphate (FBP) aldolase and a novel protein in the main root, and osmotin in the lateral root (Tada and Kashimura, 2009). In barley proteomic analysis, proteins involved in the glutathione-based detoxification of ROS were more abundant in the tolerant genotype, while proteins involved in iron uptake were expressed at a higher level in the sensitive one (Witzel et al., 2009).

In response to chilling stress in rice, ~1000 protein spots were reproducibly detected on each gel, 31 protein spots were down-regulated and 65 were up-regulated. The identified proteins are involved in several processes, that is, signal transduction, RNA processing, translation, protein processing, redox homeostasis, photosynthesis, photorespiration and metabolisms of carbon, nitrogen, sulphur and energy (Lee et al., 2009; Yan et al., 2006). These proteomic studies can help to establish stress-responsive networks revealing protein modifications and protein–protein interactions. However, gel-based methods have several limitations and often are unable to detect proteins of low abundance such as regulatory proteins and receptors. Therefore, non-gel-based proteomics approaches including multidimensional protein identification technology (MudPIT) and isotope-coded affinity tagging (ICAT) have been proposed (Glinski and Weckwerth, 2006). Nonetheless, proteomics analysis when integrated with transcriptome analyses can result in comprehensive understanding of the stress response.

7.3 REVERSE GENETICS STRATEGIES FOR THE IDENTIFICATION OF ABIOTIC STRESS-RESISTANCE GENES

7.3.1 Targeted Induced Local Lesions in Genomes (TILLING), T-DNA Insertion Mutants and RNAi

With the constantly expanding sequencing databases, reverse genetics by gene disruption is one of the preferred functional genomics methods to investigate gene function. Among reverse-genetic approaches, two insertional mutagenesis strategies, transferred DNA (T-DNA, Krysan et al., 1999) and transposon tagging (Parinov et al., 1999), resulted in several large, publicly available collections of *Arabidopsis* insertion mutants (http://signal.salk.edu/cgi-bin/tdnaexpress). However, insertion mutants have several limitations including its low frequency per genome, and due to its inability to investigate the functions of duplicated genes, their usefulness is limited to very few plant species.

Different techniques have been employed to screen stress-related gene regulation mutants in *Arabidopsis*, rice, maize and barley, in order to show the putative functions of abiotic stress-responsive genes (Xiong et al., 1999). Non-transgenic reverse genetic strategy, TILLING requires prior DNA sequence information and combines the high density of point mutations provided by traditional chemical mutagenesis, with rapid mutational screening and is appropriate for both small- and large-scale screening. It can provide an allelic series of silent, missense, nonsense and splice site mutations for examining the effect of various mutations in a gene. (Barkley and Wang, 2008; Henikoff et al., 2004; McCallum et al., 2000) (Figure 7.2). Recently, the deletion TILLING (De-TILLING) method using fast neutron mutagenesis and a sensitive polymerase chain reaction-based detection in *Medicago truncatula* plants was used for the detection of mutants. As this De-TILLING reverse genetic strategy is independent of tissue culture and transformation, it can be applicable for crop improvement with relatively few background mutations and no exogenous DNA (Gady et al., 2009). Studies on *Arabidopsis* SOS mutants showing hypersensitiveness to growth inhibition by NaCl stress confirmed the role of regulatory genes, *SOS1*, *SOS2* and *SOS3*, controlling K + nutrition essential for ion homeostasis in plants under stress (Zhu, 2001, 2002). However,

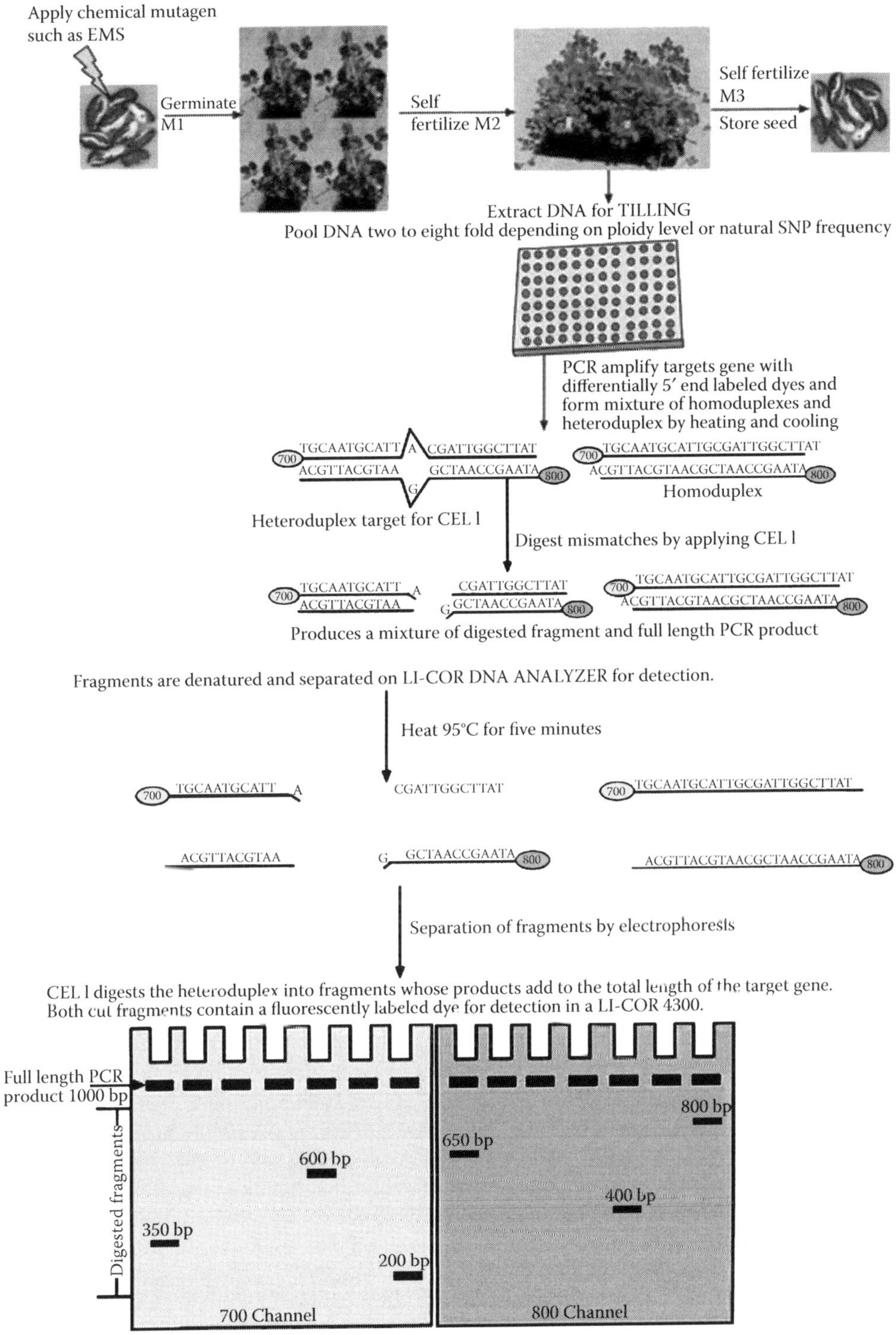

Figure 7.2 Diagrammatic representation of the TILLING method in which seeds are mutagenized with a chemical mutagen and germinated to produce M1 plants. (From Barkley, N. A. and M. L. Wang. 2008. *Curr. Genomics* 9:212–226. With permission.)

the genetic screening for identifying loci associated with abiotic stress responses in signalling can have limitations because of its inability of finding major visible phenotypes.

Post-transcriptional gene silencing via RNA interference or RNAi (Hilson et al., 2004) can reveal information on gene function by showing the effects of a drastic reduction in expression of the target gene, due to the degradation of transcripts by the Dicer/RISK pathway. Posttranscriptional gene silencing is known to be helpful for the study of gene function in *Arabidopsis* and in many other plant and metazoan species. The RNAi technology (Waterhouse and Helliwell, 2003) can relate the transcript levels of stress-responsive genes to the level of stress tolerance and provides a potential in quantitative abiotic stress studies. A collection of gene-specific sequence tags were subcloned on a large scale, into vectors designed for gene silencing, and hairpin RNA expressing lines were constructed for 21,500 *Arabidopsis* genes (Hilson et al., 2004). This study showed that in plants expression of gene-specific sequence tag hairpin RNA resulted in the expected phenotypes, in silenced *Arabidopsis* lines, providing novel and powerful tools for functional genomics.

Reverse genetic studies have been used to study the role of several genes in abiotic stress resistance. For example, plant glutathione transferases (GSTs) are induced by diverse abiotic stimuli and individual GST genes have been linked to particular stress stimuli. When a group of most highly expressed GST genes were knocked out using RNAi, no phenotypic effect was observed during stress exposure suggesting a high degree of functional redundancy within the *Arabidopsis* GST family (Sappl et al., 2009). *Cys2/His2*-type (*C2H2*) zinc-finger proteins that contain the ERF-associated amphiphilic repression domain play an important role in regulating resistance of *Arabidopsis* to abiotic stress conditions. Knockout and RNAi plants for the zinc-finger protein, *ZAT10*, were found to have enhanced tolerance to osmotic and salinity stress suggesting that *Zat10* plays a key role as both a positive and a negative regulator of plant defense (Mittler et al., 2006). Analysis of *ZAT7* RNAi lines suggested that *Zat7* functions to suppress a repressor of defense responses (Ciftci-Yilmaz et al., 2007). All these studies provided evidence that the reverse genetics approach can be used, in combination with other functional genomics approach, to improve our fragmentary knowledge about the abiotic stress response.

7.4 ROS GENE NETWORK AND ABIOTIC STRESSES

ROS play a dual role in plant biology as key regulators of biological processes, such as growth, cell cycle, programmed cell death, hormone signalling and development and defense pathways, in response to biotic and abiotic stress (Foyer and Noctor, 2005; Fujita et al., 2006; Mittler, 2002; Mittler et al., 2004; Miller and Mittler, 2006; Miller et al., 2009). A low steady-state level of ROS is maintained in the cells during the basic cycle of ROS metabolism. Different abiotic stresses, for example, drought, salinity, flooding, heat and cold stresses, result in enhanced production of highly reactive and toxic ROS. These include hydrogen peroxide (H_2O_2), superoxide anion (O_2^-), hydroxyl radicals (·OH), singlet oxygen (O_2) and nitric oxide (NO.); H_2O_2, O_2^- and OH can convert to one another and can lead to the oxidative destruction of cells (Georgiou, 2002; Mittler et al., 2004; Quan et al., 2008).

ROS signalling is modulated by a balance between different ROS-producing and ROS-scavenging pathways of the cell. A large network of genes, the 'ROS gene network', with over 152 genes in *Arabidopsis*, is responsible for tightly regulating ROS production and scavenging, to maintain the dual role for ROS in plant biology (Miller et al., 2009; Mittler et al., 2004). Furthermore, antioxidants such as ascorbate, glutathione and tocopherol also play an important role in the regulation of the cellular ROS homeostasis, by influencing gene expression associated with abiotic stresses (Figure 7.3), in addition to the different proteins and enzymes that detoxify ROS (Apel and Hurt, 2004; Foyer and Noctor, 2005).

A major source of ROS production in plant cells can be the organelles such as chloroplast and mitochondria with oxidizing metabolic activity due to a range of oxidases, pigments and electron

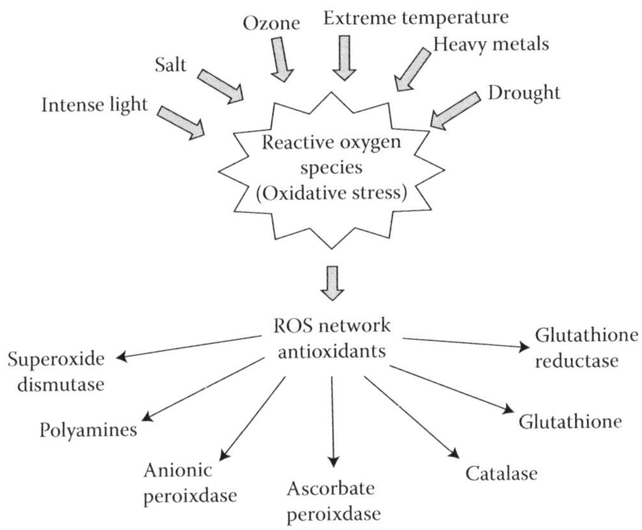

Figure 7.3 Regulation of the cellular ROS homeostasis associated with abiotic stresses.

transport chains (Apel and Hurt, 2004). Some of the major ROS-scavenging enzymes of plants include superoxide dismutase (SOD), ascorbate peroxidase (APX), catalase (CAT), glutathione peroxidase and peroxiredoxin (PrxR) and the antioxidants such as ascorbic acid and glutathione during abiotic stress exposure (Foyer and Noctor, 2005). Different redox response TFs such as *HsfA4a*, working upstream to zinc-finger family (Zat family) *WRKY* TFs, are involved in sensing the overaccumulation of ROS in the cytosol, in response to abiotic stress (Miller et al., 2009).

7.4.1 Role of TFs in Oxidative Stress and Abiotic Stress

Transcriptional regulators play a key role in modulating defense responses during abiotic stress exposure in plants (Ciftci-Yilmaz and Mittler, 2008; Mittler et al., 2006; Miller et al., 2009). Several studies involving microarray experiments and loss-of-function mutants lacking ROS-scavenging enzymes have been instrumental in the identification of the function of ROS, such as O_2, O_2^- and H_2O_2, as signalling molecules and the regulation of gene expression by these signals (Davletova et al., 2005; Lee et al., 2007; Vanderauwera et al., 2005). Several TFs such as heat-shock transcription factors (Hsfs) including members of the zinc-finger protein Zat family (*Zat12*, *Zat10* and *Zat7*), different members of the WRKY TF family, transcription coactivators such as multiprotein bridging factor 1c (*MBF1c*) have been found to be involved in ROS signalling (Miller et al., 2009; Suzuki et al., 2005).

A large family of Hsfs, with a high degree of redundancy, is known to play a role in regulating the ROS gene network in response to several abiotic stresses besides heat stress (Kotak et al., 2007). Oxidative stress-responsive Hsfs include *HsfA2*, *HsfA4a* and *HsfA8*; class An Hsfs that act as H_2O_2 sensors by gene activation has been investigated in knockdown mutants of chloroplastic CuZn-SOD *CSD2* (*csd2*) and *APX1* (*apx1*) (Miller and Mittler, 2006). Other studies have shown increased expression of *HsfA5* (a strong repressor of *HsfA4* activity) transcripts during oxidative stress, ozone treatment, salinity, heat and cold stresses (von Koskull-Doring et al., 2007; Miller and Mittler, 2006). However, further studies on double or triple mutants lacking specific Hsfs are required to investigate the different roles of specific Hsfs in regulating ROS and abiotic stress responses in plants.

Another family of transcriptional regulators, Cys2/His2-type plant-specific zinc-finger proteins (Zat family, *Zat 12*, *Zat10* and *Zat7*), are potentially involved in ROS signalling and response to abiotic stress (Miller et al., 2009; Mittler et al., 2006). Enhanced expression of *Zat12* and *Zat10* has been observed in response to osmotic, drought, salinity, temperature, oxidative, high-light stress and wounding, with *Zat12* controlling a regulon of 12 genes (Mittler et al., 2006). Increased sensitivity of knockout *Zat12 Arabidopsis* seedlings to salinity, osmotic, oxidative and heat stress (Vogel et al., 2005) was observed. The expression of *Zat7* is more specific to salt stress and heat as compared to *Zat12* and *Zat10* (Ciftci-Yilmaz et al., 2007), suggesting a high degree of complexity and specificity of the different members of the Zat family, in response to abiotic and oxidative stress.

The WRKY TFs gene family containing the invariable WRKY amino acid signature is one of the largest plant TF families, suggested to play a key role in the response of plants to abiotic and oxidative stresses (Eulgem and Somssich, 2007; Miller et al., 2009). There are three major WRKY TF proteins based on the number of WRKY domains and zinc-finger motifs (Ulker and Somssich, 2004). In *Arabidopsis*, *WRKY6* and *WRKY75* showed more than fivefold expression in response to oxidative stress (Gadjev et al., 2006). Several recent studies suggest the role of different members of WRKY TFs including *WRKY70*, *WRKY 40*, *WRKY 33*, *WRKY 25* and *WRKY 18*, as key regulators in abiotic stress defense responses and ROS stress functioning downstream or upstream to *Zat* proteins (Miller et al., 2009). Although recent studies have unraveled some of the key players in the ROS network with several key TFs and proteins, yet many questions related to its mode of regulation, during stress response remain unanswered.

7.4.2 Aluminium Resistance/Tolerance and Its Relation to ROS

Aluminium is the most limiting factor for plant productivity in acidic soils with over 50% of the world's potential arable land surface composed of acid soils (Kochian et al., 2005). The primary target of Al toxicity is the root apex, leading to poor uptake of water nutrients resulting in the reduction of root biomass (Kochian, 1995). Exudation of a variety of organic anions such as malate, citrate or oxalate upon exposure to Al has been reported which can reduce the toxic effect of Al by chelating Al in the rhizosphere (Basu et al., 1994; Kidd et al., 2001).

The accumulation of ROS during Al exposure was observed in maize, soybean and wheat (Houde and Diallo, 2008). Aluminium can inhibit a redox reaction associated with root growth and Al tolerance, leading to the accumulation of ROS and increasing the expression of antioxidant enzymes (Foyer and Noctor, 2005; Maltais and Houde, 2002). Investigations at the enzyme and gene expression levels have shown positive correlations between ROS generation and Al stress in several crop species (Kochian et al., 2005). Several genes related to oxidative stress have been found to be up-regulated during Al exposure using various molecular approaches (Basu et al., 2001, 2004; Houde and Diallo, 2008; Richards et al., 1998; Watt, 2003).

In the first large-scale, transcriptomic analysis of root responses to Al in *Arabidopsis* using a microarray, several genes related to the oxidative stress pathway and Al-responsive TFs including *AP2/EREBP*, *MYB* and *bHLH* showed increased transcript abundance (Kumari et al., 2008). Elevated levels of antioxidant systems such as catalase and APX were observed in Populus plants under increasing Al exposure (Naik et al., 2009). Another large-scale expression profiling study (Houde and Diallo, 2008) using the Affymetrix wheat genome array allowed the screening of 55,052 transcripts to identify 83 genes associated with Al stress. Of these, 25 transcripts were found to be associated with tolerance including several genes involved in the detoxification of Al and ROS (Houde and Diallo, 2008).

Genes encoding peroxidases and germin-like proteins, which are potential ROS generators, and a number of antioxidant genes including blue copper-binding proteins, glutathione reductases and glutathione *S*-transferases were found to be highly up-regulated (2.0- to 13-fold) in transcript profiling studies on root tips of an Al-sensitive genotype of *Medicago truncatula*. Besides, a gene encoding a carbohydrate oxidase, ROS generator and genes encoding a quinone oxidoreductase and a thioredoxin-like protein, which are ROS scavenging proteins, were also up-regulated in this study, suggesting that these genes may be involved in the oxidative burst response following Al stress (Chandran et al., 2008). A larger number of oxidative stress-related genes including SODs, glutathione-*S*-transferases and thioredoxin *h* were up-regulated by Al in the Al-sensitive genotype L53 than in the Al-tolerant C100-6 of maize roots suggesting that the up-regulation of oxidative stress-responsive genes could be a consequence of Al toxicity and not a component of Al-tolerant response (Maron et al., 2008). In recent studies on comparative transcriptomic analysis with Al in *Arabidopsis thaliana*, several ROS-responsive genes including ROS-scavenging enzymes (three glutathione transferases and two peroxidases), as well as *MYB15* and *tolB*-related proteins that are involved in the signal transduction pathway for ROS responses were up-regulated (Zhao et al., 2009b).

All these studies indicate a complex relationship between Al stress and different ROS signals that may work in an integrated fashion resulting in an additive effect. A new intracellular mechanism was proposed (Figure 7.4) for Al toxicity in plants cells which is associated with ion pump inhibition, lipid peroxidation and plasma membrane blebbing allowing the entry of Al into the cytosol creating respiratory dysfunction and ROS production. These ROS species lead to the opening of membrane permeability transition pore causing the loss of inner membrane potential and imbalance of Ca uptake resulting in an irreversible dysfunction of mitochondria with subsequent programmed cell in tobacco cells (Panda et al., 2008).

Continued genomics, proteomics and metabolomics efforts and other emerging technologies will be able to provide a more detailed picture of the networks involved in different ROS-related plant processes. Furthermore, these new insights into the ROS gene network and their response to

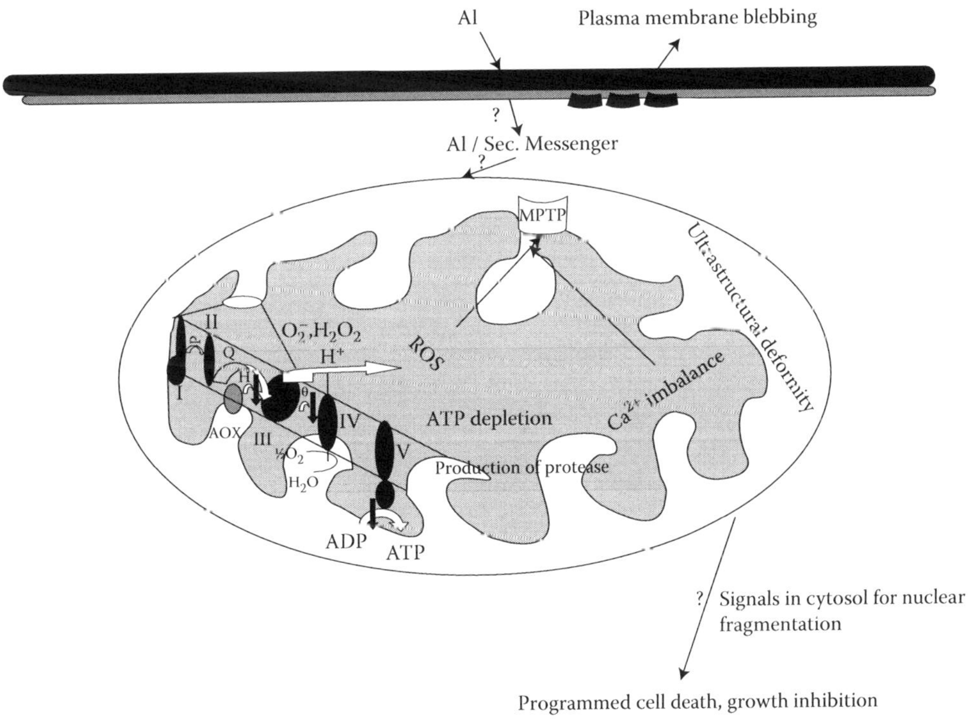

Figure 7.4 Putative model for the intracellular mechanism of aluminium toxicity in tobacco cells via mitochondria. (From Panda S. J. et al. 2008. *C.R. Biol* 331:597–610. With permission.)

various abiotic stresses might lead to the identification of genes that can ultimately be exploited to modulate ROS-related plant processes resulting in stress-tolerant plants.

7.5 PLANT MICRORNA AND ABIOTIC STRESSES

Endogenous small non-coding regulatory RNAs play a central role in mediating many biological processes, and are classified into two classes including the microRNAs (miRs) and small interfering RNAs (siRNAs) in plants (He and Hannon, 2004). Several recent studies showed important functions of miRNAs, in response to adverse abiotic stresses such as salinity, cold, drought and UV light (Floris et al., 2009; Lu and Huang, 2008; Phillips et al., 2007; Sunkar et al., 2007; Yamasaki et al., 2007; Zhao et al., 2007, 2009a). In *Arabidopsis*, abiotic stress down-regulation of miR398 caused by high light, heavy metals and oxidative stress results in an increased expression of Cu-ZnSOD (CSD) genes, *CSD1* and *CSD2* (Figure 7.5). Furthermore, transgenic *Arabidopsis* plants overexpressing miR398-resistant *CSD2* (*mCSD2*) showed more resistance to oxidative stress as compared to *CSD2* transgenic plants (Sunkar et al., 2006). On the contrary, copper deficiency up-regulates miR398 which subsequently prevents the synthesis of *CSD1* and *CSD2* by increased degradation of these transcripts. *UBC24* (*At2g33770*) encoding ubiquitin conjugating enzyme 24 is one of the large numbers of genes involved in the phosphate metabolism pathway by regulating Pi trans-

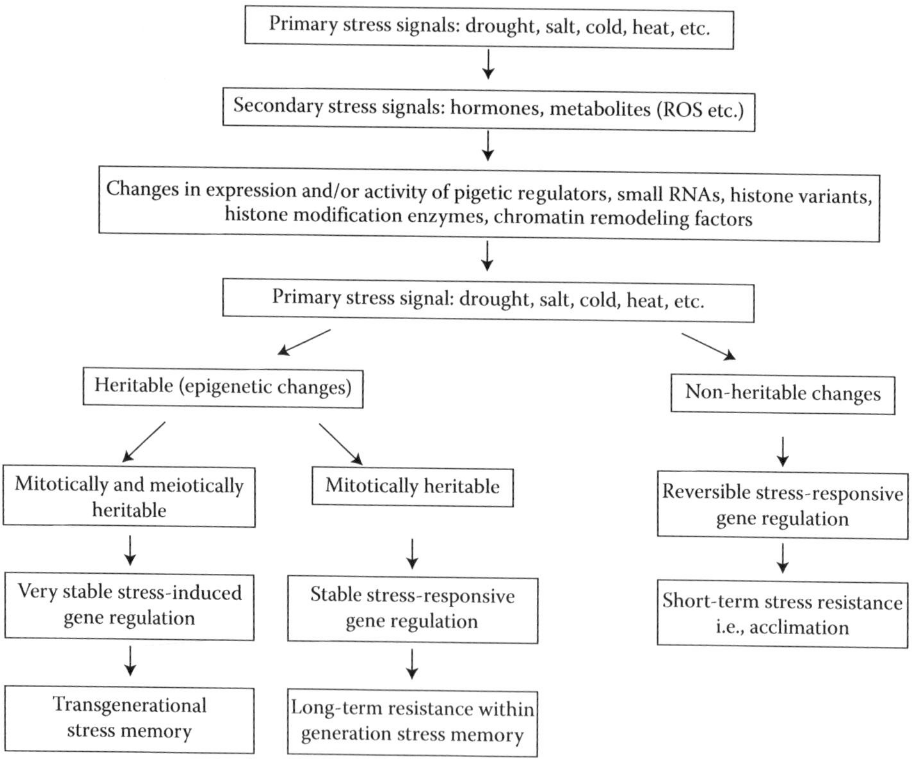

Figure 7.5 Epigenetic regulation of stress tolerance. Primary and secondary stress signals induce changes in the expression of epigenetic regulators that modify histone variants, histone modifications and DNA methylation. Some of these are heritable epigenetic modifications that provide within-generation and transgenerational stress memory. (From *Curr Opin Plant Biol.*, 12, Chinnusamy, V. and J. K. Zhu, Epigenetic regulation of stress responses in plants, 133–139, Copyright (2009), with permission from Elsevier.)

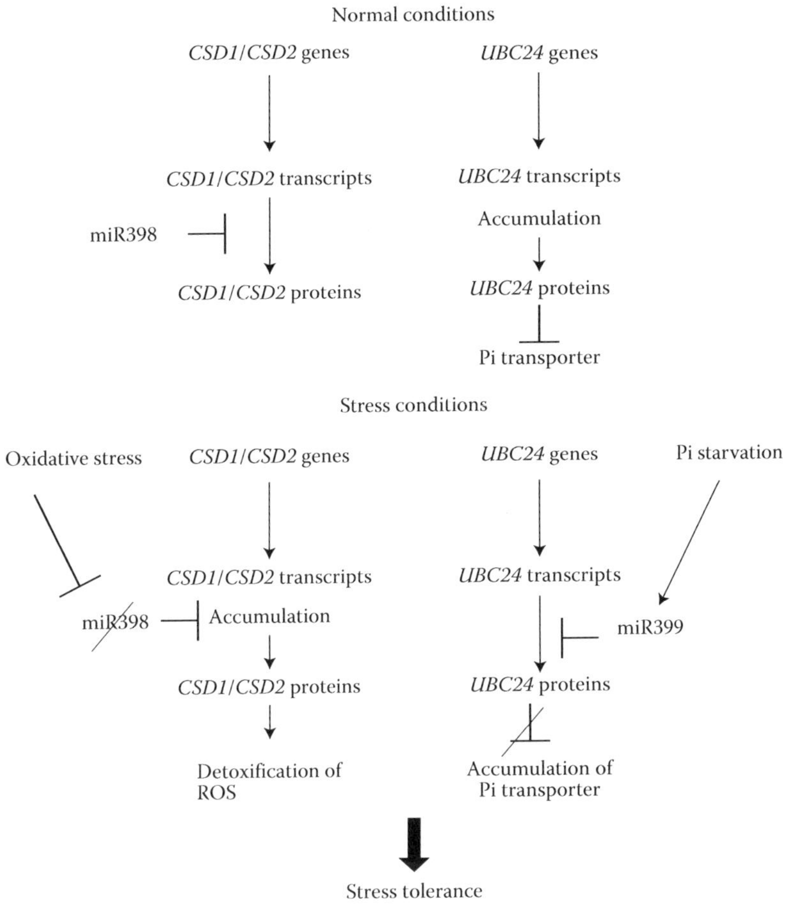

Figure 7.6 Stress conditions can induce opposite regulations of miRNAs accumulation. (From Floris, M. et al. 2009. *Int J Mol Sci* 10:3168–3185. With permission.)

porters and preventing nutrient overloading. An miR399 has five potential target sites in the 5′-untranslated region (5′UTR) of the *UBC24* transcript and induced transcription of miR399 results in reduction of the *UBC24* transcript under Pi deficiency (Figure 7.6). Overexpression of miR399 in transgenic plants did not accumulate *UBC24* mRNA under high Pi, confirming the negative regulation of *UBC24* expression (Fujii et al., 2005).

Hormonal and abiotic stress-induced microRNA, miR159 and miR160 probably function through targeting the MYB TFs, *MYB33*, *MYB65*, *MYB101* and *MYB104* (Liu et al., 2007; Reyes and Chua, 2007). Some studies have suggested that overexpression of miR159, and miR159-resistant *MYB33* and *MYB101*, resulted in ABA hypersensitivity indicating the role of miR159 in homeostasis of *MYB33* and *MYB101* mRNA levels (Reyes and Chua, 2007). Another recent study showed the up-regulation of miR417 by osmotic stress and hormones and transgenic plants overexpressing miR417 were hypersensitive to ABA; however, the target genes for miR417 still remain unclear (Jung and Kang, 2007).

In rice, drought stress up-regulated miR169g and miR393 levels and two dehydration-responsive *cis*-elements of the promoter region of miR169g, increased miR393 levels during drought. Recently, Zhao et al. (2009a) found that miR169g and miR169n were found to be induced by high salinity and identified NF-YA TF, *Os03g29760*, as a target gene of these salt-inducible miR169 members. Sanan-Mishra et al. (2009) cloned and sequenced 40 small RNAs of rice using the conventional approach, explored the expression patterns of these miRNAs under stress conditions and miR29 and miR47 are were found to be up-regulated by salt stress. MicroRNA cloning and computational EST analysis identified 72 putative miRNA sequences in *Populus euphratica* and expression analysis indicated that five miRNAs were induced by dehydration stress targeting TFs, post-translational modification factors, intracellular signal transduction factors and other proteins involved in diverse biological processes (Li et al., 2009).

All these studies provide preliminary evidence relating to cross-talk between stress signalling pathways involving TFs, miRNAs, hormones and ROS. However, the exact mechanism of all miRNAs function is still unclear, for example, the target genes for miRNAs, the cofactors involved, whether miRNAs function through translational repression or mRNA degradation. Further characterization of different miRNAs and the understanding of their role in plant responses to diverse stresses may provide a promising tool to improve plant yields by developing stress-resistant plants.

7.6 CONCLUSIONS AND FUTURE DIRECTIONS

Abiotic stresses are the primary cause of crop loss worldwide. Activation of cascades of molecular networks involved in stress perception, signal transduction and the expression of specific stress-related genes and metabolites are involved in plant adaptation to environmental stresses. The genomics era has provided us with the tools and capabilities to study abiotic stress biology with a great magnitude. Identification and engineering of genes that protect and maintain the function and structure of cellular components can provide ways to enhance resistance to stress. Further identification of stress-associated genes and interconnected pathways will enable detailed molecular dissection of abiotic stress-tolerance mechanisms in crop plants. Additionally, comprehensive profiling of stress-associated metabolites seems to be a relevant approach to the successful molecular breeding of stress-resistant plants. Recent studies integrating metabolic network analysis using metabolomic techniques, with quantitative non-targeted protein profiling have begun to add a whole new dimension for protein function analysis, offer an improved method for distinguishing among phenotypes (Wienkoop et al., 2008). Furthermore, the high-dimensional data integration from several replicates and combined with statistical analysis is a unique way to reveal responses to different abiotic stress conditions and search correlative sets of biomarkers. The genetic dissection of the quantitative traits controlling the adaptive response of crops to abiotic stress will lead to the application of genomics-based approaches to breeding programs for improving the sustainability and stability of yield under adverse conditions.

Through different genetic studies, we now understand that transcriptional regulatory network and several other molecular components and processes are very crucial to phenotype manifestation in studies on abiotic stress resistance. Recently, Glycine-rich RBPs characterized by RNA recognition motif and the K-homology domain have emerged as the regulatory factors of RNA processing in stress responses. These RBPs could potentially function by increasing the expression of RNA or help in preserving the stability for mRNA expressed during stress conditions (Lorkovic, 2009). Besides, abiotic stress-induced heritable epigenetic changes might provide better adaptation to different stresses, by providing within-generation and transgenerational stress memory (Chinnusamy and Zhu, 2009; Figure 7.5). The functional genomics studies will be able to provide connective studies on phenotypic and physiological changes using transgenics and gene inactivation techniques, comparative transcriptomics and transcript regulation, and the role of different proteins, protein complexes and metabolites in association with the phenotype analysis. A systems biology holistic approach by integrating genomics, proteomics and metabolomics with statistics, and bioinformatics will be necessary to advance the understanding of plant stress responses and resistance mechanisms.

REFERENCES

Aghaei, K., A. A. Ehsanpour and S. Komatsu. 2008. Proteome analysis of potato under salt stress. *J Proteome Res* 7:4858–4868.

Aghaei, K., A. A. Ehsanpour and S. Komatsu. 2009. Proteome analysis of soybean hypocotyl and root under salt stress. *Amino Acids* 36:91–98.

Amtmann, A., H. J. Bohnert and R. A. Bressan. 2005. Abiotic stress and plant genome evolution. Search for new models. *Plant Physiol* 138:127–130.

Amtmann, A. 2009. Learning from evolution: *Thellungiella* generates new knowledge on essential and critical components of abiotic stress tolerance in plants. *Mol Plant* 2:3–12.

Apel, K. and H. Hirt. 2004. Reactive oxygen species: Metabolism, oxidative stress, and signal transduction. *Annu Rev Plant Biol* 55:373–399.

Barkley, N. A. and M. L. Wang. 2008. Application of TILLING and EcoTILLING as reverse genetic approaches to elucidate the function of genes in plants and animals. *Curr Genomics* 9:212–226.

Bartels, D. and R. Sunkar. 2005. Drought and salt tolerance in plants. *Crit Rev Plant Sci* 24:23–58.

Basu, U., D. Goldbold and G. J. Taylor. 1994. Aluminum resistance in *Triticum aestivum* associated with enhanced exudation of malate. *J Plant Physiol* 144:747–753.

Basu, U., A. G. Good and G. J. Taylor. 2001. Transgenic *Brassica napus* plants overexpressing aluminium-induced mitochondrial manganese superoxide dismutase cDNA are resistant to aluminium. *Plant Cell Environ* 24:1269–278.

Basu, U., J. L. Southron, J. L. Stephens and G. J. Taylor. 2004. Reverse genetic analysis of the glutathione metabolic pathway suggests a novel role of PHGPX and URE2 genes in aluminum resistance in *Saccharomyces cerevisiae. Mol Genet Genomics* 271:627–637.

Behnam, B., A. Kikuchi, F. Celebi-Toprak et al. 2006. The *Arabidopsis* DREB1A gene driven by the stress-inducible rd29A promoter increases salt-stress tolerance in proportion to its copy number in tetrasomic tetraploid potato (*Solanum tuberosum*). *Plant Biotech* 23:169–177.

Bhatnagar-Mathur, P., V. Vadez and K. K. Sharma. 2008. Transgenic approaches for abiotic stress tolerance in plants: Retrospects and prospects. *Plant Cell Rep* 27:411–424.

Blum, A. 2004. The physiological foundation of crop breeding for stress environments. *In Proceedings of World Rice Research Conference*, pp. 456–458, Tsukuba, Japan, November 2004. Manila, The Philippines: International Rice Research Institute.

Bohnert, H. J., Q. Gong, P. Li and S. Ma. 2006. Unraveling abiotic stress tolerance mechanisms-getting genomics going. *Curr Opin Plant Biol* 9:180–188.

Bräutigam, M., A. Lindlöf, S. Zakhrabekova et al. 2005. Generation and analysis of 9792 EST sequences from cold acclimated oat, *Avena sativa*. *BMC Plant Biol* 5:18.

Bray, E. A., J. Bailey-Serres and E. Weretilnyk. 2000. Responses to abiotic stresses. In *Biochemistry and Molecular Biology of Plants*, eds. B. B. Buchanan, W. Gruissem and R. L. Jones, pp. 1158–1203, American Society of Plant Physiologists: Rockville, MD.

Bressan, R., H. Bohnert and J. K. Zhu. 2009. Abiotic stress tolerance: From gene discovery in model organisms to crop improvement. *Mol Plant* 2:1–2.

Brosché, M., B. Basia Vinocur, E. R. Alatalo et al. 2005. Gene expression and metabolite profiling of *Populus euphratica* growing in the Negev desert. *Genome Biol* 6:R101.

Chandran, D., N. Sharopova, S. Ivashuta et al. 2008. Transcriptome profiling identified novel genes associated with aluminum toxicity, resistance and tolerance in *Medicago truncatula*. *Planta* 228:151–166.

Chen, W., N. J. Provart, J. Glazebrook et al. 2002. Expression profile matrix of *Arabidopsis* transcription factor genes suggests their putative functions in response to environmental stresses. *Plant Cell* 14:559–574.

Chen, W. Q. J. and T. Zhu. 2004. Networks of transcription factors with roles in environmental stress response. *Trends Plant Sci* 9:591–596.

Cheng, W. H., A. Endo, L. Zhou et al. 2002. A unique short-chain dehydrogenase/reductase in *Arabidopsis* glucose signaling and abscisic acid biosynthesis and functions. *Plant Cell* 14:2723–2743.

Cheong, Y., H. Chang, R. Gupta et al. 2002. Transcriptional profiling reveals novel interactions between wounding, pathogen, abiotic stress, and hormonal response in *Arabidopsis. Plant Physiol* 129:661–677.

Chinnusamy, V., A. Jagendorf and J. K. Zhu. 2005. Understanding and improving salt tolerance in plants. *Crop Sci* 45:437–448.

Chinnusamy, V. and J. K. Zhu. 2009. Epigenetic regulation of stress responses in plants. *Curr Opin Plant Biol* 12:133–139.

Ciftci-Yilmaz, S. and R. Mittler. 2008. The zinc finger network of plants. *Cell Mol Life Sci* 65:1150–1160.

Ciftci-Yilmaz, S., M. R. Morsy, L. Song et al. 2007. The EAR-motif of the Cys2/His2-type zinc finger protein Zat 7 plays a key role in the defense response of *Arabidopsis* to salinity stress. *J Biol Chem* 282:9260–9268.

Clarke, J. D. and Y. Zhu. 2006. Microarray analysis of the transcriptome as a stepping stone towards understanding biological systems: Practical considerations and perspectives. *Plant J* 45:630–650.

Close, T. J., S. I. Wanamaker, R. A. Caldo et al. 2004. A new resource for cereal genomics: 22K barley GeneChip comes of age. *Plant Physiol* 134:960–968.

Collins, N. C., F. Tardieu and R. Tuberosa. 2008. Quantitative trait loci and crop performance under abiotic stress: Where do we stand? *Plant Physiol* 147:469–486.

Cortina, C. and F. Culiáñez-Maciá. 2005. Tomato abiotic stress enhanced tolerance by trehalose biosynthesis. *Plant Sci* 169:75–82.

Cushman, J. C. and H. J. Bohnert. 2000. Genomic approaches to plant stress tolerance. *Curr Opin Plant Biol* 3:117–124.

Davletova, S., K. Schlauch, J. Coutu and R. Mittler. 2005. The zinc-finger protein Zat12 plays a central role in reactive oxygen and abiotic stress signaling in *Arabidopsis. Plant Physiol* 139:847–856.

Degand, H., A. M. Faber, N. Dauchot et al. 2009. Proteomic analysis of chicory root identifies proteins typically involved in cold acclimation. *Proteomics* 9:2903–2907.

Deng, X., J. Phillips, A. H. Meijer, F. Salamini and D. Bartels. 2002. Characterization of five novel dehydration-responsive homeodomain leucine zipper genes from the resurrection plant *Craterostigma plantagineum. Plant Mol Biol* 49:601–610.

Eulgem, T. and I. E. Somssich. 2007. Networks of WRKY transcription factors in defense signaling. *Curr Opin Plant Biol* 10:366–371.

Floris, M., H. Mahgoub, E. Lanet, C. Robaglia and B. Menand. 2009. Post-transcriptional regulation of gene expression in plants during abiotic Stress. *Int J Mol Sci* 10:3168–3185.

Foyer, C. H. and G. Noctor, 2005. Redox homeostasis and antioxidant signaling: A metabolic interface between stress perception and physiological responses. *Plant Cell* 17:1866–1875.

Fujii, H., T. J. Chiou, S. I. Lin, K. Aung and J. K. Zhu. 2005. A miRNA involved in phosphate-starvation response in *Arabidopsis. Curr Biol* 15:2038–2043.

Fujita, M., Y. Fujita, Y. Noutoshi et al. 2006. Crosstalk between abiotic and biotic stress responses: A current view from the points of convergence in the stress signaling networks. *Curr Opin Plant Biol* 9:436–442.

Gadjev, I., S. Vanderauwera, T. S. Gechev et al. 2006. Transcriptomic footprints disclose specificity of reactive oxygen species signaling in *Arabidopsis. Plant Physiol* 141:436–445.

Gady, A. L. F., F. W. K. Hermans, M. H. B. J. van de Wal et al. 2009. Implementation of two high throughput techniques in a novel application: Detecting point mutations in large EMS mutated plant populations. *Plant Methods* 5:13–27.

Garg, A. K., J. K. Kim, T. G. Owens et al. 2002. Trehalose accumulation in rice plants confers high tolerance levels to different abiotic stresses. *Proc Natl Acad Sci USA* 99:15898–15903.

Georgiou, G. 2002. How to flip the (redox) switch. *Cell* 111:607–610.

Glinski, M. and W. Weckwerth. 2006. The role of mass spectrometry in plant systems biology. *Mass Spectrom Rev* 25:173–214.

Gong, W., Y. P. Shen, L. G. Ma et al. 2004. Genome-wide ORFeome cloning and analysis of *Arabidopsis* transcription factor genes. *Plant Physiol* 135:773–782.

Goyal, K., L. J. Walton and A. Tunnacliffe. 2005. LEA proteins prevent protein aggregation due to water stress. *Biochem J* 388:151–157.

Griffith, M., M. Timonin, C. E. Wong et al. 2007 *Thellungiella*: An *Arabidopsis*-related model plant adapted to cold temperatures. *Plant Cell Environ* 30:529–538.

Gruber, V., S. Blanchet, A. Diet et al. 2009. Identification of transcription factors involved in root apex responses to salt stress in *Medicago truncatula*. *Mol Genet Genomics* 281:55–66.

Gygi, S. P., Y. Rochon, B. R. Franza and R. Aebersold. 1999. Correlation between protein and mRNA abundance in yeast. *Mol Cell Biol* 19:1720–1730.

Hasegawa, P. M., R. A. Bressan, J. K. Zhu and H. J. Bohnert. 2000. Plant cellular and molecular responses to high salinity. *Annu Rev Plant Physiol Plant Mol Biol* 51:463–499.

He, L. and G. J. Hannon. 2004. MicroRNAs: Small RNAs with a big role in gene regulation. *Nat Rev Genet* 5:522–531.

Henikoff, S., B. J. Till and L. Comai. 2004. TILLING. Traditional mutagenesis meets functional genomics. *Plant Physiol* 135:630–636.

Hilson, P., J. Allemeersch, T. Altmann et al. 2004. Versatile gene-specific sequence tags for *Arabidopsis* functional genomics: Transcript profiling and reverse genetics applications. *Genome Res* 14:2176–2189.

Holmstrom, K. O., S. Somersalo, A. Mandal, E. T. Palva and B. Welin. 2000. Improved tolerance to salinity and low temperature in transgenic tobacco producing glycine betaine. *J Exp Bot* 51:177–185.

Houde, M. and A. O. Diallo. 2008. Identification of genes and pathways associated with aluminum stress and tolerance using transcriptome profiling of wheat near-isogenic lines. *BMC Genomics* 9:400–412.

Houde, M., M. Belcaid, F. Ouellet et al. 2006. Wheat EST resources for functional genomics of abiotic stress. *BMC Genomics* 7:149–171.

Inan, G., Q. Zhang, P. Li et al. 2004. Salt cress. A halophyte and cryophyte *Arabidopsis* relative model system and its applicability to molecular genetic analyses of growth and development of extremophiles. *Plant Physiol* 135:1718–1737.

Ishitani, M., L. Xiong, B. Stevenson and J. K. Zhu. 1997. Genetic analysis of osmotic and cold stress signal transduction in *Arabidopsis*: Interactions and convergence of abscisic acid-dependent and abscisic acid-independent pathways. *Plant Cell* 9:1935–1949.

Jaglo, K. R., S. Kleff, K. L. Amundsen et al. 2001. Components of the *Arabidopsis* C-repeat/dehydration-responsive element binding factor cold-response pathway are conserved in *Brassica napus* and other plant species. *Plant Physiol* 127:910–917.

Jaglo-Ottosen, K. R., S. J. Gilmour, D. G. Zarka, O. Schabenberger and M. F. Thomashow. 1998. *Arabidopsis* CBF1 overexpression induces *cor* genes and enhances freezing tolerance. *Science* 280:104–106.

Jiang, Y. and M. K. Deyholos. 2009. Functional characterization of *Arabidopsis* NaCl-inducible *WRKY25* and *WRKY33* transcription factors in abiotic stresses. *Plant Mol Biol* 69:91–105.

Jiang, Y., B. Yang, N. S. Harris and M. K. Deyholos. 2007. Comparative proteomic analysis of NaCl stress-responsive proteins in *Arabidopsis* roots. *J Exp Bot* 58:3591–3607.

Jung, H. J. and H. Kang. 2007. Expression and functional analyses of microRNA417 in *Arabidopsis thaliana* under stress conditions. *Plant Physiol Biochem* 45:805–811.

Kagaya, Y., T. Hobo, M. Murata, A. Ban and T. Hattori. 2002. Abscisic acid induced transcription is mediated by phosphorylation of an abscisic acid response element binding factor, TRAB1. *Plant Cell* 14:3177–3189.

Kahvejian, A., J. Quackenbush and J. F. Thompson. 2008. What would you do if you could sequence everything? *Nat Biotech* 26:1125–1133.

Kang, J. Y., H. I. Choi, M. Y. Im and S. Y. Kim. 2002. *Arabidopsis* basic leucine zipper proteins that mediate stress-responsive abscisic acid signaling. *Plant Cell* 14:343–357.

Kant P., M. Gordon, S. Kant et al. 2008a. Functional-genomics-based identification of genes that regulate *Arabidopsis* responses to multiple abiotic stresses. *Plant Cell Environ* 31:697–714.

Kant, S., P. Kant, E. Raveh and S. Barak. 2006. Evidence that differential gene expression between the halophyte, *Thellungiella halophila*, and *Arabidopsis thaliana* is responsible for higher levels of the compatible osmolyte proline and tight control of Na^+ uptake in *T. halophila*. *Plant Cell Environ* 29:1220–1234.

Kant, S., Y. Bi, E. Weretilnyk, S. Barak and S. J. Rothstein. 2008b. The *Arabidopsis* halophytic relative *Thellungiella halophila* tolerates nitrogen-limiting conditions by maintaining growth, nitrogen uptake, and assimilation. *Plant Physiol* 147:1168–1180.

Kasuga, M., S. Miura, K. Shinozaki and K. Yamaguchi-Shinozaki. 2004. A combination of the *Arabidopsis* DREB1A gene and stress inducible rd29A promoter improved drought- and low-temperature stress tolerance in tobacco by gene transfer. *Plant Cell Physiol* 45:346–350.

Kav, N. N. V., S. Srivastava, L. Goonewardene and S. F. Blade. 2004. Proteome-level changes in the roots of *Pisum sativum* in response to salinity. *Ann Appl Biol* 145:217–230.

Kidd, P. S., M. Llugany, C. Poschenrieder, B. Gunse and J. Barcelo. 2001. The role of root exudates in aluminium resistance and silicon-induced amelioration of aluminium toxicity in three varieties of maize (*Zea mays* L.). *J Exp Bot* 52:1339–1352.

Kilian, J., D. Whitehead, J. Horak et al. 2007. The AtGenExpress global stress expression data set: Protocols, evaluation and model data analysis of UV-B light, drought and cold stress responses. *Plant J* 50:347–363.

Kobayashi, F., E. Maeta, A. Terashima et al. 2008. Development of abiotic stress tolerance via bZIP-type transcription factor LIP19 in common wheat. *J Exp Bot* 59:891–905.

Kochian, L. V. 1995. Cellular mechanisms of aluminum toxicity and resistance in plants. *Annu Rev Plant Physiol Plant Mol Biol* 46:237–260.

Kochian, L. V., M. A. Pineros and O. A. Hoekenga. 2005. The physiology, genetics and molecular biology of plant aluminum resistance and toxicity. *Plant Soil* 274:175–195.

Kotak, S., J. Larkindale, J., U. Lee et al. 2007. Complexity of the heat stress response in plants. *Curr Opin Plant Biol* 10:310–316.

Kreps, J. A., Y. J. Wu, H. S. Chang et al. 2002.Transcriptome changes for *Arabidopsis* in response to salt, osmotic, and cold stress. *Plant Physiol* 130:2129–2141.

Krysan, P. J., J. C. Young and M. R. Sussman. 1999. T-DNA as an insertional mutagen in *Arabidopsis. Plant Cell* 11:2283–2290.

Kumari, M., G. J. Taylor and M. K. Deyholos. 2008. Transcriptomic responses to aluminum stress in roots of *Arabidopsis thaliana. Mol Genet Genomics* 279:339–357.

Lee, K. P., C. Kim, F. Landgraf and K. Apel. 2007. EXECUTER1- and EXECUTER2-dependent transfer of stress-related signals from the plastid to the nucleus of *Arabidopsis thaliana. Proc Natl Acad Sci USA* 104:10270–10275.

Lee, D. G., N. Ahsan, S. H. Lee et al. 2009. Chilling stress-induced proteomic changes in rice roots. *J Plant Physiol* 166:1–11.

Leung, H. 2008. Stressed genomics—Bringing relief to rice fields. *Curr Opin Plant Biol* 11:201–208.

Li, B., W. Yin and X. Xia. 2009. Identification of microRNAs and their targets from *Populus euphratica. Biochem Biophys Res Commun* 388:272–277.

Liu, F., T. Vantoai, L. P. Antoai et al. 2005. Global transcription profiling reveals comprehensive insight into hypoxic response in *Arabidopsis. Plant Physiol* 137:1115–1129.

Liu, P. P., T. A. Montgomery, N. Fahlgren et al. 2007. Repression of AUXIN RESPONSE FACTOR10 by microRNA160 is critical for seed germination and post-germination stages. *Plant J* 52:133–146.

Lorkovic, Z. J. 2009. Role of plant RNA-binding proteins in development, stress response and genome organization. *Trends Plant Sci* 14:229–236.

Lu, X. Y. and X. L. Huang. 2008. Plant miRNAs and abiotic stress responses. *Biochem Biophys Res Commun* 368:458–462.

Maltais, K. M. and M. Houde. 2002. A new biochemical marker for aluminium tolerance in plants. *Physiol Plant* 115:81–86.

Mantri, N. L., R. Ford, T. Coram and E. Pang. 2007. Transcriptional profiling of chickpea genes differentially regulated in response to high-salinity, cold and drought. *BMC Genomics* 8:303–317.

Maron, L. G., M. Kirst, C. Mao et al. 2008. Transcriptional profiling of aluminum toxicity and tolerance responses in maize roots. *New Phytol* 179:116–128.

Maruyama, K., Y. Sakuma, M. Kasuga et al. 2004. Identification of cold-inducible downstream genes of the *Arabidopsis* DREB1A/CBF3 transcriptional factor using two microarray systems. *Plant J* 38:982–993.

McCallum, C. M., L. Comai, E. A. Greene and S. Henikoff. 2000. Targeting Induced Local Lesion IN genomes (TILLING) for plant functional genomics. *Plant Physiol* 123:439–442.

Mian, M. R., Y. Zhang, Z. Wang et al. 2008. Analysis of tall fescue ESTs representing different abiotic stresses, tissue types and developmental stages. *BMC Plant Biol* 8:27–39.

Miller, G. and R. Mittler. 2006. Could heat shock transcription factors function as hydrogen peroxide sensors in plants? *Ann Bot* 98:279–288.

Miller, G., N. Suzuki, S. Ciftci-Yilmaz and R. Mittler. 2009. Reactive oxygen species homeostasis and signaling during drought and salinity stresses. *Plant Cell Environ* DOI: 10.1111/j.1365–3040.2009.02041.x

Mittler, R. 2002. Oxidative stress, antioxidants and stress tolerance. *Trends Plant Sci* 7:405–410.

Mittler, R. 2006. Abiotic stress, the field environment and stress combination. *Trends Plant Sci* 11:15–19.

Mittler, R., S. Vanderauwera, M. Gollery and F. Van Breusegem. 2004. Reactive oxygen gene network of plants. *Trends Plant Sci* 9:490–498.

Mittler, R., Y. Kim, L. Song et al. 2006. Gain- and loss-of-function mutations in Zat10 enhance the tolerance of plants to abiotic stress. *FEBS Lett* 580:6537–6542.

Naik, D., E. Smith and J. R. Cumming. 2009. Rhizosphere carbon deposition, oxidative stress and nutritional changes in two poplar species exposed to aluminum. *Tree Physiol* 29:423–436.

Nakashima, K., Y. Ito and K. Yamaguchi-Shinozaki. 2009. Transcriptional regulatory networks in response to abiotic stresses in *Arabidopsis* and Grasses. *Plant Physiol* 149:88–95.

Oh, S. J., S. I. Song, Y. S. Kim et al. 2005. *Arabidopsis* CBF3/DREB1A and ABF3 in transgenic rice increased tolerance to abiotic stress without stunting growth. *Plant Physiol* 138:341–351.

Panda, S. J., Y. Yamamoto, H. Kondo and H. Matsumoto. 2008. Mitochondrial alterations related to programmed cell death in tobacco cells under aluminium stress. *C R Biol* 331:597–610.

Parinov, S., M. Sevugan, D. Ye et al. 1999. Analysis of flanking sequences from Dissociation insertion lines: A database for reverse genetics in *Arabidopsis. Plant Cell* 11:2263–2270.

Park, J. M., C. J. Park, S. B. Lee et al. 2001. Overexpression of the tobacco Tsi1 gene encoding an EREBP/AP2-type transcription factor enhances resistance against pathogen attack and osmotic stress in tobacco. *Plant Cell* 13:35–1046.

Pellegrineschi, A., M. Reynolds and M. Pacheco. 2004. Stress-induced expression in wheat of the *Arabidopsis thaliana* DREB1A gene delays water stress symptoms under greenhouse conditions. *Genome* 47:493–500.

Phillips, J. R., T. Dalmay and D. Bartels. 2007. The role of small RNAs in abiotic stress. *FEBS Lett* 581:3592–3597.

Quan, L. J., B. Zhang, W. W. Shi and H. Y. Li. 2008. Hydrogen peroxide in plants: A versatile molecule of the reactive oxygen species network. *J Integr Plant Biol* 50:2–18.

Reyes. J. L. and N. H. Chua. 2007. ABA induction of miR159 controls transcript levels of two MYB factors during *Arabidopsis* seed germination. *Plant J* 49:592–606.

Richards, K. D., E. J. Schott, Y. K. Sharma, K. R. Davis and R. C. Gardner. 1998. Aluminum induces oxidative stress genes in *Arabidopsis thaliana. Plant Physiol* 116:409–418.

Robinson, S. J. and I. A. P. Parkin. 2008. Differential SAGE analysis in *Arabidopsis* uncovers increased transcriptome complexity in response to low temperature. *BMC Genomics* 9:434–451.

Rose, J., S. Bashir, J. J. Giovannoni and R. S. Saravanan. 2004. Tackling the plant proteome: Practical approaches, hurdles and experimental tools. *Plant J* 39:715–733.

Sakamoto, A., R. A. Valverde, T. H. Chen and N. Murata. 2000. Tranformation of *Arabidopsis* with the codA gene for choline oxidase enhances freezing tolerance of plants. *Plant J* 22:449–453.

Sakuma, Y., K. Maruyama, Y. Osakabe et al. 2006. Functional analysis of an *Arabidopsis* transcription factor, DREB2A, involved in drought-responsive gene expression. *Plant Cell* 16:6616–6628.

Sanan-Mishra, N., V. Kumar, S. K. Sopory and S. K. Mukherjee. 2009. Cloning and validation of novel miRNA from basmati rice indicates cross talk between abiotic and biotic stresses. *Mol Genet Genomics* 282:463–474.

Sappl, P. G., A. J. Carroll, R. Clifton et al. 2009. The *Arabidopsis* glutathione transferase gene family displays complex stress regulation and co-silencing multiple genes results in altered metabolic sensitivity of oxidative stress. *Plant J* 58:53–68.

Seki, A., K. Kamei, K.Yamaguchi-Shinozaki and K.Shinozaki. 2003. Molecular responses to drought, salinity and frost: Common and different paths for plant protection. *Curr Opin Biotech* 14:194–199.

Seki, M., M. Narusaka, J. Ishida et al. 2002.Monitoring the expression profiles of 7,000 Arabidopsis genes under drought, cold and high-salinity stresses using a full-length cDNA microarray. *Plant J* 31:279–292.

Shendure, J. and H. Ji. 2008. Next generation DNA sequencing. *Nat Biotech* 26:1135–1145.

Shinozaki, K., K. Yamaguchi-Shinozaki and M. Seki. 2003. Regulatory network of gene expression in the drought and cold stress responses. *Curr Opin Plant Biol* 6:410–417.

Shou, H., P. Bordallo and K. Wang. 2004. Expression of the *Nicotiana* protein kinase (NPK1) enhanced drought tolerance in transgenic maize. *J Exp Bot* 55:1013–1019.

Sreenivasulu, N., S. K. Sopory and P. B. Kavi Kishor. 2007. Deciphering the regulatory mechanisms of abiotic stress tolerance in plants by genomic approaches. *Gene* 388:1–13.

Sunkar, R., V. Chinnusamy, J. Zhu and J. K. Zhu. 2007. Small RNAs as big players in plant abiotic stress responses and nutrient deprivation. *Trends Plant Sci* 12:301–309.

Sunkar, R., A. Kapoor and J. K. Zhu. 2006. Posttranscriptional induction of two Cu/Zn superoxide dismutase genes in *Arabidopsis* is mediated by down regulation of miR398 and important for oxidative stress tolerance. *Plant Cell* 18:2051–2065.

Suzuki, N., L. Rizhsky, H. Liang et al. 2005. Enhanced tolerance to environmental stress in transgenic plants expressing the transcriptional co-activator multiprotein bridging factor 1c. *Plant Physiol* 139:1313–1322.

Swindell, W. R. 2006. The association among gene expression responses to nine abiotic stress treatments in *Arabidopsis thaliana. Genetics* 174:1811–1824.

Tada, Y. and T. Kashimura. 2009. Proteomic analysis of salt-responsive proteins in the mangrove plant, *Bruguiera gymnorhiza. Plant Cell Physiol* 50:439–446.

Takahashi, S., M. Seki, J. Ishida et al. 2004. Monitoring the expression profiles of genes induced by hyperosmotic, high salinity, and oxidative stress and abscisic acid treatment in *Arabidopsis* cell culture using a full-length cDNA microarray. *Plant Mol Biol* 56:29–55.

Ulker, B. and I. E. Somssich. 2004. WRKY transcription factors: From DNA binding towards biological function. *Curr Opin Plant Biol* 7:491–498.

Umezawa, T., M. Fujita, Y. Fujita, K. Yamaguchi-Shinozaki and K. Shinozaki. 2006. Engineering drought tolerance in plants: Discovering and tailoring genes to unlock the future. *Curr Opin Plant Biotech* 17:113–122.

Umezawa, T., R. Yoshida, K. Maruyama, K. Yamaguchi-Shinozaki and K. Shinozaki. 2004. SRK2C, a SNF1-related protein kinase 2, improves drought tolerance by controlling stress-responsive gene expression in *Arabidopsis thaliana. Proc Natl Acad Sci USA* 101:17306–17311.

Uno, Y., T. Furihata, H. Abe et al. 2000. *Arabidopsis* basic leucine zipper transcription factors involved in an abscisic acid-dependent signal transduction pathway under drought and high-salinity conditions. *Proc Natl Acad Sci USA* 97:11632–11637.

Urnov, F. D. 2003. Chromatin remodeling as a guide to transcriptional regulatory networks in mammals. *J Cell Biochem* 88:684–694.

Valliyodan, B. and H. T. Nguyen. 2006. Understanding regulatory networks and engineering for enhanced drought tolerance in plants. *Curr Opin Plant Biotech* 9:189–195.

Vanderauwera, S., P. Zimmermann, S. Rombauts et al. 2005. Genome-wide analysis of hydrogen peroxide-regulated gene expression in *Arabidopsis* reveals a high light-induced transcriptional cluster involved in anthocyanin biosynthesis. *Plant Physiol* 139:806–821.

Vincour, B. and A. Altman. 2005. Recent advances in engineering plant tolerance to abiotic stress: Achievements and limitations. *Curr Opin Biotech* 16:123–132.

Vogel, J. T., D. G. Zarka, H. A. Van Buskirk, S. G. Fowler and M. F. Thomashow. 2005. Roles of the CBF2 and ZAT12 transcription factors in configuring the low temperature transcriptome of *Arabidopsis. Plant J* 41:195–211.

Von Koskull-Doring, P., K. D. Scharf and L. Nover. 2007. The diversity of plant heat stress transcription factors. *Trends Plant Sci* 12:452–457.

Wang, H., Z. Huang, Q. Chen et al. 2004a. Ectopic overexpression of tomato JERF3 in tobacco activates downstream gene expression and enhances salt tolerance. *Plant Mol Biol* 55:183–192.

Wang, Z. I., P. H. Li, M. Fredricksen et al. 2004b. Expressed sequence tags from *Thellungiella halophila*, a new model to study plant salt-tolerance. *Plant Sci* 166:609–616.

Wang, W., Y. Wu and Y. Li. 2009a. A large insert *Thellungiella halophila* BIBAC library for genomics and identification of stress tolerance genes. *Plant Mol Biol* DOI: 10.1007/s11103–009–9553–3.

Wang, X., P. F. Yang, L. Liu et al. 2009b. Exploring the mechanism of *Physcomitrella patens* desiccation tolerance through a proteomic strategy. *Plant Physiol* 149:1739–1750.

Waterhouse, P. M. and C. A. Helliwell. 2003. Exploring plant genomes by RNA-induced gene silencing. *Nat Rev Genet* 4:29–38.

Watt, D. A. 2003. Aluminium-responsive genes in sugarcane: Identification and analysis of expression under oxidative stress. *J Exp Bot* 54:1163–1174.

Weltmeier, F., F. Rahmani, A. Ehlert et al. 2009. Expression patterns within the *Arabidopsis* C/S1 bZIP transcription factor network: Availability of heterodimerization partners controls gene expression during stress response and development. *Plant Mol Biol* 69:107–119.

Wienkoop, S., K. Morgenthal, F. Wolschin et al. 2008. Integration of metabolomic and proteomic phenotypes: Analysis of data covariance dissects starch and RFO metabolism from low and high temperature compensation response in *Arabidopsis thaliana. Mol Cell Proteomics* 7:1725–1736.

Witcombe, J. R., P. A. Hollington, C. J. Howarth, S. Reader and K. A. Steele. 2008. Breeding for abiotic stresses for sustainable agriculture. *Philos T Royal Soc B* 363:703–716.

Witzel, K., A. Weidner, G. Surabhi, A. Borner and H. P. Mock. 2009. Salt stress-induced alterations in the root proteome of barley genotypes with contrasting response towards salinity. *J Exp Bot* 60:3545–3557.

Wong, C. E., Y. Li, A. Labbe et al. 2006. Transcriptional profiling implicates novel interactions between abiotic stress and hormonal responses in *Thellungiella*, a close relative of *Arabidopsis. Plant Physiol* 140:1437–1450.

Wong, C. E., Y. Li, B. R. Whitty et al. 2005. Expressed sequence tags from the Yukon ecotype of *Thellungiella* reveal that gene expression in response to cold, drought and salinity shows little overlap. *Plant Mol Biol* 58:561–574.

Xiang, Y., N. Tang, H. Du, H. Ye and L. Xiong. 2008. Characterization of OsbZIP23 as a key player of the basic Leucine Zipper transcription factor family for conferring abscisic acid sensitivity and salinity and drought tolerance in rice. *Plant Physiol* 148:1938–1952.

Xiong, L. M., M. Ishitani and J. K. Zhu. 1999. Interaction of osmotic stress, temperature, and abscisic acid in the regulation of gene expression in *Arabidopsis. Plant Physiol* 119:205–211.

Yamada, M., H. Morishita, K. Urano et al. 2005. Effects of free proline accumulation in petunias under drought stress. *J Exp Bot* 56:1975–1981.

Yamaguchi-Shinozaki, K. and K. Shinozaki. 2006. Transcriptional regulatory networks in cellular responses and tolerance to dehydration and cold stresses. *Ann Rev Plant Biol* 57:781–803.

Yamasaki, H., S. E. Abdel-Ghany, C. M. Cohu et al. 2007. Regulation of copper homeostasis by micro-RNA in Arabidopsis. *J Biol Chem* 282:16369–16378.

Yan, S., Z. Tang, W. Su and W. Sun. 2005. Proteomic analysis of salt stress-responsive proteins in rice root. *Proteomics* 5:235–244.

Yan, S. P., Q. Y. Zhang, Z. C. Tang, W. A. Su and W. N. 2006. Comparative proteomic analysis provides new insights into chilling stress responses in rice. *Mol Cell Proteomics* 5:484–496.

Yang, L., R. Tang, J. Zhu et al. 2008. Enhancement of stress tolerance in transgenic tobacco plants constitutively expressing AtIpk2β, an inositol polyphosphate 6-/3-kinase from *Arabidopsis thaliana. Plant Mol Biol* 66:329–343.

Yi, Z. and S. Hong-Bo. 2008. The responding relationship between plants and environment is the essential principle for agricultural sustainable development on the globe. *C R Biol* 331:321–328.

Zhao, B., R. Liang, L. Ge et al. 2007. Identification of drought-induced microRNAs in rice. *Biochem Biophys Res Commun* 354:585–590.

Zhao, B., L. Ge, R. Liang et al. 2009a. Members of miR-169 family are induced by high salinity and transiently inhibit the NF-YA transcription factor. *BMC Mol Biol* 10:29–38.

Zhao, C. R., T. Ikka, Y. Sawaki et al. 2009b. Comparative transcriptomic characterization of aluminum, sodium chloride, cadmium and copper rhizotoxicities in *Arabidopsis thaliana. BMC Plant Biol* 9:32–47.

Zhu, J. K. 2001. Cell signaling under salt, water and cold stresses. *Curr Opin Plant Biol* 4:401–406.

Zhu, J. K. 2002. Salt and drought stress signal transduction in plants. *Annu Rev Plant Biol* 53:247–273.

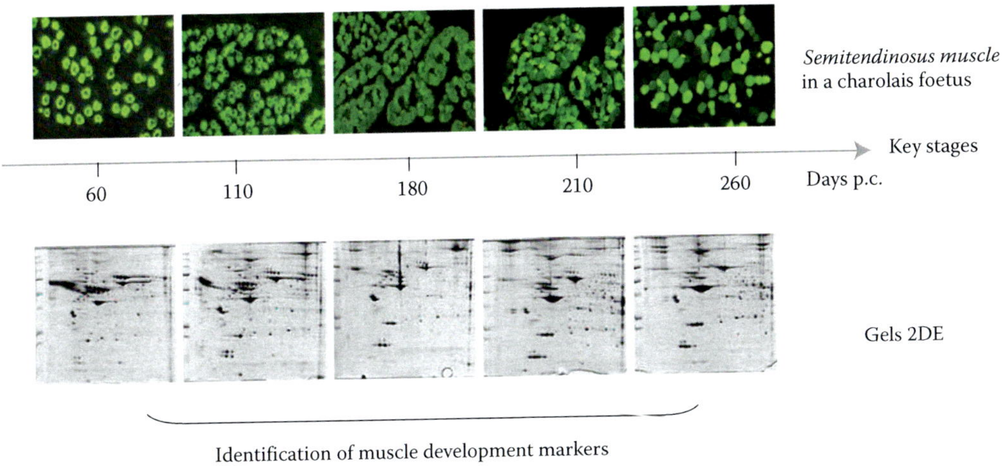

Figure 5.2 Strategy for the identification of muscle development in bovines at key stages in myogenesis by analysis of the changes in muscle proteome using two-dimensional electrophoresis (2DE) at different stages in days post-conception (p.c.).

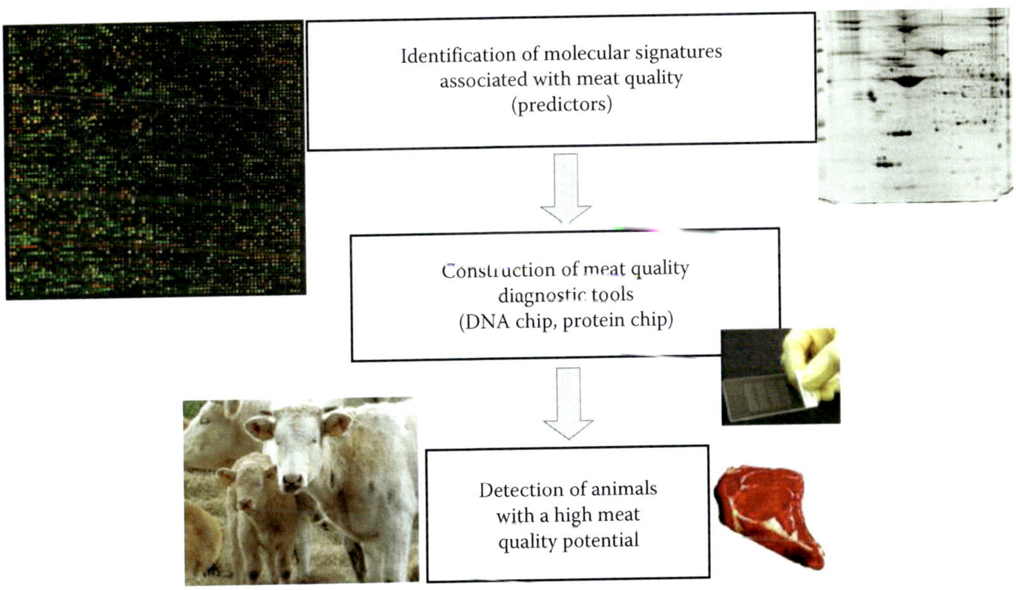

Figure 5.7 Use of genomic techniques to identify beef sensorial quality markers. It is necessary to integrate the results obtained by proteomics and transcriptomics in order to identify a combination of quality predictors. This should lead in the short term to the development of diagnostic tools allowing the identification of live animals whose muscle molecular phenotype is liable to give good-quality meat.

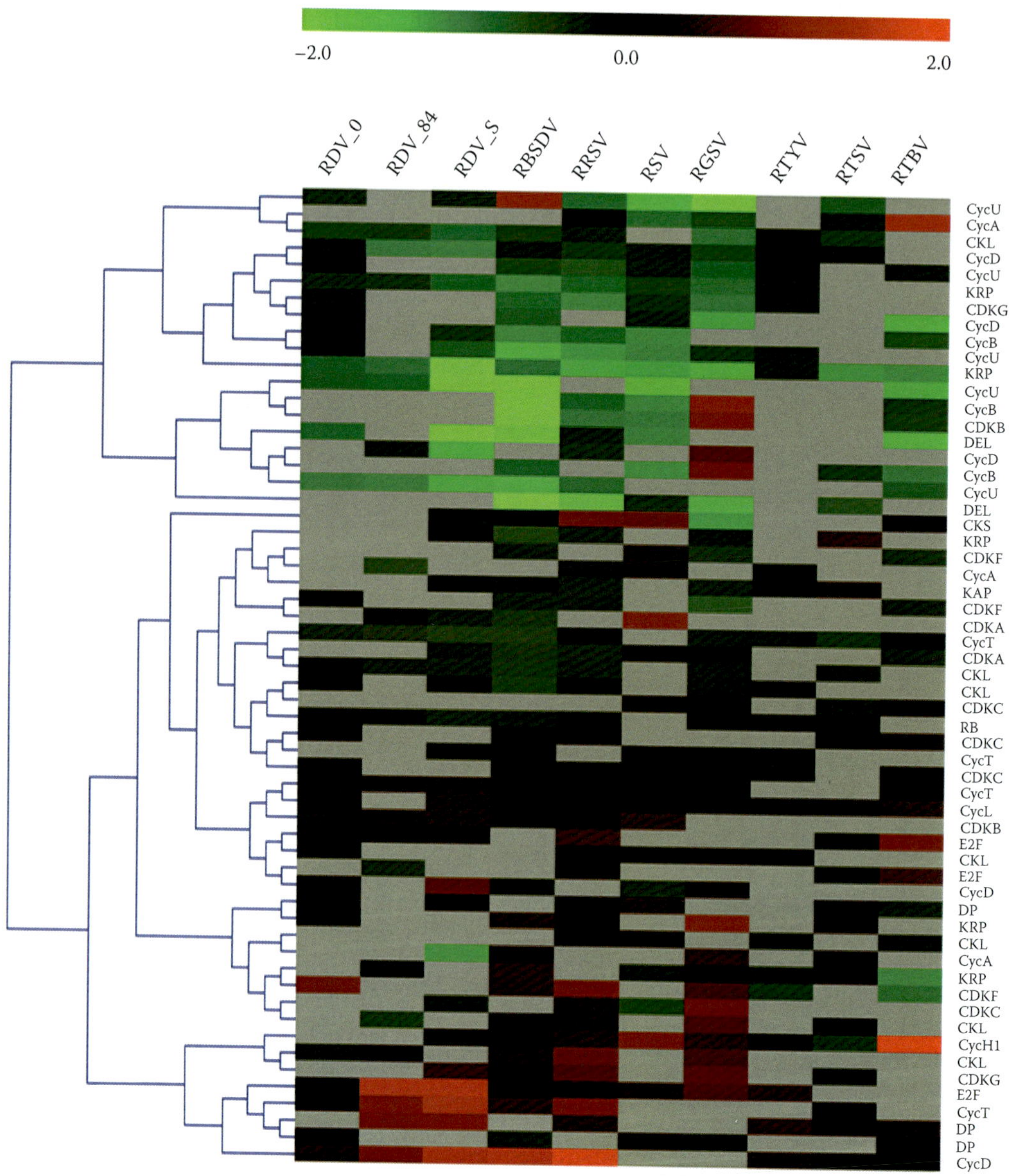

Figure 17.2 Heat-map clustering analysis of virus transcriptome data. Transcriptome data of representatively shown 58 cell-cycle-related genes from leaves of rice plants infected with eight different rice viruses were clustered on the basis of their gene-expression profiles. Red cells show upregulation and green cells show downregulation compared with mock-infected or untreated plants. Grey cells indicate low-intensity data.

CHAPTER 8

Rhizosphere Metabolomics
A Study of Biochemical Processes

Kalyan Chakravarthy Mynampati, Sheela Reuben and Sanjay Swarup

CONTENTS

8.1 INTRODUCTION

The biochemical environment of a rhizosphere system is composed of root exudations, and secretions from rhizobacteria, fungi and other soil organisms, arising from the dynamic interactions among plant roots, microbial organisms and other microfauna. These compounds directly or indirectly influence the microbial growth in the rhizosphere. In turn, the rhizosphere microbes serve as plant growth promoters or acts as pathogens inhibiting plant growth. Microbial degradation of natural or synthetic compounds (pesticides or plastics), nitrification by autotrophic bacteria or reactions of nitrate and sulphate also take place in the rhizosphere system. The rhizosphere metabolism is thus the result of complex interactions between microflora and microfauna yielding exudations, lysates, chelators, antibiotics, phytostimulators and extracellular enzymes, making it distinctive from the rest of the plant system. Hence, a systemic understanding of the rhizosphere biochemical environment is necessary to implement suitable biotechnological methods for sustainable agriculture.

Metabolomics provides an unbiased approach to investigate the underlying physiological, pathological and symbiotic interactions among the elements of the rhizosphere. Metabolomics is the comprehensive, qualitative and quantitative study of all the small molecules in an organism (i.e., the molecules that are less than or equal to about 1500 Da). The study of metabolomics therefore excludes polymers of amino acids and sugars. Instead, the focus is on intermediary metabolites used to form the macromolecular structures and other small molecules participating in important primary and secondary metabolic functions. Thus, metabolomics represents the interface between genetic predisposition and environmental influence. Metabolomics is useful to understand the function of genes, allowing us to control or design novel organisms that may benefit the rhizosphere biochemistry.

Metabolomics provides a comprehensive approach to survey the biochemical environment of the rhizosphere. The analytical tools within metabolomics—including chromatography, mass spectrometry (MS) and nuclear magnetic resonance (NMR)—profile, identify and estimate the relative abundance of metabolites in the entire system at a given time, yielding massive data sets. They can rapidly profile the impact of time, stress and environmental perturbation on hundreds of metabolites simultaneously. When used in combination with high-throughput assays such as genomics, proteomics or transcriptomics, metabolomics leads to a deeper understanding of the system's function. This generates a better picture of the rhizosphere composition than traditional plant biochemistry and natural products approaches.

This chapter focuses on how metabolomics is useful to investigate the biochemical processes taking place in the rhizosphere system. A brief overview of rhizosphere composition and its biochemistry is presented followed by a description of rhizosphere systems in soil and water environments. Rhizosphere metabolomics is comprehensively described, including the methods used to identify the compounds and the subsequent data-handling approaches to analyse and interpret the data. Case studies on rhizosphere metabolomics related to rhizoremediation and sustainable agriculture are presented in this chapter.

8.2 RHIZOSPHERE

Plants release chemicals into the rhizosphere that can positively or negatively regulate growth and development of the bacterial activity around the roots. This makes the rhizosphere a complex, dynamic and interactive microenvironment. In this section, we provide a brief description of the rhizosphere components especially in relation to the nature of metabolites and their effects on the surrounding biota.

8.2.1 Composition and Biochemistry of the Rhizosphere

8.2.1.1 Root Exudates

Root exudates play a key role in the rhizosphere, as their composition and abundance affect the growth and characteristics of the organisms thriving in the rhizosphere. Root exudates contain released ions, inorganic acids, oxygen and water and carbon-based compounds. The organic compounds include both low-molecular-weight compounds, such as amino acids, organic acids, sugars, phenolics and an array of secondary metabolites, and high-molecular-weight compounds, such as mucilage and proteins. Excellent reviews and books are available on this topic where details of exudate composition are provided (Bais et al., 2006; Mukerji et al., 2006; Pinton et al., 2000).

Exudates vary with respect to signals of biotic or abiotic origins. Allelochemicals in the root exudates govern the type of organisms that grow in the region. Some of these allelochemicals include tannins, cyanogenic glycosides, benzoquinones, flavonoids and phenolic acids.

8.2.1.2 Rhizobacteria

Rhizobacteria form an integral part of the rhizosphere. They include the microorganisms that are both beneficial as well as pathogenic. Several beneficial microorganisms are known to cause breakdown of natural products or even degrade them to simple sugars that are recycled for other anabolic reactions in the rhizosphere biota. Rhizobacteria utilize the end products of biosynthetic pathways for energy generation. In the case of the phenylpropanoid compounds, they are released in the rhizosphere and some microbes degrade these compounds through specific metabolic pathways, such as the phenylpropanoid catabolic pathway in plant growth promoting rhizobacteria strains of *Pseudomonas putida* (Pillai and Swarup, 2002) and the fluorophenol degradation pathway in different species of *Rhodococcus* (Boersma et al., 2001).

8.2.1.3 Soil Fungi

Soil fungi such as mycorrhizae are either pathogenic or symbiotic. Fungal development is often stimulated in the presence of roots especially owing to the nitrogen released by the roots. The presence of rhizobacteria may cause the inhibition of fungal growth. For example, the growth of ectomycorrhizae is inhibited by the presence of selected isolates of *Pseudomonas* and *Serratia* in the early infection stage of the fungi (Bending et al., 2002). Such growth inhibition is mediated by the secretion of antibiotics or antimicrobial compounds in the rhizosphere such as phenazines or selected flavonoids.

8.2.1.4 Soil Nematodes

Soil nematodes play an important part in the rhizosphere. Nematodes influence the nature of root exudation, which affects the physiological functioning of microorganisms in the rhizosphere. These exudates may serve as signal molecules for nematode antagonists and parasites (Kerry, 2000). These nematodes are often difficult to identify. High-throughput techniques such as matrix-assisted laser desorption ionization (MALDI) time-of-flight (TOF) analysis have been used to detect the metabolites from plant nematodes (Perera et al., 2005).

8.2.2 Aquatic versus Soil Rhizosphere

The classical definition of rhizosphere refers to the region of soil surrounding plant roots with microbial activity. In case of aquatic plants, this definition is less clear because of diffusion of nutrients in water. However, there exists a zone of influence by plant roots in this aquatic environment. The chemical conditions in this zone differ from those of the surrounding environment because of the processes that are induced either directly by the activity of plant roots or by the activity of rhizosphere microflora. According to Blotnick et al. (1980), the aquatic rhizosphere is an area of higher microbial populations and enhanced activity for nitrogen-fixing bacteria. Toyama et al. (2006) also report increased nitrification in aquatic rhizosphere along with accelerated mineralization of surfactants such as linear alkylbenze sulphonate, linear alcohol ethoxylate and mixed amino acids.

Aquatic rhizosphere is also characterized by the presence of sediments. Sediments are complex chemical and microbiological environments. The presence of anaerobic conditions, accompanied by a low oxidation–reduction potential and often by toxic constituents, places stresses on plants using anaerobic sediments as a rooting medium. Plant roots have adapted to survive in anaerobic sediments, which is the result of interactions between the surrounding sediments and the rhizosphere microflora. Stout and Nusslein (2010) have reviewed the biotechnological potential of aquatic rhizosphere in terms of remediation of metals though plant–microbial interactions in water.

8.3 RHIZOSPHERE METABOLOMICS

'Metabolome' refers to the survey of all the metabolites present in an organism. Metabolites are the small molecules that are the end products of enzymatic reactions. 'Metabolomics' is unbiased identification, quantification and analyses of all the metabolites present in the organism (Bhalla et al., 2005). Metabolomics encompasses an identification of metabolites (to understand the range of metabolites produced by the organism), their quantitation (to detect the abundance of metabolites), comparisons (to understand the differences arising from perturbations in metabolic pathways), data analysis and development of metabolic models. 'Metabolic profiling', a commonly found term in the literature, refers to obtaining a listing of the entire range of the metabolites present in the organism.

Metabolomics has applications in understanding the enzyme fluxes, uncovering novel metabolic pathways, unraveling cryptic pathways, identifying biomarkers and metabolic engineering of novel products of industrial and pharmaceutical importance. Metabolomics, in conjunction with other 'omics' such as functional genomics, proteomics and transcriptomics, aids in understanding the biological systems better. Integration of data from various fields helps in providing a holistic picture of the biological system using the systems approach.

The rhizosphere is a constantly changing microenvironment, where there is a flux of energy, nutrients and molecular signals between the plant roots and microbes that affects their mutual interactions. Metabolites exuded from plants as well as the metabolites released or secreted by the microbiota present in the rhizosphere have a considerable effect on this microenvironment. Hence, metabolic profiling constitutes a powerful technique to understand the underlying phenomenon of such exudations and the effects of metabolites on soil ecological relationships, plant–microbe interactions and other soil organisms.

8.3.1 Analytical Techniques for Rhizosphere Metabolomics

The root exudate profiling studies rely on a highly sophisticated suite of analytical techniques. Dunn et al. (2005) have extensively reviewed the use of analytical techniques in metabolomics. In this section, the major groups of techniques used for rhizosphere metabolomics are briefly discussed.

8.3.1.1 Chromatography Techniques

The following chromatography techniques help in separation and analysis of the metabolites to understand the nature of root exudation and interactions in the rhizosphere:

- Thin-layer chromatography (TLC). This technique involves the separation of metabolites on the basis of differential partitioning between the components of a mixture and the stationary solid phase. This is a simple and inexpensive analytical method. Reverse-phase TLC, along with some other techniques, has been useful in understanding fungal–bacterial interactions in the rhizosphere. The rhizobacteria *Pseudomonas chlororaphis* PCL1391 produces an antifungal metabolite phenazine-l-carboxamide, which is a crucial trait in its competition with the phytopathogenic fungus *Fusarium oxysporum* f. sp. *radicis-lycopersici* in the rhizosphere (Chin-A-Woeng et al., 2005). In this study, TLC was used to identify autoinducer compounds that were released during the expression of sigma factor psrA in different quorum-sensing gene mutants. In another application, TLC was used in studying the nodulation signalling metabolites that are secreted into the growth medium produced due to the *nod ABC* genes of *Rhizobium* and *Bradyrhizobium* strains (Spaink et al., 1992). TLC can be used to separate polar metabolites and fatty acids as well as to test the purity of compounds.

- Reverse-phase high-performance liquid chromatography (HPLC). In this technique, the metabolites are separated on the basis of their hydrophobicity and are identified by comparing the retention times with those of standard compounds. This approach is used to compare the root exudates from different cultivars. For example, root exudates from seven accessions were evaluated using HPLC (Czarnota et al., 2003). HPLC was also used to quantify the amount of sorgoleone, a photosynthetic inhibitor in the rhizosphere of sorghum plants (Weidenhamer, 2005). Polydimethylsiloxane (PDMS) was used for the study. The amounts of sorgoleone retained on the PDMS increased with time, which could be shown using HPLC methods.
- Anion-exchange chromatography. This technique is based on charge-to-charge interactions between the target compounds and the charges immobilized on the column resin. In anion-exchange chromatography, the binding ions are negative and the immobilized functional group is positive. It was used to determine the composition of soluble carbohydrates in plant tissues such as olive roots (Cataldi et al., 2000). This technique showed efficient separation of carbohydrates. Such studies can be extended to understand the movement of sugars and the types of sugars that are available in the soil for the rhizobacteria.

Chromatography techniques are powerful tools when used in conjunction with other techniques such as mass spectrometry (MS). Liquid chromatography (LC) and gas chromatography (GC) techniques have been used with different types of mass spectrometers in the context of rhizosphere metabolomics.

8.3.1.2 MS Techniques

In a mass spectrometer, the samples are ionized by different methods. This is usually done in the source part of the mass spectrometer. There are different ionization methods, such as electron impact, chemical ionization, electron spray ionization (ESI), fast atom bombardment, field ionization, field desorption and laser desorption. In electron impact ionization, the samples are ionized by the bombardment of electrons. The ionization is caused by the interaction of the fields of the bombarded electron and the molecule, resulting in the emission of an electron.

In an ESI mass spectrometer, the sample is sprayed as a fine liquid aerosol. A strong electric field is applied under atmospheric pressure to the liquid passing through a capillary tube, which induces charge accumulation at the liquid surface, which then breaks up to form highly charged droplets. As the solvent evaporates, the droplets explode to give ions. The spectra obtained are usually those of multiply charged molecular ions owing to protonation.

In laser desorption ionization, a laser pulse is focused onto the surface of the sample, some part of the compound gets desorbed and reactions among the molecules in the vapour-phase region result in ions. An extension of this method is the MALDI method. In this technique, samples are mixed with a suitable matrix and allowed to crystallize on grid surfaces. Samples are then irradiated with laser pulses to induce ionization. Most of the energy of the laser pulse is absorbed by the matrix, and so unwanted fragmentation of the biomolecule is avoided.

Chemical ionization is 'soft ionization' technique as the number of fragment ions produced is less. In this method, a reactant gas such as methane is passed through the sample and the interaction of the ions with neutral molecules produces new ions.

The ions formed by any of the above methods are accelerated through a column and deflected in a magnetic field. In the mass analyser, the ions are separated according to their mass-to-charge ratio (*m*/*z*) and finally they are detected by an ion detector. In a triple–quadrupole mass analyser, the ions from the source are passed between four parallel rods. The motion of the ions depends on the electric field, which allows only ions of the same *m*/*z* to be in resonance and to the detector at the same time. Triple–quadrupole MS is most often used for quantification purposes. In the case of an ion-trap mass analyser, the ions are focused using an electrostatic lensing system into the ion trap. An ion will be stably trapped depending on the values for the mass and the charge of the ion. A mass spectrum is obtained by changing the electrode voltages to eject the ions from the trap. In a TOF detector the molecules are detected on the basis of the time that each molecule takes to reach the detector.

Chromatographic separations followed by mass-spectral analysis provide additional separation. This is because the metabolites are first separated on a chromatographic column, which partitions the metabolites into different fractions, and each of the fractions is further analysed by a mass spectrometer. The separation of the metabolites into fractions helps in reducing the ion suppression effect and enhances detection and therefore more metabolites are analysed from samples. The metabolites can be fragmented for identification purposes using a tandem mass spectrometer. Tandem mass spectrometer helps in fragmenting targeted ions that give rise to daughter ions. These daughter ions form a fingerprint that can then be compared with fingerprints of standards or databases.

Mass spectrometers have been used in conjunction with various chromatographic methods. The two commonly used chromatographic methods are LC–MS and GC–MS. These analytical techniques are reviewed in Dunn and Ellis (2005) and Sumner et al. (2003).

- *Liquid Chromatography–Mass Spectrometry.* Although numerous metabolites can be identified in a single run using GC–MS, the technique may not always prove useful especially in the case of metabolites that are sequestered in compartments or are degraded in high-temperature regimes. Such metabolites are difficult to derivatize. In such cases, LC–MS is the technique of choice. This technique is very commonly used as it is a very convenient platform especially when used in with ESI-based MS. Nearly 13–20 isomeric isoflavone conjugates have been identified from roots of lupine species using ESI–MS (Kachlicki et al., 2005). In this study, a comparative analysis of triple–quadruple and ion-trap analysers was conducted. The study highlighted the utility of these techniques in analysing metabolites in biological samples. Such techniques can be used to study the role of metabolite conjugations in root–microbe interactions.
- *Gas Chromatography–Mass Spectrometry.* This technique is mostly used to study volatile compounds. As GC–MS relies on the hard ionization methods, ion spectra are highly uniform and reproducible between experiments. Owing to this advantage, standard databases can be created and shared between laboratories. The GC–MS technique has been used to study the differences in plants of different developmental stages with respect to their day length (Jonsson et al., 2004). GC–MS has been useful in identifying molecules such as those involved in signalling during ectomycorrhizae formation (Menotta et al., 2004). These molecules are exuded during the presymbiotic interaction between *Tuber borchii* (ectomycorrhiza) and the host plant *Tilia americana*. Seventy-three volatile organic compounds (VOCs) could be identified and 29 of these were produced during interaction between the fungi and the host and therefore, they could possibly be signalling molecules. The technique thus assists in increasing our understanding of rhizosphere signalling.

GC–combustion–isotope-ratio MS (GC–C–IRMS) is another useful technique that has been adopted in rhizosphere metabolomics. A summary of the application of this MS in rhizosphere metabolomics is provided in Rueben et al. (2008).

- *Matrix-Assisted Laser Desorption Ionization Time-of-Flight.* This is a very sensitive method and quantities as low as 10–15–10–18 mol can be detected. This method has been useful in the study of aconitum alkaloids from aconite roots (Sun et al., 1998). This kind of analysis often leads to the identification of new metabolites. MALDI-TOF is most useful for determining the mass accurately.
- *Proton Transfer Reaction Mass Spectrometry.* This new technology allows rapid and real-time analysis of most biogenic VOCs. Compounds are ionized by a chemical ionization method using H_3O^+ ions. The < H_3O^+ ions transfer their protons to the VOCs, which have higher proton affinities than water. The process is referred to as 'soft ionization' as it avoids excessive fragmentation of the biomolecules and allows real-time analysis. The detection limit in proton transfer reaction MS is as low as a few parts per trillion. This technique has been used to study rhizosphere.
- *VOCs and Their Induction by Biotic Stresses.* The VOCs can be analysed without previous separation by chromatography (Steeghs et al., 2004). In the above-mentioned study, compatible interactions of *Pseudomonas syringae* DC3000 and *Diuraphis noxia* with *Arabidopsis* roots showed rapid release of 1,8-cineole, a monoterpene detecting using proton transfer reaction MS. This detection was not previously reported in *Arabidopsis*.

8.3.1.3 Spectroscopy Techniques

Spectroscopic techniques have been increasingly used in studying and identifying metabolites. Fourier transform infrared (FTIR) spectroscopy is a technique that is useful in identifying organic and inorganic chemicals. The chemical bonds in a molecule can be determined by interpreting the IR absorption spectrum. Molecular bonds vibrate at various frequencies depending on the elements and the type of bonds and therefore give a specific absorption spectrum. FTIR spectra of pure compounds are so unique that they are like a molecular 'fingerprint'. Raman spectroscopy is another technique where the observed spectrum is based on the vibration of a scattering molecule. When a photon is incident on a molecule, it interacts with the electric dipole of the molecule. The interaction can be viewed as a perturbation of the molecule's electric field. Both FTIR and Raman spectroscopy are effective in the rapid identification of bacteria and fungi (Goodacre et al., 2000). UV resonance Raman spectroscopy was used in the characterization of *Bacillus* and *Brevibacillus* strains (López-Díez and Goodacre, 2004).

Nuclear magnetic resonance (NMR) spectroscopy is a less sensitive technique than MS. However, it is highly powerful in identifying small molecules and is one of the most used forms of spectroscopy. It helps in accurately determining the structure of a metabolite. Any molecule containing one or more atoms with nonzero moment is detectable by NMR. Biologically important atoms such as 1H, 13C, 14N, 15N and 31P are all detectable by NMR. All biologically important metabolites provide NMR signals. NMR spectra are characterized by the chemical shifts, intensity and fine structure of the signals. These signals help in identification and quantification of the metabolites. The use of NMR in metabolic fingerprinting and profiling of plants with respect to the effect of stress on wild-type, mutant and transgenic plants is reviewed by Krishnan et al. (2005). NMR has also been used in investigating the operation of networks in plants. Labelling with isotopes in conjunction with metabolite analysis by NMR can help in building flux maps that can be useful in metabolic network modeling (Ratcliffe and Shachar-Hill, 2005). Quantitative NMR is used for quantitative measurements. Solid-state NMR has been used in studying plant nitrogen metabolism (Mesnard and Ratcliffe, 2005). Solid-state 1H NMR is useful in elucidating the structure as well as the dynamic nature in the solid phase. NMR can be linked with Tandem MS to study root exudations. Structural elucidation of montecristin, a key metabolite in the biogenesis of acetogenins from the roots of *Annona muricata*, was performed using tandem MS and NMR (Gleye et al., 1997).

Combination of HPLC and NMR spectroscopy has led to the identification and quantification of a number of metabolites in the root exudates of *Arabidopsis thaliana* (Walker et al., 2003). The authors conducted a time-course study of root exudates from plants treated with salicylic acid, jasmonic acid, chitosan and two fungal cell wall elicitors. Plants treated with salicylic acid had the maximum number of compounds in their exudates, while elicitation with jasmonic acid had the least effect on exudates. This method of root exudates profiling could identify differences in root exudation with respect to plant stress.

Root exudates profiling in graminaceous plants was conducted using multinuclear and two-dimensional NMR with GC–MS and coupled with high-resolution MS for metabolite identification to understand the acquisition of metal ions from soil (Fan et al., 2001). The root profiling method was used to examine the role of exudate metal ion ligands (MILs) in the acquisition of Cadmium (Cd) and transition metals in barley and wheat. The change in the root exudate profile was studied in wheat, barley and rice grown on Fe- and Cd-deficient soils. MILs such as 3-epihydroxymugineic acid, mugineic acid, 2-deoxymugineic acid and malate in barley were elevated in Fe-deficient conditions, which in turn increased the Fe-mobilizing substances. The results suggest that enhanced exudation of murigenic acids and malate may be involved in acquisition of transition metals but not of Cd, and also that the mechanisms of acquisition for essential and toxic metal ions may be different.

8.3.2 Metabolomics Data Handling and Analysis

Metabolomic studies result in huge data sets that need to be analysed and stored. In typical LC–MS study, ions can be detected per fraction in positive mode and a slightly lesser number in the negative mode. Hence, one sample from a single injection can yield 10,000–12,000 data points. With replications and various samples, the number increases considerably, generating several hundred megabytes of data per experiment. Databases to store such data, as well as for the identification of the metabolites, are therefore essential. Raw data obtained from the instruments have to be preprocessed by several methods to minimize effects of machine variations or experimental errors during weighing of samples or injections into the instruments. Examples of preprocessing include data normalization, baseline correction and alignment of spectra. Data analysis follows preprocessing to extract useful biological information. Such analysis can lead to verification of an original hypothesis or the discovery of new associations.

Recently, metabolic modelling is being used to lead the development of metabolic networks. Such networks help in understanding the biochemical behaviour at the whole-cell level. Evolutionary computation-based methods such as genetic algorithms and genetic programming are ideal strategies for mining such high-dimensional data to generate useful relationships, rules and predictions (Goodacre, 2005).

Data obtained from metabolomic studies often are very complex as multiple dimensions may be involved. For example, various dimensions due to different treatments such as doses or time-point-based data may need to be handled. Specific types of biostatistical tools are required to make meaningful conclusions from such data. One such statistical tool is ANOVA-Simultaneous component analysis of variance—simultaneous component analysis (Smilde et al., 2005). It is a direct generalization of analysis of variance from univariate to multivariate data. This tool helps in analysing complex data generated by LC–MS involving different parameters such as time and dose factors. Other multivariate data analysis techniques such as principal component analysis (PCA) and partial least-squares regression can also be used to analyse metabolomics data. This technique can reduce the number of dimensions to two or three, which can be represented graphically. These representations allow the user to visualize the patterns or clusters in the data sets as hierarchical plots or scatter plots. PCA helps in visualizing the data in a simplified way and helps in extracting meaningful biological interpretations. A variation of PCA is the weighted PCA where spectra of repeated measurements are converted to weights describing the experimental error and it adds interpretation to the metabolomics data (Jansen et al., 2004). Multivariate data analysis in metabolomics is reviewed by van der Werf et al. (2005).

Web-based and offline data analysis tools are being developed by the research community for the preprocessing, pretreatment, processing and visualization of data. Examples include MZmine (Katajamaa and Oresic, 2005), which is useful for analysis of LC–MS studies. It contains algorithms useful in data preprocessing such as spectral filtration, peak detection, alignment and normalization. The visualization tools enable comparative viewing of the data across multiple samples and peak lists can be exported to other tools for data analysis.

MSFACT is another metabolite data analysis tool, and consists of spectral formatting, alignment and conversion tools (Duran et al., 2003). This tool helps in reformatting, alignment and export of large chromatographic data sets to allow more rapid visualization and interrogation of metabolomics data. Applications of the tool were illustrated using GC–MS profiles from *Medicago truncatula*. Metabolites from various tissues such as roots, stem and leaves from the same plant were easily differentiated on the basis of metabolite profiles. The tool uses hierarchical clustering, two-dimensional PCA and three-dimensional PCA as visualization tools.

XCMS, an open source software developed by the Scripps Center for Metabolomics (Smith et al., 2006), offers feature detection and matching, non-linear retention time alignment and statistics to process LC–MS data. It dynamically identifies hundreds of endogenous metabolites for use as standards. Following retention time correction, the relative metabolite ion intensities are directly compared to identify changes in specific endogenous metabolites, such as potential biomarkers.

A more recent and comprehensive online tool for MS data analysis is Metabolomics Data Analysis Tool (MetDAT) (http://www.sdwa.nus.edu.sg/METDAT2/). MetDAT performs alignment, baseline correction and normalization of data using a number of algorithms. It can calculate log ratios of different treatments with respect to a reference data set. It also allows generation of Venn diagrams that identify common as well as unique molecules in two or three data sets. MetDAT includes supervised and unsupervised clustering methods, such as PCA and PLS as well as biostatistical methods, such as analysis of variance at each hierarchical level. Users are able to upload their data, analyse and store the data, which can later be recalled (Biswas et al., 2010).

Upon metabolite data analysis, novel compounds and metabolic pathway features are frequently discovered. It then becomes essential to mine the literature for reports on such compounds or their enzymes or pathways. Searching databases can often lead to a large number of publications. For example, a simple search for plant sugars yields 43,926 abstracts in the PubMed database. Analysis and integration of knowledge from such abstracts becomes highly cumbersome to an extent such that it takes enormous effort, manpower and time to assimilate and interpret the information. This problem can be minimized by using knowledge mining tools such as Dragon Plant Biology Explorer (Bajic et al., 2005). This tool allows plant biologists to mine the existing literature and visualize the interconnectedness. It is a system that integrates information on genes from PubMed abstracts with gene functions based on standard gene ontologies and biochemical entity vocabularies, and presents the associations as interactive networks. It complements the existing biological resources for systems biology by identifying potentially novel associations using text analysis between cellular entities based on genome annotation terms. One of the most useful aspects of this tool to biologists is that it condenses information from a large volume of documents for easy inspection and analysis, thus making it feasible for individual users. Two modules of the explorer, the Metabolome Explorer and the Pharmacology Explorer, are especially relevant to metabolomics researchers (Bhalla et al., 2005). Interconnections among cellular entities such as metabolites, enzymes, genes, mutants, plant anatomical features, cellular components and function can be visualized as networks. Nodes in the networks are hyperlinked to the original abstracts in colour-coded forms. Hence, this tool can be used to interpret novel information generated from metabolomics projects as well as to research new topics by beginners or experienced scientists.

Different kinds of databases are required to efficiently analyse metabolomics data. Some such databases are outlined here briefly:

1. Databases for storing experimental data. These help to recall data for comparison or different types of analyses and usually such databases are created in-house. Data are stored as flat files for smaller data sets, while for larger complex data, relational databases are used. Relational databases also have functions 'built in' that help them to retrieve, sort and edit the data in many different ways.
2. Databases for comparing with other standard or data sets. One of the biggest challenges in MS is to generate reproducible fragmentation patterns especially using soft ionization methods such as ESI–MS. Databases then have to include parent, precursor and daughter ion information. Currently, there is very little understanding of the fragmentation patterns of various metabolites; hence, individual laboratories have to generate their own databases based on their methods and instrument settings. Spectral data are available in the public domain for only some metabolites. For example, the National Institute of Standards and Technology, USA, has a chemistry Web book that provides information on the molecular weight, formula, structure as well as mass spectra. Chemical information on several parameters is available for over 40,000 compounds and mass spectra of 15,000 compounds are available (http://webbook.nist.gov/). Another freely available database for drugs and metabolites is provided by the Mass Spectrometry Database Committee (http://www.ualberta.ca/~gjones/mslib.htm). In addition, a number of commercial databases are available for users. Some of these are the NIST/EPA/NIH Mass Spectral Library for electron impact (ESI) spectra (http://www.nist.gov/srd/nist1a.htm) and the Wiley Registry of Mass Spectral Data (http://www.wileyregistry.com/).
3. Databases of biochemical reactions and pathways. To understand the role of various metabolites in biological processes, it is imperative to understand the biochemical reactions in which such metabolites are involved and the pathways to which they belong. This helps to predict the molecular mechanisms that govern the various processes taking place in the cell as a whole. Consequently, several attempts are under way to create large-scale databases on gene-regulatory and biochemical networks. Such databases provide a comprehensive coverage of the chemical reactions. Such databases can therefore be helpful in deducing the reactions that are affected upon treatment or in transgenic, mutant, knockout or knockdown RNA interference plants. A summary of some of the available databases is described below.
   - MetaCyc. This is a database of experimentally elucidated pathways. It has more than 800 pathways from 600 organisms. It stores pathways involved in primary metabolism (including photosynthesis), secondary metabolism, as well as associated compounds, enzymes and genes. It is available at http://metacyc.org/.
   - AraCyc. This is a database containing biochemical pathways of *Arabidopsis* developed at The *Arabidopsis* Information Resource (http://www.arabidopsis.org) with the aim of representing *Arabidopsis* metabolism using a web-based interface (Mueller et al., 2003). This database now contains more than 250 pathways that include information on compounds, metabolic intermediates, cofactors, reactions, genes, proteins and protein subcellular locations. The web site also has an 'omics viewer' that allows users to upload the data onto a pathway chart and then visualize the variations in the data and map the pathway changes in a visual form.
   - The Kyoto Encyclopedia of Genes and Genomes (KEGG) (http://www.genome.ad.jp/kegg/) is an integrated bioinformatics database resource (Kanehsia et al., 2004). It includes 16 databases on chemical compounds, reactions, genes, proteins and their molecular interactions at pathway level, thus leading to a systems-level information. KEGG pathways for metabolism are used for genome-scale metabolic reconstruction (Kanehsia et al., 2006).
   - BioPathAt. This database allows the knowledge-based analysis of genome-scale data by integrating biochemical pathway maps (BioPathAtMAPS module) with a manually scrutinized gene-function database (BioPathAtDB) for the model plant *Arabidopsis thaliana* (Lange and Ghassemian, 2005).

8.4 APPLICATIONS OF RHIZOSPHERE METABOLOMICS

8.4.1 Rhizoremediation

Rhizoremediation involves the use of plants as well as rhizobacteria to clean up contaminated soil and water. Two processes have been described to constitute rhizoremediation, namely, *phytoremediation* and *bioaugmentation* (Kuiper et al., 2004). 'Bioaugmentation' refers to enhanced availability of a substrate using specific microbes. Microbial degradation of the pollutants is enhanced owing to stimulation of root exudates. The root system of plants aid in spreading the rhizobacteria through soil and help to penetrate otherwise impermeable soil layers. Pollutant degrading bacteria can be inoculated on plant seeds to improve the efficiency of phytoremediation or bioaugmentation.

Phytoremediation of polyaromatic hydrocarbons (PAHs) is driven by root–microbe interactions (Rugh et al., 2005). Bacterial degradation has been shown to be the dominant pathway for environmental PAH dissipation. The authors tested various plant species and the efficacy in degrading PAHs. It was found that in soils that were planted, there was an increase in heterotrophic and biodegradative cell numbers compared with the situation in unplanted soils. The study showed that the expanded metabolic range of the rhizosphere bacterial community would contribute more to effective degradation of PAHs.

Degradation of environmental pollutants can be enhanced in the rhizosphere by microorganisms that can utilize root exudates as carbon source. For example, biodegradation of polychlorinated biphenyls (PCBs) can be enhanced by the growth of PCB-degrading rhizobacteria along with plants that can exude phenylpropanoids. The rhizobacteria are able to utilize phenylpropanoids and hence are able to grow better in the rhizosphere as there is less competition for compounds such as phenylpropanoids from other microorganisms. Since they are able to grow better, the efficiency of PCB degradation is increased (Narasimhan et al., 2003). This method has the advantage that it does not rely on genetic engineering of plants as in the case of the previous methods.

Plant-assisted rhizoremediation in the long run will turn into an effective mode for rhizoremediation of toxic organic pollutants. At petroleum hydrocarbon contaminated sites, two genes encoding hydrocarbon degradation, alkane monooxygenase (*alkB*) and naphthalene dioxygenase (*ndoB*), were two and four times more prevalent in bacteria extracted from the root interior (endophytic) than from the bulk soil and sediment, respectively (Siliciano et al., 2001). These results indicate that the enrichment of catabolic genotypes in the root interior is both plant dependent and contaminant dependent.

8.4.2 Sustainable Agriculture

The guiding principle behind sustainable agriculture is to meet the needs of the present without compromising the ability of the future generations to meet their own needs. A systems perspective is essential to achieve this objective. A systems approach gives us the tools to explore the interconnections between agriculture and other aspects of our environment. New developments in biotechnology, including metabolomics, can contribute to achieving such systemic strategies for sustainable agriculture. The contributions include (1) cultivation of crops that are better able to tolerate biotic stresses (pests, diseases and weeds) and abiotic stresses (drought, salinity and temperature stress); (2) growing suitable plants (and microbes) to mitigate the effects of industrial pollution (bioremediation) by increasing their ability to remove and/or break down toxic compounds in the soil and water and (3) delivering more nutritionally beneficial staple crops, with higher content of essential vitamins and minerals, and plants for production of products for industrial purposes such as biodegradable plastics and industrial strength fibres.

The future development of new crops with improved tolerance to abiotic factors and the advent of crops that are used to produce vaccines and industrial products may change crop management practices. These new crops may either increase or decrease demand for arable land in the long term. They may also put further pressure on natural biodiversity when crop cultivation extends into presently marginal lands, or into areas not presently used for agriculture. For example, salt-tolerant rice may be able to be cultivated in coastal areas where mangroves presently grow, with resulting ecological changes in land and water use and associated plant and marine life.

High-throughput 'omics' technology can also contribute to the characterization and conservation of biodiversity. The use of pesticides (including herbicides) over the past 20 years has been a major cause of the decline in farmland birds, arable wild plants and insects in several countries. This raises the possibility of losing the food chain links between native species and the crop systems. This link is vital to preserve the early warning function of biodiversity, whereby damage to feeding species such as birds signal warning of dangers in food crops or the chemicals used to manage them. The literature suggests that use of technology-enabled genetically modified crops with insect resistance is reducing the volume and frequency of pesticide use on cotton, corn and soybean (Carpenter et al., 2002). Increasing the productivity of crops thus reduces pressure on biodiversity by reducing the need for agriculture to move into forests and marginal lands.

8.5 CONCLUSION

Metabolomics, along with genomics, proteomics and transcriptomics, provides a directed functional analysis to investigate plant–microbe interactions at full biological hierarchy, including extracellular enzyme activities and biochemical processes active within the rhizosphere. Such understanding of the plant–microbe system has broad applications in bioremediation and sustainable agriculture.

The future of rhizhosphere metabolomics will be shaped by the following:

1. Improved exudation extraction techniques. Currently, different collection procedures are required to extract root exudates. These are collected by growing plants hydroponically, aeroponically or even in soil; hence, specialized techniques and care are required for extraction.
2. Advances in separation procedures. In the present form, it is difficult to separate the exudations from various interacting biotic agents in the rhizosphere such as plants and their associated microbes.
3. Development in analytical techniques customized to root exudation profiling.
4. Expanded and comprehensive databases and data analysis tools.

ACKNOWLEDGEMENTS

The authors gratefully acknowledge the support and contributions of the Singapore-Delft Water Alliance (SDWA). This work was carried out as part of the SDWA's 'ASC—Pandan' research programme (R-264-001-002-272).

REFERENCES

Bajic, V. B., M. Veronika, P. S. Veladandi et al. 2005. Dragon Plant Biology Explorer. A text-mining tool for integrating associations between genetic and biochemical entities with genome annotation and biochemical terms lists. *Plant Physiol* 138:1914–1925.

Bais, H. P., T. L. Weir, L. G. Perry et al. 2006. The role of root exudates in rhizosphere interactions with plants and other organisms. *Annu Rev Plant Biol* 57:233–266.

Bending, G. D., E. J. Poole, J. M. Whipps et al. 2002. Characterisation of bacteria from *Pinus sylvestris–Suillus luteus* mycorrhizas and their effects on root–fungus interactions and plant growth. *FEMS Microbiol Ecol* 39:219–227.

Bhalla, R., K. Narasimhan and S. Swarup. 2005. Metabolomics and its role in understanding cellular responses in plants. *Plant Cell Rep* 24:562–571.

Biswas, A., K. C. Mynampati, S. Umashankar et al. 2010. MetDAT: A modular and workflow-based free online pipeline for mass spectrometry data processing, analysis and interpretation. *Bioinformatics* 26:2639–2640.

Boersma, M. G., I. P. Solyanikova, W. J. Van Berkel et al. 2001. 19F NMR metabolomics for the elucidation of microbial degradation pathways of fluorophenols. *J Ind Microbiol Biotechnol* 26:22–34.

Cataldi, T. R. I., G. Margiotta, L. Iasi et al. 2000. Determination of sugar compounds in olive plant extracts by anion-exchange chromatography with pulsed amperometric detection. *Anal Chem* 72:3902–3907.

Chin-A-Woeng, T. F. C., D. van den Broek, B. J. J. Lugtenberg et al. 2005. The *Pseudomonas chlororaphis* PCL1391 sigma regulator psrA represses the production of the antifungal metabolite phenazine-1-carboxamide. *Mol Plant Microbe Interact* 18:244–253.

Czarnota, M. A., A. M. Rimando and L. A. Weston. 2003. Evaluation of root exudates of seven sorghum accessions. *J Chem Ecol* 29:2073–2083.

Dunn, W. B. and D. I. Ellis. 2005. Metabolomics: Current analytical platforms and methodologies. *Trends Anal Chem* 24:28–94.

Dunn, W. B., N. J. Bailey and H. E. Johnson. 2005. Measuring the metabolome: Current analytical technologies. *Analyst* 130:606–625.

Duran, A. L., J. Yang, L. Wang et al. 2003. Metabolomics spectral formatting, alignment and conversion tools (MSFACTs). *Bioinformatics* 19:2283–2293.

Fan, T. W. M., A. N. Lane, M. Shenker et al. 2001. Comprehensive chemical profiling of gramineous plant root exudates using high resolution NMR and MS. *Phytochemistry* 57:209–221.

Gleye, C., A. Laurens, R. Hocquemiller et al. 1997. Isolation of montecristin, a key metabolite in biogenesis of acetogenins from *Annona muricata* and its structure elucidation by using tandem mass spectrometry. *J Org Chem* 62:510–513.

Goodacre, R. 2005. Making sense of the metabolome using evolutionary computation: Seeing the wood with the trees. *J Exp Bot* 56:245–254.

Goodacre, R., B. Shann, R. J. Gilbert et al. 2000. Detection of the dipicolinic acid biomarker in *Bacillus* spores using Curie-point pyrolysis mass spectrometry and Fourier transform infrared spectroscopy. *Anal Chem* 72:119–127.

Jansen, J. J., H. C. J. Hoefsloot, H. F. M. Boelens et al. 2004. Analysis of longitudinal metabolomics data. *Bioinformatics*. 20:2438–2446.

Jonsson, P., J. Gullberg, A. Nordstrom et al. 2004. A strategy for identifying differences in large series of metabolomic samples analyzed by GC/MS. *Anal Chem* 76:1738–1745.

Kachlicki, P., L. Marczak, L. Kerhoas et al. 2005. Profiling isoflavone conjugates in root extracts of lupine species with LC/ESI/MSn systems. *J Mass Spectrom* 40:1088–1103.

Kanehisa, M., S. Goto, M. Hattori et al. 2006. From genomics to chemical genomics: New developments in KEGG. *Nucleic Acids Res* 34:D354–D357.

Kanehisa, M., S. Goto, S. Kawashima et al. 2004. The KEGG resource for deciphering the genome. *Nucleic Acids Res* 32:D277–D280.

Katajamaa, M. and M. Oresic M. 2005. Processing methods for differential analysis of LC/MS profile data. *BMC Bioinformatics* 6:179.

Kerry, B. R. 2000. Rhizosphere interactions and the exploitation of microbial agents for the biological control of plant parasitic nematodes. *Annu Rev Phytopathol* 38:423–441.

Krishnan, P., N. J. Kruger and R. G. Ratcliffe. 2005. Metabolite fingerprinting and profiling in plants using NMR. *J Exp Bot* 56:255–265.

Kuiper, I., E. L. Lagendijk, G. V. Bloemberg et al. 2004. Rhizoremediation: A beneficial plant–microbe interaction. *Mol Plant Microbe Interact* 1:6–15.

Lange, B. M. and M. Ghassemian. 2005. Comprehensive post-genomic data analysis approaches integrating biochemical pathway maps. *Phytochemistry* 66:413–451.

López-Díez, E. C. and R. Goodacre. 2004. Characterization of microorganisms using UV resonance Raman spectroscopy and chemometrics. *Anal Chem* 76:585–591.

Menotta, M., A. M. Gioacchini, A. Amicucci et al. 2004. Headspace solid-phase microextraction with gas chromatography and mass spectrometry in the investigation of volatile organic compounds in an ectomycorrhizae synthesis system. *Rapid Commun Mass Spectrom* 18:206–210.

Mesnard, F. and R. G. Ratcliffe. 2005. NMR analysis of plant nitrogen metabolism. *Photosynth Res* 83:163–180.

Mueller, L. A., P. Zhang and S. Y. Rhee. 2003. AraCyc: A biochemical pathway database for *Arabidopsis.Plant Physiol* 132:453–460.

Mukerji, K. G., C. Manoharachary and J. Singh. 2006. *Soil Biology: Microbial Activity in the Rhizosphere*, Vol. 7. Heidelberg: Springer-Verlag.

Narasimhan, K., C. Basheer, V. B. Bajic et al. 2003. Enhancement of plant–microbe interactions using a rhizosphere metabolomics-driven approach and its application in the removal of polychlorinated biphenyls. *Plant Physiol* 132:146–153.

Perera, M. R., V. A. Vanstone and M. G. K. Jones. 2005. A novel approach to identify plant parasitic nematodes using matrix-assisted laser desorption/ionization time-of-flight mass spectrometry. *Rapid Commun Mass Spectrom* 19:1454–1460.

Pillai, B. V. S. and S. Swarup. 2002. Elucidation of the flavonoid catabolism pathway in *Pseudomonas putida* PML2 strain by comparative metabolic profiling. *Appl Environ Microbiol* 68:143–151.

Pinton, R., Z. Varanni, P. Nannipier et al. 2000. *The Rhizosphere: Biochemistry and Organic Dubstance at the Soil–Plant Interface*. New York, NY: Marcel Dekker.

Ratcliffe, R. G. and Y. Shachar-Hill. 2005. Revealing metabolic phenotypes in plants: Inputs from NMR analysis. *Biol Rev Camb Philos Soc* 80:27–43.

Reuben, S., V. S. Bhinu and S. Swarup. 2008. Rhizosphere metabolomics: Methods and applications. *Secondary Metabolites in Soil Ecology*. Berlin/Heidelberg: Springer.

Rugh, C. L., E. Susilawati, A. N. Kravchenko et al. 2005. Biodegrader metabolic expansion during polyaromatic hydrocarbons rhizoremediation. *Z Naturforsch C* 60:331–339.

Smilde, A. K., J. J. Jansen, H. C. J. Hoefsloot et al. 2005. ANOVA-simultaneous component analysis (ASCA): A new tool for analyzing designed metabolomics data. *Bioinformatics* 21:3043–3048.

Smith, C. A., E. J. Want, G. C. Tong et al. 2006. XCMS: Processing mass spectrometry data for metabolite profiling using nonlinear peak alignment, matching, and identification. *Analytical Chemistry* 78:779–787.

CHAPTER 9

Microbial Functionality and Diversity in Agroecosystems

A Soil Quality Perspective

Felipe Bastida, César Nicolás, José Luis Moreno, Teresa Hernández and Carlos García

CONTENTS

9.1 INTRODUCTION

Soil microbial communities and their enzymatic activities play an important role in soil fertility and quality because most organic matter and nutrient transformations are mediated by microbial activity. In this chapter, we review the last knowledge of the effects of agriculture management on soil quality from a microbiological perspective.

It has been generally stated that better management practices can increase soil carbon incorporation into soils in comparison with traditional practices. Thus, conservation tillage, in contrast to traditional or conventional tillage, has as its major aim reducing soil erosion, water conservation and the enhancement of soil quality using as organic input the crop residue. However, several regions, mainly located under arid or semiarid climate, have low soil organic matter (SOM) contents. Precisely, some of these regions (Israel, California, Greece, South-East of Spain, etc.) have developed highly productive agriculture systems. However, overpopulation and inadequate agricultural practices threaten soil quality, initiating degradation and desertification phenomena, which in turn may affect the agricultural purposes. Organic amendments are an ecological way to fight against these processes which can negatively affect the main substrate of agriculture: the soil. Conservation management practices that increases C input or decrease organic matter oxidation increase soil C pool in agricultural lands and favours microbial activity which in turn release substrate for vegetal development.

Frequently, land-use changes that imply changes in plant communities and management systems could have deep effects on soil microbial communities and nutrient turnover in soils. For instance, plant diversity, plant genotype, management practices, soil type and season are factors controlling the dynamics of the microbial community, both in terms of diversity and function. Functional methods such as enzyme activities, community-level physiological profile (CLPP) or substrate-induced respiration (SIR) provide information regarding the functional capabilities of the soil microbiota related to the mineralization of specific compounds, while others based on molecular techniques or fatty acid profile aim to trace changes in the microbial community structure.

There is a lack of integration of these methods in a holistic way which helps to trace soil quality changes. All studies on soil quality indexes (SQIs) point to the complexity of the subject since a diversity of physical, chemical, microbiological and biochemical properties need to be integrated to establish such quality. Thus, several SQIs are available for assessing different agroecosystems but they are not suitable on larger scales, or even in similar climatological or agronomic conditions probably due to the natural complexity of soil ecosystem. For the same reason, more studies regarding the effects of agricultural management on soil microbial activity and microbial composition are needed. They will provide a better understanding of the interaction between soil, microorganisms, agriculture and climate through spatial and temporal scales.

9.2 STARTING POINT: AGRICULTURE AND SOIL SUSTAINABILITY

Over the past century, large strides have been made in increasing food production to satisfy human needs. Rapidly expanding population and increasing demand for a higher standard of living have combined to tax our research compatibilities and natural resources to sustain growth of food production potential. An awareness of the impact of these food production activities on the environment has arisen during the last few decades in developed countries, particularly on natural resources as soil. Environment decline aggravates scarcity of resources—as degraded ecosystems diminished yields and biodiversity.

Soil is a non-renewable (on a human timescale at least) natural resource. Thus, we must preserve and, if possible, improve its quality and productive capacity by applying constructive measures such as preventing erosion, improving root development and replacing the nutrients removed by crops and fauna. The soil has three basic roles: (1) a medium for growing plants since it acts as physical support and reservoir of moisture and essential nutrients; (2) regulator of the flow of water and (3) a system endowed with a certain capacity to lessen the harmful effects of contaminants through physical, chemical and biological processes.

The concept of sustainable agriculture is a widespread term. But a nagging question remains. If sustainability is very self-evident and sensible, why are sustainable practices not yet adopted by farmers? Close examination reveals that inherent conflicts occur between agriculture and the other segments of the society (production/conservation, short-term/long-term objectives). Sustainable agriculture also requires some minimal economic and education expenses which the farmers cannot afford.

Research on sustainable agriculture is frequently directed to provide better solutions to resources management and predict the impact of technologies. Consequently, any assessment of sustainability should include a time dimension, because the intent is to compare a new state with a reference state of the system. Sustainability is thus a dynamic concept and is also relative. From this point of view, Lynam and Herdt (1989) defined 'a sustainable system as one with a non-negative trend in measured output' which also is the non-negative trend in total factor productivity of the system.

In this chapter, we review the last knowledge of the effects of agriculture management on soil sustainability and soil quality from a biological perspective. As we will state, microbial populations play a key role on the sustainability of soil and hence we will stress the importance of soil microorganisms in soil quality from an agricultural point of view.

9.3 SOIL QUALITY, MICROBIAL ACTIVITY AND THE NEW FIELD OF PROTEOMICS

Soil microbial communities and their enzymatic activities play an important role in soil fertility and quality because most organic matter and nutrient transformations are mediated by such activities (Nannipieri et al., 1990). Intracellular and extracellular enzymes contribute to the overall enzymatic activity of soil (Burns, 1978). The same enzyme can be present in different locations inside the cell or in the extracellular matrix (Burns, 1982). Soil enzyme activities are controlled mainly by four kinds of enzymes: hydrolases, oxidoreductases, lyases and tranferases (Gianfreda and Bollag, 1996), the first two being the most important and studied from the point of view of agricultural soil functionality. The hydrolytic function of extracellular enzymes is important in decomposition and mineralization of organic matter, because these extracellular enzymes allow microbial cells to access biologically unavailable C and nutrients in SOM. For example, the hydrolytic activity of enzymes such as urease, proteases and glucosidases are essential in the N, P and C cycles, respectively. They catalyse transformation of the organic forms of the aforementioned elements to form more readily available to microorganisms and plants. Oxido-reductase enzymatic activities, as dehydrogenase and *o*-diphenol oxidase activities, have an important role in the organic matter oxidation and formation of humic substance precursors, respectively (Moreno et al., 2007). Majority of these enzymatic processes involves the production of the nutrients essential for microorganisms and plants, and some others are important in the detoxification of xenobiotic compounds (Gianfreda and Bollag, 2002). Thus, soil enzymatic activities have an important function both at an ecological level and at an agronomical level.

The enzymatic activity in soils can be measured in a simple form (spectrophotometrically) and provide us a valuable information of soil health. However, the most important limitation of the standard soil enzyme lab assay is that it measures potential enzyme activity in optimum conditions (temperature, pH and substrate concentration) which may reflect the actual *in situ* activity. The actual enzymatic activity in soil is the result of synthesis, persistence, stabilization, regulation and catalytic behaviour of the enzymatic proteins, present in the soil environment at the moment of the assay (Gianfreda and Ruggiero, 2006). The persistence and synthesis of a soil enzyme depends on environmental natural factors (temperature, substrates and water content) and on anthropogenic activities (e.g., agriculture, management practices, pollution). In most cases it is difficult to determine what factors influence the soil enzymatic activity if a suitable control is not selected. Soil enzymology has important challenges improving the knowledge of important aspects related with extracellular enzymes such as: synthesis and regulation, stabilization and turnover and role of temperature in controlling *in situ* enzyme activity (Wallenstein and Weintraub, 2008). Microbial synthesis of extracellular enzymes is dependent on C, N and energy requirements and, as a consequence, microorganisms should only produce extracellular enzymes when nutrients and soluble C are scarce. The understanding of microbial regulation of enzyme production will be improved by using genomic, transcriptomic and proteomic methods. After extracellular enzymes are released by microbes, the stabilization of enzymes is produced through interactions with clay minerals and humic matrix (Nannipieri et al., 1996). Stabilization may protect soil enzymes against proteolysis and other denaturing agents. However, the estimation of the stabilized enzymatic activity without disturbing soil is difficult with the standard enzymes methods. The explanation of this fact would be that the measured enzyme consists of different isoenzymes changing through time. Studies of these isoenzymes using proteomic tools will help us to explain the seasonal changes and inter-site differences in enzyme temperature sensitivity.

Microbial and enzyme activity is linked to the term of soil quality at a biological level. A variety of definitions of soil quality can be found in the literature, although the most general definition would be *the soil's capacity to function* (Karlen et al., 1997). Nowadays, the soil quality must be evaluated not only by its fertility or productivity but also by its capacity to develop its principal functions (Andrews et al., 2004). It is known that microbial activity, such as it was mentioned above, has an important role in that soil functionality. More specifically, the soil functionality depends on a great variety of enzymatic reactions that occur in its matrix. It is possible with the soil enzymatic activity assay to determine its functional diversity. Functional diversity of soil enzymes can be determined from: (1) its substrate specificity; (2) distinguishing different reaction mechanisms within a given enzyme function and (3) the possible sources of soil enzyme activities (Caldwell, 2005). Human activity, especially agriculture, has a strong influence on soil quality. Several authors proposed a set of microbial and biochemical parameters to estimate the effects of the anthropic actions on soil quality (Trasar-Cepeda et al., 1998; Pascual et al., 2000; Bastida et al., 2006b). One or few enzymatic activities should not be taken as indicator of a complex metabolic picture of soil functioning. Specifically some authors have used several microbiological and biochemical parameters to compare the microbiological soil quality of different agricultural management systems in semiarid regions (Moreno et al., 2008).

Metaproteomics studies the collective proteins from all the microorganisms in an ecosystem (Wilmes and Bond, 2006) and provides information about the actual functionality in relation to metabolic pathways and regulation mechanisms, more specifically than do functional genes and the corresponding ribonucleic acids (RNAs) (Benndorf et al., 2007). Alterations of gene expression cannot necessarily be interpreted to mean that an equivalent increase or decrease in enzyme production rates has occurred. One of the most important potential of metaproteomics is to help us know the changes of enzyme diversity (protein expression) as a consequence of soil degradation by human actions (agriculture and pollution), and moreover, it can be used to elucidate the source of the extracellular enzymes in soil. Several authors have demonstrated this potential (Singleton et al., 2003; Schulze et al., 2005). Proteomics has also been used to characterize unknown extracellular enzymes through extracellular separation protein on a native (non-denaturating) gel electrophoresis, and specific enzyme activities can be detected using fluorometric substrates. However, the challenges in microbial community proteomics, including representative protein extraction, separation, purification in complex samples (environmental samples) and the scarcity of genomics sequences for the microorganisms in these environmental communities, limit the development of this approach. The workflow to obtain information of the environmental microbial proteome is the shotgun proteomics. This method consists of several steps: (1) the extraction of a representative amount of proteins, (2) digestion of these proteins with a protease (trypsin), (3) purification of the peptide mixture obtained during the digestion, (4) separation of peptides by means of liquid chromatography, (5) acquisition of mass spectrum of these peptides, (6) fragmentation of the individual molecular ion peptides to obtain MS–MS spectrum and (7) database searching to obtain amino acid sequences from comparison of the mass spectrometry (MS) data with predicted data included in database and obtained of genomic sequence of microorganisms. Nowadays, the development of new MS technology allows a better resolution during the separation and detection of peptides from soil.

One attractive research area regarding proteomics is the interphase soil–plant, rhizosphere. In principle, the probability of protein identification in soil from these microenvironments is higher since a most abundant microbial biomass is developed under water-soluble carbon derived from vegetal growth and hence protein extraction could be simpler than in other no-plant soil systems. The physical location of proteins within the apoplasm/rhizosphere at the root–soil interface positions play a strategic role in plant response to biotic and abiotic stresses and the identification of these proteins may help to understand process of adaptation to the microedaphic changes. In addition, the identification of both mycorrhizal and rhizobial symbioses-related proteins (Bestel-Corre et al., 2004) will allow us to figure out the mechanism beyond these symbiotic relationships at a functional level.

9.4 APPLICATIONS OF ORGANIC AMENDMENTS IN SEMIARID SOILS AND CARBON SEQUESTRATION IN SOIL–PLANT SYSTEMS

Several regions located in arid and semiarid areas (Israel, California, Greece, South-East of Spain, etc.) have developed high-productive agricultural systems. However, the overpopulation and inadequate agricultural practices threaten soils located in these areas, initiating degradation and desertification phenomena, which in turn may affect the agricultural productivity. As we develop in this section, organic amendments are an ecological way to fight against these processes which can negatively affect the main substrate of agriculture: the soil.

Semiarid lands are extended along the tropical meridians and appear in different regions all around the world: from the steppes in Kenya and Central Valley in California to Mediterranean drylands in south-east Spain and in great part of the Outback in Australia. Semiarid climate appears in the surrounding of desserts, preceding the arid land. After Köppen land classification, this kind of climate is classified as BS (steppe climate) characterized by high evapotranspiration and low precipitation (150–500 mm).

Soils under semiarid climate conditions have in many cases low organic carbon content. Organic amendment constitutes a source of organic matter and supply the scarce input from plant debris in natural soils and from crop residues in agroecosystems in semiarid soils. The function of organic matter in soil is important: (1) promotion of soil aggregates; (2) source of plant nutrients; (3) reduction of loss of water content and other beneficial functions. The application of organic amendment improves soil physical, chemical and biological properties (Ros et al., 2003; Tejada et al., 2009). Human actions also play an important role in soil management under semiarid climate. Soil degradation and C loss are higher under dry and warm climates than under cooler and wetter conditions (Lal, 2009). Many studies have proven their beneficial use in soil functioning and quality (García et al., 1992; Bastida et al., 2007b), but also some studies have demonstrated negative effects (Plaza et al., 2004). In addition, many studies about the evolution of organic amendments in semiarid soils have been carried out at lab (Pascual et al., 1997), but few long-term studies (more than 5 years) have studied amendments addition in experimental field (Kanchikerimath and Singh, 2001; Walton et al., 2001; Bastida et al., 2007a).

The amendment addition to soils initially causes the degradation of easily degradable compounds, followed by the microbial consumption of the more recalcitrant compounds (Pascual et al., 1999). Different types of organic amendments have been applied in semiarid soils: crops residues, pig slurry, farm yard manure, municipal solid waste, olive mill waste, sewage sludge and so on. Some of these residues have to be adequate through composting processes in order to stabilize the high organic matter content and to control pathogens.

Surface application of amendments in degraded soils are not very effective (Rostagno and Sosebee, 2001; Walter et al., 2006), and ploughing in order to mix organic amendments is not always applicable to restore degraded lands (economic and natural barriers). Rostagno and Sosebee (2001) used biosolids on surface to combat soil erosion in soils from Texas. However, surface application did not produce a significant change in soil after 18 months. The application of amendment on soil on surface is not much effective in semiarid climate in order to improve soil qualities. Although these solutions have succeeded in agricultural soils with crop residue addition and no-till (Blanco-Canqui and Lal, 2007), application on soil surface is not always the best solution in land restoration.

Other studies have shown that not all kinds of residues are adequate to use directly into soils under semiarid climate. Plaza et al. (2004) observed that pig slurry was not appropriate to be applied to agricultural soils, because of the increase of microbial population, which used soil organic carbon (SOC) as the main source of carbon and diminish for the highest dose. In an N-rich environment, where C is limited, soil often responds to this perturbation with further increases in mineralization of indigenous SOM (White et al., 1997). Tejada et al. (2007) reported that the application of fresh beet vinasse has diminished soil properties in physical and biological due to the presence of sodium ions (Tejada et al., 2007).

The rate of application is another fundamental key. A sole dose in restoration tasks produces the recuperation of the plant cover (Walter et al., 2006; Bastida et al., 2007b) and amended soils maintain higher organic carbon content after a period. Some studies using organic amendment reported that only the highest doses maintain significant differences in respect to control whereas others maintained plant cover but do not show higher organic carbon levels. The use of organic amendments supply nutrients and improve soil structure promoting plant growth. The effects of the amendment diminish along time and plant cover must contribute to C inputs after a sole dose application of amendments. In dryland agroecosystems, the application of organic amendments must be accompanied with conservational management practices, in order to maintain C inputs and soil structure (Lal, 2004).

Other important influences of organic amendments constitute a source of feed to microorganisms. Reactivation of microbial community has beneficial effects in bulk soil, due to increased availability and conversion of compounds. This beneficial effect in agroecosystems allows the degradation and conversion of residues into available forms for plants and increases the microbial population.

In conclusion, the use of organic amendments in soils with low carbon content under semiarid climate allows a higher productivity and the restoration of degraded areas.

9.4.1 Carbon Sequestration in Soil–Plant Systems

The surface in world with natural plant cover is 30%, whereas agricultural is 40–50% (IPPC, 2007). In a sustainable soil system (soil carbon levels at steady state) outputs from mineralization, run-off and leaching are balanced with inputs from plant debris and fauna residues. The C pool retained in soil is around 2500 Pg (Lal, 2004) (Figure 9.1). Among ecosystems, boreal forest is the major contributor to global carbon stocks in the soil–plant system, followed by tropical forests, tropical savannas and grasslands, temperate grasslands and shrublands, wetlands, temperate forests, deserts, croplands and tundra (Janzen, 2004). In agroecosystems, the carbon level is sustainable by means of crop residues which are returned after the harvest and organic amendments, as indicated in other sections of this chapter. In this sense, agricultural practices are essential in the future behaviour of the SOM. Atmospheric CO_2 is immobilized via photosynthesis by plants, but a half of this C pool is returned immediately via respiration whereas almost all the other half return through heterotrophic respiration via microorganisms (Janzen, 2004) (Figure 9.2).

Chemical and biochemical transformation in soil from plant depends on composition of the residue. The lignin/N ratio has been proposed as better indicator of chemical stability of plant residues,

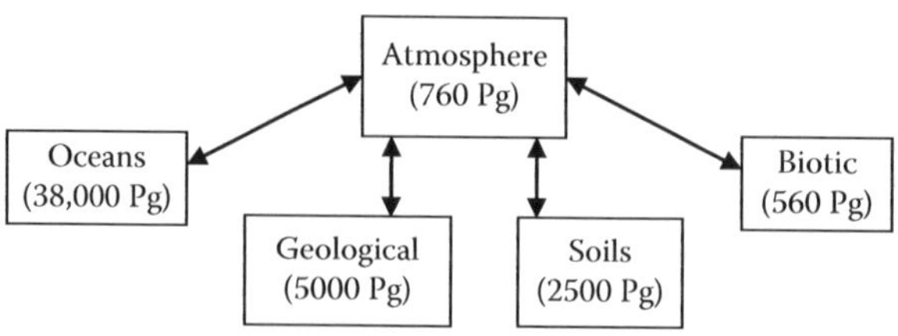

Figure 9.1 Global balance of carbon. (Modified from Lal, R. 2004. *Environ Manage* 33:528–544.)

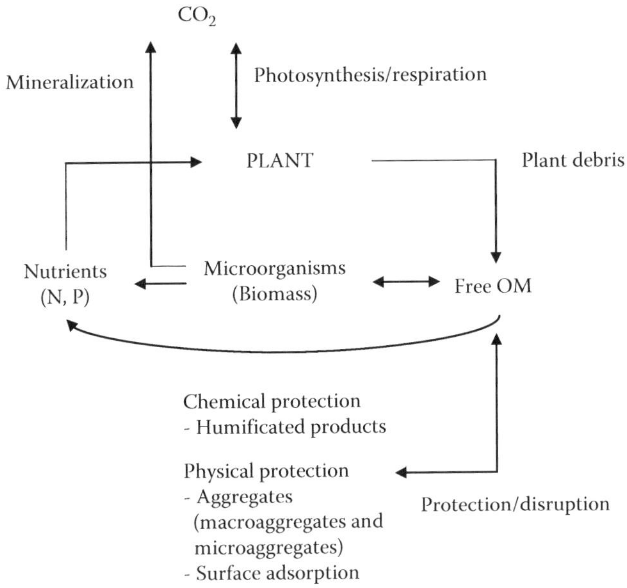

Figure 9.2 Carbon cycle in a soil–plant system. Main C inputs from vegetal debris are used by microorganisms to increase the population or to obtain energy producing CO_2. Physical and chemical protection mechanisms avoid microbial attack and increase C mean residence time.

where those difficult to degrade present high ratios whereas those easily decomposable present low ratios (Paustian et al., 1992). Nevertheless, the degradation rate of a molecular structure is not necessarily linked to residence time in soil (Gleixner and Poirier, 2001). Physical protection and biochemical protection can enlarge the mean residence time of a residue in soil (Figure 9.2). Rasse et al. (2005) also remarked the role of vegetal input, but focused on root carbon as major component in carbon sequestration in steady soil–plant systems (Rasse et al., 2005). The characteristics of the soil are also important. Hassink et al. (1997) found that the content of C and N was limited for fine soil fractions (silt and clay) in uncultivated soils, whereas the contribution of C and N in arable soils was higher (Hassink and Whitmore, 1997).

Lal (2004) suggested that better management practices can increase soil carbon incorporation into soils in comparison with traditional practices (Table 9.1). Conservation tillage, in contrast to traditional or conventional tillage, has as principal aim to reduce soil erosion, water conservation and enhancement of soil quality using as organic input the crop residue. Several types of conservational tillage have been described: no-till/strip-till, ridge-till and mulch till (Table 9.2). All types inside conservation tillage, keep at least 30% of the soil with crop residue after harvesting. In contrast, reduced-till and conventional tillage leave 15–30% and less than 15% of the crop residue, respectively. Soil carbon sequestration is a win–win strategy, because it improves soil quality and sequester atmospheric CO_2. Conservation management practices that increases C input or decrease organic matter oxidation increase soil C pool in agricultural lands (Paustian et al., 2000). Recommended management practices are then an important tool to enhance beneficial effects in agroecosystems. Although the conservation tillage is one of the best tools in the major kind of climate, the use of supplemental input through crop residue or organic amendment addition can help to increase SOC levels up to a high carbon state in equilibrium (Stewart et al., 2008). Follett (2001) estimated the C sequestration potential of selected practices in the United States, such as conservation tillage, crop residue/biomass management and including conservative reserve land programme in 30–105 million of Mg C per year. Smith (2004) also estimated the carbon sequestration potential

Table 9.1 Differences between Traditional and Recommend Management Practices

Traditional Tillage	Recommended Management Practices
Biomass burning and residue removal	Residue returned as surface mulch
Conventional tillage	Conservation tillage, no-till and mulch farming
Bare fallow	Growing cover crops during the off-season
Continous monoculture	Crop rotations with high diversity
Low-input subsistence farming and soil fertility mining	Judicious use of off-farm input
Intensive cropping	Integrated nutrient management with compost, biosolids and nutrient cycling
Surface flood irrigation	Drip, furrow or sub-irrigation
Indiscriminate use of pesticides	Integrated pest management
Cultivating marginal soils	Restoration of degraded soils through land-use change

Source: Data from Lal, R. 1997. *Soil Till Res* 43:81–107.

of European cropland (Smith, 2004). The managements studied included no-till cropland, zero-tillage, set-aside, perennial crops and deep-rooting crops, more efficient use of organic amendments (animal manure (AM), sewage sludge, cereal straw and compost), improved rotations, irrigation, bioenergy crops, extensification, organic farming and conversion of arable land to grassland or woodland. No-till and zero tillage agroecosystems were potentially able to increase carbon content about 2.4 Mt C per year, organic farming 3.9 Mt C per year and conversion to woodland 4.5 Mt C per year. Although the conservation tillage is one of the best tools in the main kind of climate, the use of supplemental input through crop residue or amendment addition can help to increase SOC levels up to a high-carbon state in equilibrium (Stewart et al., 2008). In addition, summer fallow is not recommendable in humid climates (where soil moisture and temperature, accelerate SOC oxidation) (Sherrod et al., 2003). Table 9.3 describes the influence of different land management in soil C sequestration in different locations around the world.

In case of the mineralization of the SOM, microorganisms play a fundamental role. Some authors pointed out the interest in the study of the role of the microorganisms in SOM transformation (Jastrow et al., 2007). Despite controversial studies on the C utilization efficiency between bacteria and fungi, soil carbon sequestration potential of fungi is supposed to be higher. Fungi cell walls are constructed with difficult degradable and persistent compounds such as chitin. In addition, mycorrhizal fungi sequester direct C through the symbiotic relationship with the plant and also their

Table 9.2 Characteristics of Some Conservational Tillage According to the Conservation Technology Information Center, USA

Types of Conservative Tillage	Soil Disturbance	Planting and Drill Mechanisms	Weed Control
No-till/strip-till	Undisturbed from harvest to planting except up to one-third of the row width	Disc openers, coulters, row cleaners, chisel or roto-tillers	Crop protection products, and cultivation as exception
Ridge-till	Undisturbed from harvest to planting except for strips up to one third of the row width	Sweeps, disk openers, coulter or row cleaners	Crop protection products and/or cultivation
Mulch-till	Full-width tillage involving one or more tillage trips which disturbs all of the soil surface and is done prior to and/or during planting	Chisels, field cultivators, disks, sweeps or blades	Crop protection products and/or cultivation

Table 9.3 Influence of Recommend Management Practices (RMPs) in SOC Sequestration in Agricultural Soils

RMPs	Location	Culture	Duration (year)	Other Characteristics	Treatment	ΔC^a. Mg C ha^{-1} yr^{-1}	Reference
Residue returned	Ohio, USA	Wheat	10	Stagnic Luvisol (0–50 cm)	NT 8 t ha^{-1} NT 16 t ha^{-1}	1.16 2.24	Blanco-Canqui and Lal (2007)
	Zimbabwe	Maize	9	Red clay Sandy	MT MT	0.25 0.28	Chivenge et al. (2007)
	Alaska, USA	Barley	17	Volkmar silt loam (0–20 cm)		0.04	Sparrow et al. (2006)
Conservation tillage	Sevilla, Spain	Corn Sunflower Legume	17	Entisol (0–20 cm)	CT	0.10	Melero et al. (2009)
(Traditional vs. conservational)	Senatobia, USA	Corn Cotton Soyabean	8	Grenada silt loam (0–15.2 cm)	NT	0.28 0.59 1.07	Rhoton (2000)
	Ohio, USA	Corn	2	Crosby silt loam	NT	0.38	Jarecki et al. (2005)
	Madrid, Spain	Grey pea Barley	5	Calcic Luvisol (0–30 cm)	NT ZoneT MinT	0.56 0.03 0.28	López-Fando and Pardo (2009)
	Rondonia, Brazil	Rice	2	(0–30 cm)	NT	0.23	Carvalho et al. (2009)
	North Dakota, USA	W–W–Su W–F	12		NT MinT	0.37 0.39	Halvorson et al. (2002)
	Illinois, USA	Maize Soya bean	12	Albic Luvisol (0–75 cm)	NT Chisel T	0.71 0.46	Olson et al. (2005)

continued

Table 9.3 (continued) Influence of Recommend Management Practices (RMPs) in SOC Sequestration in Agricultural Soils

RMPs	Location	Culture	Duration (year)	Other Characteristics	Treatment	ΔC[a]. Mg C ha^{-1} yr^{-1}	Reference
Crop rotations	Colorado, USA[b]	W–C–M–F	12	Pachic and Aridic	Site 1	—	Sherrod et al. (2003)
	Saskatchewan, Canada[b]	W–C–F	11	Argiustoll. Aridic	Site 2	0.07	McConkey et al. (2003)
		CC	12	Ustochrept. and	Site 3	—	
		W–C–M–F	11	Ustollic Haplargids	Site 1	0.21	
		W–C–F		(0–20 cm)	Site 2	0.18	
		CC		Haplic Kastanozen	Site 3	0.20	
		W–C–M–F		Haplic Kastanozen		0.25	
		W–C–F		Vertic Kastanozen		0.29	
		CC		(0–15 cm)		0.11	
		CC				0.13	
		CC				0.24	
		CC				0.10	
						0.22	
						0.03	
Organic addition[c]		C–W	-	(0–15 cm)	Biosolids application	4,19	Spargo et al. (2008)
		Soya bean	30	Aquic Hapludults	FYM + NPK	4.85	Bhattacharyya et al. (2007)
		Wheat		(0–45 cm)	Fertiliser	0.90	Sainju et al. (2008)
		Rye/cotton/ Corn	10	Acrisols (0–20 cm)	Poultry litter	0.51	
					Fertiliser	0.13	

Note: W: wheat; F: fallow; C: corn; Su: sunflower, M: millet; CC: continous cropping; M: maize. TT: traditional tillage; CT: conservational tillage; ZT: zone tillage; MinT: minimum tillage; T: tillage.

[a] Increases with respect to traditional tillage management.

[b] Increases with respect to W–F.

[c] Increases with respect to control soil.

hyphaes and segregated glomalin improve soil aggregate formation and stabilization (Rillig, 2004). In soil–plant systems, the rhizosphere stimulates and improves the formation of soil aggregates and roots, which are less decomposable, and could act as carbon source (Rasse et al., 2005).

Dyson (1977) made the first proposal of CO_2 atmospheric mitigation through soil and plant sequestration. Since then, the emission rate of CO_2 during the last three decades has been rising and greenhouses gases (GHG) emissions have increased by an average of 1.6% per year. The atmospheric CO_2 has increased due to the fuels burning and changes in land use. The atmospheric CO_2 concentration has increased from 280 ppmv before the industrial revolution in 1750 up to around 379 at present (IPCC, 2007). The perspectives in atmospheric emissions are expected to increase, all above in developing countries. In sectors, the principal source of GHG is fuel burning (energy supply and power for electricity and road transport). In 2004, CO_2 emissions were derived from energy supply 26%, about 19% from industry, 14% from agriculture, 17% from land use and land-use change, 13% from transport, 8% from residential, commercial and service sectors and 3% from waste. In agriculture, the principal GHG are N_2O and CH_4, whereas the CO_2 net flux in agricultural land remains unclear. Lal (2004) pointed out that SOC potential sequestration through land-use management is around 0.9 ± 0.3 Pg C y^{-1} (between one-third to one-fourth the CO_2 annual increase) and the cumulative potential of SOC sequestration is 30–60 Pg C (Lal, 2004). However, the limitations to consider in soil carbon sequestration have to answer how much carbon can contain in equilibrium and the quantity that can be preserved along time in equilibrium with the climate or inside the land management.

9.5 DO AGRICULTURAL PRACTICES AFFECT SOIL MICROBIAL DIVERSITY AND ACTIVITY?

Soil microorganisms are essential to the environment due to their role in many reactions, such as in cycling mineral compounds, decomposing organic materials and promoting or suppressing plant growth and various soil biophysical processes. Soil microbial community shows natural seasonal dynamics and their biodiversity and functions are influenced by various factors including vegetation and agricultural management. Moreover, diversity of microorganisms is a key parameter for soil structure, fertility and microbial metabolism. Metabolic activity of microorganisms in soils depends on the quality and nature of SOM rather than the quantity present. The high diversity of species ensures a great potential and soil resilience in the degradation processes.

Agricultural practices can affect soil quality, via changes in the functionality of the microbial communities. However, despite the importance of these changes in soil functionality beyond soil management, it is not still clear as to whether these changes course with a change of the microbial community structure responsible for these processes. In this section, we will focus on the last knowledge about these changes on diversity and microbial activity and the importance for the soil processes.

Frequent land-use changes which imply changes in plant communities and management systems might have profound effects on soil microbial communities and nutrient turnover in soils. Plant diversity (Carney and Matson, 2005), plant species composition (Kourtev et al., 2003), plant genotype (El Arab et al., 2001), management practices (Welbaum et al., 2004), soil type (Garbeva et al., 2004) and temperature (Waldrop and Firestone, 2004) are known to influence the soil microbial community structure. However, multiple interactions of these factors makes it difficult to predict changes.

Acosta-Martínez et al. (2003) compared the activity and microbial community structure of soils of different texture and different water management and tillage under continuous cotton rotation with peanut (*Arachis hypogaea* L.), sorghum (*Sorghum bicolor* L.), rye (*Secale cereale*) or wheat (*Triticum aestivum* L.) (Acosta-Martinez et al., 2003). In general, crop rotations and conservation tillage increased the enzyme activities in comparison to continuous cotton and conventional tillage, being correlated to soil organic C. The authors evaluated changes in the microbial community structure by fatty acid profile and suggested that the changes in microbial activity in soils with different texture are related to changes in the pattern of fatty acids, so probably with a change in the microbial community structure, as was suggested by other authors (Schutter et al., 2001). In parallel, Larkin et al. (2003) proved that different rotations of the same crop, potato, have a different effect of microbial functionality and community structure of the soil (Larkin, 2003). As these authors concluded, these types of projects are necessary to understand which type of management will optimize the soil microenvironment for reduced disease, increased production and long-term sustainability.

Trasar-Cepeda et al. (2008) analysed the role of hydrolytic enzymes as quality indicators of agricultural and forest soils and develop a critical work about its applicability (Trasar-Cepeda et al., 2008). Soil preparation and tillage, fertilization, the use of machinery and harvesting all affect the soil environment and can cause a progressive loss of soil quality. At present, the quality of a soil is defined not only in terms of its productive capacity, but also takes into consideration that the soil forms an integral part of the ecosystem. Consequently, even when soil productivity has to be maximized, environmental effects must be kept to a minimum (Doran and Parkin, 1994). For evaluation of these changes, sensitive indicators are needed. Trasar-Cepeda et al. (2008) analysed the contradictory results of different authors about the rates of enzyme activity. Various authors have pointed out that ploughing causes an increase in biochemical activity in agricultural soils due to the exposure of new surfaces as soil aggregates are broken up while others have reported a decrease in enzyme activities due to the decrease in organic matter content as a result of the mixing of horizons by ploughing.

The conclusions of the paper by Trasar-Cepeda et al. (2008) help in understanding of the role of enzyme activity. Soil use causes a clear decrease in organic matter content, but enzyme activity does not always follow the same pattern and may either increase or decrease. However, the values of enzyme activity per unit of carbon were always higher in soils subject to human interference than in oak soils, which indicate relative enrichment in enzymatic activity in agricultural and forest soils. Furthermore, the process of stabilization becomes more intense the greater the loss of organic matter suffered by the soil, and all enzymes are not affected in the same way, with some being more affected than others. Land use not only generates higher values of enzyme activity per unit of C than in oak soils, but also greater increases in the specific activity as the C content decreases, which suggests a process of enrichment of the hydrolytic enzymes in organic matter, thereby maintaining a high metabolic activity. They concluded that although land use modifies the enzyme activity of soils, the use of enzymes as indicators of the change in soil quality is not particularly useful because of the complexity of the observed behaviour. However, these handicaps, coming from the high sensitivity of enzyme to soil changes, can be solved when they are integrated together with other indicators in multi-parametrical indexes of soil quality (Bastida et al., 2008), as we develop in other sections of this chapter.

Since microorganisms are the main contributors in the conversion of residues into soil carbon and the release of CO_2 back to the atmosphere during residue decomposition and soil carbon mineralization, it would be appropriate to conceive that soil and cropping system management would not only affect the microbial populations but also their diversity and subsequent soil carbon storage (Asuming-Brempong et al., 2008). The following text is dedicated to an evaluation of the agricultural organic land uses and its effects on the microbial community.

Asuming-Brempong et al. (2008) assessed the effects of different management practices in the microbial structure of Haplic Lixiosols, in relation to changes in C sequestration and microbial biomass carbon (Asuming-Brempong et al., 2008). These authors realized a complete statistical study on the relationship of different microbiological indicators of biomass and activity in soils under different fallow management treatments. These authors conclude that those land management that lead a higher C sequestration did not involve a change in the microbial community structure.

Monokrousos et al. (2006) analysed the impacts of organic and conventional systems of *Asparagus officinalis* in chemical, biological, enzyme activities and CLPPs in soil, such as Biolog. They concluded that the most important indicators discriminating the systems were the microbial biomass C and N. The enzyme activities of amidohydrolases (l-asparaginase, l-glutaminase, urease) and phosphatases (alkaline and acid phosphatase) were higher in organic areas than in the conventional one, but in general the potential functional diversity was not different, except for the carbohydrate consumption. It is well known that carbohydrates constitute a large proportion of root exudates and they can shape soil microbial communities (Bardgett and Shine, 1999). This apparent scarcity of differences in the microbial community, in terms of metabolic potential, is in agreement with Asuming-Brempong et al. (2008).

Tu et al. (2006) evaluated the impact of different tomato organic farming systems in the microbial biomass and its activity of a sandy soil (Tu et al., 2006). The organic substrates used included composted cotton gin trash (CGT), AM and rye/vetch green manure (RV). Additions of CGT and AM enhanced microbial biomass and activity, and N supplies for plants. Surface mulching was effective in sustaining soil microbial biomass and activity. These results indicate that the amounts and quality of organic C inputs can profoundly impact microbial properties and N availability for plants, highlighting the needs for effective residue management in organic tomato farming systems. The effect of organic management in the microbial functionality is stressed in the following text.

Soil carbon sequestration is proposed as one of the major ways of capturing atmospheric carbon for safekeeping, thus reducing the ever-increasing atmospheric carbon load. Carbon is primarily stored in soil as SOM. The majority of the accumulation of soil organic C due to additional C inputs is preferentially sequestered in microaggregates or within small macroaggregates (Kong et al., 2005). Carbon stored can remain in soils for millennia or be quickly released back into the atmosphere depending on soil management and cropping system. The use of cover crops in soil management plays an important role in sequestering carbon in soil and also increases the total microbial biomass by shifting the microbial community structure towards a fungal-dominated community, thereby enhancing the accumulation of microbial-derived organic matter (Six et al., 2006). Growing leguminous cover crops also enhances biodiversity, through the quality of residue input and soil organic pool (Fullen, 1998).

Extensive research has been conducted and reported on different groundcover management systems (GMS), including the use of mulches (biomass, plastic and geotextiles), cover crops and tillage (Yao et al., 2005). GMS are important in managing fruit-tree orchards because of their effects on soil conditions, nutrient availability, tree growth and yields (Yao et al., 2005). These authors studied the long-term effects of different GMS on biotic and abiotic factors in an apple orchard soil. Four GMS treatments—Pre-emergence residual herbicides (Pre-H), post-emergence herbicide (Post-H), mowed-sod (Grass) and hardwood bark mulch (Mulch)—were evaluated. This was a complete study which studies different indicators related to microbial abundance, changes in bacterial and fungal diversity, microbial activity nematode numbers and vegetal growth and yield productivity. Soil respiration was higher in the Mulch than in the other treatments. The treatment with pre-emergence residual herbicides showed the lower culturable bacteria while the Grass treatment showed the largest population of culturable fungi. The observed differences in the microbial composition and the relative abundance of bacteria, fungi and nematodes appear to be within the range to which apple roots can adapt over time. Differences in microbial community composition and activity do not necessarily lead to strong changes in soil conditions for the apple roots.

Iovieno et al. (2009) studied changes in the microbial activity of horticultural soil under different doses of organic and mineral fertilizers during 3 years of repeated treatment. Results suggest that in Mediterranean soils repeated annual compost addition is an advisable practice to restore or to preserve soil fertility. Other authors showed that the addition 18 years ago of different dose of the organic fraction of solid wastes lead to a threshold of microbial activity (Bastida et al., 2007a). Perhaps repeated-addition of smaller doses may increase the maximum level of microbial activity in soils.

Crecchio et al. (2007) studied the long-term impact of crop residue management on the microbial biomass and on the activity and community structure of soil bacteria in a clay soil of Southern Italy, where a monoculture of durum wheat (*Triticum durum* Desf.) was grown in semiarid conditions, and burning or incorporation of post-harvest plant residues were typical practices (Crecchio et al., 2007). They did not find changes on the bacterial community due to the crop management, concluding that this long-term management is a suitable way to manage crop residues. Surprisingly, these authors did not find changes in the microbial biomass C and ATP in soils under crop incorporation. This practice supposes an increase in organic carbon in soil and this, even in long-term experiments, has supposed an increase in microbial biomass and its activity under the same climate in soils under natural conditions without agricultural management (Bastida et al., 2007a).

Roldan et al. (2005) suggested that conservation tillage practices have benefits for soil microbiological quality of sorghum cultivation under subtropical climate. They compared the effects of conservation practices (no tillage or reduced tillage) and conventional tillage practices (muldboard ploughing) on physical and biochemical indicators. Conservational practices allowed a long residence time of vegetal rest in soil and increases organic carbon and enzyme activities. No tillage systems improved soil quality regarding physical and biochemical indicators, contributing a long-term sustainability of this crop system under subtropical climate conditions.

While conservational agricultural practices are globally beneficial to soil quality, reduced tillage and residue conservation may trigger negative environmental side effects, such as increased N_2O emissions (Baggs et al., 2003). Reduced availability of oxygen and air space in soils, together with the decomposition of mulched crop residues in the superficial soil layers, are likely to favour anaerobic processes such as denitrification (Baggs et al., 2003). Baudoin et al. (2009) evaluated the effect of direct-seeding mulch-based cropping systems versus conventional tillage on the abundance and activity of denitrifiers microbial community (Baudoin et al., 2009). These authors found that denitrification activity was increased by seeding mulch cropping system and it was related to the C and N content. In addition, they found that this system had a strong effect on the microbial community structure. Previously, other authors have found that alternative management (such us direct-seeding mulch-based cropping systems) may benefit chemical, biological and structural properties of soil (Paustian et al., 2000) mainly via an increase in the amount of organic carbon. Baudoin et al. (2009) conclude that management has a stronger influence than fertilization dose on the denitrifier community in the Highlands of Madagascar. These results partially agree with the ones of Graham and Haynes (2005). These authors analyse the catabolic diversity of soils under different land management (kikuyu pasture (*Pennisetum clandestinum*), annual ryegrass pasture (*Lolium multiflorum*), maize (*Zea mays*), preharvest burnt sugarcane (*Saccharum* spp.) and undisturbed native grassland on the catabolic diversity of soil microorganisms by Biolog plates and SIR in South-African soils. CLPPs (Biolog) results indicated large differences in catabolic capability between sugarcane soil under trashing and pre-harvest burning but there were little effect of fertilizer applications. However, they found differences for fertilization treatments with SIR and in general, those treatments that increased organic carbon content, also enhanced catabolic diversity. Conversion from pre-harvest burning to green cane harvesting greatly increased catabolic evenness and richness and therefore presumably it also tended to increase the resilience of the soil to stress and disturbance, particularly in relation to decomposition functions. These considerations about the catabolic diversity of a system are probably more realistic than the simple description of species on soil (Garland and Mills, 1991). The absence or presence of different species does not ensure the increase or decrease in a particular soil function, which is called redundancy. Functional methods such as Biolog or SIR provide information regarding to the functional capabilities of the soil microbiota related to the mineralization of specific compounds.

According to Steenwerth et al. (2005) agricultural intensification (fertilizer, pesticide and herbicide application, irrigation) decreased microbial diversity and reduced the resistance of the microbial community to changes after perturbation (Steenwerth et al., 2005). In contrast, Stark et al. (2007) detected only short-term changes in the microbial community structure after fertilizer application.

Some authors evaluated changes in the microbial community structure of tropical soils from Kenya under different management (Bossio et al., 2005). They found a change in the structure of the microbial community by phospholipid fatty acid analysis (PLFA) and analysis of 16S rRNA genes. Molecular analysis concluded that soil is the primary factor in the development of the microbial community, but also land management (agricultural versus wooded soils) affects.

Nogueira et al. (2006) evaluated the effects of soil land-use change on soil microbiological quality and diversity based on 24 indicators and DGGE patterns (Nogueira et al., 2006). The sites differed significantly in many parameters, attributable to differences in the present vegetation and history of soil use, which affected—quantitatively and qualitatively—the organic matter inputs and, consequently, soil chemical and biological properties. In general, sites under native forest showed higher microbiological activities, followed by reforested sites, except eucalyptus. These sites usually contrasted with fallow, wheat-cropped, and for some variables, the eucalyptus sites. Fifty years of agricultural use after forest removal showed declines in some soil biological activities and processes mediated by soil microorganisms.

Similar patterns have been reported for other tropical soils (Islam and Weil, 2000; Lemenih et al., 2005). Nogueira et al. (2006) suggested that all these changes are mediated by changes in the moisture. Total organic carbon, microbial biomass C and N and ammonification rate were the indicators which drastically changes with the soil use. The change of an Atlantic forest to Eucalyptus plantations or wheat crop decreased microbial diversity evaluated by DGGE. Although water content in the soil samples collected from the naturally regenerated secondary forest was as low as at the fallow site, biomass C and N and basal respiration rates were comparable to those of the native forest soil, which can be attributed to organic matter inputs during 20 years of natural restoration after about 10 years of pasture. Clusters of microbial diversity and microbial activity and biomass were quite similar. By means of these two techniques, it is clear that the reforestation strategies containing native species tended to return the soil to its native forest condition. On the contrary, the strategy that employed an exotic species tended to alter soil chemical and microbial characteristics, as did agricultural systems. Fallow, sometimes used as a strategy for soil management, proved to be very negative with regard to soil microbiological and chemical properties.

Changes in moisture can be related to seasonal effects. Spedding et al. (2004) demonstrated that seasonal effects should be considered when soil quality under different agriculture uses wants to be achieved (Spedding et al., 2004). These authors analysed the microbial community dynamics, in terms of biomass and structure. PLFA showed a better distinction through seasons than treatments. Soil microbial biomass C increased when crop residue from maize are not removed probably because it offered a pool of organic compounds able to stimulate microbial growth and activity.

Hamer et al. (2008) reported a decrease on carbon and nitrogen content of a Cambisol in NE Germany after six years of intensive management. The use of substrates and the turnover of the microbial biomass were more efficient under land management than in the fallow treatment. However, the structure of the microbial community evaluated by PLFA was more influenced by seasonal changes than soil management, but still has some influenced, as it was described by Hedlund (2002). Intensive land management increased gross N mineralization and qCO_2, which was not related to an increase in the microbial biomass (no increase in total PLFA was observed under land management compared to fallow treatment).

Differences in microbial community composition as a result of past land use may have important implications for ecosystem function (Fraterrigo et al., 2006). Changes in soil microbial community composition and belowground ecosystem functions such as decomposition and nutrient transfer may create feedbacks that affect plant growth and the aboveground plant community composition and diversity (Schloter et al., 2003). However, management practices could not only affect microbial community structure and its activity, but it also affects the cultivar genotypes. El Arab et al. (2001) indicated that under controlled environmental conditions, wheat cultivars of different genotypes exhibit distinct microbial colonization in their rhizosphere by PLFA.

Other important issue is the location and mode of microbial activity in soil. Micro-scale investigation allows a deeper understanding of processes underlying the relationships between organic matter, microorganisms and functions in soil, and therefore bolsters the interpretation of modifications induced by soil management. In fact, agricultural management could change the distribution of microbial biomass and its activity. Lagomarsino et al. (2009) studied microbial biomass and enzyme activity of a soil from central Italy under conventional tillage and no tillage in a rotation of *Triticum durum* and *Zea mays* (Lagomarsino et al., 2009). These authors found that no-tillage treatment significantly increased organic input, mainly in the coarser fractions, enhancing enzyme activities and the functional diversity of the microbial community. This in agreement with Peigné et al. (2007) who indicated that tillage management plays a key role in SOM turnover, aggregation and stability, as well as in microbial abundance and carbon storage (Peigne et al., 2007). Under no-tillage management the amount of water-stable macroaggregates is generally greater and the rate of macroaggregates turnover is approximately half the rate of that under conventional tillage (Six et al., 2000).

The distribution of microbial biomass and activity in particle-size fractions may reflect changes of substrate availability and quality within fractions (Lagomarsino et al., 2009). Several studies have shown that soil enzymes are heterogeneously distributed across particle size classes: in the coarse and fine sand (Kandeler et al., 1999; Stemmer et al., 1999; Marhan et al., 2007), in the silt (Powlson and Jenkinson, 1981) and in the clay fraction (Marx et al., 2005), or bimodally distributed between the sand and clay fractions (Marx et al., 2005). Moreover, the distribution pattern also depended on the type of enzyme: xylanase was more related to the coarse sand fraction (Kandeler et al., 1999; Stemmer et al., 1999; Marhan et al., 2007). In contrast, Poll et al. (2003) found this enzyme to be mainly present in the clay and silt fractions.

Another aspect to note is the effect of soil microbial diversity as a factor controlling plant diseases. Researchers aimed to study the effects of different soil management (organic, sustainable and conventional) in physical, chemical and biological soil quality properties and the incidence of *Sclerotium rolfsii*, causal agent of Southern blight (Liu et al., 2007). Summarizing, at a physical level, soil porosity and water content were higher in organic farms, and from a biological point of view, microbial biomass, respiration and net mineralized nitrogen were higher in organic and sustainable farms than in conventional ones. In addition, bacterial functional diversity and species diversity were lower in conventional than in organic and sustainable managements. In reference to the biocontrol effect, soils under organic management showed less Southern blight. Even when it was difficult to correlate with other indicators, soil microbial communities may have been suppressive to disease (Garbeva et al., 2004).

The structure of microbial communities in a given soil, especially in the rhizosphere soil, can change with infection of soilborne pathogens such as *Phytophthora cinnamomi* (Yang et al., 2001) and *Rhizoctonia solani* (van Elsas et al., 2002). Shishido et al. (2008) analysed the response of soil microbial community in apple orchard in response to season, land use and infection by violet root rot (Shishido et al., 2008). Their results of the CLPP, DGGE and fatty acids profile suggested qualitative and quantitative effects of season and land use over the soil microbial community, but the effect of infestation was qualitative and mainly most pronounced in the adjacent bush soil and not in the orchard.

9.6 INDEXES OF SOIL QUALITY IN AGROECOSYSTEMS

As we described in different sections of this chapter, agricultural practices exert an influence in the microbial community, even when this effect may depend on the soil type and seasons. The main question is: does it exist a way to quantify the changes of the soil microbial quality in agroecosystems?

A variety of definitions were proposed for the term soil quality during the 1990s (Doran and Parkin, 1994; Harris et al., 1996; Karlen et al., 1997), ranging from a purely agricultural point of view to a more environmental perspective. In 1995, Soil Science Society of America (SSSA) suggested soil quality as *the capacity of a specific kind of soil to function, within natural or managed ecosystem boundaries, to sustain plant and animal production, maintain or enhance water quality and support human health and habitation.* From this moment, *soil quality* emerges as a branch of environmental studies since soil is a critically important component of earth's biosphere. Nowadays, the definition of soil quality still lacks a well-defined term and individual investigators use the term soil quality arbitrarily depending on their research, including those parameters of interest for each scientist.

However, despite the difficulty involved in providing a definition, the maintenance of soil quality is critical for ensuring the sustainability of the environment and the biosphere (Arshad and Martin, 2002), although this is a complex theme due to the importance of the climate, soil characteristics, plants, anthropic factors and their interactions. Indeed, soils are subject to natural or environmental degradation, often accompanied by erosion, leaching, even without human intervention (Popp et al., 2000). *An SQI could be defined as the minimum set of parameters that, when interrelated, provides numerical data on the capacity of a soil to carry out one or more functions. A soil quality indicator is a measurable property that influences the capacity of a soil to carry out a given function* (Acton and Padbury, 1993). All studies on SQIs point to the complexity of the subject since a diversity of physical, chemical, microbiological and biochemical properties need to be integrated to establish such quality (García et al., 1994; Halvorson et al., 1996).

Bastida et al. (2008b) reviewed the concept of soil quality under a biological perspective by the description of different indexes established for natural and agricultural soils (Bastida et al., 2008). Probably the most straightforward index used in the literature is the metabolic quotient (qCO_2, respiration to microbial biomass ratio). Physiologically, this index describes the substrate mineralized per unit of microbial biomass carbon. The metabolic quotient conceptually based on Odum's theory of ecosystem succession has been widely used as an ecosystem development (during which it supposedly declines) (Insam and Haselwandter, 1989) and disturbance index (supposedly increases), and also acting as an indicator of ecosystem maturity (Anderson and Domsch, 1985). Anderson and Domsch (1990) observed a decrease on qCO_2 in soils under monoculture comparing to soils under continuous crop rotations, suggesting that richness of organic C from different cultures benefits respiration (Anderson and Domsch, 1990). In addition, this ratio has been widely used as a good indicator of the alterations that take place in soil due to heavy-metal contamination (Liao and Xie, 2007), deforestation (Bastida et al., 2006a), temperature (Joergensen et al., 1990) or changes in soil management practices (Dilly et al., 2003). However, the index has also been criticized for its insensitivity to certain disturbances and to the ecosystem's development whenever stress increases along successional gradients (Wardle and Ghani, 1995).

The microbial biomass C to total organic C ratio (Anderson and Domsch, 1990) has been proposed as a more sensitive index of soil changes than total organic C, since the microbial biomass of a soil responds more rapidly to changes than organic matter (Powlson and Jenkinson, 1981). This means that if a soil is in a degradation process, this degradation could be primarily detected by microbial changes whereas changes in organic matter would not be detected at an early degradation state. Sparling (1997) proposed that this ratio could be used to compare soils with different organic matter fractions. Jenkinson and Ladd (1981) also proposed that for cultivated soils, a value of 2.2 reflects a good equilibrium between both C fractions (microbial biomass C and total organic carbon). Nevertheless, changes in this ratio due to different manuring or cropping practices could be masked by climatic factors that affect this ratio (Insam et al., 1989). As we noted in the previous section, season may affect soil microbial activity and microbial community composition.

Ratios between a given enzymatic activity and microbial biomass size might be useful as SQIs (Landi et al., 2000). Increases in enzymatic activity per unit of microbial biomass C may be due to both increased enzyme production by microorganisms, which are released and immobilized in clays or humic colloids (Kandeler and Eder, 1993), and an increase in the concentration of substrates available for enzymatic activity.

Although they are easy to apply, the use of two parameters in an SQI has almost the same limitations as the use of one parameter: the lack of information. Therefore, to obtain indexes that provide and integrate more information on the quality of a soil, multiparametric indexes have been developed for agro-ecosystems and for non-agricultural soils. Regarding soil enzyme activities, Trasar-Cepeda et al. (2008) stated single soil enzyme indicators are not sufficient for the establishment of soil quality since they are quite sensitive and their response can be quite ambiguous depending on many factors (Schloter et al., 2003). Thus, in the last years a variety of multiparametric indexes of soil quality under agricultural management have been developed. They tried to combine the sensitivity of microbiological indicators but in a coherent and holistic way.

The first multiparametric index for soil quality was probably that established by Karlen et al. (1994a). These authors used the framework established by Karlen and Stott (1994), based on the utilization of normalized scoring functions for evaluating a production system's effect on soil quality in Lancaster (WI, USA). This framework uses selected soil functions, which are weighted and integrated according to the following expression:

$$\text{Soil quality} = q_{\text{we}}(\text{wt}) + q_{\text{wma}}(\text{wt}) + q_{\text{rd}}(\text{wt}) + q_{\text{fqp}}(\text{wt}), \tag{9.1}$$

where q_{we} is the rating for the soil's ability to accommodate water entry, q_{wma} is the rating for the soil's ability to facilitate water transfer and absorption, q_{rd} is the rating for the soils's ability to resist degradation, q_{fqp} is the rating for the soil's ability to sustain plant growth and wt is the numerical weight for each soil function. Thus, overall soil quality score is given by the sum of all function scores.

Although the systems based in Karlen and Stott (1994) have been widely developed (Karlen et al., 1994a,b; Glover et al., 2000), their major handicap resides in the fact that weighting is subjective and not depending on any mathematical or statistical method. These weights are provided attending to the importance of a soil function in fulfilling the overall goals of maintaining soil quality under specific conditions or purposes.

Using Karlen and Stott's system and the same functions, Karlen et al. (1994a) evaluated the effect of different applications of residues on the long-term soil quality under *Zea mays* culture, showing that crop residue had a positive effect on soil quality. These authors made a brief but clear justification of selected parameters, indicating the importance and meaning of each indicator. Thus, porosity was selected as an indicator of the soil capacity to accommodate water entry, earthworm population was chosen as a surrogate for macropore number and water stability aggregates were chosen to reflect the ability of soil to resist degradation. Karlen et al. (1994a) gave great importance to these water relations, using a variety of physical, chemical and biological parameters. The values of parameters were standardized between 0 and 1 using standardized scoring functions, but weighted scores were subjective and authors did not use any mathematical approach. As these authors stated, one objective of this chapter was to demonstrate the methodology for computing an SQI, rather than provide a definitive tool for environmental evaluation. In this sense, this work has been successful due to its wide utilization, being the reference guide for other indexes. Karlen et al. (1994b) applied this index to evaluate the effects of tillage on soil quality showing that the absence of tillage improved soil quality.

Using the method developed by Karlen and Stott (1994) and Karlen et al. (1994a) and similar quality functions, Glover et al. (2000) established an index that reflected the effect of apple orchard management on soil quality. These authors showed that the integrated production system led the highest score of soil quality. With a similar model to Karlen's and the function resistance to biochemical degradation, Masto et al. (2007) established an SQI with the aim to evaluate and quantify the long-term effect of different fertilizer and farm yard manure treatments in a rotation system with *Zea mays*, *Pennisetum americanun*, *Triticum aestivum* and *Vigna unguiculata in* New Delhi (Masto et al., 2007). This SQI showed that soils subjected to inorganic fertilization and organic amendment treatments had higher SQI values than soils which were not fertilized or amended. A novelty of this study is that the authors also calculate SQI without weighting factors, indicating that the use of these factors did not affect the relative ranks of individual treatments, as was found previously by Andrews et al. (2002a).

Andrews et al. (2002a) were the first to compare multiparametric methods designed to establish the quality of agricultural soils, focusing on indicator selection methods: from opinion of experts (Doran and Parkin, 1994) to different mathematical–statistical systems. This mathematical development was later applied by different authors (Sharma et al., 2005; Bastida et al., 2006b). The objective of this work (Andrews et al., 2002a) was to examine the relative effectiveness of several soil quality indexing methods using assessment of complex vegetation production systems in northern California. This work may be considered the mathematical base which many authors have since followed to establish indexes in agricultural ecosystems. In addition, Andrews et al. (2002a) also focussed on the methodology for indicators standardization and transformation, from linear to non-linear methods, indicating that the latter requires depth knowledge about each indicator's behaviour and function within the system and that non-linear transformation was more representative of system function (Andrews and Carroll, 2001). Andrews et al. (2002a) concluded that both the expert opinion methods and statistical methods (mainly principal component analysis) to select a minimum set of indicators is valid to establish the soil quality in vegetable production systems (Andrews et al., 2002a). The disadvantage of statistical methods resides on the need for a wide set of data, but, conversely, once the indicators are selected, there may be no need to assess soil quality over time. However, system and selected indicators could be more difficult to interpret by statistical approaches than expert opinion systems. Indicators selected from expert opinion (soluble phosphorus, pH, electrical conductivity, sodium absorption ratio and SOM) differed from those selected by principal component analysis (soluble phosphorus, pH, calcium, sodium and total nitrogen). For most of the SQIs established by these authors, organic systems received a higher SQI value than conventional or low-input treatments.

Using the same framework based in mathematical methods, Andrews et al. (2002b) evaluated the effects of different soil management practices involving the application of amendments on soil quality under different crops (tomato, garlic and cotton) in Northern Carolina and, by means of their index, observed an increase in quality in organic cultivation systems (Andrews et al., 2002b). The indicators selected in this case were chemical and physical–chemical in nature (Table 9.1), which can be considered as more useful for practical purposes since the parameters were more accessible. Indeed, the method proposed by Andrews and co-workers in this work has been widely used by many authors (Sharma et al., 2005; Rezaei et al., 2006). Indicators selected by factor analysis were bulk density, Zn, water-stable aggregates, pH, electrical conductivity and SOM content, being electrical conductivity and organic matter the indicators which most importance has in SQI. The equation proposed by Andrews et al. (2002b) is

$$SQI = \sum_{i=1}^{n} 0.61xS_{SQMi} + 0.61xS_{EGi} + 0.16xS_{pHi} + 0.16xS_{WSAi} + 0.15xS_{Zni} + 0.09xS_{BDi} \quad (9.2)$$

where S is the score for the subscripted variable and the coefficients are the weighting factors derived from the principal component analysis.

Sharma et al. (2005) used the same methodology of Andrews et al. (2002a) based on multivariate analysis to compare and choose different cultivation systems according to their effect on dryland soil quality; these authors developed a wide mathematical model to choose parameters in a thorough work that refers to numerous samples and sampling sites. These authors concluded that soil under conventional tillage system with *Gliricidia* loppings and highest application of nitrogen obtained the highest SQI value. However, the range of variation of index is really low (from 0.90 to 1.27) and for this the differences between treatments could be very difficult to achieve.

Later, following the same methodology as described above, Andrews et al. (2004), established a computer-based mathematical model (SMAF, *Soil Management Assessment Framework*) to evaluate soil functions and quality (Andrews et al., 2004). SMAF represents an additive non-linear standardization tool for evaluating a soil's quality from its functions: (1) microorganism biodiversity and habitat, (2) filtration of contaminants, (3) nutrient cycling, (4) physical stability, (5) resistance to degradation and resilience and (6) water relations. In this work the environmental concept of soil quality takes on more importance. One or more indicators are used for each function, giving a final pool of parameters in which chemical, physical and physico-chemical indicators predominate, such as in the previous contributions by Hussain et al. (1999) and Rezaei et al. (2006). This tool was subsequently used by Cambardella et al. (2004) and Wienhold et al. (2006) to objectively evaluate the effect of agricultural management practices on soil quality, using food production and nutrient cycling as quality functions. Erkossa et al. (2007), using methodology of Andrews et al. (2004) and different scoring functions, proposed to evaluate soil quality attending to different parameters (Table 9.1) (Erkossa et al., 2007). The aim of this work was to compare the effect of four land preparation methods on soil quality: broad bed and furrows, ridge and furrow, green manure and reduced tillage. However, no significant differences were observed for soil quality values between the different treatments. Conversely, Lee et al. (2006) established an index for evaluating the effects of applying manure compost to soil under crops, which showed that quality increased significantly with the amendment dose (Lee et al., 2006).

Some quality functions have been largely overlooked when establishing quality indexes in agroecosystems. One such function is the regulation of greenhouse effect gases by using indicators such as gas flow or C sequestration (Liebig et al., 2001). This omission seems surprising, especially now that the soil is regarded as a sink of C and, especially, that plants are CO_2 capturers, at a time when the greenhouse effect has become a matter of public concern, as previously stated. Another factor that has not been widely used for establishing SQIs is climate, since this is, if not an intrinsic part of the soil, is a very important factor in its formation and development.

Despite the steps proposed on SQIs of Karlen et al. (1994) and Andrews et al. (2002) that have been the most widely used in establishing indexes at both agronomic and environmental levels, some authors have looked at other alternatives. Kang et al. (2005) used a trigonometric approach based on three subindexes (nutritional, microbiological and crop related) to establish a *Sustainability Index* in soils under wheat culture amended with manures in Punjab (India), noting that the quality increased with amendment (Kang et al., 2005). This work used a wide variety of chemical, physical and biological parameters, while Mohanty et al. (2007) established a physical index of soil quality under wheat and rice cultivation based on four parameters (real density, organic matter content, resistance to root penetration and aggregate density) using multiple regressions (Mohanty et al., 2007). They concluded that the application of plant residues will maintain soil quality for long periods of time and that this method could be extended to other soils under different cropping and climatic conditions.

While physical and chemical indicators have been widely used to elaborate indexes, only a few indexes have an enzymatic nature despite the fact that enzymes have been proposed as sensitive indicators of changes in soil (Nannipieri et al., 1990). For example, Puglisi et al. (2006) proposed three indexes of soil alteration, using different enzymatic activities to establish an index of soil degradation due to agricultural practices, including crop density and the application of organic fertilizers in different locations of Italy. The first index (AI 1) is defined by seven enzyme activities (arylsulphatase, β-glucosidase, phosphatase, urease, invertase, dehydrogenase and phenoloxidase).

The second index (AI 2) is constructed with β-glucosidase, phosphatase, urease and invertase activities. Finally, the third index (AI 3) use just three enzyme activities (β-glucosidase, phosphatase and urease) and was established and validated by mean values from literature. This validation approach may be considered a valid way to give to an index a greater spatial relevance, although one of the problems is that the methods used should be identical since small changes in methodology may lead to great variations in the results obtained. In addition, a careful conversion of units between the different previous works should have been carried out.

Other eminently enzymatic indexes are the EAN (*Enzymatic Activity Number*) proposed by Beck (1984) and the BIF (*Biological Index of Fertility*) of Stefanic et al. (1984). Beck (1984) reported values ranging from 1 to 4 for cultivated soils and from 2 to 8 for pastures and forest soils. Both have been used by other authors to show the effect of soil management on its quality (Saviozzi et al., 2001; Riffaldi et al., 2002). Saviozzi et al. (2001) also used these indexes, observing lower values for cultivated soils than for forest and native grassland soils. This fact would suggest that a strong decrease in microbial activity occurs when virgin soils are intensively cultivated. In the same way, Riffaldi et al. (2002) observed an increase in EAN and BIF in untilled management systems (natural grassland and orange groves) compared to tilled systems in southeastern Sicily (Italy).

Moreno et al. (2008) aimed to study the soil microbiological quality under different agricultural management of broccoli crop (ecological, integrated, conventional and conventional plus pig slurry addition) in semiarid climate (Moreno et al., 2008). These authors elaborated a microbiological degradation index (MDI), based on the previous investigations of Bastida et al. (2006), and concluded that ecological management during eight years (amendment with sheep and goat manure, as organic treatment) enhanced soil quality under broccoli cropping, but after a fallow period, no differences existed with regard to microbiological quality. The index was composed by five indicators: ATP, microbial biomass C, diphenol oxidase activity, urease activity and water-soluble carbohydrates.

9.7 PERSPECTIVES AND CONCLUSIONS

As we stated through the chapter, agriculture and land-use influence have a great impact on soil microbial communities, both in terms of structure and activity. In general applications of exogenous organic matter or mulches enhance microbial activity and may affect the structure of microbial community, which is especially in arid and semiarid locations where SOM content is particularly low. Microbial activity methodologies, such as enzymes, are useful as bio-indicators of soil quality, but since they are clearly influenced for several factors, they should be integrated in multi-parametrical SQIs. A variety of SQIs has been defined for different agroecosystems. However, they are not used on larger scales, nor even in similar climatologically or agronomic conditions. The lack of applicability of SQIs resides on (1) poor standardization of some methodologies, (2) some methods are out of reach in some parts of the world, (3) spatial scale problems (soil heterogeneity), (4) poor definition of soil natural conditions (climate and vegetation) and (5) poor definition of soil function to be tested for soil quality.

Changes on the soil microbial community and its activity are clearly dependent on the SOM content, moisture and season. In our opinion, despite knowledge regarding relationships soil microbial diversity and disease suppression being available, there is a lack of information related to the key players of soil microbial communities which may enhance vegetal productivity in soils. Owing to the natural complexity and heterogeneity of soil, more studies regarding the effects of agricultural management on soil microbial activity and microbial composition are needed. They will provide a better understanding of the interaction between soil, microorganisms, agriculture and climate though spatial and temporal scales and also will help to minimize global change impacts such greenhouse gas emissions.

REFERENCES

Acosta-Martínez, V., T. M. Zobeck, T. E. Gill and A. C. Kennedy. 2003. Enzyme activities and microbial community structure in semiarid agricultural soils. *Biol Fertil Soils* 38:216–227.

Acton, D. F. and G. A. Padbury. 1993. A conceptual framework for soil quality assessment and monitoring. In *Soil Quality Evaluation Summary. A Program to Assess and Monitor Soil Quality in Canada*, ed. D. F. Acton. Ottawa, Ontario: Research Branch Agriculture.

Anderson, T. H. and K. H. Domsch. 1985. Determination of ecophysiological maintenance carbon requirements of soil microorganisms in a dormant state. *Biol Fertil Soils* 1:81–89.

Anderson, T. H. and K. H. Domsch. 1990. Application of ecophysiological quotients (qCO_2 and Qd) on microbial biomasses from soils of different cropping histories. *Soil Biol Biochem* 22:251–255.

Andrews, S. S. and C. R. Carroll. 2001. Designing a soil quality assessment tool for sustainable agroecosystem management. *Ecol Appl* 11:1573–1585.

Andrews, S. S., D. L. Karlen and C. A. Cambardella. 2004. The soil management assessment framework: A quantitative soil quality evaluation method. *Soil Sci Soc Am J* 68:1945–1962.

Andrews, S. S., D. L. Karlen and J. P. Mitchell. 2002a. A comparison of soil quality indexing methods for vegetable production systems in Northern California. *Agr Ecosyst Environ* 90:25–45.

Andrews, S. S., J. P. Mitchell, R. Mancinelli, D. L. Karlen, T. K. Hartz, W. R. Horwath, G. S. Pettygrove, K. M. Scow and D. S. Munk. 2002b. On-farm assessment of soil quality in California's central valley. *Agron J* 94:12–23.

Arshad, M. A. and S. Martin. 2002. Identifying critical limits for soil quality indicators in agro-ecosystems. *Agr Ecosyst Environ* 88:153–160.

Asuming-Brempong, S., S. Gantner, S. G. K. Adiku, G. Archer, V. Edusei and J. M. Tiedje. 2008. Changes in the biodiversity of microbial populations in tropical soils under different fallow treatments. *Soil Biol Biochem* 40:2811–2818.

Baggs, E. M., M. Stevenson, M. Pihlatie, A. Regar, H. Cook and G. Cadisch. 2003. Nitrous oxide emissions following application of residues and fertiliser under zero and conventional tillage. *Plant Soil* 254:361–370.

Bardgett, R. D. and A. Shine. 1999. Linkages between plant litter diversity, soil microbial biomass and ecosystem function in temperate grasslands. *Soil Biol Biochem* 31:317–321.

Bastida, F., A. Zsolnay, T. Hernandez and C. Garcia. 2008. Past, present and future of soil quality indices: A biological perspective. *Geoderma* 147:159–171.

Bastida, F., J. L. Moreno, C. Garcia and T. Hernandez. 2007a. Addition of urban waste to semiarid degraded soil: Long-term effect. *Pedosphere* 17:557–567.

Bastida, F., J. L. Moreno, T. Hernandez and C. Garcia. 2006a. Microbiological activity in a soil 15 years after its devegetation. *Soil Biol Biochem* 38:2503–2507.

Bastida, F., J. L. Moreno, T. Hernandez and C. Garcia. 2006b. Microbiological degradation index of soils in a semiarid climate. *Soil Biol Biochem* 38:3463–3473.

Bastida, F., J. L. Moreno, T. Hernandez and C. Garcia. 2007b. The long-term effects of the management of a forest soil on its carbon content, microbial biomass and activity under a semi-arid climate. *Appl Soil Ecol* 37:53–62.

Baudoin, E., L. Philippot, D. Cheneby, L. Chapuis-Lardy, N. Fromin, D. Bru, B. Rabary and A. Brauman. 2009. Direct seeding mulch-based cropping increases both the activity and the abundance of denitrifier communities in a tropical soil. *Soil Biol Biochem* 41:1703–1709.

Beck, T. H. 1984. Methods and application of soil microbiological analysis at the Landensanstalt fur Bodenkultur und Pflanzenbau (LBB) for determination of some aspects of soil fertility. In *Proceedings of the Fifth Symposium on Soil Biology*, pp. 13–20. Rumanian National Society of Soil Science, Bucharest.

Benndorf, D., G. U. Balcke, H. Harms and M. von Bergen. 2007. Functional metaproteome analysis of protein extracts from contaminated soil and groundwater. *ISME J* 1:224–234.

Bestel-Corre, G., E. Dumas-Gaudot and S. Gianinazzi. 2004. Proteomics as a tool to monitor plant–microbe endosymbioses in the rhizosphere. *Mycorrhiza* 14:1–10.

Bhattacharyya, R., S. Chandra, R. D. Singh, S. Kundu, A. K. Srivatva and H. S. Gupta. 2007. Long-term farmyard manure application effects on properties of a silty clay loam soil under irrigated wheat-soybean rotation. *Soil Till Res* 94:386–396.

Blanco-Canqui, H. and R. Lal. 2007. Soil structure and organic carbon relationships following 10 years of wheat straw management in no-till. *Soil Till Res* 95:240–254.

Bossio, D. A., M. S. Girvan, L. Verchot, J. Bullimore, T. Borelli, A. Albrecht, K. M. Scow, A. S. Ball, J. N. Pretty and A. M. Osborn. 2005. Soil microbial community response to land use change in an agricultural landscape of Western Kenya. *Microb Ecol* 49:50–62.

Burns, R. G. 1978. Enzyme activity in soil. Some theoretical and practical considerations. In *Soil Enzymes*, ed. R. G. Burns, pp. 295–339. New York, NY: Academic Press.

Burns, R. G. 1982. Enzyme activity in soil: Location and a possible role in microbial ecology. *Soil Biology & Biochemistry* 14:423–427.

Caldwell, B. A. 2005. Enzyme activities as a component of soil biodiversity: A review. *Pedobiologia* 49:637–644.

Cambardella, C. A., T. B. Moorman, S. S. Andrews and D. L. Karlen. 2004. Watershed-scale assessment of soil quality in the loess hills of southwest Iowa. *Soil Till Res* 78:237–247.

Carney, K. M. and P. A. Matson. 2005. Plant communities, soil microorganisms and soil carbon cycling: Does altering the world belowground matter to ecosystem functioning? *Ecosystems* 8:928–940.

Carvalho, J. L. N., C. E. P. Cerri, B. J. Feigl, M. C. Píccolo, V. P. Godinho, U. Herpin and C. C. Cerri. 2009. Carbon sequestration in agricultural soils in the Cerrado region of the Brazilian Amazon. *Soil Till Res* 103:342–349.

Chivenge, P. P., H. K. Murwira, K. E. Giller, P. Mapfumo and J. Six. 2007. Long-term impact of reduced tillage and residue management on soil carbon stabilization: Implication for conservation agriculture on contrasting soils. *Soil Till Res* 94:328–337.

Crecchio, C., M. Curci, A. Pellegrino, P. Ricciuti, N. Tursi and P. Ruggiero. 2007. Soil microbial dynamics and genetic diversity in soil under monoculture wheat grown in different long-term management systems. *Soil Biol Biochem* 39:1391–1400.

Dilly, O., H. P. Blume, U. Sehy, M. Jimenez and J. C. Munch. 2003. Variation of stabilised, microbial and biologically active carbon and nitrogen in soil under contrasting land use and agricultural management practices. *Chemosphere* 52:557–569.

Doran, J. W. and T. B. Parkin. 1994. Defining and assessing soil quality. In *Defining Soil Quality for Sustainable Environment*, eds. J. W. Doran, D. C. Coleman, D. F. Beedecek and B. A. Stewart, pp. 3–21. Madison: SSSA.

Dyson, F. J. 1977. Can we control the carbon dioxide in the atmosphere? *Energ J* 2:287–291.

El Arab, H. G. D., V. Vilich and R. A. Sikora. 2001. The use of phospholipid fatty acids (PL-FA) in the determination of rhizosphere specific microbial communities (RSMC) of two wheat cultivars. *Plant Soil* 228:291–297.

Erkossa, T., F. Itanna and K. Stahr. 2007. Indexing soil quality: A new paradigm in soil science research. *Aus J Soil Res* 45:129–137.

Fraterrigo, J. M., M. G. Turner and S. M. Pearson. 2006. Interactions between past land use, life-history traits and understory spatial heterogeneity. *Landscape Ecol* 21:777–790.

Fullen, M. A. 1998. Effects of grass ley set-aside on runoff, erosion and organic matter levels in sandy soils in east Shropshire, UK. *Soil Till Res* 46:41–49.

Garbeva, P., J. A. van Veen and J. D. van Elsas. 2004. Microbial diversity in soil: Selection of microbial populations by plant and soil type and implications for disease suppressiveness. *Annu Rev Phytopathol* 42:243–270.

García, C., T. Hernández and F. Costa. 1992. Variation in some chemical-parameters and organic-matter in soils regenerated by the addition of municipal solid-waste. *Environ Manage* 16:763–768.

García, C., T. Hernández and F. Costa. 1994. Microbial activity in soils under Mediterranean environmental-conditions. *Soil Biol Biochem* 26:1185–1191.

Garland, J. L. and A. L. Mills. 1991. Classification and characterization of heterotrophic microbial communities on the basis of patterns of community-level sole-carbon-source utilization. *Appl Environ Microb* 57:2351–2359.

Gianfreda, L. and J. M. Bollag 1996. Influence of natural and anthropogenic factors on enzyme activity in soil. In *Soil Biochemistry*, eds. G. Stotzky and J. M. Bollag, pp. 123–193. New York, NY: Marcel Dekker.

Gianfreda, L. and J. M. Bollag 2002. Isolated enzymes for the transformation and detoxification of organic pollutants. In *Enzymes in the Environment: Activity, Ecology and Applications*, eds. R. G. Burns and R. P. Dick, pp. 495–538. New York, NY: Marcel Dekker.

Gianfreda, L. and P. Ruggiero. 2006. Enzyme activities in soil. In *Nucleic Acids and Proteins in Soil*, eds. P. Nannipieri and K. Smalla, pp. 257–311. Berlin: Springer.

Gleixner, G. and A. Poirier. 2001. Molecular turnover rates of soil organic matter. *Abstr Papers Am Chem Soc* 221:27-GEOC.

Glover, J. D., J. P. Reganold and P. K. Andrews. 2000. Systematic method for rating soil quality of conventional, organic and integrated apple orchards in Washington State. *Agr Ecosyst Environ* 80:29–45.

Graham, M. H. and R. J. Haynes. 2005. Catabolic diversity of soil microbial communities under sugarcane and other land uses estimated by Biolog and substrate-induced respiration methods. *Appl Soil Ecol* 29:155–164.

Halvorson, J. J., J. L. Smith and R. I. Papendick. 1996. Integration of multiple soil parameters to evaluate soil quality: A field example. *Biol Fertil Soils* 21:207–214.

Halvorson, A. D., B. J. Wiendhold and A. L. Black. 2002. Tillage, nitrogen, and cropping system effects on soil carbon sequestration. *Soil Sci Soc Am J* 66:906–912.

Hamer, U., F. MakesChin, J. Stadler and S. Klotz. 2008. Soil organic matter and microbial community structure in set-aside and intensively managed arable soils in NE-Saxony, Germany. *Appl Soil Ecol* 40:465–475.

Harris, R. F., D. L. Karlen and D. J. Mulla. 1996. A conceptual framework for assessment and management of soil quality and health. In *Methods for Assessing Soil Quality*, eds. J. W. Doran and A. J. Jones, pp. 61–82. Madison: SSSA.

Hassink, J. and A. P. Whitmore. 1997. A model of the physical protection of organic matter in soils. *Soil Sci Soc Am J* 61:131–139.

Hedlund, K. 2002. Soil microbial community structure in relation to vegetation management on former agricultural land. *Soil Biol Biochem* 34:1299–1307.

Hussain, I., K. R. Olson, M. M. Wander and D. L. Karlen. 1999. Adaptation of soil quality indices and application to three tillage systems in southern Illinois. *Soil Till Res* 50:237–249.

Insam, H. and K. Haselwandter. 1989. Metabolic quotient of the soil microflora in relation to plant succession. *Oecologia* 79:174–178.

Insam, H., D. Parkinson and K. H. Domsch. 1989. Influence of macroclimate on soil microbial biomass. *Soil Biol Biochem* 21:211–221.

Iovieno, P., L. Morra, A. Leone, L. Pagano and A. Alfani. 2009. Effect of organic and mineral fertilizers on soil respiration and enzyme activities of two Mediterranean horticultural soils. *Biol Fertil Soils* 45:555–561.

Islam, K. R. and R. R. Weil. 2000. Soil quality indicator properties in mid-Atlantic soils as influenced by conservation management. *J Soil Water Conserv* 55:69–78.

Janzen, H. H. 2004. Carbon cycling in earth systems—A soil science perspective. *Agr Ecosyst Environ* 104:399–417.

Jarecki, M. K., R. Lal and R. James. 2005. Crop Management effects on soil carbon sequestration on selected farmers' fields in Northern Ohio. *Soil Till Res* 81:265–276.

Jastrow, J. D., J. E. Amonette and V. L. Bailey. 2007. Mechanisms controlling soil carbon turnover and their potential application for enhancing carbon sequestration. *Climatic Change* 80:5–23.

Jenkinson, D. S. and J. N. Ladd. 1981. Microbial biomass in soil: Measurement and turnover: In *Soil Biochemistry*, eds. E. A. Paul and J. N. Ladd, pp. 415–471. NewYork: Marcel Dekker.

Joergensen, R. G., P. C. Brookes and D. S. Jenkinson. 1990. Survival of the soil microbial biomass at elevated-temperatures. *Soil Biol Biochem* 22:1129–1136.

Kanchikerimath, M. and D. Singh. 2001. Soil organic matter and biological properties after 26 years of maize-wheat-cowpea cropping as affected by manure and fertilization in a Cambisol in semiarid region of India. *Agr Ecosyst Environ* 86:155–162.

Kandeler, E. and G. Eder. 1993. Effect of cattle slurry in grassland on microbial biomass and on activities of various enzymes. *Biol Fertil Soils* 16:249–254.

Kandeler, E., M. Stemmer and E. M. Klimanek. 1999. Response of soil microbial biomass, urease and xylanase within particle size fractions to long-term soil management. *Soil Biol Biochem* 31:261–273.

Kang, G. S., V. Beri, B. S. Sidhu and O. P. Rupela. 2005. A new index to assess soil quality and sustainability of wheat-based cropping systems. *Biol Fertil Soils* 41:389–398.

Karlen, D. L. and D. E. Stott. 1994. A framework for evaluating physical and chemical indicators of soil quality. In *Defining Soil Quality for a Sustainable Environment*. SSSA Special Publ. 34, eds. J. W. Doran, D. C. Coleman, D. F. Bezdicek and B. A. Stewart. Madison, pp. 53–72, Wisconsin: Soil Science Society of America.

Karlen, D. L., M. J. Mausbach, J. W. Doran, R. G. Cline, R. F. Harris and G. E. Schuman. 1997. Soil quality: A concept, definition, and framework for evaluation. *Soil Sci Soc Am J* 61:4–10.

Karlen, D. L., N. C. Wollenhaupt, D. C. Erbach, E. C. Berry, J. B. Swan, N. S. Eash and J. L. Jordahl. 1994a. Crop residue effects on soil quality following 10-years of no-till corn. *Soil Till Res* 31:149–167.

Karlen, D. L., N. C. Wollenhaupt, D. C. Erbach, E. C. Berry, J. B. Swan, N. S. Eash and J. L. Jordahl. 1994b. Long-term tillage effects on soil quality. *Soil Till Res* 32:313–327.

Kong, A. Y. Y., J. Six, D. C. Bryant, R. F. Denison and C. van Kessel. 2005. The relationship between carbon input, aggregation, and soil organic carbon stabilization in sustainable cropping systems. *Soil Sci Soc Am J* 69:1078–1085.

Kourtev, P. S., J. G. Ehrenfeld and M. Haggblom. 2003. Experimental analysis of the effect of exotic and native plant species on the structure and function of soil microbial communities. *Soil Biol Biochem* 35:895–905.

Lagomarsino, A., S. Grego, S. Marhan, M. C. Moscatelli and E. Kandeler. 2009. Soil management modifies micro-scale abundance and function of soil microorganisms in a Mediterranean ecosystem. *Eur J Soil Sci* 60:2–12.

Lal, R. 1997. Residue management, conservation tillage and soil restoration for mitigating greenhouse effect by CO_2-enrichment. *Soil Till Res* 43:81–107.

Lal, R. 2004. Carbon sequestration in dryland ecosystems. *Environ Manage* 33:528–544.

Lal, R. 2009. Sequestering carbon in soils of arid ecosystems. *Land Degrad Dev* 20:441–454.

Landi, L., G. Renella, J. L. Moreno, L. Falchini and P. Nannipieri. 2000. Influence of cadmium on the metabolic quotient, L-: D-glutamic acid respiration ratio and enzyme activity: Microbial biomass ratio under laboratory conditions. *Biol Fertil Soils* 32:8–16.

Larkin, R. P. 2003. Characterization of soil microbial communities under different potato cropping systems by microbial population dynamics, substrate utilization, and fatty acid profiles. *Soil Biol Biochem* 35:1451–1466.

Lee, C. H., M. Y. Wu, V. B. Asio and Z. S. Chen. 2006. Using a soil quality index to assess the effects of applying swine manure compost on soil quality under a crop rotation system in Taiwan. *Soil Sci* 171:210–222.

Lemenih, M., E. Karltun and M. Olsson. 2005. Soil organic matter dynamics after deforestation along a farm field chronosequence in southern highlands of Ethiopia. *Agr Ecosyst Environ* 109:9–19.

Liao, M. and X. M. Xie. 2007. Effect of heavy metals on substrate utilization pattern, biomass, and activity of microbial communities in a reclaimed mining wasteland of red soil area. *Ecotox Environ Safe* 66:217–223.

Liebig, M. A., G. Varvel and J. Doran. 2001. A simple performance-based index for assessing multiple agroecosystem functions. *Agron J* 93:313–318.

Liu, B., C. Tu, S. J. Hu, M. Gumpertz and J. B. Ristaino. 2007. Effect of organic, sustainable, and conventional management strategies in grower fields on soil physical, chemical, and biological factors and the incidence of Southern blight. *Appl Soil Ecol* 37:202–214.

Lynam, J. K. and R. W. Herdt. 1989. Sense and sustainability: Sustainability as an objective in international agricultural research. *Agr Econ* 3:381–398.

López-Fando, C. and M. T. Pardo. 2009. Changes in soil chemicals characteristics with different tillage practices in a semi-arid environment. *Soil Till Res* 104:278–284.

Marhan, S., E. Kandeler and S. Scheu. 2007. Phospholipid fatty acid profiles and xylanase activity in particle size fractions of forest soil and casts of *Lumbricus terrestris* L. (Oligochaeta, Lumbricidae). *Appl Soil Ecol* 35:412–422.

Marx, M. C., E. Kandeler, M. Wood, N. Wermbter and S. C. Jarvis. 2005. Exploring the enzymatic landscape: Distribution and kinetics of hydrolytic enzymes in soil particle-size fractions. *Soil Biol Biochem* 37:35–48.

Masto, R. E., P. K. Chhonkar, D. Singh and A. K. Patra. 2007. Soil quality response to long-term nutrient and crop management on a semi-arid Inceptisol. *Agr Ecosyst Environ* 118:130–142.

McConkey, B. G., B. C. Liang, C. A. Campbell, D. Curtin, A. Moulin, S. A. Brandt and G. P. Lafond. 2003. Crop rotation and tillage impact on carbon sequestration in Canadian prairie soils. *Soil Till Res* 74:81–90.

Melero, S., R. López-Garrido, J. M. Murillo and F. Moreno. 2009. Conservation tillage: Short- and long-term effects on soil carbon fractions and enzymatic activities under Mediterranean conditions. *Soil Till Res* 104:292–298.

Mohanty, M., D. K. Painuli, A. K. Misra and P. K. Ghosh. 2007. Soil quality effects of tillage and residue under rice-wheat cropping on a Vertisol in India. *Soil Till Res* 92:243–250.

Monokrousos, N., E. M. Papatheodorou, J. D. Diamantopoulos and G. P. Stamou. 2006. Soil quality variables in organically and conventionally cultivated field sites. *Soil Biol Biochem* 38:1282–1289.

Moreno, J. L., F. Bastida, T. Hernández and C. García. 2008. Relationship between the agricultural management of a semi-arid soil and microbiological quality. *Commun Soil Sci Plan* 39:421–439.

Moreno, J. L., K. Jindo, T. Hernandez and C. Garcia. 2007. Total and immobilized enzymatic activity of organic materials before and after composting. *Compost Sci Util* 15:93–100.

Nannipieri, P., P. Sequi and P. Fusi. 1996. Humus and enzyme activity. In *Humic Substances in Terrestrial Ecosystems*, ed. A. Piccolo, pp. 293–327. Amsterdam: Elsevier.

Nannipieri, P., S. Grego and B. Ceccanti. 1990. Ecological significance of the biological activity in soils. In *Soil Biochemistry*, eds. J. M. Bollag and G. Stotzky, pp. 293–355. New York, NY: Marcel Dekker.

Nogueira, M. A., U. B. Albino, O. Brandao-Junior, G. Braun, M. F. Cruz, B. A. Dias, R. T. D. Duarte, et al. 2006. Promising indicators for assessment of agroecosystems alteration among natural, reforested and agricultural land use in southern Brazil. *Agr Ecosyst Environ* 115:237–247.

Olson, K. R., J. M. Lang and S. A. Ebelhar. 2005. Soil organic carbon changes after 12 years of no-tillage and tillage of Grantsburg soils in southern Illinois. *Soil Till Res* 81:217–225.

Pascual, J. A., C. Garcia, T. Hernandez, J. L. Moreno and M. Ros. 2000. Soil microbial activity as a biomarker of degradation and remediation processes. *Soil Biol Biochem* 32:1877–1883.

Pascual, J. A., C. García and T. Hernández. 1999. Comparison of fresh and composted organic waste in their efficacy for the improvement of arid soil quality. *Bioresource Technol* 68:255–264.

Pascual, J. A., C. García, T. Hernández and M. Ayuso. 1997. Changes in the microbial activity of an arid soil amended with urban organic wastes. *Biol Fertil Soils* 24:429–434.

Paustian, K., J. Six, E. T. Elliott and H. W. Hunt. 2000. Management options for reducing CO_2 emissions from agricultural soils. *Biogeochemistry* 48:147–163.

Paustian, K., W. J. Parton and J. Persson. 1992. Modeling soil organic-matter in organic-amended and nitrogen-fertilized long-term plots. *Soil Sci Soc Am J* 56:476–488.

Peigne, J., B. C. Ball, J. Roger-Estrade and C. David. 2007. Is conservation tillage suitable for organic farming? A review. *Soil Use Manage* 23:129–144.

Plaza, C., D. Hernandez, J. C. Garcia-Gil and A. Polo. 2004. Microbial activity in pig slurry-amended soils under semiarid conditions. *Soil Biol Biochem* 36:1577–1585.

Poll, C., A. Thiede, N. Wermbter, A. Sessitsch and E. Kandeler. 2003. Micro-scale distribution of microorganisms and microbial enzyme activities in a soil with long-term organic amendment. *Eur J Soil Sci* 54:715–724.

Powlson, D. S. and D. S. Jenkinson. 1981. A comparison of the organic-matter, biomass, adenosine-triphosphate and mineralizable nitrogen contents of ploughed and direct-drilled soils. *J Agr Sci* 97:713–721.

Puglisi, E., A. A. M. Del Re, M. A. Rao and L. Gianfreda. 2006. Development and validation of numerical indexes integrating enzyme activities of soils. *Soil Biol Biochem* 38:1673–1681.

Rasse, D. P., C. Rumpel and M. F. Dignac. 2005. Is soil carbon mostly root carbon? Mechanisms for a specific stabilisation. *Plant Soil* 269:341–356.

Rezaei, S. A., R. J. Gilkes and S. S. Andrews. 2006. A minimum data set for assessing soil quality in rangelands. *Geoderma* 136:229–234.

Rhoton, F. E. 2000. Influence of time on soil response to no-till practices. *Soil Sci Soc Am J* 64:700–709.

Riffaldi, R., A. Saviozzi, R. Levi-Minzi and R. Cardelli. 2002. Biochemical properties of a Mediterranean soil as affected by long-term crop management systems. *Soil Till Res* 67:109–114.

Rillig, M. C. 2004. Arbuscular mycorrhizae, glomalin and soil aggregation. *Can J Soil Sci* 84:355–363.

Roldan, A., J. R. Salinas-Garcia, M. M. Alguacil, E. Diaz and F. Caravaca. 2005. Soil enzyme activities suggest advantages of conservation tillage practices in sorghum cultivation under subtropical conditions. *Geoderma* 129:178–185.

Ros, M., M. T. Hernández and C. García. 2003. Soil microbial activity after restoration of a semiarid soil by organic amendments. *Soil Biol Biochem* 35:463–469.

Rostagno, C. M. and R. B. Sosebee. 2001. Surface application of biosolids in the Chihuahuan Desert: Effects on soil physical properties. *Arid Land Res Manage* 15:233–244.

Sainju, U. M., Z. N. Senwo, E. Z. Nyakatawa, L. A. Tazisong and K. C. Reddy. 2008. Soil carbon and nitrogen sequestration as affected by long-term tillage, cropping systems, and nitrogen fertilizer sources. *Agr Ecosyst Environ* 127:234–240.

Saviozzi, A., R. Levi-Minzi, R. Cardelli and R. Riffaldi. 2001. A comparison of soil quality in adjacent cultivated, forest and native grassland soils. *Plant Soil* 233:251–259.

Schloter, M., O. Dilly and J. C. Munch. 2003. Indicators for evaluating soil quality. *Agr Ecosyst Environ* 98:255–262.

Schulze, W. X., G. Gleixner, K. Kaiser, G. Guggenberger, M. Mann and E. D. Schulze. 2005. A proteomic fingerprint of dissolved organic carbon and of soil particles. *Oecologia* 142:335–343.

Schutter, M. E., J. M. Sandeno and R. P. Dick. 2001. Seasonal, soil type, and alternative management influences on microbial communities of vegetable cropping systems. *Biol Fertil Soils* 34:397–410.

Sharma, K. L., U. K. Mandal, K. Srinivas, K. P. R. Vittal, B. Mandal, J. K. Grace and V. Ramesh. 2005. Long-term soil management effects on crop yields and soil quality in a dryland Alfisol. *Soil Till Res* 83:246–259.

Sherrod, L. A., G. A. Peterson, D. G. Westfall and L. R. Ahuja. 2003. Cropping intensity enhances soil organic carbon and nitrogen in a no-till agroecosystem. *Soil Sci Soc Am J* 67:1533–1543.

Shishido, M., K. Sakamoto, H. Yokoyama, N. Momma and S. I. Miyashita. 2008. Changes in microbial communities in an apple orchard and its adjacent bush soil in response to season, land-use, and violet root rot infestation. *Soil Biol Biochem* 40:1460–1473.

Singleton, I., G. Merrington, S. Colvan and J. S. Delahunty. 2003. The potential of soil protein-based methods to indicate metal contamination. *Appl Soil Ecol* 23:25–32.

Six, J., E. T. Elliott and K. Paustian. 2000. Soil macroaggregate turnover and microaggregate formation: A mechanism for C sequestration under no-tillage agriculture. *Soil Biol Biochem* 32:2099–2103.

Six, J., S. D. Frey, R. K. Thiet and K. M. Batten. 2006. Bacterial and fungal contributions to carbon sequestration in agroecosystems. *Soil Sci Soc Am J* 70:555–569.

Smith, P. 2004. Carbon sequestration in croplands: The potential in Europe and the global context. *Eur J Agron* 20:229–236.

Spargo, J. T., M. M. Alley, R. F. Follet and J. V. Wallace. 2008. Soil carbon sequestration with continuous no-till management of grain cropping systems in the Virginia coastal plain. *Soil Till Res* 100:133–140.

Sparling, G. P. 1997. Soil microbial biomass, activity and nutrient cycling as indicators of soil health. In *Biological Indicators of Soil Health*, eds. C. Pankhurst, B. M. Doube and V. V. S. R. Gupta, pp. 97–119. Wallingford: CAB International.

Sparrow, S. D., C. E. Lewis and C. W. Knight. 2006. Soil quality response to tillage and crop residue removal under subarctic conditions. *Soil Till Res* 91:15–21.

Spedding, T. A., C. Hamel, G. R. Mehuys and C. A. Madramootoo. 2004. Soil microbial dynamics in maize-growing soil under different tillage and residue management systems. *Soil Biol Biochem* 36:499–512.

Stark, C., L. M. Condron, A. Stewart, H. J. Di and M. O'Callaghan. 2007. Effects of past and current crop management on soil microbial biomass and activity. *Biol Fertil Soils* 43:531–540.

Steenwerth, K. L., L. E. Jackson, F. J. Calderon, K. M. Scow and D. E. Rolston. 2005. Response of microbial community composition and activity in agricultural and grassland soils after a simulated rainfall. *Soil Biol Biochem* 37:2249–2262.

Stefanic, F., G. Ellade and J. Chirnageanu. 1984. Researches concerning a biological index of soil fertility. In *Fifth Symposium on Soil Biology*, eds. M. P. Nemes, S. Kiss, P. Papacostea, C. Stefanic, C. and M. Rusan, pp. 35–45. Bucharest: Romanian National Society of Soil Science.

Stemmer, M., M. H. Gerzabek and E. Kandeler. 1999. Invertase and xylanase activity of bulk soil and particle-size fractions during maize straw decomposition. *Soil Biol Biochem* 31:9–18.

Stewart, C. E., A. F. Plante, K. Paustian, R. T. Conant and J. Six. 2008. Soil carbon saturation: Linking concept and measurable carbon pools. *Soil Sci Soc Am J* 72:379–392.

Tejada, M., J. L. Moreno, M. T. Hernandez and C. Garcia. 2007. Application of two beet vinasse forms in soil restoration: Effects on soil properties in an environment in southern Spain. *Agr Ecosyst Environ* 119:289–298.

Tejada, M., M. T. Hernandez and C. Garcia. 2009. Soil restoration using composted plant residues: Effects on soil properties. *Soil Till Res* 102:109–117.

Trasar-Cepeda, C., C. Leiros, F. Gil-Sotres and S. Seoane. 1998. Towards a biochemical quality index for soils: An expression relating several biological and biochemical properties. *Biol Fertil Soils* 26:100–106.

Trasar-Cepeda, C., M. C. Leiros and F. Gil-Sotres. 2008. Hydrolytic enzyme activities in agricultural and forest soils. Some implications for their use as indicators of soil quality. *Soil Biol Biochem* 40:2146–2155.

Tu, C., J. B. Ristaino and S. J. Hu. 2006. Soil microbial biomass and activity in organic tomato farming systems: Effects of organic inputs and straw mulching. *Soil Biol Biochem* 38:247–255.

Waldrop, M. P. and M. K. Firestone. 2004. Altered utilization patterns of young and old soil C by microorganisms caused by temperature shifts and N additions. *Biogeochemistry* 67:235–248.

Wallenstein, M. D. and M. N. Weintraub. 2008. Emerging tools for measuring and modeling the *in situ* activity of soil extracellular enzymes. *Soil Biol Biochem* 40:2098–2106.

Walter, I., F. Martinez and G. Cuevas. 2006. Plant and soil responses to the application of composted MSW in a degraded, semiarid shrubland in central Spain. *Compost Sci Util* 14:147–154.

Walton, M., J. E. Herrick, R. P. Gibbens and M. D. Remmenga. 2001. Persistence of municipal biosolids in a Chihuahuan Desert rangeland 18 years after application. *Arid Land Res Manage* 15:223–232.

Wardle, D. A. and A. Ghani. 1995. A critique of the microbial metabolic quotient (qCO(2)) as a bioindicator of disturbance and ecosystem development. *Soil Biol Biochem* 27:1601–1610.

Welbaum, G. E., A. V. Sturz, Z. M. Dong and J. Nowak. 2004. Managing soil microorganisms to improve productivity of agro-ecosystems. *Crit Rev Plant Sci* 23:175–193.

White, C. S., S. R. Loftin and R. Aguilar. 1997. Application of biosolids to degraded semiarid rangeland: Nine-year responses. *J Envion Qual* 26:1663–1671.

Wienhold, B. J., J. L. Pikul, M. A. Liebig, M. M. Mikha, G. E. Varvel, J. W. Doran and S. S. Andrews. 2006. Cropping system effects on soil quality in the Great Plains: Synthesis from a regional project. *Renew Agr Food Syst* 21:49–59.

Wilmes, P. and P. L. Bond. 2006. Metaproteomics: Studying functional gene expression in microbial ecosystems. *Trends Microbiol* 14:92–97.

Yang, C. H., D. E. Crowley and J. A. Menge. 2001. 16S rDNA fingerprinting of rhizosphere bacterial communities associated with healthy and Phytophthora infected avocado roots. *FEMS Microbiol Ecol* 35:129–136.

Yao, S., I. A. Merwin, G. W. Bird, G. S. Abawi and J. E. Thies. 2005. Orchard floor management practices that maintain vegetative or biomass groundcover stimulate soil microbial activity and alter soil microbial community composition. *Plant Soil* 271:377–389.

van Elsas, J. D., P. Garbeva and J. Salles. 2002. Effects of agronomical measures on the microbial diversity of soils as related to the suppression of soil-borne plant pathogens. *Biodegradation* 13:29–40.

CHAPTER 10

Survey in Plant Root Proteomics

To Know the Unknown

Sophie Alvarez and Leslie M. Hicks

CONTENTS

10.1 INTRODUCTION

An in-depth understanding of the biology and physiology of the plant root system and its chemical and physical interactions with the surrounding soil environment remains a challenge for scientists, but needs to be elucidated in order to improve crops yields and culturing techniques. Crop yield increases depend on how roots are able to effectively utilize soil to absorb the water and nutrients necessary for plant growth. Many previous root system studies have elucidated mechanisms induced by environmental stress to identify gene products of agronomic relevance using functional genomics. More recently, the use of proteomics—defined as the systematic analysis of the protein population in a tissue, cell, or subcellular compartment—has also shown a large impact on plant biology. Because qualitative and quantitative changes in protein and enzyme patterns are manifestations of the activity of genes controlling the metabolic activity of cells, the interest in proteomics and the number of publications in plant proteomics has increased dramatically over the past 10 years (Figure 10.1). The use of proteomics tools to take snapshots of protein-expression profiles of plants has expanded due to the vast improvement of protein gel- and liquid chromatography (LC)-based separation techniques and mass spectrometry instrumentation for protein identification. Furthermore, correlation between gene expression and protein amount and activity is not always straightforward due to posttranscriptional and posttranslational regulation. It has been inferred that 5% of the *Arabidopsis thaliana* genome encodes kinases and phosphatases, representing over 1000 enzymes

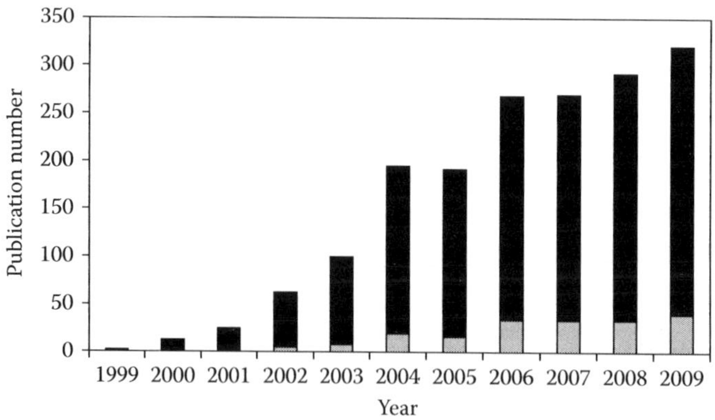

Figure 10.1 Manifestation of the terms "plant proteomics" (black bars) and "root proteomics" (gray bars) in scientific literature. The medline database was examined year by year for occurrence of the terms "plant proteomics" and "root proteomics" in the topic search field (title, keywords, abstracts). The number of hits is plotted against the respective year.

controlling the phosphorylation status of thousands of proteins (Kerk et al., 2002). Hence, the study of posttranslational modifications such as phosphorylation, methylation, acetylation, glycosylation, ubiquitination, protein degradation, using mass spectrometry has become one of the main focuses of the advancements in the proteomics field.

The plant field is still underrepresented when compared with the advancement of proteomics studies achieved in animal and medical studies. Furthermore, compared to total plant proteomics studies there has been only a slightly increased interest over the past decade in mapping proteins present in roots to understand the various synthetic functions (chemical signals, storage proteins, symbiosis structures) and secondary transformations important for normal functioning of the whole plant (Figure 10.1). One of the major factors limiting root proteomics progress is the low amount of tissue and protein. The majority of root proteomics studies were published not only on model organisms such as *Arabidopsis* and *Medicago truncatula* (Jiang et al., 2007; Larrainzar et al., 2007), but also on economical crops such as maize (Prinsi et al., 2009), rice (Koller et al., 2002), soybean (Komatsu et al., 2010), cotton (Coumans et al., 2009), and tomato (Zhou et al., 2009). Some of the aims were to study root growth and gravitropism (Di Michele et al., 2006; Kamada et al., 2005; Tian et al., 2009), plant–microbe interaction in mycorrhizal and rhizobial symbiosis (Bestel-Corre et al., 2004; Mathesius, 2009), root–pathogen interactions (Mehta et al., 2008), and to identify proteins affected by abiotic stresses. The abiotic stress studies include the effect of nutrient availability (Li et al., 2007; Prinsi et al., 2009), soil contaminants (Ahsan et al., 2008; Aloui et al., 2009; Bona et al., 2007; Li et al., 2009; Yang et al., 2007; Zhou et al., 2009), salt and water stresses (Cheng et al., 2009; Jiang et al., 2007; Komatsu et al., 2010; Yan et al., 2005; Yoshimura et al., 2008; Zhu et al., 2007), and temperature changes (Lee et al., 2009; Xu and Huang, 2008). Most of the studies describe the identification of total soluble proteins, but some focus on specific subcellular localization such as microsomal fractions (Lefebvre et al., 2007; Santoni et al., 2003; Valot et al., 2005) and cell wall proteins (Zhu et al., 2006). Other proteomics studies targeted specific cells, for example, pericycle cells (Dembinsky et al., 2007), or specific classes of proteins, for example, RNase-like major storage protein (Kim et al., 2004), copper-binding proteins (Kung et al., 2006), and reactive oxygen species (ROS)-related proteins (Kim et al., 2008). This review attempts to summarize the current findings in the comprehensive analyses of plant roots proteomics, with an emphasis on the importance of findings for sustainable agriculture.

10.2 PROTEIN SURVEY OF SPECIFIC ROOT STRUCTURES

Roots are intrinsically complex tissues to explore and only a few studies have tried to understand the different mechanisms governing root growth. The research has mainly been focused on discovering the mechanisms set during root interactions with plant pathogens and when roots are affected by environmental changes. Here we briefly describe the components of the root architecture and their functions and introduce achievements in mapping specific root proteins using proteomics and highlight the components and mechanisms that are still largely unknown.

10.2.1 Root Structure

Roots can be classified into two main systems: the taproot system and the fibrous root system. The taproot system is characterized by one main root from which smaller branch roots emerge. When a seed germinates, the primary root will develop into the taproot, and most dicots and the conifers contain this type of system. The second main system is the fibrous root system and is characterized by a mass of similarly sized roots. The first root to emerge will be replaced by adventitious roots in most of the monocots. From the outgrowing primary root up to the outermost branches of the root system, the root tip contains young tissues that are divided into the root cap, quiescent

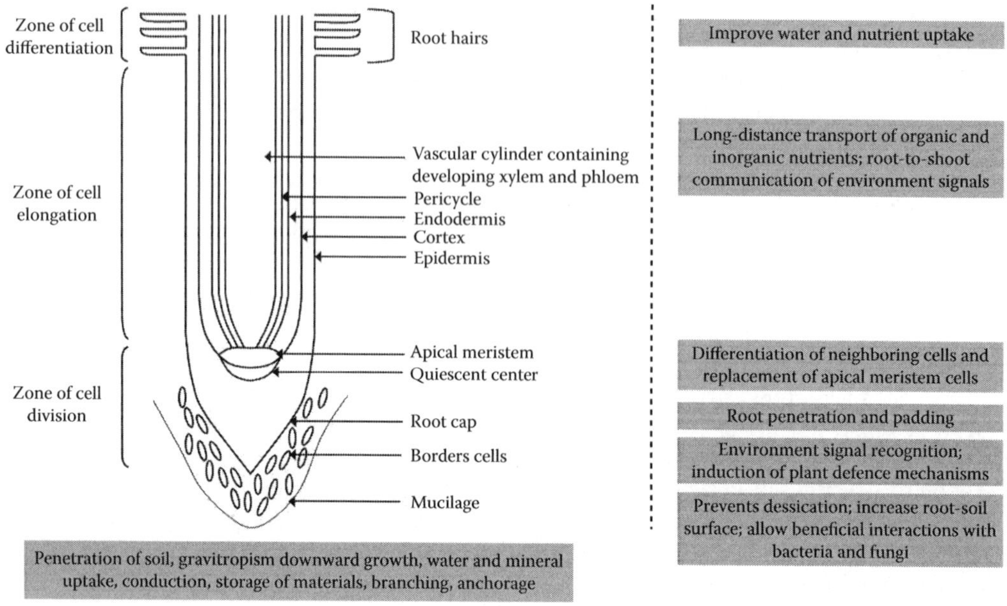

Figure 10.2 Root structure and function.

center (QC), and subapical region (Figure 10.2). The root cap protects the apical meristem and opens a passage into soil for growing roots by producing root cap mucilage (McCully, 1999), a hydrated polysaccharide formed in the dictyosomes that contain sugars, organic acids, vitamins, enzymes, and amino acids. This mucilage also aids in protection of the root by preventing desiccation and in water and nutrient absorption by increasing the soil-to-root contact. Nutrients in mucilage are used in the establishment of mycorrhizae and symbiotic bacteria. The study of root–pathogen interactions has been one of the main focuses in root proteomics research and will be discussed in more detail later (Section 10.4.1). The processes and mechanisms involved in the environmental perception and signal transmission by the root cap are still under investigation. Border cells are synthesized from the root cap meristem and programmed to separate from the root tips. They play an important role in mucilage secretion, sensing of gravity, and in specific recognition and responsiveness to soilborne microflora (bacteria, fungi, nematodes) and aluminum (Hawes et al., 2000), and therefore act in defense mechanisms. However, so far there are no published proteomics studies investigating the importance of border cells in environmental signal recognition and the plant defense response. The QC, located behind the root cap, is a population of slowly dividing cells that regulate differentiation of neighboring cells and replace the merismatic cells of the root cap meristem (Kerk and Feldman, 1995). Appearance of the QC is species specific and dependent on the rate of development. These cells differ from other root meristem cells structurally, physiologically, and biochemically, and are also characterized by a low protein content. The subapical region is divided into three zones: the cell divison zone where cells are derived from the apical meristem; the cellular elongation zone where cells increase in length; and the cellular maturation zone where cell differentiation starts. Root hairs, which function to increase water and nutrient absorption, are modified epidermal cells from the cell differentiation zone. The development of root hairs is under genetic and hormonal control (Bibikova and Gilroy, 2003) and at the same time influenced by the surrounding environment (Bates and Lynch, 1996). The proteomic map of root hair cells was recently established from soybean roots (Brechenmacher et al., 2009). They identified 1492 proteins by combining gel and LC-based proteomics tools. The proteins were mainly involved in protein synthesis and processing (31.4%) and primary metabolism (22.3%). Proteins with functions more specific to root hair cells, including water and nutrient uptake, vesicle trafficking, and hormone and secondary metabolism, were also identified.

10.2.2 Root Architecture

Roots play a pivotal role in anchoring plants in the ground. To perform this function, root systems modulate growth patterns in response to a variety of external stimuli such as gravity, touch, wind, soil density and compaction, and grazing. These factors can induce changes in plant growth and development and is called "thigmomorphogenesis." In the case where plants have to adjust to changes in gravity, it was shown that vertically growing plants placed in a horizontal orientation would result in the upward and downward growth of the apical shoot and root tip, respectively (Massa and Gilroy, 2003); this phenomenon is called gravitropism. The cells responsible for the perception of gravity in roots are the columella cells (Fukaki et al., 1998). A proteomics study focused on studying the gravity response mechanisms using the plant model *Arabidopsis* identified several proteins from the cytoskeleton decreasing by gravitational stimulation (Kamada et al., 2005), correlating well with results observed by microarray analysis (Kimbrough et al., 2004) and indicating regulation at the gene level. However, microarray and proteomics results did not correlate well for proteins such as glutathione *S*-transferase, annexin, lipase/acylhydrolase, and malate dehydrogenase, suggesting a more complex regulation of the gravitropism stimulation. This complexity seems also to be attributable to differential posttranslational modifications of proteins as shown by the observation of molecular weight shifts of proteins during the stimulus (Kamada et al., 2005). Another mechanical stimulus, the slope, was studied during root growth of *Spartium junceum* (Di Michele et al., 2006). Plants growing on slopes undergo morphological and architectural changes related to the plant's need to improve its anchorage (Chiatante et al., 2003). Several differentially expressed proteins affected by slope in *S. junceum* were identified. Among them, some were proteins involved in diverse environmental stresses and were related to alterations of cell wall properties and composition such as peroxidases (Riccardi et al., 1998). Additional proteins involved in the organization and composition of the cell wall such as annexin, tubulin, and actin were overexpressed in roots of plants grown on a slope. This suggests that mechanical stresses were first perceived at the cell wall and that cell wall changes are essential to enhance the resistance of the plant to mechanical constraints (De Jaegher and Boyer, 1987). The direction of the root growth can also change in response to a wide range of stimuli and ignore gravitropic cues. The root can grow toward nutrients and water and grow away from toxic metals (Hawes et al., 2000; Takahashi, 1997), making the root growth mechanisms more complex to understand.

Proteomic maps of total soluble proteins from rice roots and white lupin were also established. The first report of the rice root proteome showed that 63.9% (862) of the proteins detected were root specific (Koller et al., 2002) when compared with leaf and seed. The largest proportion of proteins was classified in the category of cell rescue, defense, cell death, and aging, with peroxidases being the most prominent representatives. Another proteomics study on rice roots, comparing the root proteome with the one from leaf and stem showed 36% (247) of proteins root specific but on a smaller number of total proteins detected (Nozu et al., 2006). They observed that root proteins function in disease and defense mechanisms. A more recent report on rice roots also confirmed that the main category of proteins identified after proteins involved in primary metabolism were ROS-related proteins (Kim et al., 2008). Treatment with ROS-quenching chemicals such as GSH (reduced glutathione), ascorbate, and diphenyleneidonium (DPI) inhibited root growth dose dependently by decreasing ROS generation. This demonstrates the importance of proteins such as glutathione *S*-transferases, ascorbate peroxidases, catalase isozymes, superoxide dismutases, GSH-dependent dehydroascorbate reductase, and glutathione reductase in maintaining cellular redox homeostasis during root growth. In white lupin, the protein map revealed the presence of proteins involved in ATP production (triosephosphate isomerase and glyceraldehyde 3-dehydrogenase), in ROS scavenging (ascorbate), and in flavonoid (chalcone synthase and dihydroflavonol 4-reductase), fatty acid (beta-ketoacyl synthase), and amino acid (serine hydroxymethyltransferase and glutamine synthetase) metabolisms (Tian et al., 2009).

10.2.3 Lateral Root Initiation

Lateral roots are initiated from pericycle cells that are no longer actively dividing in the differentiation zone. Developing main roots constantly proliferate new lateral roots. Signaling between main roots and lateral roots plays a pivotal role for the formation of the mature root stock (Lynch, 1995). Proteomics research helped elucidate the changes in protein composition during primary root growth and lateral root initiation in maize (Hochholdinger et al., 2004; Liu et al., 2006). By using mutants impaired in lateral root initiation (*lrt1*) and in the initiation of the seminal root formation (*rum1*), they were able to identify that proteins involved in lignification accumulated in *ltr1* and *rum1* mutants. This indicates that control of lignin biosynthesis might not only be crucial after the emergence of lateral roots but even before the initiation of these roots. Pericycle cells were isolated from maize root tips using laser capture microdissection before their first division and lateral root initiation to study gene and protein expression using transcriptomics and proteomics (Dembinsky et al., 2007). They established here the only cell-type-specific proteome for monocot roots. This study showed that the competence of pericycle cells to reenter the cell cycle after leaving the meristematic zone was related to the expression of proteins involved in protein synthesis, transcription, and signal transduction.

10.2.4 Plasma Membrane Root Proteins

Membrane proteins, which are known to play biological roles in signal transduction, endocytosis, trafficking, apoptosis, and actin cytoskeleton organization, are poorly documented in plants. Subcellular compartment purification represents a challenge in root proteomics and thus, only a few proteomics efforts on root plasma membrane proteins are reported. Several studies on rhizobial and mychorrizal symbiosis interactions focused on specific compartments such as the peribacteroid membranes (PBMs) (Catalano et al., 2004; Panter et al., 2000; Saalbach et al., 2002; Wienkoop and Saalbach, 2003) and the periarbuscular membranes (PMs) (Benabdellah et al., 2000; Valot et al., 2006; Valot et al., 2005) will be reviewed later.

A proteome map of microsomal proteins from *Medicago* roots was achieved using 2D-gel electrophoresis (Valot et al., 2004). Microsomes were isolated using differential ultracentrifugation and membrane-associated proteins were extracted using a chloroform–methanol mixture. A total of 250 proteins were resolved and 96 proteins were identified by mass spectrometry. The most abundant class of proteins identified was membrane transport proteins with 28%, followed by primary metabolism with 25%. Although the proteome map of the microsomal proteins was compared with the cytosolic and total fractions to discriminate possible contaminants, a high number of proteins were not reported as membrane proteins or membrane-associated proteins. Improvements in sample preparation and plasma membrane protein isolation still need to be optimized. In white lupin, membrane proteins were extracted from developing roots, separated by 1D-SDS-PAGE, and identified using LC–MS/MS (Tian et al., 2009). Mitochondrial and vacuolar ATPases were highly abundant in the membrane fraction. Additionally, endomembrane and plasma membrane proteins, including glycoproteins involved in cell signaling and recognition, ribosomal proteins, and heat-shock proteins (HSPs) involved in protein synthesis were also abundantly identified. Plasma membrane proteins were investigated in rice seedlings (Hashimoto et al., 2009). Using an optimized aqueous two-phase partitioning method based on the membrane surface properties and SDS-PAGE and 2D-PAGE to separate the proteins, they isolated 82 proteins predicted with at least one transmembrane helix and 46 transmembrane proteins. Using SDS-PAGE, the proteins identified were mainly involved in protein synthesis, signal transduction, and energy metabolism. With 2D-PAGE, proteins were mainly involved in energy metabolism, cell growth division, and defense.

Only one group focused on the identification of lipids and protein composition of lipid rafts in plants (Lefebvre et al., 2007). They applied aqueous two-phase partitioning for microsomal isolation of root tissue from *M. truncatula* followed by purification of plasma membranes. The detergent insoluble membranes were then isolated using a sucrose gradient and the lipid rafts were analyzed for their lipid and protein composition. They identified 270 proteins associated with *Medicago* lipid rafts, but only 120 were distinct proteins—indicating that several isoforms of a subfamily or several subunits of a complex were identified. Proteins identified included several signaling and trafficking proteins (aquaporins, ABC transporters, RLKs, remorin, prohibitin, H^+-ATPases) that were also previously identified in lipid rafts studies from other plant tissues (Borner et al., 2005; Morel et al., 2006; Shahollari et al., 2004). The most interesting result from this study was the identification of many proteins belonging to the plasma membrane redox system including NADH-ubiquinone oxidoreductases, cytochrome b561, apoplasm-facing L-AO, as well as peroxidases, whose sequences were mainly or exclusively found in root ESTs libraries.

Aquaporins are channel proteins that facilitate the diffusion of water across cell membranes and play an important role in plant–water relations. A proteomics approach was used to identify the diversity of aquaporin forms expressed in the plasma membrane of *Arabidospis* roots (Santoni et al., 2003). In this study aimed at detecting aquaporin isoforms, root plasma membranes preparations were stripped with urea/alkali treatment to enrich the intrinsic proteins and antibodies were used to probe for the specific expression of aquaporins. However, because aquaporins protein sequences are highly homogenous, the use of MS/MS was critical to characterize specific peptides of different isoforms. Five out of the 13 known aquaporin isoforms were identified. Using a methodology for the solubilization and recovery of hydrophobic proteins on 2D-gels (Chevallet et al., 1998; Santoni et al., 1999), they also used this approach to resolve aquaporins isoforms and showed that there is a higher diversity of aquaporin isoforms present than are possible from only differential phosphorylation.

Root hair membrane proteins were more recently reported in soybean (Brechenmacher et al., 2009). Prepared microsomal proteins were analyzed by both 1D-SDS-PAGE followed by LC–MS/MS and shotgun proteomics. They identified aquaporins, transporters of phosphorus and calcium, ABC transporters, and lipid transfer proteins involved in the uptake of water and nutrients. Additional proteins such as adenine nucleotide translocator and dicarboxylate/tricarboxylate carrier were identified and the authors suggested their function in energy generation required for nutrient uptake. Vesicle trafficking proteins identified as associated membrane proteins, such as ADP-ribosylation factors, clathrins, coatomer subunits, dynamins, and GTP-binding proteins, may be involved in root hair polar growth due to their role in adding new plasma membrane and cell wall components to the growing tip.

These plasma membrane proteomic studies indicate that the protein composition of the plasma membrane is complex and diverse, and it is likely that many plasma membrane proteins are expressed at specific developmental stages or under particular environmental conditions.

10.2.5 Cell Wall-Associated Root Proteins

Cell wall proteins play a crucial role in cell wall structure and architecture, cell wall metabolism, cell enlargement, signaling, and response to biotic and abiotic stresses (Cassab et al., 1988; Cosgrove, 1999; Showalter, 1993; Somerville et al., 2004). Genome-scale assessments have predicted the existence of many hundreds of extracellular proteins in *Arabidopsis*. Several proteomics approaches were used to identify cell wall proteins from leaves and suspension cells (Borderies et al., 2003; Chivasa et al., 2002; Dani et al., 2005; Watson et al., 2003), but only one study focused its interest on maize roots (Zhu et al., 2006). In this study, a vacuum infiltration–centrifugation technique was used to extract cell wall proteins. This technique allows the extraction of only water-soluble and loosely ionically bound cell wall proteins. Forty percent of the proteins identified have an N-terminal signal peptide, thus indicating the effectiveness of the extraction method. The largest proportion of the cell wall proteins was categorized in carbohydrate metabolism. The majority of the signal peptide-carrying proteins are related to cell wall structure, metabolism, and modification—including beta-xylosidase, beta-D-glucan exohydrolase, alpha-l-arabinofuranosidase, beta-galactosidase, alpha-galactosidase, endoxyloglucan transferase, chitinase, endo-1,3;1,4-beta-D-glucanase, and peroxidases. The specific contributions to wall modification or mechanisms are still unclear but their existence emphasizes the complexity of wall restructuring in roots. Some of these proteins also provide an effective barrier to most microbial pathogens to elaborate extracellular defenses such as the chitinases and endo-beta-1,3-glucanases (York et al., 2004).

10.2.6 Secondary Structures and Root-to-Shoot Communication

In the mature root, the primary, dermal, ground, and vascular tissues begin to form just behind the zone of cellular maturation in the root tip. While the epidermis and the cortex of roots have a role in protection, water and mineral uptake, and storage of photosynthetic products, the vascular tissue consists of xylem and phloem vessels that allow the transport of water and nutrients and is responsible for cellular communication between organs (Figure 10.2). Secondary structures, found in roots of gymnosperms and most dicotyledons, consist of secondary xylem and phloem cambium. Xylem and phloem are the two specialized long-distance transport systems for organic and inorganic nutrients. While phloem transports predominantly organic compounds from the site of synthesis in the leaves to the site of use, the xylem stream transports water and dissolved minerals entering the roots from the soil to the aerial plant parts. Xylem sap also contains hormones, other secondary metabolites, amino acids, and proteins that may play an important role in root to shoot communication. Proteins have been detected in xylem sap using one-dimensional electrophoresis separations in watermelon, apple, peach, and pear (Biles et al., 1989), squash (Satoh et al., 1992), cucumber, broccoli, rape, and pumpkin (Buhtz et al., 2004), tomato (Rep et al., 2002), and soybean (Djordjevic et al., 2007), but only a few proteins were identified. More recently the xylem sap proteome map of *Brassica napus* and maize were characterized using 2D-gel electrophoresis (Alvarez et al., 2006; Kehr et al., 2005). Among the 69 and 154 proteins identified in *B. napus* and maize, respectively, peroxidases, proteases, and pathogenesis-related (PR) proteins were found in both species. Many proteins involved in primary and secondary cell wall loosening and extension, such as hydrolases, were only identified in maize xylem sap. An interesting observation was that almost half of the proteins identified were present in several isoforms and that most of the isoforms were identified as glycoproteins based on gel staining techniques. Proteins secreted into the xylem stream act in cell wall extension, lignification of xylem vessels, and in the programmed cell death essential for the differentiation of the xylary elements. Our proteomic study on maize also revealed that the xylem sap contains constitutively expressed defense-related proteins that actively inhibit pathogen growth (Alvarez et al., 2006). Although the location of synthesis of those proteins secreted into the xylem stream is unknown, protein composition of the xylem sap seems to be affected when the surrounding environment of the roots is changed, as was shown in maize undergoing drought stress or in soybean response to pathogenic and symbiotic microbe interactions (Alvarez et al., 2008; Djordjevic et al., 2007; Subramanian et al., 2009). This suggests that signals (hydraulic or chemical) synthesized in roots in response to environmental changes might affect protein synthesis in xylem sap to adjust plant growth and induce plant defense mechanisms.

To summarize, primary root growth is due to the activity of apical meristems, and elongation growth prevails within it. Secondary growth leading to thickening of roots is ensured by lateral meristems and cambium will assure root-to-shoot communication. While penetrating through the soil, roots are exposed to the effects of numerous minor and major stresses induced by water deficit, inappropriate nutrient supply, unsuitable temperatures, mechanical effects, and so on.

10.3 PROTEIN NETWORK REARRANGEMENT DURING SYMBIOTIC ASSOCIATIONS

The symbiotic interactions between members of the genera *Rhizobium*, *Bradyrhizobium*, *Mesorhizobium*, *Sinorhizobium* and *Azorhizobium* and plants of the Leguminoceae family result in the formation of nitrogen-fixing root nodules. The arbuscular mycorrhizal (AM) symbiosis occurs between more than 80% of terrestrial plant families and fungi of the order *Glomales*. The most studied relationships established between plants and specific bacteria and fungi are those involving rhizobial bacteria and mycorrhizal fungi, which trigger the host genetic program to allow localized infection and controlled growth. These microorganisms improve host plant growth by providing nutrients and in turn benefit from carbohydrates synthesized by the plant. The use of specific soil microbes to improve crop yield is one strategy for an affordable sustainable agriculture. Some of the major questions that have been addressed by root proteomics studies are the mechanisms involved in signal exchange and perception between microbe and plant, suppression of plant defense mechanisms, the exchange of nutrients between partners, and the structural changes induced by microbe invasion (Table 10.1).

10.3.1 Legume–Rhizobia Interactions

10.3.1.1 Root–Rhizobial Recognition

The primary signaling event in legume–rhizhobia interactions is the recognition by the plant of bacterially synthesized lipo chitin oligosaccharide Nod factors that trigger reactions in the plant such as root hair curling and meristem induction. The production of Nod factors is induced by flavonoids released by the legume roots (Redmond et al., 1986). Nod factors are perceived by plant transmembrane receptors such as the Ser/Thr kinases whose extracellular regions contain Lys M domains (Geurts et al., 2005). The colonization of the bacteria induces meristematic activity in the inner cortical cells, which leads to the formation of nodules. The mechanisms for nutrient exchange and corresponding regulation mechanisms are then established between the partners. Rhizhobia convert the nitrogen from air to ammonia that is then transferred to the plant for amino acid biosynthesis in exchange for carbon supplied as dicarboxylic acids. To facilitate bidirectional nutrient exchange, a plant perisymbiotic membrane called peribacteroid surrounds the intracellular microsymbionts *M. truncatula* is the first legume model plant used to study rhizobial symbiosis. Signal recognition was well studied at both the metabolite level and at the protein level. A proteomics study showed an increase of secreted proteins in the rhizosphere during the rhizobia–legume interaction (De-la-Pena et al., 2008). The proteins with an increased abundance during the interaction between *M. truncatula* and *Sinorhizobium meliloti* were identified as plant PR proteins: three chitinases and one thaumatin-like protein. It was shown that Nod factors are substrates of chitinases (Perret et al., 2000), suggesting that the increase of chitinase abundance in the rhizosphere is necessary to limit the amount of active Nod factors after their perception by the host plant (Staehelin et al., 1995).

10.3.1.2 Nodulation

Another proteomics study looking at the development of nodulation on a subterranean clover cultivar Woogenellup interacting with *Rhizobium leguminosarum* biovar *trifolil* showed the upregulation of ethylene-induced proteins after 24 and 48 h postinoculation (Morris and Djordjevic, 2001). However, in this study, they were not able to show that the enhancement of ethylene levels after inoculation was likely to favor or hinder nodulation. The same observation and conclusion are supported in a proteomics study from the Mathesius group using an ethylene-insensitive mutant of *M. trunculata* (*skl*) (Prayitno et al., 2006). Here, it was shown that during early nodule development,

Table 10.1 Summary of the Proteomics Studies on Rhizobial and AM Symbiosis

Type	Interacting Partners	Reference	Biological Event	Number of Proteins Identified in Plants	Methods Used[a]
Rhyzobial symbiosis	*Medicago truncatula* and *Sinorhizobium meliloti*	De-la-Pena et al. (2008)	Root–rhizobial recognition	72	2DE, LC–MS/MS
		Schenkluhn et al. (2010)	Early association events	22	DIGE, MALDI-TOF
		Prayitno et al. (2006)	Nodulation	17	2DE, MALDI-TOF, LC–MS/MS
		Van Noorden et al. (2007)	Nodulation	170	DIGE, MALDI-TOF/TOF
		Bestel-Corre et al. (2002)	Protein metabolism in root nodules	7	2DE, MALDI-TOF, ESI–MS/MS
		Catalano et al. (2004)	Symbiosis biogenesis	51	2DE, LC–MS/MS
	Trifolium subterraneum and *Rhizobium leguminosarum*	Morris and Djordjevic (2001)	Nodulation	18	2DE, N-terminal sequencing
	Glycine max and *B. japonicum*	Wan et al. (2005)	Nodulation in root hairs	47	2DE, MALDI-TOF, ESI–MS/MS
		Oehrle et al. (2008)	Protein metabolism in root nodules	69	2DE, MALDI-TOF

		Hao et al. (2004)	Protein metabolism in mitochondrial root nodules	74	2DE, N-terminal sequencing, MALDI-TOF, ESI–MS/MS
		Panter et al. (2000)	Symbiosis biogenesis	17	2DE, N-terminal sequencing
	Melilotus alba and *S. meliloti*	Natera et al. (2000)	Protein metabolism in root nodules	18	2DE, N-terminal sequencing, MALDI-TOF
	Pisum sativum and *R. leguminosarum*	Saalbach et al. (2002)	Symbiosis biogenesis	46	2DE, ESI–MS/MS
	Lotus japonicus and *Mesorhizobium loti*	Wienkoop and Saalbach (2003)	Symbiosis biogenesis	94	BN-PAGE. SDS-PAGE, LC–MS/MS
Mychorrizal symbiosis	*M. truncatula* and *Glomus mosseae*	Bestel-Corre et al. (2002)	Colonization	7	2DE, MALDI-TOF, ESI–MS/MS
	M. truncatula and *G. intraradices*	Amiour et al. (2006)	Appresorium formation	11	2DE, MALDI-TOF
		Schenkluhn et al. (2010)	Early association events	28	DIGE, MALDI-TOF
		Valot et al. (2005)	Nutrient exchange in PM in response to phosphate	25	2DE, MALDI-TOF, ESI–MS/MS
		Valot et al. (2006)	Nutrient exchange in PM	78	2DE, 2D-LC–MS/MS
	Lycopersicon esculentum and *G. mosseae*	Benabdellah et al. (2000)	Colonization in root plasma membranes	2	2DE, N-terminal sequencing

[a] 2DE: two-dimensional electrophoresis; DIGE: differential gel electrophoresis; MALDI-TOF: matrix-assisted laser desorption ionization time of flight; ESI–MS/MS: electrospray ionization–tandem mass spectrometry; LC–MS/MS: liquid chromatography–tandem mass spectrometry.

the expression of proteins such as aminocyclopropane carboxylic acid (ACC) oxidase, Kunitz proteinase inhibitor, and pprg-2 were not induced by the inoculation of *S. meliloti*, as could be observed in the wild type. The same protein-expression pattern was observed in mutants and wild-type roots treated with ACC, a precursor of ethylene biosynthesis, suggesting that the ethylene biosynthetic pathway is induced by the rhizobia and results in an increase in the number of nodules in the mutant *skl* defective in the ethylene pathway. Using the same proteomics strategy but comparing the overlap between auxin treatment and *S. meliloti* inoculation on *M. trunculata* roots, the same group identified proteins specifically induced by auxin involved in the early stages of nodulation by *S. meliloti* (van Noorden et al., 2007). Most of the proteins were involved in energy metabolism and protein processing, indicating that preparation for cell division is required after *S. meliloti* infection. No specific auxin-response proteins were identified, but interestingly other proteins involved in jasmonate biosynthesis (12-oxophytodienoic acid 10,11-reductase, OPR), redox-related proteins, and pathogen-related proteins were induced by auxin treatment. Additionally, jasmonic acid (JA) may play a role in the regulation of the nodule number (Nakagawa and Kawaguchi, 2006). Jasmonate biosynthesis may be induced by auxin in the early stages of nodulation and may be one of the components of the autoregulation mechanism that controls the number of nodules formed. The early steps of nodulation seem to be under the control of several pathways involving hormonal crosstalk between ethylene, auxin, and jasmonate as previously suggested (Morris and Djordjevic, 2001) for the plant to allow the establishment of bacteroid invasion but simultaneously limit the number of nodules formed by the rhizobia. Protein changes induced during nodulation were also studied at the root hair level (Wan et al., 2005). Root hairs in legumes are the preferred site of infection by rhizobia. In this study, they used soybean root hairs infected by *Bradyrhizobium japonicum* to study the protein changes using gel-based proteomics. They found specific proteins (phospholipase D and phosphoglucomutase) induced in root hairs that were not previously reported by any other rhizhobia–legume study. Phospholipase D functions in lipid signaling pathways (den Hartog et al., 2001) and may play a role during root hair deformation in microtubule reorganization. Phosphoglucomutase is involved in the partitioning of stored carbon (Harrison et al., 2000) and may play an important role in the availability of phosphoglucose for synthesis of required polysaccharides by rhizobia during nodulation (Lepek et al., 2002).

10.3.1.3 Protein Metabolism in Root Nodules

Other proteomics studies looked at the protein expression in the nodules formed within the plant roots. A proteomic comparison of the protein-expression profiles of the bacteroid cells, the nodule tissue, and the root tissue of *Melilotus alba* (white sweetclover) inoculated by *S. meliloti* indicated that a high number of proteins upregulated in the nodule were from the rhizobia (Natera et al., 2000), while the rest were newly synthesized proteins and potential new nodulins. At the same time, most of the proteins were downregulated in the bacteroid cells, including proteins involved in nitrogen acquisition. Later, a time course proteomics study that was performed on *M. truncatula* roots inoculated with *S. meliloti* also showed the induction of new protein synthesis along with inoculation as well as bacteroid proteins present in the nodules (Bestel-Corre et al., 2002). At the early stages, an elongation factor Tu, involved in protein synthesis, was identified as newly synthesized but with bacterial origin. In addition, several proteins involved in nitrogen fixation and carbon metabolism were induced by rhizobia interaction. In a more recent study on *Glycine max* root nodules, the largest number of proteins identified (40%) were in the categories of carbon and nitrogen metabolism (Oehrle et al., 2008). The following most abundant categories were then oxygen protection (12%) and, most importantly, proteins involved in protein trafficking (11%), suggesting a very active exchange of macromolecules as described by (De-la-Pena et al., 2008). The specific metabolisms induced in the nodules and the plant roots were also studied at the mitochondrial level. In soybean nodules interacting with *B. japonicum*, phosphoserine aminotransferase and glycine dehydrogenase were specifically identified in the nodule mitochondria and it was suggested that they may participate in the regulation of fixed nitrogen in the form of ureides (Hao et al., 2004).

10.3.1.4 Symbiosome Biogenesis

The nutrient exchange localized in the peribacteroid space (PS) has been studied using biochemical and biophysical techniques to identify transporters involved in the bidirectional transport of nutrients. P-type H^+-ATPase pumps were identified to be active in the transport from cytoplasm to PS (Blumwald et al., 1985), as well as ammonium transporters GmSAT1 (Kaiser et al., 1998) and an integral membrane protein, nodulin 26, functioning as a water channel in PBMs from *Glycine max* root nodules (Fortin et al., 1987). Proteomics studies also specifically targeted the PS and PBM. In the rhizobia symbiosis of soybean (Panter et al., 2000) and pea (Saalbach et al., 2002), in addition to V-type ATPases already identified in the PS and PMB, BiP (Immunoglobulin heavy-chain-binding protein), protein disulfide isomerase (PDI), and an integral membrane protein known from COPI-coated (Coatomer protein) vesicles were also identified. These proteins are typically found in the ER and Golgi membranes and participate in protein import and folding of secreted proteins in organelles. The authors of this study suggest that the presence of these proteins in the PBM may explain the biogenesis of the symbiosome supplied by the ER membranes. A proteomic study on the PBM from *Lotus japonicus* root nodules identified many more membrane proteins (Wienkoop and Saalbach, 2003). In particular, transporters for sugar and sulfate, ATPases and aquaporins, endomembrane-associated proteins, and vesicle receptors were identified. This study also identified several proteins involved in signaling transduction such as 14.3.3 proteins, receptors kinases, calmodulin, and pathogen-related proteins. The ATPases identified were V-type as identified in the pea symbiosome, described earlier, which support the presence of the V-type ATPases in addition to the P-type ATPases identified by immunological analysis (Fedorova et al., 1999). The receptor kinases and 14.3.3 proteins indicate possible cell-to-cell signaling between rhizhobia and the plant host to establish and regulate the symbiosis. To summarize, four pathways are involved in symbiosome biogenesis as suggested by (Catalano et al., 2004) using previous proteomics studies and their own biochemical characterization of the symbiosome from *M. truncula*ta nodules roots after interaction with *S. meliloti.* First, proteins synthesized by free ribosomes may be targeted using N-terminal signal peptides to the PBM. Second, the presence of endomembrane proteins supports that proteins transported and modified in the Golgi may be secreted via vesicles to the symbiosome. Third, proteins may transit to the symbiosome directly from ER-derived membranes. Finally, identification of bacteroid proteins in the nodules suggests that the symbiosome membrane may also be derived from the bacteroid.

10.3.2 AM Symbiosis

10.3.2.1 Appresoria Formation and Signal Transduction

The role of vesicular–AM fungi in yield augmentation is well established, and yet little is understood about their growth, biology, and ecology. AM symbioses are the most common type of mycorrhizae and involve a majority of plants. Mycorrhizae formation significantly alters the physiology and/or morphology of roots and plants in general, starting from the formation of the appressoria at the root surface up to the fungal mycelium proliferation within the cortical parenchyma (Gianinazzi-Pearson et al., 2000). The fungus differentiates arbuscules in which the fungal cell wall is completely surrounded by a modified root cortical cell plasma membrane defined as the PM (Harrison, 1999). In this bilateral exchange, plants benefit from improved mineral nutrient uptake from the soil (mainly phosphorus), and in turn the AM fungi are supplied with the organic carbon forms essential for achieving their full life cycle. In the symbiosis between *M. truncula*ta and *Glomus mosseae*, a proteomics study performed during the time course of the inoculation reported proteins involved in redox-related defense responses such as peroxidase and glutathione-*S*-transferase (Bestel-Corre et al., 2002). In addition to ROS metabolism induction, inhibitors of polygalacturonases, which are produced by the fungi to progress through the plant cell walls (Perotto, 1995), were identified in the early stages of inoculation. In contrast to the rhizobia–legume interaction involving metabolites as recognition signals, the recognition event inducing appressorium formation was physical contact between the fungi and host root (Nagahashi and Douds, 1997). The appresorium formation step and the early events were investigated using a comparative proteomics study of wild-type *M. truncatula* and two mutants, mycorrhiza-defective (*dmi3*) and autoregulation-defective (*sunn*), during their interaction with *Glomus intraradices* (Amiour et al., 2006). DMI3 encodes a calcium and calmodulin-dependent protein kinase and its mutation in *M. truncatula* results in the loss of ability to develop a functional interaction with AM fungi (Levy et al., 2004). SUNN encodes a leucine-rich repeat receptor-like kinase which when mutated is responsible for an increase in mychorrhizal colonization intensity or an increase in nodule number when inoculated with AM fungi and rhizobia, respectively (Sagan et al., 1995). This last gene is required in the autoregulation of mychorrhization. While two glutathione-*S*-transferase abundances were downregulated in the wild type during the appresorium formation, several proteins (GSH-dependent dehydroascorbate reductase, cyclophilin, and actin depolymerization factor) described as acting in transduction signaling were upregulated. Those same proteins were found to be upregulated in *dmi3* only, suggesting a DMI3-dependent response to *G. intraradices* with DMI3 acting as repressor of signaling pathways leading to the restriction of fungal penetration by failing to develop the early cytoskeleton rearrangements necessary for AM fungus entry (Genre et al., 2005; Sanchez et al., 2005). A chalcone synthase was found to increase in abundance in wild type but not in *sunn*. Additionally, an auxin-responsive protein was also defective in the *sunn* mutant. Those elements, when compared to the early events in the nodulation of legume by rhizobia, also suggest the idea of common early regulatory pathways shared between the two symbioses.

10.3.2.2 Nutrient Exchange in Periarbuscular Space

The events of nutrient exchange occurring after the establishment of symbiosis were also analyzed with proteomics tools. A proteomics study targeting membrane proteins of *M. truncatula* roots inoculated or not by *Glomus intraradices* looked at the effect of phosphate availability in soil in the replacement of fungi symbiosis (Valot et al., 2005). Seven membrane proteins were identified with similar downregulation of protein-expression patterns in fungal colonization and phosphate supply conditions. Two of them related to ATP synthesis activated only in phosphorus-limiting conditions (Hammond et al., 2004). Two were ferritin, directly linked to iron uptake, the increase of which is correlated with phosphate-starved plant roots (Shen et al., 2004). Interestingly, a 53-kDa nodulin was also isolated from *M. truncatula* in response to AM symbiosis (Valot et al., 2005). This nodulin was earlier identified in the soybean symbiosome membrane with *B. japonicum* (Winzer et al., 1999). An earlier report also shows the upregulation of gene expression of the nodulin 26-like protein in *M. truncatula* root interactions with *G. mosseae* and *S. meliloti* (Brechenmacher et al., 2004). This supports the idea that mycorrhizal and nitrogen-fixing symbioses share common molecular pathways. The membrane proteins from the PM formed in the interaction of *M. truncatula—G. intaradices* were also investigated using a proteomics approach to study the nutrient transfer mechanisms (Valot et al., 2006). In addition to the vacuolar ATPase previously identified in PM mycorrhizal tomato roots (Benabdellah et al., 2000), and two H^+-ATPase and phosphate transporters in *M. truncatula* (Gianinazzi-Pearson et al., 2000; Harrison et al., 2002), in this study they were able to identify proteins possibly involved in nutrient transport. Two proteins, Mtha1 and MtBcp1, were specifically identified in the PM only after invasion by the fungi. Mtha1 corresponds to a proton-efflux P-type ATPase. It was suggested that ATPase proteins are responsible for generating the driving force necessary for the uptake and efflux of solutes across the membrane (Ferrol et al., 2002). MtBcp1 is a blue copper-binding protein with a domain homologue to a GPI anchored protein belonging to the plastocyanin-like family and implicated in electron transfer reactions (Garrett et al., 1984).

10.3.3 Common Symbiotic Protein Induction

Only one proteomics study recently focused its research on the identification of common pathways in signal recognition of mycorrhizal and nitrogen-fixing symbiosis in *M. truncatula* (Schenkluhn et al., 2010). They identified the early induction of a calcium-binding regulatory protein (CaM) that could be part of the calcium signaling occurring in the early pathways of myc- and nod–symbiotic associations, activating a cascade that converges on the nucleus inducing downstream gene expression. In accordance with this hypothesis, 14.3.3 proteins and nucleus signal transducers (Ran-binding proteins) were identified, which can affect plant hormone homeostasis and signaling as mentioned previously. Additionally, proteins related to energy metabolism (ATP synthase, nucleoside diphosphate kinase, malate dehydrogenase) and antioxidant defense mechanisms (ascorbate peroxidase, peroxidase, cationic peroxidase, patatin-like protein) were identified after nodulation and mycorrhization.

10.4 PROTEINS INVOLVED IN PHYSIOLOGICAL ALTERATIONS UNDER UNFAVORABLE CONDITIONS

When threaten plants are subjected to a stress situation, they translate as a deviation from the optimum which can threaten the integrity of survival of the whole system. The capacity of the plant to adapt to unfavorable environments in which the plant responds by gradual changes in its metabolism and structures responsible for long term species survival is called adaptation. Plants can also show short-term responses called acclimation in which rapid phenotypic adjustments to transient changes in environmental conditions are observed. The study of the mechanisms induced in both short- and long-term plasticity is of great importance to improve crop resistance.

10.4.1 Pathogen–Root Interactions

Root–microbe interactions include colonization of the rhizosphere by bacteria, fungi, and nematodes. These interactions can be beneficial (symbiotic) as described previously or detrimental (pathogenic) to the plant host and therefore can greatly influence not only the performance of the root growth, but also the overall plant growth. When plants are challenged by a pathogen infection, early local defense reactions (hypersensitive response) and delayed systemic responses (systemic acquired resistance) are activated in order to counteract the pathogen attack. This defense strategy is based on pathogen recognition and cell-to-cell communication in the tissue adjacent to the site of infection. When there is specific pathogen recognition and induction of hypersensitive response in the plant host which inhibits pathogen growth, the plant–pathogen interaction is called an incompatible interaction. If the pathogen causes disease symptoms on the plant host which is susceptible to the pathogen because of lack of recognition and induction of plant defense mechanisms, the interaction is called a compatible interaction. An understanding of how plants and pathogens recognize each other and differentiate to establish either a successful or an unsuccessful relationship is crucial in this field of investigation. Evidence of target proteins involved in signal recognition and plant defense events have been provided by proteomics studies (Benschop et al., 2007; Campo et al., 2004; Chivasa et al., 2006; Jones et al., 2006; Kim et al., 2003; Mahmood et al., 2006; Ndimba et al., 2003; Peck et al., 2001). However, most of them focus on pathogen infections causing damage to aerial parts of the plants. Only a few proteomics studies have focused on root pathogens, although an understanding of interactions between pathogens and the root system will help in the selection of

Table 10.2 Summary of the Proteomics Studies on Root–Pathogens Interactions

Type	Interacting Root Pathogens	Reference	Number of Proteins Identified in Plants	Proteins and Mechanisms Highlighted	Methods Used[a]
Fungi	*Medicago truncatula* and *Aphanomyces euteiches*	Colditz et al. (2004)	12	PR-10 involvement in various stress adaptations	2DE, MALDI-TOF
		Schenkluhn et al. (2010)	49	PR-10 as pathogen-specific induced	DIGE, MALDI-TOF
	M. truncatula in symbiosis with Sinorhizobium meliloti and *A. euteiches*	Colditz et al. (2005)	14	Proteasome subunits associated with disease-tolerant PR-10 as ABA-responsive proteins associated with disease susceptibility	2DE, MALDI-TOF
		Schenkluhn et al. (2010)	26	Defense mechanisms induction (ROS-related enzymes)	DIGE, MALDI-TOF
	M. truncatula in symbiosis with Glomus intraradices and *A. euteiches*	Schenkluhn et al. (2010)	22	Defense mechanisms induction (ROS-related enzymes)	DIGE, MALDI-TOF
	Gossypium hirsutum and *Thielaviopsis basicola*	Coumans et al. (2009)	57	PR-10 involved in disease resistance	2DE, LC–MS/MS
	Pisum sativum and *Orabanche crenata*	Castillejo et al. (2004)	22	PR-10 and defense mechanisms proteins associated with the disease resistance	2DE, MALDI-TOF
	Arabidopsis thaliana and *Plasmodiophora brassicae*	Devos et al. (2006)	46	Proteins involved in auxin synthesis to trigger gall development	2DE, MALDI-TOF/TOF
	Brassica napus and *Plasmodiophora brassicae*	Cao et al. (2007)	20	Proteins involved in cytokinins homestasis ROS-related proteins involved in detoxification and cell wall fortification	2DE, LC–MS/MS
	Fagus sylvatica and *Phytophtoa citricola*	Valcu et al. (2009)	68	ROS-related proteins involved in cell wall fortification	2DE, LC–MS/MS
Bacteria	*Oryza sativa* and *Azoarcus* sp	Miche et al. (2006)	20	PR associated with plant defense mechanisms inducing bacterial colonization resistance, via JA signaling pathway	2DE, MALDI-TOF, LC–MS/MS
Nematode	Gossypium hirsutum	Callahan et al. (1997)	0	Protein associated to the resistance response to root-knot nematode	2DE, N-terminal sequencing

[a] Refer to Table 10.1 for definitions.

appropriate soil chemical sanitization strategy for sustainable agriculture (Table 10.2). In compatible interactions, the degree of plant susceptibility after interaction with a virulent pathogen is dependent on the strain of the pathogen and the variety of the plant. Therefore, proteomics studies have focused more on the understanding of the mechanisms induced during compatible interactions using cultivars with different susceptibilities.

10.4.1.1 Fungi–Root Interactions

Compared to the study of root–mycorrhizal symbiosis, sparse proteomic information is available regarding biochemical and physiological interactions in root–fungi. One of the most important diseases in pea is from *Aphanomyces euteiches* which belongs to the class of Oomycete of the eukaryotic protists. Although this soilborne pathogen is an interesting parasite to study plant–oomycete interactions because of its pathogenicity on the model *M. truncatula*, so far there is no efficient chemical control or resistant cultivar available. Genomic approaches allowed the discovery of resistance quantitative trait loci that will improve legume resistance using classical breeding. But physiological and biochemical processes involved in *A. euteiches* pathogenicity still have to be investigated. Several proteomics studies in the past five years attempted to answer some of these questions. The protein expression pattern in response to *A. euteiches* infection was studied in two lines of *M. truncatula* with different susceptibility to the parasite (Colditz et al., 2005). Among the proteins induced in response to the infection in the decreased susceptible line of *M. truncatula*, and therefore involved in mechanisms that lead to an improved disease resistance, proteasome subunits involved in proteolysis pathways were identified. More interestingly, they extended the proteomics comparison using mycorrhizal plants previously demonstrated to enhance plant resistance (Kjoller, 1996). A group of PR-10 proteins were found with reduced expression in the mycorrhizal *M. truncatula* infected by *A. euteiches* compared to the infected nonmycorrhizal plants. Additionally, the two proteasome subunits induced in the more resistant line also show an induction in the infected mycorrhizal plants. Exogenous application of ABA was also tested in the same study to investigate the mechanisms induced by ABA leading to increased susceptibility in plant–oomycete interactions (Audenaert et al., 2002). They identified the induction of the PR-10 group as ABA-responsive proteins, as previously identified (Colditz et al., 2004). It was suggested that ABA plays a crucial role in the susceptibility of *M. truncatula* to root pathogen *A. euteiches* by affecting the expression of PR-10 ABA-responsive proteins. However, these data are contradictory to findings from other proteomics studies. In a more recent study on the proteins induced after *A. euteiches* in *M. truncatula* using differential gel electrophoresis (DIGE) (Schenkluhn et al., 2010), the PR-10 proteins were only induced in *A. euteiches* infected plants and not in mixed infections with mycorrhizal or nitrogen-fixing symbionts. A comparison of different genotypes showing different susceptibilities to pathogen infection was achieved by a proteomics approach in the interaction of *Pisum sativum* with crenate broomrape (*Orobanche crenata*) (Castillejo et al., 2004). In this study, they found that the most resistant genotype had higher levels of several PR proteins, irrespective of *O. crenata* infection, indicating that priming defense reactions against pathogens more rapidly in this genotype may help increase the plant's resistance to infection. The same group of PR-10 was also identified more recently as upregulated in cotton roots after inoculation with the black root rot fungus *Thielaviopsis basicola* (Coumans et al., 2009). PR-10 proteins were also found to be the most abundant proteins detected in noninfected *M. truncatula* roots suggesting that they might be constitutively expressed to provide roots with more efficient plant defense mechanisms (Mathesius et al., 2001). It is possible then to hypothesize that the PR-10 group proteins may have a different function than the PR proteins.

Hormonal involvement in the early events of plant–pathogen interactions was also investigated in *Arabidopsis* infected with *Plasmodiophora brassicae* in combination with a proteomics study (Devos et al., 2006). This protist pathogen is responsible for clubroot, an infectious disease affecting members of *Brassicaceae*. Auxin transport was suggested to be essential in the induction of gall development (Vandenbussche et al., 2003). In the proteomics study, this hypothesis was supported by an increase in the abundance of dehydrogenases after 4 days of infection, which acts in the biosynthesis of precursors of IAA–amino conjugates (LeClere et al., 2004). Another source of IAA was also suggested to come from the degradation of glucosinolates as supported by the increasing abundance of myrosinases. The susceptible reaction of *Arabidopsis* toward *P. brassicae* leads to cell division, cell expansion, and cell degradation (Devos et al., 2006). In the same study, an increase in isopentenyl-type cytokinins originating from *P. brassicae* was detected in the early stages of infection, indicating an involvement of cytokinins in the primary development of clubroot. This hypothesis was supported by the decrease in abundance of adenosine kinase (ADK) involved in cytokinin homeostasis observed in the proteomics study of *B. napus* infected with *P. brassicae* (Cao et al., 2007). ADK may control the interconversion of cytokinin bases and ribosides, which regulates the amount of the active form of cytokinins. Among the other proteins differentially expressed in response to *P. brassicae* infection, proteins involved in ROS detoxification (copper zinc superoxide dismutase (CuZnSOD), cytochrome *c* oxidase) as well as proteins indirectly involved in ROS detoxification (nucleoside diphosphate kinase, flavodoxin-like quinone reductase) were induced. The induced plant defense mechanisms involved cell wall lignification with the downregulation of caffeoyl-CoA *O*-methyltransferase and *S*-adenosine methionine synthase, and synthesis of spermidine with the upregulation of spermidine synthase. Although the plant can induce plant defense mechanisms to reduce the spread of infection, the compatible interaction induces plant disease. Additional ROS proteins were also identified during the interaction between *Fagus sylvatica* and the root pathogen *Phytophthora citricola* (Valcu et al., 2009). A catalase was identified to be upregulated locally and suggested that ROS might be locally generated to prevent pathogen invasion by inducing cross-linking of the cell wall glycoproteins and contributing to cell wall fortification. Similar observations of induction of ROS-related enzymes were also found in *A. euteiches* infected *M. truncatula* that were also colonized by either *G. intraradices* or *S. meliloti* (Schenkluhn et al., 2010).

10.4.1.2 Bacteria–Root Interactions

Azoarcus sp., a non-nodule-inducing bacterium recognized as endophytic of Gramineae and in which infection occurs at the emergence points of lateral roots, was also the focus of investigations to understand plant defense mechanisms induced by bacterial colonization (Miche et al., 2006). Two different cultivars of rice showing differences in the compatible interaction with an *Azoarcus* sp. strain were used to identify the mechanisms responsible for bacterial colonization resistance and the involvement of JA in plant defense responses. External application of JA on the roots of the less compatible rice cultivar showed the same morphological changes when this cultivar is infected by an *Azoarcus* sp. strain, suggesting that JA plays a role in the signaling pathway that leads to the induction of plant defense responses. After a comparison of the root proteomes of the two cultivars inoculated with the bacteria or after JA application, this study found that two proteins from the PR-10 group were induced in the susceptible cultivar after bacterial inoculation, but not after JA treatment. Those same two proteins were induced by the JA treatment in the less compatible cultivar in addition to more PR proteins. Similar PR-10 proteins to what was identified after fungal–root interaction seems to be associated with plant defense mechanisms inducing bacterial colonization resistance via the JA signaling pathway. It is still unknown what the specific function of those proteins is, but it was suggested that PR-10 may act as brassinolide steroid carriers that were found to induce resistance to a broad range of disease in tobacco and rice (Liu and Ekramoddoullah, 2006; Nakashita et al., 2003). Another hypothesis is the involvement of the PR-10 with RNAase activity acting directly on the pathogens (Liu and Ekramoddoullah, 2006).

10.4.1.3 Nematode–Root Interactions

Another parasite–plant interaction that causes tremendous losses in agriculture worldwide is the root-knot (*Meloidogyne* sp.) and cyst (*Heterodera* sp.) nematode. Root-knot and cyst nematodes are soilborne and feed on roots. These organisms invade plant roots as juvenile larvae (J2) and develop into adult forms that reproduce in repeated cycles. The feeding site is composed of five to seven hypertrophied multinucleated cells named giant cells. These cells serve as a permanent source of nutrients for the nematode and the stylet secretions, produced by two subventral and one dorsal oesophageal gland cells, were suggested to play a role in the feeding site induction and maintenance (Bird, 1962; Hussey, 1989). For root-knot nematode, galls will typically develop on the root. Upon becoming adults, root-knot nematodes will begin to lay eggs that are contained in a gelatinous matrix at the posterior end of the body. In contrast, cyst nematodes maintained most of the eggs laid within the body of the female, which typically breaks through the root surface and is first visible as a white, pin-head size spot or cyst on the root. The identification of proteins secreted by the nematode has been the first focus of investigations in order to understand the mechanisms of parasitism. Proteomics studies of the plant defense mechanism induced from root–nematode interactions are still in their infancy. Only one report investigated the proteins involved in the root–nematode interaction in cotton. A cotton-resistant germplasm (Auburn 634) able to arrest nematode growth and reproduction but not its penetration was studied to identify the proteins that were induced compared to a susceptible germplasm (Callahan et al., 1997). A major protein of 14 kDa was identified as correlating with the resistance response to root-knot nematode but the function was unknown. The protein expression was localized to the gall tissue but it originated from plant tissue.

10.4.2 Water Stress and Temperature Changes

There is much interest in drought and low-temperature resistance of plants because one-third of the soil resources on our planet is affected either by drought or low temperature and they are responsible for considerable losses in plant production (Boyer, 1982). Flooding and high soil temperatures are stresses that have been less well studied in general, although they can be detrimental and lethal to the whole plant (Figure 10.3).

10.4.2.1 Drought Tolerance of Roots

There are several strategies for plants to develop drought resistance: drought escape, which involves reducing the life cycle (Araus et al., 2002), drought avoidance, which includes enhancing water uptake and reducing water loss (Chaves et al., 2002; Price et al., 2002), and finally drought tolerance, which consists of osmotic adjustment and enhancement of antioxidative capacities (Chaves and Oliveira, 2004). Under drought stress, cell division and elongation of the root is inhibited and root growth is impaired (Gonzalez-Bernaldez, 1968). It was shown previously that roots were able to develop stress defense mechanisms, as well as structural adaptations in response to drought (Malamy, 2005; Spollen et al., 2000). Mechanisms of drought tolerance were studied at the proteomic level in a C3 xerophyte, *Citrullus lanatus* sp. (wild watermelon) (Yoshimura et al., 2008), which exhibits high drought resistance due to its ability to extend the root system into deep soil layers and use the roots for water storage. They observed in this study the induction of the different sets of mechanisms induced at different stages of drought stress. At the early stage of drought root growth is maintained to allow water uptake from deep soil layers. This drought avoidance event is associated with protein abundance changes in proteins involved in root morphogenesis and growth such as Ran GTPases activating cell division and lignin synthesis-related proteins increasing cell wall strength. At the later stage of drought stress, molecular chaperones are induced to protect denaturation of proteins and associated with root growth suppression and

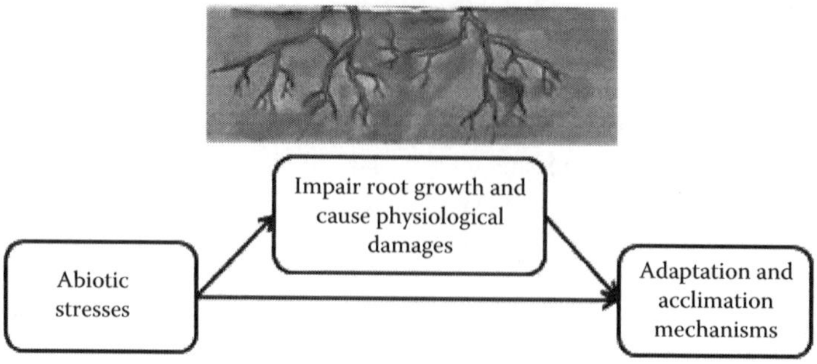

Abiotic stress	Plant	Mechanism
Drought	Maize	Cell wall loosening induced by ROS accumulation to allow root elongation (Zhu et al., 2007).
	Soybean	Isoflavone biosynthesis involved in apical root elongation and iron sequestration by ferritin involved in protection against oxidative damage (Yamaguchi et al., 2010).
	Watermelon	Root morphogenesis adaptation to allow root growth and cell wall fortification (Yoshimira et al., 2008).
Flooding	Soybean	Oxygen transport proteins induced in early stages and fermentation processes induced in later stages to maintain energy production (Komatsu et al., 2009b). Antioxidative system and mechanisms of protection against protein degradation (Komatsu et al., 2009c).
Low temperatures	Rice	Glyoxylase pathway induced for detoxification of methylglyoxal (Lee et al., 2009). Regulation of ion channel activity involved in osmotic homeostasis (Hashimoto et al., 2010).
Heat	Grass	ROS detoxification enzymes induced (Xu et al., 2008).
Salt	Arabidopsis	ROS scavenging enzymes to reduce ROS accumulation, cell wall remodeling to allow cell wall stabilization, and defense mechanisms induction (Jiang et al., 2007).
	Pea	ROS scavenging enzymes to reduce ROS accumulation and transporter of sodium into vacuole for ion nomeostasis (Kav et al., 2004).
	Rice	ROS scavenging enzymes to reduce ROS accumulation (Yan et al., 2005; Cheng et al., 2009) Cell wall remodeling to allow cell wall stabilization (Cheng et al., 2009).
Nitrogen deprivation	Maize	Protein degradation for remobilization of nitrogen (Prinsi et al., 2009).
Phosphorus deprivation	Maize	Carbon metabolism induced to allocate more carbon for root growth, cytoskeleton rearrangement to allow root growth and induced chaperones for protein folding (Li et al., 2007).
Copper contamination	*Elshotzia splendens*	Scavenging proteins induced to regulate cell redox state, cell wall remodeling to allow Cu transport and accumulation, Cu detoxification mediated by germin-like proteins (Li et al., 2009).
Cadmium contamination	*Brassica juncea*	Sulfur metabolism related proteins for the xenobiotic detoxification of Cd (Alvarez et al., 2009).
	Medicago truncatula	Mycorrhizal inoculation alleviate the oxidative stress induced (Repetto et al., 2003).
Aluminum contamination	Rice	Sulfur and ethylene metabolisms induced for antioxidant and detoxification mechanisms (Yang et al., 2007).
	Tomato	Redox regulation by induction of ROS scavenging enzymes (Zhou et al., 2009).
Arsenic contamination	Maize	Antioxidant enzymes induced to regulate cellular homeostasis (Requejo et al., 2005).
	Rice	Sulfur metabolism induced to reduce oxidative stress (Ashan et al., 2008).

Figure 10.3 Adaptation and acclimation mechanisms induced by abiotic stresses identified using the proteomics approach.

drought tolerance mechanisms. In maize, the protein changes of the root cell wall were studied by proteomics (Zhu et al., 2007) to understand the induced adaptation mechanisms responsible for maintaining cell elongation during drought as shown previously (Sharp et al., 1988). The largest group of proteins induced in water-stressed roots was involved in ROS metabolism and it was associated with an increase in hydrogen peroxide in the apical region of the elongation zone. In this study, they concluded that the increase in apoplastic ROS production might contribute to the enhancement of cell wall loosening in this same region via polysaccharide cleavage (Fry, 1998). In a more recent study, a similar proteomics approach was used on soluble proteins of soybean roots to understand the mechanisms for maintaining root elongation in the apical root tip (1–4 mm) during water stress (Yamaguchi et al., 2010). Based on the proteins identified as showing region-specific differences in response to water stress (chalcone synthase 7 and caffeyl-CoA *O*-methyltransferase), the authors suggested that an increase in isoflavone synthesis in the apical root tip might explain the maintenance of root elongation, while the increase in lignin biosynthesis in the following root region (4–8 mm) might be responsible for the inhibition of root elongation. They confirmed this hypothesis by monitoring a significant accumulation of isoflavones in the root tip and showing an enhancement of cell wall lignification in the following region. Ferritin proteins were also identified as increasing in abundance in response to water stress. Because accumulation of free iron can produce hydroxyl radicals via the Fenton reaction, they hypothesized that ferritin prevent iron accumulation in the root elongation zone during water stress and they showed that iron levels were higher with increasing distance from the apex. In nodulated legumes, symbiotic nitrogen fixation is one of the physiological processes to first show stress responses (Zahran, 1999). Plants of *M. truncatula* in symbiosis with *S. meliloti* were drought stressed and the root nodule response in the nitrogen fixation activity was studied at the proteomic level (Larrainzar et al., 2007). They showed that abundance of enzymes involved in methionine (methionine synthase) and asparagine (asparagine synthetase) biosynthesis were downregulated in response to drought stress in nodules. Methionine synthase is also involved in the regeneration of the methyl group S-adenosyl methionine (SAM) that acts as a methyl donor group and functions as a precursor for metabolites such as ethylene, which has been shown to act as an intermediate in nodule autoregulation (Wood, 2001). Asparagine synthetase in the nodule acts in the conversion of fixed nitrogen into asparagine that will be transported to the rest of the plant. The downregulation of protein abundance of those proteins suggests that drought stress may be responsible for a decrease in the nitrogen fixation localized in the nodules. It is, however, still unknown how this affects the whole plant. In pea nodulated with *R. leguminosarum*, it was shown using enzymatic assays that the decrease in nitrogen fixation in drought-stressed nodulated roots was localized and not systemic (Marino et al., 2007).

10.4.2.2 Flooding Stress on Roots

Although flooding stress has been less studied than water deprivation, partial to complete flooding is detrimental by hampering plant growth and/or causing premature death because most of the crops are not selected to cope with flooding stress (Setter and Waters, 2003). A major constraint resulting from excess water is an inadequate supply of oxygen to submerged tissues. Flooding leads to reduced gas exchange between the plant tissue and the atmosphere. In addition to the threat of oxygen deficiency, excess water also leads to other changes in the soil that influence plants such as levels of the plant hormone ethylene (Smith and Russel, 1969) and the accumulation of products of anaerobic metabolism by soil microorganisms (e.g., Mn^{2+}, Fe^{2+}, S^{2-}, H_2S, and carboxylic acids) (McKee and McKevlin, 1993). Most studies on flooding stress have focused on relatively flood-tolerant species from *Oryza*, *Rumex*, and *Echinochloa* genera. Single species studies are valuable for an understanding of the regulation of various acclimations but less meaningful toward gaining an ecological perspective. One group focused on understanding the regulatory events induced in response to flooding stress in soybean roots using proteomics (Komatsu et al., 2009a, 2010). After 12 h of flooding stress, ROS scavengers (ascorbate peroxidase and 1-Cys-peroxiredoxin) and protein chaperones (calreticulin-1 and PDI) were all downregulated (Komatsu et al., 2009a). After applying 2 days of flooding stress, they observed that enzymes involved in glycolysis or oxygen transport were induced as rescue events from flooding stress (Komatsu et al., 2010). Enzymes involved in alternative pathways of energy production such as fermentation were induced in the later stage of flooding stress in order to cope with the limited oxygen supply which impairs energy production from phostosynthesis. By comparing the flooding effect with low-oxygen supply, they also identified that some protein abundances were specifically altered by the flooding stress only, indicating that there are signaling cascades independent from low oxygen-sensing signal (Komatsu et al., 2010). In a different experiment of flooding stress in soybean roots, they also studied the differences in membrane proteins. Using gel and LC-based methods for protein separation and identification, they identified different sets of membrane proteins up- and downregulated by 1 day of flooding. Associated membrane proteins such as 14.3.3 and Ser/Thr protein kinases (Nek5) were identified as upregulated after flooding. While the involvement of Nek5 during flooding stress is unclear, authors suggested that 14.3.3 proteins might be associated with the regulation of the ATPase alpha subunit to maintain homeostasis during flooding stress. Other associated membrane proteins identified in this study, such as a SOD and HSPs, suggest that an antioxidative system and mechanisms of protection against protein degradation must be induced by flooding stress. However, only a limited number of integral membrane proteins were identified (Komatsu et al., 2009b).

10.4.2.3 Low-Temperature Effects on Root Function

Root growth and nutrient uptake are affected by low soil temperatures. Most research focuses on cold and chilling stress adaptation mechanisms on the aboveground organs (Amme et al., 2006; Bae et al., 2003; Hashimoto and Komatsu, 2007; Imin et al., 2004; Renaut, 2003), and only two studies focused on the chilling effect on rice roots specifically (Hashimoto et al., 2009; Lee et al., 2009). They observed that protein abundance changes of several proteins were previously identified in proteomics studies from aerial parts of plants, indicating that similar mechanisms are induced by chilling stress in leaves and roots. In the first study (Lee et al., 2009), proteins identifed as unique to the rice root proteomics study were involved in energy production and metabolism (acetyl transferase, pyruvate orthophosphate dikinase precursor, phosphogluconate dehydrogenase, NADP-specific isocitrate dehydrogenase, fructokinase), regulatory proteins (alpha-soluble *N*-ethylmaleimide-sensitive factor (NSF) attachment protein) and detoxification (glyoxalase 1). NSF is essential for vesicular trafficking; it is associated with membranes of the cytosol and subcellular organelles, and has been recently associated with stress responses proteins (Mazel et al., 2004). Glyoxalase 1 is the key enzyme of the glyoxalase pathway in the detoxification of methylglyoxal (MG) produced in response to several abiotic stresses (Espartero et al., 1995). It was proposed that this gene could be used to generate chilling tolerant transgenic plants. The second study focused on membrane proteins cold stress responsiveness (Haag-Kerwer, 1999). Only a few proteins were identified with significant changes in abundance. Among them, several ATPases were downregulated while a hypersensitive-induced response showed more than a 3× increase. The latter protein belongs to the hypersensitive induced reaction (HIR) family, including prohibitins and stomatins, which are involved in the regulation of ion channel activity related to the regulation of cell division and osmotic homeostasis (Nadimpalli et al., 2000).

10.4.2.4 Thermotolerance of Roots

Similar to the effect of low temperatures on roots using proteomics, the effect of increased temperatures is also not well studied at the root level. However, various studies demonstrated that roots are more sensitive to heat stress and therefore high soil temperatures are more detrimental than high air temperatures for whole plant growth (Xu and Huang, 2001). The temperature increase in soil associated with global warming has been an increasing concern, especially for temperate species. Only one study looked at the mechanisms induced in roots of grass and tried to identify the events associated with thermotolerance acquisition by comparing two grass species, *Agrostis stolonifera* L. (heat sensitive) and *Agrostis scabra* (thermotolerant) showing different heat responses (Xu and Huang, 2008). They found that the thermotolerance of *A. scabra* was associated with high levels of proteins involved in ROS detoxification. They also showed that differential phosphorylation of different isoforms of fructose aldolase was also associated with the thermotolerance of *A. scabra*. Surprisingly, no HSPs were identified in this study as upregulated. HSP expression was correlated with thermotolerance in several studies focused on different parts of the plant or different stages of plant development (Lin et al., 2005; Majoul et al., 2003; Puigderrajols et al., 2002; Wang and Luthe, 2003).

To summarize, a diverse number of root proteome responses were identified in water stress and temperature changes experiments using proteomics approachs; however, the integrated interpretation of the large number of induced proteins is still difficult.

10.4.3 Soil Composition Changes

Agriculture research is focused on how to make soils more productive. Roots are the site of perception of changes in soil composition and, as a result, the study of root development and its relationship to soil composition is essential to ascertain optimal conditions to improve crop productivity. Several elements in soil composition must be controlled for sustainable agriculture including the soil salinity and the nutrient concentrations (Figure 10.3).

10.4.3.1 Salinity Stress

Soil salinity is an abiotic stress that limits the productivity and distribution of plants. Natural phenomena or human practices can cause salts to accumulate in soils (Wiebe et al., 2007). Excess salt interferes with mineral nutrition and water uptake causing not only nutrient deprivation stress and osmotic shock, but also ionic stress by accumulating toxic ions in the plant. The protein changes induced in response to soil salinity were studied in an early report using 2D-gels, but confident protein identifications of the protein changes was still a challenge at this time with the use of N-sequencing techniques (Majoul et al., 2000). Later, salinity stress response showed the protein expression of antioxidant enzymes (SODs) in pea roots (Kav et al., 2004), which were also previously reported in response to salinity stress in leaves (Hernandez et al., 2000). They also reported additional salinity response proteins such as a vacuolar ATP synthase that may act as the proton-motive force for Na^+ transport into the vacuole by an Na^+/H^+ antiporter, and a nucleoside diphosphate kinase which regulates ROS accumulation through a MAP kinase cascade (Moon et al., 2003). More interestingly, the PR-10 group was also found upregulated in response to salinity in roots. These proteins were found accumulating in response to fungal and bacterial infection, as described above. The function of the PR-10 proteins is still unclear and controversial. It was previously shown that specific members of PR-10 protein expression were induced by hormones such as ABA in pea (Iturriaga et al., 1994), cytokinins in periwinkle callus (Carpin et al., 1998), and jasmonate and salicylate in tomato (Ding et al., 2002). These results suggest that PR-10 expression is regulated with the action of hormones that may be specifically induced by different biotic or abiotic stresses. Also PR-10 expression seems to be under the control of MAP kinase cascades which negatively regulate PR-10 expression while increasing tolerance to drought, salinity, and cold temperatures in trangenics plants (Xiong and Yang, 2003). In rice roots, additional proteins were identified to be responsive to salinity stress (Yan et al., 2005). Several enzymes involved in basic metabolism showed abundance changes in response to salt stress (enolase, triosephosphate isomerase, UGPase, cytochrome c oxidase). One peroxidase was also identified as upregulated, which suggests the involvement of ROS-scavenging enzymes in response to salinity in order to reduce ROS accumulation. Several other proteins involved in different mechanisms were identified as upregulated after salt stress—an mRNA splicing factor in RNA processing and an APB in actin cytoskeleton remodeling—while others were downregulated such as GS in nitrogen assimilation, SAMS in transmethylation reactions, and a nascent polypeptide-associated complex (NAC) in protein translocation. In 2007, a phosphoproteomics study was also conducted in rice roots under salinity stress (Chitteti and Peng, 2007). This study used multiplex staining of 2D-gels to detect phosphoproteins differentially regulated. The proteins were first stained with Pro-Q Diamond phosphoprotein stain and then with Sypro Ruby to detect all the proteins expressed. Most of the proteins identified were specifically differentially phosphorylated or differentially expressed. Different biological processes were determined to be dominantly changed under salt stress using the two different stains. With Pro-Q Diamond, proteins differentially phosphorylated were involved in transport and DNA metabolism, while with Sypro Ruby, proteins differentially expressed were involved in primary metabolism and response to biotic stimulus. A more recent study focused on proteins associated with membranes in rice roots under salt stress (Cheng et al., 2009). They found that peroxidase abundance and activity were correlated with the susceptibility to salt with a reduction in the salt-sensitive rice variety which could lead to an increase in H_2O_2 and to more severe damage of membranes. Additional proteins associated with the membranes were identified, such as remorin, NSF attachment protein, and a myosin heavy-chain-related protein. These proteins are also associated with the cytoskeleton and might be involved in cell wall membrane stabilization under salt stress. Several components of signal transduction pathways were also identified such as a 14.3.3 protein which the authors suggested may be a positive regulator of the H^+-ATPase activity involved in ion transport across vacuolar membranes and essential for ion homeostasis under salt stress. One member of the LRR-RLKs, OsRPK1, was identified as a salt-responding protein. The authors suggested that OsRPK1 was induced in rice by the osmotic shock from the salt stress, and because of its localization in the root cortex, it may play an important role in the perception of second messengers like ABA. A recent proteomics study in soybean roots identified additional components of salt stress response (Aghaei et al., 2009). Late embryogenesis-abundant proteins already known to be induced in response to several stresses were upregulated under salt stress. bHLH transcription factors were also induced, indicating regulation of transcriptional networks specific to salt stress. A proteomics study of *Arabidopsis* roots in response to salt stress identified hundreds of protein changes under their conditions (Jiang et al., 2007). Among the proteins identified, previously reported mechanisms were also observed such as the induction of antioxidant enzymes (ascorbate peroxidase, glutathione peroxidase, glutathione *S*-transferase and superoxide dismutase), the cytoskeleton rearrangement (actin, tubulin beta-chain) and cell wall remodeling (glycosyl hydrolase), the reduction of nitrogen assimilation (glutamate dehydrogenase and glutamine synthetase), and the protein translocation (NAC). They also identified additional events induced by salt stress, such as induction of ethylene and jasmonate production (1-aminocyclopropane-1-carboxylic acid oxidase and SAM), reduction of protein translation, processing and degradation (ribosomal proteins, eukaryotic translation initiation factor, PDI), and induction of defense proteins (glycine-rich proteins and germine-like proteins).

10.4.3.2 Nitrogen, Potassium, and Phosphorus Deprivation

Adaptation to short- and long-term changes in soil fertility is critical for crop productivity and nutrient capture. Improved nutrient capture reduces the need for fertilizer inputs, leading to reduced fertilizer runoff, decreased water contamination, and increased yield when soil fertility is low. Nitrogen (N), potassium (K), and phosphorus (P) are essential nutrients required in large quantities by plants. When nutrients are deficient in the soil, roots employ specialized strategies to ensure that plants obtain sufficient amounts of minerals for growth. In root cells, the uptake of mineral nutrients involves inducible and constitutive transport systems (Orsel et al., 2002). During nitrate assimilation, which occurs in both leaves and roots, nitrate is first reduced into ammonia by nitrate and nitrite reductase enzymes. The transfer of the ammonia to alpha-ketoglutaric acid is dependent on the action of glutamine synthetase and glutamate synthase (Oaks and Hirel, 1985). Nitrate assimilation appears to be tightly linked to carbon metabolism in order to control the assimilation from roots and leaves according to the nutritional status of the plant and the environmental stimuli (Crawford, 1995; Miller and Cramer, 2004; Paul and Foyer, 2001). Under high and low levels of nitrogen, a proteomics study of two varieties of wheat roots showed that differences in protein expression between the two species were correlated with their reactions to the N nutrition levels (Bahrman et al., 2005). Unfortunately, however, the proteins in this study were not identified. Another proteomics study investigated the proteins induced in maize roots after nitrate treatment (Prinsi et al., 2009). They found that enzymes mainly involved in nitrogen (nitrite reductase and glutamine synthetase) and carbon (phosphoglycerate mutase, glucose-6-phosphate dehydrogenase and 6-phosphogluconate dehydrogenase) metabolisms were induced by the nitrate treatment, indicating the activation of the ammonia assimilation. Besides those proteins, other proteins such as nonsymbiotic hemoglobin and a monodehydroascorbate reductase were also induced in response to nitrate treatment. These two proteins are known to be involved in the scavenging of NO that might be produced after exposure to nitrate (Zhao et al., 2007). Among the proteins induced in the nontreated plants were proteases, suggesting that remobilization of nitrogen by protein degradation occurs when nitrogen is not available in the soil. Most interestingly, the authors suggested based on their proteomics data that a phosphoenolpyruvate carboxylase known to increase during nitrate assimilation (Britto and Kronzucker, 2005) could be modulated by monoubiquitination in roots while it is modulated by phosphorylation in leaves. Posttranslational events seem to be essential to modulate pivotal enzymes in plant metabolic response to mineral nutrients. During potassium deprivation, one sucrose nonfermenting (SNF1)-related protein kinase SnRK2.8 was identified in *Arabidopsis* roots as downregulated and associated with a strong reduction in the growth of Arabidopsis (Shin et al., 2007). Using SnRK2.8 knockout and overexpression lines, we showed that this kinase modulates the overall growth of the plants. We used a phosphoproteomics approach to identify targets of SnRK2.8 responsible for maintaining plant growth in controlled conditions. Among the phosphorylated proteins, several 14.3.3 proteins (regulatory proteins), ribose-5-phosphate isomerase (Calvin cycle enzyme), and glyoxylase 1 (detoxification of MG) were confirmed to be targets of SnRK2.8, but their biological activity during potassium starvation is still unknown. Under low phosphorus availability, root development is dramatically changed with an increase in root-to-shoot ratio, an increase in the total root length and lateral root number and length, and a decrease in root diameter. These changes are associated with P-absorption and use efficiency enhancement (Linkohr et al., 2002). A proteomics investigation of the effect of P-starvation in maize roots (Li et al., 2007) found that a large number of proteins participating in carbon metabolism (tricarboxylic acid cycle, pentose phosphate, amino acid metabolism, polysaccharide mucilage synthesis) were regulated under P-stress, indicating an allocation of more carbon to roots to allow increased root growth for increased P-uptake. Associated to the root growth, proteins controlling cell shape and cell division such as tubulin and actin were also identified as upregulated under P-starvation. Additionally, they identified several chaperone proteins induced under P-stress to allow correct folding and assembly of proteins. Other proteins for which the function in P-stress still needs to be explored were identified in this study, including proteins involved in lignin, flavonoid, and cytokinin synthesis, as well as proteins involved in gene-expression/regulation and in RNA and protein transport.

10.4.3.3 Glycine as a Nitrogen Supply

Plant roots have the potential to acquire amino acids directly as a source of N (Nasholm et al., 1998) and soluble organic N is a quantitatively important form of N in some soils (Kielland, 1995).

Glycine is abundant in many ecosystems (Kielland, 1995) and is less readily utilized by soil microbes than other amino acids (Lipson et al., 1999). Using proteomics, researchers investigated the differences in the biochemical partitioning of acquired glycine relative to utilizing mineral N only in *Lolium perenne* roots (Thornton et al., 2007). Incorporation of ^{15}N-labeled ammonium and ^{15}N-labeled glycine showed that organic N was assimilated into glutamine, aspartate, and glutamate via the GOGAT cycle while N from glycine was metabolized into serine and then cysteine but not into methionine. It was also partly incorporated into the root amino acid pool via the GOGAT cycle. By looking at the amino acid composition, the authors suggested that with nitrate as an N-source, the rate of protein synthesis was increased to a greater extent than the rate of minor amino acids synthesis, and the reverse situation must be considered when switching the N-supply from nitrate to glycine. The difference in protein expression showed that one specific isozyme of methionine adenosyltransferase and one ADK were positively correlated with glycine supply while glyceraldehyde-3-phosphate dehydrogenase was associated with nitrate supply.

10.4.4 Metal Contamination-Responsive Proteins

Soil acts as a natural sink for toxic materials present in municipal and industrial wastes and sludges. Once the metal is introduced into the soil environment, crops may assimilate it. In developing countries that use raw sewage (consisting of mixtures of municipal sewage and industrial effluent) for irrigating crops, soils are gradually accumulating heavy metals and consequently, the agricultural produce has elevated metal content (Cheng, 2003) that could pose serious health problems. Heavy-metal contamination can also impact soil ecosystems sufficiently to result in significant losses in soil quality. The negative impacts of heavy metals result from their toxicity to biological processes. Some plants have developed strategies to tolerate and accumulate heavy metals present in soils and have been shown to be metal tolerant or hyperaccumulators. The mechanisms involved in toxicity tolerance, uptake, and transport of heavy metals in those plants have been investigated because of their economical importance for the remediation of polluted soils (Figure 10.3).

10.4.4.1 Copper

Copper (Cu) is an essential micronutrient for plant growth and development but is highly toxic to plants at elevated concentrations. The Cu-catalyzed HaberWeiss–Fenton reactions can generate ROS which can damage cellular components (proteins, amino acids, nucleic acids, membrane lipids), interfere with cellular transport processes, and produce changes in the concentrations of essential metabolites (Hartley-Whitaker et al., 2001; Stohs and Bagchi, 1995). To reduce the phytotoxicity of Cu, certain plants have developed detoxification mechanisms such as exclusion, chelation, sequestration, biotransformation, and repair (Hall, 2000). Those plants are mostly annual herbs with low or null economic value. *Elshotzia splendens* is an example of an annual herb in which the roots have been shown to accumulate high levels of copper compared to the leaves (Lou et al., 2004). A proteomics comparison was performed on *E. splendens* leaves and roots and demonstrated that *E. splendens* roots can endure Cu up to 100 μM (Li et al., 2009). Among the Cu-responsive proteins that were identified, RNA-binding proteins were upregulated by Cu-stress suggesting a role in the regulation of protein synthesis. Scavenging proteins such as ascorbate oxidase and short-chain dehydrogenase/reductase abundance were also affected by Cu-stress indicating regulation of the cell redox state in increasing Cu-tolerance. Authors also suggested that Cu-transport and accumulation in roots may be explained by an alteration of the cell wall composition to one that favors Cu-deposition and by a decrease in plant vascularization which reduces the Cu-transfer upward. Authors also suggested that detoxification may be mediated by the regulation of germin-like proteins and vacuolar membrane proteins. A better plant choice for phytoremediation is to employ crops of commercial interest. *Cannabis sativa*, which provides raw material for the production of diverse products such as rope, oil, and paper, is able to tolerate high concentrations of cadmium (Cd) and nickel (Ni) (Citterio et al., 2005) and was shown to tolerate and accumulate Cu (Angelova et al., 2004). Because *C. sativa* can develop 1 m deep roots that grow fast and easily in dense stands (Citterio et al., 2003), there has been increased interest in it as a phytoremediation facilitator, and one proteomics study investigated Cu-effects on its roots (Bona et al., 2007). Bona et al. showed that roots accumulated high levels of Cu and root growth was affected by Cu-stress at 150 ppm. However, no specific Cu-responsive proteins were identified that could be used as candidates to improve Cu-accumulation using *C. sativa*. In *Phaseolus vulgaris* L., chosen in this study for showing indicator-type characteristics in heavy metal uptake (Baker, 1981), dose–response effects of Cu-exposure were examined using a proteomics approach (Cuypers et al., 2005). Several members of the PR-10 family (homologous to PvPR1 and PvPR2) were identified in roots after Cu-uptake with different patterns of expression related to the concentration of Cu-application. They also identified two new members of the PR-10 group of proteins, which seem to be induced not only in biotic stresses but also in several abiotic stresses indicating that they share similar signaling pathways.

10.4.4.2 Cadmium

Cadmium (Cd) is another important soil contaminant that is introduced by industrial activities and application of sewage sludge. Cd introduced in the food chain can cause severe health damage in humans. For remediation of Cd-contaminated soils, it was shown that *Brassica juncea* L. can accumulate over 400 μg Cd/g dry weight in leaves over a 96-h period, although with some deleterious effects on plant growth (Haag-Kerwer, 1999), and thus is called a hyperaccumulator. Because an increase in Cd-tolerance can reduce Cd-accumulation in plants as shown previously (Ebbs et al., 2002; Gasic and Korban, 2007), it is essential to study the Cd-stress on a hyperaccumulator to identify the mechanisms involved in uptake and translocation capacity. A proteomics study from our lab using gel- and LC-based approaches identified that in response to Cd-stress in *B. juncea* (Alvarez et al., 2009) sulfur metabolism was induced indicating the xenobiotic detoxification by gluthatione and phytochelatins known to reduce Cd-toxicity by complexation and subsequent sequestration in the vacuole (Speiser et al., 1992). Generation of internal sinks for reduced sulfur was also previously found in *Arabidospis* roots upon exposure to Cd (Roth et al., 2006). We also found in our proteomics study of *B. juncea* that the energy necessary for Cd-sequestration was made available by a reduction in housekeeping metabolism, and in particular in protein synthesis and processing metabolism. In legumes, the AM symbiosis belongs to the strategies developed to cope with heavy-metal contamination such as Cd. The proposed mechanisms known are metal precipitation, metal adsorption, or metal chelating in fungal cells (Joner and Leyval, 1997; Leyval et al., 1997). Proteomic studies on *M. truncatula* colonized by *G. intracadices* and *Pisum sativum* colonized by *G. mosseae* were performed to understand the mechanisms under which mycorrhizal bioprotection was conferred to plants grown in the presence of Cd (Aloui et al., 2009; Repetto 2003). Noninoculated *M. truncatula* plants were compared with the colonized plants in the absence or presence of Cd. This study identified only minor changes in plants colonized in response to Cd-stress. The authors suggested that the mechanisms recruited during symbiosis also counteract Cd. The functions of the proteins induced during the symbiosis (cyclophilin, guanine nucleotide-binding protein, ubiquitin carboxylterminal hydrolase, thiazole biosynthetic enzyme, annexin, GST, SAM) are putatively involved in alleviating oxidative damage. The study on pea additionally compared the protein changes in response to Cd-exposure in Cd-sensitive and -tolerant cultivars and they observed that the Cd-induced growth inhibition effect was reduced when the Cd-sensitive cultivar was colonized with *G. mosseae*. Two proteins induced by Cd-treatment were repressed in mycorrhizal roots—a short-chain alcohol dehydrogenase and a UTP-1-phosphate uridylyltransferase. This repression may reflect a plant response counteracting or preventing Cd-toxicity effects.

10.4.4.3 Aluminum

In acidic soils, toxic forms of aluminum (Al) solubilized in the soil solution and the concentration of the toxic forms inhibit plant development. Plants have evolved resistance mechanisms that enable them to resist toxic levels of Al. Rice is one plant that showed natural occurrence of Al-tolerant genotypes. A proteomics study investigated the effect of Al-stress in rice roots using two different systems to take into account the complexity of interactions between the micronutrients present in soil (Yang et al., 2007). The protein changes observed in response to Al-stress were different if the soil solution only contained Al^{3+} and Ca^{2+} or Al^{3+} in a nutrient solution. Although the proteins identified were different, they were involved in similar biological processes. Sulfur and ethylene metabolism proteins (CS, GST, SAMS, and ACC oxidase) were up-regulated suggesting that Al-stress induces antioxidation and detoxification mechanisms. Physiological, cellular, and biochemical effects of Al have been well described in tomato plants (Kochian, 1995; Yamamoto et al., 2002). Al-stress on tomato root development is divided into two steps with an immediate inhibition of root cell elongation followed by a reduction in root tip cell proliferation (Doncheva et al., 2005). The mechanisms induced in tomato roots (Zhou et al., 2009) in the regulation of antioxidation and detoxification were different than those described in rice. In this study, oxalate oxidases were suppressed and enzymes involved in the glutathione–ascorbate cycle were induced (catalase, monohydroascorbate reductase, glutathione reductase). However, the induction of SAMS was also reported in tomato roots in response to Al-stress, along with two other enzymes, quercetin 3-*O*-methyltransferase and AdoHcyase, which are essential to maintaining methyl cycling to avoid the accumulation of oxidized proteins.

10.4.4.4 Arsenic

Arsenic (As) is a nonessential naturally occurring element found in soils and water. Human activities are one of the main reasons for As-mobilization in the environment (Meharg, 2004). Efforts have been made to understand As-uptake, translocation and transformation, extrusion, and sequestration in roots during As-stress (Abedin et al., 2002; Meharg and Jardine, 2003; Pickering et al., 2000; Raab et al., 2005). At the proteomic level, several studies were also performed on maize and rice to gain a better understanding of the molecular responses in roots during As-stress and the identification of As-responsive proteins which will be beneficial for phytoremediation or generation of As-resistant plants. In maize roots, a proteomics investigation of plant arsenic toxicity revealed that As-stress induced antioxidant enzymes (superoxide dismutases, glutathione peroxidases, peroxiredoxin, and ρ-benzoquinone reductase) to regulate cellular homeostasis which is perturbed by As-stress (Requejo and Tena, 2005). In rice, several SAMS and tau class GSTs, as well as CS, were identified as upregulated, while enzymes in energy and primary metabolism were downregulated (Ahsan et al., 2008). These results suggest that As-stress-induced detoxification enzymes reduce oxidative damage responsible for the inhibition of primary metabolism during As-stress. Additional proteins were also upregulated, but their function in response to As is yet to be discovered.

10.5 CHALLENGES AND PROSPECTS IN ROOT PROTEOMICS

We have described the usefulness of proteomics to investigate the molecular mechanisms taking place during root growth and root interactions with the surrounding environment. While proteomics research is relatively advanced in bacterial and animal studies, plant proteomics, and especially root proteomics has not yet reached a sufficient level of complexity to facilitate a deep understanding of the mechanisms involved during plant growth and interactions. One significant reason for this delay is explained by the early availability of complete sequences of bacterial and animal organisms. Furthermore, it is not always possible to apply proteomics to the study of plants and especially specific parts of the root due to limiting factors such as sample amount and low overall intrinsic protein amount. In general, protein concentrations in roots are lower than in aerial parts and hence, more samples are needed for root studies compared to other organs. It is a significant limitation for plants with a small developed root system such as *Arabidopsis* and it makes it even more difficult to do proteomics on subcellular root compartments. Root sampling and sample preparation are also two crucial steps for root proteomics studies. When plants are grown in soil, roots sampled have to be quickly cleaned and flash frozen to avoid protein content alteration and degradation. There were no studies evaluating the importance of this sampling step, but one study used proteomics tools to assess the dynamics of protein turnover in yeast proteins (Pratt et al., 2002). They observed that degradation rates were heterogeneous among proteins, which raises important issues on the effects of experimental variation on the proteome. An increasingly popular experimental method is to use substituted soils such as vermiculite or perlite as an alternative to soil. However, although the vermiculite-based system is more reproducible and easier for sampling roots, changes in the physiological responses might be different when roots are grown in soil and can make the biological response irrelevant to the study. For some species the rigidity of root cell walls during sample preparation can be troublesome as well as the accumulation of large quantities of secondary compounds, particularly in lignified tissues (Canovas et al., 2004; Granier, 1988). The presence of polyphenols, polysaccharides, lipids, and secondary metabolites usually hinder efficient protein extraction (Tsugita and Kamo, 1999). Protein precipitation is a crucial step to remove most of the contaminants. One of the preferred methods used for plant proteomics is using trichloroacetic acid(TCA)/acetone solution to precipitate proteins and then resolubilize the proteins using a chaotropic agent (Damerval et al., 1986). However, when using this precipitation the plant material is also pelleted and the resolubilzation of proteins can be limited. Another protein extraction method is to use phenol and then precipitate the proteins with ammonium acetate in methanol (Hurkman and Tanaka, 1986). This method was compared with the TCA/acetone extraction in tomato roots and showed a higher number of extracted (Saravanan and Rose, 2004). It was also showed that using different protein extraction methods leads to different proteomic profiles and that the most efficient protein extraction method was dependent on the tissue and plant type (Carpentier et al., 2005). Another main challenge in sample preparation is in targeting specific subcellular compartments. The difficulty is to capture the most comprehensive representation of the protein complement, while minimizing contamination with proteins from other subcellular locations. Plasma membranes and cell wall proteomes present some formidable challenges in this regard and there are only a few reports on them as previously reviewed.

2DE coupled with mass spectrometry for separating and identifying complex mixtures of proteins remains the core technology in plant proteomics. A labeling technique based on multiple fluorescent dyes to label the protein samples prior the 2D-PAGE, known as differential gel electrophoresis (DIGE), was introduced to improve the reproducibility and accuracy of the separation and therefore to increase the confidence of the 2D results (Chakravarti et al., 2005). It allows multiple samples to be co-separated and visualized on one single 2D gel. Normalization of the spot intensity across gels using the internal standard allows the ratio of relative expression of the same protein in the one gel to be compared directly, eliminating gel-to-gel variation. However, the use of 2D-gel electrophoresis in general has several limitations including unsuitability to identify less abundant proteins and/or to characterize membrane proteins. These limitations are due to stain sensitivity limits which allow detection down to the nanogram level of protein using SyproRuby and because transmembranes proteins do not resolve on denaturing 2DE gels even when using specific detergent of mixtures of chaotropic agents (Santoni et al., 1999). New approaches for protein separation have attempted to overcome these limitations. Gel-free techniques using LC-separations coupled to mass spectrometry now offer methods to simultaneously identify and quantify multiple proteins simultaneously in a high-throughput fashion and with low sample volume. The main technology used in gel-free proteomics is called MudPIT (multidimensional protein identification technology) analysis or "shotgun" proteomics. Prefractionation of the digested proteomes can improve the dynamic range. This procedure incorporates ion-exchange and reverse-phase capillary columns directly linked to the nanospray mass spectrometer for peptide analysis. The use of this technique demonstrated an increase in the coverage of both membrane-associated and low-abundance proteins (Alvarez et al., 2009; Koller et al., 2002). Because 2D-gels and MudPIT use distinct separation methods, the use of both as complementary techniques enables improved proteome coverage (Alvarez et al., 2009; Brechenmacher et al., 2009; Koller et al., 2002; Komatsu et al., 2009b) and is essential to get a more comprehensive view of the proteome. With the development of non-gel-based quantitative proteomics, several stable isotope-labeling techniques using LC-based tools were introduced in the recent years. These include isotope-coded affinity tag, stable isotope labeling by amino acids in cell culture, metabolic labeling, enzymatic labeling, isotope-coded protein labeling, tandem mass tag (TMT), and isobaric tags for relative and absolute quantification (iTRAQ). However, most labeling-based quantification approaches also have potential limitations, such as complexity of sample preparation, high cost of the reagents, incomplete labeling, and requirement for specific quantification software. Moreover, only TMT and iTRAQ allow the comparison of multiple (up to eight) samples at the same time while the other labeling methods can only compare the relative quantity of a protein between two and three different samples. Therefore, interest in label-free shotgun proteomics techniques has increased in order to address some of the issues of labeling methods but it requires high reproducibility of peptide LC-separation. In this review, only a few proteomics projects used gel-free techniques such as MudPIT (Brechenmacher et al., 2009; Koller et al., 2002; Komatsu et al., 2009b; Valot et al., 2006) or iTRAQ labeling (Alvarez et al., 2009). This indicates how much work can still be done in the plant field to understand the diverse molecular mechanisms taking place in roots.

Although the improved mass accuracy, mass resolution, and sensitivity of mass spectrometers allows increased protein identification efficiency, the lack of complete genomic cDNA sequences for many plants is still a limiting factor (van Wijk, 2001). The genome of *Arabidopsis* was the first plant genome that was completely sequenced and publicly available (Initiative, 2000). However, *Arabidopsis* is a dicotyledon and is not an economical crop, legume, or a tree and therefore shows clear limitations for the understanding of many biological processes. Since then, rice was the first crop for which a completely sequenced genome became available (Goff et al., 2002; Yu et al., 2002). In the past year, the B73 maize genome was unveiled (Schnable et al., 2009) and is already accessible online http://www.maizesequence.org/index.html. However, genome sequencing projects are currently in progress for other major crops such as wheat, tobacco, potato, barley, and soybean. The current generation of MS data search engines can search against protein sequence data, ESTs, and annotated genomic sequences. The use of EST/cDNA assemblies has shown to increase the protein coverage in proteomics studies in plants. Transcript assemblies of 254 plant species are available at the TIGR web site: http://plantta.jcvi.org/. The genomic sequences are available in fasta format to be downloaded and used in database searches for protein identification. However, they are not annotated and the sequences have to be blasted in order to identify a matching sequence to get some information on function. Over 200 ESTs assemblies are also available at the PlantGDB web site: http://www.plantgdb.org/prj/ESTCluster/. Only about half of the plant species represented overlap between the two web sites. One last aspect to be considered is validation of the proteins identified using alternatives tools such as Western blotting and enzyme activity assays. It is also essential to correctly assign the protein function due to the multiple roles a protein can play in various processes.

Proteins in plants and other organisms undergo numerous posttranslational modifications that help to regulate protein function and alter protein localization. Introduction of specific stains for phosphoproteins and glycosylated proteins helped the identification of these specific posttranslational modifications (PTMs) directly on 2D-gels. It has been shown that the use of 2D-gels is the best tool to resolve isoforms of the same proteins, characteristically identified on the gel as a train of spots (Hunzinger et al., 2006). Isoforms may reflect different gene products or may result from various PTMs. Therefore, 2D-gels have been the main tool used for studying PTM changes globally in the proteome. Phosphoproteomics in plants has been studied using different phosphoenrichment procedures prior to protein separation by 2D-gels as reviewed recently (Kersten et al., 2009), but not in roots. In the past few years, mass spectrometry has become a powerful tool for the characterization of targeted posttranslational modifications, as well as protein–protein interactions and global structural determination of protein complexes. Among the most studied PTMs are phosphorylation, glycosylation, acetylation, myristoylation, palmitoylation, methylation, sulfation, prenylation, and ubiquitination. The emerging top-down MS proteomics approach (Kelleher et al., 1999) in particular can provide a better view of the protein forms present and provide a complete characterization of their primary structures. Top-down mass spectrometry strives to preserve the posttranslationally modified forms of proteins present *in vivo* by measuring them intact, rather than measuring peptides produced from them by proteolysis in typical bottom-up proteomics experiments.

Redox-regulated proteins in response to oxidative stress are also investigated using proteomics tools. In plants, changing the redox state of protein thiol groups was shown to serve as regulatory switches of protein function in carbon storage, photosynthesis, and leaf senescence (Buchanan and Balmer, 2005; Geigenberger et al., 2005; Vanacker et al., 2006). Several methods were developed using gel- or LC-based labeling techniques (Alkhalfioui et al., 2007; Rouhier et al., 2005; Yano et al., 2002), but they essentially focused their interest on the identification of direct protein targets of thioredoxin and glutaredoxin function responsible for the cellular protein redox cycling. Whereas the thioredoxin and glutaredoxin systems seem to functions mainly in the reduction of disulfide bonds, other thiol/disulfide containing proteins (e.g., oxidoreductases) and ROS act to oxidize thiol groups. New approaches to study *in vivo* changes of the redox protein status were developed in mammals (Hurd et al., 2007; Leichert et al., 2008). Using *Arabidopsis*, we optimized a gel-based proteomics method using monobromobimane labeling (Alvarez et al., 2009) that we applied in the study of *Arabidopsis* roots in response to methyl jasmonate response, which is a component that can mimic pathogen attack (Alvarez et al., 2009). A large number of efficient functional proteomics tools are now available to analyze plant-specific processes and the responses to various biotic and abiotic factors. Despite obvious limitations in root sample preparation and in the genome sequence availability for crops, functional proteomics and the mining of protein-expression profiles certainly constitute the next challenge for the coming years in subcellular root plant proteomics.

It also should be emphasized that the integration of genomics, transcriptomics, and metabolomics with proteomics data will lead to substantial information about the regulation of protein accumulation and activity and contribute to an essential part of understanding biological functions.

REFERENCES

Abedin, M. J., J. Feldmann, and A. A. Meharg. 2002. Uptake kinetics of arsenic species in rice plants. *Plant Physiol* 128:1120–1128.

Aghaei, K., A. A. Ehsanpour, A. H. Shah, and S. Komatsu. 2009. Proteome analysis of soybean hypocotyl and root under salt stress. *Amino Acids* 36:91–98.

Ahsan, N., D. G. Lee, I. Alam et al. 2008. Comparative proteomic study of arsenic-induced differentially expressed proteins in rice roots reveals glutathione plays a central role during As stress. *Proteomics* 8:3561–3576.

Alkhalfioui, F., M. Renard, W. H. Vensel et al. 2007. Thioredoxin-linked proteins are reduced during germination of *Medicago truncatula* seeds. *Plant Physiol* 144:1559–1579.

Aloui, A., G. Recorbet, A. Gollotte et al. 2009. On the mechanisms of cadmium stress alleviation in *Medicago truncatula* by arbuscular mycorrhizal symbiosis: A root proteomic study. *Proteomics* 9:420–433.

Alvarez, S., B. M. Berla, J. Sheffield, R. E. Cahoon, J. M. Jez, and L. M. Hicks. 2009. Comprehensive analysis of the *Brassica juncea* root proteome in response to cadmium exposure by complementary proteomic approaches. *Proteomics* 9:2419–2431.

Alvarez, S., E. L. Marsh, S. G. Schroeder, and D. P. Schachtman. 2008. Metabolomic and proteomic changes in the xylem sap of maize under drought. *Plant Cell Environ* 31:325–340.

Alvarez, S., G. H. Wilson, and S. Chen. 2009. Determination of *in vivo* disulfide-bonded proteins in Arabidopsis. *J Chromatogr B Analyt Technol Biomed Life Sci* 877:101–104.

Alvarez, S., J. Q. Goodger, E. L. Marsh, S. Chen, V. S. Asirvatham, and D. P. Schachtman. 2006. Characterization of the maize xylem sap proteome. *J Proteome Res* 5:963–972.

Alvarez, S., M. Zhu, and S. Chen. 2009. Proteomics of *Arabidopsis* redox proteins in response to methyl jasmonate. *J Proteomics* 73:30–40.

Amiour, N., G. Recorbet, F. Robert, S. Gianinazzi, and E. Dumas-Gaudot. 2006. Mutations in DMI3 and SUNN modify the appressorium-responsive root proteome in arbuscular mycorrhiza. *Mol Plant Microbe Interact* 19:988–997.

Amme, S., A. Matros, B. Schlesier, and H. P. Mock. 2006. Proteome analysis of cold stress response in *Arabidopsis thaliana* using DIGE-technology. *J Exp Bot* 57:1537–1546.

Angelova, V., R. Ivanova, V. Delibatova, and K. Ivanov. 2004. Bio-accumulation and distribution of heavy metals in fibre crops (flax, cotton and hemp). *Ind Crop Prod* 19:197–205.

Araus, J. L., G. A. Slafer, M. P. Reynolds, and C. Royo. 2002. Plant breeding and drought in C3 cereals: What should we breed for? *Ann Bot* 89(Spec No):925–940.

Audenaert, K., G. B. De Meyer, and M. M. Hofte. 2002. Abscisic acid determines basal susceptibility of tomato to *Botrytis cinerea* and suppresses salicylic acid-dependent signaling mechanisms. *Plant Physiol* 128:491–501.

Bae, M. S., E. J. Cho, E. Y. Choi, and O. K. Park. 2003. Analysis of the *Arabidopsis* nuclear proteome and its response to cold stress. *Plant J* 36:652–663.

Bahrman, N., A. Gouy, F. Devienne-Barret, B. Hirel, F. Vedele, and J. Le Gouis. 2005. Differential change in root protein patterns of two wheat varieties under high and low nitrogen nutrition levels. *Plant Sci* 168:81–87.

Baker, A. J. M. 1981. Accumulators and excluders-strategies in the response of plants to heavy metals. *J Plant Nutr* 3:643–654.

Bates, T. R. and J. P. Lynch. 1996. Stimulation of root hair elongation in *Arabidopsis thaliana* by low phosphorus availability. *Plant Cell Environ* 19:529–538.

Benabdellah, K., C. Azcon-Aguilar, and N. Ferrol. 2000. Alterations in the plasma membrane polypeptide pattern of tomato roots (*Lycopersicon esculentum*) during the development of arbuscular mycorrhiza. *J Exp Bot* 51:747–754.

Benschop, J. J., S. Mohammed, M. O'Flaherty, A. J. Heck, M. Slijper, and F. L. Menke. 2007. Quantitative phosphoproteomics of early elicitor signaling in Arabidopsis. *Mol Cell Proteomics* 6:1198–1214.

Bestel-Corre, G., E. Dumas-Gaudot, and S. Gianinazzi. 2004. Proteomics as a tool to monitor plant–microbe endosymbioses in the rhizosphere. *Mycorrhiza* 14:1–10.

Bestel-Corre, G., E. Dumas-Gaudot, V. Poinsot et al. 2002. Proteome analysis and identification of symbiosis-related proteins from *Medicago truncatula* Gaertn. by two-dimensional electrophoresis and mass spectrometry. *Electrophoresis* 23:122–137.

Bibikova, T. and S. Gilroy. 2003. Root hair development. *J Plant Growth Regul* 21:383–415.

Biles, C. I., R. D. Martyn, and H. D. Wilson. 1989. Isozymes and general proteins from various watermelon cultivars and tissue types. *HortScience* 24:810–812.

Bird, A. 1962. The inducement of giant cells by *Meloidogyne javanica. Nematologica* 8:1–10.

Blumwald, E., M. G. Fortin, P. A. Rea, D. P. Verma, and R. J. Poole. 1985. Presence of host-plasma membrane type H-ATPase in the membrane envelope enclosing the bacteroids in soybean root nodules. *Plant Physiol* 78:665–672.

Bona, E., F. Marsano, M. Cavaletto, and G. Berta. 2007. Proteomic characterization of copper stress response in *Cannabis sativa* roots. *Proteomics* 7:1121–1130.

Borderies, G., E. Jamet, C. Lafitte et al. 2003. Proteomics of loosely bound cell wall proteins of *Arabidopsis thaliana* cell suspension cultures: A critical analysis. *Electrophoresis* 24:3421–3432.

Borner, G. H., D. J. Sherrier, T. Weimar et al. 2005. Analysis of detergent-resistant membranes in *Arabidopsis*. Evidence for plasma membrane lipid rafts. *Plant Physiol* 137:104–116.

Boyer, J. S. 1982. Plant productivity and environment. *Science* 218:443–448.

Brechenmacher, L., J. Lee, S. Sachdev et al. 2009. Establishment of a protein reference map for soybean root hair cells. *Plant Physiol* 149:670–682.

Brechenmacher, L., S. Weidmann, D. van Tuinen et al. 2004. Expression profiling of up-regulated plant and fungal genes in early and late stages of *Medicago truncatula-Glomus mosseae* interactions. *Mycorrhiza* 14:253–262.

Britto, D. T. and H. J. Kronzucker. 2005. Nitrogen acquisition, PEP carboxylase, and cellular pH homeostasis: New views on old paradigms. *Plant Cell Environ* 28:1396–1409.

Buchanan, B. B. and Y. Balmer. 2005. Redox regulation: A broadening horizon. *Annu Rev Plant Biol* 56:187–220.

Buhtz, A., A. Kolasa, K. Arlt, C. Walz, and J. Kehr. 2004. Xylem sap protein composition is conserved among different plant species. *Planta* 219:610–618.

Callahan, F. E., J. N. Jenkins, R. G. Creech, and G. W. Lawrence. 1997. Changes in cotton root proteins correlated with resistance to root knot nematode development. *J Cot Sci* 1:38–47.

Campo, S., M. Carrascal, M. Coca, J. Abian, and B. San Segundo. 2004. The defense response of germinating maize embryos against fungal infection: A proteomics approach. *Proteomics* 4:383–396.

Canovas, F. M., E. Dumas-Gaudot, G. Recorbet, J. Jorrin, H. P. Mock, and M. Rossignol. 2004. Plant proteome analysis. *Proteomics* 4:285–298.

Cao, T., S. Srivastava, M. H. Rahman et al. 2007. Proteome-level changes in the roots of *Brassica napus* as a result of *Plasmodiphora brassicae* infection. *Plant Sci* 174:97–115.

Carpentier, S. C., E. Witters, K. Laukens, P. Deckers, R. Swennen, and B. Panis. 2005. Preparation of protein extracts from recalcitrant plant tissues: An evaluation of different methods for two-dimensional gel electrophoresis analysis. *Proteomics* 5:2497–2507.

Carpin, S., S. Laffer, F. Schoentgen et al. 1998. Molecular characterization of a cytokinin-inducible periwinkle protein showing sequence homology with pathogenesis-related proteins and the Bet v 1 allergen family. *Plant Mol Biol* 36:791–798.

Cassab, G. I., J. J. Lin, L. S. Lin, and J. E. Varner. 1988. Ethylene effect on extensin and peroxidase distribution in the subapical region of pea epicotyls. *Plant Physiol* 88:522–524.

Castillejo, M. A., N. Amiour, E. Dumas-Gaudot, D. Rubiales, and J. V. Jorrín. 2004. A proteomic approach to studying plant response to crenate broomrape (*Orobanche crenata*) in pea (*Pisum sativum*). *Phytochemistry* 65:1817–1828.

Catalano, C. M., W. S. Lane, and D. J. Sherrier. 2004. Biochemical characterization of symbiosome membrane proteins from *Medicago truncatula* root nodules. *Electrophoresis* 25:519–531.

Chakravarti, B., S. R. Gallagher, and D. N. Chakravarti. 2005. Difference gel electrophoresis (DIGE) using CyDye DIGE fluor minimal dyes. *Curr Protoc Mol Biol* 10:10–23.

Chaves, M. M., J. S. Pereira, J. Maroco et al. 2002. How plants cope with water stress in the field. Photosynthesis and growth. *Ann Bot* 89:907–916.

Chaves, M. M. and M. M. Oliveira. 2004. Mechanisms underlying plant resilience to water deficits: Prospects for water-saving agriculture. *J Exp Bot* 55:2365–2384.

Cheng, S. 2003. Heavy metal pollution in China: Origin, pattern and control. *Environ Sci Pollut Res Int* 10:192–198.

Cheng, Y., Y. Qi, Q. Zhu et al. 2009. New changes in the plasma-membrane-associated proteome of rice roots under salt stress. *Proteomics* 9:3100–3114.

Chevallet, M., V. Santoni, A. Poinas et al. 1998. New zwitterionic detergents improve the analysis of membrane proteins by two-dimensional electrophoresis. *Electrophoresis* 19:1901–1909.

Chiatante, D., G. S. Scippa, A. Di Iorio, and M. Sarnataro. 2003. The influence of steep slopes on root system development. *J Plant Growth Regul* 21:247–260.

Chitteti, B. R. and Z. Peng. 2007. Proteome and phosphoproteome differential expression under salinity stress in rice (*Oryza sativa*) roots. *J Proteome Res* 6:1718–1727.

Chivasa, S., J. M. Hamilton, R. S. Pringle et al. 2006. Proteomic analysis of differentially expressed proteins in fungal elicitor-treated *Arabidopsis* cell cultures. *J Exp Bot* 57:1553–1562.

Chivasa, S., B. K. Ndimba, W. J. Simon et al. 2002. Proteomic analysis of the *Arabidopsis thaliana* cell wall. *Electrophoresis* 23:1754–1765.

Citterio, S., A. Santagostino, P. Fumagalli, N. Prato, P. Ranalli, and S. Sgorbati. 2003. Heavy metal tolerance and accumulation of Cd, Cr and Ni by *Cannabis sativa* L. *Plant Soil* 256:243–252.

Citterio, S., N. Prato, P. Fumagalli et al. 2005. The arbuscular mycorrhizal fungus *Glomus mosseae* induces growth and metal accumulation changes in *Cannabis sativa* L. *Chemosphere* 59:21–29.

Colditz, F., H. P. Braun, C. Jacquet, K. Niehaus, and F. Krajinski. 2005. Proteomic profiling unravels insights into the molecular background underlying increased *Aphanomyces euteiches*—Tolerance of *Medicago truncatula. Plant Mol Biol* 59:387–406.

Colditz, F., O. Nyamsuren, K. Niehaus, H. Eubel, H.-P. Braun, and F. Krajinski. 2004. Proteomic approach: Identification of *Medicago truncatula* proteins induced in roots after infection with the pathogenic oomycete *Aphanomyces euteiches. Plant Mol Biol* 55:109–120.

Cosgrove, D. J. 1999. Enzymes and other agents that enhance cell wall extensibility. *Annu Rev Plant Physiol Plant Mol Biol* 50:391–417.

Coumans, J. V., A. Poljak, M. J. Raftery, D. Backhouse, and L. Pereg-Gerk. 2009. Analysis of cotton (*Gossypium hirsutum*) root proteomes during a compatible interaction with the black root rot fungus *Thielaviopsis basicola. Proteomics* 9:335–349.

Crawford, N. M. 1995. Nitrate: Nutrient and signal for plant growth. *Plant Cell* 7:859–868.

Cuypers, A., K. M. Koistinen, H. Kokko, S. Karenlampi, S. Auriola, and J. Vangronsveld. 2005. Analysis of bean (*Phaseolus vulgaris* L.) proteins affected by copper stress. *J Plant Physiol* 162:383–392.

Damerval, C., D. de Vienne, M. Zivy, and H. Thiellement. 1986. Technical improvements in two-dimensional electrophoresis increase the level of genetic variation detected in wheat-seedling proteins. *Electrophoresis* 7:52–54.

Dani, V., W. J. Simon, M. Duranti, and R. R. Croy. 2005. Changes in the tobacco leaf apoplast proteome in response to salt stress. *Proteomics* 5:737–745.

De-la-Pena, C., Z. Lei, B. S. Watson, L. W. Sumner, and J. M. Vivanco. 2008. Root–microbe communication through protein secretion. *J Biol Chem* 283:25247–25255.

De Jaegher, G. and N. Boyer. 1987. Specific inhibition of lignification in *Bryonia dioica*: Effects on thigmomorphogenesis. *Plant Physiol* 84:10–11.

Dembinsky, D., K. Woll, M. Saleem et al. 2007. Transcriptomic and proteomic analyses of pericycle cells of the maize primary root. *Plant Physiol* 145:575–588.

den Hartog, D., A. Musgrave, and T. Munnik. 2001. Nod factor-induced phosphatidic acid and diacylglycerol pyrophosphate formation, a role for phospholipase C and D in root hair deformation. *Plant J* 25:55–66.

Devos, S., K. Laukens, P. Deckers et al. 2006. A hormone and proteome approach to picturing the initial metabolic events during *Plasmodiophora brassicae* infection on *Arabidopsis. Mol Plant Microbe Interact* 19:1431–1443.

Di Michele, M., D. Chiatante, C. Plomion, and G. S. Scippa. 2006. A proteomic analysis of Spanish broom (*Spartium junceum* L.) root growing on a slope condition. *Plant Sci* 170:926–935.

Ding, C. K., C. Y. Wang, K. C. Gross, and D. L. Smith. 2002. Jasmonate and salicylate induce the expression of pathogenesis-related-protein genes and increase resistance to chilling injury in tomato fruit. *Planta* 214:895–901.

Djordjevic, M. A., M. Oakes, D. X. Li, C. H. Hwang, C. H. Hocart, and P. M. Gresshoff. 2007. The *Glycine max* xylem sap and apoplast proteome. *J Proteome Res* 6:3771–379.

Doncheva, S., M. Amenos, C. Poschenrieder, and J. Barcelo. 2005. Root cell patterning: A primary target for aluminium toxicity in *maize. J Exp Bot* 56:1213–1220.

Ebbs, S., I. Lau, B. Ahner, and L. Kochian. 2002. Phytochelatin synthesis is not responsible for Cd tolerance in the Zn/Cd hyperaccumulator *Thlaspi caerulescens* (J. & C. Presl). *Planta* 214:635–640.

Espartero, J., I. S. Aguayo, and J. M. Pardo. 1995. Molecular characterization of glyoxalase I from a higher plant; upregulation by stress. *Plant Mol Biol* 29:1223–1233.

Fedorova, E., R. Thomson, L. F. Whitehead, O. Maudoux, M. K. Udvardi, and D. A. Day. 1999. Localization of H + -ATPase in soybean root nodules. *Planta* 209:25–32.

Ferrol, N., M.J. Pozo, M. Antelo, and C. Azcon-Aguilar. 2002. Arbuscular mycorrhizal symbiosis regulates plasma membrane H^+-ATPase gene expression in tomato plants. *J Exp Bot* 53:1683–1687.

Fortin, M. G., N. A. Morrison, and D. P. Verma. 1987. Nodulin-26, a peribacteroid membrane nodulin is expressed independently of the development of the peribacteroid compartment. *Nucleic Acids Res* 15:813–824.

Fry, S. C. 1998. Oxidative scission of plant cell wall polysaccharides by ascorbate-induced hydroxyl radicals. *Biochem J* 332 (Pt 2):507–515.

Fukaki, H., J. Wysocka-Diller, T. Kato, H. Fujisawa, P. N. Benfey, and M. Tasaka. 1998. Genetic evidence that the endodermis is essential for shoot gravitropism in *Arabidopsis thaliana. Plant J* 14:425–430.

Garrett, T. P., D. J. Clingeleffer, J. M. Guss, S. J. Rogers, and H. C. Freeman. 1984. The crystal structure of poplar apoplastocyanin at 1.8A resolution. The geometry of the copper-binding site is created by the polypeptide. *J Biol Chem* 159:2822–2825.

Gasic, K. and S. S. Korban. 2007. Transgenic Indian mustard (*Brassica juncea*) plants expressing an *Arabidopsis* phytochelatin synthase (AtPCS1) exhibit enhanced As and Cd tolerance. *Plant Mol Biol* 64:361–369.

Geigenberger, P., A. Kolbe, and A. Tiessen. 2005. Redox regulation of carbon storage and partitioning in response to light and sugars. *J Exp Bot* 56:1469–1479.

Genre, A., M. Chabaud, T. Timmers, P. Bonfante, and D. G. Barker. 2005. Arbuscular mycorrhizal fungi elicit a novel intracellular apparatus in *Medicago truncatula* root epidermal cells before infection. *Plant Cell* 17:3489–3499.

Geurts, R., E. Fedorova, and T. Bisseling. 2005. Nod factor signaling genes and their function in the early stages of Rhizobium infection. *Curr Opin Plant Biol* 8:346–352.

Gianinazzi-Pearson, V., C. Arnould, M. Oufattole, M. Arango, and S. Gianinazzi. 2000. Differential activation of H + -ATPase genes by an arbuscular mycorrhizal fungus in root cells of transgenic tobacco. *Planta* 211:609–613.

Goff, S. A., D. Ricke, T. H. Lan et al. 2002. A draft sequence of the rice genome (*Oryza sativa* L. ssp. *japonica*). *Science* 296:92–100.

Gonzalez-Bernaldez, F. 1968. Effects of osmotic pressure on root growth, cell cycle and cell elongation. *Protoplasma* 65:255–262.

Granier, F. 1988. Extraction of plant proteins for two-dimensional electrophoresis. *Electrophoresis* 9:712–718.

Haag-Kerwer, A. 1999. Cadmium exposure in *Brassica juncea* causes a decline in transpiration rate and leaf expansion without effect on photosynthesis. *J Exp Bot* 50:1827–1835.

Hall, J. L. 2000. Cellular mechanisms for heavy metal. *J Exp Bot* 53:1–11.

Hammond, J. P., M. R. Broadley, and P. J. White. 2004. Genetic responses to phosphorus deficiency. *Ann Bot* 94:323–332.

Hao, L. T., M. Nomura, H. Kajiwara, D. A. Day, and S. Tajima. 2004. Proteomic analysis on symbiotic differentiation of mitochondria in soybean nodules. *Plant Cell Physiol* 45:300–308.

Harrison, C. J., R. M. Mould, M. J. Leech et al. 2000. The rug3 locus of pea encodes plastidial phosphoglucomutase. *Plant Physiol* 122:1187–1192.

Harrison, M. J. 1999. Molecular and cellular aspects of the arbuscular mycorrhizal symbiosis. *Ann Rev Plant Physiol* 50:361–389.

Harrison, M. J., G. R. Dewbre, and J. Liu. 2002. A phosphate transporter from *Medicago truncatula* involved in the acquisition of phosphate released by arbuscular mycorrhizal fungi. *Plant Cell* 14:2413–2429.

Hartley-Whitaker, J., G. Ainsworth, and A. A. Meharg. 2001. Copper- and arsenate-induced oxidative stress in *Holcus lanatus* L. clones with differential sensitivity. *Plant Cell Environ* 24:713–722.

Hashimoto, M. and S. Komatsu. 2007. Proteomic analysis of rice seedlings during cold stress. *Proteomics* 7:1293–1302.

Hashimoto, M., M. Toorchi, K. Matsushita, Y. Iwasaki, and S. Komatsu. 2009. Proteome analysis of rice root plasma membrane and detection of cold stress responsive proteins. *Protein Pept Lett* 16(6):685–697.

Hawes, M. C., U. Gunawardena, S. Miyasaka, and X. Zhao. 2000. The role of root border cells in plant defense. *Trends Plant Sci* 5:128–133.

Hernandez, J. A., A. Jiménez, P. Mullineaux, and F. Sevilla. 2000. Tolerance of pea (*Pisum sativum* L.) to long-term salt stress is associated with induction of antioxidant defences. *Plant Cell Environ* 23:853–862.

Hochholdinger, F., L. Guo, and P. S. Schnable. 2004. Lateral roots affect the proteome of the primary root of maize (*Zea mays* L.). *Plant Mol Biol* 56:397–412.

Hunzinger, C., A. Schrattenholz, S. Poznanovic, G. P. Schwall, and W. Stegmann. 2006. Comparison of different separation technologies for proteome analyses: Isoform resolution as a prerequisite for the definition of protein biomarkers on the level of posttranslational modifications. *J Chromatogr A* 1123:170–181.

Hurd, T. R., T. A. Prime, M. E. Harbour, K. S. Lilley, and M. P. Murphy. 2007. Detection of reactive oxygen species-sensitive thiol proteins by redox difference gel electrophoresis: Implications for mitochondrial redox signaling. *J Biol Chem* 282:22040–22051.

Hurkman, W. J. and C. K. Tanaka. 1986. Solubilization of plant membrane proteins for analysis by two-dimensional gel electrophoresis. *Plant Physiol* 81:802–806.

Hussey, R. 1989. Disease-inducing secretions of plant-parasitic nematodes. *Annu Rev Phytopathol* 27: 123–141.

Imin, N., T. Kerim, B. G. Rolfe, and J. J. Weinman. 2004. Effect of early cold stress on the maturation of rice anthers. *Proteomics* 4:1873–1882.

Initiative. 2000. Analysis of the genome sequence of the flowering plant *Arabidopsis thaliana. Nature* 408:796–815.

Iturriaga, E. A., M. J. Leech, D. H. Barratt, and T. L. Wang. 1994. Two ABA-responsive proteins from pea (*Pisum sativum* L.) are closely related to intracellular pathogenesis-related proteins. *Plant Mol Biol* 24:235–240.

Jiang, Y., B. Yang, N. S. Harris, and M. K. Deyholos. 2007. Comparative proteomic analysis of NaCl stress-responsive proteins in *Arabidopsis* roots. *J Exp Bot* 58:3591–3607.

Joner, E. J. and C. Leyval. 1997. Uptake of 109 Cd by roots and hyphae of a *Glomus mosseae/Trifolium subterraneum* mycorrhiza from soil amended with high and low concentrations of cadmium. *New Phytol* 135:353–360.

Jones, A. M., V. Thomas, M. H. Bennett, J. Mansfield, and M. Grant. 2006. Modifications to the *Arabidopsis* defense proteome occur prior to significant transcriptional change in response to inoculation with *Pseudomonas syringae. Plant Physiol* 142:1603–1620.

Kaiser, B. N., P. M. Finnegan, S. D. Tyerman et al. 1998. Characterization of an ammonium transport protein from the peribacteroid membrane of soybean nodules. *Science* 281:1202–1206.

Kamada, M., A. Higashitani, and N. Ishioka. 2005. Proteomic analysis of *Arabidopsis* root gravitropism. *Biol Sci in Space* 19:148–154.

Kav, N. N., S. Srivastava, L. Goonewardene, and S. F. Blade. 2004. Proteome-level changes in the roots of *Pisum sativum* in response to salinity. *Ann Appl Biol* 145:217–230.

Kehr, J., A. Buhtz, and P. Giavalisco. 2005. Analysis of xylem sap proteins from *Brassica napus. BMC Plant Biol* 5:11.

Kelleher, N. L., H. Y. Lin, G. A. Valaskovic, D. J. Aaserud, E. K. Fridriksson, and F. W. McLafferty. 1999. Top down versus bottom up protein characterization by tandem high-resolution mass spectrometry. *J Am Chem Soc* 121:806–812.

Kerk, D., J. Bulgrien, D. W. Smith, B. Barsam, S. Veretnik, and M. Gribskov. 2002. The complement of protein phosphatase catalytic subunits encoded in the genome of *Arabidopsis. Plant Physiol* 129:908–925.

Kerk, N. M. and L. J. Feldman. 1995. A biochemical model for the initiation and maintenance of the quiescent center: Implications for organization of root meristems. *Development* 121:2825–2833.

Kersten, B., G. K. Agrawal, P. Durek et al. 2009. Plant phosphoproteomics: An update. *Proteomics* 9:964–988.

Kielland, K. 1995. Landscape patterns of free amino acids in artic tundra soils. *Biogeochemistry* 31:85–98.

Kim, S. G., S. T. Kim, S. Y. Kang, Y. Wang, W. Kim, and K. Y. Kang. 2008. Proteomic analysis of reactive oxygen species (ROS)-related proteins in rice roots. *Plant Cell Rep* 27:363–375.

Kim, S. I., S. M. Kweon, E. A. Kim et al. 2004. Characterization of RNase-like major storage protein from the ginseng root by proteomic approach. *J Plant Physiol* 161:837–845.

Kim, S. T., K. S. Cho, S. Yu et al. 2003. Proteomic analysis of differentially expressed proteins induced by rice blast fungus and elicitor in suspension-cultured rice cells. *Proteomics* 3:2368–2378.

Kimbrough, J. M., R. Salinas-Mondragon, W. F. Boss, C. S. Brown, and H. W. Sederoff. 2004. The fast and transient transcriptional network of gravity and mechanical stimulation in the *Arabidopsis* root apex. *Plant Physiol* 136:2790–2805.

Kjoller, R. 1996. The presence of the arbuscular mycorrhizal fungus *Glomus intraradices* influences enzymatic activities of the root pathogen *Aphanomyces euteiches* in pea roots. *Mycorrhiza* 6:487–491.

Kochian, L. 1995. Cellular mechanisms of aluminum toxicity and resistance in plants. *Annu Rev Plant Physiol Plant Mol Biol* 46:237–260.

Koller, A., M. P. Washburn, B. M. Lange et al. 2002. Proteomic survey of metabolic pathways in rice. *Proc Natl Acad Sci USA* 99:11969–11974.

Komatsu, S., R. Yamamoto, Y. Nanjo, Y. Mikami, H. Yunokawa, and K. Sakata. 2009a. A comprehensive analysis of the soybean genes and proteins expressed under flooding stress using transcriptome and proteome techniques. *J Proteome Res* 8:4766–4778.

Komatsu, S., T. Wada, Y. Abalea et al. 2009b. Analysis of plasma membrane proteome in soybean and application to flooding stress response. *J Proteome Res* 8:4487–4499.

Komatsu, S., T. Sugimoto, T. Hoshino, Y. Nanjo, and K. Furukawa. 2010. Identification of flooding stress responsible cascades in root and hypocotyl of soybean using proteome analysis. *Amino Acids* 38:729–738.

Kung, C. C., W. N. Huang, Y. C. Huang, and K. C. Yeh. 2006. Proteomic survey of copper-binding proteins in *Arabidopsis* roots by immobilized metal affinity chromatography and mass spectrometry. *Proteomics* 6:2746–2758.

Larrainzar, E., S. Wienkoop, W. Weckwerth, R. Ladrera, C. Arrese-Igor, and E. M. Gonzalez. 2007. *Medicago truncatula* root nodule proteome analysis reveals differential plant and bacteroid responses to drought stress. *Plant Physiol* 144:1495–1507.

LeClere, S., R. A. Rampey, and B. Bartel. 2004. *IAR4*, a gene required for auxin conjugate sensitivity in *Arabidopsis*, encodes a pyruvate dehydrogenase E1alpha homolog. *Plant Physiol* 135:989–999.

Lee, D. G., N. Ahsan, S. H. Lee et al. 2009. Chilling stress-induced proteomic changes in rice roots. *J Plant Physiol* 166:1–11.

Lefebvre, B., F. Furt, M. A. Hartmann et al. 2007. Characterization of lipid rafts from *Medicago truncatula* root plasma membranes: A proteomic study reveals the presence of a raft-associated redox system. *Plant Physiol* 144:402–418.

Leichert, L. I., F. Gehrke, H. V. Gudiseva et al. 2008. Quantifying changes in the thiol redox proteome upon oxidative stress *in vivo*. *Proc Natl Acad Sci USA* 105:8197–8202.

Lepek, V. C., A. L. D'Antuono, P. E. Tomatis, J. E. Ugalde, S. Giambiagi, and R. A. Ugalde. 2002. Analysis of *Mesorhizobium loti* glycogen operon: Effect of phosphoglucomutase (pgm) and glycogen synthase (g/gA) null mutants on nodulation of *Lotus tenuis*. *Mol Plant Microbe Interact* 15:368–375.

Levy, J., C. Bres, R. Geurts et al. 2004. A putative Ca^{2+} and calmodulin-dependent protein kinase required for baterial and fungal symbioses. *Science* 303:1361–1364.

Leyval, C., K. Turnau, and K. Haselwandter. 1997. Effect of heavy metal pollution on mycorrhizal colonization and function: Physiological, ecological and applied aspects. *Mycorrhiza* 7:139–153.

Li, F., J. Shi, C. Shen, G. Chen, S. Hu, and Y. Chen. 2009. Proteomic characterization of copper stress response in *Elsholtzia splendens* roots and leaves. *Plant Mol Biol* 71:251–263.

Li, K., C. Xu, K. Zhang, A. Yang, and J. Zhang. 2007. Proteomic analysis of roots growth and metabolic changes under phosphorus deficit in maize (*Zea mays* L.) plants. *Proteomics* 7:1501–1512.

Lin, S. K., M. C. Chang, Y. G. Tsai, and H. S. Lur. 2005. Proteomic analysis of the expression of proteins related to rice quality during caryopsis development and the effect of high temperature on expression. *Proteomics* 5:2140–2156.

Linkohr, B. I., L. C. Williamson, A. H. Fitter, and H. M. Leyser. 2002. Nitrate and phosphate availability and distribution have different effects on root system architecture of *Arabidopsis*. *Plant J* 29:751–760.

Lipson, D. A., T. K. Raab, S. K. Schmidt, and R. K. Monson. 1999. Variation in competitive abilities of plants and microbes for specific amino acids. *Biol Fert Soils* 29:257–261.

Liu, J. and A. K. M. Ekramoddoullah. 2006. The family 10 of plant pathogenesis-related proteins: Their structure, regulation, and function in response to biotic and abiotic stresses. *Physiol Mol Plant P* 68:3–13.

Liu, Y., T. Lamkemeyer, A. Jakob et al. 2006. Comparative proteome analyses of maize (*Zea mays* L.) primary roots prior to lateral root initiation reveal differential protein expression in the lateral root initiation mutant *rum1*. *Proteomics* 6:4300–4308.

Lou, L., Z. Shen, and X. Li. 2004. The copper tolerance mechanisms of *Elsholtzia haichowensis*, a plant from copper-enriched soils. *Environ Exp Bot* 51:111–120.

Lynch, J. 1995. Root architecture and plant productivity. *Plant Physiol* 109:7–13.

Mahmood, T., A. Jan, M. Kakishima, and S. Komatsu. 2006. Proteomic analysis of bacterial-blight defense-responsive proteins in rice leaf blades. *Proteomics* 6:6053–6065.

Majoul, T., E. Bancel, E. Triboi, J. Ben Hamida, and G. Branlard. 2003. Proteomic analysis of the effect of heat stress on hexaploid wheat grain: Characterization of heat-responsive proteins from total endosperm. *Proteomics* 3:175–183.

Majoul, T., K. Chahed, E. Zamiti, L. Ouelhazi, and R. Ghrir. 2000. Analysis by two-dimensional electrophoresis of the effect of salt stress on the polypeptide patterns in roots of a salt-tolerant and a salt-sensitive cultivar of wheat. *Electrophoresis* 21:2562–2565.

Malamy, J. E. 2005. Intrinsic and environmental response pathways that regulate root system architecture. *Plant Cell Environ* 28:67–77.

Marino, D., P. Frendo, R. Ladrera et al. 2007. Nitrogen fixation control under drought stress. Localized or systemic? *Plant Physiol* 143:1968–1974.

Massa, G. D. and S. Gilroy. 2003. Touch modulates gravity sensing to regulate the growth of primary roots of *Arabidopsis thaliana*. *Plant J* 33:435–445.

Mathesius, U. 2009. Comparative proteomic studies of root-microbe interactions. *J Proteom* 72:353–366.

Mathesius, U., G. Keijzers, S. H. Natera, J. J. Weinman, M. A. Djordjevic, and B. G. Rolfe. 2001. Establishment of a root proteome reference map for the model legume *Medicago truncatula* using the expressed sequence tag database for peptide mass fingerprinting. *Proteomics* 1:1424–1440.

Mazel, A., Y. Leshem, B. S. Tiwari, and A. Levine. 2004. Induction of salt and osmotic stress tolerance by overexpression of an intracellular vesicle trafficking protein AtRab7 (AtRabG3e). *Plant Physiol* 134:118–128.

McCully, M. E. 1999. ROOTS IN SOIL: Unearthing the complexities of roots and their rhizospheres. *Annu Rev Plant Physiol Plant Mol Biol* 50:695–718.

McKee, W. H. and M. R. McKevlin. 1993. Geochemical processes and nutrient uptake by plants in hydric soils. *Environ Toxicol Chem* 12:2197–2207.

Meharg, A. A. 2004. Arsenic in rice—Understanding a new disaster for South-East Asia. *Trends Plant Sci* 9:415–417.

Meharg, A. A. and L. Jardine. 2003. Arsenite transport into paddy rice (*Oryza sativa*) roots. *New Phytol* 157:39–44.

Mehta, A., A. C. Brasileiro, D. S. Souza et al. 2008. Plant-pathogen interactions: What is proteomics telling us? *Febs J* 275:3731–3746.

Miche, L., F. Battistoni, S. Gemmer, M. Belghazi, and B. Reinhold-Hurek. 2006. Upregulation of jasmonate-inducible defense proteins and differential colonization of roots of *Oryza sativa* cultivars with the endophyte *Azoarcus* sp. *Mol Plant Microbe Interact* 19:502–511.

Miller, A. J. and M. D. Cramer. 2004. Root nitrogen acquisition and assimilation. *Plant and Soil* 247:1–36.

Moon, H., B. Lee, G. Choi et al. 2003. NDP kinase 2 interacts with two oxidative stress-activated MAPKs to regulate cellular redox state and enhances multiple stress tolerance in transgenic plants. *Proc Natl Acad Sci USA* 100:358–363.

Morel, J., S. Claverol, S. Mongrand et al. 2006. Proteomics of plant detergent-resistant membranes. *Mol Cell Proteomics* 5:1396–1411.

Morris, A. C. and M. A. Djordjevic. 2001. Proteome analysis of cultivar-specific interactions between *Rhizobium leguminosarum biovar trifolii* and subterranean clover cultivar Woogenellup. *Electrophoresis* 22:586–598.

Nadimpalli, R., N. Yalpani, G. S. Johal, and C. R. Simmons. 2000. Prohibitins, stomatins, and plant disease response genes compose a protein superfamily that controls cell proliferation, ion channel regulation, and death. *J Biol Chem* 275:29579–29586.

Nagahashi, G. and J. D. D. Douds. 1997. Appressorium formation by AM fungi on isolated cell walls of carrot roots. *New Phytol* 136:299–304.

Nakagawa, T. and M. Kawaguchi. 2006. Shoot-applied MeJA suppresses root nodulation in *Lotus japonicus*. *Plant Cell Physiol* 47:176–180.

Nakashita, H., M. Yasuda, T. Nitta et al. 2003. Brassinosteroid functions in a broad range of disease resistance in tobacco and rice. *Plant J* 33:887–898.

Nasholm, T., A. Ekblad, A. Nordin, R. Giesler, M. Högberg, and P. Högberg. 1998. Boreal forest plants take up organic nitrogen. *Nature* 392:914–916.

Natera, S. H., N. Guerreiro, and M. A. Djordjevic. 2000. Proteome analysis of differentially displayed proteins as a tool for the investigation of symbiosis. *Mol Plant Microbe Interact* 13:995–1009.

Ndimba, B. K., S. Chivasa, J. M. Hamilton, W. J. Simon, and A. R. Slabas. 2003. Proteomic analysis of changes in the extracellular matrix of *Arabidopsis* cell suspension cultures induced by fungal elicitors. *Proteomics* 3:1047–1059.

Nozu, Y., A. Tsugita, and K. Kamijo. 2006. Proteomic analysis of rice leaf, stem and root tissues during growth course. *Proteomics* 6:3665–3670.

Oaks, A. and B. Hirel. 1985. Nitrogen metabolism in roots. *Ann Rev Plant Physio* 36:345–365.

Oehrle, N. W., A. D. Sarma, J. K. Waters, and D. W. Emerich. 2008. Proteomic analysis of soybean nodule cytosol. *Phytochemistry* 69:2426–2438.

Orsel, M., S. Filleur, V. Fraisier, and F. Daniel-Vedele. 2002. Nitrate transport in plants: Which gene and which control? *J Exp Bot* 53:825–833.

Panter, S., R. Thomson, G. de Bruxelles, D. Laver, B. Trevaskis, and M. Udvardi. 2000. Identification with proteomics of novel proteins associated with the peribacteroid membrane of soybean root nodules. *Mol Plant Microbe Interact* 13:325–333.

Paul, M. J. and C. H. Foyer. 2001. Sink regulation of photosynthesis. *J Exp Bot* 52:1383–1400.

Peck, S. C., T. S. Nuhse, D. Hess, A. Iglesias, F. Meins, and T. Boller. 2001. Directed proteomics identifies a plant-specific protein rapidly phosphorylated in response to bacterial and fungal elicitors. *Plant Cell* 13:1467–1475.

Perotto, S. 1995. Ericoid mycorrhizal fungi: Cellular and molecular bases of ther interactions with the host plant. *Can J Bot* 73:S557–S568.

Perret, X., C. Staehelin, and W. J. Broughton. 2000. Molecular basis of symbiotic promiscuity *Microbiol Mol Biol Rev* 64:180–201.

Pickering, I. J., R. C. Prince, M. J. George, R. D. Smith, G. N. George, and D. E. Salt. 2000. Reduction and coordination of arsenic in Indian mustard. *Plant Physiol* 122:1171–1177.

Pratt, J. M., J. Petty, I. Riba-Garcia et al. 2002. Dynamics of protein turnover, a missing dimension in proteomics. *Mol Cell Proteomics* 1:579–591.

Prayitno, J., N. Imin, B. G. Rolfe, and U. Mathesius. 2006. Identification of ethylene-mediated protein changes during nodulation in *Medicago truncatula* using proteome analysis. *J Proteome Res* 5: 3084–3095.

Price, A. H., J. E. Cairns, P. Horton, H. G. Jones, and H. Griffiths. 2002. Linking drought-resistance mechanisms to drought avoidance in upland rice using a QTL approach: Progress and new opportunities to integrate stomatal and mesophyll responses. *J Exp Bot* 53:989–1004.

Prinsi, B., A. S. Negri, P. Pesaresi, M. Cocucci, and L. Espen. 2009. Evaluation of protein pattern changes in roots and leaves of *Zea mays* plants in response to nitrate availability by two-dimensional gel electrophoresis analysis. *BMC Plant Biol* 9:113.

Puigderrajols, P., A. Jofre, G. Mir et al. 2002. Developmentally and stress-induced small heat shock proteins in cork oak somatic embryos. *J Exp Bot* 53:1445–1452.

Raab, A., H. Schat, A. A. Meharg, and J. Feldmann. 2005. Uptake, translocation and transformation of arsenate and arsenite in sunflower (*Helianthus annuus*): formation of arsenic-phytochelatin complexes during exposure to high arsenic concentrations. *New Phytol* 168:551–558.

Redmond, J. W., M. Batley, M. A. Djordjevic, R. W. Innes, P. L. Kuempel, and B. G. Rolfe. 1986. Flavones induce expression of nod genes in Rhizobium. *Nature* 323:632–635.

Renaut, J. 2003. Responses of poplar to chilling temperatures: Proteomic and physiological aspects. *Plant Biol* 5:81–90.

Rep, M., H. L. Dekker, J. H. Vossen et al. 2002. Mass spectrometric identification of isoforms of PR proteins in xylem sap of fungus-infected tomato. *Plant Physiol* 130:904–917.

Repetto, O. 2003. Targeted proteomics to identify cadmium-induced protein modifications in *Glomus mosseae*-inoculated pea roots. *New Phytol* 157:555–567.

Requejo, R. and M. Tena. 2005. Proteome analysis of maize roots reveals that oxidative stress is a main contributing factor to plant arsenic toxicity. *Phytochemistry* 66:1519–1528.

Riccardi, F., P. Gazeau, D. de Vienne, and M. Zivy. 1998. Protein changes in response to progressive water deficit in maize. Quantitative variation and polypeptide identification. *Plant Physiol* 117:1253–1263.

Roth, U., E. von Roepenack-Lahaye, and S. Clemens. 2006. Proteome changes in *Arabidopsis thaliana* roots upon exposure to Cd2+. *J Exp Bot* 57:4003–4013.

Rouhier, N., A. Villarejo, M. Srivastava et al. 2005. Identification of plant glutaredoxin targets. *Antioxid Redox Signal* 7:919–929.

Sagan, M., D. Morandi, E. Tarenghi, and G. Duc. 1995. Selection of nodulation and mycorrhizal mutants in the model plant *Medicago truncatula* after γ-ray mutagenesis. *Plant Sci* 111:63–71.

Sanchez, L., S. Weidmann, C. Arnould, A. R. Bernard, S. Gianinazzi, and V. Gianinazzi-Pearson. 2005. *Pseudomonas fluorescens* and *Glomus mosseae* trigger DMI3-dependent activation of genes related to a signal transduction pathway in roots of *Medicago truncatula. Plant Physiol* 139:1065–1077.

Santoni, V., J. Vinh, D. Pflieger, N. Sommerer, and C. Maurel. 2003. A proteomic study reveals novel insights into the diversity of aquaporin forms expressed in the plasma membrane of plant roots. *Biochem J* 373:289–296.

Santoni, V., T. Rabilloud, P. Doumas et al. 1999. Towards the recovery of hydrophobic proteins on two-dimensional electrophoresis gels. *Electrophoresis* 20:705–711.

Saravanan, R. S. and J. K. Rose. 2004. A critical evaluation of sample extraction techniques for enhanced proteomic analysis of recalcitrant plant tissues. *Proteomics* 4:2522–2532.

Satoh, S., C. Iizuka, A. Kikuchi, N. Nakamura, and T. Fufii. 1992. Proteins and carbohydrates in xylem sap from squash root *Plant Cell Physiol* 33:841–847.

Saalbach, G., P. Erik, and S. Wienkoop. 2002. Characterisation by proteomics of peribacteroid space and peribacteroid membrane preparations from pea (*Pisum sativum*) symbiosomes. *Proteomics* 2:325–337.

Schenkluhn, L., N. Hohnjec, K. Niehaus, U. Schmitz, and F. Colditz. 2010. Differential gel electrophoresis (DIGE) to quantitatively monitor early symbiosis- and pathogenesis-induced changes of the *Medicago truncatula* root proteome. *J Proteomics* 73:753–768.

Schnable, P. S., D. Ware, R. S. Fulton et al. 2009. The B73 maize genome: Complexity, diversity, and dynamics. *Science* 326:1112–1115.

Setter, T. L. and I. Waters. 2003. Review of prospects for germplasm improvement for waterlogging tolerance in wheat, barley and oats. *Plant Soil* 253:1–34.

Shahollari, B., T. Peskan-Berghöfer, and R. Oelmuller. 2004. Receptor kinases with leucine-rich repeats are enriched in Triton X-100 insoluble plasma membrane microdomains from plants. *Physiol Plantarum* 122:397–403.

Sharp, R. E., W. K. Silk, and T. C. Hsiao. 1988. Growth of the maize primary root at low water potentials: I. spatial distribution of expansive growth. *Plant Physiol* 87:50–57.

Shen, J., C. Tang, Z. Rengel, and F. Zhang. 2004. Root-induced acidification and excess cation uptake by N_2-fixing Lupinus albus grown in phosphorus-deficient soil. *Plant Soil* 260:69–77.

Shin, R., S. Alvarez, A. Y. Burch, J. M. Jez, and D. P. Schachtman. 2007. Phosphoproteomic identification of targets of the *Arabidopsis* sucrose nonfermenting-like kinase SnRK2.8 reveals a connection to metabolic processes. *Proc Natl Acad Sci USA* 104:6460–6465.

Showalter, A. M. 1993. Structure and function of plant cell wall proteins. *Plant Cell* 5:9–23.

Smith, K. A. and R. S. Russell. 1969. Occurence of ethylene and its significance in anaerobiz soil. *Nature* 222:769–771.

Somerville, C., S. Bauer, G. Brininstool et al. 2004. Toward a systems approach to understanding plant cell walls. *Science* 306:2206–2211.

Speiser, D. M., S. L. Abrahamson, G. Banuelos, and D. W. Ow. 1992. *Brassica juncea* produces a phytochelatin-cadmium-sulfide complex. *Plant Physiol* 99:817–821.

Spollen, W. G., M. E. LeNoble, T. D. Samuels, N. Bernstein, and R. E. Sharp. 2000. Abscisic acid accumulation maintains maize primary root elongation at low water potentials by restricting ethylene production. *Plant Physiol* 122:967–976.

Staehelin, C., M. Schultze, E. Kondorosi, and A. Kondorosi. 1995. Lipo-chitooligosaccharide nodulation signals from *Rhizobium meliloti* induce their rapid degradation by the host plant alfalfa. *Plant Physiol* 108:1607–1614.

Stohs, S. J. and D. Bagchi. 1995. Oxidative mechanisms in the toxicity of metal ions. *Free Radic Biol Med* 18:321–336.

Subramanian, S., U. H. Cho, C. Keyes, and O. Yu. 2009. Distinct changes in soybean xylem sap proteome in response to pathogenic and symbiotic microbe interactions. *BMC Plant Biol* 9:119.

Takahashi, H. 1997. Hydrotropism: The current state of our knowledge. *J Plant Res* 110:163–169.

Thornton, B., S. M. Osborne, E. Paterson, and P. Cash. 2007. A proteomic and targeted metabolomic approach to investigate change in *Lolium perenne* roots when challenged with glycine. *J Exp Bot* 58:1581–1590.

Tian, L., G. J. Peel, Z. Lei et al. 2009. Transcript and proteomic analysis of developing white lupin (*Lupinus albus* L.) roots. *BMC Plant Biol* 9:1.

Tsugita, A. and M. Kamo. 1999. 2-D electrophoresis of plant proteins. *Methods Mol Biol* 112:95–97.

Valcu, C. M, M. Junqueira, A. Shevchenko, and K. Schlink. 2009. Comparative proteomic analysis of responses to pathogen infection and wounding in *Fagus sylvatica. J Proteome Res* 8:4077–4091.

Valot, B., L. Negroni, M. Zivy, S. Gianinazzi, and E. Dumas-Gaudot. 2006. A mass spectrometric approach to identify arbuscular mycorrhiza-related proteins in root plasma membrane fractions. *Proteomics* 6 Suppl 1:S145–S155.

Valot, B., M. Dieu, G. Recorbet, M. Raes, S. Gianinazzi, and E. Dumas-Gaudot. 2005. Identification of membrane-associated proteins regulated by the arbuscular mycorrhizal symbiosis. *Plant Mol Biol* 59: 565–580.

Valot, B., S. Gianinazzi, and D. G. Eliane. 2004. Sub-cellular proteomic analysis of a *Medicago truncatula* root microsomal fraction. *Phytochemistry* 65:1721–1732.

van Noorden, G. E., T. Kerim, N. Goffard et al. 2007. Overlap of proteome changes in *Medicago truncatula* in response to auxin and *sinorhizobium meliloti. Plant Physiol* 144:1115–1131.

van Wijk, K. J. 2001. Challenges and prospects of plant proteomics. *Plant Physiol* 126:501–508.

Vanacker, H., L. Sandalio, A. Jimenez et al. 2006. Roles for redox regulation in leaf senescence of pea plants grown on different sources of nitrogen nutrition. *J Exp Bot* 57:1735–1745.

Vandenbussche, F., J. Smalle, J. Le et al. 2003. The *Arabidopsis* mutant *alh1* illustrates a cross talk between ethylene and auxin. *Plant Physiol* 131:1228–1238.

Wan, J., M. Torres, A. Ganapathy et al. 2005. Proteomic analysis of soybean root hairs after infection by *Bradyrhizobium japonicum. Mol Plant Microbe Interact* 18:458–467.

Wang, D. and D. S. Luthe. 2003. Heat sensitivity in a bentgrass variant. Failure to accumulate a chloroplast heat shock protein isoform implicated in heat tolerance. *Plant Physiol* 133:319–327.

Watson, B. S., V. S. Asirvatham, L. Wang, and L. W. Sumner. 2003. Mapping the proteome of barrel medic (*Medicago truncatula*). *Plant Physiol* 131:1104–1123.

Wiebe, B. H., R. G. Eilers, W. D. Eilers, and J. A. Brierley. 2007. Application of a risk indicator for assessing trends in dryland salinization risk on the Canadian prairies. *Can J Soil Sci* 87:213–224.

Wienkoop, S. and G. Saalbach. 2003. Proteome analysis. Novel proteins identified at the peribacteroid membrane from *Lotus japonicus* root nodules. *Plant Physiol* 131:1080–1090.

Winzer, T., A. Bairl, M. Linder, D. Linder, D. Werner, and P. Müller. 1999. A novel 53 kDa nodulin of the symbiosome membrane of soybean nodules, controlled by *Bradyrhizobium japonicum. Mol Plant Microbe Interact* 12:218–226.

Wood, N. T. 2001. Nodulation by numbers: The role of ethylene in symbiotic nitrogen fixation. *Trends Plant Sci* 6:501–502.

Xiong, L. and Y. Yang. 2003. Disease resistance and abiotic stress tolerance in rice are inversely modulated by an abscisic acid-inducible mitogen-activated protein kinase. *Plant Cell* 15:745–759.

Xu, C. and B. Huang. 2008. Root proteomic responses to heat stress in two *Agrostis* grass species contrasting in heat tolerance. *J Exp Bot* 59:4183–4194.

Xu, Q. and B. Huang. 2001. Lowering soil temperatures improves creeping bentgrass growth under heat stress. *Crop Sci* 41:1878–1883.

Yamaguchi, M., B. Valliyodan, J. Zhang et al. 2010. Regulation of growth response to water stress in the soybean primary root. I. Proteomics analysis reveals region-specific regulation of phenylpropanoid metabolism and control of free iron in the elongation zone. *Plant Cell Environ* 33:223–243.

Yamamoto, Y., Y. Kobayashi, S. R. Devi, S. Rikiishi, and H. Matsumoto. 2002. Aluminum toxicity is associated with mitochondrial dysfunction and the production of reactive oxygen species in plant cells. *Plant Physiol* 128:63–72.

Yan, S., Z. Tang, W. Su, and W. Sun. 2005. Proteomic analysis of salt stress-responsive proteins in rice root. *Proteomics* 5:235–244.

Yang, Q., Y. Wang, J. Zhang, W. Shi, C. Qian, and X. Peng. 2007. Identification of aluminum-responsive proteins in rice roots by a proteomic approach: Cysteine synthase as a key player in Al response. *Proteomics* 7:737–749.

Yano, H., S. Kuroda, and B. B. Buchanan. 2002. Disulfide proteome in the analysis of protein function and structure. *Proteomics* 2:1090–1096.

York, W. S., Q. Qin, and J. K. Rose. 2004. Proteinaceous inhibitors of endo-h-glucanases. *BBA- Proteins Proteomics* 1696:223–233.

Yoshimura, K., A. Masuda, M. Kuwano, A. Yokota, and K. Akashi. 2008. Programmed proteome response for drought avoidance/tolerance in the root of a C(3) xerophyte (wild watermelon) under water deficits. *Plant Cell Physiol* 49:226–241.

Yu, J., S. Hu, J. Wang et al. 2002. A draft sequence of the rice genome (*Oryza sativa* L. ssp. *indica*). *Science* 296:79–92.

Zahran, H. H. 1999. Rhizobium–legume symbiosis and nitrogen fixation under severe conditions and in an arid climate. *Microbiol Mol Biol Rev* 63:968–989, table of contents.

Zhao, D. Y., Q. Y. Tian, L. H. Li, and W. H. Zhang. 2007. Nitric oxide is involved in nitrate-induced inhibition of root elongation in *Zea mays*. *Ann Bot* 100:497–503.

Zhou, S., R. Sauve, and T. W. Thannhauser. 2009. Proteome changes induced by aluminium stress in tomato roots. *J Exp Bot* 60:1849–1857.

Zhu, J., S. Chen, S. Alvarez et al. 2006. Cell wall proteome in the maize primary root elongation zone. I. Extraction and identification of water-soluble and lightly ionically bound proteins. *Plant Physiol* 140:311–325.

Zhu, J., S. Alvarez, E. L. Marsh et al. 2007. Cell wall proteome in the maize primary root elongation zone. II. Region-specific changes in water soluble and lightly ionically bound proteins under water deficit. *Plant Physiol* 145:1533–1548.

CHAPTER 11

Applications of Agricultural and Medicinal Biotechnology in Functional Foods

Kandan Aravindaram and Ning-Sun Yang

CONTENTS

11.1 INTRODUCTION

Biotechnology is a broad discipline that studies the potential use of natural and modified organisms and systems in agriculture, medicine, environment and many other fields. It uses a wide range of techniques, from relatively simple breeding to highly sophisticated molecular and cellular manipulations to produce specific desired traits in plants, animals or microorganisms, often requiring extensive knowledge of the genetics of the target organisms. Agricultural and medicinal biotechnologies can use these tools for purposes as diverse as the genetic improvement of crops or farm animals to increase yields by enhancing growth efficiency, functional genetic characterization and conservation of specific genetic resources (e.g., endangered species), animal and human vaccine development, and plant, animal and human disease diagnosis, among many others. Some of these technologies may also have applications in the food and agricultural industries, such as the use of molecular markers in the development of improved characteristics in crop plants, farm animals and poultry, for example. Plant tissue culture manipulation technologies, such as micropropagation, the culture of embryos, organs and meristem, anther/haploid cultures, ploidy manipulation and *in vitro* cell production and genetic engineering technologies including gene transformation, plant regeneration, transgenic plants, variant selection of genetically modified organisms (GMOs) and genetic improvement of plants are responsible for many recent breakthroughs in plant biotechnology. DNA-based diagnostics for the rapid, accurate identification of microorganisms in food, and pests and diseases of crop plants, livestock, poultry and even fish, as well as other agricultural or medicinal products have added to the contributions made to the society by biotechnology.

The use of GMOs as crops can be categorized as having either first-, second- or third-generation biotechnology characteristics (Glover et al., 2005) Some GMO crop plants with first-generation characteristics were developed for resistance or tolerance to environmental (abiotic) stresses (such as salinity, drought, acid soils or temperature extremes), and for biotic or pest and disease control. Second-generation products have focused on increased nutrition, enhanced quality or better processing characteristics of food and feed crops (such as omega-3 oil production, starch modification, improved sugar content, altered sugar metabolism, improved color and increased antioxidant content). Third-generation GM plants are those where GM plants are used as a factory to produce industrial (biofuels, biodegradable plastics, enzymes or lubricant oils) or pharmaceutical compounds (animal or human hormones, vaccines, antibodies and formulation ingredients). Currently, an area of emphasis in research in functional foods or nutraceuticals is the rapid evolution and application of the 'omics' platforms (including but not limited to genomics, proteomics, metabolomics and transcriptomics). However, this technological revolution in biotechnology is still in its infancy, and will need much concerted effort from both academic and industrial sectors to develop useful and implementable systems.

The new biotechnological approaches to human and animal health care and management are making a significant contribution to attempts to improve the quality of life, food and nutrition, health care and the environment. Advances in health care include a spectrum of biotech-derived medicines and diagnostic tools, advances in quality and preservation of foods and new foods enriched with specific nutrients or with fewer calories. The first generations of health-related products in the food and nutrition industry were achieved through genetic and molecular biology approaches. These include the use of specific bioactive ingredients to combat common illnesses, such as defined antioxidants to fight cancer, vitamin B and niacin to combat heart disease and polyphenols and fatty acids to fight cardiovascular ailments. Additionally, functional food products, such as products enriched with enzymes to metabolize sugars in the stomach before absorption and coenzymes to stimulate metabolism, have been developed to combat obesity and weight gain. The second generation of health-related products in the food category comes from 'nutrigenomics' research, in which foods are developed to cater to specific groups of people according to differences in life style, developmental stage, social culture or genetic makeup. For instance, foods infused with vitamin B and folates were developed for children and adolescents to support the rapid growth and development of this population. Such products are often clearly labelled as being fortified with nutritional supplements for a specific age group and vigorously marketed.

In this chapter, we intend to focus mainly on different agricultural and medicinal biotechnology studies related to functional foods for human health care, and particularly the latest 'omics-based' approaches used in functional food research and development.

11.2 APPLICATIONS OF BIOTECHNOLOGY TO FOOD/FEED CROP IMPROVEMENT

In today's world, food is not only intended to satisfy hunger and provide necessary nutrients to humans, but is being designed to improve physical and mental well-being, and to prevent nutrition-related diseases (Menrad, 2003). Research in plant biotechnology has previously focused primarily on agronomic characteristics to improve resistance or tolerance to biotic and abiotic stresses in particular crop plants. While this effort has been relatively successful, new products that can meet the demands for increased yield and quality are limited, although recent efforts are showing some promise. These include improvement of the nutritional quality of food for human and livestock health, and the development of ingredients with superior properties for food manufacturing and processing. Systematic efforts to improve the quality and quantity of carbohydrates, proteins, lipid, vitamins and minerals in staple cereal or vegetable crops have made encouraging progress and led to the development of new approaches. We discuss here the improvement of staple crops in different parts of the world, namely rice, wheat, potato, corn and cassava.

11.2.1 Rice

Rice is the most important staple food in Asia and many parts of Africa and South America, and is now also becoming increasingly popular in Northern America and Europe. It is a very important primary source of energy for billions of people in today's world, but scientists are now trying to enhance the nutritional value of this staple of more than half of the world's population. A systematic research effort worldwide has focused on improving the nutritional quality of rice, especially the levels of total protein, and specifically lysine, as well as vitamin A and iron. The protein content of rice ranges from 7% to 8.5% of the dry mass, depending on the particular variety. The most recent success in protein improvement comes from Dr. William Folk and co-workers with their genetic modification of rice to give increased levels of the amino acid lysine (Wu et al., 2003). In Asia and other parts of the developing world, vitamin A deficiency is closely associated with the poverty-related predominant consumption of rice, since rice grain is known to lack pro-vitamin A. A notable achievement in genetic engineering of rice was the creation of the so-called 'golden rice', which involved the transfer of genes required for the synthesis of vitamin A precursor and beta-carotene (Datta et al., 2003). Paine et al. (2005) further improved the nutritional quality of golden rice by additional transformation of a phytoene synthase (*psy*)-encoding gene, with carotenoid content almost 23 times higher than the original golden rice strain (maximum 37 μg/g) and the preferential accumulation of beta-carotene. Research efforts to develop such improved golden rice varieties are continuing and their commercialization is being given a high priority.

11.2.2 Wheat

The ability to make bread and a range of other processed foods (cakes, noodles, biscuits, etc.) from wheat flour is determined mainly by the unique properties of specific wheat grain storage proteins, particularly high-molecular-weight (HMW) glutenin. Appreciable changes in glutenin subunit composition as a useful trait can now be genetically engineered without affecting the agronomic performance of wheat (Bregitzer et al., 2006). More recently, transgenic wheat with elevated levels of HMW subunit 1D × 5 and/or 1Dy10 were shown to allow doughs with increased mixing strength and tolerance, a desirable trait for bakery (Blechl et al., 2007). Apart from this approach, modifying specific protein quality is also exemplified by complementation of a null *pina* allele with the wild-type *Pina* gene sequence, shown to restore a soft phenotype in transgenic wheat (Martin et al., 2006). In another practical example of the benefit of the biotechnology genetic approaches, field evaluation of transgenic wheat expressing a modified maize ADP-glucose pyrophosphorylase large subunit (Sh2r6hs) suggested that limited abiotic stress may subsequently limit the Sh2r6hs-associated yield enhancement (Meyer et al., 2007).

11.2.3 Cassava

Little known in the West, cassava is in fact a staple food for more than 300 million people in various tropical and subtropical parts of the world. The edible part of the cassava root has, however, less than 1% protein and is thus of very poor nutritional value. Zhang et al. (2003) successfully enhanced the quality of protein in cassava tuber by transforming a cDNA gene responsible for the storage protein *ASP1*, rich in essential amino acids. Importantly, another problem in cassava consumption is the presence, even after extensive washing, of residual cyanogens (linamarin or acetone cyanohydrin), especially in incompletely processed cassava roots, as it can cause cyanide poisoning. Hydroxynitrile lyase (HNL), which catalyses the conversion of acetone cyanohydrin to cyanide, is expressed predominantly in the cell walls and laticifers of leaves and at low levels in roots. Siritunga and co-workers reported that over-expression of HNL under the control of a double 35S CaMV promoter in transgenic cassava plants increased HNL activity more than twofold in leaves and 13 times in roots of engineered transgenic plants, relative to wild-type plants. Elevated HNL levels were correlated with substantially reduced acetone cyanohydrin levels and increased cyanide volatilization in processed or homogenized roots, thus providing a safer food product (Siritunga et al., 2004).

11.2.4 Potato

Another dietary staple crop grown throughout various parts of Asia, Europe, Africa and America is the potato. Potatoes contain 0.1% fat and 2% protein, levels that are not nutritional for those people unable to source other food types. Attempts to increase the protein content of potato by genetic engineering via transformation of the seed albumin gene from *Amaranthus hypochondriacus* to potato (Chakraborty et al., 2000) have been moderately successful. Klaus et al. (2004) demonstrated increased fatty acid synthesis after transfer of the acetyl-CoA carboxylase gene, and in the resulting transgenic potatoes over-expression of fatty acids was observed. Genetic transformation of potato with a sucrose synthase (*SuSy*) gene, which encodes a highly regulated cytosolic enzyme that catalyses the conversion of sucrose and a nucleoside diphosphate into the corresponding nucleoside diphosphate glucose and fructose, has also been successful (Baroja-Fernández et al., 2009). The resultant transgenic potato exhibited appreciable increases in starch, UDP-glucose and ADP-glucose levels compared to normal plants.

11.2.5 Corn

Corn is one of the most important staple foods and feed crops in the world, and is widely grown in the America, East Asia and Africa. The nutritional quality of corn protein in high-yield hybrid corn is in general good, but corn remains unsuitable as a single food due to a relative deficiency of two essential amino acids, especially lysine. High lysine corn has thus been a key research goal for decades, using both conventional breeding and molecular biology tools.

Advances in agricultural biotechnology are still being sought to improve the nutritional quality of corn protein, especially lysine levels. Two different approaches have been evaluated recently for enhancing the lysine content in corn. First, RNAi has been used to specifically suppress α-zein production in transgenic corn, resulting in a doubling of the lysine content of corn grains from 2400 to 4800 ppm (Huang et al., 2006). However, quite similar to the original opaque mutant strains, the soft and chalky kernel phenotype displayed by these transgenic corn lines remains a deterrent factor to their commercialization. Second, lysine, along with methionine, threonine and isoleucine, is derived from aspartate, with dihydrodipicolinate synthase (DHDPS) catalysing the first step of lysine biosynthesis. A bifunctional enzyme, lysine-ketoglutarate reductase/saccharopine dehydrogenase (LKR/SDH), is responsible for lysine catabolism. The free lysine level in plant cells is believed to be regulated by lysine feedback inhibition of DHDPS and feed-forward activation of LKR/SDH. Indeed, the transgenic expression of a feedback-insensitive DHDPS transgene from *Corynebacterium glutamicum*, CordapA, as well as suppression of the LKR/SDH gene have resulted in a transgenic corn line with high levels of free lysine content (Huang et al., 2005; Houmard et al., 2007).

Recently, Christou's research team demonstrated, using elite inbred South African transgenic corn plants, that the levels of three vitamins were significantly increased specifically in the endosperm through simultaneous molecular genetic modification of three separate metabolic pathways. The transgenic kernels contained 169 times the normal amount of beta-carotene, six times as much ascorbate and twice the normal amount of folate (Naqvi et al., 2009). This sophisticated genetic engineering approach is a bright example of the potential routes to the future development of nutritionally complete or supplemented cereals to benefit the health of the world's poor populations.

11.3 FUNCTIONAL FOODS FOR APPLICATION TO HUMAN HEALTH CARE AND/OR DISEASE PREVENTION

Functional foods or nutritional supplements can benefit human health beyond the effect of baseline nutritional supply alone. The term 'functional foods' may be more widely applied to either single or combined dietary components or those sold as supplements. In these contexts, the term can be applied to various proteins, vitamins, minerals, polysaccharide fibres and other food components, and whole foods (Table 11.1), such as probiotics, prebiotics, oily fish (containing high levels of n-3 polyunsaturated fatty acids (PUFAs) and green tea (containing high levels of catechins, such as epigallocatechin-3 gallate (EGCG)) (Ghosh et al., 2006). Recent studies on these nutritional agents, substances or formulations are briefly discussed using the following examples.

11.3.1 Vitamins

Vitamins are essential biochemicals required for normal physiology and homeostasis that humans, and other animals, have lost the ability to synthesize on their own. Vitamins are often co-enzymes or the prosthetic group of specific enzymes. Depending on their biochemical properties, the various vitamins are generally classified into two groups, either lipid-soluble (A, D, E and K1) or water-soluble (B1, B2, B6, B12, C and H) vitamins. Vitamins were the first substances to be identified and employed as nutraceuticals, and are now routinely added into various foods, beverages and dietary supplements. Plants and microorganisms are capable of producing specific vitamins in their tissues (this is always the source of the vitamins taken up in a natural diet), and this has promoted the idea of using engineered plants and microbes as factories for the production of some of the above-mentioned functional foods.

Table 11.1 Some Functional Food Products for Human Health

Product Name	Description	Company
	Probiotics	
Actimel	Probiotic drinking yogurt with *Lactobacillus casei* cultures	Danone, France
Attune wellness bars	Contains three strains of bacteria to balance digestion and support immunity	Attune Foods, USA
Activia	Creamy yogurt containing Bifidus	Danone, France
Gefilus	A wide range of Lactobacillus GG (LGG) products	Valio, Finland
Jovita Probiotisch	Blend of fruit, cereals and probiotic yogurt	H&J Bruggen, Germany
Proviva	Yogurt containing *Lactobacillus plantarum*	Skåne mejerier, Sweden
Revival Active	Yogurt	Olma, Czech Republic
Soytreat	Kefir-type product with six probiotics	Lifeway, USA
Vifit	Yogurt with LGG, vitamins and minerals	Campina, The Netherlands
	Food Ingredients	
Omega-3 soya bean oil	Heart-healthy eicasopentanoic acid	Monsanto, USA
High oleic oil	For frying and baking goods	Dupont, USA
Visitive 1	Oil from soya bean produces no trans fats while cooking.	Monsanto, USA
Visitive 3	Reduced saturated fat with heart-healthy profile	Monsanto, USA
WG0401	Functional food derived from tea to control inflammation	WellGen, USA
WG0301	Restricts obesity development	WellGen, USA
	Whole Foods	
High-protein cassava	Transgenic cassava producing 2–4% dry weight protein, more than conventional variety	Biocassava Plus International Consortium
Sugar beet	Soluble fibre and low-calorie sweetener	Florimond Desprez, France
High-folic acid tomato	Twenty-five times higher folic acid in the ripened fruit, for treating folic acid deficiency	University of Florida, USA
	Dietary Supplements	
DHA omega-3	Essential for fetal development	Natrol, USA
Olivenol	Antioxidant properties	CreAgri, USA
CetylPure®	Contains cetyl myristoleate, a naturally occurring fatty acid. For joint health and comfort	Natrol, USA
Sunthi	Carminative and gastrointestinal tract stimulant, for indigestion and nausea	Himalaya Healthcare, India

11.3.1.1 Vitamin A and Other Carotenoids

Vitamin A precursor compounds, chiefly beta-carotene and retinol, are present in various animal and plant tissues. The animal-based substance known as retinol is ready for use for human health care measures. The recommended dietary allowance (RDA) for vitamin A is 1000 retinol equivalents, equal to about 6 mg beta-carotene per day. Several experimental approaches have been evaluated for enhancing the levels of beta-carotene in different plant tissues. Rosati et al. (2000) demonstrated the production of beta-carotene at levels of up to 60 μg/g fresh weight in tomato fruits, reflecting the values that would reach the RDA with an intake of 100 g fresh tomatoes per day.

Other nutritionally important carotenoids are lutein and zeaxanthin, which play major roles in protection of the retina and reduce the risk of age-related macular degeneration (Gale et al., 2003). Romer et al. (2002) genetically manipulated potato plants to block the conversion of zeaxanthin to another carotenoid, and thereby producing levels of zeaxanthin between 4 and 130 times higher than normal. Introduction of phytoene synthase and phytoene desaturase activities in the absence of heterologous beta-cyclase was also found to generate appreciable levels of provitamin A (Ye et al., 2000). Another carotenoid, astaxanthin, is a contributor to yellow color in plants, and is a powerful antioxidant. Research by Mann et al. (2000) demonstrated the production of enhanced levels of astaxanthin in transgenic tobacco plants, which resulted in a change of color in engineered plants. These elegant works pave the way for future development of provitamin A plants with a high potential for benefiting human health.

11.3.1.2 Vitamin E

The Vitamin E group comprises eight different hydrophobic derivatives; among them alpha-tocopherol is physiologically the most important in human health. Vitamin E is present in vegetable oils, wheat germ, soya bean, egg yolk and other food sources. Major efforts have been made to increase the content of vitamin E in food crops, particularly in cereals and grains. One approach undertaken by plant genetic engineers is altering the metabolic biosynthetic pathways of plants using different precursor substances. By over-expressing genes encoding specific enzymes that play a major role in vitamin E biosynthetic pathways, Estevez et al. (2001) developed transgenic Arabidopsis over-expressing the enzyme 1-deoxy-d-xylulose-5-phosphate synthase, and those plants produced increased levels of tocopherols, abscisic acid and gibberellic acid. Over-expression of an *A. thaliana* homogentisate prenyltransferase driven by the seed-specific napin promoter resulted in a doubling of vitamin E levels in seeds of transgenic *A. thaliana* plants (Savidge et al., 2002). Van Eenennaam et al. (2003) later demonstrated over-expression of vitamin E (alpha-tocopherol) in transgenic soya bean plants, with levels five times as high as untransformed (control) plants. Further improvements in experimental approaches and techniques for generating transgenic plants for commercialization are highly desirable. Such ‘nutraceutical’ plants may also assist in the elucidation of different underlying mechanisms of control and fine-tuning of vitamin E ‘super-production’ in plants. Functional genomics studies (see below) may prove to be useful approaches in this effort.

11.3.1.3 Vitamin C

Vitamin C (l-ascorbic acid) is an important water-soluble antioxidant. It is crucial for the maintenance of homeostasis and a healthy immune system in humans, and is also required for the synthesis of collagen, neurotransmitters, carnitine and other molecules. Plants and most animals have the ability to synthesize vitamin C by different metabolic pathways, whereas primates (among other groups) have lost the ability to synthesize ascorbic acid due to mutations in the enzyme l-gulono-γ-lactone oxidase. Jain and Nessler (2000) successfully developed transgenic tobacco and lettuce plants expressing a rat cDNA encoding the GLO enzyme, and this resulted in vitamin C levels four to seven times higher in leaves of transgenic lettuce and tobacco plants than the non-transformed controls. Chen et al. (2003) subsequently developed transgenic corn and tobacco plants with two to four times higher levels of vitamin C. Engineering of d-galacturonate reductase gene into strawberry plants resulted in increased expression of vitamin C in transgenic plants compared to their normal parents (Agius et al., 2003). These studies demonstrate the potential for genetically manipulating fruits and vegetables to produce high levels of vitamin C for dietary use in promoting human health.

11.3.2 Minerals

11.3.2.1 Iron

Iron deficiency is one of the most damaging and widespread nutritional deficiencies in humans, with up to 30% of people worldwide showing signs of iron-deficiency anaemia. Increasing the amount of iron provided to people in their staple diet would greatly reduce this statistic. A biochemical tweak to improve iron release from phytase activity is one possible strategy to increase iron uptake and availability to plants. Uptake of iron into roots and its transport to and from different vegetative parts of specific plants may vary considerably; for example, in wheat only 20% and in rice only 5% of the available levels of iron in the plant may reach the grain. Thus, another alternative approach is to increase the amounts of other iron-binding compounds in the seeds. Ferritin is one such iron-binding protein that could provide a storage reserve of iron in plants, animals and bacteria (Theil, 1987). The expression of ferritin genes from soya bean (Goto et al., 1999) and *Phaseolus* (Lucca et al., 2001) in developing transgenic rice plants resulted in two- to threefold increase in the iron content of the grains. Qu et al. (2005) further demonstrated that tissue-specific expression of soya bean ferritin in transgenic rice plants resulted in increased iron accumulation only in the grains but not in rice leaves. Drakakaki et al. (2005) showed that gene-pyramiding of fungal phytase and soya bean ferritin genes in transgenic maize plants resulted in 20–70% increases in the iron content of maize seeds. These findings show the possibilities offered by genetic engineering of cereals to enhance iron content and benefit human health globally.

11.3.2.2 Zinc

Zinc is a necessary element for plant growth, and it is also an important trace element for physiology and development of the human body. Zinc deficiency in animals causes symptoms such as growth retardation, delayed sexual maturity and reduced activity of sex glands. It is also involved in other diseases including diarrhoea, pneumonia, malaria and some pathogen infections (Brown et al., 1998). An effective increase in the zinc content in cereals and grains is one way to improve human nutrition. Transgenic barley plants over-expressing genes encoding specific zinc transporter proteins have been developed by Ramesh et al. (2004). They reported that the transporter proteins were degraded when plant zinc reached adequate levels, suggesting that increasing production of a zinc transporter may be one approach to increasing zinc content in cereal grains. Tomato plants have also been transformed with metallothionein (*mt-I*) gene constructs encoding the mouse MT-I protein, resulting in increased levels of zinc in engineered plants. The average zinc content in transgenic tomato leaves was 32.7 mg/100 g FW, about 1.6 times higher than that in the wild type (Sheng et al., 2007). These findings provide useful starting points for increasing the availability of zinc in foods.

11.3.2.3 Selenium

Selenium is an essential nutritional element for animals and microorganisms and is thus also a factor in human health (Birringer et al., 2002). Selenium deficiency can lead to heart disease, hypothyroidism and a weakened immune system (Combs, 2000). During the last decade it has become increasingly evident that selenium may have other potential health benefits: specific organic forms of selenium have anti-carcinogenic activities against breast, liver, prostate and colorectal cancers in experimental model systems (Vadgama et al., 2000; Whanger et al., 2000; Dong et al., 2001). Plants genetically modified to absorb above-average quantities of selenium could be used to improve animal and human nutrition. Scientists have genetically manipulated *Arabidopsis thaliana* and *Brassica juncea* plants to better absorb selenium from soil and convert it to the non-toxic form methylselenocysteine (Ellis et al., 2004; LeDuc et al., 2004). These studies indicate that selenium nutrition could be improved by genetic engineering for protection against various cardiac and malignant diseases.

11.4 OTHER FUNCTIONAL FOOD PRODUCTS

11.4.1 Probiotics

Probiotics are live microbial dietary supplements that benefit human health through effects in the intestinal tract (Stanton et al., 2005). The most well-known probiotic bacteria are lactobacilli and bifidobacteria, often manufactured in fermented dairy products and available in forms such as freeze-dried cultures or yogurt. Probiotics have been shown to reduce milk intolerance symptoms by improving lactose digestion (Sanders 1993), to enhance immune system activity (Schiffrin et al., 1995), to confer a hypocholesterolemic effect (Jackson et al., 1999) and to decrease the level of certain fecal enzymes such as beta-glucuronidase, beta-glucosidase and urease, which are known to be involved in the metabolic activation of mutagens and carcinogens. Probiotics are also used to treat inflammatory bowel disease by generating a more host-friendly gut flora; the selective use of probiotic bacteria can also stimulate immune system functions and reduce intestinal inflammation (Guarner et al., 2002). Recently, extensive biotechnological research and development of probiotics has resulted in many new diary products (Table 11.1) (Szakàly, 2007). On the other hand, the sensitivity of probiotics to both physical and chemical stress (e.g., heat and acidity) makes development of their inclusion in other types of food products more challenging (Mattila-Sandholm et al., 2002). Recently, encapsulation technology was evaluated for its ability to countervent the acid sensitivity of probiotics (Clair, 2007). Further research in genetic engineering is required to strengthen the effect of specific or mutant bacterial strains or examine the utility of new microbial colonies.

11.4.2 Prebiotics

Prebiotics are generally referred to as a non-digestible food ingredient that can beneficially affect the host by selectively stimulating the growth and/or activity of one or a limited number of bacterial populations in the colon (Stanton et al., 2005). The only prebiotics currently being used as functional food ingredients are the inulin-type fructans, which include native inulin, enzymatically hydrolysed inulin or oligofructose and the synthetic fructo-oligosaccharides (Roberfroid et al., 1998). Only one inulin-containing plant species (chicory, *Cichorium intybus*) is used to produce inulin industrially and native inulin is also processed by the food industry to produce either short-chain fructans, specifically oligofructose as a result of partial enzymatic hydrolysis, or long-chain fructans by applying an industrial physical separation technique. Prebiotics are used for selective stimulation of the growth of the bifidobacterial population in the gastrointestinal tract. They also enhance mineral absorption and balance in the large bowel, particularly by raising the concentration of Ca^{2+} and Mg^{2+} ions, a physiological condition that favours passive diffusion, suppression of diarrhoea, constipation relief and the reduction of risk for osteoporosis, atherosclerotic cardiovascular disease, obesity and Type 2 diabetes (Roberfroid, 2000). Changes in the composition of specific colonic microbiota, modulation of insulinemia, monitoring of the metabolism of triacylglycerols and improvement of bioavailability of calcium are key areas for future research into the improvement of prebiotics to enhance human health.

11.4.3 Essential Fatty Acids

Fat is an energy-dense nutrient, and high fat intakes may lead to obesity, which is now known to be one of the main causes of various chronic diseases (Holmes and Willett, 2004). However, not all fats are equal, especially with regard to their effects on health.

In one approach that addresses the need for a 'healthy' fat, scientists have made modifications to soya bean oil. Suppressing the enzyme that converts oleic acid to linoleic acid allowed GM soya beans to accumulate the healthier oleic acid to up to 80% of the total oil content, whereas the oleic acid content of their non-GM parent plants was only 20% (Kinney, 1996). Oleic acid is quite stable at high temperatures and oil from these GM soya beans is used for industrial purposes.

A specific requirement in the human diet is the need for essential fatty acids including the n-6 linoleic acid and the n-3 α-linolenic acid. The n-6 PUFAs are found in a wide variety of foods, including fish oils from salmon, tuna and sardines. In terms of plant vegetables oils, oils from soya bean, cottonseed, safflower, corn, sunflower, walnut and flax are good sources of n-3 PUFAs. Scientists have attempted to increase the proportion of n-3 PUFAs in dietary lipids, as they are believed to be protective against some chronic diseases (Sayanova and Napier, 2004).

A major approach in plant biotechnology is to identify and employ the genes that encode the specific enzymes that are responsible for the biosynthesis of particular biochemicals, and then engineer them into target crop plants. A transgenic primrose with over-expression of a cDNA encoding a borage Δ^6-desaturase had enhanced γ-linolenic and octadecatetraenoic fatty acid content compared to non-transformed control plants (Mendoza et al., 2004). Other scientists have focused on attempting to increase the saturated fatty acid content. Soya beans have been genetically modified to produce oil enriched in stearic acid, a saturated fatty acid that is believed to not raise serum cholesterol levels (Hawkins and Kridl, 1998). They also transformed canola, another popular oil crop, with three thioesterase genes from mangosteen. Interestingly, these transgenic plants accumulated up to 22% stearate in the seed oil, which is more than 10 times higher than in non-transformed canola varieties (Hawkins and Kridl, 1998).

11.4.4 Green Tea

Green tea (extract of unfermented *Camellia sinensis* leaf) is arguably the most popular drink worldwide, particularly in Asia. The regular intake of green tea is now recognized to improve blood flow, increase resistance to disease, help excrete alcohol and toxins, eliminate or reduce fat and clarify urine and improve its flow (Nakachi et al., 2003). Drinking green tea as the main source of fluid in the diet can suppress colon carcinogenesis in the Apc (min) mouse model system (Orner et al., 2003). EGCG, the major polyphenol component, has been shown to have anti-tumor effects in various human and animal models, including cancers of breast, prostate, stomach, oesophagus, colon, pancreas, skin and lung (Beltz et al., 2006, Khan et al., 2006). Mechanistic studies of EGCG have shown that it can act via different signalling pathways to regulate cancer cell growth, survival, angiogenesis and metastasis (Chen et al., 2004; Khan et al., 2006). EGCG has a relatively low toxicity and is convenient to administer due to its oral bioavailability (Chow et al., 2005). EGCG has been tested in clinical trials (Ahn et al., 2003) and seems to be a potentially useful anti-tumor agent (Bettuzzi et al., 2006).

EGCG is also a potent antioxidant and can play a major role in scavenging reactive oxygen species, metal chelation, inhibition of nitric oxide (NO) synthase, cyclooxygenase-2 (COX-2) and activator protein 1 (AP-1), induction of superoxide dismutase and detoxification of glutathione *S*-transferase (Navarro-Peran et al., 2005). Green tea is the first tea product that was allowed to claim a health benefit through an effect on body fat reduction. Fujiki et al. (2002) developed green tea capsules for use as a cancer-prevention drug, claiming that healthy volunteers can take orally the equivalent of 30–40 cups of green tea per day without apparent adverse effects, although the clinical beneficial effects have not yet been proven. Along these lines, the US Food and Drug Administration has recently approved a new drug application for Veregen, a tea leaf extract containing phytocompound mixtures of sinecatechins, for topical treatment of perianal and genital condyloma (Chen et al., 2008).

11.5 OMICS APPROACHES FOR FUNCTIONAL FOODS

Recent advances in their development mean that the biotechnology platforms of genomics, proteomics and metabolomics have started playing a major role in phytomedicine research. The so-called '-omics' approach is part of a systems biology view that looks at effects on the global or overall systems level, for example, at the genome, cell and organism levels. Functional genomics has evolved into a science that deals with temporal and spatial differences in gene expression at the RNA level, for example, as transcripts of mRNA and microRNA, resulting from a specific biological activity, drug or physiological response. The proteomics approach explores the differential expression of all detectable proteins, their biochemical identities (e.g., specific enzymes) or functions, their organellar distribution, signalling pathway or network involvement and the interactions among different groups of proteins. Metabolomics is a systems study of the overall metabolite profiles or 'chemical fingerprint' of a given biological system. Unlike the transcriptome or the proteome, many technical limitations currently do not allow the simultaneous 'whole spectrum display' of the entire metabolome (Shyur and Yang, 2008). Hopefully, in the future we will be capable of investigating a much broader spectrum of the relatively low-molecular-weight metabolites, with the eventual goal of a complete snapshot of all molecules in a cell. The 'omics' technologies (genomics, proteomics and metabolomics) have much to offer human health by improving our understanding of the cellular and molecular events associated with the cancer-protective effects of various functional foods and by unravelling the pathophysiology of various nutritional disorders.

11.5.1 Genomics and Functional Food

Nutritional genomics (nutrigenomics) studies consider the influence of the genome, as a population or an individual, on nutritional needs or activities. Hopefully, such studies can provide tailor-made nutrition advice to populations or individuals with different genetic and social backgrounds. It is important to recognize that what is appropriate dietary advice for one individual may be inappropriate, or actually harmful, to another, and such considerations hopefully could also be usefully applied to functional foods. Kreps et al. (2002) have demonstrated that various plants are in a continuous state of change and adaptation. Hence, careful experimental plans are required to reduce spatial and temporal variations in gene-expression profiles related to developmental stage, environment stress and growth conditions. These findings imply that very little functional disturbance to the genomes of transgenic plants follows the insertion of simple T-DNA constructs. This seems to contrast with the fact that expression levels of large number of genes are changed in wild-type or transgenic plantlets in response to abiotic stresses (heat, cold, salt, drought, etc.). This in turn implies that the functional differences related to specific inserted DNA sequences that are functional to the genome originated at the site of T-DNA insertion can be revealed by functional genomics assay experiments (El Ouakfaoui and Miki, 2005).

DNA microarray technologies and functional gene-expression profiling are a valuable strategy for the evaluation of the safety or effects of functional foods or newly developed food substances. Among a limited number of such studies to date is one by Kamakura et al. (2005), which looked at the safety of deteriorated royal jelly. Narasaka et al. (2006) were able to distinguish gene-expression profiles in the liver and intestine between rats fed a diet based on enzyme-treated, hypoallergenic wheat flour and rats fed with normal flour. The application of these nutrigenomics approaches in the evaluation of food safety and effect will undoubtedly be greater in the future, although some issues such as standardization of methods have to be resolved first. We believe that it is essential that the various nutritional and health care professions realize the importance of the field of nutritional genomics in human health, and call attention to the need for appropriate study design and funding to lay the proper foundations upon which clinical and scientific applications can be based. Sound interaction between food and nutrition professionals and the food industry can help promote good research and development (R&D) in nutritional science, and justify health claims made in functional food production and development. Only then will the dietetics profession be reassured that they can effectively and safely recommend functional foods to their clients.

11.5.2 Proteomics and Functional Food

Proteomics involves the large-scale study of the entire or 'global' cellular array of proteins, particularly their biochemical identity, structure and function. The term 'proteome' was originally proposed by Wilkins et al. (1995). For transgenic effects in tissues or organisms, a major concern is the possibility of unintended effects caused, for example, by the site of transgene integration (e.g., interruption of important open reading frames or gene-expression-regulatory sequences), which could result in modified metabolism, novel fusion proteins or other pleiotropic effects that could compromise study aims and safety (Kuiper et al., 2001; Cellini et al., 2004). This may include the production of new allergens, toxins or other undesirable biologics. Arguably, unintended effects are less likely to be detected with the conventional targeted analyses of a relatively limited number of molecules (Millstone et al., 1999) compared to non-targeted methods such as RNA, protein and metabolite profiling. While transcriptomics may provide, at least for specific plant species, the most complete coverage or description of potential unintended effects (Meyers et al., 2004), whole genome arrays are presently not yet available for most of the important food crops, and such assays are so far not cost effective. Lehesranta and his colleagues described potato proteome diversity in a large selection of potato varieties and assessed the significant changes in the proteome of a transgenic potato (Lehesranta et al., 2005).

Proteomics has the potential to quantify the levels of a spectrum of allergens, most of which are proteins, and detect possible post-translational modifications. Several studies have demonstrated the capacity of two-dimensional electrophoresis (2DE) to characterize and distinguish varieties and genotypes and even to identify single-nucleotide polymorphism mutations with multiple effects (Canovas et al., 2004). There is still a significant technical hurdle in proteomics, where many proteins of low abundance are still not able to be readily displayed in such 2D assays. Generally speaking, however, proteomic profiling by 2DE is a promising tool for screening purposes and, although the number of proteins that can be analysed by 2DE is still limited with respect to the predicted number of proteins present in the entire proteome of plants or mammalian cells, it remains the most widely used tool for high-resolution protein separation and quantification.

As the expression levels shown by transcriptomes and proteomes often do not correlate (Gygi et al., 1999), reliance on any one profiling technology for safety assessment could also be questionable. We have recently characterized the changes in the proteome in human dendritic cells (DCs; part of the immune system) in response to treatment with an extract of *Echinacea purpurea*, a common food and health supplement in Europe, America and Asia (Wang et al., 2008). The proteins Mn-SOD and cofilin, which are involved in human DC maturation and tissue migration, were sensitive to this treatment. On the other hand, various cytokines and chemokines, which were seen in our transcriptome study of the same system to be significantly affected, did not show any major changes at the protein level in 2D gel profiles, probably due to the relatively low protein abundance of cytokines and chemokine gene products. Hence, cytokine/chemokine protein arrays or panel ELISAs are likely needed for further detailed studies. Like other profiling methods, proteomics screening is not yet in routine use when assessing the safety of GM plant products, but it does have the potential to reduce uncertainty by providing more global information on the protein profile of engineered crops than analysis of targeted proteins alone.

11.5.3 Metabolomics and Functional Food

The term 'metabolome' was first coined by Oliver et al. (1998) to describe the whole set of metabolites synthesized by a specific organism under study, in a fashion analogous to that employed for the genome and proteome. The more precise definition is 'the quantitative complement of all of the low molecular weight molecules present in cells in a particular physiological or developmental state' (Kell et al., 2005). 'It is estimated that the metabolome extends over 7–9 magnitudes of concentration (pmol-mmol)' (Dunn and Ellis, 2005), and the number of metabolites in a plant system alone is estimated to exceed 200,000 (Hartman et al., 2005). Although the metabolome can be defined conceptually, it is at present not technically possible to analyse by a single analytical method (e.g., via a single 2D exhibition mode) the entire metabolome of even a crude plant extract (Shyur and Yang, 2008).

Investigation of the functions of a specific gene can also be achieved through metabolite profiling, especially for specific genetically altered organisms. These metabolite profiles may then be compared to that of a control organism to yield information about the targeted effect or possible side effects and metabolic consequences of a genome alteration in engineered plants (Trethewey, 2001), and ultimately help to assign the function of targeted transgene(s). There is currently some debate over whether genetically modified (GM) plants might suffer unexpected, potentially undesirable changes in overall metabolite composition. Roessner et al. (2000) analysed tubers from several transgenic potato lines, modified in either sugar or starch metabolism. The authors deemed metabolite profiling to be an important tool for characterizing the metabolic status of a plant with respect to specific environmental, developmental or genetic factors. Catchpole et al. (2005) described a comprehensive comparison of the total metabolites in potato tubers from field-grown GM versus conventional plants, using a hierarchical analysis for rapid metabolome fingerprinting, aiming to guide a detailed profiling of plant metabolites where a significant difference might be suspected. They reported that the metabolite composition of the field-grown inulin-producing potatoes was within the natural metabolite range of classical cultivars and was, in fact, very similar to that of the progenitor line Désirée, apart from the impact of the introduced transgenes and the predicted up-regulation of fructans and their derivatives. These omics data may then be integrated with other omics data and modelled computationally for a more comprehensive understanding of the metabolism of the biology of a system.

With further development and application of validated metabolome, proteome and transcriptome approaches in plant biology, we may reasonably expect that such technologies will become fundamental contributors to studies on various biological systems, with particular benefits for research into nutrition and food safety.

11.6 BIOTECHNOLOGY AGAINST FOOD ALLERGIES

It is believed that 2–3% of the human population is affected by food allergies of one kind or another, particularly children less than 3 years of age. Some specific food proteins can trigger drastic immune responses and provoke a wide spectrum of allergic reactions. Allergenic foods may have effects in the gastrointestinal tract, skin and respiratory systems via inhalation or tissue contact with the allergens. Many proteins in foods associated with food allergies have been identified by food chemists or biotechnologists, but still many remain unknown. For example, more than 28 different soya bean proteins have been found to bind IgE, an antibody type involved in allergic responses, suggesting that these soya bean proteins may have allergenic properties (Wilson et al., 2005). Herman et al. (2003) successfully demonstrated that silencing the gene expression of P34 or knocking down P34 in soya beans in partially reduced soya bean allergy.

Rice is generally considered to be a hypoallergenic food; nonetheless, some people are allergic to certain rice proteins. The major dietary allergens in rice grain are 14–16 kD protein inhibitors of human α-amylase. Engineered expression of an antisense strand to the 16 kD protein gene resulted in an 80% reduction in the sense transcripts, equivalent to a decrease in the total allergen content from about 300 μg/seed to about 60–70 μg/seed (Tada et al., 1996). Antisense technology has also been used to down-regulate the production of Lol p 1, Lol p 2 and Lol p 3 allergens in pollen of ryegrass, which are the major source of grass pollen allergy in cool temperate climates (Petrovska et al., 2004). In apple, allergy is dominated by IgE antibodies against the protein Mal d 1, and scientists have successfully used a gene-silencing approach to suppress the expression of the gene encoding Mal d 1 in transgenic apple plants. Significant reduction of Mal d 1 expression was confirmed by immunoblotting and *in vivo* allergenicity assays (Gilissen et al., 2005). A number of allergy-causing proteins in peanut have been identified, such as Ara h 1, Ara h 2 and Ara h 3. Of these, the protein Ara h 2, a Kunitz trypsin inhibitor, is believed to be the most potent allergen. Although heat-inactivated recombinant allergens from bacteria have been used as a vaccine (Li et al., 2003; Koppelman et al., 2004), ample scope exists for the elimination of these allergens from peanuts themselves by genetic engineering or another biotechnology approach, and thus providing relief from one of the more serious food allergies. Beyond the practical genetic engineering problems in removing multiple proteins from rice, peanut, soya bean and other crop species, it is also necessary to consider the potential effects of engineering specific functional properties into the modified crop plants.

11.7 CONCLUSIONS

In both developed and developing countries, functional foods are the most important, promising and dynamically developing segment of the food industry. A continuous and focused research effort into the engineering and evaluation of various crop plants to improve their production and nutritional value with mutual benefits to the consumer and the environment is urgently needed. Incorporation of the latest technologies offered by the nutritional omics is an important approach for the improvement of food and nutrition, especially in less developed countries where crop yields are limited by specific biotic and abiotic stresses. We believe that it is vital for individuals from all relevant professions, especially the nutrition specialist, plant breeder, biotechnologist and system biologist, to come together to foster the relevant biotechnology research and development, and to integrate this wealth of new knowledge into the most appropriate functional food improvement programs.

REFERENCES

Agius, F., R. González-Lamothe, J. L. Caballero, J. Muñoz-Blanco, M. A. Botella and V. Valpuesta. 2003. Engineering increased vitamin C levels in plants by overexpression of a d-galacturonic acid reductase. *Nat Biotechnol* 21:177–181.

Ahn, W. S., J. Yoo, S. W. Huh et al. 2003. Protective effects of green tea extracts (polyphenon E and EGCG) on human cervical lesions. *Eur J Cancer Prev* 12:383–390.

Baroja-Fernández, E., F. J. Muñoz, M. Montero et al. 2009. Enhancing sucrose synthase activity in transgenic potato (*Solanum tuberosum* L.) tubers results in increased levels of starch, ADP-glucose and UDP-glucose and total yield. *Plant Cell Physiol* 50:1651–1662.

Beltz, L. A., D. K. Bayer, A. L. Moss and I. M. Simet. 2006. Mechanisms of cancer prevention by green and black tea polyphenols. *Anticancer Agents Med Chem* 6:389–406.

Bettuzzi, S., M. Brausi, F. Rizzi, G. Castagnetti, G. Peracchia and A. Corti. 2006. Chemoprevention of human prostate cancer by oral administration of green tea catechins in volunteers with high-grade prostate intra-epithelial neoplasia: A preliminary report from a one-year proof of principle study. *Cancer Res* 66:1234–1240.

Birringer, M., S. Pilawa and L. Flohe. 2002. Trends in selenium biochemistry. *Nat Prod Rep* 19:693–718.

Blechl, A., J. Lin, S. Nguyen et al. 2007. Transgenic wheats with elevated levels of Dx5 and/or Dy10 high-molecular weight glutenin subunits yield doughs with increased mixing strength and tolerance. *J Cereal Sci* 45:172–183.

Bregitzer, P., A. E. Blechl, D. Fiedler et al. 2006. Changes in high molecular weight glutenin subunit composition can be genetically engineered without affecting wheat agronomic performance. *Crop Sci* 46:1553–1563.

Brown, K. H., J. M. Peerson and L. H. Allen. 1998. Effect of zinc supplementation on children's growth: A meta-analysis of intervention trials. In *Role of Trace Elements for Health Promotion and Disease Prevention*, eds. B. Sandstrom and P. Walter, pp. 76–83. Davis, CA: University of California Press.

Canovas, F. M., E. Dumas-Gaudot, G. Recorbet, J. Jorrin, H. P. Mock and M. Rossignol. 2004. Plant proteome analysis. *Proteomics* 4:285–298.

Catchpole, G. S., M. Beckmann, D. P. Enot et al. 2005. Hierarchical metabolomics demonstrates substantial compositional similarity between genetically modified and conventional potato crops. *Proc Natl Acad Sci USA* 102:14458–14462.

Cellini, F., A. Chesson, I. Colquhoun et al. 2004. Unintended effects and their detection in genetically modified crops. *Food Chem Toxicol* 42:1089–1125.

Chakraborty, S., N. Chakraborty and A. Datta. 2000. Increased nutritive value of transgenic potato by expressing a nonallergenic seed albumin gene from *Amaranthus hypochondriacus. P Natl Acad Sci USA* 97:3724–3729.

Chen, D., K. G. Daniel, D. J. Kuhn et al. 2004. Green tea and tea polyphenols in cancer prevention. *Front Biosci* 9:2618–2631.

Chen, S. T., J. Dou, R. Temple, R. Agarwal, K. M. Wu and S. Walker. 2008. New therapies from old medicines. *Nat Biotechnol* 26:1077–1083.

Chen, Z., T. E. Young, J. Ling, S. C. Chang and D. R. Gallie. 2003. Increasing vitamin C content of plants through enhanced ascorbate recycling. *Proc Natl Acad Sci USA* 100:3525–3530.

Chow, H. H., I. A. Hakim, D. R. Vining et al. 2005. Effects of dosing condition on the oral bioavailability of green tea catechins after single-dose administration of polyphenon E in healthy individuals. *Clin Cancer Res* 11:4627–4633.

Clair, A. L. 2007. New challenging food applications of probiotics. Paper presented at *Proceedings of the Fourth International FFNet Meeting on Functional Foods*, Hungary.

Combs, G. F. 2000. Food system-based approaches to improving micronutrient nutrition: The case for selenium. *BioFactors* 12:39–43.

Datta, K., N. Baisakh, N. Oliva et al. 2003. Bioengineered 'golden' indica rice cultivars with beta-carotene metabolism in the endosperm with hygromycin and mannose selection systems. *Plant Biotechnol J* 1:81–90.

Dong, Y., D. Lisk, E. Block and C. Ip. 2001. Characterization of the biological activity of gamma-glutamyl-Se-methylselenocysteine: A novel, naturally occurring anticancer agent from garlic. *Cancer Res* 61:2923–2928.

Drakakaki, G., S. Marcel, R. P. Glahn et al. 2005. Endosperm-specific co-expression of recombinant soybean ferritin and *Aspergillus* phytase in maize results in significant increases in the levels of bioavailable iron. *Plant Mol Biol* 59:869–880.

Dunn, W. B. and D. I. Ellis. 2005. Metabolomics: Current analytical platforms and methodologies. *Trend Anal Chem* 24:285–293.

El Ouakfaoui, S. and B. Miki. 2005. The stability of the Arabidopsis transcriptome in transgenic plants expressing the marker genes *nptII* and *uidA*. *Plant J* 41:791–800.

Ellis, D. R., T. G. Sors, D. G. Brunk et al. 2004. Production of Se-methylselenocysteine in transgenic plants expressing selenocysteine methyltransferase. *BMC Plant Biol* 4:1–11. http://www.biomedcentral.com/content/pdf/1471-2229-4-1.pdf.

Estevez, J. M., A. Cantero, A. Reindl, S. Reichler and P. Leon. 2001. 1-Deoxy-d-xylulose-5-phosphate synthase, a limiting enzyme for plastidic isoprenoid biosynthesis in plants. *J Biol Chem* 276: 22901–22909.

Fujiki, H., M. Suganuma, K. Imai and K. Nakachi. 2002. Green tea: Cancer preventive beverage and/or drug. *Cancer Lett* 188:9–13.

Gale, C. R., N. F. Hall, D. I. Phillips and C. N. Martyn. 2003. Lutein and zeaxanthin status and risk of age related macular degeneration. *Invest Ophthalmol Vis Sci* 44:2461–2465.

Ghosh, D., M. Skinner and L. R. Ferguson. 2006. The role of the Therapeutic Goods Administration and the Medicine and Medical Devices Safety Authority in evaluating complementary and alternative medicines in Australia and New Zealand. *Toxicology* 221:88–94.

Gilissen, L., S. T. H. Bolhaar, C. I. Matos et al. 2005. Silencing the major apple allergen Mal d 1 by using the RNA interference approach. *J Allergy Clin Immun* 115:364–369.

Glover, J., O. Mewett, D. Cunningham and K. Ritman. 2005. Genetically modified crops in Australia, the next generation. Summary report from Department of Agriculture, Fishery and Forestry, Australia.

Goto, F., T. Yoshihara, N. Shigemoto, S. Toki and F. Takaiwa. 1999. Iron fortification of rice seed by the soybean ferritin gene. *Nat Biotechnol* 17:282–286.

Guarner, F., F. Casellas, N. Borruel et al. 2002. Role of microecology in chronic inflammatory bowel diseases. *Eur J Clin Nutr* 56 (Suppl. 4):S34–S38.

Gygi, S. P., Y. Rochon, B. R. Franza and R. Aebersold. 1999. Correlation between protein and mRNA abundance in yeast. *Mol Cell Biol* 19:1720–1730.

Hartman, T., T. M. Kutchan and D. Strack. 2005. Evolution of metabolic diversity. *Phytochemistry* 66:1198–1199.

Hawkins, D. J. and J. C. Kridl. 1998. Characterization of acyl-ACP thioesterases of mangosteen (*Garcinia mangostana*) seed and high levels of stearate production in transgenic canola. *Plant J* 13:743–752.

Herman, E. M., R. M. Helm, R. Jung and A. J. Kinney. 2003. Genetic modification removes an immunodominant allergen from soybean. *Plant Physiol* 132:36–43.

Holmes, M. D. and W. C. Willett. 2004. Does diet affect breast cancer risk? *Breast Cancer Res* 6:170–178.

Houmard, N. M., J. L. Mainville, C. P. Bonin, S. Huang, M. H. Luethy and T. M. Malvar. 2007. High-lysine corn generated by endosperm-specific suppression of lysine catabolism using RNAi. *Plant Biotechnol J* 5:605–614.

Huang, S., A. Frizzi, C. A. Florida, D. E. Kruger and M. H. Luethy. 2006. High lysine and high tryptophan transgenic maize resulting from the reduction of both 19- and 22-kDa-zeins. *Plant Mol Biol* 61:525–535.

Huang, S., D. E. Kruger, A. Frizzi et al. 2005. High-lysine corn produced by the combination of enhanced lysine biosynthesis and reduced zein accumulation. *Plant Biotechnol J* 3:555–569.

Jackson, T. G., G. R. J. Taylor, A. M. Clohessy and C. M. Williams. 1999. The effects of the daily intake of inulin on fasting lipid, insulin and glucose concentrations in middle-aged men and women. *Br J Nutr* 89:23–30.

Jain, A. K. and C. L. Nessler. 2000. Metabolic engineering of an alternative pathway for ascorbic acid biosynthesis in plants. *Mol Breed* 6:73–78.

Kamakura, M., M. Maebuchi, S. Ozasa, et al. 2005. Influence of royal jelly on mouse hepatic gene expression and safety assessment with a DNA microarray. *J Nutr Sci Vitaminol* (Tokyo) 51:148–155.

Kell, B. D., M. Brown, H. M. Davey, W. B. Dunn, I. Spasic, and S. G. Oliver. 2005. Metabolic footprinting and systems biology: The medium is the message. *Nature Rev Microbiol* 3:557–565.

Khan, N., F. Afaq, M. Saleem, N. Ahmad and H. Mukhtar. 2006. Targeting multiple signaling pathways by green tea polyphenol (-)-epigallocatechin-3-gallate. *Cancer Res* 66:2500–2505.

Kinney, A. J. 1996. Development of genetically engineered soybean oils for food applications. *J Food Lipids* 3:273–292.

Klaus, D., J. B. Ohlrogge, H. E. Neuhaus and P. Dormann. 2004. Increased fatty acid production in potato by engineering of acetyl-CoA carboxylase. *Planta* 219:389–396.

Koppelman, S. J., M. Wensing, M. Ertmann, A. C. Knulst and E. F. Knol. 2004. Relevance of Ara h1, Ara h2 and Ara h3 in peanut-allergic patients, as determined by immunoglobulin E Western blotting, basophil-histamine release and intracutaneous testing: Ara h2 is the most important peanut allergen. *Clin Exp Allergy* 34:583–590.

Kreps, J. A., Y. Wu, H. S. Chang, T. Zhu, X. Wang and J. F. Harper. 2002. Transcriptome changes for Arabidopsis in response to salt, osmotic and cold stress. *Plant Physiol* 130:2129–2141.

Kuiper, H. A., G. A. Kleter, H. P. J. M. Noteborn and E. J. Kok. 2001. Assessment of the food safety issues related to genetically modified foods. *Plant J* 27:503–528.

LeDuc, D. L., A. S. Tarun, M. Montes-Bayon et al. 2004. Overexpression of selenocysteine methyltransferase in Arabidopsis and Indian mustard increases selenium tolerance and accumulation. *Plant Physiol* 135:377–383.

Lehesranta, S. J., H. V. Davies, L. V. T. Shepherd et al. 2005. Comparison of tuber proteomes of potato varieties, landraces and genetically modified lines. *Plant Physiol* 138:1690–1699.

Li, X. M., K. Srisvastava, A. Grishin et al. 2003. Persistent protective effect of heat killed *Escherichia coli* producing 'engineered', recombinant peanut proteins in a murine model of peanut allergy. *J Allergy Clin Immun* 112:159–167.

Lucca, P., R. Hurrell and I. Potrykus. 2001. Genetic engineering approaches to improve the bioavailability of the level of iron in rice grains. *Theor Appl Genet* 102:392–397.

Mann, V., M. Harker, I. Pecker and J. Hirschberg. 2000. Metabolic engineering of astaxanthin production in tobacco flowers. *Nat Biotechnol* 18:888–892.

Martin, J. M., F. D. Meyer, E. D. Smidansky et al. 2006. Complementation of the *pina* (null) allele with the wild type *Pina* sequence restores a soft phenotype in transgenic wheat. *Theor Appl Genet* 113:1563–1570.

Mattila-Sandholm, T., P. Myllärinen, R. Crittenden, G. Mogensen, R. Fondén and M. Saarela. 2002. Technological challenges for future probiotic foods. *Int Dairy J* 12:173–182.

Mendoza de Gyves, E., C. Sparks, O. Sayanova, P. Lazzeri, J. A. Napier and H. D. Jones. 2004. Genetic manipulation of gamma-linolenic acid (GLA) synthesis in commercial varieties of evening primrose (*Oenothera* spp.). *Plant Biotechnol J* 2:351–357.

Menrad, K. 2003. Market and marketing of functional food in Europe. *J Food Eng* 56:181–188.

Meyer, F. D., L. E. Talbert, J. M. Martin, S. P. Lanning, T. W. Greene and M. J. Giroux. 2007. Field evaluation of transgenic wheat expressing a modified ADP-glucose pyrophosphorylase large subunit. *Crop Sci* 47:336–342.

Meyers, B. C., D. W. Galbraith, T. Nelson and V. Agrawal. 2004. Methods for transcriptional profiling in plants. Be fruitful and replicate. *Plant Physiol* 135:637–652.

Millstone, E., E. Brunner and S. Mayer. 1999. Beyond 'substantial equivalence'. *Nature* 401:525–526.

Nakachi, K., H. Eguchi and K. Imai. 2003. Can teatime increase one's lifetime? *Ageing Res Rev* 2:1–10.

Naqvi, S., C. Zhua, G. Farrea et al. 2009. Transgenic multivitamin corn through biofortification of endosperm with three vitamins representing three distinct metabolic pathways. *Proc Natl Acad Sci USA* 106:7762–7767.

Narasaka, S., Y. Endo, Z. W. Fu et al. 2006. Safety evaluation of hypoallergenic wheat flour using a DNA microarray. *Biosci Biotechnol Biochem* 70:1464–1470.

Navarro-Peran, E., J. Cabezas-Herrera, F. Garcia-Canovas, M. C. Durrant, R. N. Thorneley and J. N. Rodriguez-Lopez. 2005. The antifolate activity of tea catechins. *Cancer Res* 65:2059–2064.

Oliver, S. G., M. K. Winson, D. B. Kell and F. Banganz. 1998. Systematic functional analysis of the yeast genome. *Trends Biotechnol* 16:373–378.

Orner, G. A., W. M. Dashwood, C. A. Blum, G. D. Diaz, Q. Li and R. H. Dashwood. 2003. Suppression of tumorigenesis in the Apc (min) mouse: Down-regulation of beta-catenin signaling by a combination of tea plus sulindac. *Carcinogenesis* 24:263–267.

Paine, J. A., C. A. Shipton, S. Chagger et al. 2005. Improving the nutritional value of golden rice through increased pro-vitamin A content. *Nat Biotechnol* 23:482–487.

Petrovska, N., X. Wu, R. Donato et al. 2004. Transgenic ryegrasses (*Lolium* spp.) with down-regulation of main pollen allergens. *Mol Breeding* 14:489–501.

Qu, L. Q., T. Yoshihara, A. Ooyama, F. Goto and F. Takaiwa. 2005. Iron accumulation does not parallel the high expression level of ferritin in transgenic rice seeds. *Planta* 222:225–233.

Ramesh, S. A., S. Choimes and D. P. Schachtman. 2004. Over-expression of an Arabidopsis zinc transporter in *Hordeum vulgare* increases short-term zinc uptake after zinc deprivation and seed zinc content. *Plant Mol Biol* 54:373–385.

Roberfroid, M. B. 2000. Concepts and strategy of functional food science: The European perspective. *Am J Clin Nutr* 71(Suppl.):1660S–1664S.

Roberfroid, M. B., J. A. E. Van Loo and G. R. Gibson. 1998. The bifidogenic nature of chicory inulin and its hydrolysis products. *J Nutr* 128:11–29.

Roessner, U., C. Wagner, J. Kopka, R. N. Trethewey and L. Willmitzer. 2000. Simultaneous analysis of metabolites in potato tuber by gas chromatography–mass spectrometry. *Plant J* 23:131–142.

Romer, S., J. Lubeck, F. Kauder, S. Steiger, C. Adomat and G. Sandmann. 2002. Genetic engineering of a zeaxanthin-rich potato by antisense inactivation and co-suppression of carotenoid epoxidation. *Metab Eng* 4:263–272.

Rosati, C., R. Aquilani, S. Dharmapuri et al. 2000. Metabolic engineering of beta-carotene and lycopene content in tomato fruit. *Plant J* 24:413–419.

Sanders, M. E. 1993. Summary of the conclusions from a consensus panel of experts on health attributes on lactic cultures: Significance to fluid milk products containing cultures. *J Diary Sci* 76:1819–1828.

Savidge, B., J. D. Weiss, Y. H. H. Wong et al. 2002. Isolation and characterization of homogentisate prenyltransferase genes from *Synechocystis* sp. PCC 6803 and *Arabidopsis*. *Plant Physiol* 129:321–332.

Sayanova, O. V. and J. A. Napier. 2004. Eicosapentaenoic acid: Biosynthetic routes and the potential for synthesis in transgenic plants. *Phytochemistry* 65:147–158.

Schiffrin, E., F. Rochat, Link-Amster H, M. Aeschlimann and A. Donnet-Hughes. 1995. Immunomodulation of blood cells following the ingestion of lactic acid bacteria. *J Dairy Sci* 78:491–497.

Sheng, J., K. Liu, B. Fan, Y. Yuan, L. Shen and B. Ru. 2007. Improving zinc content and antioxidant activity in transgenic tomato plants with expression of mouse metallothionein-I by *mt-I* gene. *J Agric Food Chem* 55:9846–9849.

Shyur, L. F. and N. S. Yang. 2008. Metabolomics for phytomedicine research and drug development. *Curr Opin Chem Biol* 12:66–71.

Siritunga, D., D. Arias-Garzon, W. White and R. T. Sayre. 2004. Over-expression of hydroxynitrile lyase in transgenic cassava roots accelerates cyanogenesis and food detoxification. *Plant Biotech J* 2:37–43.

Stanton, C., R. P. Ross, G. F. Fitzgerald and D. Van Sinderen. 2005. Fermented functional foods based on probiotics and their biogenic metabolites. *Curr Opin Biotech* 16:198–203.

Szakàly, S. 2007. Development and distribution of functional dairy products in Hungary. Paper presented at *Proceedings of the Fourth International FFNet Meeting on Functional Foods*, Hungary.

Tada, Y., M. Nakase, T. Adachi et al. 1996. Reduction of 14–16-kDa allergenic proteins in transgenic rice plants by antisense gene. *FEBS Lett* 391:341–345.

Theil, E. C. 1987. Ferritin: Structure, gene regulation and cellular function in animals, plants, and microorganisms. *Annu Rev Biochem* 56:289–315.

Trethewey, R. N. 2001. Gene discovery via metabolic profiling. *Curr Opin Biotech* 12:135–138.

Vadgama, J. V., Y. Wu, D. Shen, S. Hsia and J. Block. 2000. Effect of selenium in combination with adriamycin or taxol on several different cancer cells. *Anticancer Res* 20:1391–1414.

Van Eenennaam, A. L., K. Lincoln, T. P. Durrett et al. 2003. Engineering vitamin E content: From *Arabidopsis* mutant to soy oil. *Plant Cell* 15:3007–3019.

Wang, C. Y., V. Staniforth, M. T. Chiao et al. 2008. Genomics and proteomics of immune modulatory effects of *Echinacea purpurea* phytocompounds in human dendritic cells. *BMC Genomics* 9:479–499.

Whanger, P. D., C. Ip, C. E. Polan, P. C. Uden and G. Welbaum. 2000. Tumorigenesis, metabolism, speciation, bioavailability, and tissue deposition of selenium in selenium-enriched ramps (*Allium tricoccum*). *J Agric Food Chem* 48:5723–5730.

Wilkins, M. R., J. C. Sanchez, Gooley, A. A. et al. 1995. Progress with proteome projects: Why all proteins expressed by a genome should be identified and how to do it. *Biotechnol Genet Eng Rev* 13:19–50.

Wilson, S., K. Blaschek and E. Gonzalez de Mejia. 2005. Allergenic proteins in soybean: Processing and reduction of P34 allergenicity. *Nutr Rev* 63:47–58.

Wu, X. R., Z. H. Chen and W. R. Folk. 2003. Enrichment of cereal protein lysine content by altered tRNAlys coding during protein synthesis. *Plant Biotech J* 1:187–194.

Ye, X., S. Al-Babili, A. Kloti et al. 2000. Engineering the provitamin A (beta-carotene) biosynthetic pathway into (carotenoid-free) rice endosperm. *Science* 287:303–305.

Zhang, P., J. M. Jaynes, I. Potrykus, W. Gruissem and J. Puonti-Kaerlas. 2003. Transfer and expression of an artificial storage protein (ASP1) gene in cassava (*Manihot esculenta* Crantz). *Transgenic Res* 12:243–250.

CHAPTER 12

Nutritional Genomics and Sustainable Agriculture

María Luisa Guillén, Mercedes Sotos-Prieto and Dolores Corella

CONTENTS

12.1 INTRODUCTION

The model of health fields proposed by Laframboise in 1973 and developed in Lalonde Report in 1974 (Lalonde, 1974) has revolutionized contemporary public health due to the approaching and explaining the health levels in the populations and therefore, the way of formulating health policies. Marc Lalonde, proposed a new 'health field' concept: 'the health field can be broken up into four broad elements: Human biology, Environment, Lifestyle, and Health Care Organization' (p. 31).

Of the four determinants of health, the one which is most susceptible to be modified is lifestyle. Apart from not smoking, the most important lifestyle determinants of good health are what we eat and how active we are. A healthy diet is one that helps to maintain or improve health. It involves consuming appropriate amounts of all nutrients, and an adequate amount of water. Nutrients can be obtained from many different foods; there are a wide variety of diets that may be considered as healthy diets. However, the diet over 800 million people does not contain sufficient macronutrients, and micronutrient deficiencies are even more prevalent. On the other hand, obesity is increasing in adults and children, and authorities view it as one of the most serious public health problems of the twenty-first century. In response to the emerging problem of nutrition-related diseases, a Joint WHO/FAO Expert Consultation on Diet, Nutrition and the Prevention of Chronic Diseases was held in 2002. In May 2004, the World Health Organization (WHO) Member States adopted the Global Strategy on Diet, Physical Activity and Health at the 57th World Health Assembly, where it was stated that 'National food and agricultural policies should be consistent with the protection and promotion of public health' (p. 8, Statement 41). In its 18th Session, the FAO Committee on Agriculture in 2004 proposed to follow up on the Technical Report Series No. 916 (WHO, 2003) and requested a thorough assessment of the links between changing food consumption patterns and non-communicable diseases, and the possible effects of and changing demand on agricultural production systems and commodity trade, as well as on supply responses through diversification, taking into account the specificities of individual countries, population groups and dietary patterns.

As human genome was sequenced and genome technology accelerated the tools to study it, nutrition science introduced the new term 'nutritional genomics' or 'nutrigenomics'. One of the first references to this term in the scientific literature is that of DellaPenna (DellaPenna, 1999) who defines nutritional genomics as the general approach to gene discovery that is currently most applicable to compounds of nutritional importance that are synthesized or accumulated by plants and other organisms. This definition, which describes research at the interface of plant biochemistry, genomics and human nutrition, has the specific goal of dissecting and manipulating micronutrient pathways in plants to improve crop nutritional quality for human health. This definition of nutritional genomics is restrictive and has been extended to incorporate new dimensions of the field. Developing new foods is one of the many applications of nutritional genomics within the umbrella of its overall goal, which studies the genome-wide influences of nutrition. Currently, nutritional genomics studies the functional interaction of food and its components with the genome at the molecular, cellular and systemic level; the goal is to use diet to prevent or treat disease (Ordovas and Corella, 2004).

In this chapter, we examine the relevance of nutritional genomics to the food industry from 'farm to fork'. First, we focus on the use of genomic technologies within the agriculture sector by plant biotechnologists to manipulate plant biosynthetic pathways to improve the nutritive value of crops. Second, we review the results of a specific diet (e.g., enrichment with nutraceutical products) on the characteristics of animal food products; in addition and related to animal produce, we also considered the efforts that led to understand the genetic basis of desirable characteristics in domesticated animal species and also the way in which these genetic factors may interact with the diet of the animals. Finally, we focus on the improvement of food processing, food safety and quality assurance as well as the emerging of functional foods and the concept of 'personalized diet'.

12.2 NUTRITIONAL GENOMICS AND CROPS

A human well-balanced diet must contain a complex mixture of both macronutrients and micronutrients. Macronutrients—carbohydrates, lipids and proteins—are used primarily as an energy supply. Micronutrients are not used for energy but they are needed to avoid nutritional disorders (essential micronutrients) or to promote good health (non-essential micronutrients).

The nutritional health of humans is directly or indirectly dependent on plant foods. They offer nutrients that are essential for human nutrition and promote good health. However, micronutrient concentrations are often low in staple crops. The diet of large numbers of people in developing countries is composed primarily of wheat, rice and corn that are poor sources of some macronutrients and many essential micronutrients. It is estimated that 250 million children are at risk for vitamin A deficiency (WHO, 2009); 2 billion people—over 30% of the world's population—are anaemic, many due to iron deficiency, and in resource-poor areas, this is frequently exacerbated by infectious diseases (WHO, 2008); and 54 countries are iodine deficient (WHO, 2007). But even in industrialized nations, the risk of micronutrient deficiencies is also excessive due to poor eating habits.

Plants also can synthesize secondary metabolites as carotenoids, glucosinolates or phytoestrogens which have health-promoting properties ('nutraceutical metabolites'). These phytochemicals are often unique to certain species or genera; lycopene in tomato, resveratrol in red wine and red grape. Diets based on these plants are associated with lower morbidity and mortality in adult life (Bertelli and Das, 2009; Karppi et al., 2009) but, because dietary preferences, the health benefits associated are not fully realized in industrialized nations. Modification of dietary preferences and/or modification of micronutrition levels in food crops will be required to obtain therapeutical levels of these phytochemicals.

The advances in the 'omics' technologies have resulted in the drive to develop plant cultivars with proposed metabolites. Genetic modification (GM) technologies are used to modify biosynthesis pathways in order to achieve GM crops lines with increased amounts of essential vitamins or nutraceuticals. This part of the chapter is focused principally on recent advances on metabolic engineering of plants for improved human health characteristics. This is both for quantity and quality of proteins, vitamins, nutraceuticals or allergens.

12.2.1 General Considerations and Concerns

The identification and isolation of the gene(s) required for the synthesis, accumulation or elimination of a target compound is the first step in the process of manipulation but careful considerations must be taken. First of all, we must have a sufficient knowledge of the target system, and how is the production, accumulation or elimination of the products of interest. We must also know the flux control points or regulatory proteins as well as the interconnections with other metabolic pathways. At present, there is an incomplete knowledge of most pathways (Davies, 2007).

Second, the DNA sequences coding of the protein of interest must be available; furthermore, we must focus on genetics variations that could impact on the efficacy of the manipulation taking into account the variability of the results we can have depending on the species. For example, a copy of the *Arabidopsis* cystathionine γ-synthase gene was introduced into alfalfa in order to obtain expression of this protein that controls the synthesis of methionine (Avraham et al., 2005). The levels of free cysteine, methionine and GSH were also increased two- to fivefold. When the native potato analogue of this gene was overexpressed in potato was unsuccessful (Kreft et al., 2005). Metabolic engineering is not always straightforward and overproduction of one or another product can result in the activation or degradation pathways.

Finally, appropriate gene transfer methods must be used. Two basic approaches may be identified: manipulation of pathway flux or introduction of novel biosynthetic activities from other organisms. These approaches can be applied successfully to modify plant metabolite production. We must select the appropriate one based on our objectives. The uses of these techniques will be described further on.

12.2.2 Increased Production of Macronutrients

Amino acids are the building blocks of proteins found in all life forms. The fact is that all animals lack the enzymatic machinery to synthesize *de novo* the 'essential amino acids' (EAAs) and they have to be supplied through diet. EAAs determine the nutritive value of the protein. A diet deficient in protein quality, although it was sufficient in protein quantity, causes deleterious effects such as poor growth, tissue wasting and indeed death. Many advances in the GM of crops in order to enhance their nutritional value are emerging (Flachowsky et al., 2005). Lysine is an EAA synthesized from aspartate by a metabolic pathway regulated by several end-products. Lysine can block the aspartate kinase (AK), the first key enzyme in the pathway common to all the aspartate-family amino acids, and also the activity of dihydrodipicolinate synthase (DHDPS). This one is even more sensitive to lysine than AK. Sévenier et al. (2002) introduced the genes encoding AK and DHDPS enzymes from bacterial origin that are 100-fold less sensitive to feedback inhibition by lysine in potato. Expression of a feedback-insensitive DHDPS enzyme resulted in a sixfold increase in lysine and the introduction of a feedback-insensitive AK enzyme in potato resulted in an eightfold increase in threonine and a twofold increase in methionine. Modify the potato gene encoding DHPS to render the enzyme feedback-insensitive more effective than to introduce bacterial genes. In this way, the lysine level reached up to 15% of the total amino acid, whereas in untransformed plants, this level is only 1%.

Nowadays, however, no crops with tailored EAA content have made it to the market, and a number of hurdles remain. DNA microarray analysis also will be used together with high-throughput protein and metabolite analysis in a risk-assessment study to determine the extent to which the transgenic plants are altered by the technique. A better understanding of protein transit, association and deposition in the many cellular compartments of plants is needed to achieve the new generation of crops with tailored EAA profiles (Beauregard and Hefford, 2006). These new crops will lead to developments in the food industry, for example, nutrigenomics application demands the elaboration of functional foods tailored to meet the needs of individuals with specific genetic traits.

Another example of increasing macronutrients in plants is the fructan production in engineered sugar beet; fructans are polymers of fructose which have industrial application. One of these fructans, inulin, in which the fructose units are joined by a $\beta(2 \rightarrow 1)$, is increasingly used in processed foods because it has unusual adaptable characteristics. Its flavour ranges from bland to subtly sweet (~10% sweetness of sugar/sucrose) It can be used to replace sugar, fat and flour. This is particularly advantageous because inulin contains a quarter to a third of the food energy of sugar or other carbohydrates and a ninth to a sixth of the food energy of fat (Roberfroid, 2005). While inulin is a versatile ingredient, it also has health benefits. Inulin increases calcium absorption (Abrams et al., 2005) while promoting the growth of intestinal bacteria. Nutritionally, it is considered as a form of soluble fibre and is sometimes categorized as a prebiotic. Inulin has a minimal effect on blood sugar, and, unlike fructose, is not insulemic and does not raise triglycerides (TGs), making it generally considered suitable for diabetics and potentially helpful in managing blood sugar-related illnesses. The consumption of large quantities (particularly by sensitive or unaccustomed individuals) can lead to gas and bloating, and products that contain inulin will sometimes include a warning to add it gradually to one's diet. Sugar beet is one of the most efficient crops; it could be engineered to produce tons of fructan that makes it 'Plant as a Factory' (Sévenier et al., 1998, p. 844). Two steps are needed to achieve this transformation: the genes responsible for fructan synthesis had to be cloned, and an efficient protocol for sugar beet transformation had to be developed. cDNAs encoding enzymes necessary to catalyse the synthesis of long-chain inulin molecules have been cloned (Van der Meer et al., 1998). For the second prerequisite, Weyens et al. (2004) introduced specific onion fructosyltransferases into sugar beet. This resulted in an efficient conversion of sucrose into complex, onion-type fructans, without the loss of storage carbohydrate content. In this way, it is possible to produce *in planta* fructan types that require minimal technological process. Now nutrigenomics would define which types of fructan (depending on chain length, profile or type of linkage) are the most useful for every person.

For fatty acids, the focus has been on the production of long-chain polyunsaturated fatty acids (LC-PUFAs), such as arachidonic acid (ARA; 20:4), eicosapentaenoic acid (EPA; 20:5) and docosahexaenoic acid (DHA; 22:6) which have health-promoting effects but are only directly available from non-plant sources. Some transgenic soya bean that accumulated non-host PUFAs has been generated (Damude and Kinney, 2007). Although most studies have focussed on these predominant omega-3 fatty acids found in fish oils, recent evidence suggests similar health benefits from their common precursor, stearidonic acid (SDA). SDA is a Delta6-unsaturated C18 omega-3 fatty acid present in a few plant species (mainly Boraginaceae and Primulaceae) reflecting the general absence of Delta6-desaturation from higher plants. Using a Delta6-desaturase from *Primula vialii*, Ruiz-López et al. (2009) generated transgenic *Arabidopsis* and linseed lines accumulating SDA in their seed lipids. The achieved levels of SDA (13.4% of triacylglycerols) are very similar to those found in the sole natural commercial plant source (*Echium* spp.) or transgenic soya bean oil (SBO). However, both the latter oils contain gamma-linolenic acid, which is not normally present in fish oils and considered undesirable for heart-healthy applications. In contrast, the SDA-enriched linseed oil is essentially devoid of this fatty acid. Moreover, the overall n-3/n-6 ratio for this modified linseed oil is also significantly higher. But, this health-promoter-modified product is not beneficial to everybody. In this way, nutrigenomics has pointed out the need of personalized diets related to the quantity and quality of PUFAs. Tai et al. (2005) examined gene–nutrient interaction in relation to plasma lipid variables in a population-based study consisting of 1003 men and 1103 women participating in the Framingham cohort and consuming their habitual diets. They found significant gene–nutrient interactions between the L162V polymorphism in peroxisome proliferator-activated receptor alpha (PPAR alpha), a nuclear transcription factor regulating multiple genes involved in lipid metabolism, and total PUFA intake. The 162V allele was associated with greater TGs and apoC-III concentrations only in subjects consuming a low-PUFA diet (below the population mean, 6% of energy). However, when PUFA intake was high, carriers of the 162V allele had lower apoC-III concentrations. This interaction was significant even when PUFA intake was considered as a continuous variable suggesting a strong dose–response effect. When PUFA intake was <4%, 162V allele carriers had ~28% higher plasma TG than did 162L homozygotes. Conversely, when PUFA intake was >8%, plasma TG in 162V allele carriers was 4% lower than in 162L homozygotes. Similar results were obtained for (n-6) and (n-3) fatty acids. These data show that the effect of L162V polymorphism on plasma TG and apoC-III concentrations depends on the dietary PUFA, with a high intake triggering lower TG in carriers of the 162V allele. Using information given by nutrigenomics, as in the example mentioned above, nutricionists will be able to define more precisely which types of PUFAs (chain length, relation n-6/n-3, etc.) are the most useful to every person, and thus to the food industry that produced this specific PUFAs in transgenic plants

12.2.3 Increasing Production of Vitamins

One of the most popular examples of a cereal lacking an essential vitamin is the 'Golden Rice'. Vitamin A deficiency affects populations in at least 26 countries. The carotenoid pigments called provitamin A in plants are converted into vitamin A (retinol) in animals. There is no rice germplasm capable of synthesizing carotenoids in the endosperm; so GM is the only option for producing high-carotenoid rice. Retinol supplementation may cause side-effects at amounts greater than fivefold RDA; for this reason the engineering of provitamin A is preferable, which are safe by at least up to 100-fold RDA. Today, the current focus is to further increase the nutritional value of Golden Rice combining with additional nutritional traits, such as the accumulation of vitamin E, iron and zinc, and to increase the deposition of high-quality proteins or essential amino acids in the grains (Al-Babili and Beyer, 2005). Although a genetically modified organism (GMO) approach is straightforward in theory, past experience has shown that there is often a trade-off in cereals; for example, enhanced protein levels can lead to diminished yields or to alterations in other nutritional or grain

Table 12.1 Examples of Crops with Vitamins and Carotenoids Improvements to Provide Health Benefits

Crop	Vitamin	Difference from No Transgenic Crop	Reference
Lettuce	Folate	5.4-fold more	Nunes et al. (2009)
Tomato	Folate	25-fold more	Díaz de la Garza et al. (2007)
Tomato	B-carotene	10-fold more	Davuluri et al. (2005)
Carrot	Ketocarotenoids	70% of total carotenoids converted	Jayaraj et al. (2008)
Maize	B-carotene	169-fold more	Naqvi et al. (2009)
Maize	Ascorbate	6-fold more	Naqvi et al. (2009)
Maize	Folate	2-fold more	Naqvi et al. (2009)
Mustard greens	Vitamin E	6-fold more	Yusuf and Sarin (2007)
Tobacco	Vitamin C	2-fold more	Badejo et al. (2009)
Nabicol	Vitamin B5	1.5–2.5-fold more	Chakauya et al. (2008)

quality components (Gibbon and Larkins, 2005). Table 12.1 shows some of the last engineered crops in which levels of different vitamins have been improved.

12.2.4 Increase Nutraceutical Compounds

Flavonoids comprise a large and diverse group of polyphenolic compounds that are involved in plant growth and development. They are thought to be beneficial to human health due to their antioxidant properties (Middleton et al., 2000). Most of the genes encoding enzymes or transcriptional activators involved in the biosynthesis of different flavonoids have been encoding, given the tools to engineer the pathway towards new flavonoid species in crop plants.

We illustrate the use of 'omics' tools to increase the amount of nutraceuticals with some recent researches. AtMYB12 was originally identified as a flavonol-specific transcriptional activator in *Arabidopsis thaliana*, when Luo et al. (2008) expressed it in a tissue-specific manner in tomato; AtMYB12 activates the flavonol biosynthetic pathway. As a result of its broad specificity for transcriptional activation in tomato, AtMYB12 can be used to produce fruit with extremely high levels of multiple polyphenolic antioxidants; polyphenols collectively reached up to 10% of the dry weight of tomato fruit. These levels may represent the upper limits for polyphenol accumulation in viable plant tissues. Levels of polyphenol accumulation were achieved in tomato without any observed effects on fruit growth or yield.

Another example to increase the quantity of nutraceutical products in plants is the molecular engineering of resveratrol. The grapevine phytoalexin resveratrol displays a wide range of biological effects related to its antioxidant property. A recent paper by Delaunois et al. (2009) presented a comprehensive review on plant molecular engineering with the stilbene synthase (STS) gene. Gene and promoter options are discussed, namely the different promoters used to drive the transgene, as well as the enhancer elements and/or heterologous promoters used to improve transcriptional activity in the transformed lines. Factors modifying transgene expression and epigenetic modifications, for instance, transgene copy number, are also presented. Resveratrol synthesis in plants, together with that of its glucoside as a result of STS expression, is described, as is the incidence of these compounds on plant metabolism and development. The ectopic production of resveratrol can lead to broad-spectrum resistance against fungi in transgenic lines, and to the enhancement of the antioxidant activities of several fruits, highlighting the potential role of this compound in health promotion and plant disease control.

12.2.5 Reduction of Anti-Nutrients or Allergens

Genetic engineering has great potential for improving the safety of plant-based foods by eliminating allergenic components. There are eight foods that account for 90% of all food-allergic reactions: peanuts, tree nuts, soya bean, wheat, milk, egg, fish and shellfish. A recent publication by Dodo et al. (2008) demonstrates that RNA interference can mediate the silencing of a major allergen protein from peanuts. The focus of allergy reduction has been on an allergen that is recognized by the serum IGE, and is one of the best characterized, Ara h 2. Three additional papers (two by Le et al., 2006 and one by Lorenz et al., 2006) have reported the removal of allergenic proteins from tomato fruits. Lyce 1 (profilin) and Lyce3 (non-specific LTP), each having two isoforms, were successfully inhibited through expression of RNA interferences vectors. However, the phenotypes of Lyce1-reduced tomato were dramatically different from the wild type, showing extreme dwarfing under greenhouse condition, as well as delayed flowering, and severe yield reductions. In contrast, no detectable levels of Lyce 3 were found in transgenics that were phenotypically normal. These researches highlights the need to characterize any transgenic produced to determine stability and changes in other proteins that affect not only on allergenicity but also on agronomic traits.

Cassava is the main staple food for more than 500 million people in the tropical and subtropical regions. The primary plant organ consumed is the root, which can contain between 15 and 1500 mg cyanogen equivalents/kgfw (ÓBrien et al., 1991). Chronic exposure to cyanogens can cause hyperthyroidism and neurological disorders; acute exposure results in permanent paralysis of the legs. Moreover, in diets with insufficient amount of cysteine, cyanide is converted to cyanate, which particularly damages the brain. Cyanogenesis is initiated by the rupture of the plant cell vacuole that releases linamarin that is hydrolysed by linamarase giving acetone cyanohydrin and glucose. Acetone cyanohydrin spontaneously decompose to produce cyanide or can also be enzymatically broken down by hydroxynitrile lyase (HNL) (White et al., 1994). Transgenics strategies to reduce cyanide levels in cassava foods use two different approaches: accelerating of cyanogen turnover and cyanide volatilization during processing by elevation of root NHL activity (Arias-Garzon and Sayre, 2000; Siritunga and Sayre, 2007). This strategy has the advantage that cyanogen levels in unprocessed roots are not altered, potentially providing protection against herbivory. To produce cultivars that promote rapid cyanide volatilization, HNL, which catalyses the last step in cyanogenesis, was overexpressed in roots. The elevated HNL activity resulted in a threefold increase in the rate of cyanogen turnover. The second approach is to inhibit the expression of two cytochrome P450s, which catalysed the first dedicated step in cyanogen synthesis from linamarin or lotaustralin, in leaves leading to a 99% reduction in root cyanogen levels, indicating that the cyanogenic glycoside, linamarin, is synthesized in leaves and transported to roots.

12.2.6 Concluding Comments and Future Directions

There are alternative approaches for developing new crop cultivars with improved human health characters without using GM technology. This alternative approach encompasses the use of DNA sequences sourced from within the target crop's gene pool. It has been termed 'intragenics' or 'cisgenis'. As stated by Rommens et al. (2007), the novel intragenic approach to genetic engineering improves existing varieties by eliminating undesirable features and activating dormant traits. It transforms plants with native expression cassettes to fine-tune the activity and/or tissue specificity of target genes. Any intragenic modification of traits could, at least in theory, also be accomplished by traditional breeding and transgenic modification. However, the new approach is unique in avoiding the transfer of unknown or foreign DNA. By consequently eliminating various potential risk factors, this method represents a relatively safe approach to crop improvement. A more simple method to transform a commercial variety without employing selectable marker genes is proposed by Weeks et al. (2008). Basically, young seedlings of alfalfa (*Medicago sativa*) are cut at the apical node, cold-treated and vigorously vortexed in an *Agrobacterium* suspension also containing sand. About 7% of treated seedlings produced progenies segregating for the T-DNA. The vortex-mediated seedling transformation method was applied to transform alfalfa with an all-native transfer DNA comprising a silencing construct for the caffeic acid *o*-methyltransferase gene. The resulting intragenic plants accumulated reduced levels of the indigestible fibre component lignin that lowered forage quality. The absence of both selectable marker genes and other foreign genetic elements may expedite the government approval process for quality-enhanced alfalfa.

What then is the future of nutrigenomics in order to lead to further developments in the engineering of the crops? Modifying the nutritional composition of plant foods by plant biotechnology will lead to create appropriate products for every person with the objective to prevent, treat disease or simply enhance his health. However, one must know the precise compound(s) to target and which crops to modify. It is evident that a deep knowledge of plant biochemistry, human physiology and food biochemistry will be required. For this reason, a strong collaboration between different disciplines such as biologists, nutritionist and food scientists will be needed.

12.3 NUTRITIONAL GENOMICS AND ANIMAL PRODUCTION

Nutritional genomics offers a way to optimize human health and the quality of life through the intake of bioactive foods that interact with our genome. Animals play some roles in developing this discipline. First, foods derived from animals are an important source of nutrients in the diet; the composition of animal-derived foods may contain products that may actively promote long-term health, and research is urgently required to fully characterize the benefits associated with the consumption of these compounds and to understand how the levels in natural foods can be enhanced. On the other hand, animals can be used as models to study different interactions between gene and diet when it is difficult, for example, to control other many factors involved in this interaction in humans. So, in this part of the chapter we make a review of the last approaches to improve the nutritive value of animal food products and secondly, we comment on some of the animal models which have been used in the field of the nutrigenomics.

12.3.1 Improving the Nutritive Value of Animal-Derived Foods

It is known that during the second half of the twentieth century there were very major changes to diet in developed and, also, in developing areas. Plant-based diets have been replaced by an increased consumption of animal products. On the other hand, the composition of the animal food products has also varied, due to a change in the composition of the animal product (often animal fats) or due to an addition of some components during the food processing. In this part of the chapter we focus on the first point; we examine the possibilities to improve the composition of animal products in order to promote human health.

Lipids are the macronutrients associated with the risk of developing cardiovascular disease (CVD) that have been more extensively investigated. Intake of high saturated fatty acids (SFAs) raises total cholesterol and LDL-cholesterol but with markedly different effects depending on chain length. For example, myristic acid (14:0) and palmitic acid (16:0) have been associated with elevated plasma LDL-cholesterol levels in humans (Temme et al., 1996) whereas stearic acid (18:0) has been shown as essentially neutral (Bonanome and Grundy, 1988). Regarding lauric acid (12:0), controversial results have been found; Zock et al. (1994) indicated a more potent effect of this SFA on plasma cholesterol than palmitic acid whereas Denke and Grundy (1992) found the opposite effect. When substituting for SFA in metabolic studies both MUFA and PUFA lower plasma total cholesterol and

LDL-cholesterol. Regarding health promotion, the most important n-3 PUFA are EPA (20: 5n-3), DHA (22:6n-3) and α-linolenic acid (18: 3n-3). The results of prospective cohort studies indicate that consuming fish or fish oil containing the n-3 fatty acids EPA and DHA is associated with decreased cardiovascular death, whereas consumption of the vegetable oil-derived n-3 fatty acid α-linolenic acid is not as effective. Randomized control trials in the context of secondary prevention also indicate that the consumption of EPA plus DHA is protective at doses <1 g/d. The therapeutic effect appears to be due to the suppression of fatal arrhythmias rather than stabilization of atherosclerotic plaques. At doses >3 g/d, EPA plus DHA can improve CVD risk factors, including decreasing plasma triacylglycerols, blood pressure, platelet aggregation and inflammation, while improving vascular reactivity (Breslow, 2006). All these data have led to a widespread belief that there should be small increases in n-3 PUFA intake. In general, fats in daily animal-derived foods as milk and meat are very poor sources of n-3 PUFA. The fatty acid composition of animal-derived foods can vary considerably in response to changes in the diet of productive animals; in this way, one of the approaches is to modify the diet of the animals to obtain lower concentrations of SFA and enhanced amounts of PUFA (DHA and EPA) without diminishing the nutritional benefit of the animal-derived food. PUFA are not synthesized in any appreciable quantities in ruminant tissues, and concentrations in milk are, therefore, essentially a reflection of the amount leaving the rumen. When using fish oil as a source of EPA and DHA, the transfer efficiency of these ones from the diet into milk is very low because of extensive biohydrogenation in the rumen and subsequent transport of absorbed n-3 PUFA in phospholipids and cholesteryl ester fractions on plasma that are poorly utilized by the mammary gland (Rymer et al., 2003). In addition, inclusion of fish oil in the diet of dairy cows reduces levels of 18:0 and markedly increases the amount of *trans*-18:1 and *trans*-18:2 in milk (Shingfield et al., 2003). For this reason, Bernal-Santos et al. (2010) utilized SDA-enhanced SBO derived from genetically modified soya beans—SDA is an n-3 fatty acid that humans are able to convert to EPA—to examine the potential to increase the n-3 fatty acid content of milk fat and to determine the efficiency of SDA uptake from the digestive tract and transfer to milk fat. Treatment also had no effect on milk production, milk protein percentage and yield or lactose percentage and yield or milk fat yield but milk fat percentage was lower for the SDA treatment via ruminal. The SDA treatment via abomasal increased n-3 fatty acids to 3.9% of total milk fatty acids, a value more than fivefold greater than that for the control. Overall, results demonstrate the potential to use SDA-enhanced SBO from genetically modified soya beans combined with proper ruminal protection to achieve impressive increases in the milk fat content of SDA and other n-3 fatty acids that are beneficial for human health.

A right intake of PUFA (derived from milk or from other animal food) could have different health promoting effects depending on genome background. As an example of these gene*diet interactions, Ordovas et al. (2002a,b) reported a genetic variant in the APOA1 lipoprotein gene that modulates HDL-cholesterol concentrations in human. Dietary fat modulates the association between the APOA1 G-A polymorphism and HDL-cholesterol concentrations. When PUFA intake was <4% of energy, G/G subjects had 14% higher HDL-cholesterol concentrations than did the carriers of the A allele. Conversely, when PUFA intake was >8%, HDL-cholesterol concentrations in carriers of the A allele were 13% higher than those of G/G subjects. No significant allelic difference was observed for subjects in the range of PUFA intake of 4–8% of energy. These interactions were not significant in humans.

Another daily animal food product is the meat. In general, the fatty acid composition in meats derived from ruminant animals (beef and lamb) is manipulated through the diet using high-forage-based diets or oil supplements rich in PUFA. But most of the PUFAs are hydrogenated in the rumen and only a small proportion of them can escape, and, once absorbed, is available for incorporation into tissue lipids. In contrast, the fatty acid-profile of non-ruminant meat is essentially a reflection of that in the diet. So, it is theoretically easier to enhance long-chain n-3 PUFA levels in non-ruminant tissue lipids using dietary sources. But the enrichment of the diet in EPA and DHA leads to the production of meat with a metallic taint, fish-like flavour and reduced shelf-life (Leskanich and Noble, 1997). Since the sequence of the genome of some ruminant and non-ruminant animals was completed, the use of molecular genetic information in selection and improvement programmes of animal production has allowed us to use alternative ways to improve the profile of animal-derived food. To illustrate this section we reviewed the novel sites of genetic variation within candidate genes implicated to be associated with economically relevant traits of meat quality and growth in pork and beef.

Owing to recent drastic increases in feed costs as the exigencies of the consumer, traits related to lean growth and meat quality have received more relative interest in selection schemes. For example, a sufficient level of intramuscular fat (IMF) is needed to enhance consumer acceptance of pork or beef products, and is currently receiving greater attention within genetic improvement programmes. Schwab et al. (2009) investigated candidate genes implicated to play a role in adipogenesis within two lines of purebred Duroc pigs in order to evaluate its potential use of genetic markers-assisted selection (MAS) for IMF deposition. These included melanocortin-4 receptor (MC4R), heat fatty acid-binding protein 3 (FABP3), muscle and heart (mammary-derived growth inhibitor), delta-like 1 homologue 1 (DLK1) (*Drosophila*) and transcription factor 7-like 2 (TCF7L2) (T-cell specific, HMG-box). The investigators described significant variation in IMF by the MC4R genotype and by a novel BsrfI single-nucleotide polymorphism (SNP) within the FABP3 gene within the control line but not in the selection line (the selection line is the result of five generations of selection for increased IMF based on a two-trait animal model that included IMF measured on the carcass and predicted via ultrasound). Researchers indicated that owing to the complexity and polygenic nature of traits such as growth and composition, one possibility for the discrepancy of results between lines may be because of background gene effects. Genetic markers for DLK1 and TCF7L2 evaluated in this population are not currently recommended for selection in Duroc swine. In conclusion, the authors pointed out that the existence of MC4R and FABP3 mutations may be useful markers in MAS aimed at IMF improvement, provided that gene effects are segregating and the presence of an association is detected within the population. However, additional work to confirm the use of the investigated genetic markers in selection programmes is needed.

Other genome polymorphisms associated with meat composition in pig have been studied. Fernández et al. (2008) carried out a study to investigate the associations between mitochondrial DNA polymorphisms and meat quality traits (IMF and protein content of the longissimus) in an Iberian porcine line. The selected pigs were assigned to six mitochondrial haplotypes (H1 to H6), based on Cytochrome b and Dloop sequences. In H3 haplotype, greater fat content and lower protein content were found. The authors determined the complete mitochondrial DNA sequence of six individuals, each carrying a different mitochondrial haplotype, to identify the causative mutation of these effects on IMF and protein contents. Two polymorphic positions were exclusively detected in H3 carriers: a synonymous transition 9104C > T in the gene-coding region of cytochrome c oxidase subunit III and a substitution 715A > G in 12S rRNA. The detected candidate substitutions are located in essential mitochondrial genes, and although they do not change the amino acid composition they could bring about potential change in the secondary structure of their corresponding mRNA. However, researchers pointed out that the usefulness of these polymorphisms as markers in selection programmes requires validation of the consistency of these results in other Iberian pig lines.

A question that arises once we have selected potential polymorphisms as markers for a specific trait is whether they would be useful also in other species. In this way, Marques et al. (2009) identified polymorphisms in genes known to affect lipid metabolism in other species different from beef cattle and to assess their association with carcass quality traits in beef. Two genes, 2,4 dienoyl CoA reductase 1 (DECR1) and core binding factor, runt domain, alpha subunit 2, translocated to 1 gene (CBFA2T1), had been previously evaluated in other species and found to contain polymorphisms influencing lipid metabolism. Another gene, fibroblast growth factor 8 (FGF8), had been linked to several quantitative trait loci (QTL) affecting obesity in mice, indicating its potential for regulating adiposity in other species. Sequencing analysis identified nine polymorphisms in DECR1, four in CBFA2T1 and four in FGF8. Researchers found associations between these genetic variants and ultrasound marbling score (CBFA2T1), ultrasound backfat (DECR1), carcass backfat (FGF8) and lean meat yield (FGF8). QTL analysis including a set of previously genotyped markers on BTA14 and BTA26 (several studies have reported the presence of carcass quality QTL on BTA14 and BTA26, with no specific genes being conclusively linked as their cause), and DECR1 and FGF8 polymorphism resulted in several significant QTL peaks: ultrasound backfat, lean meat yield, carcass grade fat and yield grade. These results suggest that polymorphisms discovered in DECR1, CBFA2T1 and FGF8 may play a role in the lipid metabolism pathway affecting carcass quality traits in beef cattle.

We have principally focused on the lipid profile of meat but other genetic variants are also being investigated to improve quality traits and/or promote human health. In this way, researchers are looking for associations between candidate genes and traits as colour, pH, conductivity, ham weight, muscle fibre number, size and lean meat in differents animal species (see, e.g., Wimmers et al., 2007; Kim et al., 2009).

The most traditional approach to improve the nutritive value of animal products is through animal diet and for this reason the role of animal nutrition in creating foods closer to the optimum composition for promoting human health (Givens, 2005) has more importance. On the other hand, the development of 'omics' disciplines has allowed us to use tools and genome information that were not available to increase the capacity to select the animals with appropriate genetic characteristics to lead to health-promoter animal foods. But we take into account that variations in gene effects between different species or in different lines in the same species indicate the existence of possible epistatic effects within different background genomes. For this reason, not only additional work is necessary to confirm the use of the investigated genetic markers; further investigations to identify causative mutation underlying animal food quality variation are also needed.

12.3.2 Animal Models in Nutrigenomics

The heterogeneous genetic and environmental conditions of human populations impart considerable difficulty for uncovering how genome and diet are related. Animal models where these variations, genetics and environments are more controlled can be of major value for both identifying mechanisms of interactions as well as validation of human data (Koch and Britton, 2008). For example, animal models provide a means of assessing the consequences of manipulating the perinatal environment in ways that cannot be done in humans. During the gestational period, maternal overnutrition resulting from high fat intake produces offspring obesity in out-bred rats (Guo and Jen, 1995). However, the results are different if the genetic background of the offspring is different. Levin and Goveck (1998) selectively bred rats prone to develop diet-induced obesity (DIO; these rats overeat and become obese over 3–4 weeks on a 31% fat diet) and prone to be diet resistant (DR; these rats quickly adjust for the increased caloric density of the diet and remain lean). Only those offspring with the DIO genotype become more obese as adults when their dams are made obese throughout gestation and lactation, even when those offspring are fed only a low-fat diet from weaning. Importantly, these obese DIO offspring have abnormal development of important monoamine pathways that influence the development in energy homeostasis (HeisLer et al., 2007). Thus, maternal obesity influences neural development specifically in animals which are genetically predisposed to become obese. But, on the other hand, this predisposition can be modified by other environmental condition: exercise; running wheels prevent DIO pups from becoming obese, even when fed a high-fat diet. Only 3 weeks of exercise is sufficient to prevent them from becoming obese for up to 10 weeks after the wheels are removed (Pattersson et al., 2008). In contrast, when sedentary DIO pups on 31% fat diet were calorically restricted to match their weights to those of the exercising pups for 6 weeks, they become more obese as adults once they were allowed to eat *ad libitum*. These studies emphasize the importance of taking prevention measures (in this case, appropriate exercise and diet) as soon as possible in those subjects with a genetic predisposition.

Animal models have provided a fundamental contribution to the historical development of understanding the basic parameters that regulate the components of our energy balance. Obesity results from prolonged imbalance of energy intake and energy expenditure. Speakman et al. (2008) carried out an in-depth review of the contribution of animal models to the study of obesity. They provide some examples of the animal work that has been performed to understand the physiological/genetic basis of obesity. Five different types of animal models have been employed in the study of the physiological and genetic basis of obesity. The first models reflect single-gene mutations that have arisen spontaneously in rodent colonies and have subsequently been characterized. The second approach is to speed up the random mutation rate artificially by treating rodents with mutagens or exposing them to radiation. The third types of models are mice and rats where a specific gene has been disrupted or overexpressed as a deliberate act. Such genetically engineered disruptions may be generated through the entire body for the entire life (global transgenic manipulations) or restricted in both time and to certain tissue or cell types. In all these genetically engineered scenarios, there are two types of situations that lead to insights: where a specific gene hypothesized to play a role in the regulation of energy balance is targeted, and where a gene is disrupted for a different purpose, but the consequence is an unexpected obese or lean phenotype. A fourth group of animal models concern experiments where selective breeding has been utilized to derive strains of rodents that differ in their degree of fatness. Finally, studies have been made of other species including non-human primates and dogs. They also review environmental factors, focussing on varied approaches for using animals, rather than aiming to be a comprehensive summary of all work in the field. Studies in this context include exploring the responses of animals to high-fat or high-fat/high-sugar diets, investigations of the effects of dietary restriction on body mass and fat loss, and studies of the impact of candidate pharmaceuticals on the components of energy balance. Despite all this work, there are many gaps in the understanding of how body composition and energy storage are regulated, and a continuing need for the development of pharmaceuticals to treat obesity. The review concludes with an assessment of the potential for reducing the use of animals to study obesity.

12.4 NUTRITIONAL GENOMICS AND FOOD PROCESSING

Given that nutrition is part of our life from birth and humans feed daily, it is of particular relevance to get safe food production since the beginning of the food chain until it reaches the consumer. Furthermore, eating outside home and the usage of partly or fully cooked food is increasing (Soriano et al., 2002). As the incidence and type of foodborne diseases are also changing taking into account the current food habits, providing the consumer with safe food is especially important in the age of globalization. Assuring safe food is the most difficult task in the food industry. The food-processing industry is faced with an ever-increasing demand for safety and minimally processed wholesome foods (Sun and Ockerman, 2005). Currently, many groups explore the use of the recently booming genomics technology in setting food-processing parameters.

Genomics projects have been and are defined in order to obtain a transparent view of the food chain with respect to microbial food spoilage and resistance development against food preservation techniques (Brul et al., 2006).

Research and practice are focusing on development, validation and harmonization of technologies and methodologies to ensure complete traceability process throughout the food chain. The International Organization for Standardization defines traceability as the 'ability to trace the history, application or location of an entity by means of recorded identifications' (Furness and Osman, 2003). In the food industry, and taking into account that thousands of new items enter in the food market every year, traceability has emerged in the last century as a key instrument of policy quality, safety and competitiveness. The industry has to cope with the increasing population and consumer demands through the development of new products that provide not only a nutritional add value but that ensure a quality end-product throughout the food chain process, from collection in the field to human consumption, known as 'from farm to fork' (Brown and van der Ouderaa, 2007, p. 1027). For food safety reasons, among other commercial reasons, we need to trace items from farm to fork. An integrated production chain control system should be able to identify and document with accuracy materials and actions applied in food processing to increase product safety, identifying the source of possible contamination, facilitate the product recall procedure, and control public health risks derived from product consumption (Raspor, 2005). Thus, the food industry applies the hazard analysis and critical control point as an approach to food safety that addresses physical, chemical and biological hazards as a means of preventive rather than finished product inspection. The development of biological identification technologies based on DNA testing enables traceability, food authenticity, the search of indicator microorganisms in food processes, the toxicological evaluation and so on. The new genetics biotechnological methods developed recently provide numerous opportunities for the food industry.

In this part of the chapter we present a general view of how the nutrigenomics technologies can be used in other parts of the food chain to improve food processing, food safety (microbial genomics and toxicogenomics evaluations) and quality assurance.

12.4.1 Genetic Process Markers

Genetics process markers of food processing are informative molecular markers that may be used to guide industrial processes or improve supply chain management (Brown and van der Ouderaa, 2007). As the food industry is aware of the quality of raw material and improving processability and quality traits, genomics could be a very helpful tool in order to facilitate the discovery of new genes for further selection by DNA markers. Generally, we need biomarkers to chase and trace the quality and safety of foodstuffs during their production or consumption and we have to select such biomarkers that can be used in many different places and different circumstances along the food chain.

First, we illustrate the use of genetic markers in the *organoleptic characteristics and physic properties in food*. As a first example of using genetic markers we present the case of the dry-cured ham. There are several genes involved in biological processes affecting dry-cured-ham production. Some researchers have determined the effect of candidate genes on the processing quality traits of US country hams. Before analysing several genes, they found two genetic markers associated with cured weight and yield (a gene from the cathepsin family (cathepsin F) and the stearoyl-CoA desaturase (delta-9-desaturase) gene, involved in lipid metabolism). They also identified other markers related to colour traits, pH and lipid percentage (Ramos et al., 2008). These markers could be used for screening and sorting of carcasses prior to ham processing and, eventually in pig improvement programmes designed to select animals possessing genotypes more suitable for the production of dry-cured hams.

Other studies have focused on studying hundreds genes involved in the determination of fruit texture, pigmentation, flavour and chilling injury. Ogundiwin et al. (2009) studied such parameters as an integrated fruit quality gene map of *Prunus*. They were able to bin–map further 158 markers to the reference map: 59 ripening candidate genes, 50 cold-responsive genes and 50 novel microsatellite markers (they are repeating sequences of 1–6 bp of DNA. Microsatellites are typically neutral and co-dominant and they are used as molecular markers in genetics and other studies) from ChillPeach (). This information regarding the fruit quality gene map is a valuable tool for dissecting the genetic architecture of fruit quality traits in *Prunus* crops.

The genetics approach has also been used in the genetic control of wheat processing of baking parameters, particularly sponge and dough baking. The QTL analysis techniques are commonly used for this purpose, that is, for identifying genetic markers. It refers to the inheritance of a phenotypic characteristic that varies in degree and can be attributed to the interactions between two or more genes and their environment. Although not necessarily genes themselves, QTLs are stretches of DNA that are closely linked to the genes that underlie the trait in question. QTLs can be molecularly identified (e.g., with amplified fragment-length polymorphism (AFLP)) to help map regions of the genome that contain genes involved in specifying a quantitative trait, in this case, for instance, in the sponge and dough baking of the wheat processing. Using this technique, Mann et al. (2009) performed a QTL analysis using a population of doubled haploid lines derived from a cross between Australian cultivars Kukri 9 Janz grown at sites across different Australian wheat production zones in order to examine the genetic control of protein content (with two loci, 3A and 7A), protein expression, dough rheology and sponge and dough baking performance, where the major effects were associated with the Glu-B1 and Glu-D1 loci. Finally, they concluded that dough rheology measurements were poor predictors of sponge and dough quality providing further evidence for the contention that wheat quality is a consequence of a network of interacting genes.

Second, genetic markers can also help in *the identification of increasing varieties* such as new potato varieties that are obtained as Plant Breeders' Rights with the use of a set of nine microsatellite markers that can differentiate over 1000 cultivars including the majority of varieties in the European Union Common Catalogue (Reid et al., 2009). Another good example is the tomato species, given that the ecological interest and for being one of the most common vegetable crops grown. An extensive collection of tomato expressed sequence tags (ESTs) is available at the SOL Genomics Network to prevent genome-wide evolutionary analyses. The ESTs is a short sub-sequence of a transcribed c-DNA sequence. They may be used to identify gene transcripts, and are instrumental in gene discovery and gene sequence determination. The identification of ESTs has proceeded rapidly, with approximately 52 million ESTs now available in public databases (e.g., Gene Bank). ESTs contain enough information to permit the design of precise probes for DNA microarrays and later on can be used to determine the gene expression.

Taking all this information together and using the available meta-information, a group of research carried out by Jiménez-Gómez and Maloof (2009) classified genes into functional categories and obtained estimations of single SNPs quality, position in the gene and effect on the encoded proteins, allowing them to perform evolutionary analyses. They developed a set of more than 10,000 between-species molecular markers optimized by sequence quality and predicted intron position. Experimental validation of 491 of these molecular markers resulted in the confirmation of 413 polymorphisms revealing genes potentially important for the evolution and domestication of tomato; interestingly, these sets were enriched with genes involved in response to the environment.

Third, another example that shows how gene markers can be used is that involved in *post-harvest performance*. In terms of supply chain management, molecular markers may prove useful for predicting and improving the shelf-life of fresh produce. For example, Broccoli, has very short life days after the harvest resulting in losing turgor. This process is similar to those seen in developmental leaf senescence. Page et al. (2001) studied which genetic and environmental factors influence post-harvest performance studying the molecular and biochemical changes in broccoli florets stored at two different temperatures after harvest. They have used a genomic approach to identify senescence-enhanced genes using previously identified senescence-enhanced genes in the model plant of Arabidopsis and newly isolated, and they found that post-harvest changes in broccoli florets show many similarities to the processes of developmental leaf senescence. The leaf senescence studies and the putative network of signalling pathways has already been used to manipulate the post-harvest characteristics of broccoli and also other vegetables such as lettuce.

12.4.2 Genomics and Food Safety

Food safety issues are triggers by various hazards, including microbiological (*E. coli*, *Salmonella*, *Listeria*, etc.), nutritional (fat consumption), environmental (pesticides, nitrates, etc.), natural (ingredients) and food composition hazards.

External contamination occurs during harvesting, processing and storage, and includes enteric bacteria, viral pathogens, bacterial derived toxins and by-products as well as residues such as herbicides, antibiotics and so on. These products could end up in the food chain and therefore interfere with the human health. Genomics in food safety concentrated on the safety evaluation of foods components and the detection of microorganisms which may cause food spoilage or that considered a risk for human health. In this part of the chapter we explain the application of genomics in toxicological and microbiological evaluation illustrating some examples to better understand it.

12.4.2.1 Genomics and Toxicological Evaluation

For the safety evaluation of food components, it is important to focus on food components that cause adverse health effects and also the level of exposure required to elicit adverse effects. A dose–response relationship is essential in toxicological testing. A safe level of human exposure can be established by determining the level of exposure at which no adverse effects occur. The margins of safety between animal and human intake are calculated from a 90-day feeding study by dividing a no-observed-adverse-effect level (NOAEL) by the anticipated mean per capita daily dietary intake by adults or sensitive groups such as pregnant women, and so on. In this issue, genomic technologies can offer a variety of opportunities through the toxicological evaluations.

Some examples related to genetic toxicology to identify the acceptable daily intake in GM food are the case of the maize. A young adult male rat weighing 250 g eats typically 25 g rodent diet/day, that is, 100 g diet/kg body weight/day. At 33% (w/w) dietary incorporation, this represents 33 g maize/kg body weight/day. Averaged over the whole study, a rat typically consumes 25 g maize/kg/day, which provides a conservative NOAEL. A typical EU theoretical maximum daily intake (TMDI) for maize as described above would be 17 g/person/day. For a 70 kg human, this equates to 0.24 g maize/kg body weight/day. This again provides an exposure margin of over 100-fold when the NOAEL is divided by the EU TMDI for maize and its derivatives (Reported by the EFSA GMO Panel Working Group on Animal Feeding Trials, 2008). On the other hand, the NOAEL is important in the design of genetic toxicology studies to at least inform the risk assessor whether or not an adverse event is likely to occur at a particular exposure level. Moreover, some studies with DNA-reactive chemicals consistently suggest that even direct-acting genotoxic substances may exhibit clear NOAEL responses to differentiate genotoxic or no genotoxic compounds because there are data to demonstrate that almost any chemical can be genotoxic under certain conditions such as sugar, salt or vegetables (Pottenger et al., 2007).

The possible toxicological consequences of intended changes in GM plants/food/feed product should be studied in comparison with its most appropriate non-GM. In each case it is necessary to determine macro- and micronutrients, anti-nutritional compounds or natural toxins (EFSA, 2004). The requirements of toxicological testing in the safety assessment of food/feed derived from GM plants should be considered on a case-by-case basis. The interest must be in the, through the insertion of a particular gene, newly expressed proteins and the consequence of any GM. A search for sequence similarity that implies homology to proteins known to cause adverse effects should be conducted. The scientific tools available for studies on the safety of GM food and feed include *in silico*, *in vitro* and *in vivo* methods. Repeated dose toxicity studies (normally a 28-day oral toxicity study in rodents) should be carried out to demonstrate the safety of the newly expressed proteins. At present, *in vitro* studies are not suitable for the study of whole foods. However, genomics and microarray technologies are quite helpful in providing genes that serve as biomarkers for a cell's response to toxins and allergens. The only thing that needs to be carried out is to ensure a high degree of reproducibility and validity of these assays (a slightly difficult aspect, taking into account the lack of information about the regulation as we have commented above).

Also it is important to consider the allergenicity of the newly expressed protein and the assessment of allergenicity of the proteins sourced from the whole GM plant or crop. A review carried out by Prescott and Hogan (2006) reported that when using the Brown Norway rat model, a model for the assessment of allergenicity in GM foods, studies showed that when fed common allergenic whole foods (cow milk and egg white) the rats developed IgE-associated allergic responses to the same proteins that are commonly allergenic in humans.

On the other hand, studies analysing the GMOs in animal models should be carried out in a 90-day rodent feeding design (as indicated by EFSA Panel opinions). Animal studies are essential in the safety assessment of many compounds including food additives, chemicals and pesticides. In most cases, the test substances of known purity are fed to laboratory animals at a range of doses to identify any potential adverse effects. It should also be highlighted that when conducting animal feeding studies, the nutritional value and balance of the diet should be maintained (Report by the Scientific Panel Working Group on Animal Feeding Trials).

The advances in molecular biology during the past two decades have resulted in an increase in gene structure and function. These advances have led to a new sub-discipline of toxicology: Toxicogenomics, a study of the relationship between the structure and activity of the genome and the adverse biological effects of exogenous agents. With the new 'omics' technologies, it is possible to study the functional activity of biochemical pathways applying transcriptomics, proteomics and metabolomics (see Section 12.5). These advances present methods to evaluate potential human and environmental toxicants and of monitoring the effects of exposures to these toxicants. For example, with the application of these technologies in the genetic toxicology, it is possible to identify polymorphisms responsible for sensitivity to toxicity from particular agents and the identification of chemical-induced genetic changes with particular diseases. One example of the influence of genetic variation and toxicity is the sensitivity to fava bean toxicity among Mediterranean populations with glucose-6-phosphate dehydrogenase deficiency (G6PD) (Weber, 1999). The potential application of toxicogenomics will improve the risk assessment and provide a more holistic understanding of the effects of chemicals on cellular alterations.

12.4.2.2 Genomics and Microbiological Evaluation

One application of genomics to food safety is the control and detection of food-borne microorganisms. The genomics in food microbiology is expected to predict models of behaviour of spoilage and pathogenic microorganisms upon and after processing. Furthermore, the identification of the organism by microarrays saves time compared to the conventional cultivation.

The classical methods to preserve and control the microbial spoilage of food are sterilization, freezing, curing, blanching and preservatives and also the use of high pressures, lyophilization techniques and so on. As organic food has increased due to practices related to sustainable agriculture, the challenge to preserve food products by the industry has increased. The genome sequences of many bacteria that cause foodborne disease are known, including *Listeria monocytogenes*, *Yersinia enterocolitica*, *Escherichia coli*, *Clostridium botulinum A*, *Campylobacter jejuni* and *Salmonella* species. The use of the genomic techniques can identify strain-specific genetic signatures that can be relevant for the detection of microbial contamination and the microbial food stability. For example, research on food spoilage *Bacillus subtilis* isolates using the AFLP and microarray technology has identified a number of genomic factors correlated to the level of spore heat resistance. Furthermore, it was shown with the sequenced *B. subtilis* laboratory strain that sporulation in the presence of some particular calcium ions in a cocktail of calcium, magnesium, iron, manganese and potassium promotes thermal resistance of developing spores (Oomes et al., 2009). The use of the DNA-Chip technology indicates molecular markers that allow the occurrence of spoilage and pathogenic bacteria and prediction of their thermal preservation stress resistance as, for example, occurs in the exposure of vegetative cells to elevated temperatures.

In other cases it is important to evaluate if the process employed to preserve the stability of the food is enough to kill a specific microorganism. For example, Guilbaud et al. (2008) investigated the effect of liquid smoke on growth, survival, proteomic pattern and haemolytic potential of *Listeria monocytogenes*. They studied gene expression of the protein of interest expressed by the microorganism and concluded that the presence of liquid smoke in a rich medium strongly affected growth and survival of *L. monocytogenes*. Brief smoke stress affected the metabolic pathways and inhibited the haemolytic activity of *L. monocytogenes*. This technique, where genomic tools are employed, is relevant if we take into account the gravity and the fact that *L. monocytogenes* is very often associated with processed foods industry, and that it is among the food-borne diseases that has greater health impact. Major outbreaks of listeriosis have been reported in North America and Europe since 1980. In connection with the previously explained, the identification of pathogen bacteria by molecular genetics can be advanced by DNA microarray technology. 16S rRNA has been reported to be a suitable target for identifying various specific microorganisms. Microarray experiments are a useful tool for phylogenetic analysis and species identification. For instance, currently, the NUTRI®Chip in a modified form is used for the detection of the presence of pathogenic bacteria. One study comparing the sensitivity and specificity of *Salmonella* detection by the conventional cultural method according to Section 35 LMBG (Food and Consumer Goods Act), Tecra® uniqueTM Salmonella ELISA, PCR (Gene Scan) and biochip-based Microarray (NUTRI-Chip) showed that the new NUTRI-Chip proved to have a higher-than-average sensitivity and specificity with artificially as well as naturally contaminated food samples for the detection of *Salmonella* in food. This highlights the advantage of the current technology in detecting several different pathogenic microorganisms at the same time.

Another genomic approach using the amplified polymorphic DNA (RAPD) analysis and enterobacterial repetitive intergenic consensus PCR (ERIC-PCR) analysis has been used, for example, for investigating the biodiversity of the enterococcal species (a major concern by their potential pathogens) present in farmhouse goat's milk cheese made in Spain to study their traits with regard to consumer safety (Martín-Platero et al. 2009). They found that most of their enterococcal isolates proved to be safe except for some *E. faecalis* strains. However, the species described free of virulence may prove to be an interesting source of new strains for use in cheese technology after their characteristics have been thoroughly examined. This technique is also a good tool for the identification and typification of microorganisms in foodstuffs. The use of combination of these technologies (AFPL, RAPD, PCR, ribotyping (a method to determine homologies and differences between bacteria at the species or sub-species (strain)), pulsed-field gel electrophoresis (PFGE, a method of separating large DNA molecules that can be used for typing microbial strains) with the new technology (DNA chips and microarrays) and other genetics-based detection of microbes offers significant benefits over the conventional detection methodology, particularly from the human health perspective.

On the other hand, microbial genomics can be applied in the total food chain in the quality control system for fermented food, development of an integrated system for tailor-made processing, microbiological control for ingredients, process-line and processed food, development of microbial detection systems for the quality control of raw materials, identification of novel (natural) antimicrobial compounds and development of novel diagnostic systems for pathogens (Brul et al., 2006). Understanding the cellular response at the level of molecular events could help us to know the microbial responses during processing, storage and transport before arriving to the consumer. For example, Lee et al. (2002) determined how most of the transcriptional regulators encoded in the eukaryote *Saccharomyces cerevisiae* associate with genes across the genome in living cells using genome-wide location analysis to discover transcriptional regulatory networks in higher eukaryotes.

Knowledge of these networks will be important for understanding human health and designing new strategies to combat disease. Other models describing the kinetics of the intracellular metabolic processes with their linked regulatory systems have been described.

12.4.3 Genomics and Quality Assurance

The authenticity of plants, animals and packaged food products by DNA identification is currently applied in the food industry for quality assurance. Quality is defined as 'the totality of characteristics of an entity that bears its ability to satisfy stated and implied needs' (ISO 8402: 1994; EN ISO 8402; BS EN ISO 8402: 1995). Therefore, in the field of food safety, traceability becomes the principal tool to both ensure the effective responsibility of food manufacturers, farmers and food operators in relation to the final product quality (Raspor, 2005) and to assess and manage risks effectively. One application of the genetics analysis in the quality assurance is as a *means of authenticating and controlling conventional animal identification systems.* One example is the National disease monitoring and eradication programmes that depend on the conventional identification programmes, usually with ear tags. As DNA is unalterable, DNA identification avoids fraudulent practices shown in traditional methods.

DNA chips are also already being used in food analysis: with the help of the CarnoCheck® it is possible to detect eight different animal species in food products and animal feed. Given the current crisis in agriculture and the food industry, the issue regarding the monitoring of an accurate food-labelling procedure becomes more urgent. Furthermore, incorrect or vague declarations with regard to foodstuffs can cause allergies. For example, CarnoCheck allows the quick and efficient identification of eight species in all (horse, ass, beef, pork, goat, sheep, turkey and chicken), even from processed animal feed and food products (Spielbauer and Stahl, 2005).

In relation to that explained above, another application of DNA identification is the *molecular authentication on food ingredients and packaged food products.* Authentication of foodstuffs may be carried out by manufacturers as a part of their quality assurance processes, because the consumer trusts in the food labelling, especially when some ingredients have been removed. Some food fraud are being detected in the regional and traditional ingredients that give to a food 'designation of origin' or products with better quality. For example, the traditional Basmati rice has a superior aroma and grain quality than cross-bred Basmati and non-Basmati rice. Woolfe and Primrose (2004) has developed a microsatellite-based technique to distinguish between different varieties of Basmati and avoid fraudulences. Similar examples can be found in this chapter regarding the virgin olive oil and tuna.

Genomics in quality assurance can be applied as well in the *analysis of GMO.* As we have explained in Section 12.2 of this chapter, GMOs derived through genetic engineering, may be used as food, feeds, seeds or forestry materials. The use of GMOs in the food industry has been ascending abruptly in the past years. The food and feed under investigation were derived mostly from GM plants with improved agronomic characteristics such as herbicide tolerance and/or insect resistance. The majority of these experiments did not indicate clinical effects or histopathological abnormalities in organs or tissues of exposed animals. Therefore, the necessity of labelling and tracing the GMOs are current issues that are considered in trade and regulation. This regulation will affect shipments of non-GM products that may contain adventitious levels of GMOs. Regulation (EC) No. 1829/2003 and No. 1830/2003 of the European Parliament implicate a credible identification and documentation system for GMOs in food and feed products. The rapid introduction of novel GMOs associated with the legal requirements in food industry necessitate the cost efficient and reliable testing of foods for the unique identifiers of GMOs. Current methodologies for the analysis of GMOs focus on the transgenic DNA inserted or the novel protein expressed in a genetic product. Real-time polymerase chain reaction (PCR) has been the most commonly used technology for quantification of GMOs.

The PRC-based GMO test can be categorized into four levels (Miraglia et al., 2004): first, *screening methods* (target DNA elements such as promoters and terminators); second, *gene-specific* methods, for example, the Bt gene coding for a toxin acting against certain insects (it consists in targeting a specific part of DNA associated with the specific GM); third, *construct specific methods* (targets DNA sequence junctions not naturally present in nature) and finally, *event-specific methods* (the target the unique junction found at the integration locus between the inserted DNA and the recipient genome). Furthermore, the high-throughput methods such as arrays technology can be also used, although they required the capability of standardization, reproducibility and sensitivity of the detection system which makes difficult the introduction of biochips as a method for the routine analysis because of the need of the exact quantity of GMOs in foods are allowed to comply with legal requirements. The microarray technology also has the theoretic potential to detect unauthorized GM varieties that have any similarity with known genetics constructs. The microarray for protein-based detection will be an interesting alternative in the near future.

12.5 NUTRITIONAL GENOMICS AND HUMAN HEALTH

As we have commented in the introduction section, diet is the most important environmental factor that may interact with our genes to increase or decrease the likelihood of developing chronic diseases. Food contains a variety of chemical substances that have multiple effects on different biochemical pathways in the organism and it is consumed by everybody as a regular basis. The current changes in the lifestyle in part promoted by the nutritional transition, described as the transition from a traditional diet based on fruits and vegetables to a western diet enriched in calories based on meat and saturated fats, have led to CVDs, cancer and other non-communicable diseases.

Nutritional health is dependent on the interaction between the environmental aspects of supply, bioavailability, consumption and co-ingestion of dietary components and the genetically controlled aspects (absorption, metabolism, excretion, etc.). A better understanding of these interactions, known as gene–diet interaction, has the potential to support disease prevention and will lead to different requirements between individuals via modification dietary recommendations.

Since the Human Genome Project has been completed, the molecular nutrition begins with the approach to better understand the mechanism involved in the gene–diet interaction and to personalize the diet. Since then, the classic candidate gene approach and the newer genome-wide association studies have identified genetic variants that predispose patients to common diseases. This research has led to the development of concepts and research on the effect of genetic variation on the interaction between diet and specific phenotypes known as *nutrigenetics*. The goal of the nutrigenetics is to generate recommendations regarding the risk and benefits to the individual of specific dietary components. It is also termed *personalized nutrition*. And, on the other hand, the study of the characterization of gene products and the physiological function and interactions of these products is known as *nutrigenomics* (Figure 12.1). The unifying term *nutritional genomics* refers to nutrigenomics and nutrigenetics. The term 'nutrigenetics' was coined for the first time by Brennan in 1975 in his book *Nutrigenetics: New concepts for Relieving Hypoglycemia* while 'nutrigenomics' was used in 1999 by DellaPenna when he studied the plant genomics field. The new technology research together with the knowledge of the human genome sequence have led to the study of the new 'omics' to better understand the molecular basis of the disease development; these disciplines are transcriptomics, proteomics and metabolomics (Table 12.2). The genomic revolution has led to the development of several new technologies that facilitated the study of gene–nutrient interactions at the cell, individual and population level.

Apart from nutrigenetics and the study of gene–diet interaction, the identification of the cellular response to a nutritional signal may provide ways to decipher the mechanism by which a nutritional signal is transduced into a given response. Due to nutritional differences, only a few of

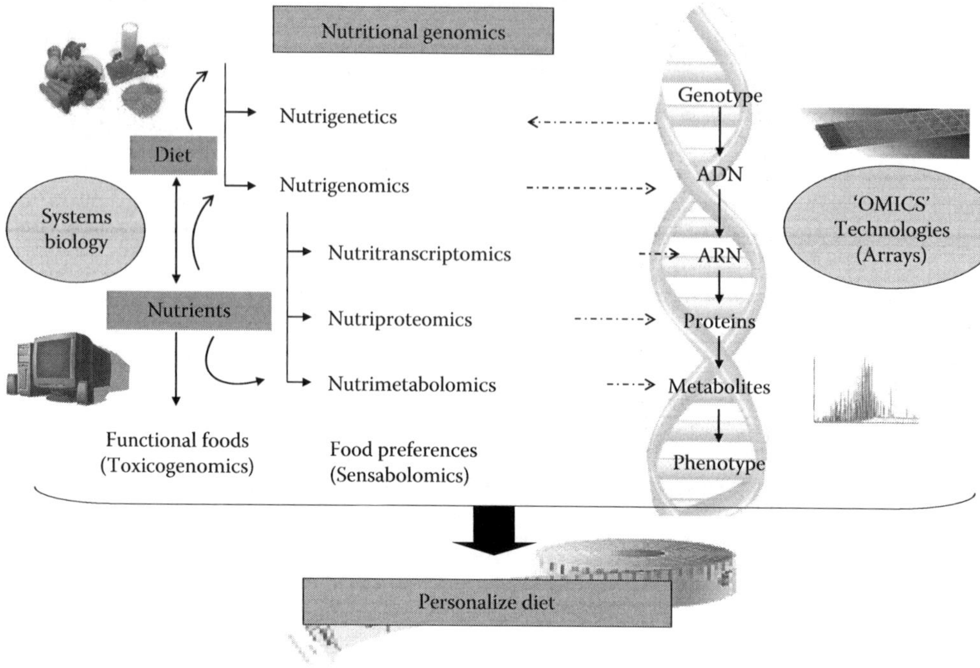

Figure 12.1 *Personalized Nutrition.* The unifying term '*nutritional genomics*' refers to nutrigenomics and nutrigenetics. The new technology research together with the knowledge of the human genome sequence have led the study of the new 'omics' to better understand the molecular basis of the disease development; they are the disciplines knows as transcriptomic, proteomic and metabolomic that allow us to know phenotypes characteristics. Sensabolomics and toxicogenomics sciences, combined with the 'omics' mentioned above, will contribute to create personalized diets.

the patients carrying the mutations will develop the disease. It is already well known that bioactive food compounds can interact with genes affecting transcription factors, protein expression and metabolite production. The determination of molecular effects of food components by DNA–microarray experiments will in future afford the development of functionalized food and personalized advice.

However, although 10 years ago the number of publications regarding gene–diet interaction was low, nowadays the research in the nutrigenomics field is rapidly increasing. The genetic variants that have been mostly studied have been the SNPs; however, there exist a lot of possibilities to study the genome variation as copy number variation, microsatellites and so on. Although several candidate genes for CVD, cancer and other diseases as well as thousands of polymorphism have been identified, this number decreased when we considered the gene–diet interactions.

12.5.1 Gene–Diet Interactions

One of the classical example of single-gene defect that responded to dietary treatment in the case of the phenylketonuria (PKU) employing a low phenylalanine-containing diet for nutrigenetic management. PKU is an autosomal recessive disorder resulting from phenylalanine (Phe) hydroxylase deficiency and characterized by mental retardation. For patients with classical PKU, adequate restriction of Phe intake ensures normal development whereas in general, no dietary restriction is required in subjects with mild hyperphenylalaninemia (Scriver and Waters, 1999).

Table 12.2 Definitions of the New 'Omics' Involved in Nutritional Genomics

Nutritional genomics	The application of high-throughput functional genomic technologies in nutrition research. These technologies can be integrated with databases of genomic sequences and inter-individual genetic variability, enabling the process of gene expression to be studied for many thousands of genes in parallel.	Elliott and Ong (2002)
Nutritranscriptomic	The genome-wide study of mRNA expression levels in one cell or in a population of biological cells for a given set of nutritional conditions.	Panagiotou and Nielsen (2009)
Nutrimetabolomic	The measurement of all metabolites to access the complete metabolic response of an organism to a nutritional stimulus.	Panagiotou et al. (2009)
Nutriproteomics	The large-scale analysis of the structure and function of proteins as well as of protein–protein interactions in a cell to identify the molecular targets of diet components.	Panagiotou et al. (2009)
Sensomics	The map of the sensometabolome and to identify the most intense taste-active metabolites in fresh and processed foods.	Hofmann (2009)
Toxicogenomics	Sub-discipline of toxicology, resulted from the natural convergence of the field of conventional toxicological research and the emergent field of functional genomics. Toxicogenomics endeavours to elucidate molecular mechanisms evolved in the expression of toxicity, and to derive molecular expression patterns (i.e., molecular biomarkers) that predict toxicity or the genetic susceptibility to it.	Mei et al. (2010)
Systems biology	The analysis of the relationships among the elements in a system in response to genetic or environmental perturbations, with the goal of understanding the system or the emergent properties on the system.	Weston and Hood (2004)

On the other hand, it is well known that the response of plasma cholesterol concentration to diet is variable in populations and it is estimated that 50% of the variant is genetically determined. Furthermore, the effect of dietary changes on plasma lipid concentrations differs significantly between individuals. While some individuals appear to be insensitive to dietary changes, others have high sensitivity. Then we can classify the people into hyporesponders, normoresponders and hyperresponders. One established example of that is the variation in the *APOE* gene. Significantly, LDL-C responses to changes in the fat content of the diet by *APOE* genotype were reported in more than 10 studies (Masson et al., 2003). Individuals with the E4/4 genotype respond with an increase in serum cholesterol, whereas those with Apo E2/2 or Apo E3/2 do not show that increase. Changes in the fat content of the diet result in greatest responses in E4 (23% of reduction in LDL-C) than in E3/E3 (14% of reduction) or E3/E2 (13%) (López-Miranda et al., 1994). Thus, specific genetic information is needed to define the optimal diet for an individual.

Another study reporting gene–diet interactions on lipid concentrations is carried out by Ordovas et al. (2002a,b) in the hepatic lipase gene. Hepatic lipase is a plasma lipolytic enzyme that participates in the metabolism of LDL-C into smaller, denser particles and in the conversion of HDL_2 into HDL_3 during reverse cholesterol transport. Several polymorphisms have been described in this gene. One gene–diet interaction was found between –514 C/T *LIPC* polymorphism and dietary fat intake, affecting HDL-related measures in the Framingham Study (1020 men and 1110 women). The T-allele was associated with significantly greater HDL-C concentration and particle size only in subjects consuming <30% of energy from fat. When the total fat intake was =30% of energy, mean HDL-C concentrations and HDL-particle size were lowest among those with the TT genotype, whereas no differences were observed between CC and CT individuals (Ordovas et al., 2002a,b). Later on, this modulation was replicated in Asian populations (Tai et al., 2003).

Four years later, the same group described other interesting interaction between dietary fat and the 1131T > C and 56C > G polymorphism in the *APOA5* gene. The researchers measured plasma fasting TGs, remnant-like particle (RLP) concentrations, and lipoprotein particle size in 1001 men and 1147 women in the Framingham Heart Study. Significant gene–diet interactions between –1131T > C polymorphism and PUFA intake were found ($P < 0.001$). The -1131C allele was associated with higher fasting TGs and RLP concentrations ($P < 0.01$) in only the subjects consuming a high-PUFA diet (>6% of the total energy). The effects were particularly true for w-6 PUFA-rich diets, which seem to be related to a more atherogenic lipid profile in these subjects. Similar gene–diet interactions were not observed in individuals with *APOA5* 56C > G polymorphism (Lai et al., 2006).

We have presented some important examples of gene–diet interaction based on observational and cross-sectional studies. We may distinguish the results derived from interventional studies, in which subjects received a controlled dietary intake providing the best approach for conducting gene–nutrient phenotype, and the observational studies that are considered with lower evidence. Nevertheless, in the first one, the level of replication has been lower probably due to difference in the study design. In the second one, the level of evidence in this particular case studying the nutritional genetics approach can be considered higher due to the Mendelian randomization (random assortment of alleles at the time of gamete formation). Therefore, replication studies are really important to increase the level of evidence and they are needed to give future personalized guidelines based on genetic susceptibility. Recently, Corella et al. (2009) have reported for the first time, a gene–diet interaction in the –265T > C *APOA2* gene influencing body mass index (BMI) and obesity that has been strongly and consistently replicated in three independent populations: the Framingham Offspring Study (1454 whites), the Genetics of Lipid Lowering Drugs and Diet Network Study (1078 whites) and Boston–Puerto Rican Centers on Population Health and Health Disparities Study (930 Hispanics of Caribbean origin). They found that the magnitude of the difference in BMI between the individuals with the CC and TT + TC genotypes differed by saturated fats. The CC genotype was significantly associated with higher obesity prevalence in all populations only in the high saturated fat stratum. These examples illustrate the importance of tailored nutritional advice to populations or individuals because what is appropriate dietary advice for one individual may be inappropriate or harmful to another.

12.5.2 Nutrigenomics and Food Intolerance

Another important field of research for food development and personalized advice is the knowledge of the molecular basis of food intolerance. Gene–diet interactions have also been described in food intolerance. For example, it is well known in lactose intolerance. More than half of the world's population is deficient in lactase (lactase non-persistence). Lactase non-persistence results in the presence of mal-digested lactose in the colon, which may cause unpleasant symptoms such as bloating, abdominal pain and diarrhoea. Genetic lactase non-persistence is a recessively inherited condition caused by a decline in the activity of lactase-phlorizin hydrolase (LPH) in the small intestine during maturation. Enattah et al. (2002) found that lactase persistence correlates strongly with the C/T-13910 polymorphism located 13.9 kb upstream from the lactase gene. Further studies have focussed on this SNP and recently Laaksonen et al. (2009) concluded that subjects with the lactase non-persistence (C/C-13910) genotype consumed less milk since childhood, but the consumption of other milk products did not differ between the genotypes. Because of the gene–nutrient interaction, subjects with specific mutations in the genes involved in LPH deficiency should avoid lactose-containing food.

12.5.3 Nutrigenomics and Food Preferences

Modification of dietary habits is not easy because apart from the difficulty to find methods for nutritional education, little is known about the factors that determine the factors that influence the food preferences. Some of the variation in taste sensitivity has a genetic origin, and many other inter-individual differences are likely to be partially or wholly determined by genetic mechanisms. The food preferences based on genetic susceptibility could help the food industry in the development of new products taking into account that consumer accepts some products and rejects others. Driven by the need to discover the key players that induce food taste, a new 'omic' has emerged, 'sensomics', to map the sensometabolome and to identify the most intense taste-active metabolites in fresh and processed foods. Recent advances in the understanding of taste at the molecular level have provided candidate genes that can be evaluated for contributions to phenotypic differences in taste abilities. One example of that is the response to bitter-tasting compounds such as phenylthiocarbamide or 6-*n*-propylthiouracil individuals can be classified as supertasters, tasters or non-tasters. Variations in the *TAS2R* gene are associated with food choice (Tepper et al., 2009) and a recent study indicated that a polymorphism in *TAS2R38* is associated with differences in ingestive behaviour (Dotson et al., 2010).

Establishing a genetic basis for food likes/dislikes may explain, in part, some of the inconsistencies among epidemiological studies relating diet to risk of chronic diseases. Identifying populations with preferences for particular flavours or foods may lead to the development of novel food products targeted to specific genotypes or ethnic populations.

In conclusion, the current evidence from nutrigenetics studies is not enough to begin implementing specific personalized information. However, there are a large number of examples of common SNPs modulating the individual response to diet as proof of concept of how gene–diet interactions can influence lipid metabolism, BMI or other disorders and it is expected that in the near future we will be able to harness the information contained in our genomes using the behavioural tools to achieve successful personalized nutrition. For achieving that, it is recommended that the future advices in nutrition are based on studies using the highest level of epidemiological evidence and supported by mechanistic studies within the Nutritional Genomics and Systems biology. In the 2007 the U.S government established the Genes and Environment Initiative (GEI) (http://www.gei.nih.gov) with the aim to analyse genetic variations in groups of patients with specific illnesses and to apply an environmental technology development programme to produce and validate the gene environmental interactions (measuring environmental toxins, dietary intake and physical activity) for the development of human diseases. For that purpose GEI uses new tools based on genomics, proteomics and metabolomics.

However, we have to take into account that for achieving successful results in personalized nutrition through the genetics variation, the health professionals need to be prepared to learn and interpret genetic knowledge about their patients.

REFERENCES

Abrams, S. I. Griffin, K. Hawthorne, L. Liang, S. Gunn, G. Darlington and K. Ellis. 2005. A combination of prebiotic short- and long-chain inulin-type fructans enhances calcium absorption and bone mineralization in young adolescents. *Am J Clin Nutr* 82:471–476.

Al-Babili, S. and P. Beyer. 2005. Golden Rice—Five years on the road—five years to go? *Trends Plant Sci* 10:565–573.

Arias-Garzon, D. I. and R. T Sayre. 2000. Genetic engineering approaches to reducing cyanide toxicity in cassava (*Manihot esculenta* Crantz). In *Proceedings of the Fourth International Scientific Meeting Cassava Biotechnology Network*, eds. L. Carvalho, A. M. Thro and A. D. Vilarinhos, pp. 213–221. EMBRAPA, Brazilia, 3–7 November, 1998, Salvador, Brazil.

Avraham, T., H. Badani, S. Galili and R. Amir. 2005. Enhanced levels of methionine and cysteine in transgenic alfalfa (*Medicago sativa* L.) plants over-expressing the Arabidopsis cystathionine γ-synthase gene. *Plant Biotechnol* 3:71–79.

Badejo, A. A., H. A. Eltelib, K. Fukunaga, Y. Fujikawa and K. Esaka. 2009. Increase in ascorbate content of transgenic tobacco plants overexpressing the acerola (*Malpighia glabra*) phosphomannomutase gene. *Plant Cell Physiol* 50:423–428.

Beauregard, M. and M. A. Hefford. 2006. Enhancement of essential amino acid contents in crops by genetic engineering and protein design. *Plant Biotechnol J* 4:561–574.

Bernal-Santos, G., A. M. O'Donnell, J. L.Vicini, G. E. Hartnell and D. E. Bauman. 2010. Hot topic: Enhancing omega-3 fatty acids in milk fat of dairy cows by using stearidonic acid-enriched soybean oil from genetically modified soybeans. *J Dairy Sci* 93:32–37.

Bertelli, A. A. and D. K. Das. 2009. Grapes, wines, resveratrol and heart health. *J Cardiovas Pharmacol* 4:468–476.

Bonanome, A. and S. M. Grundy. 1988. Effect of dietary stearic acid on plasma cholesterol and lipoprotein. *N Engl J Med* 318:1244–1248.

Brennan, R. O. 1975. *Nutrigenetics: New Concepts for Relieving Hypoglycemia.* New York, NY: M. Evans and Co.

Breslow, J. L. 2006. n-3 fatty acids and cardiovascular disease. *Am J Clin Nutr* 83:1477S–1482S.

Brown, L. and F. van der Ouderaa. 2007. Nutritional genomics: Food industry applications from farm to fork. *Br J Nutr* 97:1027–1035.

Brul, S., F. Schuren, R. Montijn, B. J. Keijser, H. van der Spek and S. J. Oomes. 2006. The impact of functional genomics on microbiological food quality and safety. *Int J Food Microbiol* 112:195–199.

Chakauya, E., K. M. Coxon, M. Wei et al. 2008. Towards engineering increased pantothenate (vitamin B(5)) levels in plants. *Plant Mol Biol* 68:493–503.

Damude, H. G. and A. J. Kinney. 2007. Engineering oilseed plants for a sustainable, land-based source of long chain polyunsaturated fatty acids. *Lipids* 42:179–185.

Davies, K. M. 2007. Genetic modification of plant metabolism for human health benefits. *Mutat Res* 622:122–137.

Davuluri, G. R., A. Van Tuinen, P. D. Fraser et al. 2005. Fruit-specific RNAi-mediated suppression of DET1 enhances carotenoid and flavonoid content in tomatoes. *Nat Biotechnol* 23:890–895.

Delaunois, B., S. Cordelier, A. Conreux, C. Clément and P. Jeandet. 2009. Molecular engineering of resveratrol in plants. *Plant Biotech J* 7:2–12.

DellaPenna, D. 1999. Nutricional genomics: Manipulating plant micronutrients to improve human health. *Science* 285:375–379.

Denke, M. A. and S. M. Grundy. 1992. Comparison of effects of lauric acid and palmitic acid on plasma lipids and lipoproteins. *Am J Clin Nutr* 56:895–898.

Díaz de la Garza, R. I., J. F. 3rd Gregory and A. D. Hanson. 2007. Folate biofortification of tomato fruit. *Proc Natl Acad Sci USA* 104:4218–4222

Dodo, H. W., K. N. Konan, F. C. Chen, M. Egnin and O. M. Viquez. 2008. Alleviating peanut allergy using genetic engineering: The silencing of the immunodominant allergen Ara h 2 leads to its significant reduction and a decrease in peanut allergenicity. *Plant Biotechnol J* 6:135–145.

Dotson, C. D., H. L. Shaw, B. D. Mitchell, S. D. Munger and N. I. Steinle. 2010. Variation in the gene TAS2R38 is associated with the eating behavior disinhibition in Old Order Amish women. *Appetite* 54:93–99.

EFSA. 2004. Food and feed safety and the use of animal http://www.efsa.europa.eu/EFSA/DocumentSet/animal_welfare_mb15_doc5_en1,0.pdf. Accessed 12 March, 2010.

EFSA GMO Panel Working Group on Animal Feeding Trials. 2008. Safety and nutritional assessment of GM plants and derived food and feed: The role of animal feeding trials. *Food Chem Toxicol* 46:S2–S70.

Elliott, R. and T. J. Ong. 2002. Nutritional genomics. *Br Med J* 324:1438–1442.

Enattah, N. S., T. Sahi, E. Savilahti, J. D. Terwilliger, L. Peltonen and I. Järvelä. 2002. Identification of a variant associated with adult-type hypolactasia. *Nat Genet* 30:233–237.

Fernández, A. I., E. Alves, A. Fernández et al. 2008. Mitochondrial genome polymorphisms associated with longissimus muscle composition in Iberian pigs. *J Anim Sci* 86:1283–1290.

Flachowsky, G., A. Chesson and K. Aulrich. 2005. Animal nutrition with feeds from genetically modified plants. *Arc Anim Nutr* 59:1–40.

Furness, A. and K. A. Osman. 2003. Developing treaceability systems across the supply chain. *Food Authenticity and Traceability*, ed. M. Lees, pp. 473–495. Cambridge: Woodhead Publishing Limited.

Gibbon, B. C. and B. A. Larkins. 2005. Molecular genetic approaches to developing quality protein maize. *Trends Genet* 21:227–233.

Givens, D. I. 2005. The role of animal nutrition in improving the nutritive value of animal-derived foods in relation to chronic disease. *Proc Nutr Soc* 64:395–402.

Gu, F. and K. L. Jen. 1995. High-fat feeding during pregnancy and lactation affects offspring metabolism in rats. *Physiol Behav* 57:681–686.

Guilbaud, M., I. Chafsey, M. F. Pilet et al. 2008. Response of *Listeria monocytogenes* to liquid smoke. *J Appl Microbiol* 104:1744–1753.

Heisler, L. K., N. Pronchuk, K. Nonogaki et al. 2007. Serotonin activates the hypothalamic–pituitary–adrenal axis via serotonin 2C receptor stimulation. *J Neurosci* 27:6956–6964.

Hofmann, T. 2009. Identification of the key bitter compounds in our daily diet is a prerequisite for the understanding of the hTAS2R gene polymorphisms affecting food choice. *Ann N Y Acad Sci* 1170:116–125.

Jiménez-Gómez, J. M. and J. N. Maloof. 2009. Sequence diversity in three tomato species: SNPs, markers, and molecular evolution. *BMC Plant Biol* 3:9–85.

Karppi, J., S. Kurl, T. Nurmi, T. H. Rissanen, E. Pukkala and K. Nyyssönen. 2009. Serum lycopene and the risk of cancer: The Kuopio Ischaemic Heart Disease Risk Factor (KIHD) study. *Ann Epidemiol* 19: 512–518.

Kim, J. M., B. D. Choi, B. C. Kim, S. S. Park and K. C. Hong. 2009. Associations of the variation in the porcine myogenin gene with muscle fibre characteristics, lean meat production and meat quality traits. *Anim Breed Genet* 126:134–141.

Koch, L. G. and S. L. Britton. 2008. Development of animal models to test the fundamental basis of gene–environment interaction. *Obesity* 16:S28–S32.

Kreft, O., R. Höfgen and H. Hesse. 2005. Functional analysis of cystathionine-synthase in genetically engineered potato plants. *Plant Physiol* 131:1843–1854.

Laaksonen, M. M., V. Mikkilä, L. Räsänen et al. 2009. Cardiovascular risk in young Finns study group. 2009. Genetic lactase non-persistence, consumption of milk products and intakes of milk nutrients in Finns from childhood to young adulthood. *Br J Nutr* 102:8–17.

Lai, C. Q., D. Corella, S. Demissie et al. 2006. Dietary intake of *n*-6 fatty acids modulates effect of apolipoprotein A5 gene on plasma fasting triglycerides, remnant lipoprotein concentrations, and lipoprotein particle size: The Framingham Heart Study. *Circulation* 13:2062–2070.

Lalonde, M. 1974. *A New Perspective on the Health of Canadians: A Working Document*. Ottawa, ON: Department of Health and Welfare.

Lang, M. L., Y. X. Zhang and T. Y. Chai. 2004. Advances in the research of genetic engineering of heavy metal resistance and accumulation in plants. *Sheng Wu Gong Cheng Xue Bao* 20:157–164.

Le, L. Q., Y. Lorenz, S. Scheurer et al. 2006. Design of tomato fruits with reduced allergenicity by dsRNAi-mediated inhibition of ns-LTP (Lyc e 3) expression. *Plant Biotechnol* 4:231–242.

Le, L. Q., V. Mahler, Y. Lorenz et al. 2006. Reduced allergenicity of tomato fruits harvested from Lyc e 1-silenced transgenic tomato plants. *J Allergy Clin Immunol* 118:1176–1183.

Lee, T. I., N. J. Rinaldi, F. Robert et al. 2002. Transcriptional regulatory networks in *Saccharomyces cerevisiae*. *Science* 298:799–804.

Leskanich, C. O. and R. C. Noble. 1997. Manipulation of the n-3 polyunsatured fatty acid composition of avian eggs and meat. *Poult Sci* 53: 155–183.

Levin, B. E. and E. Goveck. 1998. Gestational obesity accentuates obesity in obesity-prone progeny. *Am J Physiol* 275:R1374–R1379.

Lopez-Miranda, J., J. M. Ordovas, P. Mata et al. 1994. Effect of apolipoprotein E phenotype on diet-induced lowering of plasma low density lipoprotein cholesterol. *J Lipid Res* 35:1965–1975.

Lorenz, Y., E. Enrique, L. LeQuynh et al. 2006. Skin prick tests reveal stable and heritable reduction of allergenic potency of gene-silenced tomato fruits. *J Allergy Clin Immunol* 118:711–718.

Luo, J., E. Butelli, L. Hill et al. 2008. AtMYB12 regulates caffeoyl quinic acid and flavonol synthesis in tomato: Expression in fruit results in very high levels of both types of polyphenol. *Plant J* 56:316–326.

Mann, G., S. Diffey, B. Cullis et al. 2009. Genetic control of wheat quality: Interactions between chromosomal regions determining protein content and composition, dough rheology, and sponge and dough baking properties. *Theor Appl Genet* 118:1519–1537.

Marques, E., J. D. Nkrumah, E. L. Sherman and S. S. Moore. 2009. Polymorphisms in positional candidate genes on BTA14 and BTA26 affect carcass quality in beef cattle. *J Anim Sci* 87:2475–2484.

Martín-Platero, A. M., E. Valdivia, M. Maqueda and M. Martínez-Bueno. 2009. Characterization and safety evaluation of enterococci isolated from Spanish goats' milk cheeses. *Int J Food Microbiol* 132:24–32.

Masson, L. F., G. McNeill and A. Avenell. 2003. Genetic variation and the lipid response to dietary intervention: A systematic review. *Am J Clin Nutr* 77:1098–1111.

Mei, N., J. C. Fuscoe, E. K. Lobenhofer and L. Guo. 2010. Application of microarray-based analysis of gene expression in the field of toxicogenomics. *Methods Mol Biol* 597:227–241.

Middleton, E., C. Kandaswami and T. C. Theoharides. 2000. The effect of plant flavonoids on mammalian cells: Implications for inflammation, heart disease and cancer. *Pharmacol Rev* 52:673–751.

Miraglia, N., G. Assennato, E. Clonfero, S. Fustinoni and N. Sannolo. 2004. Biologically effective dose biomarkers. *G Ital Med Lav Ergon* 26:298–301.

Naqvi, S., C. Zhu, G. Farre et al. 2009. Transgenic multivitamin corn through biofortification of endosperm with three vitamins representing three distinct metabolic pathways. *Proc Natl Acad Sci USA* 106: 7762–7767.

Nunes, A. C., D. C. Kalkmann and F. J. Aragão. 2009. Folate biofortification of lettuce by expression of a codon optimized chicken GTP cyclohydrolase I gene. *Transgenic Res* 18: 661–667.

O'Brien, G. M., A. J. Taylor and N. H. Poulter. 1991. Improved enzymatic assay for cyanogens in fresh and processed cassava. *J Sci Food Agric* 56:277–289.

Ogundiwin, E. A., C. P. Peace, T. M. Gradziel, D. E. Parfitt, F. A. Bliss and C. H. Crisosto. 2009. A fruit quality gene map of *Prunus. BMC Genom* 10:587.

Oomes, S. J., M. J. Jonker, F. R. Wittink, J. O. Hehenkamp, T. M. Breit and S. Brul. 2009. The effect of calcium on the transcriptome of sporulating *B. subtilis* cells. *Int J Food Microbiol* 133:234–242.

Ordovas, J. M. and D. Corella. 2004. Nutritional genomics. *Annu Rev Genomics Hum Genet* 5:71–118.

Ordovas, J. M., D. Corella, L. A. Cupples et al. 2002a. Polyunsaturated fatty acids modulate the effects of the APOA1 G-A polymorphism on HDL-cholesterol concentrations in a sex-specific manner: The Framingham Study. *Am J Clin Nutr* 75:38–46.

Ordovas, J. M., D. Corella, S. Demissie et al. 2002b. Dietary fat intake determines the effect of a common polymorphism in the hepatic lipase gene promoter on high-density lipoprotein metabolism: Evidence of a strong dose effect in this gene–nutrient interaction in the Framingham Study. *Circulation* 106:2315–2321.

Page, T., G. Griffiths and V. Buchanan-Wollaston. 2001. Molecular and biochemical characterization of post-harvest senescence in broccoli. *Plant Physiol* 125:718–727.

Panagiotou, G. and J. Nielsen. 2009. Nutritional systems biology: Definitions and approaches. *Annu Rev Nutr* 29:329–339.

Patterson, C. M., S. G. Bouret, A. A. Dunn-Meynell and B. E. Levin. 2008. Three weeks of postweaning exercise in DIO rats produces prolonged increases in central leptin sensitivity and signaling. *Am J Physiol Regul Integr Comp Physiol* 296:R537–R548.

Pottenger, L. H., J. S. Bus and B. B. Gollapudi. 2007. Genetic toxicity assessment: Employing the best science for human safety evaluation part VI: When salt and sugar and vegetables are positive, how can genotoxicity data serve to inform risk assessment? *Toxicol Sci* 98:327–331.

Prescott, V. E. and S. P. Hogan. 2006. Genetically modified plants and food hypersensitivity diseases: Usage and implications of experimental models for risk assessment. *Pharmacol Ther* 111:374–383.

Ramos, A. M., K. L. Glenn, T. V. Serenius, K. J. Stalder and M. F. Rothschild. 2008. Genetic markers for the production of US country hams. *J Anim Breed Genet* 125:248–257.

Raspor, P. 2005. Bio-markers: Traceability in food safety issues. *Acta Biochim Pol* 52:659–664.

Reid, A., L. Hof, D. Esselink and B. Vosman. 2009. Potato cultivar genome analysis. *Methods Mol Biol* 508:295–308.

Roberfroid, M. 2005. Introducing inulin-type fructans. *Br J Nutr* 93:S13–S25.

Rommens, C. M., M. A. Haring, K. Swords, H. V. Davies and W. R. Belknap. 2007. The intragenic approach as a new extension to traditional plant breeding. *Trends Plant Sci* 12:397–403.

Ruiz-López, N., R. P. Haslam, M. Venegas-Calerón et al. 2009. The synthesis and accumulation of stearidonic acid in transgenic plants: A novel source of 'heart-healthy' omega-3 fatty acids. *Plant Biotech J* 7:704–716.

Rymer, C., D. I. Givens and K. W. J. Wahle. 2003. Dietary strategies for increasing docosahexaenoic acid (DHA) and eicosopentaenoic acid (EPA) concentrations in bovine milk: A review. *Nutr Abst Rev* 73B:9R–25R.

Schwab, C. R., B. E. Mote, Z. Q. Du, R. Amoako, T. J. Baas and M. F. Rothschild. 2009. An evaluation of four candidate genes for use in selection programmes aimed at increased intramuscular fat in Duroc swine. *J Anim Breed Genet* 126:228–236.

Scriver, C. R. and P. J. Waters. 1999. Monogenic traits are not simple: Lessons from phenylketonuria. *Trends Genet* 15:267–272.

Sevenier, R., R. D. Hall, I. M. van der Meer, H. J. C. Hakkert, A. J. van Tunen and A. J. Koops. 1998. High level fructan accumulation in a transgenic sugar beet. *Nature Biotechnol* 16:843–846.

Sevenier, R., I. M. Van der Meer, R. Bino and A. Koops. 2002. Increased production of nutriments by genetically engineered crops. *J Am Coll Nutr* 21:199S–204S.

Shingfield, K. J., S. Ahvenjärvi, V. Toivonen et al. 2003. Effect of fish oil on biohydrogenation of fatty acids and milk fatty acid content in cows. *J Anim Sci* 77:165–179.

Siritunga, D. and R. T. Sayre. 2007. Transgenic approaches for cyanogen reduction in cassava. *J AOAC Int* 90:1450–1455.

Soriano, J. M., H. Rico, J. C. Moltó and J. Mañes. 2002. Effect of introduction of HACCP on the microbiological quality of some restaurant meals. *Food Control* 13:253–261.

Speakman, J., C. Hambly, S. Mitchell and E. Król. 2008. The contribution of animal models to the study of obesity. *Lab Anim* 42:413–432.

Spielbauer, B. and F. Stahl. 2005. Impact of microarray technology in nutrition and food research. *Mol Nutr Food Res* 49:908–917.

Sun, Y. M. and H. W. Ockerman. 2005. A review of the needs and current applications of hazard analysis and critical control point (HACCP) system in foodservice areas. *Food Control* 16:325–332.

Tai, E. S., D. Corella, S. Demissie et al. 2005. Framingham Heart Study. Polyunsaturated fatty acids interact with the PPARA-L162V polymorphism to affect plasma triglyceride and apolipoprotein C-III concentrations in the Framingham Heart Study. *J Nutr* 135:397–403.

Tai, E. S., D. Corella, M. Deurenberg-Yap et al. 2003. Dietary fat interacts with the -514C > T polymorphism in the hepatic lipase gene promoter on plasma lipid profiles in a multiethnic Asian population: The 1998 Singapore National Health Survey. *J Nutr* 133:3399–3408.

Temme, E. H. M., R. P. Mensink and G. Hornstra. 1996. Comparison of the effects of diets enriched in lauric, palmitic, or oleic acids on serum lipids and lipoproteins in healthy women and men. *Am J Clin Nutr* 63:897–903.

Tepper, B. J., E. A. White, Y. Koelliker, C. Lanzara, P. d'Adamo and P. Gasparini. 2009. Genetic variation in taste sensitivity to 6-*n*-propylthiouracil and its relationship to taste perception and food selection. *Ann N Y Acad Sci* 1170:126–139.

Van der Meer, I., A. J. Koops, J. C. Hakkert and A. J. van Tunen. 1998. Cloning of the fructan biosynthesis pathway of Jerusalem artichoke. *Plant J* 15:489–500.

Weber, W. W. 1999. Populations and genetic polymorphisms. *Mol Diagn* 4:299–307.

Weeks, J. T., J. Ye and C. M. Rommens. 2008. Development of an in planta method for transformation of alfalfa (*Medicago sativa*). *Transgenic Res* 17:587–597.

Weston, A. D. and L. Hood. 2004. Systems biology, proteomics, and the future of health care: Toward predictive, preventative, and personalized medicine. *J Proteome Res* 3:179–196.

Weyens, G., T. Ritsema, K. Van Dun et al. 2004. Production of tailor-made fructans in sugar beet by expression of onion fructosyltransferase genes. *Plant Biotech J* 2: 321–327.

White, W. L. B., J. M. McMahon and R. T. Sayre. 1994. Regulation of cyanogenesis in cassava. *Acta Hort* 375:69–78.

WHO. World Health Organization. 2003. Diet, nutrition and the prevention of chronic diseases. Report of a Joint WHO/FAO Expert Consultation, Geneva, Technical Report Series, No. 916.

WHO. World Health Organization. 2007. Assessment of iodine deficiency disorders and monitoring their elimination. A guide for programme managers. WHO, Geneva.

WHO. World Health Organization. 2008. WHO Global database on anaemia. Worldwide prevalence of anaemia 1993–2005. WHO, Geneva.

WHO. World Health Organization. 2009. Global prevalance of vitamin A deficiency in populations at risk 1995–2005. WHO, Geneva.

Wimmers, K., E. Murani, M. F. Te Pas et al. 2007. Associations of functional candidate genes derived from gene-expression profiles of prenatal porcine muscle tissue with meat quality and muscle deposition. *Anim Genet* 38:474–484.

Woolfe, M. and S. Primrose. 2004. Food forensics: Using DNA technology to combat misdescription and fraud. *Trends Biotechnol* 22:222–226.

Yusuf, M. A. and N. B. Sarin. 2007. Antioxidant value addition in human diets: Genetic transformation of *Brassica juncea* with gamma-TMT gene for increased alpha-tocopherol content. *Transgenic Res* 16:109–113.

Zock, P. L., J. H. de Vries and M. B. Katan. 1994. Impact of myristic acid versus palmitic acid on serum lipid and lipoprotein levels in healthy women and men. *Arterioscler Thromb* 14:567–575.

CHAPTER 13

Metabolomics

Current View on Fruit Quality in Relation to Human Health

Ilian Badjakov, Violeta Kondakova and Atanas Atanassov

CONTENTS

13.1 INTRODUCTION

The technologies that aimed at giving general information about the total genome sequence or the complete transcriptional analysis of an organism are already expanding research knowledge over the past years.

According to Hall (2006), the useful working definitions of metabolomics can be formulated as the technology geared towards providing an essentially unbiased, comprehensive qualitative and quantitative overview of the metabolites present in an organism. During the recent years, plant metabolomics has been a valuable technology applied in global research achievement on the molecular organization of multicellular organisms. Considerable advances have been made in metabolomics applications in the past years.

Metabolomics technology plays a significant role in bridging the phenotype–genotype gap. The increasing number of publications in this domain demonstrate that metabolomics is not just a new 'omics' but a valuable emerging tool to study phenotype and changes in phenotype caused by environmental influences, disease or changes in genotype (Bum and Withell, 1941; Desbrosses et al., 2005; Dettmer and Hammock, 2004; Patel et al., 2004). More detailed information on the biological and biochemical composition of plant tissues has great potential value in a wide range of scientific and applied fields (Fiehn, 2002; Methora and Mendes, 2006; Ross et al., 2007). The quality of plants is a direct function of their metabolite content (the existence of definite metabolic profiles, even in single analyses) and indicates their commercial value (Hakkinen et al., 1999a,b, 2000; Hung et al., 2004; Ozarowski and Jaroniewski, 1989; Puupponen-Pimia et al., 2004). According to Mazza and

Miniati (1993), the total number of metabolites that are produced by plants vary considerably and they are in the range between 100,000 and 200,000. This complexity can be used to define plants at every level of genotype, phenotype, tissue and cell. The secondary metabolites can be found and each species may contain its own phenotypic expression pattern. Substantial quantitative and qualitative variation in metabolite composition is observed in plant species (Gullberg et al., 2004; Keurentjes et al., 2006).

Stitt and Fernie (2003) reports that in contrast to transcriptomics and proteomics, which rely to a great extent on genome information, metabolomics is mainly species independent, which means that it can be applied to widely diverse species with relatively little time required for reoptimizing protocols for a new species (Kopka et al., 2004).

The potential of metabolomics technologies in the study of the genetics of metabolite accumulation and metabolic regulation is in progress too. The first report in this aspect has been published (Hall, 2006; Schauer and Fernie, 2006) suggesting that the combination of metabolite profiling and marker-assisted selection could prove highly informative in understanding and influencing the chemical composition of crop species (Fernie, 2003).

Nowadays many reviews and books with fundamental principles, technical aspects of metabolite profiling (Fernie et al., 2004; Harrigan and Goodacre, 2003; Methora and Mendes, 2006; Sumner et al., 2003) and metabolite profiles in spectral libraries (Kopka et al., 2005; Schauer et al., 2005) are being published, and are formulating a series of reporting standards (Bino et al., 2004; Brazma et al., 2001; Hall, 2006). The study of metabolomics is a major challenge to analytical chemistry and metabolomics analysis.

The antioxidant effect of fruits phenolics towards inhibiting lipid and protein oxidation is a well-recognized factor in improving the quality of food. The phenolics have been related to the onset of many diseases, including cardiovascular diseases and cancer. Accordingly, many studies concerning the relevance of polyphenols to humans, ingestion of cranberry juice, black currants and red wine have been reported to increase plasma antioxidant activity (Williamson and Manach, 2005). Recently, many investigations have shown that some fruits phenolics components have antiproliferative effects on cultured cancer cell lines.

The interest of research community is to find evidences for the effects of polyphenol consumption on health. It is very important to specify which out of the hundreds of existing polyphenols are likely to provide the greatest protection in the context of preventive nutrition. Such knowledge will allow the evaluation of polyphenol intake and enable epidemiologic analysis that will in turn provide an understanding of the relation between the intake of these substances and the risk of development of several diseases.

13.2 METABOLITE PROFILING

Currently, two complementary approaches are used for metabolomics investigations: metabolic profiling and metabolic fingerprinting (Schauer et al., 2005; Urbanczyk-Wochniak and Fernie, 2005). Metabolic profiling focuses on the analysis of a group of metabolites related either to a specific metabolic pathway or a class of compounds. An even more directed approach is target analysis that aims at the defining biomarkers of disease, toxicant exposure or substrates and products of enzymatic reactions (Hamzehzarghani et al., 2005). Another approach is the metabolic fingerprinting. This approach is intended to compare patterns or 'fingerprints' of metabolites that change in response to disease, toxin exposure and environmental or genetic alterations. Both metabolic fingerprinting and profiling can be used while searching for new biomarkers. Having more detailed information on the biochemical composition of plant tissues has great potential value in a wide range of both scientific and applied fields (Brosche et al., 2005; Robinson et al., 2005). When previously unidentified metabolites covary with known metabolites, this can be used to build testable hypotheses about their biosynthetic pathway. Such compounds can then be used to elucidate metabolite synthesis in pathways associated with key phenotypic features such as taste or other food quality.

Metabolite profiling is a fast-growing technology and is useful for phenotyping and diagnostic analyses of plants. It is also rapidly becoming a key tool in functional annotation of genes and in the comprehensive understanding of the cellular response to biological conditions. It has been performed on a diverse array of plant species. In its earliest applications, metabolite profiling used metabolite composition as a diagnostic tool to ascertain change in different aspects such as the metabolic response to herbicide (Ott et al., 2003; Sauter et al., 1988), the equivalence of genetically modified organisms, conventional crops (Catchpole et al., 2005; Defernez et al., 2004) and the classification of plant genotypes (Fiehn et al., 2000; Roessner et al., 2000, 2001a,b; Tikunov et al., 2005). More recently, it has been much used to ascertain the response of plants to a wide range of biotic or abiotic stresses (Cook et al., 2004; Hirai et al., 2004, 2005; Kaplan et al., 2004; Nikiforova et al., 2005a,b; Urbanczyk-Wochniak et al., 2005). At the same time an important work is being carried out in these areas, where metabolite profiling has also been used increasingly to decipher gene function (Goossens et al., 2003; Hirai et al., 2004; Morikawa et al., 2006; Suzuki et al., 2005), investigate metabolic regulation (Junker et al., 2004; Morino et al., 2005; Roessner et al., 2001a,b), and as part of integrative analyses of the systemic response to environmental or genetic varying (Sweetlove and Fernie, 2005). One of the recent applications of metabolite profiling is as a diagnostic aid in relation with the determination of various herbicides using gas chromatography coupled with gas chromatography–mass spectrometry (GC–MS).

A similar GC–MS-based approach to that used by Sauter et al. (1988) has also been used to assess the utility of metabolite profiling as a genotyping tool. He showed that metabolite profiling could be used in connection with statistics, not only as a genotyping tool, but also for identification of the most important compounds underlying the separation of genotypes. Metabolite profiling has also been useful for diagnosis in plants in taxonomic evaluation (Merchant et al., 2006).

The metabolic response of plants to various kinds of stresses—biotic (Broeckling et al., 2005; Desbrosses et al., 2005; Hamzehzarghani et al., 2005) or abiotic (Ishizaki et al., 2006, 2005; Kaplan et al., 2004; Merchant et al., 2006; Nikiforova et al., 2005a,b; Urbanczyk-Wochniak et al., 2005) is currently receiving increasing attention. This is driven partly by the increasing losses that these processes impose on agriculture and partly by the development of tools to measure such responses (Bohnert et al., 2006). The majority of such studies are still dominated by analysis at the level of the transcript, including measurements of steady-state metabolite levels. However, much targeted research has been carried out in the past, studying metabolites at the level of single metabolic pathways that are thought by the researchers to be important within a given process (Fernie, 2003). The stress that has been best characterized using metabolite profiling is the response and acclimation to temperature stress, which has been the subject of several investigations (Cook et al., 2004; Kaplan et al., 2004).

The use of metabolite profiling for the characterization of different kinds of stresses has resulted in a range of novel observations that could enable the understanding of the complex responses of such stresses (Stitt and Fernie, 2003). Although many studies were focused on a relatively limited number of metabolites (Ferrario-Mery et al., 2002; Noctor et al., 2002; Stitt et al., 2002), mass spectrometry has recently been applied to tomato plants exposed to nitrogen limitations (Urbanczyk-Wochniak and Fernie, 2005).

In summary, specific metabolites have been associated with stress for decades, metabolite profiling of these stresses and of diurnal alterations of metabolite abundance in leaves (Thimm et al., 2004) have created a far greater number of responses (Zimmermann et al., 2004) .

Metabolite profiling provides direct functional information on metabolic phenotypes and indirect functional information on a range of phenotypes that are determined by small molecules and it has great potential as a tool for functional genomics (Broeckling et al., 2005). This research aim received more attention than any other in the metabolomics field in recent years (Achnine et al., 2005; Hirai et al., 2004; Mercke et al., 2004; Morikawa et al., 2006; Suzuki et al., 2005; Tohge et al., 2005).

Early studies attempted gene annotation by biochemical assay, but this approach has not been commonly applied, because of the need for more time for labour experience (Martzen et al., 1999). Recent studies have found that analysis of the metabolite composition of mutants can aid the assignment of functions to genes (Raamsdonk et al., 2001). In plants, several genes have been commented upon on the basis of correlations between transcript and metabolite levels or simply of altered metabolite profiles. In these examples, altered metabolite profiles could be correlated to specific genes.

Integrated genomics approaches are increasing including metabolite-based approaches (Alba et al., 2005; Andersson-Gunneras et al., 2006; Kristensen et al., 2005; Nikiforova et al., 2005a,b; Suzuki et al., 2005; Tohge et al., 2005; Villas-Boas et al., 2005). The testing of several candidate genes by using reverse genetic approaches has recently confirmed that manipulation of at least one of the genes identified by this method resulted in the anticipated change in metabolite content.

The majority of cultivated crops carry only a small fraction of the genetic variation available in related wild species and land races (Fernie et al., 2006). It has therefore become a goal of modern plant breeding to screen wild genetic resources that could be introduced into modern varieties to improve specific traits. Multiple traits were analysed in one population enabling an assessment of which metabolite accumulation traits were associated with changes in fruit yield; this should prove to be a highly useful resource because it enables the selection of genotypes that have beneficial traits introgressed from the wild species without detrimental effects on other important agronomic parameters. Such integrated approaches therefore offer great potential in the commercialization of crops selected for nutritional fortification.

The studies that give the answer for the influence of genetic factors on metabolic accumulation and metabolite regulation will be of great potential for identifying right strategies for breeding and also for fundamental understanding of metabolism. In recent years, many groups have taken up the challenge of integrating metabolite profiling within broader experimental analyses for understanding metabolic regulation.

The intense changes in the environment, extreme temperatures, drought periods and frequent storms have a negative influence on the growth and development of fruit plants. The skin of fruits is composed of different substances which have an important environmental function. They give colour to the fruits, attract insects and other animals that are important for the pollination and spreading of the seeds, protect plants from harmful ultraviolet radiation and perform a screening role from diverse forms of stress. Among the secondary products, the group of phenolic substances is extremely important. The secondary metabolites and especially phenols are also very important in the plant defense mechanisms. Many scientists claim that phenols and their rapid synthesis enhance plant resistance to pathogens.

Recent research has indicated that the majority of these secondary substances have a beneficial impact on human health and their increased concentrations in fruits are positive. They have an antioxidative role and can even prevent the development of some chronic diseases.

13.3 FRUIT QUALITY RELATION WITH HUMAN HEALTH

Metabolomics assay as a new dimension in studies and practice focuses its attention on the biochemical contents of cells and tissues, and has a rapidly increasing knowledge of fruits importance for human health. General fruits are very rich sources of bioactive compounds such as phenolics and organic acid. Bioactive fruits compounds, their characterization and utilization in functional foods and clinical assessment of antimicrobial properties for human health are among the major targets of contemporary research. Phenolic compounds inhibit the growth of human pathogens. Especially small fruits—raspberry, strawberry, cranberry and crowberry—showed evidence of antimicrobial effects against bacterial pathogens as *Salmonella* and *Staphylococcus*. The evaluation of fruit resources for the presence of bioactive compounds and their properties as natural agents is significant and will be of great benefit for breeders, food and pharmaceutical industry.

The beneficial health effects of fruits are indisputable and largely discussed and proved (Goodacre, 2005; Gullberg et al., 2004; Hancock et al., 2005; Kahkonen et al., 2001; Puupponen-Pimia et al., 2004, 2005a,b; Verhoeven et al., 2006). Berry fruits, wild or cultivated, are proved as a traditional and rich source of bioactive compounds, possessing important biological activities (Dunn and Ellis, 2005; Facchini et al., 2004). Studies of Dettmer et al. (2007) on biochemical profiles by high-performance liquid chromatography (HPLC), revealed the presence of flavonoids (kaempferol, quercetin, myricetin), phenolic acids (galic, *p*-coumaric, caffeic, ferulic) and ellagic acid polymers (ellagitannins).

During the past years, many research projects were targeted at studying bioactive fruit compounds, their characterization and utilization in functional foods and clinical assessment of antimicrobial properties for human health (Lee and Lee, 2006; Maas et al., 1991; Michels et al., 2006; Misciagna et al., 2000; Moyer et al., 2002a,b; Mullen et al., 2002). In recent documents, the World Health Organization emphasized on the importance of antioxidants' activity of flavonoids, especially from fruits, for prevention of most important health problems as cardiovascular and diabetes diseases (Arts and Hollman, 2005).

As a matter of fact, phenolic acids and flavonoids are widely distributed in higher plants and form part of human diet. High antioxidant content in foods provides potential health benefits such as reduction of coronary heart disease, antiviral and anticancer activity. The earlier reports have listed standard cultivars of dark-fruited berries (raspberry, blueberry, gooseberry, blackberry) having high natural antioxidant content relative to vegetables or other foods (Hall et al., 2005; Puupponen-Pimia et al., 2004; Stewart et al., 2007; Wang et al.,1996, 1997). In addition to vitamins and minerals, extracts of raspberries are also rich in anthocyanins, other flavonoids and phenolic acids.

Anthocyanins in fruits comprise a large group of water-soluble pigments. In the fruit, they are found mainly in the external layers of the hypodermis (the skin). In the cells, they are present in vacuoles in the form of various-sized granules; meanwhile cell walls and flesh tissue practically do not contain anthocyanins.

The determination of the range of anthocyanin content, phenolic content and antioxidant capacity in wild species (*Rubus* L. and *Ribes* L.) and cultivar germplasm of dark fruits is very impressive (Memelink, 2004; Rojas-Vera et al., 2002; Scalbert et al., 2005). The berry species belonging to the *Rubus* and *Ribes* genus have very high amounts of antioxidant compounds. For example, the quantity of anthocyanin content in black currant cultivars (*Ribes nigrum* L.) ranges between 128 and 420 mg/100 g fruit; for blackberries (*Rubus hybrid*) to 250 mg/100 g fruit and for black raspberry to 630 mg/100 g fruit.

The earlier studies confirm that the antioxidant ability of raspberry fruit is derived from the contribution of phenolic compounds in raspberries (Tsiotou et al., 2005; Wang et al., 2002).

For example, the antioxidant capacity of raspberry fruit is of course not determined by a single component. Different growing conditions influence the flavonoid content and antioxidant activity of strawberry and raspberry cultivars (Trethewey, 2004).

The dominant antioxidants in small fruits could be classified as being vitamin C, several anthocyanins, ellagitannins and some minor proanthocyanidins-like tannins. Vitamin C is quite abundant in many fruit and vegetable species. It is not specific only for berries, but nevertheless the fruit provides about 20–30 mg vitamin C per 100 g fruits. Vitamin C can make up about 20% of the total antioxidant capacity of raspberry fruits (De Ancos et al., 2000). Anthocyanins contribute about 25% to the antioxidant capacity of red raspberry fruit. They are often involved in the pigmentation of fruits and flowers. As in other red fruits, the average content of anthocyanins in raspberry is 200–300 mg per 100 g dry weight. Some other berries, such as bilberries (*Vaccinium myrtillus*) accumulate even more—between 2 and 3 g anthocyanin per 100 g (Beekwilder et al., 2005).

On a quantitative basis, anthocyanin contents of fruits vary considerably with concentrations ranging from 0.25 mg/100 g FW in pear peel to >200 mg/100 g FW for black fruits rich in anthocyanins (Macheix et al., 1990). However, much higher anthocyanin contents have been reported; anthocyanins contribute 10.7% to the dry weight of fruits of the Mediterranean shrub, *Coriaria myrtifolia* (Escribano-Bailón et al., 2002).

Fruits of some cultivated crops (e.g., apple, mango and peach) may even be bicoloured due to a light requirement for anthocyanin synthesis. This light requirement results in fruits having a red sun-exposed side and a green shaded side that may turn yellow when they ripen (Awad et al., 2000; Hetherington, 1997).

The biggest contribution to the antioxidant capacity in raspberry is of ellagitannins. Among different fruits species, ellagitannins are only represented in cloudberry and raspberry (between 1 and 2 g/100 g dry weight and to a minor extent in strawberry (around 100 mg/100 g FW) (Daniel et al., 1989; Hancock et al., 2005).

Among the naturally occurring pigments in fruits, anthocyanins are arguably the best understood and most studied group. Research into their occurrence, inheritance and industrial use encompasses hundreds of years of human history, and many volumes are dedicated to describing the prevalence, type and biosynthesis of anthocyanins.

Only recently have studies begun to explain the reasons for the accumulation of these red pigments in various tissues of plants. With the fundamental advances in understanding the functional attributes of anthocyanins *in-planta*, the potential of anthocyanins as compounds of industrial importance, as pigments, has been realized. With the increasing knowledge of the biosynthesis of anthocyanins in plant systems it has become feasible to engineer microbial species to contain a functional anthocyanin pathway.

The various functions are considered and discussed with reference to the prevalence of different fruit colours and the contribution of anthocyanins to the rate as well as anthocyanin accumulation in response to environmental factors and fruit quality parameters. Blue, purple, black and most red fruits derive their colour from anthocyanin accumulating during ripening.

No differences in antioxidant capacity have been found between berry cultivars. For most raspberry cultivars defined by HPLC analysis in Europe, nine different anthocyanin peaks were detected. Significant differences were reported within the cultivars with respect to the relative amounts of each of these individual components. The ellagitannins were always the dominant antioxidants in all wild fruits and cultivars. The value of pink fruit, compared to fully ripe, red fruit, can be up to 50% lower. These differences would seem to be very relevant for determining the best harvest time. This indicates that growth conditions, including stress, may affect antioxidants and thus might be used in the future to manipulate antioxidant levels at the time of harvest (Hall et al., 2005). Fruit development from a flower to a ripe fruit is a complex process that involves modification of cellular compartments, loss of cell wall structure causing softening and accumulation of carbohydrates (Brady, 1987). The production of secondary metabolites during the ripening process is an essential phenomenon for the contribution of seed dispersal of the plant in the form of accumulation of pigments and flavour compounds.

The antioxidant capacity of anthocyanins is one of their most significant biological and human health properties. The prevalence of black fruits could relate to the powerful antioxidant ability of anthocyanins. On the other hand, blackness also correlates with fruit maturity and quality (Whelan and Willson, 1990). The presence of anthocyanins in immature fruits and its regulation by environmental factors could relate to the photoprotective ability of anthocyanins. Since anthocyanins are able to fulfil a range of functions in plants, their adaptive value in fruits should be interpreted against a background of interaction with dispersers, genotype and environment. Anthocyanins also accumulate in vegetative tissues where they are considered to confer protection against various biotic and abiotic stresses (Gould and Lister, 2006). Anthocyanins are able to protect tissues from photoinhibition caused by high levels of visible light (Smillie and Hetherington, 1999) and from oxidative damage (Neill and Gould, 2003). Many wild and cultivated plant species display stable polymorphisms in fruit colour (Traveset et al., 2001; Whitney and Lister, 2004; Willson, 1986). This means that plants of the same species carry fruits that differ in colour. Comparison between species does not indicate a consistent preference of dispersers.

Some temperate *Rosaceous* fruits, that is, apple, pear, peaches, nectarines and apricots, are considered to have an absolute light requirement for anthocyanin synthesis in the skin (Allen, 1932). Other fruits including strawberries, blackberries, grapes, cherries and plums are considered to be able to develop colour, albeit to a lesser extent, in the absence of light (Allen, 1932). An absolute light requirement has also been documented for ripening figs (Puech et al., 1976). It can also be inferred, for at least some mango and pomegranate cultivars, from the absence of anthocyanins from shaded skin (Gil et al., 1995; Hetherington, 1997). Anthocyanin levels in cranberry also increase with light exposure (Zhou and Singh, 2002).

The effect of light levels on anthocyanin levels in red-fleshed fruits (e.g., strawberry, blood orange and some apple, pear, plum and peach genotypes) has generally not been recorded (Traveset et al., 2004).

Second to light, the most important environmental factor that influences anthocyanin synthesis is temperature. Low temperatures either before harvest and/or during storage generally favour anthocyanin synthesis (reported for apple, pear, grape, blackberry and cranberry) (Curry, 1997; Hall and Stark, 1972; Kliewer, 1970; Naumann and Wittenburg, 1980; Steyn et al., 2004a). In contrast, high temperatures are associated with anthocyanin degradation and the preharvest loss of red colour in pears (Steyn et al., 2004b). Consequently, the colour of some pear cultivars has been found to fluctuate between red and green in response to the passing of cold fronts and intermittent warmer conditions (Steyn et al., 2004b). In addition to inductive low temperatures at night, anthocyanin synthesis in mature apples requires mild day temperatures between 20°C and 25°C (Curry, 1997).

However, anthocyanin synthesis in immature apple fruit requires lower temperatures (Faragher, 1983). Differential cultivar responsiveness to low temperatures has been recorded in grape (Kliewer and Schultz, 1973) and pear (Steyn et al., 2004a). Anthocyanins are synthesized via the phenylpropanoid pathway. Anthocyanin biosynthesis has been extensively studied in several plant species and therefore, detailed information of the course of reactions is available. Two classes of genes are required for anthocyanin biosynthesis: the structural genes encoding the enzymes that directly participate in the formation of anthocyanins and other flavanoids and regulatory genes that controlled the transcription of structural genes (Holmstrom et al., 2000).

Anthocyanins are a large group of phenolic compounds in the human diet and the daily intake of anthocyanins in human has been estimated to be as much as 180–215 mg/day in the United States (Kuhnau, 1976).

Comparison of the phenolic content of different berries is difficult because of the various analytical methods used. Berries, especially of family *Rosaceous*, genus *Rubus* (strawberry, red raspberry and cloudberry), are rich in ellagitannins (51–88%) (Facchini et al., 2004; Hakkinen et al., 1999a,b; Joshipura et al., 1999).

The presence of phenolic acids has been confirmed mainly by chromatographic analysis and further studies have revealed the presence of quercetin, kaempferol and ellagic acid. Ellagic acid is a naturally occurring phenolic constituent of many plant species (Bruno et al., 2006) and has shown promising antimutagenic and anticarcinogenic activity against chemical-induced cancers (Hall, 2005; Kikuchi et al., 2004).

The analyses on berry fruits indicate that they are rich source of flavonoids, ellagic acid and tannins which may be used for the quality assessment of *Rubus* species leaves and may suggest that some leaves could be of equal value to those which have been characterized as having medicinal properties (Dettmer et al., 2007).

The polyphenols found in fruits *in vivo*, can generally be attributed to several distinct base structures. These encompass the anthocyanins, flavonols, flavanals, isoflavones, phenolic acids, catechins and ellagitannins. Ellagic acid is a compound that is known to have significant contribution to human health. It is formed by oxidation, and dimerization of gallic acid. Ellagic acid is known to occur in strawberry and significant variation is anticipated.

To guarantee a sufficient scientific challenge, for both health aspects and flavour properties, two groups of compounds are interesting targets to study with respect to their biosynthetic pathways.

There are fundamental differences between intervention studies and dietary assessment. These observations may indicate that only particular antioxidants, such as specific flavonoids, have a beneficial effect.

Current scientific evidence does not allow ascribing strong protective effects to specific compounds. For more detailed study, it is required not only to determine the total antioxidant capacity of foods, but also to identify the antioxidants involved (Bruno et al., 2006).

13.4 FRUIT PHENOLICS AND HUMAN HEALTH

The modern consumers are increasingly interested in their personal health, and expect the foods to be tasty, attractive and also safe and healthy. Phenolic compounds are one of the most diverse groups of secondary metabolites in edible plants. Plant phenols have many potential biological properties and extensive studies are being carried out at present on their effects on human health (Nakaishi et al., 2000; Shanmuganayagam et al., 2007). Interest has been focused on two large groups of phenolics–flavonoids and phytoestrogens. Flavonoids are found in many food products with plant origin such as vegetables, fruits, berries, tea and wine. Flavonols (quercetin and kaempferol) and flavones (apigenin and luteolin) are abundantly found in various plant-based foods. In addition, flavonoids exhibit various physiological activities including antiallergic, anticarcinogenic, antiarthritic and antimicrobial activities. It is known that the leaves of raspberry (*R. idaeus* L.) have been commonly used in traditional medicines to treat a variety of ailments including diseases of the alimentary canal, air passage, heart and the cardiovascular system (Mazza and Miniati, 1993; Torronen et al., 1997). They may also be applied externally as antibacterial, anti-inflammatory, sudorific, diuretic and choleretic agents (Beekwilder et al., 2005; Kresty et al., 2006; Simpson et al., 2001). Raspberry leaf extracts have been reported to have relaxant effect, particularly on uterine muscles (Bazzano, 2005; Mcdougall et al., 2005; Michels et al., 2006). Beneficial effects of using raspberry leaves during pregnancy have been noticed (Okuda et al., 1989; Valsta, 1999; Viberg et al., 1997). Generally, berries are good sources of various phenolic compounds, especially flavonoids. Strawberry, raspberry and cloudberry contain few flavonols, but they are rich in ellagitannins that are polymers of ellagic acid (Gudej and Tommczyk, 2004). The study on antibacterial activities of phenolic compounds has proved that the widest bactericidical activity of berries belongs to genus *Rubus* (raspberry and cloudberry) (Goodacre, 2005). In general, berry extracts inhibited the growth of Gram-negative bacterial species but not Gram-positive lactobacillus species. The experiments show that raspberry and strawberry extracts are strong inhibitors of *Salmonella, Escherichia* and *Staphylococcus* strains. The antimicrobial activities of the pure phenolic compounds are widely studied. However, there is very little information about antimicrobial activity of the berries, which contain a very complex mixture of phenolics compounds, specific for each berry species (Ozarowski and Jaroniewski, 1989).

It can be hypothesized that ellagitannins could be one of the components in cloudberries, raspberries and strawberries causing the inhibition against *Salmonella. Escherichia coli* strain 50 (Okuda et al., 1989; Valsta, 1999; Viberg et al., 1997). Ellagitannins are not found in any other common foods; so, the berries remain the most important sources of them.

In 107 blueberry, blackberry and blackcurrant genotypes, the total anthocyanins were found to correlate significantly ($r = 0.57$–0.93) to the total phenolics and to the average 34% of the total phenolics (Moyer et al., 2002a,b). It was proved that antioxidant capacity correlated better to the total phenolics than to anthocyanin levels. This is because fruits containing low levels of anthocyanins, but high levels of total phenolics, may still display high antioxidant capacity (Deighton et al., 2000). The beneficial effects of anthocyanins may be modified by complex interactions with various other fruit constituents (Lila, 2004). Much uncertainty still exists regarding the extent of uptake of anthocyanins from the human body (Manach, 2004).

Interactions between anthocyanin pigments and other flavonoids or other phytochemicals accumulating within a plant contribute significantly to the ability of natural plant extracts to protect human health or mitigate disease damage.

Anthocyanin pigments are an excellent example of a stress-induced secondary compound that confers protection to the host plant.

Interactions between flavonoids, including anthocyanins and other flavonoid classes, are increasingly cited as responsible for more intense potency of natural mixtures (as compared to purified compounds) in both *in vitro* and *in vivo* bioactivity trials.

Seeram et al. (2004) determined phytochemical constituents in the American cranberry; although individually efficacious against human carcinogenesis, they provided maximum protection only when coadministered in natural mixtures.

Combinations of two polyphenolic compounds from grapes (resveratrol and quercetin) demonstrated synergistic ability to induce apoptosis (activating caspase-3) in a human pancreatic carcinoma cell line (Mouria et al., 2002).

Similarly, a mixed polyphenolic extract from red wine demonstrated stronger inhibition of DNA synthesis in oral squamous carcinoma cells than individual compounds, even when the concentrations of individually administered quercetin or resveratrol were higher than those in the mixed extract (Elattar and Virji, 1999). Similarly, interactions between catechins in tea were responsible for protecting cells against damage induced by exposure to lead (Chen et al., 2002). Mertens-Talcott et al. (2003) evaluated the interactions between two common grape polyflavonoid, as well as potential interactions between anthocyanins and lipophilic fruit constituents, which appear to account for differences in the chemopreventive value of European and American elderberry fruits (Thole et al., 2006). Phenolics, quercetin and ellagic acid, at low, physiologically relevant concentrations, were found to inhibit the human leukemia cell line. Combination of these two compounds greatly reduced proliferation and viability, and induced apoptosis. The multiplicity of health benefits associated with consumption of anthocyanin-rich pomegranate fruits and juices have been largely attributed to flavonoid content; however, the direct contribution of individual components or mixtures have not been established (Aviram et al., 2000, 2002; Hora et al., 2003). In the same way, genotypic differences in the complement of interacting flavonoid components in blackberries from discrete geographical regions were determined to account for distinct differences in their antioxidant capacities, as confirming two complementary antioxidant assays (Reyes-Carmona et al., 2005).

The accumulated research experience, knowledge and practical methodology applications during the past years concerning bioactive berry compounds, in particular phenolic compounds, have advanced a lot. Future work is supposed to be focused on treatments of fruit promoting bioavailability and also on more determined confirmation of the effects of antioxidant compounds from berries on consumer health.

Unfortunately, data on the health benefits of ingesting anthocyanins are almost exclusive to humans. Anthocyanins are potent radical scavengers and antioxidants with proven health benefits and activity against a range of chronic human diseases (Lila, 2004). Various studies have shown highly pigmented berry fruits to possess considerable antioxidant capacity, at least partially due to their high anthocyanin concentrations (Deighton et al., 2000; Kähkönen et al., 2003; Moyer et al., 2002a,b; Reyes-Carmona et al., 2005).

The biosynthetic capacity of the whole plants will be evaluated and used. The potential value of secondary metabolite profiling in the field of fruits quality is in relation with the development of breeding strategies for plant improvement.

The utilization of antimicrobial activity of berry phenolic compounds as natural antimicrobial agents may offer many opportunities for use in food industry and medicine. The metabolic profiling approaches are highly relevant to study the interface between plant breeding for food and human nutrition. The development of alternative approaches, by implementing berry compounds for the prevention and control of infections caused by bacteria resistant to antibiotics will also be very important issue for definite research priorities in the future.

It is well documented in the literature that flavonoids possess a wide variety of biological properties (as antioxidants, enzyme inhibitors, induction of detoxification enzymes, enhanced membrane stability, induction of apoptosis, arrest of cell cycle, etc.), which may account for the ability for a mixture of interacting flavonoids to provide enhanced synergistic therapeutic or protective action, through multiple pathways of intervention. The anticancer potential of berries has been related, at least in part, to a multitude of bioactive phytochemicals including polyphenols (flavonoids, proanthocyanidins, ellagitannins, gallotannins, phenolic acids), stilbenoids, lignans and triterpenoids. Studies show that the anticancer effects of berry bioactives are partially mediated through their abilities to counteract, reduce and also repair damage resulting from oxidative stress and inflammation.

In addition, berry bioactives also regulate carcinogen and xenobiotic metabolizing enzymes, various transcription and growth factors, inflammatory cytokines and subcellular signalling pathways of cancer cell proliferation, apoptosis and tumour angiogenesis.

Various mechanisms of action of flavonoids on cancer cell growth or other therapeutic targets may be complementary and a combination of these actions may be responsible for the overall efficacy of ingested natural phytochemicals. It will be critical to identify the flavonoids in natural mixtures that interact together to fortify biological activity for human heath.

If programmed therapies involving key functional foods are to be considered for human health maintenance regimes, it is important to appreciate flavonoid interactions and how effective dosages can be prescribed.

The rapidly developing studies suggest that berry fruits may have immense potential for cancer prevention and therapy, but there are still important gaps in our knowledge. In this direction, our understanding of some of the potential mechanisms of the action of berry phytochemicals in cancer prevention has increased over the past decade, and research efforts should continue to focus on the elucidation of mechanisms of action at the cellular and molecular levels.

Research focus on nutrigenomics (effects of nutrients on the genome, proteome and metabolome) and nutrigenetics (effects of genetic variation on the interaction between diet and disease) will be essential.

Whether the chemopreventive potential of berry bioactives is increased by complex interactions of multiple substances within the natural food matrix of berry fruits, and/or in combination with phytochemicals from other foods, should be investigated.

Finally, interdisciplinary research is highly recommended so that basic and preclinical studies can lead to translational research (from the laboratory to bedside).

It is strongly recommended that this area of research for berry fruits continue to be explored, as this will lay the foundation for the development of diet-based strategies for the prevention and therapy of several types of human cancers (Seeram, 2008).

REFERENCES

Achnine, L., D. V. Huhman, M. A. Farag and L. W. Sumner. 2005. Genomics-based selection and functional characterization of triterpene glycosyltransferases from the model legume *Medicago truncatula*. *Plant J* 41:875–887.

Alba, R., P. Payton, Z. Fei and R. McQuinn. 2005. Transcriptome and selected metabolite analyses reveal multiple points of ethylene control during tomato fruit development. *Plant Cell* 17:2954–2965.

Allen, F. W. 1932. Physical and chemical changes in the ripening of deciduous fruits. *Hilgardia* 6:381–441.

Andersson-Gunneras, S., E. J. Mellerrowicz, J. Love, B. Segerman et al. 2006. Biosynthesis of cellulose enriched tension wood in Populus: Global analysis of transcripts and metabolites identifies biochemical and developmental regulators in secondary wall biosynthesis. *Plant J* 45:144–165.

Arts I. C. and P. C. Hollman. 2005. Polyphenols and disease risk in epidemiologic studies. *Am J Clin Nutr* 81:317–325.

Aviram, M., L. Dornfeld, M. Kaplan et al. 2002. Pomegranate juice flavonoids inhibit low-density lipoprotein oxidation and cardiovascular diseases: Studies in atherosclerotic mice and in humans. *Drugs Exp Clin Res* 28:40–62.

Aviram, M., L. Dornfeld, M. Rosenblat et al. 2000. Pomegranate juice consumption reduces oxidative stress, atherogenic modifications to LDL, and platelet aggregation: Studies in humans and in atherosclerotic apolipoprotein E-deficient mice. *Am J Clin Nutr* 71:1062–1076.

Awad, M. A., A. De Jager and L. M. Van Westing. 2000. Flavonoid and chlorogenic acid levels in apple fruit: Characterization of variation. *Sc Hort* 83:249–263.

Bazzano, L. A., Y. Song, V. Bubes et al. 2005. Dietary intake of whole and refined grain breakfast cereals and weight gain in men. *Obes Res* 13:1952–1960.

Beekwilder J., R. D. Hall and C. H. Ric de Vos. 2005. Identification and dietary relevance of antioxidants from raspberry. *BioFactors* 23:197–205.

Bino, R. J., R. D. Hall, O. Fiehn et al. 2004. Potential of metabolomics as a functional genomics tool. *Trends Plant Sci* 9:418–425.

Bohnert, H. J., Q. Gong, P. Li et al. 2006. Unraveling abiotic stress tolerance mechanisms—Getting genomics going. *Curr Opin Plant Biol* 9:180–188.

Brady, C. J. 1987. Fruit ripening. *Annu Rev Plant Physiol* 38:155–178.

Brazma, A., P. Hingamp, J. Quackenbush et al. 2001. Minimum information about a microarray experiment (MIAME)—Toward standards for microarray data. *Nat Genet* 29:365–371.

Broeckling, C. D. et al. 2005. Metabolic profiling of *Medicago truncatula* cell cultures reveals the effects of biotic and abiotic elicitors on metabolism. *J Exp Bot* 56:323–336.

Brosche, M. B. Vinocur, E. R. Alatalo, A. Lamminmaki et al. 2005. Gene expression and metabolite profiling of *Populus euphratica* growing in the Negev desert. *Genome Biol* 6:R101.

Bruno, E. J. J., T. N. Ziegenfuss and J. Landis. 2006. Vitamin C: Research update. *Curr Sports Med Rep* 5:177–181.

Bum, J. H. and E. R. Withell. 1941. A principle in raspberry leaves which relaxes uterine muscle. *Lancet* 5:1–3.

Catchpole, G. S., M. Beckmann, D. P. Enot et al. 2005. Hierarchical metabolomics demonstrates substantial compositional similarity between genetically modified and conventional potato crops. *Proc Natl Acad Sci USA* 102:14458–14462.

Chen, L., X. Yang, H. Jiao and B. Zhao. 2002. Tea catechins protect against lead-induced cytotoxicity, lipid peroxidation, and membrane fluidity in HepG2 cells. *Toxicol Sci* 69:149–156.

Cook, D., S. Fowler, O. Fiehn et al. 2004. A prominent role for the CBF cold response pathway in configuring the low-temperature metabolome of Arabidopsis. *Proc Natl Acad Sci USA*. 101:15243–15248.

Curry, E. A. 1997. Temperatures for optimal anthocyanin accumulation in apple tissue. *J Hort Sci* 72:723–729.

Daniel E. M., A. S. Krupnick, Y. H. Heur, J. A. Blinzler, R. W. Nims and G. D. Stoner. 1989. Extraction, stability, and quantitation of ellagic acid in various fruits and nuts. *J Food Comp Anal* 2:338–349.

De Ancos, B. E. M. Gonzalez and M. P. Cano. 2000. Ellagic acid, vitamin C, and total phenolic contents and radical scavenging capacity affected by freezing and frozen storage in raspberry fruit. *J Agric Food Chem* 48:4565–4570.

Defernez, M., Y. M. Gunning, A. J. Parr, L. V. Shepherd et al. 2004. NMR and HPLC–UV profiling of potatoes with genetic modifications to metabolic pathways. *J Agric Food Chem* 52:6075–6085.

Deighton, N., R. Brennan, C. Finn and H. V. Davies. 2000. Antioxidant properties of domesticated and wild *Rubus* species. *J. Sci. Food Agric* 80:1307–1313.

Desbrosses, G. G., J. Kopka and M. K. Udvardi. 2005. *Lotus japonicus* metabolic profiling. Development of gas chromatography–mass spectrometry resources for the study of plant–microbe interactions. *Plant Physiol* 137:1302–1318.

Dettmer K. and B. D. Hammock. 2004. Metabolomics: A new exciting field within the 'omics' sciences.*Environ Health Perspect* 112:396–397.

Dettmer K., P. A. Aronov and B. D. Hammock. 2007. Mass spectrometry-based metabolomics. *Mass Spectrom Rev* 26:51–78.

Dunn W. B. and D. I. Ellis. 2005. Metabolomics: Current analytical platforms and methodologies. *Trends Anal Chem* 24:285–294.

Elattar, T. and A. Virji. 1999. The effect of red wine and its components on growth and proliferation of human oral squamous carcinoma cells. *Anticancer Res* 19:5407–5414.

Escribano-Bailón, M. T., C. Santos-Buelga, G. L. Alonso and M. R. Salinas. 2002. Anthocyanin composition of the fruit of *Coriaria myrtifolia* L. *Phytochem Anal* 13:354–357.

Facchini P. J., D. A. Bird and B. St-Pierre. 2004. Can Arabidopsis make complex alkaloids? *Trends Plant Sci* 9:116–122.

Faragher, J. D. 1983. Temperature regulation of anthocyanin accumulation in apple skin. *J Exp Bot* 34: 1291–1298.

Fernie, A. R., 2003. Metabolome characterisation in plant system analysis. *Funct Plant Biol* 30:111–120.

Fernie, A. R., R. N. Trethewey, A. J. Krotzky et al. 2004. Innovation—Metabolite profiling: From diagnostics to systems biology. *Nat Rev Mol Cell Biol* 5:763–769.

Fernie, A. R. Y. Tadmor and D. Zamir. 2006. Natural genetic variation for improving crop quality. *Curr Opin Plant Biol* 9:196–202.

Ferrario-Mery, S. et al. 2002. Diurnal changes in ammonia assimilation in transformed tobacco plants expressing ferredoxindependent glutamate synthase mRNA in the antisense orientation. *Plant Sci* (Shannon) 163:59–67.

Fiehn, O., J. Kopka, P. Dörmann et al. 2000. Metabolite profiling for plant functional genomics. *Nat Biotechnol* 18:1157–1161.

Fiehn O. 2002. Metabolomics—The link between genotype and phenotype. *Plant Mol Biol* 48:155–171.

Gil, M. A., C. García-Viguera, F. Artés and F. A. Tomás-Barberán. 1995. Changes in pomegranate juice pigmentation during ripening. *J Sci Food Agric* 68:77–81.

Goodacre, R. 2005. Making sense of the metabolome using evolutionary computation: Seeing the wood with the trees. *J Experimental Botany* 56:245–254.

Goossens, A., S. T. Hakkinen, I. Laakso et al. 2003. A functional genomics approach toward the understanding of secondary metabolism in plant cells. *Proc Natl Acad Sci USA* 100:8595–8600.

Gould, K. S. and C. Lister. 2006. Flavonoid functions in plants. In *Flavonoids: Chemistry, Biochemistry, and Applications*, eds. Ø. M. Andersen and K. R. Markham, pp. 397–442. CRC Press, Taylor & Francis Group: Boca Raton, FL.

Gudej, J. and M. Tommczyk. 2004. Determination of flavonoids, tannins and ellagic acid in leaves from *Rubus* L. sp. *Arch Pharm Res* 27:1114–1119.

Gullberg, J., P. Jonsson, A. Nordstrom, M. Sjostrom and T. Moritz. 2004. Design of experiments: An efficient strategy to identify factors influencing extraction and derivatization of *Arabidopsis thaliana* samples in metabolomic studies with gas chromatography/mass spectrometry. *Anal Biochem* 331:283–295.

Hakkinen, S. H., M. Heinonen and S. Karenlampi. 1999a. Screening of selected flavonoids and phenolic acids in 19 berries. *Food Res Int* 32:345–353.

Hakkinen, S. H., S. Karenlampi, M. I. Heinonen, H. M. Mykkanen and R. A. Torronen. 1999b. Content of the flavonols quercetin, myricetin, and kaempferol in 25 edible berries. *J Agric Food Chem* 47:2274–2279.

Hakkinen, S., S. Karenlampi, H. Mykkanen, M. I. Heinonen and R. Torronen. 2000. Influence of domestic processing and storage on flavonol contents in berries. *Eur Food Res Technol* 212:75–80.

Hall, I. V. and R. Stark. 1972. Anthocyanin production in cranberry leaves and fruit, related to cool temperatures at a low light intensity. *Hort Res* 12:183–186.

Hall, R. D., C. H. R. Vos, H. A. Verhoeven and R. J. Bino. 2005. Metabolomics for the assessment of functional diversity and quality traits in plants. In *Metabolome Analyses: Strategies for Systems Biology*, eds. S. Vaidyanathan, G. G. Harrigan and R. Goodaece. Springer: New York, NY, USA.

Hall, R. D. 2006. Plant metabolomics: From holistic hope, to hype, to hot topic. *New Phytol* 169:453–468.

Hamzehzarghani, H., A. C. Kushalappa, Y. Dion, S. Rioux et al. 2005. Metabolic profiling and factor analysis to discriminate quantitative resistance in wheat cultivars against *Fusarium* head blight. *Physiol Mol Plant Pathol* 66:119–133.

Hancock R. D. and R. Viola. 2005. Improving the nutritional value of crops through enhancement of L-ascorbic acid (vitamin C) content: Rationale and biotechnological opportunities. *J Agric Food Chem* 53:5248–5257.

Harrigan, G. G. and Goodacre, R. 2003. *Metabolic Profiling: Its Role in Biomarker Discovery and Gene Function Analysis*. Kluwer Academic Publishers: Boston, MA.

Hetherington, S. E. 1997. Profiling photosynthetic competence in mango fruit. *J Hort Sci* 72:755–763.

Hirai, M. Y. and Saito, K. 2004. Post-genomics approaches for the elucidation of plant adaptive mechanisms to sulphur deficiency. *J Exp Bot* 55:1871–1879.

Hirai, M. Y., M. Yano, D. B. Goodenowe et al. 2004. Integration of transcriptomics and metabolomics for understanding of global responses to nutritional stresses in *Arabidopsis thaliana. Proc Natl Acad Sci USA*. 101:10205–10210.

Hirai, M. Y. et al. 2005. Elucidation of gene-to-gene and metabolite-to-gene networks in Arabidopsis by integration of metabolomics and transcriptomics. *J Biol Chem* 280:25590–25595.

Holmstrom, K. O. et al. 2000. Improved tolerance to salinity and low temperature in transgenic tobacco producing glycine betaine. *J Exp Bot* 51:177–185.

Hora, J., E. Maydew, E. Lansky and C. Dwivedi. 2003. Chemopreventive effects of pomegranate seed oil on skin tumor development in CD1 mice. *J Med Food* 6:157–161.

Hung, H. C., K. J. Joshipura, R. Jiang et al. 2004. Fruit and vegetable intake and risk of major chronic disease. *J Natl Cancer Inst* 96:1577–1584.

Ishizaki, K. et al. 2005. The critical role of Arabidopsis electron transfer flavoprotein: Ubiquinone oxidoreductase during dark induced starvation. *Plant Cell* 17:2587–2600.

Ishizaki, K. et al. 2006. The mitochondrial electron transfer flavoprotein complex is essential for survival of Arabidopsis in extended darkness. *Plant J* 47:751–760.

Joshipura, K. J., A. Ascherio, J. E. Manson et al.1999. Fruit and vegetable intake in relation to risk of ischemic stroke. *JAMA* 282:1233–1239.

Junker, B. H. et al. 2004. Temporally regulated expression of a yeast invertase in potato tubers allows dissection of the complex metabolic phenotype obtained following its constitutive expression. *Plant Mol Biol* 56:91–110.

Kähkönen, M. P., J. Heinämäki, V. Ollilainen and M. Heinonen. 2003. Berry anthocyanins: Isolation, identification and antioxidant activities. *J Sci Food Agric* 83:1403–1411.

Kahkonen, M. P., A. I. Hopia and M. Heinonen. 2001. Antioxidant activity of plant extracts containing phenolic compounds. *J Agric Food Chem* 49:4076–4082.

Kaplan, F., J. Kopka, D. Haskel et al. 2004. Exploring the temperature–stress metabolome of Arabidopsis. *Plant Physiol* 136:4159–4168.

Keurentjes, J. B., F. Jingyuan C. H. Ric de Vos et al. 2006. The genetics of plant metabolism. *Nat Genet* 38:842–849.

Kikuchi, J., K. Shinozaki and T. Hirayama. 2004. Stable isotope labeling of *Arabidopsis thaliana* for an NMR-based metabolomics approach. *Plant Cell Physiol* 45:1099–1104.

Kliewer, W. M. 1970. Effect of day temperature and light intensity on coloration of *Vitis vinifera* L. grapes. *J Am Soc Hort Sci* 95:693–697.

Kliewer, W. M. and H. B. Schultz. 1973. Effect of sprinkler cooling of grapevines on fruit growth and composition. *Am J Enol Viticult* 24:17–26.

Kopka, J., A. Fernie, W. Weckwerth et al. 2004. Metabolite profiling in plant biology: Platforms and destinations. *Genome Biol* 5:109.

Kopka, J., N. Schauer, S. Krueger et al. 2005. GMD@CSB DB: The Golm Metabolome Database. *Bioinformatics* 21:1635–1638.

Kresty, L. A., W. L. Frankel, C. D. Hammond et al. 2006. Transitioning from preclinical to clinical chemopreventive assessments of lyophilized black raspberries: Interim results show berries modulate markers of oxidative stress in Barrett's esophagus patients. *Nutr Cancer* 54:148–156.

Kristensen, C., M. Morant, C. Olsen et al. 2005. Metabolic engineering of dhurrin in transgenic Arabidopsis plants with marginal inadvertent effects on themetabolome and transcriptome. *Proc Natl Acad Sci USA*. 102:1779–1784.

Kuhnau, J. 1976. The flavonoids. A class of semi-essential food components: Their role in human nutrition. *World Rev Nutr Diet* 24:117–191.

Lee, K. W. and H. J. Lee. 2006. The roles of polyphenols in cancer chemoprevention. *Biofactors* 26: 105–121.

Lila, M. A. 2004. Plant pigments and human health. In *Plant Pigments and Their Manipulation*, ed. K. M. Davies, pp. 248–274. *Annu Plant Rev.* Blackwell Publishing/CRC Press: Boca Raton, FL.

Maas J. L., G. J. Galletta and G. D. Stoner. 1991. Ellagic acid, an anticarcinogen in fruits, especially in strawberries: A review. *Hort Science* 26:10–14.

Macheix, J.-J., A. Fleuriet and J. Billot. 1990. *Fruit Phenolics*. CRC Press: Boca Raton, FL.

Manach, C. 2004. Polyphenols: Food sources and bioavailability. *Am J Clin Nutr* 79:727–747.

Martzen, M. R., S. M. McCraith, S. L. Spinelli et al. 1999. A biochemical genomics approach for identifying genes by the activity of their products. *Science* 286:1153–1155.

Mazza, G. and E. Miniati. 1993. *Anthocyanins in Fruits, Vegetables, and Grains*. CRC Press: Boca Raton, FL.

McDougall, G. J., P. Dobson, P. Smith, A. Blake and D. Stewart. 2005. Assessing potential bioavailability of raspberry anthocyanins using an *in vitro* digestion system. *J Agric Food Chem* 53:5896–5904.

Mehrotra, B. and P. Mendes K., R. 2006. Bioinformatics approaches to integrate metabolomics and other systems biology data. In *Plant Metabolomics*, eds. K. Saito, R. A. Dixon and L.Willmitzer, pp. 105–116. Springer-Verlag: Berlin, Heidelberg.

Memelink, J. 2004. Tailoring the plant metabolome without a loose stitch. *Trends Plant Sci* 7:305–307.

Merchant, A., A. Richter, M. Popp and M. Adams. 2006. Targeted metabolite profiling provides a functional link among eucalypt taxonomy, physiology and evolution. *Phytochemistry* 67:402–408.

Mercke, P., I. F. Kappers, F. W. Verstappen et al. 2004. Combined transcript and metabolite analysis reveals genes involved in spider mite induced volatile formation in cucumber plants. *Plant Physiol* 135:2012–2024.

Mertens-Talcott S., S. Talcott and S. Percival. 2003. Low concentrations of quercetin and ellagic acid synergistically influence proliferation, cytotoxicity and apoptosis in MOLT-4 human leukemia cells. *J Nutr* 133:2669–2674.

Michels K. B., E. Giovannucci, A. T. Chan, R. Singhania, C. S Fuchs and W. C. Willett. 2006. Fruit and vegetable consumption and colorectal adenomas in the nurses' health study. *Cancer Res* 66:3942–3953.

Misciagna, G., A. M. Cisternino and J. Freudenheim. 2000. Diet and duodenal ulcer. *Dig Liver Dis* 32:468–472.

Morikawa, T., M. Mizutani, N. Aoki et al. 2006. Cytochrome P450 CYP710A encodes the sterol C-22 desaturase in Arabidopsis and tomato. *Plant Cell* 18:1008–1022.

Morino, K., F. Matsuda, H. Miyazawa et al. 2005. Metabolic profiling of tryptophan overproducing rice calli that express a feedback-insensitive alpha subunit of anthranilate synthase. *Plant Cell Physiol* 46:514–521.

Mouria, M., A. Gukovskaya, Y. Jung et al. 2002. Food-derived polyphenols inhibit pancreatic cancer growth through mitochondrial cytochrome C release and apoptosis. *Int J Cancer* 98:761–769.

Moyer, R. A., K. E. Hummer, C. E. Finn, B. Frei and R. E. Wrolstad. 2002a. Anthocyanins, phenolics, and antioxidant capacity in diverse small fruits: *Vaccinium*, *Rubus*, and *Ribes*. *J Agric Food Chem* 50:519–525.

Moyer R., K. Hummerand and R. E. Wrolstad. 2002b. Anthocyanins, phenolics, and antioxidant capacity in diverse small fruits: *Vaccinium*, *Rubus*, and *Ribes*. *Acta Hort* 585:501–505.

Mullen, W., A. J. Stewart, M. E. J. Lean, P. Gardner, G. G. Duthie and A. Crozier. 2002. Effect of freezing and storage on the phenolics, ellagitannins, flavonoids, and antioxidant capacity of red raspberries. *J Agric Food Chem* 50:5197–5201.

Nakaishi, H., H. Matsumoto, S. Tominaga and M. Hirayama. 2000. Effects of black current anthocyanoside intake on dark adaptation and VDT work-induced transient refractive alteration in healthy humans. *Altern Med Rev* 5:553–562.

Naumann, W. D. and U. Wittenburg. 1980. Anthocyanins, soluble solids, and titratable acidity in blackberries as influenced by preharvest temperatures. *Acta Hort* 112:183–190.

Neill, S. and K. S. Gould. 2003. Anthocyanins in leaves: Light attenuators or antioxidants? *Funct Plant Biol* 30:865–873.

Nikiforova, V. J., J. Kopka, V. Tolstikov, O. Feihn et al. 2005a. Systems rebalancing of metabolism in response to sulfur deprivation, as revealed by metabolome analysis of Arabidopsis plants. *Plant Physiol* 138:304–318.

Nikiforova, V. J., C. Daub, H. Hesse et al. 2005b. Integrative gene–metabolite network with implemented causality deciphers informational fluxes of sulphur stress response. *J Exp Bot* 56:1887–1896.

Noctor, G., L. Gomez, H. Vanacker and C. Foyer. 2002. Interactions between biosynthesis, compartmentation and transport in the control of glutathione homeostasis and signalling. *J Exp Bot* 53:1283–1304.

Okuda, T., T. Yoshida and T. Hatano. 1989. Ellagitannins as active constituents of medicinal plants. *Planta Med* 55:117–122.

Ott, K. H., N. Aranibar, B. J. Singh and G. W. Stockton. 2003. Metabolomics classifies pathways affected by bioactive compounds. Artificial neural network classification of NMR spectra of plant extracts. *Phytochemistry* 62:971–985.

Ozarowski, A. and W. Jaroniewski.1989. Medicinal plants and their practical use (in Polish), *IWZZ, Warszawa*, 184:243.

Patel, A. V., J. Rojas-Vera and C. G. Dacke. 2004. Therapeutic constituents and actions of *Rubus* species. *Curr Med Chem* 11:1501–1512.

Puech, A. A., C. A. Rebeiz and J. C. Crane. 1976. Pigment changes associated with the application Ethephon ((2-chloroethyl)phosphonic acid) to fig (*Ficus carica* L.) fruits. *Plant Physiol* 57:504–509.

Puupponen-Pimiä, R., L. Nohynek, H. L. Alakomi and K. M. Oksman-Caldentey. 2004. Bioactive berry compounds—novel tools against human pathogens. *Appl Microbiol Biotechnol* 23:11–24

Puupponen-Pimia, R., L. Nohynek, H. L. Alakomi, K. Oksman-Caldentey. 2005a. The action of berry phenolics against human intestinal pathogens. *BioFactors* 23:243–251.

Puupponen-Pimia R., L. Nohynek, S. Schmidlin et al. 2005b. Berry phenolics selectively inhibit the growth of intestinal pathogens. *J Appl Microbiol* 98:991–1000.

Raamsdonk, L. M., B. Teusink, D. Broadhurst et al. 2001. A functional genomics strategy that used metabolome data to reveal the phenotype of silent mutations. *Nat Biotechnol* 19:45–50.

Reyes-Carmona, J., G. G. Yousef, R. A. Martínez-Peniche and M. A. Lila. 2005. Antioxidant capacity of fruit extracts of blackberry (*Rubus* sp.) produced in different climatic regions. *J Food Sci* 70:497–503.

Robinson, A. R., R. Gheneim, R. A. Kozak et al. 2005. The potential of metabolite profiling as a selection tool for genotype discrimination in Populus. *J Exp Bot* 56:2807–2819.

Roessner, U. A., Luedemann, D. Brust, O. Fiehn et al. 2001a. Metabolic profiling allows comprehensive phenotyping of genetically or environmentally modified plant systems. *Plant Cell* 13:11–29.

Roessner, U., C. Wagner, J. Kopka et al. 2000. Simultaneous analysis of metabolites in potato tuber by gas chromatography–mass spectrometry. *Plant J* 23:131–142.

Roessner, U., L. Willmitzer and A. R. Fernie. 2001b. High-resolution metabolic phenotyping of genetically and environmentally diverse potato tuber systems. Identification of phenocopies. *Plant Physiol* 127:749–764.

Rojas-Vera J., A. V. Patel and C. G. Dacke. 2002. Relaxant activity of raspberry (*Rubus idaeus*) leaf extract in guinea-pig ileum in vitro. *Phytother Res* 16:665–668.

Ross H. A., G. J. McDougall and D. Stewart. 2007. Antiproliferative activity is prominently associated with ellagitannins in raspberry extracts. *JSci Food Agric* 70:133–150.

Sauter, H., M. Lauer and H. Fritsch. 1988. Metabolic profiling of plants—A new diagnostic-technique. Abstract. Paper. *Am Chem Soc.* 195:129.

Scalbert A., I. T. Johnson and M. Saltmarsh. 2005. Polyphenols: Antioxidants and beyond. *Am J Clin Nutr* 81:215–217.

Schauer, N. and A. R. Fernie. 2006. Plant metabolomics: Towards biological function and mechanism. *Trends Plant Sci* 11:508–516.

Schauer, N, Steinhauser D, Strelkov S, et al. 2005. GC–MS libraries for the rapid identification of metabolites in complex biological samples. *FEBS Lett* 579:1332–1337.

Schauer, N., D. Zamir and A. Fernie. 2005. Metabolic profiling of leaves and fruit of wild species tomato: A survey of the *Solanum lycopersicum* complex. *J Exp Bot* 56:297–307.

Seeram, N. P. 2008. Berry fruits for cancer prevention: Current status and future prospects. *J Agric Food Chem* 56:630–635.

Seeram, N. P., L. S. Adams, M. I. Hardy and D. Heber. 2004. Total cranberry extract versus its phytochemical constituents: Antiproliferative and synergistic effects against human tumor cell lines. *J Agric Food Chem* 52:2512–2517.

Shanmuganayagam, D., T. F. Warner, C. G. Krueger, J. D. Reed and J. D. Folts. 2007. Concord grape juice attenuates platelet aggregation, serum cholesterol and development of atheroma in hypercholesterolemic rabbits. *Atherosclerosis* 190:135–142.

Simpson, M., M. Parsons, J. Greenwood and K. Wade. 2001. Raspberry leaf in pregnancy: Its safety and efficacy in labor. *J. Midwifery Women's Health* 46:51–59.

Smillie, R. M. and S. E. Hetherington. 1999. Photoabatement by anthocyanin shields photosynthetic systems from light stress. *Photosynthetica* 36:451–463.

Stewart D., G. L. McDougall, J. Sungurtas, S. Verrall, J. Graham and I. Martinussen. 2007. Metabolomic approach to identifying bioactive compounds in berries: Advances toward fruit nutritional enhancement. *Mol Nutr Food Res* 51(6):645–651.

Steyn, W. J., D. M. Holcroft, S. J. E. Wand and G. Jacobs. 2004a. Regulation of pear color development in relation to activity of flavonoid enzymes. *J Amer Soc Hort Sci* 129:6–12.

Steyn, W. J., D. M. Holcroft, S. J. E. Wand and G. Jacobs. 2004b. Anthocyanin degradation in detached pome fruit with reference to preharvest red color loss and pigmentation patterns of blushed and fully red pears. *J Amer Soc Hort Sci* 129:13–19.

Stitt, M. and A. R. Fernie. 2003. From measurements of metabolites to metabolomics: An 'on the fly' perspective illustrated by recent studies of carbon–nitrogen interactions. *Curr Opin Biotechnol* 14:136–144.

Stitt, M., C. Müller, P. Matt et al. 2002. Steps towards an integrated view of nitrogen metabolism. *J Exp Bot* 53:959–970.

Sumner, L. W., P. Mendes and R. A. Dixon. 2003. Plant metabolomics: Large-scale phytochemistry in the functional genomics era. *Phytochemistry* 62:817–836.

Suzuki, H., M. S. Reddy, M. Naomkina et al. 2005. Methyl jasmonate and yeast elicitor induce differential transcriptional and metabolic re-programming in cell suspension cultures of the model legume *Medicago truncatula. Planta* 220:696–707.

Sweetlove, L. J. and A. R. Fernie. 2005. Regulation of metabolic networks: Understanding metabolic complexity in the systems biology era. *New Phytol* 168:9–24.

Thimm, O., O. Blasing, Y. Gibon et al. 2004. MAPMAN: A user-driven tool to display genomics data sets onto diagrams of metabolic pathways and other biological processes. *Plant J* 37:914–939.

Thole, J., T. Kraft, L. Sueiro, Kang et al. 2006. A comparative evaluation of the anticancer properties of European and American elderberry fruits. *J Med Food* 9:498–504.

Tikunov, Y., A. Lommen, C. H. R. de Vos, R. D. Hall et al. 2005. A novel approach for nontargeted data analysis for metabolomics. Large-scale profiling of tomato fruit volatiles. *Plant Physiol* 139:1125–1137.

Tohge, T., Y. Nishiyana and M. Y. Hirai. 2005.Functional genomics by integrated analysis of metabolome and transcriptome of Arabidopsis plants over-expressing an MYB transcription factor. *Plant J* 42:218–235.

Torronen, R., S. Hakkinen, S. Karenlampi and H. Mykkanen. 1997. Flavonoids and phenolic acids in selected berries. *Cancer Lett* 114:191–192.

Traveset, A., N. Riera and R. E. Mas. 2001. Ecology of fruit-colour polymorphism in *Myrtus communis* and differential effects of birds and mammals on seed germination and seedling growth. *J Ecol* 89:749–760.

Traveset, A., M. F. Willson and M. Verdú. 2004. Characteristics of fleshy fruits in southeast Alaska: Phylogenetic comparison with fruits from Illinois. *Ecography* 27:41–48.

Trethewey R. N. 2004. Metabolite profiling as an aid to metabolic engineering in plants. *Curr Opin Plant Biol* 7:196–201.

Tsiotou A. G., G. Sakorafas, G. Anagnostopoulos and J. Bramis. 2005. Septic shock; current pathogenetic concepts from a clinical perspective. *Med Sci Monit* 11:76–85.

Valsta, L. M., 1999. Food-based dietary guidelines for Finland—A staged approach. *Br J Nutr* 81:49–55.

Verhoeven H. A., C. H. R. de Vos, R. J. Bino and R. D. Hall. 2006. Plant metabolomics strategies based upon quadrupole time of flight mass spectrometry (QTOF–MS). In *Biotechnology and Forestry 57 Plant Metabolomics*, eds. Saito, K., Dixon, R. A. and Willmitzer, L., pp. 33–46. Springer-Verlag: Berlin, Heidelberg.

Viberg U., G. Ekstrom, K. Fredlund, R. E. Oste and I. Sjoholm. 1997. A study of some important vitamins and antioxidants in a blackcurrant jam with low sugar content and without additives. *Int J Food Sci Nutr* 48:57–66.

Villas-Boas, S. G., J. Moxley, M. Akesson et al. 2005. High-throughput metabolic state analysis: The missing link in integrated functional genomics of yeasts. *Biochem J* 388:669–677.

Urbanczyk-Wochniak, E. and A. R. Fernie. 2005. Metabolic profiling reveals altered nitrogen nutrient regimes have diverse effects on the metabolism of hydroponically-grown tomato (*Solanum lycopersicum*) plants. *Exp Bot* 56:309–321.

Wang H., G. Cao and R. L. Prior. 1996. Total antioxidant capacity of fruits. *J Agric Food Chem* 44:701–705.

Wang H., G. Cao and R. L Prior. 1997. Oxygen radical absorbing capacity of anthocyanins. *J Agric Food Chem* 45:304–309.

Wang S. Y., W. Zheng and G. J. Galletta. 2002. Cultural system affects fruit quality and antioxidant capacity in strawberries. *J Agric Food Chem* 50:6534–6542.

Williamson, G. and C. Manach. 2005. Bioavailability and bioefficacy of polyphenols in humans. II.Review of 93 intervention studies. *Am J Clin Nutr* 81:243S–255S.

Willson, M. F. 1986. Avian frugivory and seed dispersal in eastern North America. *Curr Ornithol* 3:223–279.

Willson, M. F. and C.J. Whelan. 1990. The evolution of fruit color in fleshy-fruited plants. *Am Nat* 136:790–809.

Whitney, K. D. and C.E. Lister. 2004. Fruit colour polymorphism in *Acacia ligulata*: Seed and seedling performance, clinal patterns, and chemical variation. *Evol Ecol* 18:165–186.

Zhou, Y. and B. R. Singh. 2002. Red light stimulates flowering and anthocyanin biosynthesis in American cranberry. *Plant Growth Regul* 38:165–171.

Zimmermann, P. et al. 2004. GENEVESTIGATOR. Arabidopsis microarray database and analysis toolbox. *Plant Physiol* 136:2621–2632.

CHAPTER 14

New Farm Management Strategy to Enhance Sustainable Rice Production in Japan and Indonesia

Masakazu Komatsuzaki and Faiz M. Syuaib

CONTENTS

14.1 INTRODUCTION

Rice is a common and essential food plant in Asia, and from the 1960s to recent years, rice yields have increased remarkably. Since the 'green revolution' programme launched in the late 1960s, application of chemical fertilizers has dramatically increased because of governmental encouragement to succeed in the food self-sufficiency goal. Chemical fertilizer consumption in the agriculture sector reached five times the level of 1975 in 1990 and increased slightly afterwards in Japan and Indonesia. In Indonesia, since the economic crisis, in 1998 the government reduced the subsidy on fertilizers and therefore the cost of agriculture input has increased; thus farmers have reduced the use of chemical fertilizer and started improving their application methods, and organic fertilizer has presently become more favourable. A similar situation occurred even in Japan: with an increase in chemical fertilizer price due to increasing oil prices, most farmers have evinced interest in using organic fertilizer. '*Bochashi*' is the traditional way of making compost using agricultural subproducts and waste in Japan, and this technique has spread widely to other Asian countries, including Indonesia.

Recent intensive research is aimed at evaluating differences in the conservation rice farming system between Indonesia and Japan and also at finding out the common aspects. To develop the ecological management of the farming system, an ecosystem approach is needed. The ecosystems between Indonesia and Japan are so different, even though we are all facing similar situations such as global warming and globalization. Through these case studies, we will discuss what is needed to achieve a sustainable ecosystem, and how we can collaborate to develop a community-based approach. These framework studies will be able to point out the appropriate technical transfer and development for each agro-ecosystem. This chapter addresses the similarity and variability of organic rice production from the viewpoints of regional and global sustainability between Japan and Indonesia.

14.2 RICE PRODUCTION AND THEIR SUSTAINABILITY IN JAPAN

14.2.1 Country Facts and Agricultural Practices at a Glance

Japan is an island country, made up of more than 3000 islands of a large strato-volcanic archipelago along the Pacific coast of Asia. Japanese agriculture is more than 2000 years old. Today, agriculture represents only 1% of Japan's gross domestic product (GDP). Approximately 73% of the land area of Japan is mountainous and about 13% is farmland (4,628,000 ha). Rice (*Oryza sativa* var, *japonica*) has been recognized as the most important crop from economic, political and cultural perspectives in Japanese agricultural history. Rice plants grow well in Japan, with its abundant rainfall and high summer temperatures. Therefore, paddy rice occupies more than half of the total cultivated land (2,516,000 ha).

Up until the 1940s, Japanese agriculture had many features of sustainability. Most food was grown in integrated, mixed farming systems with a closed-loop nutrient cycle. Traditionally, Japanese farming used to be an environmentally friendly system because most crop rotations included both summer cash crops and winter grain crops, such as wheat and barley. However, recent statistical data have revealed that the annual ratio of crop planting area in cultivated fields was 94.4% in 2004, representing a dramatic decrease over the past few decades (Statistics of Agriculture, Forestry and Fisheries, 2005). This suggests that Japanese crop rotation systems are being destroyed and intensive farming and single-cropping systems are now widely being employed. These new farming systems can degrade the quality and health of soil through the increase of synthetic chemical inputs and leaching of soil residual nutrients into groundwater (Kusaba, 2001).

Today, agricultural technologies that are overly dependent on chemical inputs might be increasing productivity to economically meet food demand, but they may also be threatening agricultural ecosystems and local environments. With the growing interest in reducing excessive synthetic chemical inputs to farming, the importance of cover crops as determinants of soil quality has been recognized in agriculture. The recent policies of the Japanese government in developing more environmentally friendly farming practices and the growing awareness of the importance of surplus reduction have led to a widespread interest in organic farming and conservation farming.

According to recent statistical data, 167,995 farms in Japan were engaged in conservation farming, accounting for 21.5% of the total cropping area in the country (Sustainable Agriculture Office, 2008). Under conservation management, traditional agronomic methods are combined with modern farming techniques, and conventional inputs such as synthetic pesticides and fertilizers are excluded or reduced. Instead of synthetic inputs, cover crops, compost and animal manure are used to build up soil fertility. Cover crops are particularly beneficial to cropland soils, as they are instrumental in supplying soil organic matter (SOM), adding biological fixed nitrogen, scavenging soil residual nutrients, suppressing weeds and breaking pest cycles (Peet, 1996; Sarrantonio, 1998; Magdoff, 1998).

14.2.2 Rice Production and Sustainable Farming System Using Cover Crops

More than 2000 years ago, farmers in China and the Mediterranean countries sowed cover crops to improve soil productivity. Chinese milk vetch (*Astragalus sinicus* L.), which was the most useful cover crop in China more than 1000 years ago, was introduced in Japan from China in about the seventh century. In 'Nougyou Zensyo', one of the corpuses of agricultural techniques of the Edo period that was published in 1697, green manure and cover crops were noted as the most important tools for improving soil fertility. Chinese milk vetch used to be planted in most Japanese paddy fields as a cover crop until the 1960s, when it was replaced by increasing use of chemical fertilizers on farmland (Yasue, 1993). Some beneficial impacts on the agro-ecosystem using cover crops in paddy fields are given below.

14.2.2.1 Soil Residual N Scavenging

Wet paddy rice cultivation is one of the traditional agricultural techniques in Asia. In Japan, half of all croplands are cultivated with paddy rice, and the total value of rice in Japan exceeds its monetary return because of its significantly high quality in the world marketplace. If a grower were to base the decision to grow rice solely on the market value of rough rice, it is doubtful that rice would be grown at all. However, when the overall value of rice and its effects on the environment, such as flooding water in the fields, storing water after intense rainfalls and the overall profitability of the land, are considered, the benefits of rice production outweigh the costs. Moreover, paddy fields have unique ways of regulating the movement, accumulation and transformation of SOM and nitrogen because paddy fields are controlled by long-term seasonal oxidation–reduction interaction. Therefore, in Japan, paddy fields play a significant role in the country's agricultural ecosystems.

In general, soil subsidence is the loss of surface elevation due to decomposition (mineralization) of organic soil. Microbial activity is the major cause of mineralization and requires the presence of oxygen. A deep water table allows a large amount of soil to be aerated, which promotes mineralization (Snyder et al., 1978). High summer temperatures accelerate the process (Bonner and Galston, 1952). Paddy rice effectively stops the subsidence of muck soil during the hot summer months, the time of year when the rate of subsidence is the highest (Snyder et al., 1978).

Wet paddy farming also has benefits for improving water quality. Nitrogen reaction in paddy fields can effectively process inorganic N through nitrification and denitrification, ammonia volatilization and plant uptake. In Japan, 55% of all land used for agriculture is paddy fields. However, paddy rice production requires abundant amounts of water, accounting for 95% of total agricultural water use (Tabuchi and Hasegawa, 1995), as well as a lot of chemical fertilizer. Therefore, paddy fields play a particularly significant role in catchment environments. Depending on the amounts of water and fertilizer that are used and how they are applied, some paddy fields remove nutrients (Takeda et al., 1997). The recycling of irrigation water may reduce the need for both irrigation water and nutrient inputs in agricultural catchments (Kudo et al., 1995).

However, these benefits occur only during the growing season, when paddies are flooded. After the rice is harvested in autumn, most paddy fields are not irrigated and are left fallow from autumn to spring. This dries out the soil, which can lead to N leaching. For example, in the area around Biwako, the largest lake in Japan, considerable N leaching from paddy fields to the lake has been observed in winter (Tanaka, 2001). However, when planted with winter cover crops, paddies have shown significant environmental benefits. Figure 14.1 illustrates the cover crop benefits in paddy fields during winter. As we can see, non-legume cover crops have particularly significant N uptake during winter in paddy fields and add organic matter to the soil (Komatsuzaki, 2009).

Paddy field rice can conserve N in the soil under flooded conditions; however, residual soil N represents a potential environmental concern when fields are no longer flooded. Komatsuzaki (2009) reported that winter annual non-legume cover crops provide an alternative means of conserving

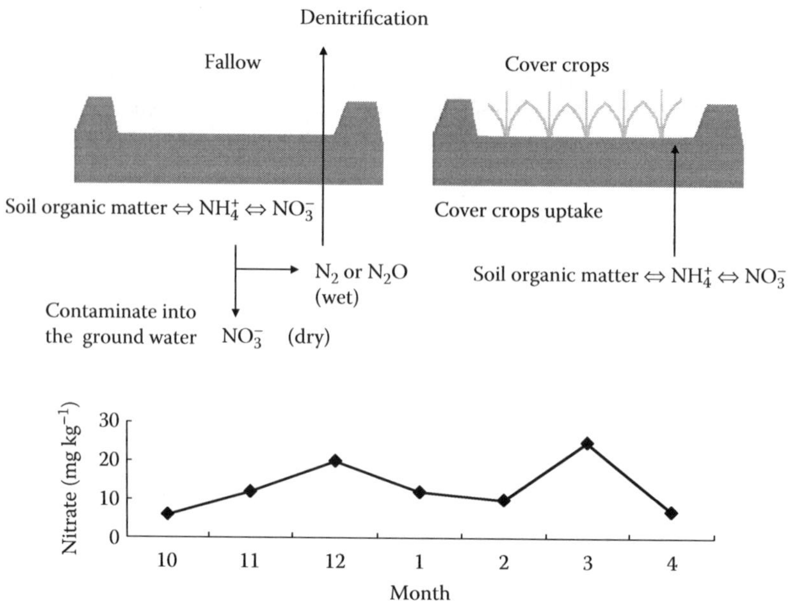

Figure 14.1 Illustration of N dynamics between fallow and cover crop fields in the non-irrigated winter season and soil nitrate concentration change during the entire non-irrigated season in the Kanto region of Japan.

residual soil N following rice harvest. A two-year field experiment was conducted at the Ibaraki University Experimental Farm to compare dry matter (DM) and N accumulation of rye (*Secale cereale* L.), oat (*Avena sativa* L.), triticale (*Triticum secale* L.), wheat (*Triticum aestivum* L.) and fallow (no cover) in relation to soil residual N levels. Figure 14.2 shows that DM and N accumulation by the following April were of the order rye > triticale > wheat = oat > fallow, whereas residual soil N levels occurred in the reverse order.

Residual soil N levels exerted the greatest influence on cover crop DM accumulation, with differences in N levels becoming more pronounced by the April sampling date. By 17 April, DM differences between the low and high residual soil N levels were 3.45 versus 6.82 Mg/ha for rye (98% increase), 1.15 versus 1.45 Mg/ha for oat (26% increase), 1.49 versus 1.99 Mg/ha for wheat (34% increase) and 1.70 versus 2.98 Mg/ha for triticale (75% increase). N accumulation for rye was greater than for oat and wheat across all planting dates, whereas triticale showed moderate ability to accumulate N (Komatsuzaki, 2009).

These results demonstrated that non-legume cover crops have great potential for conserving soil residual N. However, additional research would be needed to determine the contribution of cover crop N to the growth of the following crop of rice.

14.2.2.2 Reducing or Eliminating Fertilizer Use

Cover crops in paddy fields also increase the nutrient use efficiency of farming systems, especially legume cover crops that do not require fertilizer. Leguminous cover crops can make a significant nitrogen contribution to the following rice crop, reducing and sometimes eliminating the need for synthetic N fertilizer (Horimoto et al., 2002). This is due to the biological fixation of atmospheric N by bacteria in nodules on the roots of most legumes. Furthermore, both legume and non-legume cover crops will take up N from the soil that would otherwise be lost to leaching or denitrification (conversion to N gases) during the winter. In field experiments conducted in Butte

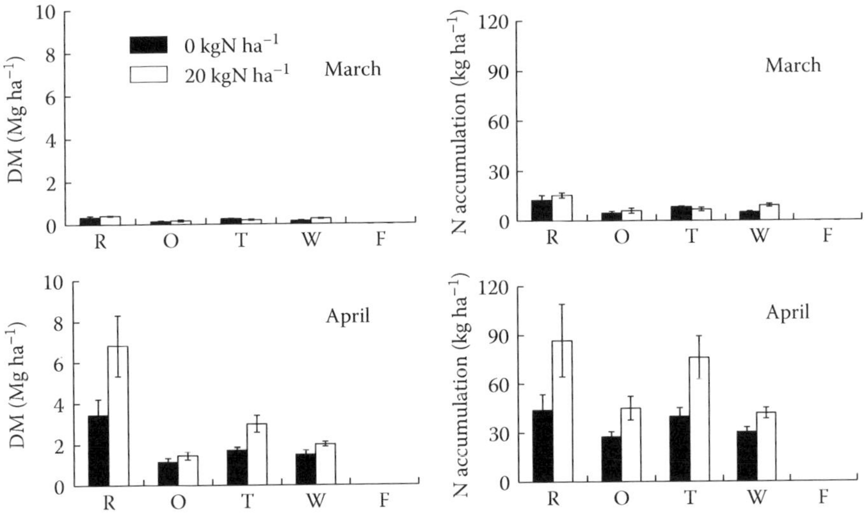

Figure 14.2 Cover crop DM accumulation and N accumulation in relation to cover crop type, soil nitrogen level and growth termination. Cover crops are indicated by letters as follows: R for rye, O for oat, T for triticale, W for wheat and F for fallow. Vertical bars indicate standard error. (Data from Komatsuzaki, M. 2009. *Jpn J Farm Work Res* 44(4):201–210.)

and Sutter counties in California, USA, cover cropping with vetch reduced the rice fertilizer N requirement by 30–100 lb N/acre, with an average over several experiments of about 50 lb N/acre (Pettygrove and Williams, 1997). Similar values have also been reported for Japanese paddy fields. For example, Asagi and Ueno (2009) reported that winter white clover, hairy vetch and crimson clover with no-fertilizer plots showed similar yields as fertilized plots, suggesting that the cover crop contributed about 80 kg N/ha to rice growth.

With the increasing costs of chemical fertilizer and energy, a new no-tillage rice planting system that uses hairy vetch as a cover crop has been developed in Japan. In field experiments, hairy vetch planted in the fall covered the land well in the spring, and suppressed weed growth. Two days before rice transplanting, fields were flooded, and a no-tillage transplanter was used for transplanting rice without tillage and pudding. In this system, hairy vetch contributed about 60 kg N/ha to rice growth (National Agricultural Research Center for the Kyushu Okinawa Region, 2003).

Legume cover crops can supply enough nutrients for rice growth because they require relatively low N input in the flooded condition. However, Asagi and Ueno (2006) reported that white clover could produce a high level of biomass and can fix a sufficient amount of nitrogen for rice plants but only a small amount of the fixed nitrogen was taken up by rice plants in the past, because of the discrepancies in N patterns between the legume supplier and N uptake by the crop.

By estimating cover crop N content shortly before incorporation, a rice grower can adjust the rate of N fertilizer in response to year-to-year variations in cover crop growth. Synchronization of cover crop N release with demand from the following rice crop is needed for efficient production. Release of N from cover crops depends on species, growth stage, farming method and climate. Irrigation is a practical method that can control the timing of the killing and decomposition of the legume and following N mineralization (Ueno, 2004).

Asagi and Ueno (2009) evaluated the effects of irrigation timing on N concentration in the soil, rice plant growth and yield, and found that delayed irrigation could retard legume decomposition and N mineralization. The rice harvested from cover crop plots that were irrigated at 30 days after transplanting had a 13.7% higher grain weight than rice from plots that had been irrigated 10 days before transplanting. It appears that the delay of N mineralization was more effective for supplying available N to the rice crop.

Soil management practices in paddy fields, such as the synchronization of cover crop N release with crop growth, still remain a challenge for rice production. Sainju et al. (2006) reported that legume and non-legume cover crop biculture improved cover crop N release patterns and also improved crop yields. Further research will be needed to optimize cover crops used in biculture in paddy fields.

14.2.2.3 Landscape Management

Cover crops also help to conserve paddy levees, which are an integral part of the traditional Japanese landscape, and this aspect is unique in the world. Japanese paddy fields are often located in mountainous areas, and levee areas occupy a total of 143,600 ha, about 5.7% of the area of paddy fields (National Agriculture and Food Research Organization, 2008). Paddy ridge or levee management is one of the most burdensome tasks for farmers. For example, most farmers usually mow levees 6 to 8 times during the growing season with a brush cutter. However, there are a number of ecological benefits: for instance, paddy levees prevent soil erosion and conserve rice traces and landscapes.

Perennial cover crops are intensively introduced to levee management in paddy fields. *Imperata cylindrical* L. is an especially effective crop for eliminating the need for weeding levees, because of its wide leaves, creeping, short plant height, early growth and early root development (Tominaga, 2007). When these ecological functions dominate levee slopes, they can suppress weeds and reduce the need to mow. For example, the National Agriculture and Food Research Organization (2008) reported that if *I. cylindrical* L. were planted at 12 hills/m, it would dominate the levee slope two years later and would reduce the amount of labour required for mowing by about 50%.

Zoysia japonica and *Ophiopogon japonicus* Gyokuryu are also effective cover crops for levee areas. They are low plants that have significant ability for ground covering, are tolerant to trampling and provide traction against slipping during levee work. Since these cover crops were originally wild plants that were domesticated in Japan, such management strategies should also help conserve local agro-ecosystems and biodiversity.

14.2.3 Sustainable Rice Production Practices in Japan

Flooding paddy rice—winter non-legume cover crops are an appropriate farming system to develop sustainable farming, because they can prevent the leaching of soil residual nutrient, add

Plate 14.1 Photo of winter cover crop field in Ushiku, Kanto, Japan. Non-legume winter cover crop can prevent the leaching of N (a), and their roots matt system also improves soil stricture (b).

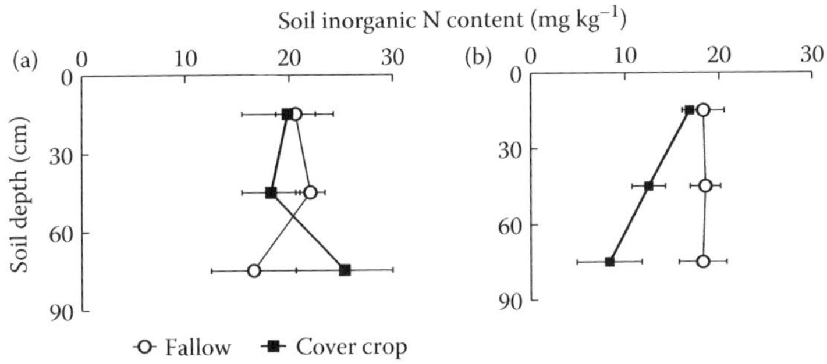

Figure 14.3 Soil inorganic N distribution between cover crop and winter fallow in a paddy rice field. Soil inorganic N was measured on 13 October 2000 when cover crop was seeded (a) and on 17 April 2001 when cover crop was terminated (b). (Data from Komatsuzaki, M. et al. 2004. *Jpn J Farm Work Res* 39:23–26.)

SOM, improve yields and eliminate fertilizer for rice growth (Plate 14.1). Italian ryegrass (*Lolium multiflorum* Lam.) was used as a non-legume cover crop in paddy fields. Komatsuzaki et al. (2004) reported that winter annual non-legume cover crops provide an alternative means of conserving residual soil N following rice harvest in Kanto, Japan.

Non-legume cover crop was grown well when sown immediately after rice harvest. The cover crop biomass was 4.4 Mg/ha and N accumulation was 90 kg N/ha. Figure 14.3 shows the soil inorganic N distribution between winter cover crop and fallow treatment in a paddy field. There were no significant differences in soil inorganic N content between cover crop and fallow treatment; however, cover crop soil showed significantly lower inorganic N content compared with fallow soil in 0–90 cm soil depth layer. Non-legume cover crop was to provide enough soil cover during the winter, and supply enough biomass and N for subsequent rice production (Komatsuzaki et al., 2004).

Because cover crop residue decomposition is difficult to manage, cover crop N release often decreases rice food quality due to increase in grain protein content. In this regard, *bochashi* is one of the alternatives for enhancing cover crop residue decomposition after incorporating with cover crop residues before puddling. Farmers use a self-produced organic fertilizer called '*bochashi*', which is composed of 60% rice bran, 20% rice chaff and 20% soybean mill (in volume). Nutrient values of this organic fertilizer are 3.7% for nitrogen, 2.3% for phosphorus and 1.6% for potassium (in dry base). *Bochashi* also has benefit in reducing the cost for nutrient in rice production, and can cut 56% of the fertilizer cost compared with chemical fertilizer.

The combination of non-legume cover crop and *bochashi* application showed good performance in paddy rice production. Table 14.1 shows rice yield response between winter cover crop and fallow treatment in a paddy field. In both treatments, *bochashi* was applied as an initial fertilizer. The results suggest that cover crop with *bochashi* treatment enhanced rice yield and rice food quality: especially, cover crop with *bochashi* treatment increased the straw–grain ratio and reduced protein

Table 14.1 Rice Yield Response between Winter Cover Crop and Fallow Treatment

Treatment	No. of Hills (hill/m^2)	Straw (g/m^2)	Grain (g/m^2)	Straw–Grain Ratio %	Food Quality	Amylose %	Protein %
Cover crop	397.0	488.6	552.6	58.3	82.0	18.4	6.0
Fallow	337.3	483.2	451.3	52.4	81.7	18.5	6.1
Significance	NS	NS	NS	*	*	NS	*

Source: Data from Komatsuzaki, M. 2008. *Bokuso to Engei* 56(6):10–14.
* Indicates significance at 5%, and NS indicates non significance.

content in the grain. Winter cover crop and *bochashi* have great potential in reducing N leaching in the river and pond from paddy fields, reducing the cost of fertilizer for rice production, and improving rice yield and rice food quality.

Cover crop with *bochashi* application is one of the new farming strategies that improves the soil as a result of adding organic matter, reducing N leaching and improving rice yield and quality; however, cover crop introduction needs an additional cost for cover crop seeds for farmers. Therefore, political assistance and incentives should be provided to encourage the introduction of cover crops that can increase SOM and control soil residual nutrients. For example, the city of Ushiku, located in the northern part of the Kanto area of Japan, implemented a cost-sharing programme in 2004 in which the city shares the cost for cover crop seed that farmers use in their crop rotation. This programme has succeeded in preventing N leaching and protecting the soil from wind erosion (Komatsuzaki and Ohta, 2007).

14.3 RICE PRODUCTIONS AND THEIR SUSTAINABILITY IN INDONESIA

14.3.1 Country Facts and Agricultural Practices at a Glance

Indonesia is the world's largest archipelagic country, which extends between two continents, Asia and Australia, and between two oceans, Pacific and Indian. The Archipelago stretches over 5500 km from east to west and 1900 km from north to south, consisting of more than 17,000 islands, an 81,000 km coastline, 1.9 million km^2 land territory and 3.2 million km^2 of sea territory. Indonesia is made up of reservoirs of rich biodiversity in tropical agro and marine ecosystems, which contain various indigenous varieties of flora and fauna, many of them are typically indigenous and are never found in other parts of the world.

Indonesia lies on the zone of tropical environment with almost no extreme changes in weather. The daily temperature range is 23–33°C in the low plains and 15–27°C in the highland areas. The east and west monsoon strongly influence the weather, which periodically brings the dry season in April–September and the wet season in October–March. The annual average rainfall in the country is about 2400 mm, but varies widely among areas, from about 1000 to 4500 mm. However, the farming system in the country is generally determined as the variable of rainfall pattern rather than temperature. Based on soil, rainfall and length of the growing period, five pragmatic agro-ecological zones might be recognized in the country: (1) dryland–dry climate, (2) dryland–wet climate, (3) highland, (4) lowland irrigation and (5) tidal swamp.

More than 90% of the land surface of Indonesia is still covered by vegetation, as tropical rainforest, woodland, mangrove, agricultural crops and grassland. Based on the land utilization features of Indonesia, more than 30% (58 million ha) is recognized as agricultural land (lowland, upland, estate plantation, grassland, pond and dike), 60% as forest (permanent and industrial forest) and 3% as housing and settlement area (Central Bureau of Statistics (CBS), 2006). A rough estimation of land utilization is shown in Table 14.2.

Indonesia is the fourth most populous country in the world, with a present population of 228 million and an annual growth rate of 1.4% (CBS, 2008). Agriculture plays a substantial role in the Indonesian economy, especially for grassroot people. Agriculture also plays a leading role in alleviating poverty and malnutrition, as well as environmental mitigation. Agriculture provides productive employment opportunities and income for the huge population residing in rural areas. The agricultural sector now contributes more than 40% of employment and 17% GDP. Over the last decades, the domestic agricultural product has grown by about 3–5% annually, and at present the growth is 4.8% (CBS, 2009).

In general, the farming system in Indonesia is diverse on commodities and ecosystems as well, which are basically categorized into four types: (1) intensive wetland (lowland–irrigated) paddy

Table 14.2 A Rough Estimation of Land Utilization in Indonesia

No.	Type of Land Utilization	Area (× 1000 ha)	% of Total Land Area
1	Permanent forest	114,192	60.0
2	Woodland/agro-forestry	9304	4.9
3	Estate plantation	18,490	9.7
4	Dryland (upland and garden)	15,585	8.2
5	Temporary fallow land	11,342	6.0
6	Wetland (rice field)	7886	4.1
7	Housing/settlement	5686	3.0
8	Swamp/marsh land	4755	2.5
9	Grassland/meadows	2432	1.3
10	Pond and dike	779	0.4
	TOTAL LAND AREA	190,457	100.0

Source: Summarized from BPS and the Ministry of Agriculture (2006).

fields, (2) upland (rainfed–dryland) secondary crop fields, (3) estate plantations (industrial crops) and (4) agro-forestry. Lowland and upland crops are predominantly practised by common or individual farmers, whereas estate plantation and agro-forestry are industrial or company-based management.

Humans and animals remain predominant power sources of farm work in Indonesia. Mechanization is a 'luxury' for most Indonesian farmers. In some 'well-developed' areas, farm mechanization has been applied; however, it still has a narrow meaning and is limited to the utilization of a hand tractor for land preparation or the utilization of a power thresher and a rice milling unit for postharvest handling. Utilization of a four-wheel tractor, power sprayer, cultivator and other machinery is merely found in bigger commercial estate plantations, such as oil palm, sugarcane or some industrial crop plantations. These facts clearly show that mechanization is only applied for <20% of total farm activities in Indonesia (Syuaib, 2006).

Food self-sufficiency has therefore always been the main goal of the agricultural development agenda of the Indonesian government. The increasing demand for food has resulted in intensive agricultural practice. Since the so-called 'green revolution' era, intensive exploitation of land and increasing application of chemical fertilizers have been practised by most farmers so as to increase the yield. Straight fertilizers (N, P, K) in accordance with the recommended composition have been used for decades for most farming systems in Indonesia, for crop farming and industrial plantations as well. In 1990, the total national consumption of fertilizers reached five times that in 1975, and increased slightly afterwards.

Since the economic crisis, in 1998 the government reduced the subsidy on fertilizers and therefore the cost of agriculture input has increased; thus, farmers reduced the use of chemical fertilizers and started improving their application methods, and organic fertilizer became more preferable. Although the economic crisis in the industrial sector has not fully recovered yet, agricultural estates have been intensified to lift up the national income; fertilizer consumption for estate plantations has therefore been increasing since 2002. Figure 14.4 shows the total domestic consumption of agricultural fertilizers and the total consumption for crop farming and estate plantation. Figure 14.5 shows the domestic retail price of four common fertilizers consumed in Indonesia. The figures show that the retail price of fertilizers doubled since the economic crisis of 1998 (Figure 14.5) when the prices of fuel and other industrial input materials were increasing. At the time the consumption of fertilizer was then decreasing for some years, and afterward it has been increasing again since 2002 (Figure 14.4). The increase of chemical fertilizers from 2002 is mainly caused by the increasing consumption of estate plantation due to planted area expansion. Meanwhile, the demand of fertilizer for food crop farming actually decreased due to the significant increase of retail price which was not affordable for common farmer.

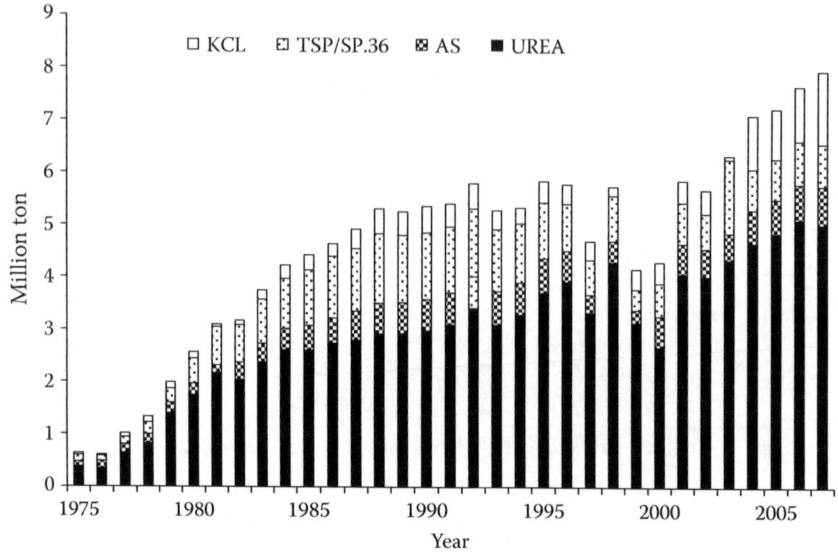

Figure 14.4 Domestic consumption of fertilizers in Indonesia.

Agriculture is the dominant land use, the largest consumer of water and one of the main contributors to groundwater and surface water pollution as well as a powerful emitter of greenhouse gases. Furthermore, increasing inputs of chemicals and fertilizers may pose a threat to the natural areas surrounding farmlands. Concerning the present condition of agricultural practices in Indonesia (as well as in many countries in Asia), we need to adjust our understanding and formulate an action strategy to develop better and sustainable agricultural practices in the future. A sustainable approach to agriculture is necessary in order to keep increasing the productivity to fulfil human needs, whereas at the same time it has to be in harmony with the environment so as to conserve resources and maintain environmental balance.

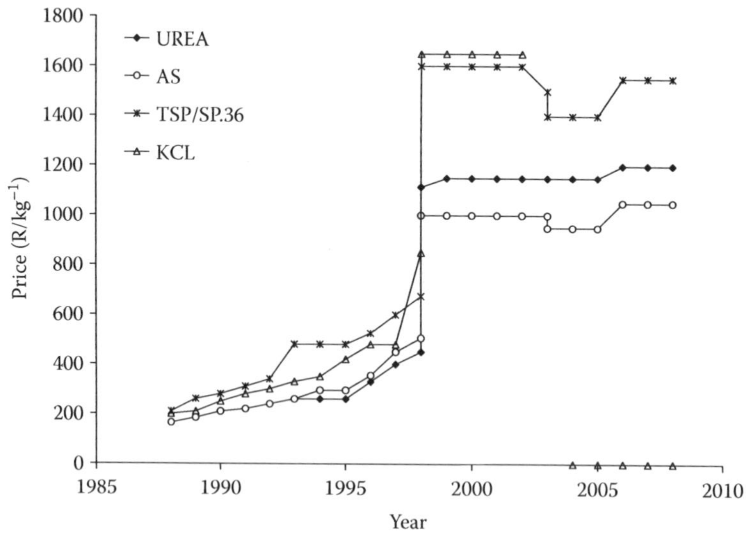

Figure 14.5 Domestic retail price of fertilizers in Indonesia.

14.3.2 Rice Production and Farming System in Indonesia

Concerning the amount of production, Indonesia is now the world's number 3 among rice producers, after China and India. However, the amount of domestic consumption is about the same as production itself, since more than 95% of the population consumed rice as the staple food, and therefore paddy is the most important crop grown in Indonesia. Paddy is grown on flat lowland up to terraced middle-range altitude. Java Island is the main area for growing paddy, which comprises about 50% of total harvested area and 60% of total rice production of the country. The other major areas of paddy fields are Bali, Lombok, west and southern parts of Sumatra and South Sulawesi (Syuaib, 2009).

Owing to the advantage of tropical climate conditions, a common feature of Indonesian farming activities is the fact that crops can be grown any time of the year. By using short growth period varieties, it is theoretically possible to grow three crops a year in Indonesia. However, the average cropping indexes of the country so far are still 1.8 and 1.0, respectively, for irrigated and non-irrigated fields.

Rice farming in Indonesia is dominated by small farms; that is, the average land ownership is now <0.3 ha per farm household. Potentially, the total paddy field area in the country is now about 10.2 million ha, which is half of its wetland-irrigated paddy field. However, the official data (Ministry of Agriculture, 2009) show that the present rice-harvested area (year 2008) is totally 12.3 million ha: 11.2 million ha is wetland field and 1.1 million ha is dryland field. Also, the productions are 57.2 million and 3.2 million tonnes of rice (unhusked) for wetland and dryland fields, respectively. Almost no increment of harvested area as well as production of rice occurred in the country between 1997 and 2002. Meanwhile, there was about a 12% increment of harvested area and 15% of production occurred in the country between 2004 and 2008. This means that there is presently an average 3% annual growth of rice production in the country. However, comparing with the population growth, Figures 14.6 and 14.7 show that rice production per capita is stagnant, whereas the harvested area per capita tends to decrease.

According to the condition of agro-ecosystem and farm infrastructures, four types of farm fields can be recognized: irrigated, rainfed, flood-prone (marshland) and upland (dry). More than 80% of irrigated field can be cropped twice or more a year, whereas nearly 90% of non-irrigated field has only one crop a year. Technically-irrigated fields are mostly planted as monoculture by paddy for the whole year and it can be managed for two or three crops annually. Semi-technically and simply irrigated fields are mostly managed as rotated farming system, one or two crops by paddy and

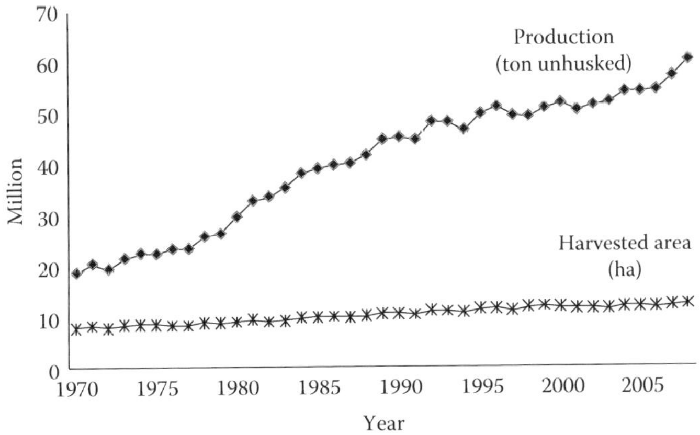

Figure 14.6 Domestic rice production and harvested area.

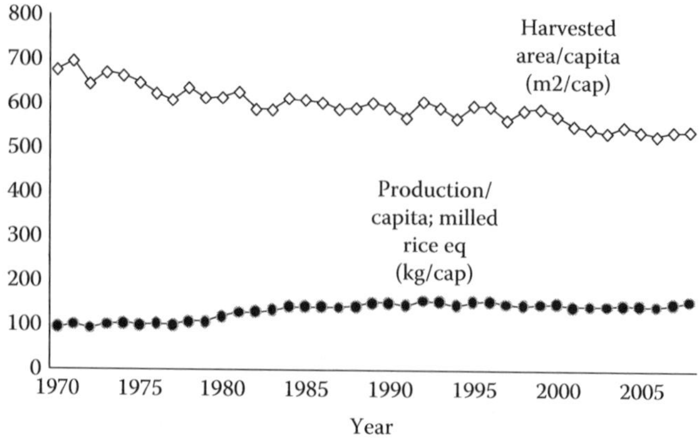

Figure 14.7 Per capita rice production and harvested area.

another one by secondary crops or vegetables. Table 14.3 shows the crop field area in Indonesia regarding the types of farming systems.

14.3.3 Organic Rice Production and Sustainable Agriculture

Despite diversity in agro-ecological conditions in Indonesian agriculture, it is useful to distinguish between agricultural problems related to intensive-irrigated areas and those related to extensive rainfed farming areas. Productivity has grown fastest in intensive-irrigated areas because of the increasing use of modern inputs such as irrigation water, chemical input of fertilizer and pesticide, and high-yielding variety and machinery. However, this system is associated with some environmental problems, that is, deterioration of water and soil quality, micronutrient deficiency and soil toxicity, weather and pest-related vulnerability, and loss of indigenous (traditional) varieties.

On the other hand, in the absence of an adequate increase of productivity in the less-productive rainfed agricultural area, farmers need to reduce fallows and expand into new areas. Sometimes they open a new fragile area with inappropriate land clearing, which causes pressure on the properties of natural resources and leads to degradation of environmental quality. Some major problems associated with extensive farming are loss of biodiversity, air pollution caused by slash and burn land opening, erosion, hillsides, drought and flood in lowland areas.

Concerning the present condition of agricultural practices in dealing with the needs to increase productivity and at the same time conserve the environment and natural resources, we need to adjust our understanding and formulate an action strategy to develop better and sustainable agricultural practices in the future. A sustainable approach to agriculture in Indonesia is necessary

Table 14.3 Areas of Farm Fields Regarding the Types of Farm Infrastructure in Indonesia

Farming Systems	Area (× 1000 ha)	%
Technically irrigated	2310	22.7
Semi-technically irrigated	1000	9.8
Simply irrigated	1760	17.3
Rainfed	2300	22.6
Flood prone and other	1610	15.8
Upland/dryland	1200	11.8
Total	**10,180**	**100.0**

Source: Summarized from BPS and the Ministry of Agriculture (2008).

to keep increasing the productivity (to fulfil human needs), whereas at the same time it has to be in harmony with the environment (to maintain environmental balance). Furthermore, a sustainable agricultural system has to always take into account three dimensions of sustainability: economy, social and ecology.

Agricultural activities can basically result in both positive and negative effects in the environment. They vary significantly as a function of the type of production and management system. Environmental issues related to agricultural production and management practices include water quality and use, use and management of agricultural inputs (nutrients, pesticides and energy), land use and management, soil quality, biodiversity, climate and air quality. The organic farming system might be one of the most reasonable approaches to be disseminated to address the above-mentioned issues.

According to the latest report of Willer and Yussefi (2006), more than 31 million ha are currently managed organically by at least 623,174 farms worldwide. In Asia, the area under organic management has been comparatively small in the past years: the total organic area in Asia is now about 4.1 million ha, managed by almost 130,000 farms. The highest reported domestic market growth, estimated to be up to 30%, is in China and an organic boom seems to be taking place in Indonesia.

Organic agriculture in Indonesia is still in its infancy, although there is some sign that the movement may be in the take-off stage. The development so far is largely in the hands of farmers and the private sector while government supports have just started recently. The majority of organic producers are family farms with small land holdings and they are organized under grower group or organic projects. The predominant organic agriculture in Indonesia is rice crops and especially some vegetables.

The area under organic management in Indonesia is comparatively small; it is approximately only 0.2% of total agricultural land. However, according to IFOAM organic net (2008), organic management in Indonesia has been significantly increasing within the last 5 years: 17,800 ha in 2004, 40,400 ha in 2005, 52,882 ha in 2006 and 66,184 ha in 2007. Figure 14.8 gives the initial impression that Indonesia is at a good pace, being a big organic farming country in Asia. This fact will be more optimistic because the price of chemical fertilizers and pesticides is increasing and it will not be affordable anymore to most grassroot farmers.

Food quality and safety are now becoming of prime concern to customers along with the aim of protecting the environment through good practices of sustainable agriculture. These facts have been stimulating the demand for organic products and eventually have become the driving force in the

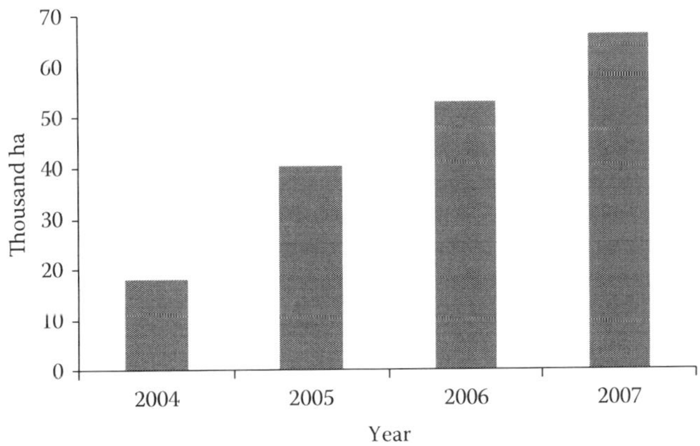

Figure 14.8 Area under organic farm in Indonesia. (Data from IFOAM Organic Net, 2008. Available at URL: http://www.organic-world.net. Accessed 15 March 2010.)

development of organic agriculture. Indonesian governments have responded by setting targets for the expansion of organic production, and new market opportunities have developed as part of the strategy to address such concerns.

With a population over 228 million people that is still growing, there is no doubt that organic agricultural products will have a market in Indonesia, now and in the future. The Indonesian government, as well as farmers' communities, has recently shown interest in organic farming with the expansion of the market for organic products and their potential for promoting sustainable agriculture. The department of agriculture established an ambitious programme, titled Go Organic, with the target of becoming one of the biggest producers of organic commodities in the world.

14.3.4 Organic Rice Production Practice in West Java, Indonesia

Organic farming provides several benefits to the farming system in Indonesia, because it can improve soil and food quality, and increase soil organic carbon (SOC) storage in the soil. For global environmental conservation, this soil management strategy has great potential to contribute to carbon sequestration, because the carbon sink capacity of the world's agricultural and degraded soil is 50–66% of the historic carbon loss of 42–72 petagrams (1Pg = 10^{15} g), although actual carbon storage in cultivated soil may be smaller if climate change leads to increasing mineralization (Lal, 2004). The importance of SOC in agricultural soil is, however, not controversial because SOC helps to sustain soil fertility and conserve soil and water quality, and these compounds play a variety of roles in nutrient, water and biological cycles.

Organic farming also has great potential in improving soil carbon storage (Pimentel et al., 1995; Marriott and Wander, 2006). However, only a few studies have been conducted in Indonesia, and there are few data for comparing soil carbon storage between organic farming and conventional farming. In addition, the organic farming system and associated farm work have not been studied in Indonesia. Therefore, this research was designed to evaluate the ability of soil carbon sequestration and make comparisons between the conventional farming system and the organic farming system for rice production on the island of Java.

The case study was observed in the city of Bogor, located in the Cisadane watershed, West Java, Indonesia. The Cisadane River flows through urban areas from Bogor to Jakarta, the capital of Indonesia, and is a major rice and vegetable production area. The soil type is Lotosole. Organic and conventional rice farmer groups for the study were selected from the Situgede district of Bogor.

For organic rice production in Bogor, farmers use a self-produced organic fertilizer called 'bochashi', which is composed of 10% rice bran, 20% rice chaff and 70% cow manure (in volume). Nutrient values of this organic fertilizer are 28.29% for carbon, 0.35% for nitrogen, 0.17% for phosphorus, 2.31% for potassium, 1.87% for calcium and 0.42% for magnesium (in dry base). Plate 14.2 illustrates the difference between conventional and organic rice production systems.

Figure 14.9 shows a comparison of soil carbon content and carbon storage in soil between organic and conventional farming. The soil in organic farming showed higher soil carbon content than conventional soils after four years of continuous organic farming; however, there were no significant differences in soil bulk density between the two farming systems. Soil carbon storage in organic farming significantly increased compared with conventional farming.

Using these data, we can estimate the ability of soil carbon sequestration. Organic farming can increase by 1.85 tonne C/ha/year compared with the conventional farming system. This value also agrees with the data that Shirato et al. (2005) obtained from paddy fields in Thailand. The rate of increase in SOC stock resulting from changes in land use and adoption of recommended farming practices follows a sigmoid curve that attains the maximum 5–20 years after adoption of recommended farming practices (Lal, 2004). In addition, the amount of organic carbon stored in paddy soils is greater than in dry field soils because of different biochemical processes and mechanisms specifically caused by the presence of floodwater in paddy soils (Katoh, 2003). These results show

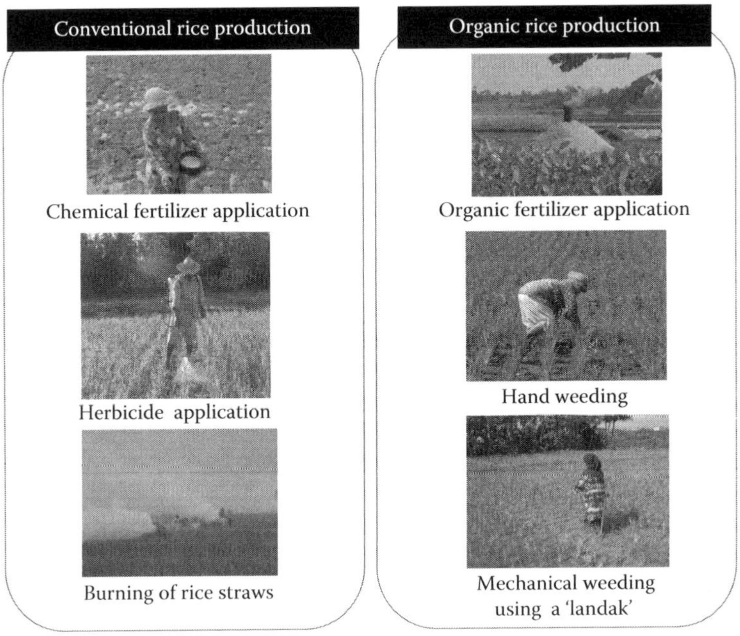

Plate 14.2 Comparison of farm work system between conventional and organic rice production in West Java, Indonesia.

that organic rice farming has a lot of potential in improving soil carbon sequestration and it may also mitigate global warming in Indonesia.

Figure 14.10 shows the costs for conventional and organic rice production. Organic farming helped in reducing the cost of rice production. For example, conventional farmers had to pay 1,410,000 Rp (rupiah, 1 US$ = 9,135 Rp) for chemical fertilizers, whereas organic farmers only had to pay 30,000 Rp for the *bochashi* organic fertilizer, with the result that organic farming could cut 90% of the total cost of rice production.

According to the latest data, the cost of farming with chemical fertilizers is on average twice as high as the use of organic products, although the production levels are the same, if not a fraction higher in the organic sector (*The Jakarta Post*, 2009) This indicates that the economic crisis helped boost the growth of Indonesia's organic farming sector.

Figure 14.10 also compares the labour inputs and methods of conventional and organic rice farming systems. The organic system required more labour to apply the organic fertilizer and weeding. The amount of organic fertilizer applied was 2 Mg/ha for each rice-growing season, which was 4 times greater than conventional farming due to the lack of appropriate technology for applying the organic fertilizer. Weeding in Indonesia is mainly done by hand, and while there are also traditional weeding tools called 'landak' (Plate 14.2), these tools still require a lot of manual labour. The total labour time for rice cultivation was 768 man hours/ha for the conventional system, whereas it was almost twice as high, 1406 man hours/ha, for the organic system.

Table 14.3 shows the gross profit and wages per working hour between the organic and conventional farming systems. The yield of organic farming was lower than in conventional farming, whereas the price of organic rice was 18% higher than conventionally grown rice, resulting in almost the same gross profits for organic and conventional farming. The wages per working hour in the organic system, however, were significantly lower, only about half those in the conventional system. The low labour productivity in organic farming is a major factor limiting the expansion of this farming system in West Java.

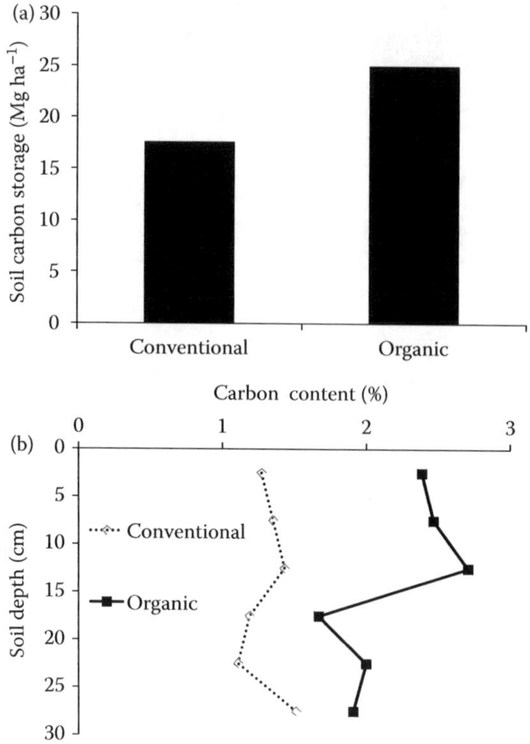

Figure 14.9 Comparison of soil carbon sequestration between organic and conventional rice fields in the top 10 cm soil depth (a) and soil carbon distribution (b). The different letters on the bar indicate the significant difference of soil carbon storage between organic and conventional rice fields ($P < 0.01$). (Data from Komatsuzaki, M. and M. F. Syuaib. 2010. *Sustainability* 2(3):833–843.)

According to the soil analysis, organic farming showed significantly higher SOC storage; hence it may help not only in mitigating global warming, but also in establishing a sustainable food system in Indonesia. Thus, organic farming will be one of the keys to establishing sustainable agriculture there. In Indonesia, organic farming for paddy rice cultivation also has a lot of potential to improve soil quality, reduce the cost of chemicals that have recently been increasing with the price of fossil fuels and increase farmers' incomes due to its higher price. However, organic farming requires intensive labour such as weeding and applying *bochashi* fertilizer to the fields.

The biggest difference was observed in the share-cropping system of organic farming in Indonesia compared with Japanese organic farmers. In the study area, profits from rice production were shared among the land owners, farmers (managers) and workers, but workers could receive only about 20% of the yield base of rice production. This suggests that by converting conventional farming to organic farming, landowners and farmers can increase their profits, whereas workers must work harder at organic farming but receive relatively little added benefit (Table 14.4). Thus, although organic farming has great potential to improve environmental quality, it also has problems regarding social justice in Indonesia.

14.4 CONCLUSIONS

Agriculture dominates land and water use like no other human enterprise, with landscape providing critical products for human sustenance. Farmers, consumers, researchers and policymakers

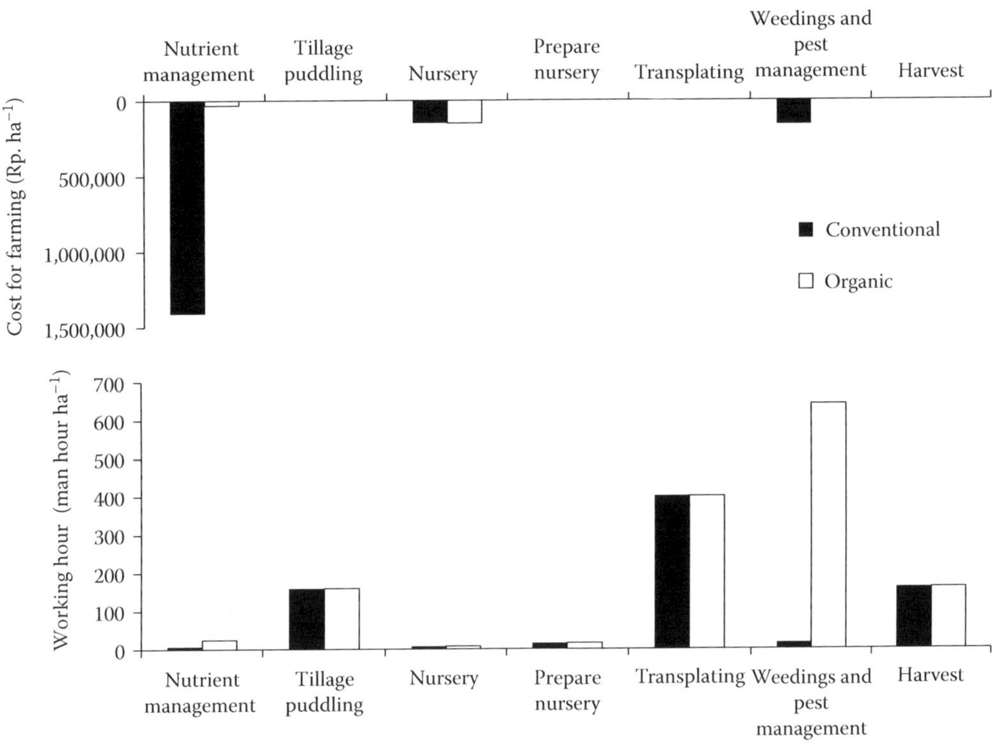

Figure 14.10 Comparison of costs for farming and man hours needed for each farming procedure between organic and conventional rice farming in West Java. (Data from Komatsuzaki, M. and M. F. Syuaib. 2010. *Sustainability* 2(3):833–843.)

in many parts of the world have begun to develop and promote sustainable agriculture. But adoption of sustainable farming is essential on a much larger scale to achieve the Millennium Development Goals on hunger, poverty and environmental sustainability in developing countries, and to sustain the ecosystem in rural economics in industrialized countries.

As local environmental quality becomes increasingly degraded by agricultural practices, the importance of protecting and restoring soil resources is being recognized by the world community (Lal, 1998, 2001; Barford et al., 2001). Sustainable management of soil received strong support at the Rio Summit in 1992 as well as at Agenda 21 (UNCED, 1992), at the UN Framework Convention on Climate Change (UNFCCC, 1992), in articles 3.3 and 3.4 of the Kyoto Protocol (UNFCCC, 1998), and elsewhere. These conventions are indicative of the recognition by the world community of the strong linkages between soil degradation and desertification on the one hand and loss of

Table 14.4 Comparison of Gross Benefit and Hourly Wages between Organic and Conventional Farming Systems

Management	Yield[a] (without husk) (Mg)	Price (Rp/kg)	Gross Benefits (Rp/ha)	Hourly Wages[b] (Rp)
Organic	3.2	6500	20,800,000	2955
Conventional	4.1	5500	22,550,000	5872

[a] Yields were obtained by quadrate sampling on September 2008.
[b] Hourly wages were calculated by taking account of the cost share of total benefit to the landowner, manager and worker.

biodiversity, threats to food security, increases in poverty, and risks of accelerated greenhouse effects and climate change on the other. This situation suggests that global support network systems between the countries are needed to conserve the local environment such as Indonesia's organic farmlands and Japanese croplands.

Therefore, new farming system development will be needed to establish a sustainable farming system in Indonesia and Japan in terms of both the local and global environmental levels. Based on these social and ecological situations and understanding, appropriate technology should be developed to conserve the ecological environments in Asia.

The use of *bochashi* organic fertilizer as an alternative soil fertility amendment in nutrient tropical paddy soils of West Java has resulted in increased SOC storage and gross benefit for farming. These techniques are also effective in eliminating chemical fertilizer and enhancing submaterials from rice production. Therefore, a new farming system for rice production may help not only in mitigating global warming due to carbon sequestration in the atmosphere, but also in establishing a sustainable food system in Japan and Indonesia.

Recent intensive research has shown that sustainable farming practices that use cover crops or bochashi can contribute to mitigating global warming, conserving biodiversity, and maintaining soil fertility and productivities. However, these farming practices often do not return enough for a farmers directory. Therefore, political and social incentives are much more important based on the common understanding that soil and agro-ecosystems are essential in developing a nature–human coexistence society in this century.

REFERENCES

Asagi, N. and H. Ueno. 2006. Nutrient release management for white clover by irrigation timing and their effects on rice growth and yield in paddy field with cover crop. In *ASA–CSSA–SSSA International Meetings*, pp. 105–158. Indianapolis: American Society of Agronomy.

Asagi, N. and H. Ueno. 2009. Nitrogen dynamics in paddy soil applied with various 15N-labelled green manures. *Plant Soil* 322:251–262.

Barford, C. C., S. C. Wofsy, M. L. Goulden et al. 2001. Factors controlling long- and short-term sequestration of atmospheric CO_2 in a mid-latitude forest. *Science* 294:1688–1691.

Bonner, J. and A. W. Galston. 1952. *Principles of Plant Physiology*. San Francisco, CA: W. H. Freeman and Co.

CBS: Central Bureau of Statistics (Biro Pusat Statistik: BPS). 2006. *Statistical Year Book of Indonesia 2005*. Jakarta, Indonesia.

CBS: Central Bureau of Statistics (Biro Pusat Statistik: BPS). 2008. *Statistical Year Book of Indonesia 2007*. Jakarta, Indonesia.

CBS: Central Bureau of Statistics (Biro Pusat Statistik: BPS). 2009. *Statistical Year Book of Indonesia 2008*. Jakarta, Indonesia.

Horimoto, S., H. Araki, M. Ishimoto, M. Ito, and Y. Fujii. 2002. Growth and yield of tomatoes grown in hairy vetch incorporated and mulched field. *Jpn Farm Work Res* 37:231–240 (In Japanese).

IFOAM Organics Net. 2008. Available at URL: http://www.organic-world.net. Accessed 15 March 2010.

Katoh, T. 2003. Carbon accumulation in soils by soil management, mainly by organic matter application—Experimental results in Aichi prefecture. *Jpn Soil Sci Plant Nutr* 73:193–201.

Komatsuzaki, M. 2008. Use of Italian ryegrass as a winter green manure in flooding paddy field for environmental friendly farming system. *Bokuso to Engei* 56(6):10–14.

Komatsuzaki, M. 2009. Nitrogen uptake by cover crops and inorganic nitrogen dynamics in Andisol paddy rice field. *Jpn J Farm Work Res* 44(4):201–210.

Komatsuzaki, M. and H. Ohta. 2007. Soil management practice for sustainable agroecosystem. *Sustain Sci* 2:103–120.

Komatsuzaki, M. and Syuaib, M. F. 2010. Comparison of the farming system and carbon sequestration between conventional and organic rice production in West Java, Indonesia. *Sustainability* 2(3):833–843.

Komatsuzaki, M., S. Moriizumi, S. Gu, S. Abe, and Y. Mu. 2004. A case study on paddy culture utilizing cover cropping. *Jpn J Farm Work Res* 39:23–26.

Kudo, A., N. Kawagoe, and S. Sasanabe, 1995. Characteristics of water management and outflow load from a paddy field in a return flow irrigation area. *J Jpn Soc Irrig Drain Reclam Eng* 63:179–184 (In Japanese).

Kusaba, T. 2001. Development of soil diagnosis research, *J Agr Sci* 56:7–12 (In Japanese).

Lal, R. 1998. Soil erosion impact on agronomic productivity and environment quality. *Crit Rev Plant Sci* 17:319–464.

Lal, R. 2004. Soil carbon sequestration impacts on global climate change and food security. *Science* 304:1623–1627.

Magdoff, F. 1998. Cover crops. In *Building Soils for Better Crops*, ed. F. Magdoff, pp. 73–89. Lincoln, NE: University of Nebraska Press.

Marriott, E. E. and M. M. Wander. 2006. Total and labile soil organic matter in organic and conventional farming systems. *Soil Sci Soc Am J* 70:950–959.

Ministry of Agriculture of Republic of Indonesia. 2009. Central of Agricultural Statistical Data Base. Available at http://database.deptan.go.id/bdsp/index.asp. Accessed 15 March 2010.

National Agricultural Research Center for Kyushu Okinawa Region. No-tillage rice transplanting machine using hairy vetch cover crop. 2003. Available from: URL: http://konarc.naro.affrc.go.jp/kyushu_seika/2003/2003621.html. Accessed 15 March 2010.

National Agriculture and Food Research Organization. Levee management manual. 2008. Available from: URL: http://wenarc.naro.affrc.go.jp/tech-i/covercrop/covercrop_200801.pdf. Accessed 15 March 2010.

Peet, M. 1996. Chapter 1: Soil management. In *Sustainable Practices for Vegetable Production in the South*, ed. M. Peet, pp. 1–28. Newburyport: Focus Publishing R. Pullins Company.

Pettygrove, G. S. and J. F. Williams. 1997. Nitrogen-fixing cover crops for California rice production. Available from: URL: http://www.plantsciences.ucdavis.edu/uccerice/covercrp/covcrop4.htm. Accessed March 15, 2010.

Pimentel, D., C. Harvey, P. Resosudarmo et al. 1995. Environmental and economic costs of soil erosion and conservation benefits. *Science* 267:1117–1122.

Sainju, U. M., W. F. Whitehead, B. P. Singh, and S. Wang. 2006. Tillage, cover crops, and nitrogen fertilization effects on soil nitrogen and cotton and sorghum yields. *Eur J Agron* 25:372–382.

Sarrantonio, M. 1998. Building fertility and tilth with cover crops. In *Managing Cover Crops Profitably*, ed. A. Clark, pp. 16–24. Beltsville, MD: Sustainable Agriculture Network.

Shirato, Y., K. Paisancharoen, P. Sangtong, C. Nakviro, M. Yokozawa, and N. Matsumoto. 2005. Testing the Rothamsted carbon model against data from long-term experiments on upland soils in Thailand. *Eur J Soil Sci* 56:179–188.

Snyder, G. H., H. W. Burdine, J. R. Crockett et al. 1978. Water table management for organic soil conservation and crop production in the Florida Everglades. Bulletin 801. Gainesville, FL: University of Florida.

Statistics of Agriculture, Forestry and Fishers. 2005. Crop land area and field use ratio in 2004–2005. Available from: URL: http://www.maff.go.jp/toukei/sokuhou/data/nobemenseki2004/nobemenseki2004.pdf (In Japanese). Accessed 15 March 2010.

Sustainable Agriculture Office. Results of certification of ecological friendly farmer. 2008. Available from: URL: http: http://www.maff.go.jp/soshiki/nousan/nousan/kanpo/ecofarmer.pdf. Accessed 15 March 2010.

Syuaib, M. F. 2006. An overview of farm work and farming system in Indonesia. In *International Symposium on Sustainable Agriculture in Asia: Challenges for Agricultural Sciences on Environmental Problems under Global Changes*. Bogor, Indonesia.

Syuaib, M. F. 2009. Perspective of sustainable agriculture in Indonesia: Keep growing in harmony with environment. In *International Symposium and Workshop on Sustainable Science Thinking the Shift and the Role of Asian Agricultural Science*. Ibaraki, Japan.

Tabuchi, T. and S. Hasegawa. 1995. *Paddy Fields in the World*. Tokyo: Japanese Society of Irrigation, Drainage and Reclamation Engineering.

Takeda, I., A. Fukushima, and R. Tanaka. 1997. Non-point pollutant reduction in a paddy field watershed using a circular irrigation system. *Water Res* 31:2685–2692.

Tanaka, Y. 2001. Challenges for reduction of water quality in Biwako Lake. *Nougyo Gijyut* 56(6):251–256 (In Japanese).

The Jakarta Post. 2009. Economic crisis helps boost growth in Indonesia's organic fertilizer sector. Available at http://www.thejakartapost.com/news/2009/03/02/economic-crisis-helps-boost-growth-indonesia039s-organic-fertilizer-sector.html. Accessed 15 March 2010.

Tominaga, T. 2007. Ecological studies on *Imperata cylindrica* and its use in Japan. *J Weed Sci Tech* 52:66–71.

Ueno, H. 2004. Use of cover crops in lowland rice cultivation. *Jpn J Farm Work Res* 39(3):165–170.

UNCED. 1992. Agenda 21: Programmed of action for sustainable development, Rio declaration on environment and development, statement of principles. In *Final Text of Agreement Negotiated by Governments at the United Nations Conference on Environment and Development,* UNCED, Rio de Janeiro, Brazil, pp. 3–14. New York, NY: UNDP.

UNFCCC. 1992. *United Nations Framework Convention on Climate Change*. Bonn, Germany: UNFCCC,

UNFCCC. 1998. Kyoto Protocol to the United Nations Framework Convention on Climate Change. Available by United Nations Framework Convention on Climate Change at www.unfccc.int/resource/docs/convkp. Accessed 15 March 2010.

Willer, H. and M. Yussefi. 2006. *The World of Organic Agriculture: Statistics and Emerging Trends 2006*. Bonn, Germany: International Federation of Organic Agriculture Movements (IFOAM), and Frick, Switzerland: Research Institute of Organic Agriculture FiBL.

Yasue, T. 1993. *Chinese Milk Vetch*. Tokyo, Japan: Nou-Bun-Kyo (In Japanese).

CHAPTER 15

Advances in Genetics and Genomics for Sustainable Peanut Production

Baozhu Guo, Charles Chen, Ye Chu, C. Corley Holbrook, Peggy Ozias-Akins, and H. Thomas Stalker

CONTENTS

15.1 INTRODUCTION

Peanuts (*Arachis hypogaea* L.), or groundnuts, are one of the major economically important legumes that are cultivated worldwide for their ability to grow in semiarid environments with relatively low inputs of chemical fertilizers. On a global basis, peanuts are also a major source of protein and vegetable oil for human nutrition, containing about 28% protein, 50% oil, and 18% carbohydrates.

India and China together produce almost two-thirds of the world's peanuts, and the United States produces about 6% (http://faostat.fao.org/). Nearly two-thirds of global production is crushed for oil and the remaining one-third is consumed as food. From 2004 to 2006, peanuts were grown on an average of 21.49 million hectares worldwide, with production totaling 32.98 million metric tons (http://www.nass.usda.gov/Publications/Ag_Statistics/2008/Chap03.pdf). During the same time, the U.S. peanut crop averaged 571 thousand hectares with production of 1.9 million metric tons. Peanut production has a significant role in sustainable agriculture in terms of global food security and nutrition, fuel and energy, sustainable fertilization, and enhanced agricultural productivity.

Genetics and genomics have substantial potential to enhance sustainable production. The major contribution of these technologies for peanut will likely be to improve disease resistance and enhance productivity. It is important to preserve the diversity of germplasm collections to have resources for maximizing the genetic potential in plant breeding and genetics programs in order to maximize desirable genetic traits in improved germplasm and to provide growers with cultivars that are locally adapted and highly productive. Genomics analyzes the complete genetic makeup of plants, through mapping, sequencing, and functional studies to identify genes regulating, controlling, and modifying traits. Together, plant breeding, genetics, and genomics are a powerful approach to enhance sustainability of agriculture.

Peanut improvement faces challenges due to low genetic diversity among cultivated genotypes and a large, polyploid genome. In spite of these obstacles, however, a wealth of information in peanut is available concerning germplasm collection and utilization, conventional breeding and cultivar improvement for disease resistance and drought tolerance, cytogenetics and genome composition, molecular genetics and genomics, molecular methods for mutation screenings (TILLING, Targeting Induced Local Lesions in Genomes) and genetic transformation. In recent years, there have been several excellent reviews with different focuses such as Holbrook and Stalker (2003) and Tillman and Stalker (2009) concentrating on germplasm issues, interspecific hybridization, and breeding methodologies, Ozias-Akins (2007) who reviewed the progress made in cultivar improvement through applications of molecular markers and transformation approaches, and Burow et al. (2008) who summarized the state of genomics for peanut particularly as related to the crop's nutritional value in oil and protein. For the last few years, particularly since the first *International Peanut Genomics and Aflatoxin Conference* held in Guangzhou CHINA in 2006 (Guo et al., 2009b) and the second *International Peanut Conference: Advances in Arachis Through Genomics and Biotechnology* in Atlanta, Georgia in 2007 (http://www.peanutbioscience.com), progress has been made within the international peanut community to foster research collaboration and the development of sustainable peanut production through the use of genetic and genomic resources and tools such as genetic mapping and marker-assisted selection (MAS). In this review, only the most pertinent aspects of peanut genetics and genomics are discussed. Research efforts will be summarized in three areas including: (1) worldwide germplasm collection and utilization; (2) genetic breeding and cultivar development; and (3) molecular genetics and genomic biotechnology.

15.2 GERMPLASM COLLECTION AND UTILIZATION

A large number of accessions of peanut exist in germplasm collections (Holbrook and Stalker, 2003) that harbor numerous alleles that have been underused for cultivar development. To better utilize the peanut germplasm, core collections have been created to reduce the efforts needed in evaluating the larger collections; and application of genomics is expected to enhance the use of these resources (Barkley et al., 2007; Burow et al., 2008; Wang et al., 2009).

There are three major peanut germplasm collections in the world. The United States maintains a collection of over 9711 accessions of *A. hypogaea* (Holbrook and Isleib, 2001). The International Crops Research Institute for the Semi-Arid Tropics (ICRISAT) maintains a larger *Arachis* germplasm collection of 14,723 accessions (Upadhyaya et al., 2001), and the Chinese Oil Crops Institute preserves 6390 accessions in Wuhan, China (Jiang et al., 2008). About half of U.S. *A. hypogaea* accessions are unimproved landraces collected from expeditions made to South America where the primary centers of origin and diversity are found. About one-third of the U.S. peanut accessions originated from Africa, an important secondary center of diversity for peanut. In contrast, in the ICRISAT collection about 35.7% of the accessions are from Asia, 32.7% from Africa, 17% from South America, and 13.6% from North America.

A diverse germplasm collection is a valuable resource for breeding new cultivars. Because of the size of these collections, it is not feasible for peanut breeders to screen an entire collection for traits that have economic value. Frankel (1984) proposed the use of core collections to provide smaller subsets of germplasm that could be used to mine valuable traits from entire germplasm collections. Thus, Holbrook et al. (1993) developed an *A. hypogaea* core collection that has been used to evaluate several of the major production-limiting diseases of peanut. This core collection was used to evaluate the core collection theory for germplasm evaluations. Examination of data for resistance to leaf spot (*Cercosporidium personatum*) (Holbrook and Anderson, 1995) and peanut root-knot nematode (*Meloidogyne arenaria*) (Holbrook et al., 2000b) clearly demonstrated that the peanut core collection can be used to improve the efficiency of identifying valuable traits in the entire germplasm collection that have led to greatly expanded efforts to mine the larger germplasm collection for agronomically important traits for cultivar development. For example, accessions in the core collection have been evaluated for resistance to tomato spotted wilt virus (TSWV) (Anderson et al., 1996), preharvest aflatoxin contamination (Holbrook et al., 2009), and *Rhizoctonia* limb rot (Franke et al., 1999). Accessions in the core collection have also been evaluated for fatty acid composition (Hammond et al., 1997), and eight aboveground and eight belowground morphological characteristics (Holbrook and Dong, 2005).

The U.S. peanut core collection has been very effective in enhancing the utilization of peanut genetic resources (Holbrook and Stalker, 2003); however, an even smaller subset of germplasm is needed for traits which are difficult and/or expensive to measure. Holbrook and Dong (2005) also developed a mini core collection for the U.S. germplasm collection and demonstrated that this subset can be used to further improve the efficiency for identifying desirable traits in the larger core and entire collections. Accessions in the mini core collection have been evaluated for seed nutrient content (Dean et al., 2009), and long-chain fatty acid content (Dean and Sanders, 2009), percentage oil variation (Wang et al., 2009), simple sequence repeat (SSR) marker variation (Barkley et al., 2007; Kottapalli et al., 2007) and Chu et al. (2007b) used a cleaved amplified polymorphic sequence (CAPS) marker to assess the frequency (31.6%) of a loss-of-function mutation in oleoyl-PC desaturase in the mini core collection. Association mapping of U.S. mini core collection is in progress using more polymorphic SSRs to identify interesting alleles and traits for breeding programs (M. Wang, personal communication).

The largest collection of *A. hypogaea* is maintained at ICRISAT (Upadhyaya et al., 2001). A core collection (Upadhyaya et al., 2003) and a mini core collection with 184 accessions (Upadhyaya et al., 2002) also have been selected for the ICRISAT collection. These subset collections have been assessed for several characteristics including early maturity (Upadhyaya et al., 2006), and tolerance to drought and low temperature (Upadhyaya et al., 2001; Upadhyaya, 2005). A core collection specific for the Valencia market-type peanut was developed and evaluated for morphological characteristics (Dwivedi et al., 2008). In China, a core collection comprising 576 accessions has recently been developed for their *A. hypogaea* collection (Jiang et al., 2008). The evaluation of this core collection and subsequent comparison to the ICRISAT mini core collection indicated that it is an important source of genetic variation for varieties *hirsuta* and *vulgaris*.

Recent efforts funded by the Generation Challenge Program have led to the assembly of a reference collection of groundnuts based on SSR marker genotyping (V. Vadez, personal communication). This collection of 264 *A. hypogaea* accessions is representative of most allelic diversity of the entire collection housed at ICRISAT, and provides a workable entry point to the peanut germplasm for the identification of useful germplasm sources. The ICRISAT mini core collection is included in this reference collection.

15.3 GENETIC BREEDING AND CULTIVAR DEVELOPMENT

Genetic improvement of cultivars has been playing an essential role in sustainable peanut production and currently used cultivars have provided economic benefits to growers and consumers. Peanut production in the United States initially began with introduced landraces originally from South America. The first cultivar selected from planned crosses between plant introductions was released in the 1940s; and the initial crosses were made in two separate programs of W. A. Carver at the University of Florida and B. B. Higgins at the University of Georgia in the early to mid-1930s (Isleib et al., 2001). Cultivars developed from the first populations formed by hybridization were used as parents to generate new populations for the second cycle of selection. The process of hybridization between elite cultivars to form new populations for selection continues in peanut-breeding programs.

In order to increase genetic diversity, artificial hybridization became the mechanism to form a population with increased genetic variability. Artificial mutagenesis was also used extensively in the late 1950s to early 1970s, but it has not been a productive method of obtaining useful genetic variability for peanut cultivar development in the United States. Thus, population formation by hybridization has dominated in the present-day peanut-breeding community. Populations used for cultivar development range from a two-parent population to complex populations involving multiple parents. However, the majority of peanut cultivars developed in the United States have been selected from two-parent populations. For example, "Georgia Green" was selected from a two-parent cross between "Southern Runner" and "Sunbelt Runner." "Tamrun OL07" was derived from a three-parent cross among "Tamrun 96" (Smith et al., 1998), breeding line Tx901639-3, and "SunOleic 95R" (Gorbet and Knauft, 1997), the donor of the high-O/L genes (Baring et al., 2006). A double-cross population of (F627B 3 × "Andru 93") × ("Sunrunner" × F435–2 HO) was used to develop cultivar "Andru II," a runner market-type peanut cultivar with high-oleic (80%) oil chemistry and good resistance to TSWV (Gorbet, 2006).

The procedure most often used in peanut to transfer genes for desirable traits from an agronomically unacceptable landrace into an elite population is referred to as backcrossing. "Georgia-04S" was selected from a backcross between an F_4 Georgia high-oleic selection derived from ("Georgia Browne" × UF 435-OL-2) × "Georgia Browne," the recurrent parent (Branch, 1994). In this case UF 435-OL-2 was used as a donor parent to contribute the high-oleic trait into recurrent parent "Georgia Browne." Molecular MAS has been used in conjunction with backcrossing to improve efficiency for converting existing cultivars into new ones with one or a few added traits. For example, only three years were needed to create a new version of high-oleic "Tifguard" (Holbrook et al., 2008a) by utilizing the backcrossing procedure with MAS for root-knot nematode resistance and the high-oleic trait (Y. Chu, P. Ozias-Akins, and C. C. Holbrook, unpublished data).

Yield is the character of primary importance to peanut production. Genetic improvement in yield by hybridization and selection has more than doubled in the past. However, yield improvements have decreased in recent years. The lower rate may reflect the difficulty in obtaining genetic improvement beyond that of already well-adapted cultivars. Another possibility that the slowdown in yield improvement is because peanut-breeding programs have been concentrating on improving disease resistance, and so more efforts have been with utilizing resistance genes in non-adapted genotypes versus elite × elite crosses, which has slowed yield improvement progress. It also may be related to the limited amount of genetic diversity among the high-yielding cultivars used as parents for hybridization. There is evidence that the allelic diversity in runner-type peanuts has significantly increased over the years and the diversity in Virginia type fluctuated greatly over the years (it is about the same now as in the 1940s) (S. R. Milla-Lewis and T. G. Isleib, personal communication). The genetic base of peanut in the United States has been extremely narrow. Parents with diverse origins have a higher probability to produce superior progenies than those of similar ancestry, but it has become increasingly difficult to find high-yielding genotypes that do not have common parentage. Isleib et al. (2001) reported that in the runner and Virginia market types, the average PI (plant inventory) ancestry of all cultivars was 17.9%. The parents of all runner-type cultivars released in the United States can be traced back to 13 PIs. Only three PIs occur in the pedigrees of Virginia-type cultivars and seven PIs are in the pedigrees of Spanish-type cultivars. Over the past 30 years, great efforts have been made to integrate unique germplasm into the breeding populations in the United States in order to expand the peanut genetic base.

Yield is a quantitative trait and genetically controlled by multiple genes. Halward and Wynne (1991) found that additive effects were to be significant for pod yield and seed weight. The genotype-by-environment (G × E) interactions for yield and market grade also exist in peanuts (for review see Knauft and Wynne, 1995). The amount of emphasis that can be placed on yield improvement also depends on the numbers of other characters that must be considered. With this in mind, the following characters are generally considered in a peanut-breeding program in the United States.

15.3.1 Resistance to Root-Knot Nematodes

The peanut root-knot nematode is one of the most important pests of peanut in the United States and globally. Using host plant resistance is the most effective way of controlling the damage caused by peanut root-knot nematode. More than 100 accessions with resistance have been identified from 3000 accessions evaluated in the U.S. collection (Holbrook and Noe, 1992; Holbrook et al., 2000a). Only moderate levels of resistance can be found in cultivated peanuts, but high levels of resistance exist in *A. batizocoi*, *A. cardenasii*, and *A. diogoi* (Baltensperger et al., 1986; Nelson et al., 1989; Holbrook and Noe, 1990). Two dominant genes are reported to control the resistance (Garcia et al., 1996), *Mag* for inhibiting root galling and *Mae* for inhibiting egg production. The reactions of resistance genes from *A. cardenasii* and *A. batizocoi* to *M. arenaria* are different. Resistance from *A. cardenasii* results in a complete inhibition of nematode development (Nelson et al., 1990). The resistance of *A. batizocoi* reduces the total number of invading nematodes reaching maturity and producing eggs, thus extending the time for *M. arenaria* to finish its life cycle. No hypersensitive reaction has been observed in *A. batizocoi* plants (Nelson et al., 1990). The resistance found from *A. cardenasii* has been intensively studied and has been introgressed into *A. hypogaea*. "TxAG-6," a complex interspecific hybrid, carries the resistance to *M. arenaria* from three nematode-resistant species, *A. batizocoi*, *A. cardenasii*, and *A. diogoi* (Simpson, 1991). The first peanut cultivar "COAN" with nematode resistance was developed from a backcross progeny of "Florunner" × TxAG-6 (Simpson and Starr, 2001). Resistance of "COAN" to *M. arenaria* was similar to that of *A. cardenasii* by reducing nematode reproduction. Owing to an extreme susceptibility to TSWV, "COAN" is difficult to adapt to southeast U.S. peanut production region where TSWV is epidemic. "Tifguard," a runner-type peanut cultivar, has high level of resistance to both root-knot nematode and TSWV (Holbrook et al., 2008a). "Tifguard" was selected from the breeding progeny of the cross between the TSWV-resistant "C-99R" and the nematode-resistant "COAN."

15.3.2 Resistance to Soil-Borne Fungal Diseases

Among important soil-borne fungal pathogens, three can cause significant yield losses in the United States, including white mold or stem rot (*Sclerotium rolfsii*), *Sclerotinia* blight (*Sclerotinia minor*), and *Cylindrocladium* black rot (CBR) (*Cylindrocladium parasiticum*). Compared to the foliar diseases, which can be adequately controlled by application of fungicides, these soil-borne diseases are less predictable in their occurrence and more difficult to manage. In heavily infested fields, the plant damage and resulting yield loss can be devastating. There are useful genetic variations for resistance to white mold, and sources of resistance have been identified for breeding programs (Branch and Csinos, 1987; Grichar and Smith, 1992). In general, runner-type cultivars have higher levels of resistance to the disease than Virginia-type or Spanish-type peanuts, while Valencia-type peanuts are the most susceptible to white mold. In a three-year study under heavy white mold disease pressure, "Tamrun 96" (Smith et al., 1998) demonstrated less disease incidence caused by white mold and higher pod yields compared to other tested commercial cultivars (Besler et al., 1997). "Ap-3" (Gorbet, 2007) and "Florida-07" (Gorbet and Tillman, 2009) express significant improvement of white mold resistance compared to "Georgia Green" in the presence (inoculated) and absence (not-inoculated) of white mold (Gorbet and Tillman, 2009).

The first *Sclerotinia* blight occurring in the U.S. peanut was reported in 1971 in Virginia and North Carolina (Porter and Beute, 1974). Yield loss due to severe disease epidemics can be as high as 50%. Development of cultivars with resistance to *Sclerotinia* blight has become an important component of U.S. breeding programs. Sources of resistance to *Sclerotinia* blight were identified (Coffelt and Porter, 1982; Akem et al., 1992; Porter et al., 1992), and several moderately resistant cultivars and germplasm were released (Coffelt et al., 1994; Kirby et al., 1998). A partially resistant cultivar, "VA 93B," was the first Virginia-type cultivar released in 1994 (Coffelt et al., 1994), and was followed by the release of "VA 98R" and "Perry" (Mozingo et al., 2000; Isleib et al., 2003). Spanish-type peanut seems to be more resistant to *Sclerotinia* blight than Valencia or Virginia genetic backgrounds (Akem et al., 1992). Inheritance of physiological resistance to *S. minor* has been studied based on the cross "TxAG-4" × "VA 93B" and its reciprocal cross in Australia (Cruickshank et al., 2002). The broad-sense heritability of the resistance was estimated at 47% based on a Generation Means Analysis on a single plant basis. An additive effect was detected for gene action from "TxAG-4," but no dominance effects were found. In addition, a small but significant reciprocal effect between "TxAG-4" and "VA 93B" was found, which indicated that a maternal factor also plays a role in the resistance (Cruickshank et al., 2002).

CBR was first reported in Georgia in 1966. It currently threatens peanut production throughout the southeastern United States. A considerable genetic variation for resistance to CBR exists in peanut; however, Spanish-type cultivars generally express higher levels of resistance than either the Valencia or Virginia types (Phipps and Beute, 1997). Among released germplasm, the breeding line NC3033 (Virginia-type), and cultivars "NC 2" (Virginia-type) and "Argentine" (Spanish-type) possess moderate to high levels of resistance and have been used as parents for CBR resistance in peanut-breeding programs (Hadley et al., 1979). The CBR resistance seems to delay the epidemics of disease progress (Culbreath et al., 1991). Recently, "Bailey" and "Sugg," Virginia-type cultivars were released with high yield, high resistance to CBR, and improved pod and seed characteristics (Isleib et al., 2010).

15.3.3 Resistance to Foliar Diseases

Early leaf spot caused by *Cercospora arachidicola* and late leaf spot caused by *C. personatum* are the most important foliar diseases affecting peanut throughout the world. Although commonly used fungicides can manage both pathogens in the production field, cultivars with resistance are more cost effective. Holbrook and Isleib (2001) found that *A. hypogaea* accessions originating from Bolivia have a high frequency of resistance to late leaf spot and are a valuable source for resistance to early leaf spot. More sources of resistance to both early and late leaf spots have been identified in cultivated peanut (Chiteka et al., 1988a,b; Anderson et al., 1993), and used to develop resistant breeding lines (Wells et al., 1994; Branch and Fletcher, 2001). "Southern Runner" was the first released cultivar that has a moderate resistance to late leaf spot in 1984. Moderate levels of resistance are also available in the cultivars "Florida MDR 98," "C-99R," and "Florida-07" (Gorbet and Tillman, 2009).

A. diogoi (syn. *A. chacoense*) and *A. cardenasii* are immune to early and late leaf spots, respectively. Ouedraogo et al. (1994) reported that resistance to both diseases was introgressed from wild species into runner-type peanut-breeding lines and several germplasm lines with high levels of resistance have been released (Stalker and Beute, 1993; Stalker et al., 2002). Only a few successes in transferring the resistances from wild *Arachis* species to cultivated peanut were reported, because of interspecific compatibility barriers, undesirable traits drag in progeny, and the long period of time required for developing stable tetraploid interspecific derivatives. With different sources in the studies, the inheritances of resistance to late leaf spot have been reported as having two recessive genes controlling the resistance (Tiwari et al., 1984) or a five-gene model (Nevill, 1981). In general, partial resistance is a function of multiple components of resistance that contribute additively to a decrease in the rate of epidemic progress (Parlevliet, 1979). Breeding programs have made a great progress to suppress both early and late leaf spots, for example in "C-99R," "Hull," "DP-1," "GA-01R," "Georganic," and "Tifrunner," all of which were developed for resistance to early and late leaf spots and/or TSWV. Cultivar "DP-1" had the best field resistance to early leaf spot among the genotypes tested by Cantonwine et al. (2006), and "Georganic" (Holbrook et al., 2008b) had an intermediate level of resistance relative to "Georgia Green" and "DP-1" (Cantonwine et al., 2006). However, high yield combined with high levels of resistance to both diseases while maintaining acceptable market traits continues to be very difficult to achieve.

15.3.4 Resistance to TSWV

TSWV was first identified as a significant plant pathogen in Australia in 1916 (Brittlebank, 1919). Since the first spotted wilt observed in 1971 in Texas, it has become a serious threat to peanut production and now causes significant yield losses in the United States. Annual losses due to TSWV on peanuts were estimated at \$40–\$100 million annually in Georgia alone (Jain et al., 1998; Culbreath et al., 2003).

TSWV is primarily transmitted by thrips, *Frankliniella fusca* (tobacco thrips) and *F. occidentalis* (western flower thrips) in the United States, with tobacco thrips playing the most important vector for secondary spread of TSWV (Lynch and Mack, 1995). Breeding efforts have been made to screen peanut germplasm accessions and improved breeding lines for TSWV resistance and resulted in the release of several TSWV-resistant cultivars including "Georgia Browne," "Georgia Green" (Branch, 1996), "C-99R," and "Florida MDR 98." Germplasm accessions with high levels of resistance to TSWV have been found in wild diploid species (Lyerly et al., 2002). The new breeding line NC94022 (a botanical variety *hirsuta*) has been identified with higher resistance to TSWV by Culbreath et al. (2005). Wang et al. (2007) identified PI 590483 originated from Brazil as having a high resistance to TSWV in the field and in greenhouse evaluation. Li et al. (2009) recently evaluated 22 genotypes from the United States and China in field trials in 2007–2008 for TSWV resistance and found that NC94022, "Georganic," C 689-6-2, "Georgia-OIR," C724-19 25, C 209-6-13, and "Tifguard" were the most resistant genotypes to TSWV. Transgenic peanut progenies with the insertion of the nucleocapsid protein (NP) gene of TSWV (Yang et al., 1998; Magbanua et al., 2000) were developed, and demonstrated a significantly lower incidence of TSWV in the field as well as inoculation under controlled environment (Li et al., 1997; Magbanua et al., 2000; Yang et al., 2004). This transgenic plant could potentially be used in a traditional breeding program to enhance host resistance to TSWV.

15.3.5 Resistance to Aflatoxin Contamination and Drought Tolerance

Aspergillus fungi can colonize seeds of several agricultural crops including peanut, and this can result in the contamination of the edible yield from these crops with toxic chemical compounds, aflatoxins (Guo et al., 2008a, 2009b). Scientists have been conducting research on aflatoxin contamination of peanut since the early 1960s. The first efforts were focused on identifying the risk factors for increased aflatoxin contamination and helped in documenting the importance of drought, high soil temperatures, and pod damage (for review see Cole et al., 1995). Later efforts were focused on the development of screening techniques (Anderson et al., 1996; Holbrook et al., 2008c) and identifying sources of resistance to *Aspergillus* colonization and/or aflatoxin contamination (Holbrook et al., 2009; Liang et al., 2009b; Nigam et al., 2009). This laid the foundation for conventional breeding programs that have resulted peanut-breeding lines with reduced *Aspergillus* colonization and/or reduced aflatoxin contamination relative to standard control cultivars (Holbrook et al., 2009; Liang et al., 2009b; Liao et al., 2009; Nigam et al., 2009). Recent research efforts include studies on the use of molecular genetic approaches to reduce aflatoxin contamination (Guo et al., 2009b). This includes the evaluation of genetically engineered peanut (Singsit et al., 1997; Ozias-Akins et al., 2002), molecular (Luo et al., 2005a, 2005c; Guo et al., 2008a, 2009b; Liao et al., 2009), and protein (Liang et al., 2005, 2006) markers.

Drought-tolerance traits in peanut may have potential for use as indirect selection criteria for resistance to preharvest aflatoxin contamination (Guo et al., 2008a). Holbrook et al. (2000c) observed lower levels of aflatoxin contamination in several improved lines that had been documented as having improved drought tolerance. Arunyanark et al. (2009) observed an association between aflatoxin contamination and the drought-tolerance traits, specifically leaf area and root length, and proposed that these might be used as selection criteria for reduced aflatoxin contamination. Girdthai et al. (2010a) also reported associations between physiological traits for drought tolerance and aflatoxin contamination in peanut genotypes subjected to terminal drought stress.

In addition to preharvest aflatoxin contamination, yield reduction by drought costs millions of dollars each year. While direct selection for yield under stressed conditions can be effective, the limitations of this approach are high resource investments and poor repeatability of results due to large G × E interactions that result in slow breeding progress (Wright et al., 1996). More rapid progress may be achieved by using physiological traits (Nigam and Aruna, 2008) such as harvest index, water-use efficiency (WUE), specific leaf area (SLA), and SPAD chlorophyll meter reading (SCMR). Although SLA and SCMR have been intensively used over the past two decades to proxy transpiration efficiency (TE), both SLA and SCMR were robust surrogate measures for TE in some studies (Upadhyaya, 2005; Sheshshayee et al., 2006), but not in others (Krishnamurthy et al., 2007). SCMR and TE are significantly and positively correlated with chlorophyll density in peanut leading to the conclusion that photosynthetic capacity has a major effect on TE (Sheshshayee et al., 2006). A major drawback in using these surrogate measures is that their relationship with TE was assessed in a very narrow range of genotypes. Recent work at ICRISAT (V. Vadez, personal communication) indicates that SLA and SCMR bear no significant relationship with TE when assessed over a large range of genotypes; therefore, their use as surrogate measures may be genotype specific. These data agree with similar data obtained in a mapping population (Krishnamurthy et al., 2007). This may also explain why a trait-based approach using surrogates SLA and SCMR has not proved more useful than a yield-based selection approach (Nigam et al., 2005).

For testing drought-tolerance traits, the ICRISAT group has developed a high-throughput method that allows the estimation of TE in a large range of genotypes in a replicated manner (Krishnamurthy et al., 2007). Stomatal regulation has been identified as a principal mechanism to explain differences in TE. Genotypes having lower canopy conductance have higher TE (Bhatnagar-Mathur et al., 2007), and lower canopy conductance is related to differences in how transpiration responds to an increased vapor pressure deficit (Devi et al., 2010). Therefore, certain peanut genotypes restrict water losses under high evaporative demand, and others maximize water losses. The former loses an opportunity to fix carbon since this trait entails at least partial stomata closure, but helps to conserve water, whereas the latter continues to fix carbon, but at a high water cost. Genotypic differences for canopy conductance appear to be large. An observation also indicates that water uptake during critical periods of the pod filling stage leads to higher pod yield and is more important than the total amount of water uptake (Ratnakumar et al., 2009).

Rucker et al. (1995) found that some peanut genotypes with large root systems had high yield under drought conditions, and they suggested that these genotypes possessed drought avoidance traits. In a study examining the root distribution of drought-resistant genotypes, Songsri et al. (2008) also proposed that drought avoidance can be an important mechanism for enhanced drought tolerance. Girdthai et al. (2010b) developed a hydroponic system that should be useful in breeding peanut for larger root systems. Songsri et al. (2009) observed different mechanisms for improved water-use efficiency in a set of drought-tolerant lines. They proposed that it may be possible to combine these mechanisms together to develop new peanut cultivars with improved water-use efficiency.

15.3.6 Improvement of Oil Quality

Oleic and linoleic acids contribute 80% of fatty acids in peanut oil. Normal peanut cultivars average 55% oleic acid and 25% linoleic acid (Knauft et al., 1993), whereas, high oleate peanut genotypes have approximately 80% oleic acid and 2% linoleic acid (Norden et al., 1987). Oleic acid also can be beneficial in decreasing blood low-density lipoprotein levels, suppressing tumorigenesis, and ameliorating inflammatory diseases. These obvious benefits drive the breeding effort toward developing high-oleic peanuts. The first high-O/L peanut, F435 with 80% oleic acid and 2% linoleic acid was identified by Norden et al. (1987). From F435, numerous high-oleic peanut cultivars such as "SunOleic 95R," "SunOleic 97R," "Tamrun OL01," "Georgia04S," "Andru II," "Florida-07," and "Hull" have been released through conventional breeding. Another two high-O/L peanut lines available for breeders were developed through chemical mutagenesis; C458, also named "Flavorunner 458" or "Mycogen-Flavorunner," was selected from ethyl methanesulfonate treated "Florunner" (U.S. patent 5948954); and M2-225 was generated from "AT-108" with diethyl sulfate mutagenesis treatment (U.S. patent 5945578). Gamma-radiation treatment also induced a high-oleic trait in peanut. "Georgia-02C" and "Georgia Hi-O/L" were derived from a high-O/L mutation that was selected from "Georgia Runner" treated with gamma-radiation (Branch, 2000, 2003). Georgia-05E is a high-O/L cultivar selected from a cross of "Georgia-01R" × "GA 942010" but the pedigree of the high-O/L advanced breeding line GA 942010 was not clear stated (Branch, 2006).

The high-oleic trait in peanut is simply inherited. Moore and Knauft (1989) found that two recessive genes control the high-oleate trait. Knauft et al. (1993) reported monogenic inheritance observed in crosses of the runner market-type cultivars and breeding lines. Isleib et al. (1996) examined five different Virginia-type cultivars and found that four were monogenic inheritance and one was digenic inheritance. Furthermore, Lopez et al. (2001) examined the inheritance of high-oleic acid in six Spanish market-type peanut cultivars and the segregation patterns indicated that two major genes were involved. Therefore, it seems that one of the recessive alleles appears to be common in runner and Virginia type but absent in Spanish-type peanuts. This was confirmed by Chu et al. (2007b) who reported that one of the recessive alleles, *ahFAD2A*, was found to be frequently present in *A. hypogaea* spp. *hypogaea* but not spp. *fastigiata*. Two recessive genes that control a high O/L ratio in peanut were named as *ahFAD2A* and *ahFAD2B* or ol_1 and ol_2, respectively.

15.4 CYTOGENETICS AND GENOME COMPOSITION

The first comprehensive cytological study of peanut was by Husted (1936) when the somatic chromosome number of the species was reported as $2n = 4x = 40$. During meiosis there are mostly bivalents, but a few multivalents can be formed, which means that genes generally behave in a diploid manner but tetraploid inheritance patterns also can occur. Chromosomes are small (1–4 μm) and difficult to karyotype, but Husted was able to distinguish one chromosome with a secondary constriction which he termed the "B chromosome" and one pair that was smaller than others in the genome and named as the "A chromosome." Thus, the cultivated peanut is believed to be an allotetraploid and the terminology of A and B genomes has persisted in the literature. Several types of secondary constrictions occur in *A. hypogaea* and Stalker and Dalmacio (1986) grouped cultivars based on distinct karyotypes among the botanical designations. At least 15 of the 20 chromosome pairs can be distinguished in somatic cells. More recent technologies such as genomic *in situ* hybridization better differentiate the peanut chromosomes and shed light on species relationships (Seijo et al., 2007). This suggests that all *A. hypogaea* have arisen from a unique allotetraploid plant population that originated from the same two diploid species, *A. duranensis* (A genome) and *A. ipaensis* (B genome). This further supports the hypothesis that *A. monticola* is either the immediate wild antecessor of *A. hypogaea* or a wild derivative of the cultivated species.

Gregory (1946) reported the first chromosome number in a related species (*A. glabrata*) in the genus as $2n = 4x = 40$. The four diploid ($2n = 2x = 20$) species *A. diogoi*, *A. marginata*, *A. prostrata*, and *A. villosulicarpa* were reported a year later by Mendes (1947). By 1982, there were only 26 *Arachis* species with associated chromosome numbers (Smartt and Stalker, 1982). Of the 80 described species of *Arachis*, five are tetraploid (*A. hypogaea* and *A. monticola* in section *Arachis* and *A. glabrata, A. pseudovillosa*, and *A. nitida* in section *Rhizomatosae*); four are aneuploid ($2n = 18$) (*A. decora*, *A. palustris*, and *A. praecox* in section *Arachis* and *A. prophycrocalyx* in section *Erectoides*), and the remaining 71 species are diploid ($2n = 2x = 20$) (Krapovickas and Gregory, 1994; Lavia, 1998; Valls and Simpson, 2005). Polyploidy arose independently in sections *Arachis* and *Rhizomatosae* and aneuploidy independently in sections *Arachis* and *Erectoides*. Discovery of aneuploids is unexpected because even though trisomics are relatively easy to find by observing plants derived from shrivelled seeds, no monosomics or nullisomics were observed after extensive investigations of *A. hypogaea* (H. T. Stalker, unpublished data).

Intra-sectional hybrids can be produced artificially whereas inter-sectional hybrids are very difficult to make (Gregory and Gregory, 1979). However, hybrids between species in sections *Rhizomatose* and *Arachis*; *Ambinervosae* and *Extranervosae*; and *Erectoides* and *Rhizomatosae*, *Arachis*, *Caulorhizae*, and *Ambinervosae* have been produced (Gregory and Gregory, 1979). Intersectional hybrids are generally sterile, although meiotic chromosome pairing has been observed among species of different groups. For example, a few bivalents will form in crosses between *A. rigoni* (section *Procumbentes*) × *A. paraguarensis* (section *Erectoides*) (Stalker, 1985), which indicates that the species are not closely related. Krapovickas and Gregory (1994) considered section *Erectoides* to be less isolated than the other sections based on hybridization data. Smartt and Stalker (1982) and Stalker (1991) designated genomes in different groups as section *Arachis* with A, B, and D genomes; section *Ambinaervosae* = Am; Section *Caulorhizae* = C, section *Erectoides* = E (which may need to be further subgrouped); section *Extranervosae* = Ex, section *Triseminalae* = T and section *Rhizomatosae*, series *Prorhizomatosae* = R1). However, because inter-sectional hybrids are very difficult to produce, cytological characterization of most species and their relationships in the genus is lacking.

Most cytological research has been done in section *Arachis*, which has the largest number of species and contains the domesticated species and cultivated peanuts. Several *Arachis* species have been karyotyped (Stalker and Dalmacio, 1981; Singh and Moss, 1982) even though mitotic chromosomes of section *Arachis* species have median centromeres and are very difficult to distinguish. Interspecific hybrids between A-genome species have mostly bivalent chromosome pairing whereas A × B, A × D, and B × D genome hybrids are highly sterile, but meiotic chromosomes have bivalents as well as many univalents, indicating homology between the A and B genomes (Stalker et al., 1991b). Stalker (1985) suggested that more appropriate terminology may be A_1 and A_2 to replace the A and B genomic designations. Meiotic chromosome pairing in hybrids among accessions of *A. duranensis* are mostly bivalents, but both quadravalents and hexavalents have been observed (Stalker et al., 1995); and hybrids between different accessions of *A. batizocoi* have quadravalents (Stalker et al., 1991a). This indicates that chromosome rearrangements, and especially translocations, have played a role in the evolution of both species and likely in other species of the genus.

Triploid interspecific hybrids between *A. hypogaea* and A- or B-genome diploids in section *Arachis* have 4–12 bivalents and 0–4 multivalents (Singh and Moss, 1984). This presents further cytological evidence for *A. hypogaea* being an allopolyploid derived from two diverse section *Arachis* species. To restore fertility, the chromosome number can be doubled to the hexaploid ($2n = 6x = 60$) level and up to 32 univalents have been observed (Spielman et al., 1979; Company et al., 1982), whereas each chromosome should have a homolog and 30 bivalents are expected. Thus, genetic regulators controlling chromosome pairing in *Arachis* are believed to be present (Stalker and Simpson, 1995). Although the hexaploids are generally stable at this chromosome level even though they have reduced fertility, the chromosome number can be reduced to 40 chromosomes either by selecting the most fertile progenies and selfing for numerous generations or by backcrossing with *A. hypogaea*. Hexaploid × diploid hybrids, or reciprocals, have not been obtained (H. T. Stalker, unpublished data). Further, introgression of genes from diploid species of section *Arachis* into the cultivated peanuts has been reported (Garcia et al., 1995; Burow et al., 2001).

15.5 MOLECULAR GENETICS AND BIOTECHNOLOGY

15.5.1 Genetic Markers

Historically, peanut cultivar development has been dominated by conventional breeding, which will continue to play a very important role in peanut improvement and production. Limited DNA polymorphisms have impeded the application of marker-assisted breeding in peanut, which has not been feasible in peanut because DNA markers, mapping information, and other resources critical for genotyping applications in cultivated × cultivated crosses have been lacking. The first reported genetic maps in intraspecific (cultivated × cultivated) crosses were only published recently by Varshney et al. (2009) and Hong et al. (2010). Both maps of the linkage groups (LGs) are incomplete and sparsely populated with SSR markers. Nevertheless, marker-assisted breeding has been applied on a limited scale (Jung et al., 2000a,b; Chu et al., 2007a,b, 2009; Nagy et al., 2010). Development and applications of molecular markers in peanut has been reviewed recently (Ozias-Akins, 2007). Since then, a large number of SSRs and other types of new markers have been developed allowing the construction of linkage maps for wild *Arachis* species, introgression lines, and cultivated × cultivated peanuts. Markers linked to simply inherited traits are also applied to molecular-assisted breeding.

In the early 1990s, peanut proteins were used as markers to identify genetic diversity among cultivated peanut and its wild diploid species; however, limited polymorphisms were detected. Three out of 25 isozymes from seeds were polymorphic among 71 peanut lines and only one isozyme isolated from flowers could document the introgression of *A. cardenasii* in an interspecific hybrid (Lacks and Stalker, 1993). Seed storage protein profiles were evaluated among various *Arachis* species. A greater level of polymorphism was found among wild species compared to cultivated peanut (Lanham et al., 1994). Liang et al. (2006) suggested that there may be an association between these polymorphic seed storage protein isoforms and peanut ssp. *fastigiata* (Spanish type) and *hypogaea* (runner type). They found that polymorphic protein peptides that were distinguished by 2-D PAGE could be used as markers for the identification of runner and Spanish peanuts. Advances in other marker types, such as randomly amplified polymorphic DNA (RAPD), restriction fragment-length polymorphism (RFLP), amplified fragment-length polymorphism (AFLP), sequence-characterized amplified region (SCAR), SSR, single-nucleotide polymorphism (SNP), and single-strand conformational polymorphism (SSCP) soon replaced the early exploration with proteins.

RAPD markers are amplified using primers of arbitrary sequence, usually 10 nucleotides long. Amplicons of RAPD markers can be multiallelic and multilocus. Initial RAPD screening of 27 peanut germplasm lines and cultivars found that there was no variation among the tested lines, but wild diploid *Arachis* species were highly polymorphic (Halward and Stalker, 1991; Halward et al.,

1992). More recently, RAPD markers were used to screen a F_2 mapping population segregating for rust resistance (Mondal et al., 2007), which was derived from an interspecific hybrid of *A. hypogaea* and *A. cardenasii*. Eleven of 160 RAPD markers were polymorphic and one marker cosegregated with the rust resistance trait.

RFLP methodology is a non-PCR based and a laborious process to detect DNA polymorphism. RFLP variability in cultivated peanut is also low (Halward and Stalker, 1991; Burow et al., 2009). Halward and Stalker (1991) used 13 restriction enzymes and 60 probes for digestion and hybridization, and found very little variation among 27 tested peanut genotypes but detected a large variation in wild species. The distinction of A and B genome came from cytogenetic evidence where the A-genome species possess a smaller pair of chromosomes than the B-genome species (Husted, 1936). Classification of diploid A-genome species by this criterion is supported by RFLP data (Gimenes et al., 2002b). However, B-genome species grouped by the absence of a pair of small chromosomes is heterogeneous (Gimenes et al., 2002b; Burow et al., 2009). Five RFLP markers for B-genome species developed from 154 cDNA probes redefined the B-genome group as seven *Arachis* species in two subgroups (Burow et al., 2009). The RFLP banding patterns of these two diploid species can reconstitute that of the cultivated peanuts (Kochert et al., 1996; Gimenes et al., 2002b; Ramos et al., 2006; Seijo et al., 2007; Burow et al., 2009).

AFLP is another PCR-based technique that involves restriction digestion of genomic DNA. Compared to RFLP and RAPD markers, AFLP markers can sample a larger number of loci, usually 10 to 50, in a single PCR-reaction. Application of AFLP markers to study genetic diversity of *Arachis* species resulted in similar conclusions as the previously mentioned marker types, such as low polymorphism among cultivated peanuts but large genetic variation among wild species. Taxonomic placement of wild *Arachis* species by AFLP polymorphisms agrees with the classification based on morphology, geographic distribution and cross compatibility (Gimenes et al., 2002a; Milla et al., 2005). AFLP markers were used to map the resistance trait to aphids, the vector for groundnut rosette disease (Herselman et al., 2004). The aphid-resistant parent was a Spanish-type breeding line that was crossed to a susceptible cultivar. A total of 308 AFLP primer sets were used to screen both the parents and F_2 bulks, and 12 polymorphic AFLP markers were mapped in the F_2 segregating population. The resistance trait to aphid was found to be controlled by a single recessive gene. A marker tightly linked to aphid susceptibility also was identified. AFLP markers are often sequenced and converted to SCAR, SNP, or CAPS markers (Agarwal et al., 2008).

SSR or microsatellite markers are a class of molecular markers based on PCR primers designed to flank a set of tandem repeats of a short DNA sequence. Insertion or deletion of these repeat units contributes to the marker polymorphism (Liang et al., 2009a). SSRs are highly polymorphic, multiallelic, codominant, ubiquitously distributed in the genome, analytically simple, and readily transferable between species such as from soybean to peanut (He et al., 2006). Amplification of SSR markers requires only a PCR reaction and the genotyping can be performed on an agarose gel if the fragment size difference is large enough or on a polyacrylamide gel to separate smaller differences (<10 bp).

Development of SSR markers is costly and three strategies have been used. First, peanut genomic libraries can be constructed specifically for SSR isolation. Early work was performed to hybridize SSR probes to the genomic library, 26 pairs of SSR primer pairs were designed and only six SSR markers detected polymorphisms in cultivated peanut (Hopkins et al., 1999). Subsequently, the efficiency of SSR marker development was improved by enriching SSR sequences in the genomic DNA library (He et al., 2003; Gimenes et al., 2007; Cuc et al., 2008). Second, expressed sequence tag (EST) libraries from various tissues of cultivated peanuts and diploid *Arachis* species were sequenced and assembled for SSR discovery (Ferguson et al., 2004; Luo et al., 2005a; Proite et al., 2007; Liang et al., 2009a; Guo et al., 2008b, 2009a; Nagy et al., 2010). EST-SSRs comprise the largest portion of available peanut SSR markers. Since EST-SSRs are derived from transcripts, they are highly conserved across related species and can serve as anchor markers for comparative mapping (Varshney et al., 2005); but they produce very low polymorphism among cultivated peanuts (Hong et al., 2010). Third, *in silico* development of SSR markers was performed using bioinformatics to mine SSR regions from nucleotide sequences deposited in a public database such as GenBank (Mace et al., 2008; Hong et al., 2010). Currently, the total number of SSRs discovered in peanut is estimated to be over 6000 (B. Guo, unpublished data). Two SSR-based genetic linkage maps for cultivated peanuts recently have been constructed (Varshney et al., 2009; Hong et al., 2010). There are also 1450 new SSRs deposited to NCBI GSS in the context of BAC end sequences data (D. Cook and G. He, personal communication) for genetic map development.

15.5.2 Other Markers

The genus *Arachis* belongs to the Leguminosae family which includes several model plants such as *Lotus japonicus* and *Medicago truncatula*. The large-scale genome sequence information from these model plants is useful for enriching the limited genomic information of peanut and its wild relatives. In order to identify syntenic relationships among different species, legume anchor markers (LEGs) were developed by aligning the EST sequences from *Arachis*, *Lotus*, and *Medicago* with the orthologs in *Arabidopsis*. The LEG primers are designed from conserved intron-spanning exon sequences that have a single copy in *Arabidopsis* (Hougaard et al., 2008). The rationale is that the unique sequence within a genome facilitates comparative mapping. A single copy in *Arabidopsis* is assumed to represent a single copy in other legumes. The marker-spanning intron region is more variable than exons allowing for polymorphism detection. LEG primers are anchored in highly conserved coding regions which increase their transferability across species (Bertioli et al., 2009). Macrosynteny and microsynteny were established among legume species and model species suggesting genome conservation across species is sufficient for comparative mapping (Choi et al., 2004; Hougaard et al., 2008). Subsequently, using anchored markers a unified genetic map or comparative map of the Papilionoids such *Arachis–Lotus–Medicago* could be established (Burow et al., 2008; Bertioli et al., 2009).

Resistance gene analogs (RGAs) were isolated from both *A. hypogaea* and wild *Arachis* species to develop disease resistance markers by Bertioli et al. (2003) and Yuksel et al. (2005). Degenerative primers were designed to amplify conserved disease resistance nucleotide binding site (NBS) domains. Seventy-eight and 250 putative resistance loci were isolated from *Arachis* species and *A. hypogaea*, respectively. SSCP markers were developed from these RGA sequences to identify markers and candidate genes for nematode resistance in peanut cultivars (Nagy et al., 2010). Cook and He (personal communication) is also focusing on the discovery of disease resistance genes in the legume family including peanut, cowpea, chickpea, pigeon pea, common bean, lupin, *Cercis*, and lentil and has discovered ~4000 disease resistance gene homologs (RGHs), including ~580 RGHs in cultivated peanut. Based on the comparison to the sequenced genomes of *M. truncatula* and *Glycine max*, they estimated that these ~580 genes represent the vast majority of RGHs in the genome of cultivated peanut.

15.5.3 Genetic Linkage Map

One of the major applications for molecular markers is for the construction of linkage maps. A genetic linkage map constructed from a population segregating for a trait of interest is required for QTL (quantitative trait loci) identification. A more detailed linkage map of all chromosomes and with sufficient markers is necessary for QTL analysis and marker-assisted breeding. Over the last few years, progress has been made in marker development from multiple sources enabling the expansion of quality resources needed for genotyping applications in cultivated × cultivated populations (Varshney et al., 2009; Hong et al., 2010; Nagy et al., 2010).

The first genetic map for *Arachis* species was constructed with RFLP markers from an F_2 population developed from the interspecific hybridization of two related diploid species with AA genome

(*A. stenosperma* × *A. cardenasii*) resulting in 11 LGs (Halward et al., 1993). Another RFLP-based linkage map was derived from a synthetic interspecific tetraploid BC_1 population {[*A. batizocoi* × (*A. cardenasii* × *A. diogoi*)]^{4x} × *A. hypogaea* ("Florunner")} with 23 LGs (Burow et al., 2001). Because of the complex pedigree, this map is complicated and difficult to use in terms of extraction of useful genetic information. The first SSR-based linkage map for *Arachis* was constructed from an F_2 population of diploid AA genome wild peanuts (*A. duranensis* and *A. stenosperma*) with 170 SSRs and 11 LGs (Moretzsohn et al., 2005), and an advanced version of same map has been published recently with 369 markers, including 188 microsatellites, 80 legume anchor markers, 46 AFLPs, 32 NBS profiling, 17 SNP, 4 RGA-RFLP, and 2 SCAR markers, mapped into 10 LGs by Leal-Bertioli et al. (2009). A diploid BB genome map was also established from a F_2 population of *A. ipaensis* × *A. magna* (Moretzsohn et al., 2009). Seven-hundred forty-five SSRs were initially screened on the parents and yielded 166 polymorphic markers with a linkage map of 10 LGs. Comparative mapping of the BB genome with the AA genome using 51 common markers revealed a high level of synteny. Foncéka et al. (2009) published a SSR-based interspecific tetraploid map using 88 individuals of the BC_1F_1 population of ["Fleur 11" × (*A. ipaënsis* × *A. duranensis*)^{4x}, and 298 loci were mapped in 21 LGs.

In addition to the linkage maps of diploid *Arachis* species, there are two intra-specific maps for cultivated peanuts published recently (Varshney et al., 2009; Hong et al., 2010). Varshney et al. (2009) used two distinct peanut cultivars, "ICGV 86031" and "TAG 24" that are drought tolerant and sensitive, respectively, for the mapping population. They screened 1145 SSRs that yielded a total of 144 polymorphic markers. A linkage map was constructed with 135 of these markers which are sparsely populated in 22 LGs. Hong et al. (2010) reported a SSR-based composite genetic linkage map of cultivated peanut genome based on three recombinant inbred lines constructed from three crosses with one common female parent. The four parents were screened with 1044 SSRs of which 901 produced PCR products. Of the 901 primer pairs, 146, 124, and 64 SSRs were polymorphic in the hybrid populations, respectively. Individual linkage maps were constructed from each of the three populations and a composite map based on 93 common loci was constructed. The composite linkage map consists of 22 composite LGs with 175 SSR markers, representing the 20 chromosomes of *A. hypogaea* (Hong et al., 2010). This composite map also consists of 47 SSRs that were used in the A–A genome diploid map (Moretzsohn et al., 2005). There was potential synteny as shown by collinear order of some markers and conservation of collinear LGs in the composite map and the A–A genome maps; for example, LG4 of the *A. hypogaea* map corresponded with Group 3 of the A–A genome diploid map. Nevertheless, the application of biotechnology for improving the allotetraploid cultivated peanut has been hampered by an inability to visualize sufficient genetic variation among paired genotypes and by lacking a high-resolution linkage map. Further development and improvement of a tetraploid genetic map *A. hypogaea* with newly developed markers such as SNPs would facilitate map resolution and coverage. The fine-maps also will be used to accelerate plant breeding efforts with peanut.

15.5.4 MAS for Nematode-Resistant Peanuts

Wild species generally harbor greater levels of resistance, and even apparent immunity, as compared to cultivated species. However, linkages of agronomically un-adapted wild alleles with disease resistance genes are inevitable. MAS has the potential to facilitate combining resistance loci from both cultivated and wild species with agronomically adapted alleles. Utilization of wild peanut species germplasm in breeding programs has been limited, but successes have occurred for several foliar pathogens, for insects, and for nematode resistance. MAS has been applied successfully in peanut-breeding programs for the two simply inherited traits of nematode resistance and high-oleic acid peanut cultivars. Nematode resistance was introgressed into cultivated peanut from *A. cardenasii* via a synthetic amphidiploid line TxAG6 [*A. batizocoi* × (*A. cardenasii* × *A. diogoi*)] that served as the nematode-resistant donor parent. By backcrossing TxAG6 to cultivar "Florunner," nematode-resistant cultivars "COAN" and "NemaTAM" were released (Simpson and Starr, 2001; Simpson et al., 2003). Additional selections were made by crossing "COAN" and the TSWV-resistant cultivar "C99R," and two near-isogenic lines, "Tifguard" (nematode-resistant cultivar) and 724-19-25 (nematode-susceptible breeding line), were produced (Holbrook et al., 2008a). A single dominant gene (*Rma*) of nematode resistance was identified (Burow et al., 1996; Choi et al., 1999).

Application of MAS for breeding nematode-resistant cultivars has been implemented from RAPD markers, RFLP probes, to more recent PCR-based markers (Burow et al., 1996; Choi et al., 1999; Chu et al., 2007a). Three RAPD markers (RKN 229, RKN 410, and RKN 440) that are linked to *M. arenaria* resistance were identified and three RFLP loci R2430E, R2545E, and S113E were reported to have association with *M. arenaria* resistance. Unfortunately when these markers were applied to additional breeding populations they were unreliable as selectable markers (Chu et al., 2007a). Based on the full-length sequence of RKN 440, a robust SCAR marker (S197) was designed whose amplicons demonstrate a distinct size difference between "COAN" and "Georgia Green" (Chu et al., 2007a). This PCR-based marker discriminates fragments amplified from the nematode-resistant and susceptible NILs by a 29 bp size difference.

However, a single dominant marker is unable to detect heterozygosity in genotypes and additional markers flanking *Rma* are needed to define the introgressed region for nematode resistance. To conduct the fine mapping of *Rma*, 2847 SSRs and 380 SSCPs were used to screen the parents of two mapping populations, F_4: "Gregory" × "Tifguard" and F_5: "NemaTAM" × GP-NC WS14 (Nagy et al., 2010). Thirteen polymorphic markers (including S197) were mapped in both populations. Severe suppression of recombination in the introgressed region was found. Comparative mapping of the *Rma*-linked markers to a diploid F_2 population derived from a cross between two *A. duranensis* accessions suggested that the introgressed region covers a third to half of the chromosome. The upper and lower borders of the interstitial alien region were also delineated by genetic markers. Flanking markers for the *Rma* introgressed region can ensure the integration of *Rma* into other elite cultivars. A codominant SSR marker GM565 was selected for MAS in breeding populations (Y. Chu, P. Ozias-Akins, and C. C. Holbrook, unpublished results).

15.5.5 MAS for High-Oleic Oil Peanuts

Oleic to linoleic acid ratios (O/L) in wild-type peanut are 1.0–4.0 whereas the O/L ratio in high oleic acid mutants is 35–40 (Norden et al., 1987). High O/L is desirable for the healthy cholesterol-lowering benefits and the oxidative stability of the oil (Wilson et al., 2006). The rate-limiting enzyme for the conversion of oleic to linoleic acid is oleoyl-PC desaturase (ahFAD2) (Ray et al., 1993). The two genes are homeologous and *ahFAD*2A belongs to the A sub-genome and *ahFAD*2B to B sub-genome of *A. hypogaea* (Jung et al., 2000a,b). From independent studies on F435 and F435-derived high-O/L cultivars, a mutant allele of *ahFAD*2A was characterized as having a 448G → A transition (Jung et al., 2000a,; Lopez et al., 2002; Patel et al., 2004). A CAPS marker targeting the mutation site has been designed to target this SNP in order to differentiate the mutant and wild-type *ahFAD*2A alleles (Chu et al., 2007b). This CAPS marker showed that 31.6% of accessions from the mini core collection of the U.S. peanut germplasm (Holbrook and Dong, 2005) carry this mutant allele (Chu et al., 2007b). Two alleles of *ahFAD*2B are found among high-O/L peanut cultivars (Chu et al., 2009). One allele is characterized by a 441_442insA mutation, found in the spontaneous high-O/L mutant line F435 and its derivatives (Jung et al., 2000b; Lopez et al., 2002). The second mutant allele was identified from a chemically mutagenized cultivar C458 (Flavorunner) which possesses a 225 bp insertion of a miniature-inverted repeat-transposable-element (MITE) at position 717 bp of the coding region (Patel et al., 2004). Another CAPS marker digested with *Hpy*188I was designed to track the 441_442insA mutation (Chu et al., 2009). Barkley et al. (2009) developed a real-time PCR genotyping assay to identify the wild-type and mutant alleles of *ahFAD*2B gene in peanuts.

As an example of MAS in a breeding program for peanut cultivar improvement, an intensive backcross schedule has been developed to pyramid the high-O/L trait with nematode resistance in the cultivar "Tifguard" (Holbrook et al., 2008a). Crosses with two high-O/L cultivars, "Georgia 02C" (GAO2C) and "Florida 07," were made with "Tifguard," the female parent. Markers S197 and GM565 were used to detect the inheritance of the introgressed segment carrying *Rma*. Both high-O/L male parents possess the 441_442insA mutation which can be identified by CAPS marker *Hpy*188I, and all three parents carry the A-genome 448G → A transitional mutation in *ahFAD*2A allele; therefore, *Hpy*188I-CAPS was the only marker necessary to track inheritance of high O/L. Since these genetic markers can identify true hybrids at each stage of backcrossing, the backcross and selection process can be accelerated by using heterozygous F_1 hybrids as donor parents. In contrast to conventional breeding, which takes 8–10 years for a new cultivar release, this MAS approach is expected to produce a high O/L "Tifguard" within 26 months (Y. Chu, P. Ozias-Akins, and C. C. Holbrook, unpublished). As more marker-tagged traits become available through genome and genetic mapping studies in peanut, marker-assisted breeding will become reality and more precise for specific trait integration in genetic improvement of peanut.

15.6 TILLING AND TRANSFORMATION

Another option for enhancing peanut genetic diversity other than utilization of wild germplasm is biotechnology. TILLING is a mutation screening tool designed for functional genomics. TILLING was first implemented in *Arabidopsis thaliana* where the genome and predicted gene sequences preceded expansive knowledge of gene function (Greene et al., 2003). Although mutation strategies to generate TILLING populations are not novel, knowledge of gene sequence is a prerequisite since this reverse genetics tool initially identifies mutations in the DNA sequence rather than a mutant phenotype. A spectrum of mutations can be recovered, most of which are silent or missense, but a few of which are nonsense. Therefore, functional analysis of genes is informed by phenotyping multiple variant alleles at a locus, and an allelic series can allow the recovery of sublethal alleles in essential genes.

Development of a TILLING population follows standard mutagenesis procedures for mutagens that have been shown to induce single-nucleotide changes (e.g., ethyl methanesulfonate, EMS) or small deletions (e.g., diepoxybutane, DEB). The novel aspect of TILLING is the screening technique whereby (1) pools of mutant and wild-type DNAs are amplified using fluorescently labeled gene-specific primers; (2) the amplicons are melted and reannealed to form homo- and heteroduplexes; (3) the heteroduplexes are recognized and cleaved by an endonuclease; and (4) the cleavage products are separated by slab gel or capillary electrophoresis and detected by the appropriate fluorescent detection system. Detailed protocols and variations of the method have been previously published (Till et al., 2006; Raghavan et al., 2007; Uauy et al., 2009). TILLING now has been used to identify mutations in genes of interest in barley (Lababidi et al., 2009), wheat (Uauy et al., 2009), rice (Suzuki et al., 2008), maize (Weil and Monde, 2007), sorghum (Xin et al., 2008), soybean (Dierking and Bilyeu, 2009), bean (Porch et al., 2009), pea (Triques et al., 2007), and the diploid peanut *A. duranensis* (Ramos et al., 2009) as well as model species (Perry et al., 2009). The mutations are predominantly G/C-to-A/T transitions for EMS-treated materials, and the mutation rate has ranged from three per Mb for *Arabidopsis* (Greene et al., 2003) to 0.4 per Mb for one barley population (Lababidi et al., 2009).

A TILLING population has been generated for cultivated peanut using the mutagen EMS (P. Ozias-Akins, unpublished data). The results suggest that the mutation density is approximately 1 per Mb. After screening more than 3000 M_2 lines for mutations in six genes, two copies each of *Ara* h 1, *Ara* h 2, and *FAD*2, more than 20 mutations have been identified including four that putatively affect the function of each gene. *Ara* h 1 and *Ara* h 2 are seed storage proteins with food allergen properties that can cause serious allergic reactions in peanut-sensitive individuals (Burks et al., 1998). A significant reduction in *Ara* h 2, even though its primary function is nutrient reservoir activity in the seed, does not affect seed viability (Chu et al., 2008b); therefore, a knockout of *Ara* h 2 by TILLING should be a feasible approach for reducing the allergenicity of peanut. One *Ara* h 2.02 knockout mutation has been recovered from the TILLING population and carried forward to homozygous M_3 plants where the corresponding protein is absent (unpublished data). A putative knockout of one copy of *Ara* h 1 also has been discovered. Interestingly, reversion of a frequently occurring mutant allele of *FAD*2A (Chu et al., 2007b) has been observed in the TILLING population (unpublished data), which may limit the use of this technology to suppress allergenicity.

An alternative reverse genetics approach for inferring gene function is RNA interference (RNAi), which exploits a natural RNA-mediated defense mechanism in plants (Baulcombe, 2004) to down-regulate endogenous gene expression upon the insertion of a homologous transgene fragment, usually as an inverted repeat. RNAi in plants relies on stable insertion of the transgene into the genome of the host plant or transient silencing such as obtained through virus-induced gene silencing (Watson et al., 2005). Stable insertion can be achieved by introducing DNA into plant cells via direct DNA delivery or *Agrobacterium* (Newell, 2000). For peanut, stable transformation is most reliably accomplished using microprojectile bombardment (Ozias-Akins et al., 1993; Ozias-Akins and Gill, 2001; Ozias-Akins, 2007). A limited number of genotypes also are amenable to transformation via *Agrobacterium* (Chenault et al., 2008).

Collectively, the above-mentioned transformation systems have resulted in the generation of transgenic peanut engineered for resistance to insects (Singsit et al., 1997; Ozias-Akins et al., 2002; Tiwari et al., 2008), viruses (Li et al., 1997; Yang et al., 1998, 2004; Magbanua et al., 2000; Sharma and Anjaiah, 2000; Chenault and Payton, 2003), fungal pathogens (Chenault et al., 2005; Livingstone et al., 2005; Niu et al., 2009), and tolerance to herbicide and drought (Bhatnagar-Mathur et al., 2007, 2009; Chu et al., 2008a). For drought, a transcription factor (*AtDREB*1A) known to be involved in environmental stress responses was introduced to peanut under the control of a stress inducible promoter. Its expression resulted in differential responses under drought stress compared to wild type with respect to the accumulation of antioxidants and proline and the mitigation of lipid peroxidation (Bhatnagar-Mathur et al., 2009). RNAi also was used as an alternative to mutation for increasing oleic to linoleic acid ratios (O/L) in peanut (Yin et al., 2007). Since mutants of the oleate desaturase FAD2 have been shown to be primarily responsible for increasing the O/L ratio in peanut, the *FAD*2 gene was targeted by RNAi with a resulting 70% increase in oleic acid content in peanut seeds. While no transgenic peanut has yet been deregulated and brought to market because of governmental regulation, transformation technologies have been developed and used to test the efficacy of a variety of genes, and will play a significant role in peanut improvement in the future.

15.7 PEANUT EXPRESSED SEQUENCE TAGS (ESTS) AND TRANSCRIPTOME ANALYSIS

Peanut genomics research can provide new tools and resources to enhance crop genetic characteristics. The current status of peanut genomic research has been documented in the book entitled, *Legume Crop Genomics* (Wilson et al., 2004), published under the auspices of the USLCGI (US Legume Crops Genome Initiative) (Gepts et al., 2005). Genomic resources such as EST sequence, microarray technologies, and whole genome sequencing provide high-throughput tools for profiling genes. In spite of decreasing costs for DNA sequencing, it is improbable that many large-plant genomes, such as peanut (2800 Mb/1C), will be sequenced in the near future. However, sequencing large numbers of expressed genes (ESTs) can deliver substantial amounts of genetic information on protein-coding sequences for comparative and functional genomic studies. ESTs provide putative genes, markers, and microarray resources to devise more effective breeding strategies to target sustainable peanut production. However, only 85,337 *A. hypogaea* ESTs were deposited in GenBank (http://www.ncbi.nlm.nih.gov/dbEST/dbEST_summary.html) as of January 22, 2010. There are more not released yet to GenBank (http://www.peanut.uga.edu). These ESTs were from different tissues, some of which had been subjected to different abiotic and biotic stresses (Luo et al., 2005a; Proite et al., 2007; Guo et al., 2008b, 2009a).

Microarray technology allows the simultaneous determination of expression levels of thousands of genes, making it possible to obtain a global view of the transcriptional state in a cell or tissue, and to associate genes with functions or specific physiological conditions. Transcriptome expression analysis in peanut to date has been limited to a relatively small set of genes (Luo et al., 2005b,c) and only recently has a significant number of ESTs been released into the public domain. Utilization of these ESTs for oligonucleotide microarrays provides a means to investigate large-scale transcript responses to a variety of developmental and environmental factors (Payton et al., 2009). Luo et al. (2005c) identified 25 ESTs that were potentially associated with drought stress or that responded to *A. parasiticus* challenge. Likewise, 56 up-regulated transcripts were identified and confirmed by real-time PCR upon infection with *C. personatum*, peanut late leaf spot pathogen. Payton et al. (2009) developed a high-density oligonucleotide microarray for peanut using 49,205 publicly available ESTs and tested the utility of this array for expression profiling and compared transcript levels in a variety of peanut tissues. The peanut oligonucleotide array represents the majority of publicly available peanut ESTs and can be used as a tool for expression profiling studies in diverse tissues.

15.8 CONCLUSION

Plant breeding, genetics, and genomics have a critical role to play in sustainable agriculture. These technologies are contributing to rapid progress in improving crop productivity, quality, and resistance to pests and diseases. The advances in genetics and genomics are opening new frontiers in peanut breeding, including rapid and targeted advances in specific traits such as nematode resistance and high-oleic acid content. Germplasm is the treasure of crop genetic resources and the three largest germplasm collections for peanut are maintained at ICRISAT in India, in the United States, and in China. Utilization of the collections has come a long way, such as development of core and mini core collections for more efficient use in breeding for new cultivars. Conventional breeding will continue to play an essential role in sustainable peanut production; however, the progress made in peanut molecular genetics and genomics has demonstrated its potential for transforming peanut breeding and cultivar development programs in the United States and globally through increased integration of molecular genetics and marker-assisted breeding.

The construction of genetic linkage maps for cultivated peanut continues to be an important research goal to facilitate QTL analysis and gene tagging for use in marker-assisted breeding. Even though several maps have been developed using diploid or interspecific tetraploid populations, recently there are two published intra-specific maps constructed from crosses of cultivated peanuts, in which only 135 and 175 SSR markers, respectively, were sparsely populated in 22 LGs representing the 20 chromosomes of tetraploid cultivated peanuts. A more detailed linkage map with sufficient markers is necessary to make QTL identification and MAS feasible on a large scale. One caution that needs to be considered in both genetic and physical maps is due to the tetraploid nature of the peanut genome, requiring distinction of polymorphisms between homologous versus homeologous (subgenome) templates. The promises of other biotechnologies such as genetic engineering have come to reality in other crops, but not in peanut yet because of government regulation. The transformation technology has been developed in peanut. Nevertheless, much progress is still needed for peanut with regard to genome sequencing, genetic and physical mappings, SNP discovery, and understanding of the influence of environmental factors, both abiotic and biotic, on gene expression in peanut. Through new biotechnologies, enhancement of the genetic diversity, preservation, and utilization of these natural treasures will lead to improved sustainable peanut production and agriculture at large as well as improved human living standards, and improved food security and safety.

REFERENCES

Agarwal, M., N. Shrivastava, and H. Padh. 2008. Advances in molecular marker techniques and their applications in plant sciences. *Plant Cell Rep* 27:617–631.

Akem, C. N., H. A. Melouk, and O. D. Smith. 1992. Field evaluation of peanut genotypes for resistance to *Sclerotinia* blight. *Crop Protec* 11:345–348.

Anderson, W. F., C. C. Holbrook, and A. K. Culbreath. 1996. Screening the peanut core collection for resistance to tomato spotted wilt virus. *Peanut Sci* 23:57–61.

Anderson, W. F., C. C. Holbrook, and T. B. Brenneman. 1993. Resistance to *Cercosporidium personatum* within peanut germplasm. *Peanut Sci* 20:53–57.

Arunyanark, A., S. Jogloy, S. Wongkaew, C. Akkasaeng, N. Vorasoot, G. C. Wright, R. C. N. Rachaputi, and A. Patanothai. 2009. Association between aflatoxin contamination and drought tolerance traits in peanut. *Field Crop Res* 114:14–22.

Baltensperger, D. D., G. M. Prine, and R. A. Dunn. 1986. Root-knot nematode resistance in *Arachis glabrata*. *Peanut Sci* 13:78–80.

Baring, M. R., C. E. Simpsonb, M. D. Burowc, M. C. Blackd, J. M. Casonb, J. Ayerse, Y. Lopeze, and H. A. Meloukf. 2006. Registration of ‘Tamrun OL07’ Peanut. *Crop Sci* 46:2721–2722.

Barkley, N. A., K. D. Chenault Chamberlin, M. L. Wang, and R. N. Pittman. 2009. Development of a real-time PCR genotyping assay to identify high oleic acid peanuts (*Arachis hypogaea* L.). *Mol Breed* 25:541–548.

Barkley, N. A., R. E. Dean, R. N. Pittman, M. L. Wang, C. C. Holbrook, and G. A. Pederson. 2007. Genetic diversity of cultivated and wild-type peanuts evaluated with M13-tailed SSR markers and sequencing. *Genet Res Cambridge* 89:93–106.

Baulcombe, D. 2004. RNA silencing in plants. *Nature* 431:356–363.

Bertioli, D. J., M. C. Moretzsohn, L. H. Madsen et al. 2009. An analysis of synteny of *Arachis* with *Lotus* and *Medicago* sheds new light on the structure, stability and evolution of legume genomes. *BMC Genomics* 10:45.

Bertioli, D. J., S. C. Leal-Bertioli, M. B. Lion, V. L. Santos, G. Pappas, Jr., S. B. Cannon, and P. M. Guimaraes. 2003. A large scale analysis of resistance gene homologues in *Arachis*. *Mol Genet Genomics* 270:34–45.

Besler, B. A., W. J. Grichar, and O. D. Smith. 1997. Reaction of selected peanut cultivars and breeding lines to southern stem rot. *Peanut Sci* 24:6–9.

Bhatnagar-Mathur, P., M. Devi, D. Reddy et al. 2007. Stress-inducible expression of *At Dreb1a* in transgenic peanut (*Arachis hypogaea* L.) increases transpiration efficiency under water-limiting conditions. *Plant Cell Rep* 26:2071–2082.

Bhatnagar-Mathur, P., M. J. Devi, V. Vadez, and K. K. Sharma. 2009. Differential antioxidative responses in transgenic peanut bear no relationship to their superior transpiration efficiency under drought stress. *J Plant Physiol* 166:1207–1217.

Branch, W. D. 1994. Registration of ‘Georgia Browne’ peanut. *Crop Sci* 34:1125–1126.

Branch, W. D. 1996. Registration of ‘Georgia Green’ peanut. *Crop Sci* 36:806.

Branch, W. D. 2000. Registration of ‘Georgia Hi-O/L’ peanut. *Crop Sci* 40:1823–1824.

Branch, W. D. 2003. Registration of ‘Georgia-02C’ peanut. *Crop Sci* 43:1883–1884.

Branch, W. D. 2006. Registration of ‘Georgia-05E’ peanut. *Crop Sci* 46:2305.

Branch, W. D. and A. S. Csinos. 1987. Evaluation of peanut cultivars for resistance to field infection by *Sclerotium rolfsii*. *Plant Dis* 71:268–270.

Branch, W. D. and S. M. Fletcher. 2001. No-pesticide preliminary yield trials in peanut. *Peanut Sci* 28:21–4.

Brittlebank, C. C. 1919. Tomato diseases. *J Agric Victoria* 27:213–235.

Burks, W., H. A. Sampson, and G. A. Bannon. 1998. Peanut allergens. *Allergy* 53:725–730.

Burow, M. D., C. E. Simpson, A. H. Paterson, and J. L. Starr. 1996. Identification of peanut (*Arachis hypogaea* L.) RAPD markers diagnostic of root-knot nematode (*Meloidogyne arenaria* (Neal) Chitwood) resistance. *Mol Breed* 2:369–379.

Burow, M. D., C. E. Simpson, J. L. Starr, and A. H. Paterson. 2001. Transmission genetics of chromatin from a synthetic amphidiploids to cultivated peanut (*Arachis hypogaea* L.): Broadening the gene pool of a monophyletic polyploid species. *Genetics* 159:823–837.

Burow, M. D., C. E. Simpson, M. W. Faries, J. L. Starr, and A. H. Paterson. 2009. Molecular biogeographic study of recently described B- and A-genome *Arachis* species, also providing new insights into the origins of cultivated peanut. *Genome* 52:107–119.

Burow, M. D., M. G. Selvaraj, H. Upadhyaya et al. 2008. Genomics of peanut, a major source of oil and protein. In *Genomics of Tropical Crop Plants*, eds. P. H. Moore, and R. Ming, pp. 421–40. New York: Springer.

Cantonwine, E. G., A. K. Culbreath, K. L. Stevenson et al. 2006. Integrated disease management of leaf spot and spotted wilt of peanut. *Plant Dis* 90:493–500.

Chenault, K., M. Gallo, P. Ozias-Akins, and P. Srivastava. 2008. Peanut. In *Compendium of Transgenic Crop Plants: Transgenic Cereals and Forage Grasses*, eds. C. Kole, and T. Hall, pp. 169–98. Oxford: J. Wiley & Sons.

Chenault, K. D., H. A. Melouk, and M. E. Payton. 2005. Field reaction to *Sclerotinia* blight among transgenic peanut lines containing antifungal genes. *Crop Sci* 45:511–515.

Chenault, K. D. and M. E. Payton. 2003. Genetic transformation of a runner-type peanut with the nucleocapsid gene of tomato spotted wilt virus. *Peanut Sci* 30:112–115.

Chiteka, Z. A., D. W. Gorbet, D. A. Knauft, F. M. Shokes, and T. A. Kucharek. 1988a. Components of resistance to late leafspot in peanut. II. Correlations among components and their significance in breeding for resistance. *Peanut Sci* 15:25–30.

Chiteka, Z. A., D. W. Gorbet, F. M. Shokes, T. A. Kucharek, and D. A. Knauft. 1988b. Components of resistance to late leafspot in peanut. I. Levels and variability-implications for selection. *Peanut Sci* 15:25–30.

Choi, H. K., J. H. Mun, D. J. Kim et al. 2004. Estimating genome conservation between crop and model legume species. *Proc Natl Acad Sci USA* 101:15289–15294.

Choi, H. K., M. D. Burow, G. T. Church, A. H. Paterson, C. E. Simpson, and J. L. Starr. 1999. Genetics and mechanisms of resistance to *Meloidogyne arenaria* in peanut germplasm. *J Nematol* 31:283–290.

Chu, Y., C. C. Holbrook, and P. Ozias-Akins. 2009. Two alleles of *ahFAD2B* control the high oleic acid trait in cultivated peanut. *Crop Sci* 49:2029–2036.

Chu, Y., C. C. Holbrook, P. Timper, and P. Ozias-Akins. 2007a. Development of a PCR-based molecular marker to select for nematode resistance in peanut. *Crop Sci* 47:841–847.

Chu, Y., L. Ramos, C. C. Holbrook, and P. Ozias-Akins. 2007b. Frequency of a loss-of-function mutation in oleoyl-pc desaturase (ahFAD2A) in the mini-core of the U.S. peanut germplasm collection. *Crop Sci* 47:2372–2378.

Chu, Y., P. Faustinelli, M. L. Ramos et al. 2008b. Reduction of IgE binding and nonpromotion of *Aspergillus flavus* fungal growth by simultaneously silencing *Ara* h 2 and *Ara* h 6 in peanut. *J Agric Food Chem* 56:11225–11233.

Chu, Y., X. Y. Deng, P. Faustinelli, and P. Ozias-Akins. 2008a. *Bcl-Xl* transformed peanut (*Arachis hypogaea* L.) exhibits paraquat tolerance. *Plant Cell Rep* 27:85–92.

Coffelt, T. A. and D. M. Porter. 1982. Screening peanuts for resistance to *Sclerotinia* blight. *Plant Dis* 66:385–387.

Coffelt, T. A., D. M. Porter, and R. W. Mozingo. 1994. Registration of 'VA 93B' peanut. *Crop Sci* 34:1126.

Cole, R. J., J. W. Dorner, and C. C. Holbrook. 1995. Advances in mycotoxin elimination and resistance. In *Advances in Peanut Science*, eds. H. E. Pattee and H. T. Stalker, pp. 456–74. Stillwater, OK: American Peanut Research and Education Society.

Company, M., H. T. Stalker, and J. C. Wynne. 1982. Cytology and leafspot resistance in *Arachis hypogaea* × wild species hybrids. *Euphytica* 31:885–893.

Cruickshank, A. W., M. Cooper, and M. J. Ryley. 2002. Peanut resistance to *Sclerotinia minor* and *S. sclerotiorum. Aust J Agri Res* 53:1105–1110.

Cuc, L. M., E. S. Mace, J. H. Crouch, V. D. Quang, T. D. Long, and R. K. Varshney. 2008. Isolation and characterization of novel microsatellite markers and their application for diversity assessment in cultivated groundnut (*Arachis hypogaea*). *BMC Plant Biol* 8:55.

Culbreath, A. K., D. W. Gorbet, N. Martinez-Ochoa et al. 2005. High levels of field resistance to tomato spotted wilt virus in peanut breeding lines derived from *hypogaea* and *hirsuta* botanical varieties. *Peanut Sci* 32:20–24.

Culbreath, A. K., J. W. Todd, and S. L. Brown, 2003. Epidemiology and management of tomato spotted wilt in peanut. *Annu Rev Phytopathology* 41:53–75.

Culbreath, A. K., M. K. Beute, and C. L. Campbell. 1991. Spatial and temporal aspects of epidemics of *Cylindrocladium* black rot resistant and susceptible peanut genotypes. *Phytopathology* 81:144–150.

Dean, L. L., K. W. Hendrix, C. C. Holbrook, and T. H. Sanders. 2009. Content of some nutrients in the core of the core of the peanut germplasm collection. *Peanut Sci* 36:104–120.

Dean, L. L. and T. H. Sanders. 2009. Hexacosanoic acid and other very long-chain fatty acids in peanut seed oil. *Plant Genet Res* 7:252–256.

Devi, J. M., T. R. Sinclair, and V. Vadez. 2010. Genotypic variation in peanut for transpiration response to vapor pressure deficit. *Crop Sci* 50:191–196.

Dierking, E. C. and K. D. Bilyeu. 2009. New sources of soybean seed meal and oil composition traits identified through TILLING. *BMC Plant Biol* 9:89.

Dwivedi, S. L., N. Puppala, H. D. Upadyaya, N. Manivannan, and S. Singh. 2008. Developing a core collection of peanut specific to Valencia market type. *Crop Sci* 48:625–632.

Ferguson, M. E., M. D. Burow, S. R. Schulze et al. 2004. Microsatellite identification and characterization in peanut (*A. hypogaea* L.). *Theor Appl Genet* 108:1064–1070.

Foncéka, D., T. Hodo-Abalo, R. Rivallan et al. 2009. Genetic mapping of wild introgressions into cultivated peanut: A way toward enlarging the genetic basis of a recent allotetraploid. *BMC Plant Biol* 9:103.

Franke, M. D., T. B. Brenneman, and C. C. Holbrook. 1999. Identification of resistance to *Rhizoctonia* limb rot in a core collection of peanut germplasm. *Plant Dis* 83:944–948.

Frankel, O. H. 1984. Genetic perspectives of germplasm conservation. In *Genetic Manipulation: Impact on Man and Society*, eds. W. K. Arber, K. Llimensee, W. J. Peacock, and P. Starlinger, pp. 161–170. Cambridge: Cambridge University Press.

Garcia, G. M., H. T. Stalker, E. Shroeder, and G. Kochert. 1996. Identification of RAPD, SCAR and RFLP markers tightly linked to nematode resistance genes introgressed from *Arachis cardenasii* to *A. hypogaea*. *Genome* 39:836–845.

Garcia, G. M., H. T. Stalker, and G. A. Kochert. 1995. Introgression analysis of an interspecific hybrid population in peanuts (*Arachis hypogaea* L.) using RFLP and RAPD markers. *Genome* 38:166–176.

Gepts, P., W. D. Beavis, E. C. Brummer et al. 2005. Legumes as a model plant family. Genomics for food and feed: Report of the Cross-Legume Advances through Genomics Conference. *Plant Physiol* 137:1228–1235.

Gimenes, M. A., A. A. Hoshino, A. V. Barbosa, D. A. Palmieri, and C. R. Lopes. 2007. Characterization and transferability of microsatellite markers of the cultivated peanut (*Arachis hypogaea*). *BMC Plant Biol* 7:9.

Gimenes, M. A., C. R. Lopes, and J. F. Valls. 2002a. Genetic relationships among *Arachis* species based on AFLP. *Genet Mol Biol* 25:349–353.

Gimenes, M. A., C. R. Lopes, L. Galgaro, J. F. Valls, and G. Kochert. 2002b. RFLP analysis of genetic variation in species of section *Arachis*, genus *Arachis* (Leguminosae). *Euphytica* 123:421–429.

Girdthai, T., S. Jogloy, T. Kesmala et al. 2010a. Associations between physiological traits for drought tolerance and aflatoxin contamination in peanut genotypes under terminal drought. *Plant Breed* 125:693–699.

Girdthai, T., S. Jogloy, T. Kesmala et al. 2010b. Relationship between root characteristics of peanut in hydroponics and pot studies. *Crop Sci* 50:159–167.

Gorbet, D. W. 2006. Registration of 'Andru II' Peanut. *Crop Sci* 46:2712–2713.

Gorbet, D. W. 2007. Registration of 'Ap-3' Peanut. *J Plant Registrations* 1:126–127.

Gorbet, D. W. and B. L. Tillman. 2009. Registration of 'Florida-07' Peanut. *J Plant Registrations* 3:14–18.

Gorbet, D. W. and D. A. Knauft. 1997. Registration of 'SunOleic 95R' peanut. *Crop Sci* 37:1392.

Greene, E. A., C. A. Codomo, N. E. Taylor et al. 2003. Spectrum of chemically induced mutations from a large-scale reverse-genetic screen in *Arabidopsis*. *Genetics* 164:731–740.

Gregory, W. C. 1946. Peanut breeding program underway. In *Research and Farming*, 69th Annual Report, pp. 42–44. Raleigh, NC: N.C. Agric. Expt. Stat.

Gregory, W. C. and M. P. Gregory. 1979. Exotic germ plasm of *Arachis* L. interspecific hybrids. *J Herd* 70:185–193.

Grichar, W. J. and O. D. Smith. 1992. Variation in yield and resistance to southern stem rot among peanut (*Arachis hypogaea* L.) lines selected for pythium pod rot resistance. *Peanut Sci* 19:55–58.

Guo, B. Z., J. Yu, C. C. Holbrook, T. E. Cleveland, W. C. Nierman, and B. T. Scully. 2009b. Strategy in prevention of preharvest aflatoxin contamination in peanuts: Aflatoxin biosynthesis, genetics and genomics. *Peanut Sci* 36:11–20.

Guo, B. Z., X. Chen, P. Dang et al. 2008b. Peanut gene expression profiling in developing seeds at different reproduction stages during *Aspergillus parasiticus* infection. *BMC Dev Biol* 8:12.

Guo, B. Z., X. Chen, Y. Hong et al. 2009a. Analysis of gene expression profiles in leaf tissues of cultivated peanut and development of EST-SSR markers and gene discovery. *Int J Plant Genomics* 2009:1–14.

Guo, B. Z., Z.-Y. Chen, R. D. Lee, and B. T. Scully. 2008a. Drought stress and preharvest aflatoxin contamination in agricultural commodity: Genetics, genomics and proteomics. *J Integr Plant Biol* 50:1281–1291.

Hadley, B. A., M. K. Beute, and J. C. Wynne. 1979. Heritability of *Cylindrocladium* black rot resistance in peanut. *Peanut Sci* 6:51–54.

Halward, T. and H. T. Stalker. 1991. Genetic variation detectable with molecular markers among unadapted germplasm resources of cultivated peanut and related wild species. *Genome* 34:1013–1020.

Halward, T., H. T. Stalker, and G. Kochert. 1993. Development of an RFLP linkage map in diploid peanut species. *Theor Appl Genet* 87:379–384.

Halward, T., T. Stalker, E. LaRue, and G. Kochert. 1992. Use of single-primer DNA amplifications in genetic studies of peanut (*Arachis hypogaea* L.). *Plant Mol Biol* 18:315–325.

Halward, T. M. and J. C. Wynne. 1991. Generation means analysis for productivity in two diverse peanut crosses. *Theor Appl Genet* 82:784–792.

Hammond, E. G., D. Duvick, T. Wang, H. Dodo, and R. N. Pittman. 1997. Survey of fatty acid composition of peanut (*Arachis hypogaea*) germplasm and characterization of their epoxy and eicosenoic acids. *J Am Oil Chem Soc* 74:1235–1239.

He, G., F. E. Woullard, I. Marong, and B. Z. Guo. 2006. Transferability of soybean SSR markers in peanut (*Arachis hypogaea* L.). *Peanut Sci* 33:22–28.

He, G., R. Meng, M. Newman, G. Gao, R. N. Pittman, and C. S. Prakash. 2003. Microsatellites as DNA markers in cultivated peanut (*Arachis hypogaea* L.). *BMC Plant Biol* 3:3.

Herselman, L., R. Thwaites, F. M. Kimmins, B. Courtois, P. J. van der Merwe, and S. E. Seal. 2004. Identification and mapping of AFLP markers linked to peanut (*Arachis hypogaea* L.) resistance to the aphid vector of groundnut rosette disease. *Theor Appl Genet* 109:1426–1433.

Holbrook, C. C., B. Z. Guo, D. M. Wilson, and P. Timper. 2009. The U.S. breeding program to develop peanut with drought tolerance and reduced aflatoxin contamination. *Peanut Sci* 36:50–53.

Holbrook, C. C., C. K. Kvien, K. S. Rucker D. M. Wilson, and J. E. Hook. 2000c. Preharvest aflatoxin contamination in drought tolerant and intolerant peanut genotypes. *Peanut Sci* 27:45–48.

Holbrook, C. C. and H. T. Stalker. 2003. Peanut breeding and genetic resources. *Plant Breed Rev* 22:297–356.

Holbrook, C. C. and J. P. Noe. 1990. Resistance to *Meloidogyne arenaria* in *Arachis* spp. and the implications on development of resistant peanut cultivars. *Peanut Sci* 17:35–38.

Holbrook, C. C. and J. P. Noe. 1992. Resistance to the peanut root-knot nematode (*Meloidogyne arenaria*) in *Arachis hypogaea. Peanut Sci* 19:35–37.

Holbrook, C. C., M. G. Stephenson, and A. W. Johnson. 2000a. Level and geographical distribution of resistance to *Meloidogyne arenaria* in the U.S. peanut germplasm collection. *Crop Sci* 40:1168–1171.

Holbrook, C. C., P. Ozias-Akins, P. Timper et al. 2008c. Research from the coastal plain experiment station, Tifton Georgia, to minimize aflatoxin contamination in peanut. *Toxin Rev* 27:391–410.

Holbrook, C. C., P. Timper, A. Culbreath, and C. K. Kvien. 2008a. Registration of 'Tifguard' peanut. *J Plant Registrations* 2:92–94.

Holbrook, C. C., P. Timper, and H. Q. Xue. 2000b. Evaluation of the core collection approach for identifying resistance to *Meloidogyne arenaria* in peanut. *Crop Sci* 40:1172–1175.

Holbrook, C. C., P. Timper, W. Dong, C. K. Kvien, and A. K. Culbreath. 2008b. Development of near-isogenic peanut lines with and without resistance to the peanut root-knot nematode. *Crop Sci* 48:194–198.

Holbrook, C. C. and T. G. Isleib. 2001. Geographical distribution of genetic diversity in *Arachis hypogaea. Peanut Sci* 28:80–84.

Holbrook, C. C. and W. Dong. 2005. Development and evaluation of a mini core collection for the U.S. peanut germplasm collection. *Crop Sci* 45:1540–1544.

Holbrook, C. C. and W. F. Anderson. 1995. Evaluation of a core collection to identify resistance to late leafspot in peanut. *Crop Sci* 35:1700–1702.

Holbrook, C. C., W. F. Anderson, and R. N. Pittman. 1993. Selection of a core collection from the U.S. germplasm collection of peanut. *Crop Sci* 33:859–861.

Hong, Y. X. Chen, X. Liang et al. 2010. A SSR-based composite genetic linkage map for the cultivated peanut (*Arachis hypogaea* L.) genome. *BMC Plant Biol* 10:17.

Hopkins, M. S., A. M. Casa, T. Wang et al. 1999. Discovery and characterization of polymorphic simple sequence repeats (SSRs) in peanut. *Crop Sci* 39:1243–1247.

Hougaard, B. K., L. H. Madsen, N. Sandal et al. 2008. Legume anchor markers link syntenic regions between *Phaseolus vulgaris*, *Lotus japonicus*, *Medicago truncatula* and *Arachis*. *Genetics* 179:2299–2312.

Husted, L. 1936. Cytological studies of the peanut *Arachis*. II. Chromosome number, morphology and behavior and their application to the origin of cultivated forms. *Cytologia* 7:396–423.

Isleib, T. G., C. C. Holbrook, and D. W. Gorbet. 2001. Use of *Arachis* spp. plant introductions in peanut cultivar development. *Peanut Sci* 28:96–113.

Isleib, T. G., C. T. Young, and D. A. Knauft. 1996. Fatty acid genotypes of five Virginia type cultivars. *Crop Sci* 36:556–558.

Isleib, T. G., P. W. Rice, R. W. Mozingo, J. E. Bailey, R. W. Mozingo, and H. E. Pattee. 2003. Registration of 'Perry' peanut. *Crop Sci* 43:739–740.

Isleib, T. G., S. R. Milla-Lewis, H. E. Pattee et al. 2011. Registration of 'Bailey' peanut. *J Plant Registration* 5:27–39.

Jain, R. K., S. S. Pappu, H. R. Pappu, A. K. Culbreath, and J. W. Todd, 1998. Molecular diagnosis of tomato spotted wilt tospovirus infection of peanut and other field and greenhouse crops. *Plant Dis* 82:900–904.

Jiang, H. F., X. P. Ren, B. S. Liao et al. 2008. Peanut core collection established in China and compared with ICRISAT mini core collection. *Acta Agron Sinica* 34:25–30.

Jung, S., D. Swift, E. Sengoku et al. 2000b. The high oleate trait in the cultivated peanut (*Arachis hypogaea* L.). I. Isolation and characterization of two genes encoding microsomal oleoyl-PC desaturases. *Mol Gen Genet* 263:796–805.

Jung, S., G. Powell, K. Moore, and A. Abbott. 2000a. The high oleate trait in the cultivated peanut (*Arachis hypogaea* L). II. Molecular basis and genetics of the trait. *Mol Gen Genet* 263:806–811.

Kirby, J. S., H. A. Melouk, T. E. Stevens et al. 1998. Registration of 'Southwest Runner' peanut. *Crop Sci* 38:545–546.

Knauft, D. A. and J. C. Wynne. 1995. Peanut breeding and genetics. *Adv Agron* 55:393–445.

Knauft, D. A., K. Moore, and D. W. Gorbet. 1993. Further studies on the inheritance of fatty acid composition in peanut. *Peanut Sci* 20:74–76.

Kochert, G., H. T. Stalker, M. A. Gimenes, L. Galgaro, C. R. Lopes, and K. Moore. 1996. RFLP and cytogenetic evidence on the origin and evolution of allotetraploid domesticated peanut, *Arachis hypogaea* (Leguminosae). *Am J Bot* 83:1282–1291.

Kottapalli, K. R., M. Burow, G. Burow, J. Burke, and N. Puppala. 2007. Molecular characterization of the U.S. peanut mini core collection using microsatellite markers. *Crop Sci* 47:1718–1727.

Krapovickas, A. and W. C. Gregory. 1994. Taxonomía del género *Arachis* (Leguminosae). *Bonplandia* 8:1–186.

Krishnamurthy, L., V. Vadez, M. J. Devi et al. 2007. Variation in transpiration efficiency and its related traits in a groundnut (*Arachis hypogaea* L.) mapping population. *Field Crops Res* 103:189–197.

Lababidi, S., N. Mejlhede, S. K. Rasmussen, G. Backes, W. Al-Said, M. Baum, and A. Jahoor. 2009. Identification of barley mutants in the cultivar 'Lux' at the *Dhn* loci through TILLING. *Plant Breed* 128:332–336.

Lacks, G. D. and H. T. Stalker. 1993. Isozyme analyses of *Arachis* species and interspecific hybrids. *Peanut Sci* 20:76–81.

Lanham, P. G., B. P. Foster, and P. McNicol. 1994. Seed storage protein variation in *Arachis* species. *Genome* 37:487–496.

Lavia, G. I. 1998. Karyotypes of *Arachis palustris* and *A. praecox* (section *Arachis*), two species with basic chromosome number $x = 9$. *Cytologia* 63:177–181.

Leal-Bertioli, S. C., A. C. Jose, D. M. Alves-Freitas et al. 2009. Identification of candidate genome regions controlling disease resistance in *Arachis*. *BMC Plant Biol* 9:112.

Li, Y, A. K. Culbreath, B. Z. Guo, S. J. Knapp, and C. C. Holbrook. 2009. Tomato spotted wilt and early leaf spot reactions in peanut genotypes from the U.S. and China. *Phytopathology* 99:S73

Li, Z. J., R. L. Jarret, and J. W. Demski. 1997. Engineered resistance to tomato spotted wilt virus in transgenic peanut expressing the viral nucleocapsid gene. *Transgen Res* 6:297–305.

Liang, X., C. C. Holbrook, R. E. Lynch, and B. Z. Guo. 2005. Beta-1,3-glucanase activity in peanut seed (*Arachis hypogaea*) is induced by inoculation with *Aspergillus flavus* and co-purifies with a conglutin-like protein. *Phytopathology* 95:506–511.

Liang, X., G. Zhou, Y. Hong, X. Chen, H. Liu, and S. Li. 2009b. Overview of research progress on peanut (*Arachis hypogaea* L.) host resistance to aflatoxin contamination and genomics at the Guangdong Academy of Agricultural Sciences. *Peanut Sci* 36:29–34.

Liang, X., X. Chen, Y. Hong, H. Liu, G. Zhou, S. Li, and B. Z. Guo. 2009a. Utility of EST-derived SSR in cultivated peanut (*Arachis hypogaea* L.) and *Arachis* wild species. *BMC Plant Biol* 9:35.

Liang, X. Q., M. Luo, C. C. Holbrook, and B. Z. Guo. 2006. Storage protein profiles in Spanish and runner market type peanuts and potential markers. *BMC Plant Biol* 6:24.

Liao, B., W. Zhuang, R. Tang et al. 2009. Peanut aflatoxin and genomics research in China: Progress and perspectives. *Peanut Sci* 36:21–28.

Livingstone, D. M., J. L. Hampton, P. M. Phipps, and E. A. Grabau. 2005. Enhancing resistance to *Sclerotinia minor* in peanut by expressing a barley oxalate oxidase gene. *Plant Physiol* 137:1354–1362.

Lopez, Y., H. L. Nadaf, O. D. Smith, C. E. Simpson, and A. K. Fritz. 2002. Expressed variants of Δ12-fatty acid desaturase for the high oleate trait in Spanish market-type peanut lines. *Mol Breed* 9:183–190.

Lopez, Y., O. D. Smith, S. A. Senseman, and W. L. Rooney. 2001. Genetic factors influencing high oleic acid content in Spanish market-type peanut cultivars. *Crop Sci* 41:51–56.

Luo, M., P. Dang, B. Z. Guo et al. 2005a. Generation of expressed sequence tags (ESTs) for gene discovery and marker development in cultivated peanut. *Crop Sci* 45:346–353.

Luo, M., P. Dang, C. C. Holbrook et al. 2005b. Identification of transcripts involved in resistance responses to leaf spot disease caused by *Cercosporidium personatum* in peanut (*Arachis hypogaea* L.). *Phytopathology* 95:381–387.

Luo, M., X. Liang, P. Dang et al. 2005c. Microarray-based screening of differentially expressed genes in peanut in response to *Aspergillus parasiticus* infection and drought stress. *Plant Sci* 169:695–703.

Lyerly, J. H., H. T. Stalker, J. W. Moyer, and K. Hoffan. 2002. Evaluation of *Arachis* species for resistance to tomato spotted wilt virus. *Peanut Sci* 29:79–84.

Lynch, R. E. and T. P. Mack. 1995. Biological and biotechnical advances for insect management in peanut. In *Advances in Peanut Science*, eds. H. E. Pattee and H. T. Stalker, pp. 95–159. Stillwater, OK: American Peanut Research and Education Society.

Mace, E. S., R. K. Varshney, V. Mahalakshmi et al. 2008. *In silico* development of simple sequence repeat markers within the aeschynomenoid/dalbergoid and genistoid clades of the Leguminosae family and their transferability to *Arachis hypogaea*, groundnut. *Plant Sci* 174:51–60.

Magbanua, Z. V., H. D. Willed, J. K. Roberts et al. 2000. Field resistance to tomato spotted wilt virus in transgenic peanut (*Arachis hypogaea* L.) expressing an antisense nucleocapsid gene sequence. *Mol Breed* 6:227–236.

Mendes, A. J. T. 1947. Estudos citologicos no gênero *Arachis*. *Bragantia* 7:257–267.

Milla, S. R., T. G. Isleib, and H. T. Stalker. 2005. Taxonomic relationships among *Arachis* sect. *Arachis* species as revealed by AFLP markers. *Genome* 48:1–11.

Mondal, S., A. M. Badigannavar, and G. S. S. Murty. 2007. RAPD markers linked to a rust resistance gene in cultivated groundnut (*Arachis hypogaea* L.). *Euphytica* 159:233–239.

Moore, K. M. and D. A. Knauft. 1989. The inheritance of high oleic acid in peanut. *J Hered* 80:252–253.

Moretzsohn, M. C., A. V. Barbosa, D. M. Alves-Freitas et al. 2009. A linkage map for the B-genome of *Arachis* (Fabaceae) and its synteny to the A-genome. *BMC Plant Biol* 9:40.

Moretzsohn, M. C., L. Leoi, K. Proite et al. 2005. A microsatellite-based, gene-rich linkage map for the AA genome of *Arachis* (Fabaceae). *Theor Appl Genet* 111:1060–1071.

Mozingo, R. W., T. A. Coffelt, and T. G. Isleib. 2000. Registration of 'VA 98R' peanut. *Crop Sci* 40:1202–1203.

Nagy, E. D., Y. Chu, Y. Guo et al. 2010. Recombination is suppressed in an alien introgression on chromosome 5A of peanut harboring *Rma*, a dominant root knot nematode resistance gene. *Mol Breed* 26:357–370.

Nelson, S. C., C. E. Simpson, and J. L. Starr. 1989. Resistance to *Meloidogyne arenaria* in *Arachis* spp. germplasm. *J Nematol* 21:654–660.

Nelson, S. C., J. L. Starr, and C. E. Simpson. 1990. Expression of resistance to *Meloidogyne arenaria* in *Arachis batizocoi* and *A. cardenasii*. *J Nematol* 22:423–425.

Nevill, D. J. 1981. Components of resistance to *Cercospora arachidicola* and *Cercosporidium personatum* in groundnuts. *Ann Appl Biol* 99:77–86.

Newell, C. A. 2000. Plant transformation technology—Developments and applications. *Mol Biotech* 16:53–65.

Nigam, S. N., F. Waliyar, R. Aruna et al. 2009. Breeding peanut for resistance to aflatoxin contamination at ICRISAT. *Peanut Sci* 36:42–49.

Nigam, S. N. and R. Aruna. 2008. Stability of soil plant analytical development (SPAD) chlorophyll meter reading (SCMR) and specific leaf area (SLA) and their association across varying soil moisture stress conditions in groundnut (*Arachis hypogaea* L.). *Euphytica* 160:111–117.

Nigam, S. N., S. Chandra, K. R. Sridevi et al. 2005. Efficiency of physiological trait-based and empirical selection approaches for drought tolerance in groundnut. *Ann Appl Biol* 146:433–439.

Niu, C., Y. Akasaka-Kennedy, P. Faustinelli et al. 2009. Antifungal activity in transgenic peanut (*Arachis hypogaea* L.) conferred by a nonheme chloroperoxidase gene. *Peanut Sci* 36:126–132.

Norden, A. J., D. W. Gorbet, D. A. Knauft, and C. T. Young. 1987. Variability in oil quality among peanut genotypes in the Florida breeding program. *Peanut Sci* 14:7–11.

Ouedraogo, M., O. D. Smith, C. E. Simpson, and D. H. Smith. 1994. Early and late leaf spot resistance and agronomic performance of nineteen interspecific derived peanut lines. *Peanut Sci* 21:99–104.

Ozias-Akins, P. 2007. Peanut. In *Biotechnology in Agriculture and Forestry 61—Transgenic Crops VI*, eds. E. C. Pua, and M. R. Davey, pp. 81–105. Heidelberg: Springer.

Ozias-Akins, P., H. Yang, R. Gill, H. Fan, and R. E. Lynch. 2002. Reduction of aflatoxin contamination in peanut: A genetic engineering approach. *ACS Symp Series* 829:151–160.

Ozias-Akins, P., J. A. Schnall, W. F. Anderson et al. 1993. Regeneration of transgenic peanut plants from stably transformed embryogenic callus. *Plant Sci* 93:185–194.

Ozias-Akins, P. and R. Gill. 2001. Progress in the development of tissue culture and transformation methods applicable to the production of transgenic peanut. *Peanut Sci* 28:123–131.

Parlevliet, J. E. 1979. Components of resistance that reduce the rate of epidemic development. *Annu Rev Phytopathol* 17:203–222.

Patel, M., S. Jung, K. Moore, G. Powell, C. Ainsworth, and A. Abbott. 2004. High-oleate peanut mutants result from a MITE insertion into the *FAD2* gene. *Theor Appl Genet* 108:1492–1502.

Payton, P., K. R. Kottapalli, D. Rowland et al. 2009. Gene expression profiling in peanut using high density oligonucleotide microarrays. *BMC Genomics* 10:265.

Perry, J., A. Brachmann, T. Welham et al. 2009. TILLING in *Lotus japonicus* identified large allelic series for symbiosis genes and revealed a bias in functionally defective ethyl methanesulfonate alleles toward glycine replacements. *Plant Physiol* 151:1281–1291.

Phipps, P. M. and M. K. Beute. 1997. Cylindrocladium black rot. In *Compendium of Peanut Diseases*, eds. N. Kokalis-Burelle, D. M. Porter, R. Rodriguez-Kabana, D. H. Smith, and P. Subrahmanyam, pp. 12–15. St. Paul, MN: Am. Phytopath. Soc. Press.

Porch, T. G., M. W. Blair, P. Lariguet, C. Galeano, C. E. Pankhurst, and W. J. Broughton. 2009. Generation of a mutant population for TILLING common bean genotype BAT 93. *J Am Soc Hort Sci* 134:348–355.

Porter, D. M. and M. K. Beute. 1974. Sclerotinia blight of peanuts. *Phytopathology* 64:263–264.

Porter, D. M., T. A. Coffelt, F. S. Wright, and R. W. Mozingo. 1992. Resistance to *Sclerotinia* blight and early leaf spot in Chinese germplasm. *Peanut Sci* 19:41–43.

Proite, K., S. C. Leal-Bertioli, D. J. Bertioli et al. 2007. ESTs from a wild *Arachis* species for gene discovery and marker development. *BMC Plant Biol* 7:7.

Raghavan, C., M. E. B. Naredo, H. H. Wang et al. 2007. Rapid method for detecting SNPs on agarose gels and its application in candidate gene mapping. *Mol Breed* 19:87–101.

Ramos, M. L., G. Fleming, Y. Chu, Y. Akiyama, M. Gallo, and P. Ozias-Akins. 2006. Chromosomal and phylogenetic context for conglutin genes in *Arachis* based on genomic sequence. *Mol Genet Genom* 275:578–592.

Ramos, M. L., J. J. Huntley, S. J. Maleki, and P. Ozias-Akins. 2009. Identification and characterization of a hypoallergenic ortholog of *Ara* h 2.01. *Plant Mol Biol* 69:325–335.

Ratnakumar, P., V. Vadez, S. N. Nigam, and L. Krishnamurthy. 2009. Assessment of transpiration efficiency in peanut (*Arachis hypogaea* L.) under drought by lysimetric system. *Plant Biol* 11:124–130.

Ray, T. K., S. P. Holly, D. A. Knauft, A. G. Abbott, and G. L. Powell. 1993. The primary defect in developing seed from the high oleate variety of peanut (*Arachis hypogaea* L.) is the absence of Δ12–desaturase activity. *Plant Sci* 91:15–21.

Rucker, K. S., C. K. Kvien, C. C. Holbrook, and J. E. Hook. 1995. Identification of peanut genotypes with improved drought avoidance traits. *Peanut Sci* 22:14–18.

Seijo, G., G. I. Lavia, A. Fernández et al. 2007. Genomic relationships between the cultivated peanut (*Arachis hypogaea* L.) and its close relatives revealed by double GISH. *Am J Bot* 94:1963–1971.

Sharma, K. K. and V. Anjaiah. 2000. An efficient method for the production of transgenic plants of peanut (*Arachis hypogaea* L.) through *Agrobacterium tumefaciens*-mediated genetic transformation. *Plant Sci* 159:7–19.

Sheshshayee, M. S., H. Bindumadhava, N. R. Rachaputi et al. 2006. Leaf chlorophyll concentration relates to transpiration efficiency in peanut. *Ann Appl Biol* 148:7–15.

Simpson, C. E. 1991. Pathways for introgression of pest resistance into *Arachis hypogaea* L. *Peanut Sci* 18:22–26.

Simpson, C. E. and J. L. Starr. 2001. Registration of 'COAN' peanut. *Crop Sci* 41:918.

Simpson, C. E., M. D. Burow, A. H. Paterson, J. L. Starr, and G. T. Church. 2003. Registration of 'NemaTAM' peanut. *Crop Sci* 43:1561.

Singh, A. K. and J. P. Moss. 1982. Utilization of wild relatives in genetic improvement of *Arachis hypogaea* L. Part 2. Chromosome complements of species in section *Arachis*. *Theor Appl Genet* 61:305–314.

Singh, A. K. and J. P. Moss. 1984. Utilization of wild relatives in genetic improvement of *Arachis hypogaea* L. V. Genome analysis section *Arachis* and its implications in gene transfer. *Theor Appl Genet* 68:355–364.

Singsit, C., M. J. Adang, R. E. Lynch et al. 1997. Expression of a *Bacillus thuringiensis* Cryla (C) gene in transgenic peanut plants and its efficacy against lesser cornstalk borer. *Transgen Res* 6:169–176.

Smartt, J. and H. T. Stalker. 1982. Speciation and cytogenetics in *Arachis*. In *Peanut Science and Technology*, eds. H. E. Pattee, and C. T. Young, pp. 21–49. Yoakum, TX: American Peanut Research and Education Society.

Smith, O. D., C. E. Simpson, M. C. Black, and B. A. Besler. 1998. Registration of 'Tamrun 96' peanut. *Crop Sci* 38:1403.

Songsri, P., S. Jogloy, C. C. Holbrook et al. 2009. Association of root, specific leaf area and SPAD chlorophyll meter reading to water use efficiency of peanut under different available soil water. *Ag Water Manage* 96:790–798.

Songsri, P., S. Jogloy, N. Vorasoot, C. Akkasaeng, A. Patanothai, and C. C. Holbrook. 2008. Root distribution of drought-resistant peanut genotypes in response to drought. *J Agron Crop Sci* 194:92–103.

Spielman, I. V., A. P. Burge, and J. P. Moss. 1979. Chromosome loss and meiotic behavior in interspecific hybrids in the genus *Arachis* L. and their implications inbreeding for disease resistance. *Z Pflanzenzuchrg* 83:236–250.

Stalker, H. T. 1985. Cytotaxonomy of *Arachis*. Paper presented in International Workshop on Cytogenetics of *Arachis*, Hyderabad, India, 65–79.

Stalker, H. T. 1991. A new species in section *Arachis* of peanuts with a D genome. *Am J Bot* 78:630–637.

Stalker, H. T. and C. E. Simpson. 1995. Genetic resources in *Arachis*. In *Advances in Peanut Science*, eds. H. E. Pattee and H. T. Stalker, pp. 14–53. Stillwater, OK: American Peanut Research and Education Society.

Stalker, H. T., J. S. Dhesi, and D. Parry. 1991a. An analysis of the B genome species *Arachis batizocoi*. *Plant Syst Evol* 174:159–169.

Stalker, H. T., J. S. Dhesi, D. Parry, and J. H. Hahn. 1991b. Cytological and interfertility relationships of *Arachis* section *Arachis*. *Am J Bot* 78:238–246.

Stalker, H. T., J. S. Dhesi, and G. D. Kochert. 1995. Variation within the species *A. duranensis*, a possible progenitor of the cultivated peanut. *Genome* 38:1201–1212.

Stalker, H. T. and M. K. Beute. 1993. Registration of four interspecific peanut germplasm lines resistant to *Cercospora arachidicola*. *Crop Sci* 33:1117.

Stalker, H. T., M. K. Beute, B. B. Shew, and T. G. Isleib. 2002. Registration of five leafspot-resistant peanut germplasm lines. *Crop Sci* 42:314–316.

Stalker, H. T. and R. D. Dalmacio. 1981. Chromosomes of *Arachis* species, section *Arachis* (Leguminosae). *J Hered* 72:403–408.

Stalker, H. T. and R. D. Dalmacio. 1986. Karyotype analysis and relationships among varieties of *Arachis hypogaea* L. *Cytologia* 58:617–629.

Suzuki, T., M. Eiguchi, T. Kumamaru et al. 2008. MNU-induced mutant pools and high performance TILLING enable finding of any gene mutation in rice. *Mol Genet Genomics* 279:213–223.

Till, B. J., T. Zerr, L. Comai, and S. Henikoff. 2006. A Protocol for TILLING and ecoTILLING in plants and animals. *Nature Protocols* 1:2465–2477.

Tillman, B. L. and H. T. Stalker. 2009. Peanut. In *Oil Crops, Handbook of Plant Breeding 4*, eds. J. Vollmann, and I. Rajcan, pp. 297–315. New York, NY: Springer Science Press.

Tiwari, S., D. K. Mishra, A. Singh, P. K. Singh, and R. Tuli. 2008. Expression of a synthetic Cry1ec gene for resistance against *Spodoptera litura* in transgenic peanut (*Arachis hypogaea* L.). *Plant Cell Rep* 27:1017–1025.

Tiwari, S. P., M. P. Ghewande, and D. P. Misra. 1984. Inheritance of resistance to rust and late leafspot in groundnut (*Arachis hypogaea* L.). *J Cytol Genet* 19:97–101.

Triques, K., B. Sturbois, S. Gallais et al. 2007. Characterization of *Arabidopsis thaliana* mismatch specific endonucleases: Application to mutation discovery by TILLING in pea. *Plant J* 51:1116–1125.

Uauy, C., F. Paraiso, P. Colasuonno et al. 2009. A modified TILLING approach to detect induced mutations in tetraploid and hexaploid wheat. *BMC Plant Biol* 9:115.

Upadhyaya, H. D. 2005. Variability for drought related traits in the mini core collection of peanut. *Crop Sci* 45:1432–1440.

Upadhyaya, H. D., L. J. Reddy, C. L. L. Gowda, and S. Singh. 2006. Identification of diverse groundnut germplasm: Sources of early maturity in a core collection. *Field Crop Res* 97:261–267.

Upadhyaya, H. D., M. E. Ferguson, and P. J. Bramel. 2001. Status of the *Arachis* germplasm collection at ICRISAT. *Peanut Sci* 28:89–96.

Upadhyaya, H. D., P. J. Bramel, R. Ortiz, and S. Singh. 2002. Developing a mini core of peanut for utilization of genetic resources. *Crop Sci* 42:599–600.

Upadhyaya, H. D., R. Ortiz, P. J. Bramel, and S. Singh. 2003. Development of a groundnut core collection using taxonomical, geographical and morphological descriptors. *Genet Res Crop Evol* 50:139–148.

Valls, J. F. M. and C. E Simpson. 2005. New species of *Arachis* (Leguminosae) from Brazil, Paraguay and Bolivia. *Bonplandia* 14:35–63.

Varshney, R. K., A. Graner, and M. E. Sorrells. 2005. Genic microsatellite markers in plants: Features and applications. *Trends Biotechnol* 23:48–55.

Varshney, R. K., D. J. Bertioli, M. C. Moretzsohn et al. 2009. The first SSR-based genetic linkage map for cultivated groundnut (*Arachis hypogaea* L.). *Theor Appl Genet* 118:729–739.

Wang, M. L., C. Y. Chen, J. Davis, B. Z. Guo, T. Stalker, and R. Pittman. 2009. Assessment of oil content and fatty acid composition variability in different peanut subspecies and botanical varieties. *Plant Genet Res* 8:71–73.

Wang, M. L., D. L. Pinnow, and R. N. Pittman. 2007. Preliminary screening of peanut germplasm in the U.S. collection for TSWV resistance and high flavonoid content. *Plant Path J* 6:219–226.

Watson, J. M., A. F. Fusaro, M. B. Wang, and P. M. Waterhouse. 2005. RNA silencing platforms in plants. *FEBS Lett* 579:5982–5987.

Weil, C. F. and R. A. Monde. 2007. Getting the point—Mutations in maize. *Crop Sci* 47:S60–S67.

Wells, M. A., W. J. Grichar, O. D. Smith, and D. H. Smith. 1994. Response of selected peanut germplasm lines to leaf spot and southern stem rot. *Oléagineux* 49:21–26.

Wilson, R. F., H. T. Stalker, and E. C. Brummer. 2004. *Legume Crop Genomics*, Champaign, IL: AOCS Press.

Wilson, T. A., D. Kritchevsky, T. Kotyla, and R. J. Nicolosi. 2006. Structured triglycerides containing caprylic (8:0) and oleic (18:1) fatty acids reduce blood cholesterol concentrations and aortic cholesterol accumulation in hamsters. *Biochim Biophys Acta* 1761:345–349.

Wright, G. C., R. C. Nageswara Rao, and M. S. Basu. 1996. A physiological approach to the understanding of genotype by environment interactions—A case study on improvement of drought adaptation in peanut. In *Plant Adaptation and Crop Improvement*, eds. M. Cooper, and G. L. Hammer, pp. 365–381. Wallingford: CAB International.

Xin, Z. G., M. L. Wang, N. A. Barkley et al. 2008. Applying genotyping (TILLING) and phenotyping analyses to elucidate gene function in a chemically induced sorghum mutant population. *BMC Plant Biol* 8:103.

Yang, H., C. Singsit, A. Wang, D. Gonsalves, and P. Ozias-Akins. 1998. Transgenic peanut plants containing a nucleocapsid protein gene of tomato spotted wilt virus show divergent levels of gene expression. *Plant Cell Rep* 17:693–699.

Yang, H., P. Ozias-Akins, A. Culbreath et al. 2004. Field evaluation of *tomato spotted wilt virus* resistance in transgenic peanut (*Arachis hypogaea*). *Plant Dis* 88:259–264.

Yin, D. M., S. Z. Deng, K. H. Zhan, and D. Q. Cui. 2007. High-oleic peanut oils produced by HpRNA-mediated gene silencing of oleate desaturase. *Plant Mol Biol Rep* 25:154–163.

Yuksel, B., J. C. Estill, S. R. Schulze, and A. H. Paterson. 2005. Organization and evolution of resistance gene analogs in peanut. *Mol Genet Genomics* 274:248–263.

CHAPTER 16

The Relevance of Compositional and Metabolite Variability in Safety Assessments of Novel Crops

George G. Harrigan, Angela Hendrickson Culler, William P. Ridley, and Kevin C. Glenn

CONTENTS

16.1 INTRODUCTION

The composition of food and feed crops has economic value, and is also an important determinant of quality for the development of diets to promote the healthy growth of people and animals. A major goal of many plant breeding programs, whether using traditional methods or modern biotechnology such as genetic modification (GM), is to maintain compositional quality while improving agronomic traits such as yield, pest resistance, or herbicide tolerance. Other programs seek to enhance the nutritional value of important crops through change in the endogenous levels of nutrients, antinutrients, vitamins, and minerals. Crop composition, however, encompasses a very broad range of natural variability in endogenous levels of different constituents. This variation is an important consideration in experimental designs and analyzes to evaluate the impact of novel traits and nutritional enhancements on compositional and metabolite profiles in new crops. We describe here how compositional studies conducted during comparative safety assessments of novel crops have led to a further understanding of the contribution of environmental and genetic factors to compositional and metabolite variation.

16.2 PHILOSOPHY OF COMPOSITIONAL ANALYSES IN COMPARATIVE SAFETY ASSESSMENTS OF NEW CROPS

The goal of comparative safety assessments of novel crops containing new traits is to provide assurance that the new variety is as safe as conventional variants of the crop when prepared or eaten according to its intended use, and that it can be consumed under the same dietary terms as foods or feed derived from conventional crops (OECD, 1998, 2002, 2006; Delaney, 2007; Kok et al., 2008). Comparative compositional studies represent a key component of these safety assessments. A task force established in 1999 by the Organization for Economic Cooperation and Development (OECD) developed science-based protocols (consensus documents) for the safety assessments of foods and feed derived through modern biotechnology. These OECD consensus documents, which are in the public domain, address compositional considerations for new crop varieties by identifying key food and feed nutrients, antinutrients, and secondary plant metabolites found in each crop species (OECD, 1998, 2002, 2006). These internationally accepted recommendations outline a comparative approach for assessing compositional equivalence of crops containing biotechnology-derived traits (Codex, 2003). The OECD documents "provide a technical tool for regulatory officials, and also for industry and other interested parties" and are meant to "promote international harmonization in biotechnology" and "increase the efficiency of the risk/safety assessment process."

Analyses conducted according to the principles outlined in the OECD consensus documents and Codex guidelines focus on quantitative measurements of key biochemical components of the "articles of commerce" (usually grain and forage) of the crop containing the new trait. The recommended analytes include major constituents such as proximates (oil, protein, ash, carbohydrate, and fiber), key nutrients such as amino acids, fatty acids, vitamins, minerals, and crop-specific antinutrients, and secondary metabolites such as phytic acid, ferulic acid, and *p*-coumaric acid in maize, isoflavones in soybean, and gossypol in cotton. The list of analytes can be modified depending on the properties of the crop and the intended trait, particularly if the trait offers nutritional enhancement through increased levels of selected vitamins or minerals. A list of selected crops and components typically analyzed is presented in Table 16.1.

The product to be assessed is usually referred to as the test substance and contains the transgene(s) that confers the desired trait. The control substance should have a history of safe use and broad commercial acceptance. It should be isogenic or near-isogenic to the test substance but should not contain the novel trait or any additional inserted DNA. Direct comparisons between test and control substances form the basis of compositional assessments. This comparative approach is also applicable to foods derived from plants that have been altered by other techniques (Codex, 2003).

There are two key considerations that are incorporated into assessments because of their importance in defining natural variability in grain and forage composition. First, composition studies utilize samples from multiple replicated field trials encompassing diverse geographic regions. Studies from multiple growing seasons are also analyzed. This allows for a comparison of the test and control substances over several unique environments and can provide an assessment of the reproducibility of any differences observed at individual sites. Second, composition studies often include a range of unique commercially available conventional crops (reference substances) grown concurrently at the same sites as the test and control substances. This provides an overview of the impact of genetic diversity on compositional variability in readily available foodstuffs. Ranges reported in the literature can also be incorporated into compositional studies to describe compositional variability inherent in conventional crops.

Component differences of sufficient magnitude and high statistical significance may be considered biologically relevant and would warrant additional safety assessments of the GM crop. Setting a universal threshold for relative magnitude of differences as a trigger for further safety assessments of GM crops has been considered. In 2000, the Nordic Council of Ministers recommended that if a

Table 16.1 Summary of Recommendations for Composition Analyses for New Varieties of Crops

Crop	Tissue	Composition	Nutrients[a]	Vitamins and Minerals	Amino/Fatty Acids	Antinutrients/Secondary Metabolites
Alfalfa	Forage	OECD standard	Proximates, fiber	Ca, P	AA	Lignin
		Optional considerations		Cu, Fe, Mg, Mn, K, Na, Zn	FA	Saponins (soyasapogenols, zanhic acid glycosides, medicagenic acid), phytoestrogens, cyanogenic glycosides
Canola	Seed	OECD standard	Proximates, fiber	Ca, P	AA	Tannins, sinapine, phytic acid
		Optional considerations		Mg, K, Na, S, Fe, Zn, Cu, Mg, vitamins (B_1, B_2, B_6, E), biotin, choline, folate, niacin, pantothenic acid, thiamin	FA in oil	Sterols in oil
Corn	Grain	OECD standard	Proximates, fiber	Na, K, Ca, P, Mg, Mn, Fe, Cu, Zn, vitamins (A, B_1, B_2, B_6, C, E), folate, niacin	AA and FA	Phytic acid, raffinose, furfural, ferulic acid, *p*-coumaric acid
		Optional considerations				
Cotton	Seed	OECD standard	Proximates, fiber	Vitamin E	FA	Gossypol, malvalic acid, sterculic acid, dihydrosterculic acid
		Optional considerations		Na, K, Ca, P, Mg, Fe, Cu, Zn, Mn, vitamins (A, B_1, B_2, B_6, C), folate, niacin	AA	
Soybean	Grain	OECD standard	Proximates, fiber		AA and FA	Phytic acid, trypsin inhibitors, lectins, isoflavones
		Optional considerations				Stachyose, raffinose

continued

Table 16.1 (continued) Summary of Recommendations for Composition Analyses for New Varieties of Crops

Crop	Tissue	Composition	Nutrients[a]	Vitamins and Minerals	Amino/Fatty Acids	Antinutrients/Secondary Metabolites
Potato	Tubers	OECD standard	Dry matter, sugars, protein	Vitamin C		Glycolalkaloids[b]
		Optional considerations	Fiber		AA	Lecithin, trypsin inhibitors, lectins
Wheat	Grain	OECD standard	Proximates	Vitamins (B_1, B_2, B_6, B_{12}, E)	AA and FA	Phytate
		Optional considerations		Carotenoids, folic acid		β-Sitosterol, campesterol, lignans, phenolics, flavanoids
Rice	Grain	OECD standard	Proximates, fiber, amylase	Ca, Mg, P, K, Si, S, Cu, Fe, Mn, Na, Zn, vitamins (A, B_1, B_2, B_6, B_{12}, E), niacin, inositol,	AA and FA	Phytic phosphorus
		Optional considerations		Panthothenic acid, biotin, choline, *p*-aminobenzoic acid (PABA)		Trypsin inhibitor, oryzacystatin, alpha-amylase subtilisin inhibitor, lectins
Sugar beet	Roots/Tops	OECD standard	Proximates,	Na, K, P	Total AA (α-amino-nitrogen)	Sucrose, pectins
		Optional considerations				

[a] Proximates include protein, fat, ash, and carbohyhdrates.
[b] A variety of traditionally bred potato containing high levels of solanine was taken off the market for high levels of anti-nutrients.

component in a GM crop differed from the conventional control by ±20% in relative magnitude, additional analyses of the GM crop were warranted (cited in Hothorn and Oberdoerfer, 2006). This concept was refined to account for the nutritional relevance of a component and the experimental precision of its measurement (Hothorn and Oberdoerfer, 2006), although, as pointed out in Chassy et al. (2004), considerations of proposed use and actual dietary intake of a specific crop product are also important. Threshold ranges for GM components were suggested as follows: 0.833–1.20 of the conventional control for "nutritionally very relevant" components (minerals, vitamins, antinutrients, bioactives, essential amino acids, and fatty acids), 0.769–1.30 for "relevant" (nonessential amino and fatty acids), and 0.667–1.50 for components of "less relevance" (proximates and fiber). In general, however, statistical approaches are more commonly adopted and may utilize statistical difference or equivalence testing approaches. Any differences, whether arithmetic or statistical, have to account for natural variability in component levels. The use of multiple geographic locations and reference substances provides insights into natural variability in crop composition because of environmental and genetic variability and provides an important context for assessing the relevance of any differences recorded between test and control substances.

16.3 OVERVIEW OF COMPOSITIONAL ASSESSMENTS OF EQUIVALENCE AND NATURAL VARIABILITY

The inclusion of compositional analysis in safety assessments is primarily to evaluate equivalence. Compositional equivalence of GM crops and their conventional (non-GM) comparators have now been demonstrated in a range of products derived from potato (Rogan et al., 2000; Shepherd et al., 2006), cotton (Berberich et al., 1996; Hamilton et al., 2004; Nida et al., 1996), soybean (Berman et al., 2009, 2010; Harrigan et al., 2007; Lundry et al., 2008; McCann et al., 2005; Padgette et al., 1996; Taylor et al., 1999), corn (Drury et al., 2008; Harrigan et al., 2009; McCann et al., 2007; Ridley et al., 2002; Sidhu et al., 2000), rice (Oberdoerfer et al., 2005), wheat (Obert et al., 2004), and alfalfa (McCann et al., 2006). Also significant is that the results from the inclusion of multiple sites, growing seasons, and different genetically distinct commercial varieties have validated the use of natural variability in providing a framework to evaluate pair-wise comparisons of test and controls. The relative impacts of biotechnology-derived agronomic traits and natural variation on crop composition based on data generated by Monsanto on several GM corn and soybean products grown in a range of geographies, under different regional agronomic practices, and over multiple seasons are presented in Table 16.2.

Table 16.2 lists GM corn and soybean products containing drought tolerance, insect resistance, and herbicide tolerance traits. Each required a compositional evaluation prior to deregulation. Consistent with OECD principles, each study utilized multiple replicated sites. Compositional data were combined across all study sites prior to statistical analysis. This is referred to as a combined site analysis. The number of tests to control comparisons is equal to the number of components analyzed.

Overall, for corn, a total of 457 (number of studies × number of compositional components) statistical comparisons between the GM products and their corresponding near-isogenic conventional controls were conducted. Of these, 91.9% were not significantly different ($P > 0.05$). In most, if not all, cases the statistical differences between the GM and conventional components represented modest differences in relative magnitude (generally <20%). Only one comparison where a significant difference ($P < 0.05$) was observed had a relative magnitude of difference greater than 20%. In this instance (see Table 16.1, insect-protected MON 88017) vitamin B_1 levels in the GM product were 2.47 mg/kg DW and in the conventional counterpart, 3.24 mg/kg DW. However, the range of values observed in the study for GM and conventional corn vitamin B_1 levels in these studies was 2.33–4.58 and 2.30–4.79 mg/kg DW, respectively, a remarkable overlap. There were no

Table 16.2 Summary of Differences between Components in Non-GM and GM Crops

GM Crop[ref]	Trait	Location (Season)	Number of Comparisons (Sites)[a]	Number of Statistical Differences ($P < 0.05$)	Number of Magnitude > 20% Differences	Number of Components Different ($P < 0.05$) at All Individual Sites
MON 87460 (Harrigan et al., 2009)	Drought tolerance	Chile A[b] (2006–2007)	52 (three sites)	2	0	0
MON 87460 (Harrigan et al., 2009)	Drought tolerance	Chile B[b] (2006–2007)	52 (three sites)	1	0	0
MON 87460 (Harrigan et al., 2009)	Drought tolerance	USA (2006)	53 (six sites)	3	0	0
MON 89034 (Drury et al., 2008)	Insect protection	Argentina (2004–2005)	52 (five sites)	5	0	0
MON 89034 (Drury et al., 2008)	Insect protection	USA (2004)	52 (five sites)	2	0	0
MON 88017 (McCann et al., 2007)	Insect protection	Argentina (2003–2004)	52 (four sites)	9	0	2[c]
MON 88017 (McCann et al., 2007)	Insect protection	USA (2002)	53 (three sites)	3	1	0
NK603 (Ridley et al., 2002)	Herbicide tolerance	Europe (1999)	44 (four sites)	6	0	1[d]
NK603 (Ridley et al., 2002)	Herbicide tolerance	USA (1998)	47 (two sites)	7	0	2[e]
All corn	—	*All*	*457*	*37 (8.1%)*	*1 (0.2%)*	
MON 87701 (Berman et al., 2009)	Insect protection	Argentina (2007–2008)	46 (five sites)	4	1	0
MON 87701 (Berman et al., 2010)	Insect protection	Brazil north (2007–2008)	46 (two sites)	7	0	4[f]
MON 87701 (Berman et al., 2010)	Insect protection	Brazil south (2007–2008)	46 (two sites)	2	2	1[g]
MON 87701 (Berman et al., 2009)	Insect protection	USA (2007)	48 (five sites)	15	1	0
MON 89788	Herbicide tolerance	Brazil north (2007–2008)	46 (two sites)	11	1	6[h]
MON 89788	Herbicide tolerance	Brazil south (2007–2008)	46 (two sites)	6	0	0
MON 89788 (Lundry et al., 2008)	Herbicide tolerance	USA (2005)	39 (five sites)	3	0	0
MON 89788 (Lundry et al., 2008)	Herbicide tolerance	Argentina (2004–2005)	39 (five sites)	0	0	0
40–3–2 (Harrigan et al., 2007)	Herbicide tolerance	Romania (2005)	42 (five sites)	3	0	0
All soy	—	*All*	*368*	*51 (13.9%)*	*5 (1.4%)*	

Source: Data from Harrigan et al. 2009. *J Agric Food Chem* 57:9754–9763; Drury et al. 2008. *J Agric Food Chem* 56:4623–4630; McCann et al. 2007. *J Agric Food Chem* 55:4034–4042; Ridley et al. 2002. *J Agric Food Chem* 50:7235–7243; Berman et al. 2010. *J Agric Food Chem* 58:6270–6276; Berman et al. 2009. *J Agric Food Chem* 57:11360–11369; Lundry et al. 2008. *J Agric Food Chem* 56:4611–4622; and Harrigan et al. 2007. *J Agric Food Chem* 55:6160–6168.

[a] All studies used multiple replicated field trials; compositional data were combined across all sites of each specific study. The number of comparisons per study is equal to the number of components statistically assessed.

[b] Chile A was conducted under well-watered conditions, Chile B was conducted under water-limited conditions. All other field trials followed normal regional agronomic practices.

[c] Oleic acid and linolenic acid.

[d] Palmitic acid.

[e] Cystine and stearic acid.

[f] Palmitoleic acid, heptadecanoic acid, linolenic acid, and raffinose.

[g] Vitamin E.

[h] Protein, proline, palmitoleic acid, heptadecanoic acid, stearic acid, raffinose, and stachyose.

other instances where significant differences between GM and control values were observed during combined site comparisons.

An analysis of the corn data also highlighted a lack of reproducibility in differences observed at individual sites; in other words, while a component may have been statistically significantly different at one site, in most instances it was not different at other sites, indicating that the difference was most likely due to natural variation resulting from different geographic environmental influences versus a result of the inserted trait. In nine field studies encompassing 35 sites, there were only five instances where a component was observed to be statistically significant ($P < 0.05$) at all sites within a study. For example, insect-protected MON 88017 grown in Argentina during the 2003/2004 growing season showed only two components that were different at all four sites. These were oleic acid and linolenic acid. Statistical differences between MON 88017 and the control were not observed during the earlier 2002 U.S. production, again indicating that the difference was most likely not a result of the inserted trait. For herbicide-tolerant NK603 grown in Europe during the 1999 growing season, one component (palmitic acid) was different at all four sites. Two components (cystine and stearic acid) were different for NK603 during the U.S. 1998 growing season. In summary, no consistent or reproducible differences between the GM and non-GM products were observed across multiple geographic sites throughout multiple growing seasons. This indicates that differences observed were due to natural variation, and are not attributed to the inserted trait.

The use of multiple sites and growing seasons therefore serves to offer a more comprehensive overview of the contribution of transgene insertion to compositional variation relative to other factors. As an illustration, Figure 16.1 presents the range of values observed for amino acids in corn evaluated in OECD consensus-based safety assessments as listed in Table 16.1. These ranges are extensive and highlight the compositional plasticity of corn. Furthermore, there is an extensive

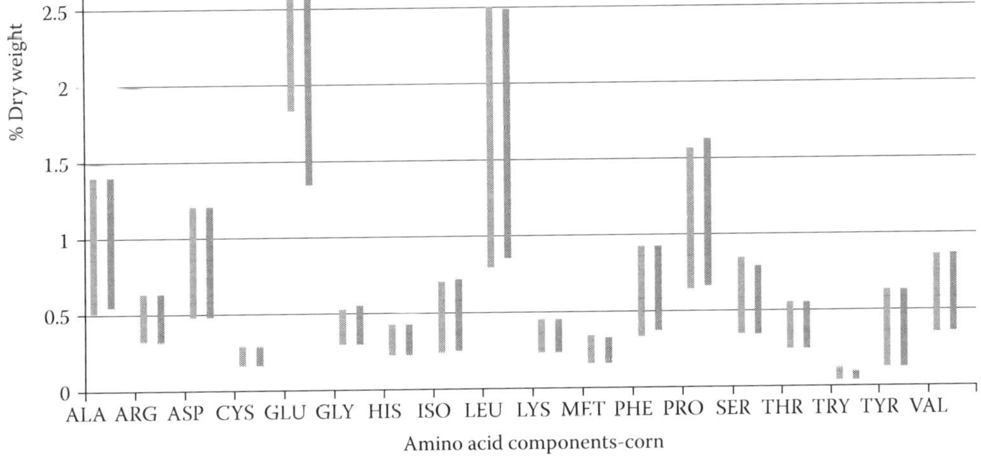

Figure 16.1 Summary of amino acid levels in non-GM and GM corn from a total of 10 growing seasons. Each vertical bar represents the range of values for the corresponding amino acids as measured in studies listed in Table 16.1.

overlap in the values for the conventional and GM components which suggests that overall, GM and conventional composition cannot be disaggregated.

For soybean, there were a total of 368 statistical comparisons between the GM products and the corresponding conventional controls (Table 16.2). Of these, 86.1% were not significantly different ($P > 0.05$). As with corn, the statistically significant differences between the GM and conventional soybean components generally represented modest differences in relative magnitude (generally <20%). In addition, there were few instances where components were observed to be different at all sites. In fact, the only instances were in studies where the combined site analysis included only two individual sites. These studies include the evaluation of MON 87701 and MON 89788 grown in distinct geographic regions of Brazil (northern and southern) (Berman et al., 2010) (Table 16.2, Figure 16.2). Along with the United States and Argentina, Brazil is one of the top three soybean producers in the world. Two of its major growing regions (northern and southern) are separated by geography and climate and require different germplasms adapted for growth in each respective region. Thus, for insect-protected MON 87701 and glyphosate-tolerant MON 89788, both biotechnology-derived traits are introgressed into both the conventional variety Monsoy 8329, which is adapted for cultivation in the northern region, and the conventional variety A5547 for cultivation in the southern region (Berman et al., 2010). As shown in Table 16.2 there are more differences between test and control substances when grown in the northern region of Brazil compared to the south. It is also notable that the differences observed between test and control substances in Brazil north are not reproduced in other countries or growing regions.

Figure 16.2 presents hierarchical cluster analysis and principal component analysis of compositional data generated on MON 87701, MON 89788, and their respective region-specific controls grown at two replicated field sites in each of the northern and southern growing regions. It is apparent that cultivation in different regions and type of control germplasm contributes more to statistically significant differences recorded in this study than GM.

Although differences in mean values of test and control fatty acids were generally statistically insignificant ($P > 0.05$) or of small relative magnitude, there was a remarkable difference in the fatty

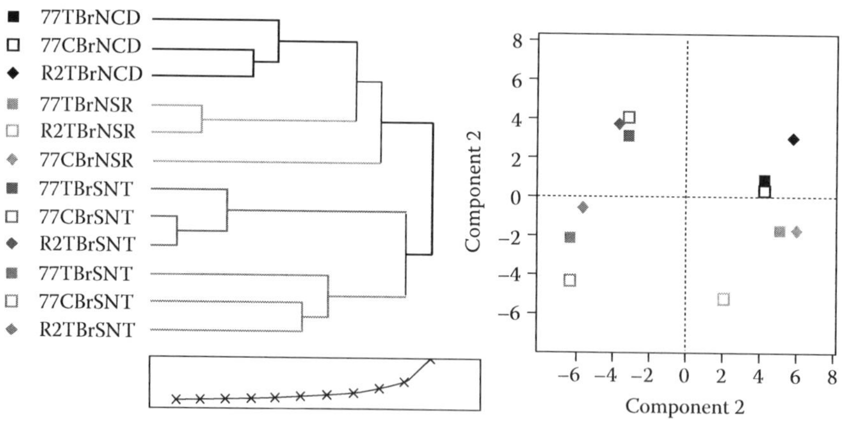

Figure 16.2 Hierarchical cluster analysis and principal component analysis of compositional data generated on the harvested seed of insect-protected MON 87701 and glyphosate-tolerant MON 89788 grown in the Northern and Southern Regions of Brazil during the 2007–2008 seasons. The sample codes are as follows. The first three digits indicate the sample; 77T = MON 87701, R2T = MON 89788, and 77C = conventional control for both MON 87701 and MON 87988. The remaining digits indicate the sites; Cachoeira Dourada, Minas Gerais (BrNCD), Sorriso, Mato Grosso (BrNSR), Nao-Me-Toque, Rio Grande do Sul (BrSNT), and Rolandia, Parana (BrSRO).

acid profiles of the two region-specific controls. This was particularly true of the most abundant fatty acids. Thus, for the northern region control (Monsoy 8329) the mean values for 18:1 oleic acid and 18:2 linoleic acid were 40.43% and 39.73% total FA, respectively; corresponding values for A5547 were 22.60% and 52.23% total FA, respectively.

There were also striking differences in values of the isoflavones from the two region-specific controls. There were 1.9-, 2.7-, and 4.4-fold differences in glycitein, genistein, and daidzein, respectively, between the conventional controls (Berman et al., 2010). Levels were much higher in the Southern region samples. At least one other study (Carrao-Panizzi et al., 2009) has highlighted variability in isoflavone levels in Brazilian soybean cultivars.

The observations are consistent with the known "tremendous oscillations" in isoflavone content. A particular genotype planted in the same location can vary up to three fold in its isoflavone content; variation is attributable to temperature, water, soil conditions, light, and biotic stresses. As stated by Gutierrez-Gonzalez et al. (2009) in a separate study in U.S. soybean "the range of values of isoflavones is overwhelming."

16.4 METABOLOMICS AND METABOLITE VARIATION IN CROPS

In many instances, the equivalence of GM crops and their conventional comparators has been independently assessed by modern profiling technologies such as transcriptomics, proteomics, and metabolomics. Although there are complexities in the interpretation of data generated through modern profiling technologies (Lay et al., 2006) including the fact that the data are not quantitative and there is no standardized framework for comparisons, the lack of variation between GM crops and their conventional comparators at the transcriptomic, proteomic, and metabolomic level has, nonetheless, been independently corroborated. These profiling evaluations extend to a wide range of plants including wheat (Baker et al., 2006; Gregersen et al., 2005; Ioset et al., 2007), potato (Catchpole et al., 2005; Defernez et al., 2004; Lehesranta et al., 2005), soybean (Cheng et al., 2008), rice (Dubouzet et al., 2007; Wakasa et al., 2006), tomato (Le Gall et al., 2003), tobacco, *Arabidopsis* (Kristensen et al., 2005), and *Gerbera* (Ainasoja et al., 2008).

As with the compositional studies reported above, results from many of the "omics studies emphasize the need to understand natural variation in levels of endogenous metabolites in providing biological context to pair-wise differences in any recorded profiles."

In Baker et al. (2006) NMR-based metabolic profiles of three GM wheat varieties and the corresponding parents were generated. The incorporated trangenes encoded high-molecular-weight subunits of the storage protein, glutenin. The wheat varieties were grown at two different sites over three different growing seasons (1999–2001). Differences between the GM and parental lines were within the same range as the differences between the control lines grown on different sites and in different years. Analogous to the approach adopted in targeted compositional analyses adopting OECD recommendations, this study emphasized the importance of data from multiple years and multiple sites and that environmental variation influences metabolome composition.

In Catchpole et al. (2005) two GM potato crops modified in fructan chemistry were grown over two different seasons (2001, 2003). Metabolic profiles of the GM and five conventional crops were generated using flow-injection MS (FIE–MS), GC–MS, and LC–MS. These demonstrated that differences between GM and conventional potatoes were due to intended metabolic changes. The study emphasized the importance of reference cultivars to establish a natural range of metabolite levels in a crop population as adopted in OECD-based compositional analyses.

Experimental designs that will both account for natural variation and have enough power to identify differences that can be attributed to transgene insertion will offer opportunities to maximize the value of omics technologies as tools in plant breeding.

16.5 IMPLICATIONS OF VARIATION FOR REGULATORY ASSESSMENTS FOR NUTRITIONALLY ENHANCED CROPS

The above compositional examples focused on agronomic traits where the goal was to enhance stress tolerance, insect resistance, or herbicide tolerance qualities while maintaining compositional quality. The modification of plants to improve nutritional and flavor qualities through changes in endogenous levels of metabolites is another major goal of modern biotechnology (Hirschi, 2009). Improved protein quality and quantity, altered fatty acid composition, increased levels of vitamins and functional metabolites (such as antioxidants), and improved mineral availability have been achieved through transgene insertion.

Such nutritionally enhanced crops will also be subject to comparative safety assessments including compositional evaluations prior to commercialization. In 2004, the task force of the International Life Sciences Institute (ILSI) International Food Biotechnology Committee published a comprehensive review providing recommendations for safety assessments of nutritionally enhanced products (Chassy et al., 2004). The overall conclusion was that the existing comparative assessment process used to assess agronomically enhanced crops was also appropriate for nutritionally enhanced crops, but that additional product-related studies could be considered on a case-by-case basis.

A major recommendation was that evaluations of nutritionally enhanced products should consider the exposure level of altered levels of metabolites and the potential impact on human and animal nutrition. This would include (1) targeted compositional analysis of the altered nutrient (or antinutrient) as well as an assessment of metabolites that may be impacted in or by the modified metabolic pathways, (2) feeding studies in animal models and targeted species, and (3) postmarket monitoring to verify target species impacts if such monitoring can be based on a valid testable hypothesis.

A particularly intriguing consideration proposed by the task force was that impacts on human and animal nutrition should be evaluated regardless of the technology used (e.g., GM, chemical or radiation mutagenesis, or conventional breeding). ILSI also recommended that the evaluation of the possible consequences of the adoption of improved nutrition crops should assess not only potential risks but also the benefits in alleviating nutritional deficiencies. The potential for more foods with higher nutritional functionality to alleviate hunger/malnutrition has spurred great efforts utilizing both conventional and GM technologies. Conventional breeding is an attractive option when genetic variation in nutrient levels is sufficient to allow exploitation of high natural levels of a nutrient found either in the edible portion of the crop of choice, or a closely related, sexually compatible species, and when genetic heritability is consistent. Crop varieties developed through conventional breeding thus far have not been required to clear the same regulatory obstacles encountered by varieties containing GM traits, which therefore allows faster time to market. Conventional breeding is a more realistic approach for many of the populations targeted by the nutritionally enhanced varieties, as it is generally thought to be more cost-effective, easily applicable, and affordable (Cakmak, 2008). The HarvestPlus project, instituted through the Consultative Group on International Agricultural Research (CGIAR), was developed to breed higher levels of essential nutrients into key staple foods. It focuses on enhancing levels of the three key micronutrients identified by the World Health Organization as most limiting in diets: iron, zinc, and vitamin A. This is to help eliminate micronutrient deficiency in humans, known as "hidden hunger," which affects more than half of the human population (Pfeiffer and McClafferty, 2007). CGIAR holds over 530,000 accessions available for screening for high endogenous levels of the key micronutrients. Utilizing core collections of crops that represented gene pool variation for the target micronutrients, seven fortified crops were targeted for release by 2013: beans (iron), cassava (provitamin A), maize (provitamin A), pearl millet (iron), rice (zinc), sweet potato (provitamin A), and wheat (zinc). Although the products are targeted to specific countries, more than 25 countries primarily in Asia and Africa are expected to benefit from spillover effects following the initial release (http://www.harvestplus.org). The biofortified crops and target levels of nutrients can be found in Table 16.3.

Table 16.3 Micronutrient-Fortified Crops Developed by HarvestPlus (http://www.harvestplus.org)

Crop	Micronutrient	Target Release	Average Baseline Content (μg/g)	Nutrient Target Level (μg/g)	Targeted Countries
Beans	Iron	2010	50	94	D.R. Congo and Rwanda
Cassava	Provitamin A	2011–2012	0.5	15.5	D.R. Congo and Nigeria
Maize	Provitamin A	2011–2012	0.5	15	Zambia
Pearl millet	Iron	2010	47	77	India
Rice	Zinc	2012–2013	16.6	24	Bangladesh and India
Sweet potato	Provitamin A	2007	2	32	Uganda and Mozambique
Wheat	Zinc	2012	25	33	India and Pakistan

Zinc is one micronutrient targeted through the HarvestPlus project. Over one-third of the world's population suffers from zinc deficiency (Hotz, 2007), which is the fifth leading cause of illness and disease in low-income countries, and 11th overall in the world. CGIAR accessions show a fourfold variation in rice grain zinc levels (14–58 mg/kg), showing significant potential to breed for increased levels (White and Broadley, 2005). However, the maximum level HarvestPlus hopes to achieve is 24 μg/g (approx. 1.5 × average baseline content, Table 16.3), possibly because of the poor (zinc-deficient) soil the crops will be cultivated in. Additional gains could be achieved through GM of rice. Transforming rice with the soybean ferritin gene (there appears to be cross-talk between iron and zinc transport pathways) (Hirschi, 2009) increased levels of zinc from 33.6 μg/g DW in the controls to over 55.5 μg/g in one transgenic line (Vasconcelos et al., 2003), which is at the high end of (but still within) the natural variation found in the CGIAR accessions.

Another example of a nutritionally enhanced crop developed using GM, rice with enriched β-carotene content, is known as Golden Rice, and was developed to combat vitamin A deficiency. Vitamin A deficiency is a major problem in developing countries, many of which depend on rice as their primary food source. β-Carotene, a carotenoid, is a precursor to vitamin A, and gives the rice endosperm its yellow color. The original Golden Rice contained approximately 1.6 μg/g DW of total carotenoids, and different lines, all containing the transgenes, presented different carotenoid profiles (Ye et al., 2000). Golden Rice 2 was created with the intention of increasing carotenoid content to meet the US-recommended daily allowance while factoring in bioavailability/bioconversion considerations. The carotenoid content increased up to 23-fold (maximum 37 μg/g DW) compared to the original Golden Rice, and β-carotene was more preferentially accumulated (Paine et al., 2005). This value is much lower than levels in commonly consumed foods such as carrots (raw: 83 μg/g fresh weight (FW), sweet potatoes (baked: 115 μg/g FW), and spinach (frozen: 72.4 μg/g FW), and expected consumption levels of the enhanced rice and resulting β-carotene consumption are orders of magnitude lower than dietary supplement levels and are not of toxicological concern (ILSI, 2007). Still, recommendations from the ILSI task force were to undertake a safety assessment (including composition analysis and carotenoid metabolite pool analysis), premarket testing that includes palatability of the rice, and risk versus benefit analysis.

16.6 CONCLUDING REMARKS

The acquisition of extensive compositional data generated during safety assessments has corroborated the need to utilize natural variability to provide a context to the evaluation of pair-wise comparisons. It has further offered a greater appreciation of the extent and sources of natural variation, and the potential to benefit from natural variation to breed for higher levels of beneficial nutrients. The current evidence supports the conclusion that incorporation of biotechnology-derived agronomic traits has little impact on natural variation in crop composition and that most compositional variation is attributable to genetic background, growing region, and agronomic practices.

This observation is also relevant to appropriate experimental design for compositional or metabolomic studies assessing differences between different but closely related crop varieties. It is also an important consideration when addressing safety evaluations for crops designed to have altered levels of metabolites.

REFERENCES

Ainasoja, M. M., L. L. Pohjala, P. S. Tammela, P. J. Somervuo, P. M. Vuorela, and T. H. Teeri. 2008. Comparison of transgenic *Gerbera hybrida* lines and traditional varieties shows no differences in cytotoxicity or metabolic fingerprints. *Transgenic Res*17:793–803.

Baker, J. M., N. D. Hawkins, J. L. Ward et al. 2006. A metabolomic study of substantial equivalence of field-grown genetically modified wheat. *Plant Biotechnol J* 4:381–392.

Berberich, S. A., J. E. Ream, T. L. Jackson et al. 1996. The composition of insect-protected cottonseed is equivalent to that of conventional cottonseed. *J Agric Food Chem* 44:365–371.

Berman, K. H., G. G. Harrigan, C. Hanson et al. 2010. Composition of forage and seed from MON 87701 and MON 89788 grown in Brazil is equivalent to that of conventional soybean. *J Agric Food Chem* 58:6270–6276.

Berman, K. H., G. G. Harrigan, M. Smith et al. 2009. Composition of seed, forage and processed fractions from insect-protected soybean MON 87701 is equivalent to that of conventional soybean. *J Agric Food Chem* 57:11360–11369.

Cakmak, I. 2008. Enrichment of cereal grains with zinc: Agronomic or genetic biofortification? *Plant Soil* 302:1–17.

Carrao-Panizzi, M. C., M. Berhow, J. M. G. Mandarino, and M. C. N. De Oliveira. 2009. Environmental and genetic variation of isoflavone content of soybean seeds grown in Brazil. *Pesqui Agropecu Bras* 44:1444–1451.

Catchpole, G. S., M. Beckmann, D. P. Enot et al. 2005. Hierarchical metabolomics demonstrates substantial compositional similarity between genetically modified and conventional potato crops. *Proc Natl Acad Sci USA* 102:14458–14462.

Chassy, B., J. J. Hlywka, G. A. Kleter et al. 2004. Nutritional and safety assessments of foods and feeds nutritionally improved through biotechnology: An executive summary. *Comp Rev Food Sci Food Saf* 3:38–104.

Cheng, K. C., J. Beaulieu, E. Iquira, F. J. Belzile, M. G. Fortin, and M. V. Stromvik. 2008. Effect of transgenes on global gene expression in soybean is within the natural range of variation of conventional cultivars. *J Agric Food Chem* 56:3057–3067.

Codex. 2003. Guideline for the conduct of food safety assessment of foods derived from recombinant DNA plants. CAC/GL 45-2003.

Defernez, M., Y. M. Gunning, A. J. Parr, L. V. Shepherd, H. V. Davies, and I. J. Colquhoun. 2004. NMR and HPLC-UV profiling of potatoes with genetic modifications to metabolic pathways. *J Agric Food Chem* 52:6075–6085.

Delaney, B. 2007. Strategies to evaluate the safety of bioengineered foods. *Int J Toxicol* 26:389–399.

Drury, S. M., T. L. Reynolds, W. P. Ridley et al. 2008. Composition of forage and grain from second-generation insect-protected corn MON 89034 is equivalent to that of conventional corn (*Zea mays* L.). *J Agric Food Chem* 56:4623–4630.

Dubouzet, J. G., A. Ishihara, F. Matsuda, H. Miyagawa, H. Iwata, and K. Wakasa. 2007. Integrated metabolomic and transcriptomic analyses of high-tryptophan rice expressing a mutant anthranilate synthase alpha subunit. *J Exp Bot* 58:3309–3321.

Gregersen, P. L., H. Brinch-Pedersen, and P. B. Holm. 2005. A microarray-based comparative analysis of gene expression profiles during grain development in transgenic and wild type wheat. *Transgenic Res* 14:887–905.

Gutierrez-Gonzalez, J. J., X. L. Wu, J. Zhang et al. 2009. Genetic control of soybean seed isoflavone content: Importance of statistical model and epistasis in complex traits. *Theor Appl Genet* 119:1069–1083.

Hamilton, K. A., P. D. Pyla, M. Breeze et al. 2004. Bollgard II cotton: Compositional analysis and feeding studies of cottonseed from insect-protected cotton (*Gossypium hirsutum* L.) producing the Cry1Ac and Cry2Ab2 proteins. *J Agric Food Chem* 52:6969–6976.

Harrigan, G. G., W. P. Ridley, K. D. Miller et al. 2009. The forage and grain of MON87460, a drought tolerant corn hybrid, are compositionally equivalent to that of conventional corn. *J Agric Food Chem* 57:9754–9763.

Harrigan, G. G., W. P. Ridley, S. G. Riordan et al. 2007. Chemical composition of glyphosate-tolerant soybean 40-3-2 grown in Europe remains equivalent with that of conventional soybean (*Glycine max* L.). *J Agric Food Chem* 55:6160–6168.

Hirschi, K. D. 2009. Nutrient biofortification of food crops. *Annu Rev Nutr* 29:401–421.

Hothorn, L. A. and R. Oberdoerfer. 2006. Statistical analysis used in the nutritional assessment of novel food using the proof of safety. *Regul Toxicol Pharmacol* 44:125–135.

Hotz, C. 2007. Dietary indicators for assessing the adequacy of population zinc intakes. *Food Nutr Bull* 28:S430–S453.

ILSI. 2007. Nutritional and safety assessments of foods and feeds nutritionally improved through biotechnology: Case studies: Executive summary of a task force report by the International Life Sciences Institute, Washington, D.C. *J Food Sci* 72:R131–R137.

Ioset, J. R., E. Urbaniak, B. Ndjoko-Ioset et al. 2007. Flavonoid profiling among wild type and related GM wheat varieties. *Plant Mol Biol* 65:645–654.

Kok, E. J., J. Keijer, G. A. Kleter, and H. A. Kuiper. 2008. Comparative safety assessment of plant-derived foods. *Regul Toxicol Pharmacol* 50:98–113.

Kristensen, C., M. Morant, C. E. Olsen et al. 2005. Metabolic engineering of dhurrin in transgenic Arabidopsis plants with marginal inadvertent effects on the metabolome and transcriptome. *Proc Natl Acad Sci USA* 102:1779–1784.

Lay, J. O., S. Borgmann, R. Liyanage, and C. L. Wilkins. 2006. Problems with the "omics." *Trac-Trends Anal Chem* 25:1046–1056.

Le Gall, G., I. J. Colquhoun, A. L. Davis, G. J. Collins, and M. E. Verhoeyen. 2003. Metabolite profiling of tomato (*Lycopersicon esculentum*) using 1H NMR spectroscopy as a tool to detect potential unintended effects following a genetic modification. *J Agric Food Chem* 51:2447–2456.

Lehesranta, S. J., H. V. Davies, L. V. Shepherd et al. 2005. Comparison of tuber proteomes of potato varieties, landraces, and genetically modified lines. *Plant Physiol* 138:1690–1699.

Lundry, D. R., W. P. Ridley, J. J. Meyer et al. 2008. Composition of grain, forage, and processed fractions from second-generation glyphosate-tolerant soybean, MON 89788, is equivalent to that of conventional soybean (*Glycine max* L.). *J Agric Food Chem* 56:4611–4622.

McCann, M. C., G. J. Rogan, S. Fitzpatrick et al. 2006. Glyphosate-tolerant alfalfa is compositionally equivalent to conventional alfalfa (*Medicago sativa* L.). *J Agric Food Chem* 54:7187–7192.

McCann, M. C., K. Liu, W. A. Trujillo, and R. C. Dobert. 2005. Glyphosate-tolerant soybeans remain compositionally equivalent to conventional soybeans (*Glycine max* L.) during three years of field testing. *J Agric Food Chem* 53:5331–5335.

McCann, M. C., W. A. Trujillo, S. G. Riordan, R. Sorbet, N. N. Bogdanova, and R. S. Sidhu. 2007. Comparison of the forage and grain composition from insect protected and glyphosate-tolerant MON 88017 corn to conventional corn (*Zea mays* L.). *J Agric Food Chem* 55:4034–4042.

Nida, D. L., S. Patzer, P. Harvey, R. Stipanovic, R. Wood, and R. L. Fuchs. 1996. Glyphosate-tolerant cotton: The composition of the cottonseed is equivalent to that of conventional cottonseed. *J Agric Food Chem* 44:1967–1974.

Oberdoerfer, R. B., R. D. Shillito, M. De Beuckeleer, and D. H. Mitten. 2005. Rice (*Oryza sativa* L.) containing the bar gene is compositionally equivalent to the nontransgenic counterpart. *J Agric Food Chem* 53:1457–1465.

Obert, J. C., W. P. Ridley, R. W. Schneider et al. 2004. The composition of grain and forage from glyphosate tolerant wheat MON 71800 is equivalent to that of conventional wheat (*Triticum aestivum* L.). *J Agric Food Chem* 52:375–384.

OECD. 1998. *Report of the OECD Workshop on the Toxicological and Nutritional Testing of Novel Foods*. Paris, France: Organisation for Economic Co-operation and Development.

OECD. 2002. *Consensus Document on Compositional Considerations for New Varieties of Maize (Zea Mays): Key Food and Feed Nutrients, Anti-Nutrients and Secondary Plant Metabolites*. Paris, France: Organization for Economic Cooperation and Development.

OECD. 2006. *An Introduction to the Food/Feed Safety Consensus Documents of the Task Force*. Paris, France: Organization for Economic Cooperation and Development.

Padgette, S. R., N. B. Taylor, D. L. Nida et al. 1996. The composition of glyphosate-tolerant soybean seeds is equivalent to that of conventional soybeans. *J Nutr* 126:702–716.

Paine, J. A., C. A. Shipton, S. Chaggar et al. 2005. Improving the nutritional value of Golden Rice through increased pro-vitamin A content. *Nat Biotech* 23:482–487.

Pfeiffer, W. H. and B. Mcclafferty. 2007. HarvestPlus: Breeding crops for better nutrition. *Crop Sci* 47:S88–S105.

Ridley, W. P., R. S. Sidhu, P. D. Pyla, M. A. Nemeth, M. L. Breeze, and J. D. Astwood. 2002. Comparison of the nutritional profile of glyphosate-tolerant corn event NK603 with that of conventional corn (*Zea mays* L.). *J Agric Food Chem* 50:7235–7243.

Rogan, G. J., J. T. Bookout, D. R. Duncan et al. 2000. Compositional analysis of tubers from insect and virus resistant potato plants. *J Agric Food Chem* 48:5936–5945.

Shepherd, L. V., J. W. McNicol, R. Razzo, M. A. Taylor, and H. V. Davies. 2006. Assessing the potential for unintended effects in genetically modified potatoes perturbed in metabolic and developmental processes. Targeted analysis of key nutrients and anti-nutrients. *Transgenic Res* 15:409–425.

Sidhu, R. S., B. G. Hammond, R. L. Fuchs et al. 2000. Glyphosate-tolerant corn: The composition and feeding value of grain from glyphosate-tolerant corn is equivalent to that of conventional corn (*Zea mays* L.). *J Agric Food Chem* 48:2305–2312.

Taylor, N. B., R. L. Fuchs, J. Macdonald, A. R. Shariff, and S. R. Padgette. 1999. Compositional analysis of glyphosate-tolerant soybeans treated with glyphosate. *J Agric Food Chem* 47:4469–4473.

Vasconcelos, M., K. Datta, N. Oliva et al. 2003. Enhanced iron and zinc accumulation in transgenic rice with the ferritin gene. *Plant Sci* 164:371–378.

Wakasa, K., H. Hasegawa, H. Nemoto et al. 2006. High-level tryptophan accumulation in seeds of transgenic rice and its limited effects on agronomic traits and seed metabolite profile. *J Exp Bot* 57:3069–3078.

White, P. J. and M. R. Broadley. 2005. Biofortifying crops with essential mineral elements. *Trends Plant Sci* 10:586–593.

Ye, X. D., S. Al-Babili, A. Kloti et al. 2000. Engineering the provitamin A (beta-carotene) biosynthetic pathway into (carotenoid-free) rice endosperm. *Science* 287:303–305.

CHAPTER 17

Gene-Expression Analysis of Cell-Cycle Regulation Genes in Virus-Infected Rice Leaves

Shoshi Kikuchi and Kouji Satoh

CONTENTS

17.1 INTRODUCTION

With the availability of the complete rice genome sequence and the massive collection of full-length complementary DNA clones, many kinds of tools for transcriptome (RNA expression) analysis have been established for functional genomics. These tools include oligomicroarrays that cover nearly all expressed genes, genome tiling arrays, the serial analysis of gene expression and massively parallel signature sequencing systems and next-generation sequencing systems. Transcriptome data on responses to various environmental stress treatments and on tissue- and stage-specific expression have been accumulated. Many kinds of transcriptome data on responses to infection of rice by bacteria or fungi have also been collected, but the data on infection by rice viruses are very limited.

In this chapter, we describe our findings on the effects of virus infection on the expression of cell-cycle genes in rice. We first review the progress of research on eight viruses that infect rice in east Asia from the perspective of how each virus interacts with the host system. Then, by focusing on genes related to cell-cycle regulation, we describe our data on gene expression following infection by each of the eight viruses. These data are compared and contrasted with data from drought-stressed plants.

17.2 CURRENT BIOLOGY OF THE EIGHT RICE VIRUSES USED IN THIS STUDY

Fifteen plant viruses are known to occur in rice: rice black-streaked dwarf virus (RBSDV), rice bunchy stunt virus (RBSV), rice dwarf virus (RDV), rice gall dwarf virus (RGDV), rice giallume virus (RGV), rice grassy stunt virus (RGSV), rice hoja blanca virus (RHBV), rice necrosis mosaic virus (RNMV), rice ragged stunt virus (RRSV), rice stripe necrosis virus (RSNV), rice stripe virus (RSV), rice transitory yellowing virus (RTYV), rice tungro bacilliform virus (RTBV), rice tungro spherical virus (RTSV) and rice yellow mottle virus (RYMV) (Hibino, 1996). Of these viruses, 11 (RBSDV, RBSV, RDV, RGDV, RGSV, RNMV, RRSV, RSV, RTYV, RTBV, RTSV) occur in Asia, two (RSNV and RYMV) in Africa, one (RGV) in Europe and one (RHBV) in North and South America. In this chapter, we have focused on eight viruses occurring mainly in East Asia: RBSDV, RDV, RGSV, RRSV, RSV, RTYV, RTSV and RTBV. Characteristics of each of these eight viruses are summarized in Table 17.1 and detailed below.

17.2.1 Rice Black-Streaked Dwarf Virus

RBSDV, a member of the fijivirus group of the family *Reoviridae*, is the causal agent of RBSK and maize rough dwarf diseases, both of which are responsible for severe yield losses in some countries (Bai et al., 2002; Fang et al., 2001; Shikata and Kitagawa, 1977). RBSDV virions consist of a capsid, a core and a nucleoprotein complex. The virus capsid is not enveloped. The capsid is isometric with icosahedral symmetry and has a diameter of 75–80 nm (in cells of infected plants and insects) or 70 nm (intact virions). The capsid shells are composed of two layers. The outer capsid is smooth, showing A-type spikes of about 11 nm length and breadth (ICTVdB Management, 2006). The genome (23,898 nucleotides (nt)) consists of 10 segments (S1–S10) of linear double-stranded (ds) RNA. S1 is fully sequenced and is 3998 nt long (ICTVdB Management, 2006; Wang et al., 2003; Zhang et al., 2001). Sequence analysis suggests that S1 encodes a putative RNA-dependent RNA polymerase (168.8 kD) with similarities to S1 of *Nilaparvata lugens* reovirus (NLRV), whereas S2 and S4 are both monocistronic and probably encode the core protein and outer-shell B-spike protein, respectively (Zhang et al., 2001). The S3-encoded protein has been identified as a possible guanylyltransferase, with similarities to VP3 of mycoreovirus 1 (Supyani et al., 2007), and the S6-encoded protein functions as a viral RNA-silencing suppressor (Zhang et al., 2005). S8 and S10 encode a core capsid and a major outer-capsid protein, respectively, whereas S7 and S9 encode nonstructural proteins (Isogai et al., 1998).

Table 17.1 Viruses Used in This Transcriptome Analysis, Structure of Genome, Symptoms of Infected Plants and Localization of Viruses

Virus	Genome	Symptom[a]	Localization of Virus	
RDV	Phytoreovirus Reoviridae	Segmented and consists of 12 segments of linear, dsRNA	Pronounced stunting, increased tillering and short leaves that are darker green in colour and with fine chlorotic specks	Vascular bundles and in parenchymatous cells in the portions of leaves that correspond to the white specks
RBSDV	Fijivirus Reoviridae	Segmented and consists of 10 segments of linear dsRNA	Pronounced stunting, darkening of leaves, twisting of leaf tips, splitting of the leaf margin, waxy white-to-black galls along the veins on the underside of leaf blades and the outer surface of sheaths and columns. The galls result from hyperplasia and hypertrophy of the phloem tissues	Phloem and gall tissues
RRSV	Oryzavirus Reoviridae	Segmented and consists of 10 segments of linear dsRNA	Stunting, abnormal leaves with serrated edges or twisted tips, and vein swelling or galls on the underside of leaf blades and outer surface of the leaf sheaths. The gall results from hyperplasia and hypertrophy of the phloem tissue. Plants infected with RRSV at the seedling stage develop new leaves with identifiable symptoms 2 weeks after inoculation and thereafter develop leaves showing milder or no definite symptoms. At the heading stage, the plants again show symptoms on upper leaves and flag leaves	Phloem and gall tissues
RSV	Tenuivirus	Segmented and consists of four segments of linear, negative-sense and ambisense, single-stranded RNA	Chlorotic stripes or mottling and necrotic streaks on leaves, and premature wilting. Chlorotic leaves are unfolded, and later droop and wilt	
RGSV	Tenuivirus	Segmented and consists of four segments of linear negative-sense and ambisense, single-stranded RNA	Pronounced stunting proliferation of short, erect and narrow leaves that are pale-green or pale-yellow in colour. Infected leaves may show mottling symptoms	

continued

Table 17.1 (continued) Viruses Used in This Transcriptome Analysis, Structure of Genome, Symptoms of Infected Plants and Localization of Viruses

Virus	Genome	Symptom[a]	Localization of Virus	
RTYV	Nucleorhabdovirus plant rhabdovirus	Not segmented and contains a single molecule of linear single-stranded RNA	Leaf yellowing, reduced tillering and mild stunting. Later, infected plants develop normal looking leaves and may appear healthy, but symptoms may reappear after this temporary recovery	Phloem tissues
RTBV	Badnavirus	A circular dsDNA of 8 kbp that encodes four proteins	Tungro symptoms, including stunting, yellow or yellow-orange discolouration and reduced tillering (61). Discoloured leaves may show irregularly shaped dark-brown blotches. The leaves, especially the younger ones, may show striping or mottling and intervernal chlorosis. RTBV alone show similar but milder tungro symptoms	Vascular bundles
RTSV	Ribotungrovirus	A single-stranded RNA genome of 12 kb that encodes a large polyprotein and possibly one or two smaller proteins	RTSV alone show no obvious symptoms except for very mild stunting	Phloem tissues

Source: Data from Hibino, H. 1996. *Annu Rev Phytopathol* 34:249–274.

[a] Symptoms induced by the infection of viruses and localization of the viruses follow the description by Hibino (1996).

S9 is ~1900 nt long and contains two large, nonoverlapping open-reading frames (ORFs; S9-1 and S9-2), which encode a 39.9-kD (P9-1) and a 24.2-kD (P9-2) polypeptide, respectively. No sequence similarity has been found between P9-1 and P9-2. Both ORF1 and ORF2 are conserved in RBSDV S9, maize rough dwarf virus S8 and NLRV S9, indicating that proteins encoded by RBSDV S9 may play a role in the virus life cycle. P9-2 is not detected in RBSDV-infected plants and insects, whereas P9-1 accumulates in intracellular viroplasms (viral inclusion bodies), suggesting that P9-1 plays an important role in viroplasm formation and viral morphogenesis (Isogai et al., 1998). P9-1 is a thermostable, helical protein with an intrinsic ability to self-interact and form homodimers *in vitro* and *in vivo* (Zhang et al., 2008). When P9-1 was transiently expressed in *Arabidopsis* protoplasts, it formed large, discrete viroplasm-like structures in the absence of infection or other RBSDV proteins. P9-1 is suggested to be the minimal viral component required for viroplasm formation, and it may play an important role in the early stages of the virus life cycle by forming intracellular viroplasms that serve as the sites of virus replication and assembly (Zhang et al., 2008).

RBSDV-infected rice plants show pronounced stunting, darkening of leaves, twisting of leaf tips, splitting of the leaf margin and waxy white-to-black galls along the veins on the underside of leaf blades and the outer surface of leaf sheaths and culms (Figure 17.1). The galls result from hyperplasia and hypertrophy of the phloem tissues. RBSDV is localized in the phloem and gall

Figure 17.1 Symptoms induced by infection with various rice viruses. (Top) Symptoms induced RBSDV, RDV and RGSV. For transcriptome analysis of RDV infection, most data were obtained from the severest strain of RDV_S infection. (Bottom) Symptoms induced by RRSV, RSV and RTD (caused by infection with RTSV and RTBV). (Adapted from Satoh, K. et al. 2010. *J Gen Virol* 91:294–305.)

tissues. It is transmitted in a persistent manner by the planthopper *Laodelphax striatellus* and two other planthopper species (Hibino, 1989; Shikata, 1974). RBSDV naturally infects rice, maize, wheat and barley, as well as *Alopecurus aequalis* and some other weeds.

17.2.2 Rice Dwarf Virus

RDV is a member of the phytoreovirus group of the family *Reoviridae*. RDV virions consist of a nonenveloped capsid, a core and a nucleoprotein complex. The capsid/nucleocapsid is isometric with icosahedral symmetry and has a diameter of 70 nm. The capsid shells are composed of two layers (Hagiwara et al., 2004; ICTVdB Management, 2006; Miyazaki et al., 2005; Nakagawa et al., 2003). The genome (25,130 nt) consists of 12 segments of linear dsRNA and has been fully sequenced. Segments S1, S2, S3, S5, S7, S8 and S9 encode structural proteins P1, P2, P3, P5, P7, P8 and P9, respectively (Boccardo and Milne, 1984; Hagiwara et al., 2003; ICTVdB Management, 2006; Ueda et al., 1997; Zhong et al., 2003; Zhou et al., 2007; Zhu et al., 1997b, 2005). P2 protein interacts with *ent*-kaurene oxidases, which play a key role in the biosynthesis of plant growth hormones (gibberellins), in infected plants (Zhu et al., 2005). P2 and P8 are components of the virion outer shell. Coexpression of P3 and P8 in transgenic rice plants or insect cells results in the formation of double-shelled virus-like particles (Hagiwara et al., 2003; Zheng et al., 2000).

The five nonstructural proteins of RDV—Pns4, Pns6, Pns10, Pns11 and Pns12—are encoded by segments S4, S6, S10, S11 and S12, respectively (Suzuki et al., 1996; Xu et al., 1998; Li et al., 2004; Cao et al., 2005). Pns4 is a phosphoprotein that is localized around the viroplasm matrix and forms minitubules (Wei et al., 2006b). Pns6 was identified as a viral movement protein (Li et al., 2004; Xu et al., 2009), and Pns10 as an RNA-silencing suppressor (Cao et al., 2005). Pns10 forms viral particle–containing tubules that move along actin-based filopodia into adjacent insect vector cells in monolayers. These tubular structures extend into the medium and are surrounded by plasma membrane (Wei et al., 2006a). The observed formation of tubular structures in nonhost Sf9 (*Spodoptera frugiperda*) tissue culture cells indicates that Pns10 has the intrinsic ability to form tubules without a requirement for other viral proteins (Wei et al., 2006a). Pns11 is a nucleic acid-binding protein (Xu et al., 1998), and Pns12 is essential for the formation of viroplasms and for nucleation of viral–assembly complexes (Wei et al., 2006c).

RDV-infected rice plants show pronounced stunting, increased tillering and short leaves that are dark green with fine chlorotic specks (Figure 17.1). RDV is distributed in vascular bundles and in parenchymatous cells in the portions of leaves that correspond to the white specks. The cytoplasm of infected cells contains large round or oval inclusion bodies consisting of a viroplasmic matrix and numerous virus particles. Tubules enclosing virus particles and paracrystalline bodies occur in or around the inclusion bodies. Starch accumulates in infected rice tissues. RDV is transmitted in a persistent manner by leafhoppers, including *Nephotettix cincticeps*, *N. nigropictus*, *Recilia dorsalis* and some other *Nephotettix* spp. (Hibino, 1989; Iida et al., 1972). RDV naturally infects rice and a few grass weeds.

Gene-expression analysis in rice plants infected with RDV revealed significant decreases in the expression of genes involved in cell wall formation, as reflected in the stunted growth of diseased plants. The expression of plastid-related genes was also suppressed, as reflected by the white chlorotic appearance of infected leaves. In contrast, the expression of defence- and stress-related genes was enhanced after viral infection (Shimizu et al., 2007). Introduction of Pns12- and Pns4-specific RNA interference (RNAi) constructs into rice caused the transgenic plants to accumulate short interfering RNAs specific to the constructs. The progeny of rice plants transformed with Pns12-specific RNAi constructs were strongly resistant to viral infection. In contrast, resistance was less apparent in rice plants transformed with Pns4-specific RNAi constructs, and delayed symptoms appeared in some plants from each line (Shimizu et al., 2009).

To identify host factors involved in RDV infection, a set of rice transposon *Tos17* insertion mutant lines was screened for reduced response to infection, and a mutant (*rim1-1*) was identified. Inoculated *rim1-1* mutant plants did not have typical disease symptoms and showed reduced accumulation of RDV capsid proteins (Yoshii et al., 2009). RIM1 functions as a transcriptional regulator of jasmonic acid (JA) signalling and is degraded in response to JA treatment via a 26S-proteasome-dependent pathway. Plants carrying *rim1* mutations show root growth inhibition. The expression profiles of the mutants were similar to those of JA-treated wild-type plants (which do not accumulate endogenous JA), indicating RIM1 functions as a component of JA signalling (Yoshii et al., 2010).

17.2.3 Rice Grassy Stunt Virus

RGSV is a member of the tenuivirus group (Hibino, 1986; Mayo et al., 2000; Toriyama, 1995). Virions consist of a nucleocapsid that is filamentous, flexuous or circular, with segments having a length proportional to the size of their RNA and a width of 6 nm. The nonenveloped capsid/nucleocapsid is elongated with helical symmetry. The genome (15,460 nt) consists of four segments of linear negative-sense and ambisense single-stranded RNA. RNA 1 is partially sequenced. RNA 2,

RNA 3 and RNA 4 have been sequenced and have lengths of ~3940, 3640 and 3490 nt, respectively (ICTVdB Management, 2006; Miranda et al., 2000; Toriyama et al., 1997, 1998). Little is known about the functions of the individual RGSV proteins, except for the 339-kD RNA-dependent RNA polymerase encoded on the complementary strand of RNA 1 (cRNA 1) and the 36-kD nucleocapsid protein encoded on cRNA 5, both of which are found along with genomic and possibly cRNAs as thin filamentous ribonucleoprotein particles. The 22-kD p5 protein encoded on the virus genomic strand of RNA 5 (virus RNA (vRNA 5)) accumulates in large amounts in both RGSV-infected rice leaves and viruliferous brown planthoppers (Chomchan et al., 2002). The p5 protein interacts with itself to form oligomeric complexes *in vitro* and *in vivo* (Chomchan et al., 2003). The 23-kD p2 protein encoded on vRNA 2 and the 21-kD p6 protein encoded on vRNA 6 are present at higher levels in infected rice leaf tissues than in viruliferous insects (Chomchan et al., 2002).

RGSV-infected rice plants show pronounced stunting and proliferation of short, erect, narrow, pale-green or pale-yellow leaves (Figure 17.1). Infected leaves may show mottling symptoms. RGSV-infected plants produce a virus-specific protein that is serologically related to a similar protein produced in RSV-infected plants. RGSV-infected rice cells contain masses of fibrils in the nucleus and cytoplasm, and membrane-bound bodies with fibrils in the cytoplasm. Tubules associated with isometric particles, 18–25 nm in diameter, can be seen in the sieve tubes. RGSV is transmitted in a persistent manner by the brown planthopper *N. lugens*, which is monophagous to rice, and by two other *Nilaparvata* spp. (Hibino 1986, 1989). RGSV naturally infects only rice.

17.2.4 Rice Ragged Stunt Virus

RRSV is a member of the oryzavirus group of the family *Reoviridae* (Milne and Ling, 1982; Uyeda et al., 1995). RRSV virions consist of a nonenveloped capsid, a core and a nucleoprotein complex. The capsid/nucleocapsid is round and exhibits icosahedral symmetry. The capsid shells of virions are composed of a single inner capsid layer (ICTVdB Management, 2006). The genome (26,660 nt) consists of 10 segments of linear dsRNA (S1–S10) and has been sequenced (Kawano, 1984; ICTVdB Management, 2006). The complete nucleotide sequences of all genomic segments of the RRSV Thailand isolate have been determined (ICTVdB Management, 2006). The sequences of S9 and S10 of isolates from the Philippines and India have also been determined by several laboratories (Li et al., 1996; Lu et al., 1997; Upadhyaya et al., 1995, 1996, 1997, 1998; Uyeda et al., 1995; Yan et al., 1995). All 10 genomic segments possess identical conserved nucleotide sequences of GAUAAA at the 5′-terminus and GUGC at the 3′-terminus of each segment (Yan et al., 1992).

The virion is composed of six major, highly immunoreactive structural proteins with estimated molecular weights of 31, 39, 43, 70, 90 and 120 kD, respectively, and several minor structural proteins (Lu et al., 1990). Three proteins of 33, 63 and 88 kD have already been identified by *in vitro* translation of RRSV genomic segments and immunoprecipitation, and they have been identified as non-structural (Lu et al., 1987). The S6 product, Pns6, is a 71 kD nonstructural protein with nucleic acid-binding activity. The protein interacts with both single and ds forms of RNA and DNA in a sequence-independent manner, although it interacts preferentially with single-stranded nucleic acids. The nucleic acid-binding domain is located in the basic region of Pns6, from aa 201 to 273. The Pns6 protein may play an important role in virus replication and assembly (Shao et al., 2004). The products of S7 and S10 are nonstructural proteins with molecular weights of 68 and 33 kD, respectively (Upadhyaya et al., 1997). S5 encodes a 91-kD structural protein (Li et al., 1996). S8 encodes a 67-kD structural protein that has both self-aggregation and self-cleavage abilities. One of the cleavage products, a 43-kD protein, may take part in virus assembly; the other product, a 26-kD protein, may act as a self-cleavage proteinase (Lu et al., 2002; Upadhyaya et al., 1996). P9 is a 39-kD viral spike protein that plays an important role in virus transmission by the insect vector (Zhou et al., 1999).

RRSV-infected rice plants show stunting, abnormal leaves with serrated edges or twisted tips, and vein swelling or galls on the undersides of leaf blades and outer surfaces of the leaf sheaths

(Figure 17.1). The galls result from hyperplasia and hypertrophy of the phloem tissue. Plants infected with RRSV at the seedling stage develop new leaves with identifiable symptoms 2 weeks after inoculation and thereafter develop leaves showing milder symptoms or none at all. At the heading stage, the plants again show symptoms on upper leaves and flag leaves. In infected plants, RRSV is localized in the phloem and gall tissues. Infected cells contain large inclusion bodies consisting of a viroplasmic matrix and numerous virus particles. RRSV is transmitted in a persistent manner by the brown planthopper, *N. lugens*, and other *Nilaparvata* spp. (Hibino, 1989; Milne and Ling, 1982). In inoculation tests using *N. lugens*, RRSV infects many grasses. However, natural infection of weeds and cereals other than rice is rare or nonexistent, as *N. lugens* is monophagous to rice (Hibino, 1979; Milne and Ling, 1982).

17.2.5 Rice Stripe Virus

RSV is a member of the tenuivirus group (Toriyama, 1983, 1995). Virions consist of a nucleocapsid, which is not enveloped, and the capsid/nucleocapsid is elongated with helical symmetry. The nucleocapsid is filamentous, flexuous or spiral, or branched, circular or tightly coiled, and segments have a length proportional to the size of their RNA and a width of 8 nm (ICTVdB Management, 2006). The genome (9000 nt) consists of four segments of linear, single-stranded RNA, and has been fully sequenced (Falk and Tsai, 1998; ICTVdB Management, 2006; Ramírez and Haenni, 1994). RNA 1 is of negative sense and encodes a putative protein with a molecular weight of 337 kD, which is considered to be part of the RNA-dependent RNA polymerase associated with the RSV filamentous ribonucleoprotein (Barbier et al., 1992; Toriyama, 1986; Toriyama et al., 1994). RNA 2–4 are ambisense, each containing two ORFs, one in the 5′ half of the viral RNA (vRNA) and the other in the 5′ half of the viral cRNA (vcRNA). The vRNA strands of RNA 2, 3 and 4 encode proteins p2, p3 and p4, respectively; the vcRNA strands encode proteins pc2, pc3 and pc4, respectively. Pc2 has stretches of weak amino acid similarity with membrane protein precursors found in members of the *Bunyaviridae*. These precursors are processed into two membrane-spanning glycoproteins; however, there is no evidence that RSV forms enveloped particles that could incorporate such glycoproteins (Takahashi et al., 1993). P3 of RSV shares 46% identity with the gene-silencing suppressor NS3 protein of the tenuivirus RHBV (Bucher et al., 2003). Pc3 is the nucleocapsid protein (CP). P4 is the major nonstructural protein, the accumulation of which correlates with symptom development in infected plants (Espinoza et al., 1993; Kakutani et al., 1990, 1991; Zhu et al., 1991).

Pc4 shares some common structures with the viral 30k superfamily of viral movement proteins (Melcher, 2000). This protein is localized predominantly near or within the cell walls, can move from cell to cell, and is able to complement movement-defective potato virus X (Xiong et al., 2008). These findings suggest that Pc4 is a movement protein of RSV. However, as it is a negative-strand RNA virus that does not seem to form intact virions, it appears that cell-to-cell movement of RSV would be a very complex process. The yeast two-hybrid (Y2H) system was used to investigate all the potential interactions between RSV-encoded proteins and host factors; Pc4 was shown to interact with a DNAJ protein and an hsp20 family protein of rice (Lu et al., 2009).

RSV-infected rice plants show chlorotic stripes or mottling and necrotic streaks on leaves, and plants wilt prematurely (Figure 17.1). The chlorotic leaves are unfolded, and they later droop and wilt. Infected plants produce relatively high concentrations of virus-specific proteins that are serologically related to similar proteins produced in RGSV-infected plants. Infected rice cells contain large masses of granular or sandy structures in the cytoplasm and nuclei, and masses of needlelike and paracrystalline structures in the cytoplasm and vacuole. RSV is transmitted in a persistent manner by *Laodelphax striatellus* and some other planthoppers (Hibino, 1989; Toriyama, 1983). RSV naturally infects rice, wheat, barley, oat, foxtail millet and some grass weeds (Toriyama, 1983).

To reveal which rice cellular systems are influenced by RSV infection, temporal changes in the transcriptome of RSV-infected plants were monitored by a rice 44k oligoarray system (platform no.

GPL7252) available at NCBI GEO (Barrett et al., 2007). Transcription of genes encoding protein-synthesis machinery and energy production in the mitochondrion were activated by RSV infection, whereas transcription of genes related to energy production in the chloroplast and synthesis of cell-structure components was suppressed. The transcription of genes related to host–defence systems under hormone signals and gene silencing were not upregulated at the early infection phase. Together with concurrent observation of virus concentration and symptom development, these transcriptome changes in RSV-infected plants suggest that different sets of host genes are regulated depending on the development of disease symptoms and the accumulation of RSV (Satoh et al., 2010). The locations and identities of some genes associated with resistance to RSV have been determined by QTL analysis (Hayano-Saito et al., 2000; Maeda et al., 2006; Wu et al., 2009).

17.2.6 Rice Transitory Yellowing Virus

RTYV is a member of the nucleorhabdovirus subgroup of the plant rhabdovirus group (Shikata, 1972; Hiraguri et al., 2010). Virus particles are bullet shaped, 180–210 nm in length and 94 nm in width. The RTYV genome is 14,029 nt in length. The overall nucleotide identity between RTYV and rice yellow stunt virus (RYSV) is 98.5%, and the deduced amino acid sequence identities between the seven genes in RTYV and RYSV ranges from 82.3% to 99.7%. This sequence information indicates that these two viruses should be categorized as members of the same species (Hiraguri et al., 2010). RTYV has a nonsegmented, negative-sense, single-stranded RNA genome, which contains a 3′-leader (le), a 5′-trailer (tr) and seven genes in the order 3′-le-N-P-3-M-G-6-L-tr-5′ (Huang et al., 2003, 2005; Luo et al., 1998; Luo and Fang, 1998; Zhu et al., 1997a).

RTYV-infected rice plants initially show leaf yellowing, reduced tillering and mild stunting. They then develop normal-looking leaves and may appear healthy, but symptoms may reappear after this temporary recovery. RTYV is localized in the phloem tissues. A mass of RTYV particles appears between two nuclear membranes in infected cells. RTYV is transmitted in a persistent manner by *N. cincticeps*, *N. nigropictus* and *Nephotettix virescens* (Hibino, 1989; Shikata, 1972). RTYV naturally infects only rice.

17.2.7 Rice Tungro Disease

Tungro (Anjaneyulu et al., 1982; Hibino, 1995) is the most important virus disease of rice in the Asian tropics. It is a composite disease caused by RTBV and RTSV.

17.2.7.1 Rice Tungro Bacilliform Virus

RTBV is a member of the badnavirus group. Virions consist of a nonenveloped capsid which is bacilliform and exhibits icosahedral symmetry. The capsid shells are composed of multiple layers with a length of 110–400 nm and a width of 30–35 nm. The genome is not segmented and contains a single ds circular DNA molecule. The complete genome (8002 nt; Hay et al., 1991; ICTVdB Management, 2006) contains four ORFs (ORFs I–IV), of which only ORF III has been functionally characterized. ORF III encodes a polyprotein containing domains of several structural and nonstructural proteins, including the coat protein (CP), protease and reverse-transcriptase RNase H (Hay et al., 1991; Qu et al., 1991). The single 37-kD CP has been mapped to nt 2427–3372 within ORF III (Marmey et al., 1999). The C-terminal portion of the ORF III-encoded polyprotein shows reverse-transcriptase RNase H activity (Laco et al., 1995), a common feature of all pararetroviruses.

RTBV has a transcriptional promoter (Yin and Beachy, 1995; Yin et al., 1997a) and encodes the transcription factors that activate its expression. The ‘E’ fragment of the promoter (nt −164 to +45) is sufficient to confer high-level, tissue-specific gene expression. The *cis*-elements include a GATA motif (nt –143 to –135), an AS1-like box (nt –98 to –79), Box II (nt –53 to –39) and Box I (nt –3 to +8) (He et al., 2000, 2001, 2002; Yin and Beachy, 1995; Yin et al., 1997a). Box II plays an essential role in the expression of genes driven by the RTBV promoter, and two host transcription factors that bind to the Box II *cis*-element have been identified and partially characterized. These factors, RF2a and RF2b, bind to and activate transcription from the RTBV promoter (Dai et al., 2003, 2004, 2006; Petruccelli et al., 2001; Yin et al., 1997b). To reduce the expression of the RTBV promoter, and to ultimately reduce virus replication, three synthetic zinc-finger protein transcription factors (ZF-TFs) were tested, each composed of six finger domains and designed to bind to promoter sequences between nt –58 and +50. Two of these ZF-TFs reduced expression from the promoter in transient assays and in transgenic *Arabidopsis*. One of the ZF-TFs had significant effects on plant regeneration, apparently as a consequence of binding to multiple sites in the *A. thaliana* genome. Expression from the RTBV promoter was reduced by ~45% in transient assays and was reduced by up to 80% in transgenic plants. Coexpression of two different ZF-TFs did not further reduce expression of the promoter (Ordiz et al., 2010).

To apply the concept of RNAi to the control of RTBV infection, transgenic rice plants were produced expressing DNA encoding ORF IV of RTBV in both sense and antisense orientation, resulting in the formation of dsRNA. RNA blot analysis of two representative lines indicated specific degradation of the transgene transcripts and the accumulation of low-molecular-weight RNA, a hallmark of RNAi (Tyagi et al., 2008). Using a more traditional approach to obtain transgenic resistance against RTBV, *indica* rice cultivar Pusa Basmati-1 was transformed to express the CP gene of an Indian isolate of RTBV. Two independent transgenic lines showed significantly lowered levels of RTBV DNA, especially towards later stages of infection, and a concomitant reduction of tungro symptoms (Ganesan et al., 2009).

17.2.7.2 Rice Tungro Spherical Virus

RTSV is a single-stranded RNA virus belonging to the family *Sequiviridae*. Virions consist of a round, nonenveloped capsid with polyhedral symmetry. The isometric capsid has a diameter of 30 nm (ICTVdB Management, 2006). The genome is monopartite: only one particle size of linear, positive-sense, single-stranded RNA has been recovered. The complete genome is 10,422 nt long and has been fully sequenced (ICTVdB Management, 2006; Isogai et al., 2000; Shen et al., 1993). The genome RNA contains a single large ORF predicted to begin ~500 nt downstream from the 5′ end. The leader sequence (5′-noncoding region) contains many AUGs (translation initiation codons), a structure often found in the 5′-noncoding regions of picornaviruses. The virus requires the translational machinery to skip the first AUGs until it reaches the appropriate initiation codon (Jang et al., 1988; Pelletier and Sonenberg, 1988). The viral polyprotein contains a putative leader protein (72 kD), three CPs (CP1–3), a nucleoside triphosphate-binding protein, a 3C-like protease and a polymerase. The CPs and protease have been characterized (Druca et al., 1996; Miranda et al., 1995; Thole and Hull, 1996, 1998), but the leader protein and polymerase have not. The characterized proteins are thought to be processed by virus-encoded or host proteases; some of the characterized proteins have been detected in RTSV-infected tissues. Additionally, the protease region has both *cis*- and *trans*-proteolytic activities (Thole and Hull, 2002). When the gene encoding the 3C-like protease from RTSV strain Vt6 was expressed *in vitro*, it appeared that the mature protease might be generated by its own proteolytic activity (Sekiguchi et al., 2005).

17.2.7.3 Interaction of RTBV and RTSV

RTSV appears to play the role of a helper virus for insect transmission of RTBV (Hibino, 1983). RTSV alone does not cause any distinctive symptoms except for mild stunting in some rice cultivars

(Cabauatan et al., 1993; Hibino, 1983). Rice plants infected with RTBV and RTSV together show tungro symptoms, including stunting, yellow or yellow-orange discolouration and reduced tillering (Hibino et al., 1978, Figure 17.1). Discoloured leaves may show irregularly shaped dark-brown blotches. The leaves, especially the younger ones, may show striping or mottling and interveinal chlorosis. Plants infected with RTBV alone show similar but milder tungro symptoms. RTBV is localized in the vascular bundles and RTSV in the phloem tissues (Sta Cruz et al., 1993). In infected cells, both RTBV and RTSV particles are scattered or aggregated in the cytoplasm. RTSV particles also occur in vacuoles. Viroplasm-like inclusions and membranous masses occur in the cytoplasm of RTSV-infected cells. RTBV and RTSV are transmitted in a semipersistent manner by leafhoppers such as *N. virescens*, *N. nigropictus*, *N. cincticeps*, *R. dorsalis* and other *Nephotettix* spp. *N. virescens* is monophagous to rice, and its density can reach high levels depending on the environment.

RTSV is likely to be a bearer of the helper activity. The helper component of RTSV has not yet been isolated or identified. One hypothesis is that the helper component is a protein coded by the RTSV genome and produced in RTSV-infected cells, similar to the helpers identified for potyviruses and caulimoviruses (Pirone, 1977). Both RTBV and RTSV are styletborne, and the helper component may be essential for specific adsorption of RTBV on the leafhopper's mouth wall.

Cultivar resistance to rice tungro disease (RTD) is an important breeding objective for rice improvement in many Asian countries (Hibino et al., 1990). Cultivars resistant to green leafhopper (GLH) were initially used to control RTD (Hibino et al., 1987), but most did not show enough durability for continued use in the field (Angeles and Khush, 2000; Dahal et al., 1990). Dozens of germplasm sources were found to be highly resistant to RTSV (Hibino et al., 1990), although a formal distinction between resistance to RTSV and to GLH is yet to be made (Sebastian et al., 1996; Shibata et al., 2007; Zenna et al., 2008). Meanwhile, germplasm sources effective for practical field control of RTBV are still very limited (Hibino et al., 1990; Zenna et al., 2006). Indonesian rice cultivar Utri Merah is highly resistant to RTSV and tolerant of RTBV (Azzam et al., 2001; Shahjahan et al., 1990). Advanced breeding lines derived from Utri Merah consistently showed low levels of infection with both RTBV and RTSV in several field trials (Cabunagan et al., 1999). Genetic studies showed that RTSV resistance in Utri Merah was determined by two recessive genes, *tsv1* and *tsv2* (Ebron et al., 1994); the specific genes necessary to confer resistance varied depending on the virus strains (Azzam et al., 2001). RTBV infects Utri Merah, but its level of accumulation is lower than in susceptible cultivars (Shahjahan et al., 1990). RTBV tolerance in Utri Merah is reportedly controlled by multiple genes (Shahjahan et al., 1990). RTD resistance of Utri Merah involves suppression of both RTSV and RTBV; however, the traits that suppress RTSV and RTBV are inherited separately (Encabo et al., 2009).

To identify the gene or genes involved in RTSV resistance of Utri Merah, the association of genotypic and phenotypic variations for RTSV resistance was examined in backcross populations derived from Utri Merah and rice germplasm with known RTSV resistance. Genetic analysis revealed that resistance to RTSV in Utri Merah was controlled by a single recessive gene (*tsv1*) that mapped between 22.05 and 22.25 Mb on chromosome 7. A gene for putative translation initiation factor 4G (eIF4G^{tsv1}) was found in the *tsv1* region. Comparison of *eIF4G*^{tsv1} gene sequences among susceptible and resistant plants suggested that RTSV resistance was associated with a single-nucleotide polymorphism (SNP) site found in exon 9 of *eIF4G*^{tsv1}. Examination of this SNP site in rice cultivars resistant and susceptible to RTSV corroborated the association of SNPs or deletions in codons for Val1060-1061 of the predicted eIF4G^{tsv1} protein sequence with RTSV resistance in rice (Lee et al., 2010).

As reviewed by Truniger and Aranda (2009), about half of the 200 known virus resistance genes in plants are recessively inherited, suggesting that this form of resistance is more common for viruses than for other plant pathogens. A growing number of recessive resistance genes have been cloned from crop species, and further analysis has shown all of them to encode translation initiation factors of the 4E (eIF4E) and 4G (eIF4G) families. Mutations in eIF4G or eIF(iso)4G have been shown to affect turnip mosaic virus, lettuce mosaic virus, plum pox virus and clover yellow vein virus multiplication in *A. thaliana* (Nicaise et al., 2007) and RYMV multiplication in rice (Albar et al., 2006).

17.3 HOST–VIRUS INTERACTIONS

After this review of progress in research on these eight rice viruses, it is clear that much information on the virus genomes, proteins and possible biochemical functions is available, but much less is known about how viruses use plant replication machinery or how host plants mount defence responses to limit infection. Some of the existing data that might be helpful for the understanding of the interaction between virus and plant host are discussed below.

Two possible viral mechanisms that might protect against host defence systems are viral proteins that function as RNA-silencing suppressors and formation of intracellular viroplasms to serve as the site of virus replication and assembly. RNA silencing is an evolutionarily conserved sequence-specific gene-inactivation system that also functions as an antiviral mechanism in higher plants and insects. To overcome antiviral RNA silencing, viruses express silencing-suppressor proteins that can counteract the host silencing-based antiviral process. After the discovery of virus-encoded silencing suppressors, it was shown that these viral proteins can target one or more key points in the silencing machinery (Csorba et al., 2009). In three of the viruses described here, proteins that function as RNA-silencing suppressors have been identified: the S6-encoded protein of RBSDV (Zhang et al., 2005); the Pns10 protein of RDV (Cao et al., 2005) and the P3 protein of RSV, which shares 46% identity with the gene-silencing suppressor NS3 protein of tenuivirus RHBV (Bucher et al., 2003).

Transgenic rice plants that expressed the Pns12-specific RNAi of RDV showed strong resistance to viral infection, and some plants that expressed Pns4-specific RNAi showed delayed appearance of symptoms (Shimizu et al., 2009). These results suggest that viral gene inactivation by RNAi could be constructed to knock down the key gene for viral multiplication (in this case, Pns12). In a similar study, transgenic rice plants expressing DNA encoding ORF IV of RTBV, in both sense and antisense orientations, resulted in the formation of dsRNA. RNA blot analysis of two representative lines indicated specific degradation of the transgene transcripts and the accumulation of low-molecular-weight RNA, a hallmark of RNAi (Tyagi et al., 2008).

The formation of viroplasms (viral inclusion bodies) is a common feature of dsRNA viruses from the family *Reoviridae*, which have genomes composed of 10–12 segments (Brookes et al., 1993; Fabbretti et al., 1999; Fukushi et al., 1962; Petrie et al., 1984; Rhim et al., 1962; Shikata and Kitagawa, 1977; Touris-Otero et al., 2004). These inclusion bodies are composed mainly of viral dsRNA, viral proteins and partially and fully assembled viral particles (Dales et al., 1965; Fabbretti et al., 1999; Isogai et al., 1998; Rhim et al., 1962; Silverstein and Schur, 1970; Touris-Otero et al., 2004; Wei et al., 2006b, c). In addition, the inclusion bodies have been shown to contain microtubules and thinner 'kinky' filaments suggested to be intermediate filaments (Dales, 1963; Dales et al., 1965; Spendlove et al., 1964). Although the mechanism of formation of these inclusions is largely unknown, the inclusions are thought to play an important role in viral infection, because they are probable sites of viral genome replication, protein synthesis and virus assembly (Fabbretti et al., 1999; Isogai et al., 1998; Petrie et al., 1984; Rhim et al., 1962; Wei et al., 2006b, c). The P9-1 protein of RBSDV has been suggested to form intracellular viroplasm without the need for other RBSDV proteins (Zhang et al., 2008). The Pns4 protein of RDV is a phosphoprotein that is localized around the viroplasm matrix and forms minitubules (Wei et al., 2006b), whereas Pns12 is essential for the formation of viroplasms and nucleation of viral–assembly complexes (Wei et al., 2006c).

Some information is available on the host factors involved in virus multiplication. For example, the plant-specific NAC (petunia NAM and Arabidopsis ATAF1, ATAF2 and CUC2) transcription factor RIM1 (RDV multiplication 1) supported the multiplication of RDV (Yoshii et al., 2009).

Further analysis showed that RIM1 functions as a transcriptional regulator of JA signalling and is degraded in response to JA treatment via a 26S-proteasome-dependent pathway (Yoshii et al., 2010). The P2 protein in RDV was also reported to interact with *ent*-kaurene oxidases, which play a key role in the biosynthesis of plant growth hormones (gibberellins) in infected plants (Zhu et al., 2005). The Pc4 protein of RSV, a putative movement protein, has been reported to interact with a DNAJ protein and an hsp20-family protein of rice (Lu et al., 2009). Two host transcription factors that bind to the Box II *cis*-element in the promoter region of RTBV have been identified and partially characterized: RF2a (Dai et al., 2003; Yin et al., 1997b) and RF2b (Dai et al., 2004), which bind to and activate transcription from the RTBV promoter (Dai et al., 2003, 2004, 2006; Petruccelli et al., 2001; Yin et al., 1997b).

As described above for RTSV, cultivar resistance to virus is an important breeding objective for rice improvement in many Asian countries (Hibino et al., 1990). Recessive mutations in translation initiation factors of the 4E (eIF4E) and 4G (eIF4G) families have conferred resistance to RYMV and RTSV (Albar et al., 2006; Lee et al., 2010).

17.4 TRANSCRIPTOME ANALYSIS OF VIRUS-INFECTED HOST PLANTS

Currently we are now performing the transcriptome analysis of eight rice viruses using a rice 44k oligoarray system as shown before. Global changes of host genes are shown in Table 17.2. The only published global gene-expression analyses in rice specific to viruses relate to infection by RDV (Shimizu et al., 2007) and RSV (Satoh et al., 2010), and proteome analysis has been reported for RYMV (Ventelon-Debout et al., 2004). This is possibly because insect vectors are needed for virus infection, and it is very difficult to set up an infection system without collaboration.

Extensive global gene-expression data from infections of geminivirus and other plant RNA viruses are available for *Arabidopsis* (Ascencio-Ibáñez et al., 2008; Espinoza et al., 2007; Golem and Culver, 2003; Marathe et al., 2004; Whitham et al., 2003; Yang et al., 2007), *Nicotiana* (Dardick, 2007; Senthil et al., 2005), *Zea mays* (Uzarowska et al., 2009) and *Vitis vinifera* (Espinoza et al., 2007). Geminiviruses represent a unique opportunity to study changes in the expression of host genes involved in cell-cycle regulation and DNA replication as well as the induction of the pathogen response. Cabbage leaf curl virus (CaLCuV), a member of the *Begomovirus* genus (Hill et al., 1998), encodes seven proteins, including two viral replication proteins designated AL1 (AC1, C1 or Rep) and AL3 (AC3, C3 or REn). CaLCuV does not encode a DNA polymerase; instead, it depends on host DNA replication machinery to amplify its small, circular genome via a combination of rolling-circle replication and recombination-mediated replication. This dependence on host machinery constitutes a barrier to infection of mature plant cells, which no longer supports DNA replication (Hanley-Bowdoin et al., 2004). To overcome this constraint, the geminivirus AL1 protein binds to the host retinoblastoma-related (RBR) protein and relieves the repression of E2F transcription factors. This in turn allows the activation of host genes required for transition into the S-phase and establishment of a DNA replication-competent environment (Desvoyes et al., 2006; Egelkrout et al., 2001, 2002).

Geminiviruses may also influence host gene expression by altering signal transduction pathways through interactions with host protein kinases. Reduced activity of a host SNF1-related kinase (SnRK1) in response to binding by geminivirus AL2 protein has been implicated in host susceptibility to infection (Hao et al., 2003). The geminivirus protein BR1 binds to host proteins NSP-INTERACTING KINASE (NIK)-1, NIK2 and NIK3, members of the Leu-rich-repeat-receptor-like kinase family, and inhibits their phosphorylation and antiviral activities (Fontes et al., 2004). In contrast, interaction and phosphorylation of BR1 by a PERK-like receptor kinase, NsAK (nuclear shuffle protein (NSP)-related kinase), is necessary for efficient infection and full symptom development (Florentino et al., 2006). AL1 binding to GRIK1 and GRIK2, host proteins that

Table 17.2 Number of Genes Upregulated and Downregulated by the Infection of Viruses on 44K Gen-Expression Oligo Arrays

	RDV-S		RRSV		RBSDV		RSV		RGSV		RTYV		RTSV		RTBV		IR64 Drought (0.5 FTSW)	
	Up	Down	Up	Down	Up	Down	Up	Down	Up	Down	Up	Down	Up	Down	Up	Down	Up	Down
	3598	3599	2619	2967	2694	3096	4250	4148	3221	3564	725	395	560	548	1743	1824	5616	5278
Days p.i.	21	28	28	12	28	28	15	18										

Note: Change of gene expression was defined by the difference in mean intensity between RSV-infected and mock-inoculated plants, which was greater than 1.56, and the significance of changes in gene expression was $P < 0.05$ by a paired t-test.

accumulate in infected cells (Kong and Hanley-Bowdoin, 2002), may modulate their proposed dual roles in controlling precursor and energy resources needed for DNA replication and activation of SnRK1 and the pathogen response (Shen and Hanley-Bowdoin, 2006). The divergent AL4 and C4 proteins from geminivirus may alter cell signalling through their interactions with two members of the *Arabidopsis* shaggy protein kinase-related family, which is involved in brassinosteroid signalling (Piroux et al., 2007). The diverse interactions and activities of the viral proteins suggest that geminiviruses modulate a variety of plant processes by altering the host gene expression.

Studies of plant RNA virus-infected plants show that viral infection is accompanied by many changes in the plant transcriptome. The transcript profiles of *Arabidopsis* ecotype Uk-4 infected with tobacco mosaic virus strain Cg and *Vitis vinifera* 'Carménère' infected with grapevine leafroll-associated virus strain 3 were analysed using *Arabidopsis* microarrays. A large number of senescence-associated genes (SAGs) exhibited altered expression during these two compatible interactions. Among the SAGs were genes that encode proteins such as proteases and lipases, proteins involved in the mobilization of nutrients and minerals, transporters, transcription factors and proteins related to translation and antioxidant enzymes, among others. Thus, part of the plant's response to virus infection appears to be the activation of the senescence programme. It was also demonstrated that several virus-induced genes are expressed at elevated levels during natural senescence in healthy plants (Espinoza et al., 2007).

17.5 EXPRESSION PROFILES OF CELL-CYCLE-RELATED GENES IN RICE LEAVES FOLLOWING VIRUS INFECTION OR DROUGHT STRESS

As the first step of our comparative transcriptome analysis, we focused on global gene expression of approximately 90 genes related to cell-cycle regulation in virus-infected cells (Table 17.3). Progression of the cell cycle is primarily controlled by a highly conserved molecular machinery in which cyclin-dependent kinases (CDKs) play a central role. CDK activity is regulated in a complex manner, including reversible phosphorylation by specific kinases or phosphatases and association with other regulatory proteins. Core cell-cycle regulators, such as CDKs, cyclins, CDK inhibitors (CKIs) and homologs of retinoblastoma (Rb) protein and E2F transcription factors (E2F), have been identified in eukaryotes. In higher plants, the cell cycle must be integrated into a complex system of histogenesis and organogenesis, because spatial and temporal regulation of cell division is essential to the development of plant form (Mironov et al., 1999; Meijer and Murray, 2001; Stals and Inzé, 2001; De Veylder et al., 2003). The basic functional characteristics of the cell-cycle proteins are highly conserved in all eukaryotes (Novak et al., 1998). The cell-cycle protein families of plants are, however, larger than those of mammals and yeast (see Table 17.3), probably on account of genome duplication events (Vandepoele et al., 2002). Plants have acquired unique regulatory mechanisms to accommodate specialized types of cell division (Sheen and Key, 2004); they have developed unique cytoskeletal structures, the preprophase band and the phragmoplast, to position the cell division plane and to construct the cell plate (Mineyuki, 1999; Van Damme and Geelen, 2008). Thus, the occurrence of plant-specific structures implies that proteins conserved between plants and animals might have acquired different functions and localizations. In Figure 17.2, the overall picture of gene-expression profiles of the cell-cycle-related genes from our microarray data are presented. The following paragraphs provide the details.

17.5.1 Cyclins

Cyclins are the factors that determine the timing of CDK activation in the cell cycle. Rice has 10 different types of cyclins (A, B, D, F, H, L, P, SDS, T and U). Among them, F-type cyclins are unique to rice and are not found in *Arabidopsis* (La et al., 2006).

Table 17.3 List of Rice Cell-Cycle-Related Genes

Class	Gene Name	ProbeName	Identifier
CycA	Orysa;CycA1;1	R5_c12_41982_R	LOC_Os12g20324
CycA	Orysa;CycA1;2	R5_c01_01054_R	LOC_Os01g13260
CycA	Orysa;CycA2;1	R5_c12_42538_R	LOC_Os12g31810
CycA	Orysa;CycA3;1	R5_c03_13041_F	LOC_Os03g41100
CycA	Orysa;CycA3;2	R5_c12_43070_F	LOC_Os12g39210
CycA	Orysa;CycA3;3	R5_c03_10995_F	LOC_Os03g11040
CycA	Orysa;CycA3;4	R5_c03_10994_F	LOC_Os03g11030
CycB	Orysa;CycB1;1	R5_c01_04277_R	LOC_Os01g59120
CycB	Orysa;CycB1;2	R5_c05_21482_F	LOC_Os05g41390
CycB	Orysa;CycB1;3	R5_c02_08594_R	LOC_Os02g41720
CycB	Orysa;CycB1;4	R5_c01_01393_R	LOC_Os01g17402
CycB	Orysa;CycB2;1	R5_c04_17815_R	LOC_Os04g47580
CycB	Orysa;CycB2;2	R5_c06_25819_R	LOC_Os06g51110
CycD	Orysa;CycD1;1	R5_c09_33520_R	LOC_Os09g21450
CycD	Orysa;CycD1;2	R5_c06_23338_R	LOC_Os06g12980
CycD	Orysa;CycD2;1	R5_c07_28712_R	LOC_Os07g42860
CycD	Orysa;CycD2;2	R5_c06_23191_F	LOC_Os06g11410
CycD	Orysa;CycD2;3	R5_c03_12321_R	LOC_Os03g27420
CycD	Orysa;CycD3;1	R5_c09_32421_R	LOC_Os09g02360
CycD	Orysa;CycD4;1	R5_c09_34108_F	LOC_Os09g29100
CycD	Orysa;CycD4;2	R5_c08_31672_F	LOC_Os08g37390
CycD	Orysa;CycD5;1	R5_c12_43111_R	LOC_Os12g39830
CycD	Orysa;CycD5;2	R5_c03_13100_R	LOC_Os03g42070
CycD	Orysa;CycD5;3,4	R5_c03_10960_R	LOC_Os03g10650
CycD	Orysa;CycD6;1	R5_c07_28216_R	LOC_Os07g37010
CycD	Orysa;CycD7;1	R5_c11_40640_R	LOC_Os11g47950
CycF	Orysa;CycF1;1	R5_c02_08445_R	LOC_Os02g39470
CycF	Orysa;CycF1;2	R5_c02_08423_F	LOC_Os02g39220
CycF	Orysa;CycF1;3	R5_c02_08425_F	LOC_Os02g39240
CycF	Orysa;CycF1;4	R5_c02_08424_F	LOC_Os02g39230
CycF	Orysa;CycF2;1	R5_c02_08443_R	LOC_Os02g39420
CycF	Orysa;CycF2;2	R5_c02_08426_F	LOC_Os02g39260
CycF	Orysa;CycF2;3	R5_c02_08388_R	LOC_Os02g38820
CycF	Orysa;CycF3;1	R5_c03_10994_F	LOC_Os03g11030
CycF	Orysa;CycF3;2	R5_c03_10995_F	LOC_Os03g11040
CycH1	Orasa;CycH1;1	R5_c03_13926_F	LOC_Os03g52750
CycL	Orysa;CycL1;1	R5_c01_02019_R	LOC_Os01g27940
CycSDS	Orysa;SDS	R5_c03_11122_R	LOC_Os03g12414
CycT	Orysa;CycT1;1,2	R5_c02_07331_F	LOC_Os02g24190
CycT	Orysa;CycT1;3	R5_c02_05867_R	LOC_Os02g04010
CycT	Orysa;CycT1;4	R5_c11_37966_F	LOC_Os11g05850
CycT	Orysa;CycT1;5,6	R5_c12_42437_F	LOC_Os12g30020
CycU	Orysa;CycU1;1	R5_c04_18317_F	LOC_Os04g53680
CycU	Orysa;CycU2;1	R5_c04_17733_F	LOC_Os04g46660
CycU	Orysa;CycU3;1	R5_c05_20827_R	LOC_Os05g33040
CycU	Orysa;CycU4;1,3	R5_c10_37408_F	LOC_Os10g41430

Table 17.3 (continued) List of Rice Cell-Cycle-Related Genes

Class	Gene Name	ProbeName	Identifier
CycU	Orysa;CycU4;2	R5_c04_17678_R	LOC_Os04g46000
CycU	Orysa;CycU4;4,5	R5_c02_08759_R	LOC_Os02g43550
CDKA	Orysa;CDKA;1	R5_c03_10288_F	LOC_Os03g02680
CDKA	Orysa;CDKA;2	R5_c02_05777_F	LOC_Os02g03060
CDKB	Orysa;CDKB;1	R5_c01_04972_R	LOC_Os01g67160
CDKB	Orysa;CDKB;2	R5_c08_31921_F	LOC_Os08g40170
CDKC	Orysa;CDKC;1	R5_c05_20767_F	LOC_Os05g32360
CDKC	Orysa;CDKC;2	R5_c01_05428_F	LOC_Os01g72790
CDKC	Orysa;CDKC;3	R5_c08_31514_R	LOC_Os08g35220
CDKD	Orysa;CDKD;1	R5_c05_20789_F	LOC_Os05g32600
CDKE	Orysa;CDKE;1	R5_c10_37539_R	LOC_Os10g42950
CDKF	Orysa;CDKF;1	R5_c06_23957_R	LOC_Os06g22820
CDKF	Orysa;CDKF;2	R5_c12_42114_F	LOC_Os12g23700
CDKF	Orysa;CDKF;3	R5_c03_14797_F	LOC_Os03g63020
CDKG	Orysa;CDKG;1	R5_c02_08405_F	LOC_Os02g39010
CDKG	Orysa;CDKG;2	R5_c04_17254_F	LOC_Os04g41100
CKL	Orysa;CKL1	R5_c12_43041_R	LOC_Os12g38860
CKL	Orysa;CKL2	R5_c07_28522_R	LOC_Os07g40550
CKL	Orysa;CKL3	R5_c01_05190_R	LOC_Os01g70130
CKL	Orysa;CKL4	R5_c01_00814_F	LOC_Os01g10430
CKL	Orysa;CKL6	R5_c07_29069_F	LOC_Os07g47180
CKL	Orysa;CKL7	R5_c11_38536_F	LOC_Os11g13860
CKL	Orysa;CKL8	R5_c12_41398_F	LOC_Os12g10184
CKL	Orysa;CKL9	R5_c02_08100_F	LOC_Os02g35300
CKL	Orysa;CKL10	R5_c08_29389_R	LOC_Os08g02050
KRP	Orysa;KRP1	R5_c02_09557_F	LOC_Os02g52480
KRP	Orysa;KRP2	R5_c06_23160_R	LOC_Os06g11050
KRP	Orysa;KRP3	R5_c11_40081_F	LOC_Os11g40030
KRP	Orysa;KRP4	R5_c10_36718_R	LOC_Os10g33310
KRP	Orysa;KRP5	R5_c03_10448_F	LOC_Os03g04490
KRP	Orysa;KRP6	R5_c09_34061_R	LOC_Os09g28580
KRP	Orysa;KRP7	R5_c01_02527_R	LOC_Os01g37740
E2F	Orysa;E2F1	R5_c02_07950_R	LOC_Os02g33430
E2F	Orysa;E2F2	R5_c12_41089_R	LOC_Os12g06200
E2F	Orysa;E2F3	R5_c04_16720_R	LOC_Os04g33950
E2F	Orysa;E2F4	R5_c04_14995_R	LOC_Os04g02140
DP	Orysa;DP1	R5_c01_03347_R	LOC_Os01g48700
DP	Orysa;DP2	R5_c10_36513_R	LOC_Os10g30420
DP	Orysa;DP3	R5_c03_10560_F	LOC_Os03g05760
DEL	Orysa;DEL1	R5_c02_09389_R	LOC_Os02g50630
DEL	Orysa;DEL2	R5_c06_23402_F	LOC_Os06g13670
RB	Orysa;Rb1	R5_c11_39567_F	LOC_Os11g32900
RB	Orysa;Rb2	R5_c08_32123_R	LOC_Os08g42600
CKS	Orysa;CKS1	R5_c03_10515_F	LOC_Os03g05300
Wee	Orysa;Wee1,Wee2	R5_c02_05886_R	LOC_Os02g04240

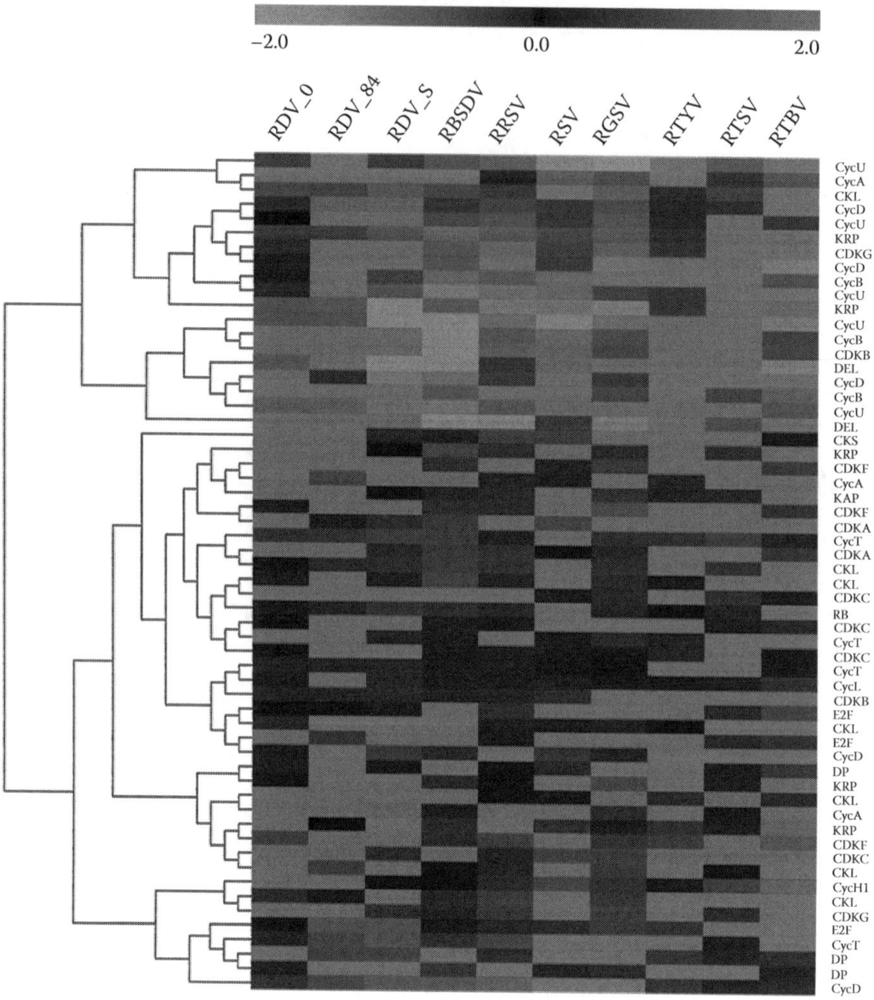

Figure 17.2 **(See color insert.)** Heat-map clustering analysis of virus transcriptome data. Transcriptome data of representatively shown 58 cell-cycle-related genes from leaves of rice plants infected with eight different rice viruses were clustered on the basis of their gene-expression profiles. Red cells show upregulation and green cells show downregulation compared with mock-infected or untreated plants. Grey cells indicate low-intensity data.

Cyclin A binds to CDK produced before the S-phase and is required for the cell to progress through the S-phase. Both B- and A-type cyclins in plants are expressed during mitosis (and hence are referred to as 'mitotic cyclins') and contribute greatly to the G2/M transition (Burssens et al., 2000; Girard et al., 1991; Lehner and O'Farrell, 1990; Obaya and Sedivy, 2002; Pagano et al., 1992; Swenson et al., 1986). In contrast to animals, plants have larger A- and B-type cyclins with extensive and complex functions. For instance, Nicta;CYCA3;2 functions as an E-type cyclin in animals, and alfalfa A2-type cyclin Medsa;CYCA2;2 is persistently expressed throughout all phases of the cell cycle (Roudier et al., 2000; Yu et al., 2003). The A-type cyclins generally appear at the onset of the S-phase (Fuerst et al., 1996; Ito et al., 1997; Reichheld et al., 1996; Setiady et al., 1995). In contrast, the B-type cyclins are induced in the late S- to G2-phase, reach their maximum levels during mitosis and are completely degraded before anaphase via the functions of the destruction boxes in their

N-termini (Genschik et al., 1998; Peters, 1998). Table 17.4a summarizes the cyclin genes analysed in this study.

In the rice genome, seven cyclin A genes (*CycA1;1*, *-A1;2*, *-A2;1*, *-A3;1*, *-A3;2*, *-A3;3* and *-A3;4*) have been identified. *CycA1;1* and *-A1;2* show constitutively high levels of expression, *CycA2;1*, *-A3;1* and *-A3;2* show intermediate levels and *CycA3;3* and *-A3;4* show low levels. Although *CycA1;1*, *-A1;2* are upregulated during drought stress treatment in roots, the only cyclin A gene upregulated during virus infection were *CycA2;1* (during RTBV infection). Specific upregulation of *CycA2;1* during infection by RTBV suggests utilization of this protein for its own DNA replication. A protein localization study (Boruc et al., 2009) revealed that A2 and A3 cyclins were localized to the nucleus in interphase BY-2 tobacco culture cells, while CycA1;1 was detected in both nuclear and cytoplasmic compartments.

Like cyclin A, cyclin B is a mitotic cyclin. In the rice genome, six cyclin B genes (*CycB1;1*, *-B1;2*, *-B1;3*, *-B1;4*, *-B2;1* and *-B2;2*) have been identified. Among these genes, *CycB2;2* shows relatively low expression level and the others show high expression. dsRNA-type viruses (RDV, RBSDV, RRSV) and RSV commonly downregulate the expression of *CycB1;1*, *-B1;4* and *-B2;1*, whereas RTBV downregulates *CycB1;2* and *-B2;1*. Drought stress treatments downregulate many cyclin B genes except for *CycB1;3*. In contrast, only *CycB1;1* was upregulated by virus infection (by RGSV). This upregulation might be correlated with the specific symptom induced by infection with RGSV: 'proliferation of short, erect and narrow leaves' (Hibino, 1996). The phenotype induced by the knockdown of *CycB1;1* suggests that *CycB1;1* expression is critical for endosperm formation via the regulation of mitotic division (Guo et al., 2010).

Cyclin D is the major cyclin expressed during G1 in higher organisms, and its activity is crucial for progression to the S-phase. Plant D-type cyclins might function as the mediators of internal and environmental stimuli to drive cell division (Riou-Khamlichi et al., 1999). The cyclin D-RBR-E2F pathway is conserved in plants (RBR gene) (de Jager et al., 2005; Francis, 2007). Of the 10 D-type cyclins present in *Arabidopsis*, *CYCD3;1* has been most extensively characterized. Overexpression of *CYCD3;1* drives cell-cycle progression in cell culture (Menges et al., 2006) and ectopic cell division in plants (Dewitte et al., 2003), and blocks endoreduplication (Schnittger et al., 2002), while loss of CYCD3 activity enhances endoreduplication (Dewitte et al., 2007). Expression of *E2Fa*, *E2Fc* or *CYCD3;1* was induced, and the effects on gene expression over time were detected by using microarrays, allowing both short- and longer-term effects to be observed. Overlap between CYCD3;1- and E2Fa-modulated genes substantiates their action in a common pathway with a key role in controlling the G1/S transition. Additional targets for CYCD3;1 were genes related to chromatin modification, and for E2Fa, genes involved in cell wall biogenesis and development. E2Fc induction led primarily to gene downregulation, but it did not antagonize E2Fa action. E2Fc appears to function outside the CYCD3-RBR pathway; it does not have a direct effect on cell-cycle genes, and promoter analysis suggests a distinct binding-site preference, that is, the sequences NNNCCCGC, NNNGGCGC or NNNCGCGC are overrepresented by E2Fa but not by E2Fc (de Jager et al., 2009). Phytohormones, particularly IAA, modified the kinase activity (CDK-A type) associated with D cyclins (D4;1 and D5) during the first few hours of germination of maize (Lara-Núñez et al., 2008).

Thirteen cyclin D genes have been identified in rice and were classified into three groups according to their expression levels. The first group, which is the most highly expressed, includes *CycD2;2*, *-D5;2*, *-D5;3.4* and *-D6;1*; the second group, which is moderately expressed, contains *CycD2;1*, *-D3;1*, *-D4;2* and *-D5;1* and the third group, which is expressed at the lowest levels, includes *CycD1;1*, *-D1;2*, *-D2;3*, *-D4;1* and *-D7;1*. Both down- and upregulation of cyclin D genes in response to virus infection have been observed. *CycD3;1* and *-5;1* were downregulated in response to RDV infection, and *CycD5;1* and *-5;2* were downregulated by RTBV. The finding that RDV downregulates the expression of *CycD3;1* and *-D5;1* suggests that RDV effectively targets the cyclin D genes important for the G1/S transition. *CycD5;2* was also downregulated by RBSDV and RGSV. Cyclin D genes *CycD2;1*, *-D3;1*, *-D5;2* and *-D5;3,4* were downregulation by drought. Conversely, *CycD6;1*

Table 17.4a Rice Cyclin Genes and Their Expression Profiles in Virus-Infected and Drought-Stressed Tissues

	RDV		RRSV		RBSDV		RSV		RGSV	
Gene Name	**Inf**	**Mock**	**Inf**	**Mock**	**Inf**	**Mock**	**Inf**	**Mock**	**Inf**	**Mock**
					Cyc A					
Orysa;CycA1;1	8.69	9.27	9.34	9.5	9.57	9.59	8.97	8.81	9.58	9.55
	0		–0.16		0		0.36		0	
Orysa;CycA1;2	8.14	8.4	8.47	8.59	8.51	8.67	7.85	7.54	8.6	8.63
	0		0		0		0.26		–5.20E–02	
Orysa;CycA2;1	7.29	6.7	6.56	6.4	6.46	6.34	4.76	5.89	6.23	6.85
	0		0		0		0		–0.68	
Orysa;CycA3;1	5.67	6.17	5.44	5.27	5.49	5.94	4.56	5.03	5.31	5.59
	0		0		0		0		0	–0.87
Orysa;CycA3;2	6.37	7.61	5.52	5.62	6.69	6.4	5.61	5.16	6.08	5.5
	–1.15	0	0	0	0	0	0.22	0	–0.13	0
Orysa;CycA3;3	4.33	3.95	4.13	4.29	4.45	3.74	3.64	3.47	4.58	3.99
	0	0	0	0	0	0	0	0	0	–0.28
Orysa;CycA3;4	1.88	1.91	1.86	1.86	1.97	2.11	1.83	1.84	1.93	1.9
	0	0	0	0	0	0	0	0	–0.05	–2.70E–02
					Cyc B					
Orysa;CycB1;1	6.12	7.38	6.75	7.64	5.49	7.54	5.51	7.37	7.17	6.2
	0	–0.79	–2.05	–1.39	1.08	0	0	–0.49	0	0
Orysa;CycB1;2	5.83	7.21	6.07	6.6	5.37	6.37	4.27	6.06	5.43	4.41
	0	0	0	0	0	0	–0.57	–0.94	0	–0.84
Orysa;CycB1;3	6.37	6.85	6.5	6.72	6.71	6.49	6.73	6.81	6.91	6.64
	–0.48	0	0	0	0	0	0	0	0	0
Orysa;CycB1;4	5.7	7.33	7.46	8.28	7.3	7.9	6.08	5.72	7.57	7.31
	–1.49	–0.95	–0.6	0	0	0	0	0	–0.34	–0.35
Orysa;CycB2;1	6.48	6.95	6.55	7.36	5.98	7.19	5.37	7.21	6.74	7.43
	–0.46	–0.84	–1.06	–1.4	0	0	0	–0.63	0	–0.79
Orysa;CycB2;2	4.08	5.17	3.92	4.26	2.01	4.07	3.09	4.2	3.6	1.92
	0	0	0	0	0	0	0	0	0	0
					Cyc D					
Orysa;CycD1;1	1.91	1.9	1.82	1.8	1.88	1.9	2.24	2.19	1.85	1.89
	0	0	0	0	0	0	0	0	0	0
Orysa;CycD1;2	3.78	3.11	4.33	3.11	2.88	3.42	2.89	3.47	2.33	4.57
	0	0	0	0	0	0	0	0	0.35	0
Orysa;CycD2;1	5.8	6.71	6.1	6.56	6.03	6.21	5.97	5.13	6.84	6.13
	0	0	0	0	0.71	0	0	0	0.76	0
Orysa;CycD2;2	8.84	9.09	9.93	9.96	10.26	10.12	9.63	7.82	10.16	10.28
	0	0	0	1.45	–0.12	0	0	0	0	0
Orysa;CycD2;3	1.85	1.84	1.9	1.9	1.98	1.9	1.87	2.5	1.9	1.77
	0	0	0	0	0	0	0	0	0	–9.10E–02
Orysa;CycD3;1	7.58	8.6	8.99	9.52	8.84	9.12	8.46	8.63	8.68	9.29
	–1.06	–0.45	–0.27	–0.4	–0.57	–9.80E–02	–0.21	0	0	–0.69
Orysa;CycD4;1	4.96	4.52	4.71	2.77	4.51	3.79	3.81	4.75	5.03	6.03
	0	0	0	0	0	0	0	0	0	0
Orysa;CycD4;2	7.03	6.88	7.84	8.13	7.48	8.13	6.45	6.21	8.4	7.33
	0	–0.32	–0.43	0	0.95	0	0	0	0	–0.7
Orysa;CycD5;1	5.26	7.17	3.44	3.79	4.22	4.77	4.8	5.07	4.22	5.07
	–1.57	0	0	0	0	0	0	–1.2	0	–0.38
Orysa;CycD5;2	10.7	11.43	9.4	9.64	9.5	10.29	8.86	9.11	9.97	11.24
	0	0	–0.71	–0.43	–1.28	0	0	–1.26	0	–0.55
Orysa;CycD5;3,4	11.67	10.95	11.14	11.23	10.96	11.18	10.8	11.2	11.13	10.84
	0.72	0	–0.19	–0.4	0.29	0	0	0	0.67	–0.39

RTYV		RTSV		RTBV		Leaf		Panicle		Root 0.2 FTSW		Root 0.5 FTSW	
Inf	Mock	Inf	Mock	Inf	Mock	D	C	D	C	D	C	D	C
						Cyc A							
9.02	9.24	10.15	10.13	9.56	9.68	10.38	10.47	9.96	9.88	10.13	9.29	10.35	9.54
−0.18		0		0		0		3.30E−02		0.78		0.81	
8.45	8.43	8.48	8.39	8.35	8.47	9.41	9.5	9.07	9	9.36	8.6	8.99	8.45
0		0		0		−0.1		0		0.74		0.55	
6.52	6.69	7.6	7.81	9.44	8.22	7.91	7.98	8.99	9.43	8.68	8.64	9.79	8.98
0		−0.27		1.22		0		0		0		0.09	
6.29	6.32	5.42	5.39	5.95	5.76	3.82	4.84	6.49	7.59	5.67	7.99	7.76	8.2
0		0		0		−1.02		−0.87		−1.29		−1.06	
7.22	7.19	6.86	6.65	5.66	5.91	5.96	6.7	7.62	7.96	6.19	6.56	7.47	7.15
−0.13	0.25												
4.79	4.2	4.61	3.98	4	1.88	4.87	4.59	4.43	4.84	5.92	5.14	3.18	5.67
0.55	−1.32												
1.87	1.87	1.91	1.93	1.93	2.08	2.49	2.53	2.38	2.42	2.8	2.79	2.83	2.72
5.00E−03	0												
						Cyc B							
7.28	7.91	7.54	7.52	7.89	8.38	6.48	7.2	9.49	10.2	3.43	9.86	9.42	10.04
−5.56	−1.13												
6.03	6.73	6.82	6.98	7.11	8.16	4.27	5.25	9.14	9.98	2.86	9.3	8.18	9.67
−5.42	−1.7												
6.55	6.88	5.62	5.59	6.31	6.27	7.1	7.11	8.34	8.89	8.21	8.62	9.47	9.18
0	0												
6.22	6.54	3.49	3.49	6.71	6.99	6.58	7.01	8.36	8.79	2.88	5.71	6.58	6.03
−2.26	0.5												
6.75	7.22	6.42	6.45	6.87	7.5	5.71	6.01	8.77	9.55	3.98	9.4	8.5	9.17
−4.42	−0.67												
4.98	5.27	4.91	5.35	5.74	5.76	2.63	3.88	7.17	7.79	2.86	8.66	7.39	8.67
−5.76	−1.28												
						Cyc D							
2.02	1.85	1.86	1.85	1.91	1.88	2.41	2.39	2.35	2.35	2.78	2.72	2.7	2.78
0	−3.30E−02												
3.89	3.26	4.96	4.54	5.35	5.62	6.81	6.54	3.7	4.03	9.24	5.46	10.55	6.15
4.3	4.29												
6.07	6.32	6.8	6.68	8.14	8.39	7.56	6.8	8.03	8.64	9.72	11.46	11.5	11.4
−1.85	1.00E−01												
9.01	9.06	7.17	7.37	8.65	8.74	10.8	10.78	9.67	9.94	8.88	9.06	9.60	9.15
0	0.42												
1.9	1.96	1.94	2.74	2.52	2.63	2.47	2.47	2.7	2.86	3.39	3.36	2.71	2.86
0	0												
8.1	8.19	8.13	8.34	8.07	8.1	8.41	9.06	8.22	9.15	4.74	6.99	8.04	7.35
−2.25	0.43												
4.92	5.27	2.42	3.14	3.99	4.59	4.9	4.17	6.32	5.84	5.09	4.34	6.53	5.33
0	0.63												
7.94	7.98	3.21	4.29	8.05	8.22	8.55	9.07	9.03	9.77	5.89	4.9	6.28	5.64
0.99	0.64												
7.08	7.16	3.8	3.1	6.89	7.62	5.78	3.71	9.31	9.64	9.92	9.77	11.68	9.81
0.31	1.84												
10.25	10.14	9.68	9.72	8.85	9.94	10.99	12	10.27	10.81	11.13	13.7	12.64	13.87
−2.83	−1.23												
11.14	11.03	9.24	9.12	10.21	9.99	9.94	9.27	7.45	7.85	7.88	9.16	8.58	8.89
−1.28	−0.52												

continued

Table 17.4a (continued) Rice Cyclin Genes and Their Expression Profiles in Virus-Infected and Drought-Stressed Tissues

	RDV		RRSV		RBSDV		RSV		RGSV	
Gene Name	**Inf**	**Mock**	**Inf**	**Mock**	**Inf**	**Mock**	**Inf**	**Mock**	**Inf**	**Mock**
Orysa;CycD6;1	12.4	11.01	11.18	9.49	11.44	10.06	12.34	11.9	11.22	10.69
	1.38	1.67	1.38	0	0	0.46	0.15	−0.1	0.39	0
Orysa;CycD7;1	2.12	1.88	1.85	1.85	2.95	1.99	1.85	1.8	1.91	1.87
	0	0	0	0	0	0	0	0	0.05	2.50E−02
Cyc F										
Orysa;CycF2;3	9.57	9.9	10.28	11.44	10.16	11.17	10.64	11.09	10.66	10.88
	−0.42	−1.16	−1.06	0	−0.26	−0.53	0	0	0	0.42
Orysa;CycF1;2	1.9	2	1.9	1.93	2	2.11	1.85	1.82	1.95	1.92
	0	0	0	0	0	0	0	0	0	0
Orysa;CycF1;4	1.97	1.89	2.59	1.83	1.84	1.92	3.48	3.28	2	2.72
	0	0	0	0	0	0	0	0	0	0
Orysa;CycF1;3	4.62	4.93	4.52	5.28	4.05	5.07	6.34	6.55	4.33	5.12
	0	0	0	0	0	0	0	0	0	0
Orysa;CycF2;2	2.43	2	1.83	1.83	1.91	1.95	1.84	1.84	1.89	1.86
	0	0	0	0	0	0	0	0	0	0
Orysa;CycF2;1	4.87	4.86	4.88	5.47	6.19	6.33	4.33	5.06	5.04	5.66
	0	0	0	0	0	0	0	0	0	0
Orysa;CycF1;1	7.72	7.35	8.85	9.01	8.44	8.54	5.33	6.05	7.64	7.38
	0	0	0	0	0.25	0	0	−0.35	0.37	0
Orysa;CycF3;1	1.88	1.91	1.86	1.86	1.97	2.11	1.83	1.84	1.93	1.9
	0	0	0	0	0	0	0	0	0	0
Orysa;CycF3;2	4.33	3.95	4.13	4.29	4.45	3.74	3.64	3.47	4.58	3.99
	0	0	0	0	0	0	0	0	0	0
Cyc H, Cyc L, Cyc SDS and Cyc T										
Orasa;CycH1;1	6.91	6.61	7.81	7.36	7.59	7.45	6.15	4.98	7.83	7.29
	0.11	0.46	9.60E−02	0	0.55	0.14	0	0	0	0.13
Orysa;CycL1;1	13.46	13.06	13.42	13.25	13.51	13.29	13.09	12.81	13.46	13.35
	0.4	0.16	0.21	0.26	0.11	5.40E−02	0.11	0.4	0	0.42
Orysa;CycSDS	6.1	5.81	6.17	6.17	5.71	6.2	6.72	7.09	6.1	6.25
	0	0	0	0	0	0	0	0	0	0.31
Orysa;CycT1;1,2	11.22	11.81	11.41	11.57	11.13	11.6	11.59	11.71	11.14	11.44
	−0.51	−0.16	−0.5	0	−0.3	−0.3	−0.46	−0.12	−0.47	−0.31
Orysa;CycT1;3	12.87	11.49	12.53	11.53	12.17	11.66	12.15	12.65	11.67	11.46
	1.32	0.88	0.51	0	0	0	0.17	0	0	0.51
Orysa;CycT1;4	13.07	13.24	13.29	13.33	13.29	13.38	13.15	12.98	13.11	13.31
	−0.15	0	−0.1	0.16	−0.2	−9.60E−02	0	0	−0.46	0
Orysa;CycT1;5,6	13	12.61	12.94	12.66	12.96	12.69	12.13	11.98	12.68	12.62
	0.37	0.22	0.22	0.15	0.05	0	0	0.13	0	0.47
Cyc U										
Orysa;CycU1;1	12.42	12.64	12.61	13.33	12.29	12.85	12.3	12.82	12.21	12.91
	0	−0.7	−0.57	−0.35	−0.8	−0.15	0	−0.27	0	−0.59
Orysa;CycU2;1	8.61	12.46	7.41	7.65	7.29	9.08	10.45	12.04	6.99	6.94
	−3.85	0	−1.85	−1.99	0	0	0	−1.2	3.34	0
Orysa;CycU3;1	6.89	6.78	7	7.22	6.77	7.25	8.54	9.21	6.47	7.26
	0	0	−0.57	0	−0.89	0	−0.23	0	−0.21	−0.23
Orysa;CycU4;1,3	6.96	8.46	6.36	7.26	6.22	7.52	7.56	8.78	6.92	6.96
	−1.4	−0.9	−1.3	0	0	0	0	−0.74	0	0
Orysa;CycU4;2	10.4	11.13	8.49	9.27	10.51	9.5	8.87	10.86	8.11	10.15
	−0.36	−0.83	1.02	−1.79	−2.04	0	−0.71	−0.23	0	0
Orysa;CycU4;4,5	10.48	11.22	11.07	12.31	11.14	12.47	10.87	12.09	11.87	12.36
	−0.75	−1.21	−1.4	−1.26	−0.49	−0.37	0	0	0	0

RTYV		RTSV		RTBV		Leaf		Panicle		Root 0.2 FTSW		Root 0.5 FTSW	
Inf	Mock	Inf	Mock	Inf	Mock	D	C	D	C	D	C	D	C
11.59	11.23	10.82	10.59	10.9	11.09	10.41	9.79	11.43	11.34	9.26	9.48	9.85	9.66
0	0.19												
1.98	1.87	1.96	1.89	1.98	1.94	2.48	2.43	2.4	2.38	2.74	2.75	2.76	2.82
−0.01	0												
Cyc F													
9.85	10.38	10.11	10.24	11.82	12.35	12.48	12.03	9.81	9.43	8.24	6.61	7.88	6.79
1.39	1.1												
2.13	1.9	2.08	1.92	2.6	2.11	2.58	2.49	2.48	2.44	2.78	2.79	2.79	2.87
0	0												
2.29	2.3	3.56	3.6	1.89	1.95	2.38	2.4	2.87	3.24	3.93	4.04	3.12	2.7
0	0												
4.44	4.58	6.8	6.85	4.12	5.94	5.21	5.84	4.62	5.14	4.93	6.23	3.14	5.86
−0.4	−2.08												
2.53	1.86	1.93	1.87	1.94	1.98	2.44	2.4	3	2.37	2.73	2.74	2.75	2.8
0	0												
5.33	5.39	5	5.43	4.56	4.9	4.78	4.92	4.14	3.71	5.93	4.98	5.09	5.18
0	0												
7.73	7.92	5.36	6.62	6.73	7.24	7.57	7.21	5.96	5.92	8.47	7.47	7.96	7.71
1.07	0.31												
1.87	1.87	1.91	1.93	1.93	2.08	2.49	2.53	2.38	2.42	2.8	2.79	2.83	2.72
0	0												
4.79	4.2	4.61	3.98	4	1.88	4.87	4.59	4.43	4.84	5.92	5.14	3.18	5.67
0	0												
Cyc H, Cyc L, Cyc SDS and Cyc T													
7.52	7.37	5.45	6.29	6.49	4.53	8.21	8.3	7.18	7.02	9.28	7.35	7.96	7.46
1.92	0.51												
13.32	13.27	13.18	12.98	13.17	12.77	13.69	13.61	13.28	12.86	14.06	13.27	13.72	13.45
0.67	0.29												
5.88	5.83	8.49	8.47	7.42	7.59	6.38	6.36	5.94	5.5	6.62	6.47	6.23	6.45
0.35	0												
11.26	11.6	11.57	12.07	11.06	11.14	11.43	11.78	11.15	11.46	11.48	12.17	11.39	12.08
−0.69	−0.53												
12.07	11.86	12.82	12.48	13.26	13.01	12.44	12.25	11.73	11.48	12.13	11.95	11.42	11.86
0	−0.31												
13.19	13.3	13.37	13.23	13.14	13.12	13.6	14.05	13.31	13.32	14.14	13.92	14.11	13.82
0.23	0.32												
12.73	12.6	12.59	12.27	12.59	12.46	13.95	14.01	14.92	14.44	15.39	14.68	15.02	14.69
0.58	0												
Cyc U													
11.54	11.69	10.16	9.56	13.5	14.07	14.07	14.01	9.73	10.47	8.14	9.99	9.02	9.96
−1.85	−0.94												
11.46	11.8	9.1	9.05	10.75	12.03	7.86	5.03	13.91	12.55	13.72	13.66	15.14	13.69
0.6	0												
6.81	6.66	9.76	10.07	6.96	6.73	7.88	8.01	7.86	8.02	9.19	10.15	9.9	10.27
−0.87	−0.37												
6.95	7.38	9.02	9.64	8.25	8.99	7.35	6.91	11.7	11.82	8.67	10.98	10.89	11.11
−2.82	−0.89												
11.23	11.2	7.78	8.83	9.56	9.78	10.17	9.66	11.14	11.45	8.48	12.35	9.99	12.72
−2.43	−2.73												
10.86	11.61	9.78	9.59	11.01	11.23	12.27	12.31	13.36	13.13	10.96	10.74	9.95	10.94
0.22	−1.24												

was commonly upregulated in plants infected by three dsRNA viruses. RGSV upregulates two cyclin D genes, *CycD2;1* and *-D4;2*. These data also suggest the grassy symptoms specifically induced by RGSV infection. Expression of cyclin D genes is relatively higher in root than other tissues, and upregulation by drought treatment is evident.

Cyclin F is a rice-specific cyclin, and nine genes have been identified (La et al., 2006). Cyclin F genes are also classified into three groups according to their expression levels: *CycF2;3* with the highest expression; *CycF1;1*, *-1;3* and *-2;1* with moderate expression and *CycF1;2*, *-1;4*, *-2;2*, *-3;1* and *-3;2* with very low expression. Only *CycF2;3* showed a change in expression caused by viral infection. RRSV and RBSDV downregulate this gene, and drought stress in root upregulates it. Drought stress also upregulates *CycF1;1* and downregulates *CycF1;3*.

Other cyclins, such as H-, L- and T-types, have been discovered in both animals and plants (Burssens et al., 2000; De Veylder et al., 1999; Umeda et al., 1999). In addition, some cyclins found only in animals or in plants, such as C-, G-, E-, P-, SDS- and K-types, are regarded as being specific to particular species, and their functions are being unravelled in depth (La et al., 2006). Genes encoding cyclins H1 and SDS show relatively high expression levels, while genes for cyclins L (*CycL1;1*) and T (*CycT1;1*, *-1;3*, *-1;4* and *-1;5*) show very high expression compared to other types of cyclin genes. Viral infection by RDV and RRSV upregulated only *CycT1;3*. During drought stress treatment, expression changes were observed only in root. *CycH1;1* and *CycHL1;1* were upregulated, and *CycT1;1,2* was downregulated.

Cyclin U (also known as UNG2) is a uracil DNA glycosylase. It removes uracil bases from the sugar backbone of genomic and mitochondrial DNA, leaving abasic sites and initiating the uracil base-excision repair pathway (Nilsen et al., 2000). UNG2 activity is crucial for the rapid removal of dUMP residues incorporated during genomic DNA replication (Kavli et al., 2002). Six cyclin U genes have been identified in the rice genome. They can be divided into two groups according to their expression levels: group 1, with highly expressed genes *CycU1;1*, *CycU4;2* and *CycU4;4,5*; and group 2, with moderately expressed genes *CycU2;1*, *CycU3;1* and *CycU4;1,3*. The transcripts of this family's members are detected only in leaves, and this family might have a major function in leaf growth and development (Guo et al., 2007). In our analysis, however, the transcripts were also abundant in panicle and root. Many of the cyclin U family members were drastically downregulated by infection with viruses and drought stress treatment.

Cyclins, especially A, B and D, are well conserved among many species and are well characterized in plants. However, our gene-expression data and differential expression profiles among virus infections and drought stress treatments raise new questions as to which transcription factors are committed to the transcriptional regulation of the cyclin genes.

17.5.2 Cyclin-Dependent Kinases

CDKs belong to a group of protein kinases originally found to be involved in the regulation of the cell cycle. CDKs are also involved in the regulation of transcription and mRNA processing. CDKs phosphorylate proteins on serine and threonine and are therefore classified as serine/threonine kinases. A CDK is upregulated by association with a cyclin, forming a CDK complex (King et al., 1996; Morgan, 1995; Rossi and Varotto, 2002). By phylogenetic analysis, 25 CDK genes have been identified in *Arabidopsis* (Menges et al., 2005; Potuschak and Doerner, 2001; Vandepoele et al., 2002). CDKA;1 is the only CDK in *Arabidopsis* to contain the canonical PSTAIRE amino acid motif required for cyclin binding in Cdk1 or Cdc2 + /Cdc28p yeast cells. CDKA;1 is also the only CDK from *Arabidopsis* known to complement yeast cdc2/cdc28 cells (deficient in CDK), while CDKB1;1 (formerly known as Cdc2bAt) did not do so (Ferreira et al., 1991; Hirayama et al., 1991; Imajuku et al., 1992). Mammalian, yeast and plant PSTAIRE kinases share key regulatory sites in the T- and P-loops (Dissmeyer et al., 2007; Harashima et al., 2007). As in other organisms, phosphorylation of the T-loop of *Arabidopsis* CDKA;1 is required for proper functioning of this enzyme (Dissmeyer et al., 2007; Harashima et al., 2007). However, it is not thought to serve a regulatory function, because no indication of a change in phosphorylation state during cell-cycle progression was found in *S. cerevisiae* Cdc28p (Hadwiger and Reed, 1988). CDKA;1 from *Arabidopsis* and CDKA from tomato (*Solanum lycopersicum*) have been found to be phosphorylated at a tyrosine residue *in vitro*, presumably within the P-loop (Gonzalez et al., 2007; Shimotohno et al., 2006). In addition, although no protein with similarities to Myt1/Mik1+ (a regulator of CDK) has been isolated in plants, a homologue of the protein kinase gene *Wee1* has been identified in different plant species (Gonzalez et al., 2004; Sorrell et al., 2002; Sun et al., 1999). In *in vitro* experiments, WEE1 was able to phosphorylate CDK and block its activity (Gonzalez et al., 2007; Shimotohno et al., 2006; Sun et al., 1999). RNAi-mediated downregulation of *WEE1* expression in tomato resulted in reduced plant growth and smaller fruit size (Gonzalez et al., 2007). In contrast, *wee1* mutants in *Arabidopsis* displayed no growth defects but were susceptible to genotoxic stress consistent with a function of Wee1 at a DNA damage checkpoint in metazoans (de Schutter et al., 2007). Furthermore, the expression of the *S. pombe* Cdc25+ phosphatase stimulates proliferation of tobacco (*Nicotiana tabacum*) BY-2 cell cultures (Orchard et al., 2005), and a potential candidate gene for a CDC25-like phosphatase in *Arabidopsis* has been identified (Landrieu et al., 2004). These findings have suggested that CDKA;1 regulation is conserved between plants and metazoans. However, dephosphomimetic and phosphomimetic mutants of CDKA;1 and mutants and overexpression lines of a putative CDC25 ortholog indicate that regulatory phosphorylation patterns are strikingly different between *Arabidopsis*, animals and yeasts (Dissmeyer et al., 2009).

In rice, 24 CDK genes have been identified (Guo et al., 2007). Twenty-three of the 24 genes could be classified into four groups based on relative expression level (CKL5 was missing from our array). The first group consists of the most highly expressed members among all tissues tested and includes *CDKA;1, -B;1, -C;1, -C;2, -D;1, -E;1, -F;1, -F;3, -G;1* and *CKL2*. The second group includes the moderately expressed genes *CDKA;2, CDKG;2, CKL1, CKL4, CKL6* and *CKL10*. The third group includes members with intermediate expression, and sometimes tissue-specific expression: *CDKB;2, CDKC;3, CKDF;2, CKL3, CKL7* and *CKL8*. The fourth group includes only *CKL9*, which has low overall expression but very high expression in panicle. Table 17.4b summarizes the CDK genes analysed in this study.

Virus infection by RRSV, RBSDV and RSV downregulated *CDKB;2*, as did the drought stress treatment in panicle and root. Only RGSV infection upregulated it. *CDKG;2* was also downregulated by virus infection (RRSV, RBSDV and RGSV), but upregulated by drought stress treatments in root. *CKL1* showed a similar phenomenon, being downregulated during infection by RDV and RGSV, but upregulated in panicle during the drought stress treatment. *CDKG;1* was commonly upregulated by infection with RDV, RRSV and RGSV, but drought stress had no effect. *CKL3* was upregulated during infection with RDV, RRSV and RGSV, and by drought stress in root. *CKL8* showed a similar response, being upregulated by RGSV infection and drought stress treatment (in leaf and root). *CKL2* was downregulated by drought stress in leaf, panicle and root, but was unaffected by virus infection. *CKL9* was also downregulated only by drought stress treatment. Although *CDKA;1* is the only CDK known to complement yeast cdc2/cdc28 cells, in rice infected by viruses or treated with drought stress, *CDKA;1* appears to be upregulated by 0.5 fraction of transpirable soil water (FTSW) drought treatment. On the other hand, *CDKA;2, CDKB;2, CDKF;2, CDKG;1, CDKG;2, CKL1, CKL2, CKL3* and *CKL6, -7* and *-8* clearly responded to the infection of viruses.

17.5.3 CDK Inhibitors

CKIs bind CDKs and inhibit cell-cycle progression. In mammals, there are two types of CKIs: INK4 and Kip/Cip. Plants do not appear to have INK4 CKIs, but proteins related to the Kip/Cip CKIs have been identified in plants and designated as Kip-related proteins (KRPs). Recently, a second group of CKIs, called SIAMESE in *Arabidopsis* and EL2 in rice, has been identified (Churchman et al., 2006; Peres et al., 2007). The SIAMESE-related genes are not present in animals and fungi,

Table 17.4b Rice CDK Family Genes and Their Expression Profiles in Virus-Infected and Drought-Stressed Tissues

	RDV		RRSV		RBSV		RSV_12D		RGSV	
Gene Name	**Inf**	**Mock**	**Inf**	**Mock**	**Inf**	**Mock**	**Inf**	**Mock**	**Inf**	**Mock**
					CDKA					
Orysa;CDKA;1	10.8	11.3	10.98	11.36	10.97	11.31	11.26	11.23	11.13	11.31
	−0.23		−0.35		−0.43		0.11		−0.18	
Orysa;CDKA;2	9.23	9.46	10.22	10.32	9.95	10.42	10.05	8.93	9.84	9.91
	−0.28		0		−0.42		**1.12**		0	
					CDKB					
Orysa;CDKB;1	11.26	10.84	11.63	11.26	11.35	11.45	10.78	10.39	11.22	11.37
	0.4		0.33		−0.12		0.55		0	
Orysa;CDKB;2	8.48	8.75	7.93	8.86	6.46	8.67	7.62	9.24	8.32	7.52
	0		**−0.98**		**−2.27**		**−1.3**		**0.94**	
					CDKC					
Orysa;CDKC;1	12.2	12.26	12.56	12.4	12.35	12.45	12.01	11.78	12.27	12.27
	0		0.17		−0.11		0		0	
Orysa;CDKC;2	13.53	13.52	13.44	13.48	13.41	13.39	13.32	13	13.34	13.53
	0		0		0		0.19		−0.19	
Orysa;CDKC;3	5.09	5.37	5.01	4.44	4.95	5.05	4.61	5.23	5.36	4.57
	0		0		0		0		0	
					CDKD					
Orysa;CDKD;1	12.36	12.43	12.62	12.68	12.74	12.81	12.2	12.04	12.67	12.59
	0		−0.08		−6.50E−02		0.16		7.90E−02	
					CDKE					
Orysa;CDKE;1	13.54	13.46	13.74	13.71	13.75	13.72	13.31	13.28	13.75	13.85
	0		3.40E−02		0		0		−9.20E−02	
					CDKF					
Orysa;CDKF;1	13.35	13.35	12.57	12.69	12.34	12.7	12.76	12.41	12.36	12.81
	0		0		−0.31		0.35		−0.45	
Orysa;CDKF;2	8.29	7.57	7.44	6.43	8.22	7.71	5.76	5.53	6.8	6.12
	0		**0.83**		0.46		0		0	
Orysa;CDKF;3	12.08	12.11	12.02	12.28	12.09	12.34	12.77	12.61	11.67	12.34
	0		−0.26		−0.32		0		**−0.68**	
					CDKG					
Orysa;CDKG;1	13.06	12.47	13.46	12.67	13.17	12.82	12.76	12.43	13.14	12.5
	0.59		**0.71**		0.36		0		**0.78**	
Orysa;CDKG;2	9.79	10.06	10.38	11.48	10.18	11.03	9.49	9.99	10.17	11.15
	0		**−1.1**		**−0.81**		−0.5		**−1.01**	
					CKL					
Orysa;CKL1	11.65	12.53	11.63	11.97	10.21	10.77	14.86	15.44	10.35	11.29
	−0.82		−0.33		−0.49		0		**−0.94**	
Orysa;CKL2	10.93	11.05	11.4	11.55	10.68	11.18	10.74	10.84	11.17	11.43
	−0.13		−0.15		−0.5		0		−0.22	
Orysa;CKL3	9.36	8.25	9.15	8.36	8.78	8.47	8.06	8.16	8.48	7.87
	0.6		**0.79**		0.4		0		**0.62**	
Orysa;CKL4	10.73	10.98	11.2	11.31	11.5	11.43	10.19	10.26	11.42	11.3
	0		−0.11		3.70E−02		0		0.16	
Orysa;CKL6	9.9	10.33	10.12	10.09	9.98	9.95	10.02	9.85	9.99	9.98
	0		0.11		0		0.22		0	
Orysa;CKL7	8.95	9.02	9.37	9.69	9.09	9.53	8.8	8.53	9.55	9.82
	−0.32		−0.34		−0.43		0		−0.26	
Orysa;CKL8	7.58	7.84	8.67	8.35	8.57	8.46	6.97	6.75	8.74	7.99
	0		0.39		0.12		0		**0.69**	
Orysa;CKL9	2.11	2.61	2.26	2.38	2.53	2.86	2.35	1.95	2.43	2.34
	0		0		0		0		0	
Orysa;CKL10	9.49	9.58	9.52	9.33	9.38	9.39	10.35	10.52	9.54	9.3
	0		0.24		0		−9.70E−02		0.24	

RTYV		RTSV		RTBV		Leaf		Panicle		Root 0.2 FTSW		Root 0.5 FTSW	
Inf	Mock	Inf	Mock	Inf	Mock	D	C	D	C	D	C	D	C
						CDKA							
10.75	10.58	11.32	11.35	11.47	11.65	11.89	11.7	12.16	11.99	12.25	12.5	13.21	12.62
;2		0		–0.31		0.15		0		0		**0.63**	
9.6	9.55	8.51	9.17	10.74	10.92	11	11.31	11.37	11.58	10.88	11.5	11.21	11.75
0		0		0		0		0		–0.31		0	
						CDKB							
11.06	11.07	11.3	11.11	10.98	11.1	11.36	11.68	11.66	12.12	10.74	11.55	10.52	11.52
–0.1		0		0		0		0		**–0.95**		**–0.7**	
8.57	9.04	9.14	9.31	10.37	10.83	7.61	8.28	11.77	12.34	8.88	11.42	10.9	11.32
0		0		–0.48		–0.47		**–0.86**		**–2.13**		**–1.11**	
						CDKC							
12.2	12	12.34	12.59	11.97	11.56	12.18	12.62	12.05	12.21	12.52	12.66	12.65	12.56
0		–6.10E–02		0.29		–0.48		–0.16		0		0.16	
13.38	13.47	13.63	14.16	13.39	13.2	13.87	14.15	13.66	13.81	14.57	14.67	14.33	14.46
0		–0.26		0.14		–0.38		–0.08		0		–0.13	
5.83	5.79	5.55	5.57	6.21	6.23	5.54	5.95	4.64	5.21	5.26	5.71	5.65	6.02
0		0		0		0		–0.43		–0.3		–0.49	
						CDKD							
12.45	12.5	12.78	12.51	12.38	12.31	13	13.33	12.82	12.87	13.77	13.47	13.53	13.46
–0.13		0		9.70E–02		–0.33		0		0.3		–5.00E–02	
						CDKE							
13.36	13.46	13.21	13.4	12.78	12.95	13.84	14.11	13.7	13.98	14.01	14.22	14.04	14.34
–0.12		–0.16		0		–0.33		–0.2		0		–0.31	
						CDKF							
12.57	12.51	12.89	13.1	13.11	13.33	13.4	14.03	14.05	14.02	14.75	15.51	14.44	15.21
0		0		–0.3		–0.51		0		0		**–0.78**	
7.84	8.74	8.37	7.96	10.17	11.05	5.99	5.91	6.31	7.08	9.35	10.1	10.84	10.36
–0.48		0		**–0.81**		0		0		**–1.03**		0.47	
11.79	11.96	12.56	12.84	12.8	13.11	12.21	12.51	12.44	12.5	13.26	12.89	13.27	13.06
0		0		–0.31		–0.28		0		0.16		0.4	
						CDKG							
12.86	12.89	13.38	13.92	13.89	13.58	12.98	13.44	12.67	12.53	13.61	13.16	13.37	13.16
0		–0.25		0		–0.35		0		0.44		8.20E–02	
10.01	10.24	9.8	9.71	10.29	10.36	10.25	10.55	10.08	9.98	9.75	8.11	9.72	8.47
–0.32		0		0		–0.11		0.1		**1.12**		**1.28**	
						CKL							
11.61	11.94	15.29	15.46	10.32	10.63	9.43	10.47	13.63	11.3	9.52	9.71	9.84	8.89
–0.2		–0.47		0		0		**1.21**		0		0	
10.84	10.92	11.86	11.99	12.1	12.09	11.6	12.43	10.73	11.34	12.42	13.1	12.47	12.82
–7.90E–02		0		0		**–0.68**		**–0.62**		**–1.04**		–0.11	
9.39	9.22	9.3	9.18	9.65	9.36	9.95	9.75	8.41	8.03	9.24	7.93	8.94	7.92
0		0		0		0		0		0		**1.32**	
11.05	11.21	9.17	9.13	10.13	10.29	11.66	11.7	11.14	11.08	11.86	12.14	12.43	12.32
0		0		0		0		0		–0.18		0	
9.73	10.01	10.74	10.88	10.52	10.81	10.47	10.57	10.23	10.08	11.45	10.72	10.83	10.5
–0.25		0		–0.23		0		0		**0.71**		0.33	
8.8	8.84	7.16	7.6	9.29	9.87	7.36	7.12	7.25	6.92	6.66	5.72	6.45	5.77
0		–0.31		0		0.4		0		**0.68**		**0.72**	
7.91	7.83	7.48	7.31	7.32	7.89	9.35	8.41	7.13	7.57	7.05	5.76	6.27	6.29
0		9.80E–02		0		**1.14**		–0.44		**1.37**		–0.38	
2.19	2.37	5.35	4.71	2.56	3.07	6.57	6.09	13.74	12.5	3.44	5.02	5.43	5.26
0		0		0		0		0.51		**–1.57**		–6.20E–02	
9.6	9.47	10.86	10.87	9.8	9.94	10.3	10.22	10.79	10.47	12.97	11.12	12.02	11.24
0.01		0		0		0		0.3		**1.58**		**0.79**	

and appear to be unique to plants. In *Arabidopsis*, seven *KRP* genes (*Arath;KRP1–7*) were identified, and their inhibitory activity against CDK was confirmed both *in vitro* and *in vivo* (De Veylder et al., 2001; Verkest et al., 2005; Wang et al., 1998; Zhou et al., 2002). The *KRP* genes also arrest the cell cycle in response to specific developmental or environmental cues. Abscisic acid and cold conditions induce *Arath;KRP1* expression (Wang et al., 1998). *Arath;KRP2* expression is negatively regulated by auxin during early lateral root initiation (Himanen et al., 2002). *Arath;KRP6* and *Arath;KRP7* are involved in the control of germline proliferation (Gusti et al., 2009; Kim et al., 2008; Liu et al., 2008). In addition to their role in blocking cell-cycle progression, KRPs have also been suggested to regulate nuclear DNA endoreduplication. In *Arabidopsis*, there is functional evidence that KRPs promote endoreduplication (Verkest et al., 2005; Weinl et al., 2005). In monocots, two maize (*Zea mays*) genes, *Zeama;KRP1* and *Zeama;KRP2*, were characterized and shown to be expressed in developing endosperm. Both proteins inhibit endosperm Cdc2-related CDK activity *in vitro*. Zeama;KRP1 was suggested to be involved in endoreduplication during the middle stage of endosperm development (Coelho et al., 2005). Seven rice KRP genes (*Orysa;KRP1–7*) were identified in the rice genome database (Barrôco et al., 2006; Guo et al., 2007). Overexpression of *Orysa;KRP1* reduced cell production and seed filling, which was accompanied by a drop in endoreduplication during the late stage of endosperm development; this suggests a role of Orysa;KRP1 in cell-cycle control at the maturation stage of seed development (Barrôco et al., 2006). KRP3 plays a key role in the cell-cycle control of syncytial endosperm. That is, the expression of *KRP3* is strongly upregulated in multinucleate syncytial endosperm and significantly reduced in cellularized endosperm. Y2H analysis and yeast mutant complementation analysis further support KRP3's role in syncytial endosperm (Mizutani et al., 2009).

Seven KRP genes in rice are classified into three groups by their expression levels. The most highly expressed group consists of *KRP1*, *-4* and *-6*. The second group contains the moderately expressed *KRP3* and *KRP5*. The third group contains the very slightly expressed *KRP2* and *KRP7*. Table 17.4c summarizes the CKI family genes analysed in this study.

Infection by virus (RDV, RRSV, RBSDV, RSV, RGSV, RTSV and RTBV) mainly downregulated the expression of KRP6. Under drought stress, the expression of *KRP6* was upregulated in leaf but downregulated in root at 50% FTSW. Drought stress upregulated the expression of *KRP5* (leaf only) and root (at 20% FTSW). Virus infection also induced the expression of *KRP1* (RGSV) and *KRP3* (RBSDV).

17.5.4 E2F/DP Transcription Factor and Rb Homologs

The E2F pathway is conserved in mammals and plants (De Veylder et al., 2003; Dewitte and Murray, 2003; Inzé, 2005). In the genome of *Arabidopsis*, three E2F genes (*E2Fa*, *E2Fb* and *E2Fc*) and two DP genes (*DPa* and *DPb*) have been identified (Vandepoele et al., 2002). E2Fa and E2Fb are potent transcriptional activators, as demonstrated by their ability to transactivate reporter genes harbouring the E2F consensus *cis*-acting element (Mariconti et al., 2002; Stevens et al., 2002). Moreover, transient overexpression of *E2Fa* and *DPa* induces nondividing mesophyll cells to reenter the S-phase (Rossignol et al., 2002), whereas their constitutive overexpression induces plant cells to undergo either ectopic cell division or enhanced DNA endoreduplication (De Veylder et al., 2002; Kosugi and Ohashi, 2003). In contrast, E2Fc, which lacks a strong activating domain, negatively regulates the E2F-responsive genes, because its ectopic expression inhibits cell division (del Pozo et al., 2002). E2F and DP each have a DNA-binding domain and a leucine zipper dimerization domain, a 'marked box' domain and an Rb-binding domain, while DP-like (DEL) protein has only two DNA-binding domains.

Four E2F genes (*E2F1–4*), three DPs (*DP1–3*), two DELs (*DEL1–2*) and two RBs (*Rb1–2*) have been identified in the rice genome (Table 17.4d). *E2F1* showed lower expression than the other three *E2F* genes in all tissues tested. *DP1* and *DP2* showed higher expression than *DP3*. Both *DEL1* and *DEL2* showed intermediate expression.

Table 17.4c Rice CKI Family Genes and Their Expression Profiles in Virus-Infected and Drought-Stressed Tissues

	RDV		RRSV		RBSV		RSV		RGSV		RTYV		RTSV		RTBV		Leaf		Panicle		Root 0.2 FTSW		Root 0.5 FTSW	
Gene Name	**Inf**	**Mock**	**Inf**	**Mock**	**Inf**	**Mock**	**Inf**	**Mock**	**Inf**	**Mock**	**Inf**	**Mock**	**Inf**	**Mock**	**Inf**	**Mock**	**D**	**C**	**D**	**C**	**D**	**C**	**D**	**C**
Orysa;KRP1	12.09	12.23	12.37	12.23	12.74	12.27	11.53	11.95	12.97	12.11	12.78	12.83	11.87	11.68	12.51	12.15	13	13.2	12.53	12.78	10.75	11.35	11.57	11.57
	0		0.11		0.47		**0**		**0.98**		0		0.25		5.70E−02		0		−0.27		**−0.61**		−8.00E−03	
Orysa;KRP2	5.09	5.89	5.72	6.56	5.3	5.42	6.87	7.3	5.31	6.28	5.34	5.48	6.71	6.78	6.64	7.39	5.68	5.97	7.92	8.3	5.5	6.6	3.94	6.75
	0		0		0		−0.72		0		0		0		0		−0.53		**0**		**−1.28**		**−2.56**	
Orysa;KRP3	8.45	8.14	6.38	6.44	7.36	6.38	9.35	9.18	6.19	6.25	9.26	9.56	11.06	10.98	8.03	7.91	8.68	7.91	6.38	6.11	8.47	7.53	7.07	6.61
	0		**0**		**0.92**		**0**		0		−0.2		**0**		**0**		0.41		**0**		**1.3**		**0.75**	
Orysa;KRP4	12.48	12.61	12.47	12.7	12.66	12.79	12.53	12.35	12.6	12.96	12.58	12.77	11.47	11.07	12.08	11.98	13.19	12.76	12.12	12.13	13.16	13.39	13.33	13.77
	−5.90E−02		−0.27		−0.18		0		−0.35		−0.18		0.37		0		0.37		0		0		−0.5	
Orysa;KRP5	8.4	8.43	8.34	8.27	8.6	8.12	7.55	8.05	8.64	8.51	7.92	8.65	8.49	8.28	8.66	9.73	9.31	8.69	9.9	10.37	11.65	12.34	12	12.53
	0		0		0.48		−0.38		0.24		−0.29		0.25		**−0.99**		**0.72**		−0.47		**−0.72**		**−0.88**	
Orysa;KRP6	10.33	12.22	10.25	11.65	10.16	10.9	11.12	12.52	9.43	10.93	11.92	12.29	8.29	9.53	11.49	12.5	11.86	9.99	13.12	12.77	11	11.74	11.01	11.87
	−2.11		**−1.34**		**−0.92**		**−1.4**		**−1.48**		−0.38		**−1.24**		**−1.01**		**1.98**		0.31		**0**		**−1.6**	
Orysa;KRP7	5.05	4.94	5.48	5.78	4.71	5.43	6.79	6.98	4.78	5.06	4.52	4.7	8.52	8.01	7.51	7.41	6.69	6.31	5.64	5.63	8.27	7.19	5.92	6.79
	0		0		0		0		0		**0**		0.54		**0**		0.34		**0**		**1.07**		**−1.03**	

Table 17.4d Rice E2F/DP and Rb Genes and Their Expression Profiles in Virus-Infected and Drought-Stressed Tissues

	RDV		RRSV		RBSV		RSV		RGSV		RTYV		RTSV		RTBV		Leaf		Panicle		Root 0.2 FTSW		Root 0.5 FTSW	
Gene Name	**Inf**	**Mock**	**Inf**	**Mock**	**Inf**	**Mock**	**Inf**	**Mock**	**Inf**	**Mock**	**Inf**	**Mock**	**Inf**	**Mock**	**Inf**	**Mock**	**D**	**C**	**D**	**C**	**D**	**C**	**D**	**C**
											E2F													
Orysa;E2F1	7.32	7.15	7.11	6.8	7	7.09	5.78	6.02	6.88	6.86	7.13	7.13	7.93	8.23	8.51	7.73	7.32	7.6	6.74	6.8	7.55	7.69	7.81	7.71
	0.21		0.49		0		0		0		0		−0.18		**0.78**		0		0		0		−0.13	
Orysa;E2F2	9.52	9.86	10.06	9.36	9.64	9.29	8.82	8.62	9.62	9.85	9.22	9.07	10.57	10.42	10.84	10.43	10.15	10.73	10.01	10.27	10.36	9.95	9.98	10.19
	0		**0.82**		0.29		0		−0.23		0		0		0		**−0.59**		−0.26		0.19		−0.57	
Orysa;E2F3	10.76	9.29	9.72	9.28	9.52	9.35	9.83	9.33	10.14	9.37	10.37	9.67	8.81	8.92	9.21	9.83	9.44	9.24	9.89	9.88	10.56	8.5	9.72	8.54
	1.47		0.33		0.16		0.51		**0.7**		0.58		**0**		**0**		**0**		7.00E−02		**2.06**		**1.19**	
Orysa;E2F4	10.55	10.7	10.7	10.44	10.53	10.52	10.24	10.27	10.6	10.71	10.71	10.73	11.04	11.11	10.89	10.37	11.62	11.47	10.87	10.83	10.93	10.84	10.94	10.95
	0		0.22		0		0		−0.11		0		−0.18		0.52		0.16		0.13		0.25		0	
											DP													
Orysa;DP1	9.61	9.24	9.8	9.76	10.05	10.15	8.91	8.45	10.17	10.2	9.1	9.14	7.86	7.44	8.31	8.64	9.6	9.22	11.07	10.95	12.22	11.1	11.98	11.34
	0.18		8.80E−02		0		0.49		0		−0.11		0.22		−0.33		0.24		0.13		**1.11**		0.45	
Orysa;DP2	10.2	10.16	10.31	10.35	10.1	10.47	9.76	9.5	10.19	10.36	10.02	9.99	10.27	10.29	9.95	9.82	9.81	10.24	10.74	10.79	10.54	10.84	11.21	10.95
	0		0		−0.31		0.27		−0.17		0		0		0.15		−0.32		0		−0.29		0	
Orysa;DP3	6.1	5.58	6.41	6.09	6.52	6.15	7.27	7.29	6.18	6.16	6.17	5.86	8.19	7.77	8.06	7.71	7.62	7.57	6.67	6.52	8.59	9.2	7.5	7.99
	0		0		0		0		0		0		0.23		0.35		0		0.15		0		−0.12	
											DEL													
Orysa;DEL1	5.14	7.19	5.94	6.3	3.4	5.51	5.21	7.05	4.8	5.17	6.62	7.16	7.58	7.97	8.09	9.93	5.72	5.91	7.42	8.43	3.93	8.57	6.55	8.75
	−2.05		0		0		**−1.45**		0		**0**		**0**		**−1.42**		**0**		**−1.04**		**−4.03**		**−2.24**	
Orysa;DEL2	4.44	4.6	5.36	7.02	4.26	7.57	4.12	4.65	6.05	7.29	5.03	5.01	7.73	8.39	6.96	6.61	3.21	4.65	5.55	6.22	7.59	8.73	9.62	8.81
	0		**−1.66**		**−3.27**		**0**		**−1.36**		**0**		**−0.66**		0		**−1.53**		0		**−1.14**		0	
											Rb													
Orysa;Rb1	11.79	12.08	12.11	12.32	12.12	12.41	12.02	11.98	12.29	12.46	11.94	11.96	11.18	11.27	10.34	10.41	10.54	10.99	10.85	10.86	10.76	11.01	10.59	10.7
	−0.38		−0.18		−0.29		0		−0.15		−2.30E−02		−0.12		0		−0.44		0		0		0	
Orysa;Rb2	4.46	5.06	2.73	3.63	2.39	4.18	2.59	3.97	2.61	5.16	5.43	5.78	6.98	7.65	8.36	5.78	3.67	4.92	6.86	7.51	4.34	8.05	8.28	8.04
	0		0		0		0		**0**		**0**		0		**2.22**		0		0		**−3.66**		9.50E−02	

Table 17.4e Rice CKS and Wee Genes and Their Expression profiles in Virus-Infected and Drought-Stressed Tissues

	RDV		RRSV		RBSV		RSV		RGSV		RTYV		RTSV		RTBV		Leaf		Panicle		Root 0.2 FTSW		Root 0.5 FTSW	
Gene Name	**Inf**	**Mock**	**Inf**	**Mock**	**Inf**	**Mock**	**Inf**	**Mock**	**Inf**	**Mock**	**Inf**	**Mock**	**Inf**	**Mock**	**Inf**	**Mock**	**D**	**C**	**D**	**C**	**D**	**C**	**D**	**C**
Orysa; CKS1	13.61	13.75	13.21	12.39	13.09	12.82	13.53	12.61	12.71	13.89	13.7	13.56	13.52	14.1	11.62	11.42	12.72	12.75	13.99	14.4	13.46	13.77	13.81	13.71
	−0.18		**0.83**		0.27		**0.86**		**−1.23**		0		0		0		0		−0.39		0		0.1	
Orysa; Wee1, Wee2	5.04	5.18	4.54	4.27	4.18	4.15	2.91	4.39	4.32	5.52	5.29	5.22	5.32	5.36	7.29	4.65	3.2	5.46	6.37	6.52	6.2	6.77	8.16	6.87
	0		0		0		0		0		0		0		**2.46**		**−1.23**		0		−0.23		**0.92**	

DEL2 was highly expressed specifically in root. *Rb1* showed high expression in all tissues tested, but *Rb2* showed low expression except in root. Virus infection specifically downregulated the expression of *DEL* genes. Among the eight viruses, RDV, RSV and RTBV specifically downregulated *DEL1*, and most of the other viruses downregulated *DEL2*. Drought stress in leaf downregulated only *DEL2*, that in panicle downregulated *DEL1* and that in root downregulated both *DEL1* and *DEL2*. Virus infection also upregulated the expression of specific E2F genes. *E2F1* was upregulated by RTBV, *E2F2* by RRSV and *E2F3* by RDV and RGSV. It was expected that upregulation of E2F would be coupled with upregulation of its heterodimeric partner DP, but the only example of this association was upregulation of both DP1 and E2F3 in root under drought stress. RTBV upregulated *Rb2* specifically, and drought stress in root drastically downregulated it. Downregulation of *DEL* genes and upregulation of *Rb* genes suggest deactivation of the cell cycle, while upregulation of *E2F* suggests activation of the cell cycle. These complicated gene-expression changes suggest that under conditions of virus infection and drought stress, cell division might be activated in some meristematic cells but suppressed in others.

17.5.5 Cell Division Cycle Kinase Subunit 1 and Wee

Cell division cycle (CDC) kinase subunit 1 (CKS) proteins have been proposed to act as docking factors that mediate the interactions between the kinase complex and the substrate (Morris et al., 2003). One gene, *CKS1*, has been identified in the rice genome (Table 17.4e). *CKS1* showed high expression in all tissues, and infections by RRSV and RSV further increased its expression, whereas infection by RGSV downregulated it. Drought stress produced no clear change in expression.

Wee protein kinase inhibits CDKs by phosphorylation of a tyrosine in the active site (Gonzalez et al., 2007; Shimotohno et al., 2006; Sun et al., 1999). Two rice wee genes (*Wee1* and *Wee2*) have been reported, but in our array analysis, one probe was used to detect the expression of both genes (Table 17.4e). Expression of *Wee* genes was upregulated by the infection of RTBV and drought stress in root, but clearly downregulated in leaf.

17.6 CONCLUSION

Transcriptome analysis of the responses of rice cell-cycle-regulator genes to virus infection and drought stress treatment has revealed complex differences among the responses to different viruses, and between the responses to virus infection and abiotic stress (Figure 17.2). It seems that the cell cycle can be stopped in a number of ways, both through inactivation of positive regulators (e.g., Cyclin, CDK, E2F/DP/DEL) and through activation of negative regulators (e.g., CKI, Rb, Wee). After a careful analysis of the transcriptome data, it is clear that virus infection and drought not only stop the cell cycle, but also induce the gene expression necessary for cell-cycle activation, probably because there are many meristematic cells that may be at different points in the cell cycle.

By comparing the gene-expression changes between virus infection and drought stress treatment, we found that virus infection generally targeted a specific member of a gene family, whereas no particular family member seemed to respond to drought stress treatment. This observation suggests that in the case of virus infection, specific protein factors act as messengers, whereas in the case of drought stress, such specific factors do not exist. In this chapter, we have shown the gene-expression profiles and changes after virus infection as an example. Similar kinds of analyses can be done for categories of genes that might be involved in response to other types of infection or stress.

REFERENCES

Albar, L., M. Bangratz-Reyser, E. Hébrand et al. 2006. Mutations in the eIF(iso)4G translation initiation factor confer high resistance to rice to rice yellow mottle virus. *Plant J* 47:417–426.

Angeles, E. R. and G. S. Khush. 2000. Genetic analysis of resistance to green leafhopper, *Nephotettix virescens* (Distant) in three varieties of rice. *Plant Breed* 119:446–448.

Anjaneyulu, A., V. D. Shukla, G. M. Rao and S. K. Singh. 1982. Experimental host range of rice tungro virus and its vectors. *Plant Dis* 66:54–56.

Ascencio-Ibáñez, J. T., R. Sozzani, T. J. Lee et al. 2008. Global analysis of *Arabidopsis* gene expression uncovers a complex array of changes impacting pathogen response and cell cycle during geminivirus infection. *Plant Physiol* 148:436–454.

Azzam, O., T. Imbe, R. Ikeda, P. D. Nath and E. Coloquio. 2001 Inheritance of resistance to Rice tungro spherical virus in a near-isogenic line derived from Utri Merah and in rice cultivar TKM6. *Euphytica* 122:91–97.

Bai, F. W., J. Yan, Z. C. Qu et al. 2002. Phylogenetic analysis reveals that a dwarfing disease on different cereal crops in China is due to rice black streaked dwarf virus (RBSDV). *Virus Genes* 25:201–206.

Barbier, P., M. Takahashi, I. Nakamura, S. Toriyama and A. Ishihama. 1992. Solubilization and promoter analysis of RNA polymerase from rice stripe virus. *J Virol* 66:6171–6174.

Barrett, T., D. B. Troup, S. E. Wilhite et al. 2007. NCBI GEO: Mining tens of millions of expression profiles—database and tools update. *Nucleic Acids Res.* 35 (Database issue):D760–765.

Barrôco, R. M., A. Peres, A.-M. Droual et al. 2006. The cyclin-dependent kinase inhibitor Orysa; KRP1 plays an important role in seed development of rice. *Plant Physiol* 142:1053–1064.

Boccardo, G. and R. G. Milne. 1984. Plant reovirus group. In *CM/AAB Descriptions of Plant Viruses*, eds. A. F. Morant and B. D. Harrison, 294pp. United Kingdom: Commonwealth Agricultural Bureaux, Farnham Royal, Slough.

Boruc, J., E. Mylle, M. Duda et al. 2009. Systematic localization of the *Arabidopsis* core cell cycle proteins reveals novel cell division complexes. *Plant Physiol* 152:553–565.

Brookes, S. M., A. D. Hyatt and B. T. Eaton. 1993. Characterization of virus inclusion bodies in bluetongue virus-infected cells. *J Gen Virol* 74:525–530.

Bucher, E., T. Sijen, P. De Haan, R. Goldbach and M. Prins. 2003. Negative-strand tospoviruses and tenuiviruses carry a gene for a suppressor of gene silencing at analogous genomic positions *J Virol* 77:1329–1336.

Burssens, S., J. D. A. Engler, T. Beeckman et al. 2000. Developmental expression of the *Arabidopsis thaliana* CycA2;1 gene. *Planta* 211:623–631.

Cabauatan, P. Q., N. Kobayashi, R. Ikeda and H. Koganezawa. 1993. *Oryza glaberrima*: An indicator plant for rice tungro spherical virus. *Int J Pest Manage* 39:273–276.

Cabunagan, R. C., E. R. Angeles, S. Villareal et al. 1999. Multilocation evaluation of advanced breeding lines for resistance to rice tungro viruses. In *Rice Tungro Disease Management*, ed. T. C. B. Chancellor, O. Azzam and K. L. Heong, pp. 45–55. Manila, Philippines: International Rice Research Institute.

Cao, X. S., P. Zhou, X. M. Zhang et al. 2005. Identification of an RNA silencing suppressor from a plant double-stranded RNA virus. *J Virol* 79:13018–13027.

Chomchan, P., S.-F. Li and Y. Shirako. 2003. Rice grassy stunt tenuivirus nonstructural protein p5 interacts with itself to form oligomeric complexes *in vitro* and *in vivo*. *J Virol* 77:769–775.

Chomchan, P., G. I. Miranda and Y. Shirako. 2002. Detection of rice grassy stunt tenuivirus nonstructural proteins p2, p5 and p6 from infected rice plants and from viruliferous brown planthoppers. *Arch Virol* 147: 291–300.

Churchman, M. L., M. L. Brown, N. Kato et al. 2006. SIAMESE, a plant-specific cell cycle regulator, controls endoreplication onset in *Arabidopsis thaliana. Plant Cell* 2006. 18:3145–3157.

Coelho, C. M., R. A. Dante, P. A. Sabelli et al. 2005. Cyclin-dependent kinase inhibitors in maize endosperm and their potential role in endoreduplication. *Plant Physiol* 138:2323–2336.

Csorba, T., V. Pantaleo and J. Burgyán. 2009. RNA silencing: An antiviral mechanism. *Adv Virus Res* 75:35–71.

Dahal, G., H. Hibino, R. C. Cabunagan et al. 1990. Changes in cultivar reaction to tungro due to changes in "virulence" of the leafhopper vector. *Phytopathology* 80:659–665.

Dai, S., Z. Zhang, S.Chen and R. N. Beachy. 2004. RF2b, a rice bZIP transcription activator, interacts with RF2a and is involved in symptom development of rice tungro disease. *Proc Natl Acad Sci USA* 101:687–692.

Dai, S., Z. Zhang, J. Bick and R. N. Beachy. 2006. Essential role of the Box II cis element and cognate host factors in regulating the promoter of Rice tungro bacilliform virus. *J Gen Virol* 87:715–722.

Dai, S., S. Petruccelli, M. I. Ordiz et al. 2003. Functional analysis of RF2a, a rice transcription factor. *J Biol Chem* 38:36396–36402.

Dales, S. 1963. Association between the spindle apparatus and reovirus. *Proc Natl Acad Sci USA* 50:268–275.

Dales, S., P. J. Gomatos and K. C. Hsu. 1965. The uptake and development of reovirus in strain L cells followed with labelled viral ribonucleic acid and ferritin–antibody complexes. *Virology* 25:193–211.

Dardick, C. 2007. Comparative expression profiling of *Nicotiana benthamiana* leaves systemically infected with three fruit tree viruses. *Mol Plant Microbe Interact* 20:1004–1017.

De Jager, S. M., Maughan, W. Dewitte, S. Scofield and J. A. H. Murray 2005. The developmental context of cell-cycle control in plants. *Semin Cell Dev Biol* 16:385–396.

De Jager, S. M., S.Scofield, R. P. Huntley, A. S. Robinson, B. G. den Boer and J. A. H. Murray. 2009. Dissecting regulatory pathways of G1/S control in *Arabidopsis*: common and distinct targets of CYCD3;1, E2Fa and E2Fc. *Plant Mol Biol* 71:345–365.

De Schutter, K., J. Joubes, T. Cools et al. 2007. Arabidopsis WEE1 kinase controls cell cycle arrest in response to activation of the DNA integrity checkpoint. *Plant Cell* 19:211–225.

De Veylder, L., T. Beeckman, G. T. Beemster et al. 2001. Functional analysis of cyclin-dependent kinase inhibitors of *Arabidopsis. Plant Cell* 13:1653–1668.

De Veylder, L., T. Beeckman, G. T. S.Beemster et al. 2002. Control of proliferation, endoreduplication and differentiation by the *Arabidopsis* E2Fa/DPa transcription factor. *EMBO J* 21:1360–1368.

De Veylder, L., J. D. A. Engler, S. Burssens et al. 1999. A new D-type cyclin of *Arabidopsis thaliana* expressed during lateral root primordial formation. *Planta* 208:453–462.

De Veylder, L., J. Joubès and D. Inzé. 2003. Plant cell cycle transitions. *Curr Opin Plant Biol* 6:536–543.

del Pozo, J. C., M. B. Boniotti and C. Gutierrez. 2002. *Arabidopsis* E2Fc functions in cell division and is degraded by the ubiquitin-SCFAtSKP2 pathway in response to light. *Plant Cell* 14:3057–3071.

Desvoyes, B., E. Ramirez-Parra, Q. Xie, N. H. Chua and C. Gutierrez. 2006. Cell type-specific role of the retinoblastoma/E2F pathway during *Arabidopsis* leaf development. *Plant Physiol* 140:67–80.

Dewitte, W. and J. A. H. Murray. 2003. The plant cell cycle. *Annu Rev Plant Biol* 54:235–264.

Dewitte, W., C. Riou-Khamlichi, S. Scofield et al. 2003. Altered cell cycle distribution, hyperplasia, and inhibited differentiation in *Arabidopsis* caused by the D-type cyclin CycD3. *Plant Cell* 15:79–92.

Dewitte, W., S. Scofield, A. A. Alcasabas et al. 2007. *Arabidopsis* CYCD3 D-type cyclins link cell proliferation and endocycles and are rate-limiting for cytokinin responses. *Proc Natl Acad Sci USA* 104:14537–14542.

Dissmeyer, N., M. K. Nowack, S. Pusch et al. 2007. T-loop phosphorylation of *Arabidopsis* CDKA;1 is required for its function and can be partially substituted by an aspartate residue. *Plant Cell* 19:972–985.

Dissmeyer, N., A. K.Weimer, S. Pusch et al. 2009. Control of cell proliferation, organ growth, and DNA damage response operate independently of dephosphorylation of the *Arabidopsis* Cdk1 homolog CDKA;1. *Plant Cell* 21:3641–3654.

Druca, A., T. Burns, S. Zang and R. Hull. 1996. Immunological characterization of rice tungro spherical virus coat proteins and differentiation of isolates from Philippines and India. *J Gen Virol* 77:1975–1983.

Ebron, L. A., R. R. Yumol, R. Ikeda and T. Imbe. 1994. Inheritance of resistance to rice tungro spherical virus in some rice cultivars. *Int Rice Res Notes* 19:10–11.

Egelkrout, E., L. Mariconti, R. Cella, D. Robertson and L. Hanley-Bowdoin. 2002. The activity of the proliferating cell nuclear antigen promoter is differentially regulated by two E2F elements during plant development. *Plant Cell* 14:3225–3236.

Egelkrout, E. M., D. Robertson and L. Hanley-Bowdoin. 2001. Proliferating cell nuclear antigen transcription is repressed through an E2F consensus element and activated by geminivirus infection in mature leaves. *Plant Cell* 13:1437–1452.

Encabo, J. R., P. Q. Cabauatan, R. C. Cabunagan et al. 2009. Suppression of two tungro viruses in rice by separable traits originating from cultivar Utri Merah. *Mol Plant Microbe Interact.* 22:1268–1281.

Espinoza, C., C. Medina, S. Somerville and P. Arce-Johnson. 2007. Senescence-associated genes induced during compatible viral interactions with grapevine and *Arabidopsis. J Exp Bot* 58:3197–3212.

Espinoza, A. M., R. Pereira, A. V. Macaya-Lizano et al. 1993. Comparative light and electron microscopic analyses of tenuivirus major noncapsid protein (NCP) inclusion bodies in infected plants, and of the NCP in vitro. *Virology* 195:156–166.

Fabbretti, E., I. Afrikanova, F. Vascotto and O. R. Burrone. 1999. Two non-structural rotavirus proteins, NSP2 and NSP5, form viroplasm-like structures in vivo. *J Gen Virol* 80:333–339.

Falk, B. W. and J. H. Tsai. 1998. Biology and molecular biology of viruses in the genus Tenuivirus. *Annu Rev Phytopathol* 36:139–163.

Fang, S., J. Yu, J. Feng et al. 2001. Identification of rice black-streaked dwarf fijivirus in maize with rough dwarf disease in China. *Arch Virol* 146:167–170.

Ferreira, P. C., A. S. Hemerly, R. Villarroel, M. van Montagu and D. Inzé. 1991. The *Arabidopsis* functional homolog of the p34cdc2 protein kinase. *Plant Cell* 3:531–540.

Florentino, L. H., A. A. Santos, M. R. Fontenelle et al. 2006. A PERK-like receptor kinase interacts with the geminivirus nuclear shuttle protein and potentiates viral infection. *J Virol* 80:6648–6656.

Fontes, E. P., A. A. Santos, D. F. Luz, A. J. Waclawovsky and J. Chory. 2004. The geminivirus nuclear shuttle protein is a virulence factor that suppresses transmembrane receptor kinase activity. *Genes Dev* 18:2545–2556.

Francis, D. 2007. The plant cell cycle—15 years on. *New Phytol* 174:261–278.

Fuerst, R. A., R. Soni, J. A. Murray and K. Lindsey. 1996. Modulation of cyclin transcript levels in cultured cells of *Arabidopsis thaliana. Plant Physiol* 112:1023–1033.

Fukushi, T., E. Shikata and I. Kimura. 1962. Some morphological characters of rice dwarf virus. *Virology* 18:192–205.

Ganesan, U., S. S. Suri, S. Rajasubramaniam, M. V. Rajam and I. Dasgupta. 2009. Transgenic expression of coat protein gene of rice tungro bacilliform virus in rice reduces the accumulation of viral DNA in inoculated plants. *Virus Genes* 39:113–119.

Genschik, P., M. C. Criqui, Y. Parmentier, A. Derevier and J. Fleck. 1998. Cell cycle-dependent proteolysis in plants: Identification of the destruction box pathway and metaphase arrest produced by the proteasome inhibitor mg132. *Plant Cell* 10:2063–2076.

Girard, F., U. Strausfeld, A. Fernandez and N. J. C. Lamb. 1991. Cyclin A is required for the onset of DNA replication in mammalian fibroblasts. *Cell* 67:1169–1179.

Golem, S. and J. N. Culver. 2003. Tobacco mosaic virus induced alterations in the gene expression profile of *Arabidopsis thaliana. Mol Plant Microbe Interact* 16:681–688.

Gonzalez, N., F. Gevaudant, M. Hernould, C. Chevalier and A. Mouras. 2007. The cell cycle-associated protein kinase WEE1 regulates cell size in relation to endoreduplication in developing tomato fruit. *Plant J* 51:642–655.

Gonzalez, N., M. Hernould, F. Delmas et al. 2004. Molecular characterization of a WEE1 gene homologue in tomato (*Lycopersicon esculentum* Mill.). *Plant Mol Biol* 56:849–861.

Guo, J., J. Song, F. Wang and X. S. Zhang. 2007. Genome-wide identification and expression analysis of rice cell cycle genes. *Plant Mol Biol* 64:349–360.

Guo, J., F. Wang, J. Song, W. Sun and X. S. Zhang. 2010. The expression of Orysa; CycB1;1 is essential for endosperm formation and causes embryo enlargement in rice. *Planta* 231:293–303.

Gusti, A., N. Baumberger, M. Nowack et al. 2009. The *Arabidopsis thaliana* F-box protein FBL17 is essential for progression through the second mitosis during pollen development. *PLoS One* 4: e4780.

Hadwiger, J. A. and S. I. Reed. 1988. Invariant phosphorylation of the *Saccharomyces cerevisiae* Cdc28 protein kinase. *Mol Cell Biol* 8:2976–2979.

Hagiwara, K., T. Higashi, K. Namba, T. Uehara-Ichiki and T. Omura. 2003. Assembly of single-shelled cores and double-shelled virus-like particles after baculovirus expression of major structural proteins P3, P7 and P8 of rice dwarf virus. *J Gen Virol* 84:981–984.

Hagiwara, K., T. Higashi, N. Miyazaki et al. 2004. The amino-terminal region of major capsid protein P3 is essential for self-assembly of single-shelled core-like particles of Rice dwarf virus. *J Virol* 78:3145–3148.

Hanley-Bowdoin, L., S. B. Settlage and D. Robertson. 2004. Reprogramming plant gene expression: A prerequisite to geminivirus DNA replication. *Mol Plant Pathol* 5:149–156.

Hao, L., H. Wang, G. Sunter and D. M. Bisaro. 2003. Geminivirus AL2 and L2 proteins interact with and inactivate SNF1 kinase. *Plant Cell* 15:1034–1048.

Harashima, H., A. Shinmyo and M. Sekine. 2007. Phosphorylation of threonine 161 in plant cyclin-dependent kinase A is required for cell division by activation of its associated kinase. *Plant J* 52:435–448.

Hay, J. M., M. C. Jones, M. L. Blakebrough et al. 1991. An analysis of the sequence of an infectious clone of rice tungro bacilliform virus, a plant pararetrovirus. *Nucleic Acids Res* 19:2615–2621.

Hayano-Saito, Y., K. Saito, S. Nakamura, S. Kawasaki and M. Iwasaki. 2000. Fine physical mapping of the rice stripe resistance gene locus, Stvb-i. *Theor Appl Genet* 101:59–63.

He, X., J. Futterer and T. Hohn. 2001. Sequence-specific and methylation-dependent and -independent binding of rice nuclear proteins to a rice tungro bacilliform virus vascular bundle expression element. *J Biol Chem* 276:2644–2651.

He, X., J. Futterer and T. Hohn. 2002. Contribution of downstream promoter elements to transcriptional regulation of the rice tungro bacilliform virus promoter. *Nucleic Acids Res* 30:497–506.

He, X., T. Hohn and J. Futterer. 2000. Transcriptional activation of the rice tungro bacilliform virus gene is critically dependent on an activator element located immediately upstream of the TATA box. *J Biol Chem* 275:11799–11808.

Hibino, H. 1979. Rice ragged stunt, a new virus disease occurring in tropical Asia. *Rev Plant Prot Res* 12:98–110.

Hibino, H. 1983. Relations of rice tungro bacilliform and rice tungro spherical viruses with their vector *Nephotettix virescens. Annu. Phytopathol Soc. Jpn* 49:45–53.

Hibino, H. 1986. Rice grassy stunt virus. *CMI/AAB Description of Plant Viruses No. 320.*

Hibino, H. 1989. Insect-borne viruses of rice. In *Advances in Disease Vector Research*, ed. K. F. Harris, Vol. 6, pp. 209–421. New York, NY: Springer-Verlag.

Hibino, H. 1995. Rice tungro-associated viruses and their relationships to host plants and vector leafhoppers. In *Pathogenesis and Host Specificity in Plant Diseases, Histopathological, Biochemical, Genetic and Molecular Bases. Viruses and Viroids*, Vol. III, eds. P. R. Singh, U. S. Singh and K. Khomoto, pp. 393–403. Oxford: Elsevier.

Hibino, H. 1996. Biology and epidemiology of rice viruses. *Annu Rev Phytopathol* 34:249–274.

Hibino, H., R. D. Daquiaog, E. M. Mesina and V. M. Aguiero. 1990. Resistances in rice to tungro-associated viruses. *Plant Dis* 74:923–926.

Hibino, H., M. Roechan and S. Sudarisman. 1978. Association of two types of virus particles with penyakit habang (tungro disease) of rice in Indonesia. *Phytopathology* 68:1412–1416.

Hibino, H., E. R. Tiongco, R. C. Cabunagan and Z. M. Flores. 1987. Resistance to rice tungro-associated viruses in rice under experimental and natural conditions. *Phytopathology* 77:871–875.

Hill, J. E., J. O. Strandberg, E. Hiebert and S. G. Lazarowitz. 1998. Asymmetric infectivity of pseudorecombinants of cabbage leaf curl virus and squash leaf curl virus: Implications for bipartite geminivirus evolution and movement. *Virology* 250:283–292.

Himanen, K., E. Boucheron, S. Vanneste et al. 2002. Auxin-mediated cell cycle activation during early lateral root initiation. *Plant Cell* 14:2339–2351.

Hiraguri, A., H. Hibino, T. Hayashi et al. 2010. Complete sequence analysis of rice transitory yellowing virus and its comparison to rice yellow stunt virus. *Arch Virol* 155: 243–245.

Hirayama, T., Y. Imajuku, T. Anai, M. Matsui and A. Oka, 1991. Identification of two cell-cycle-controlling cdc2 gene homologs in *Arabidopsis thaliana. Gene* 105:159–165.

Huang, Y. W., Y. F. Geng, X. B. Ying, X. Y. Chen and R. X. Fang. 2005. Identification of a movement protein of rice yellow stunt rhabdovirus. *J Virol* 79:2108–2114.

Huang, Y., H. Zhao, Z. Luo, X.Chen and R. X. Fang. 2003. Novel structure of the genome of rice yellow stunt virus: Identification of the gene 6-encoded virion protein. *J Gen Virol* 84:2259–2264.

ICTVdB Management. 2006. *ICTVdB—The Universal Virus Database, Version 4.* ed. C. Büchen-Osmond, New York, NY, USA: Columbia University.

Iida, T. T., A. Shinkai and I. Kimura. 1972. Rice dwarf virus. *CMI/AAB Description of Plant Viruses No. 172.*

Imajuku, Y., T. Hirayama, H. Endoh and A. Oka. 1992. Exon-intron organization of the *Arabidopsis thaliana* protein kinase genes CDC2a and CDC2b. *FEBS Lett.* 304: 73–77.

Inzé, D. 2005. Green light for the cell cycle. *EMBO J* 24:657–662.

Isogai, M., I. Uyeda and B. C. Lee. 1998. Detection and assignment of proteins encoded by rice black streaked dwarf fijivirus S7, S8, S9 and S10. *J Gen Virol* 79:1487–1494.

Isogai, M., P. Q. Cabauatan, C. Masuta, I. Uyeda and O. Azzam. 2000. Complete nucleotide sequence of the rice tungro spherical virus genome of the highly virulent strain Vt6. *Virus Genes* 20:79–85.

Ito, M., M.-C. Criqui, M. Sakabe et al. 1997. Cell-cycle regulated transcription of A- and B-type plant cyclin genes in synchronous cultures. *Plant J* 11:983–992.

Jang, S. K., H.-G. Krausslich, M. Nicklin et al. 1988. A segment of the 5' nontranslated region of encephalomyocarditis virus RNA directs internal entry of ribosomes. *J Virol* 62:2636–2643.

Kakutani, T., Y. Hayano, T. Hayashi and Y. Minobe. 1990. Ambisense segment 4 of rice stripe virus: Possible evolutionary relationship with phleboviruses and uukuviruses (*Bunyaviridae*) *J Gen Virol* 71:1427–1432.

Kakutani, T., Y. Hayano, T. Hayashi and Y. Minobe. 1991. Ambisense segment 3 of rice stripe virus: The first instance of a virus containing two ambisense segments *J Gen Virol* 72:465–468.

Kavli, B., O. Sundheim, M. Akbari et al. 2002. hUNG2 is the major repair enzyme for removal of uracil from U:A matches, U:G mismatches, and U in single-stranded DNA, with hSMUG1 as a broad specificity backup. *J Biol Chem* 277:9926–9936.

Kawano, S. 1984. Particle structure and double stranded RNA of rice ragged stunt virus. *J Fac Agric Hokkaido Univ* 61:408–418.

Kim, H. J., S. A. Oh, L. Brownfield et al. 2008. Control of plant germline proliferation by SCF(FBL17) degradation of cell cycle inhibitors. *Nature* 455:1134–1137.

King, R. W., R. J. Deshaies, J. M. Peters and M. W.Kirschner. 1996. How proteolysis drives the cell cycle. *Science* 274:1652–1659.

Kong, L. and L. Hanley-Bowdoin. 2002. A geminivirus replication protein interacts with a protein kinase and a motor protein that display different expression patterns during plant development and infection. *Plant Cell* 14:1817–1832.

Kosugi, S. and Y. Ohashi. 2003. Constitutive E2F expression in tobacco plants exhibits altered cell cycle control and morphological change in a cell type-specific manner. *Plant Physiol* 132:2012–2022.

La, H., J. Li, Z. Ji et al. 2006. Genome-wide analysis of cyclin family in rice (*Oryza sativa* L.). *Mol Gen Genomics* 275:374–386.

Laco, G. S., S. B. H. Kent and R. N. Beachy. 1995. Analysis of the proteolytic processing and activation of the rice tungro bacilliform virus reverse transcriptase. *Virology* 208:207–214.

Landrieu, I., M. da Costa, L. De Veylder et al. 2004. A small CDC25 dual-specificity tyrosine-phosphatase isoform in *Arabidopsis thaliana. Proc Natl Acad Sci USA* 101:13380–13385.

Lara-Núñez, A., N. de Jesús and J. M. Vázquez-Ramos. 2008. Maize D4;1 and D5 cyclin proteins in germinating maize. Associated kinase activity and regulation by phytohormones. *Physiol Plant* 132:79–88.

Lee, J. H., M. Muhsin, G. A. Atienza et al. 2010. Single nucleotide polymorphisms in a gene for translation initiation factor (eIF4G) of rice (*Oryza sativa*) associated with resistance to Rice tungro spherical virus. *Mol Plant Microbe Interact.* 23:29–38.

Lehner, C. F. and P. H. O'Farrell. 1990. The roles of *Drosophila* cyclins A and B in mitotic control. *Cell* 61:535–547.

Li, Z., N. M. Upadhyaya, W. Kositratana, A. J. Gibbs and P. M. Waterhouse. 1996. Genome segment 5 of rice ragged stunt virus encodes a virion protein. *J Gen Virol* 77:3155–3160.

Li, Y., Y. M. Bao, C. H. Wei et al. 2004. Rice dwarf phytoreovirus segment S6-encoded nonstructural protein has a cell-to-cell movement function. *J Virol* 78:5382–5389.

Liu, J., Y. Zhang, G. Qin et al. 2008. Targeted degradation of the cyclin-dependent kinase inhibitor ICK4/KRP6 by RING-type E3 ligases is essential for mitotic cell cycle progression during *Arabidopsis* gametogenesis. *Plant Cell* 20:1538–1554.

Lu, H. H., Z. X. Gong and T. Q. Cao. 1990. Studies on the genomic coding assignments of rice ragged stunt virus. *Chin J Virol* 6:167–172.

Lu, H. J., C. G. Shao and Z. X. Gong. 2002. Self-aggregation of the structural protein encoded by rice ragged stunt Oryzavirus genome segment 8. *Acta Biochim Biophys Sin* 34:565–570.

Lu, H. H., J. H. Wu, Z. X. Gong and T. Q. Cao. 1987. *In vitro* translation of 10 segments of the dsRNA of rice ragged stunt virus (RRSV). *Acta Biochim Biophys Sin* 19:354–359.

Lu, X. B., G. Y. Zhou, J. H. Wu and Z. X. Gong. 1997. Sequence analysis of segment 9 of RRSV and its expression in *E. coli* DE3. *Chin J Virol* 13:68–74.

Lu, L., Z. Du, M. Qin et al. 2009. Pc4, a putative movement protein of rice stripe virus, interacts with a type I DnaJ protein and a small Hsp of rice. *Virus Genes* 38:320–327.

Luo, Z. L. and R. X. Fang. 1998. Structure analysis of the rice yellow stunt rhabdovirus glycoprotein gene and its mRNA. *Arch Virol* 143:2453–2459.

Luo, Z., X. Chen, D. Gao and R. Fang. 1998. The gene 4 of rice yellow stunt rhabdovirus encodes the matrix protein. *Virus Genes* 16:277–280.

Maeda, H., K. Matsushita, S. Iida and Y. Sunohara. 2006. Characterization of two QTLs controlling resistance to rice stripe virus detected in a Japanese upland rice line Kanto 72. *Breed Sci* 56:59–64.

Marathe, R., Z. Guan, R. Anandalakshmi, H. Zhao and S. P. Dinesh-Kumar. 2004. Study of *Arabidopsis thaliana* resistome in response to cucumber mosaic virus infection using whole genome microarray. *Plant Mol Biol* 55:501–520.

Mariconti, L., B. Pellegrini, R. Cantoni et al. 2002. The E2F family of transcription factors from *Arabidopsis thaliana*. Novel and conserved components of the retinoblastoma/E2F pathway in plants. *J Biol Chem* 277:9911–9919.

Marmey, P., B. Bothner, E. Jacquot et al. 1999. Rice tungro bacilliform virus open reading frame 3 encodes a single 37-kDa coat protein. *Virology* 253:319–926.

Mayo, M. A., J. R. De Miranda, B. W. Falk, R. Goldbach, A.-L. Haenni and S. Toriyama. 2000. Genus Tenuivirus. In *Virus Taxonomy*, eds. M. H. V. van Regenmortel, C. M. Fauquet, D. H. L. Bishop, E. B. Carstens, M. K. Estes, S. M. Lemon, J. Maniloff et al. pp. 904–908. New York, NY: Academic Press.

Meijer, M. and J. A. H. Murray. 2001. Cell cycle controls and the development of plant form. *Curr Opin Plant Biol* 4:44–49.

Melcher, U. 2000. The '30K' superfamily of viral movement proteins. *J Gen Virol* 81:257–266.

Menges, M., S. M. de Jager, W. Gruissem and J. A. Murray. 2005. Global analysis of the core cell cycle regulators of *Arabidopsis* identifies novel genes, reveals multiple and highly specific profiles of expression and provides a coherent model for plant cell cycle control. *Plant J* 41:546–566.

Menges, M., A. K. Samland, S. Planchais and J. A. H. Murray. 2006. The D-type cyclin CYCD3;1 is limiting for the G1-to-S-phase transition in *Arabidopsis. Plant Cell* 18:893–906.

Milne, R. G. and K. C. Ling. 1982. Rice ragged stunt virus. *CMI/AAB Description of Plant Viruses No. 248.*

Mineyuki, Y. 1999. The preprophase band of microtubules: Its function as a cytokinetic apparatus in higher plants. *Int Rev Cytol* 187:1–50.

Miranda, G. J., O. Azzam and Y. Shirako. 2000. Comparison of nucleotide sequences between northern and southern Philippine isolates of rice grassy stunt virus indicates occurrence of natural genetic reassortment. *Virology* 266:26–32.

Miranda, G., M. L. M. Yambao, P. Q. Cabauatan, S. R. Venkitesh and H. Koganezawa. 1995. Degradation of a coat protein of rice tungro spherical virus during purification. *Ann Phytopathol Soc Jpn* 61:41–43.

Mironov, V., L. De Veylder, M. van Montagu and D. Inzé. 1999. Cyclin-dependent kinases and cell division in higher plants—The nexus. *Plant Cell* 11:509–521.

Miyazaki, N., K. Hagiwara, H. Naitow et al. 2005. Transcapsidation and the conserved interactions of two major structural proteins of a pair of phytoreoviruses confirm the mechanism of assembly of the outer capsid layer. *J Mol Biol* 345:229–237.

Mizutani, M., T. Naganuma, K. Tsutsumi and Y. Saitoh. 2009. The syncytium-specific expression of the Orysa;KRP3 CDK inhibitor: Implication of its involvement in the cell cycle control in the rice (*Oryza sativa* L.) syncytial endosperm. *Exp Bot* 61:91–98.

Morgan, D. O. 1995. Principles of CDK regulation. *Nature* 374:131–134.

Morris, M. C., P. Kaiser, S. Rudyak et al. 2003. Cks1-dependent proteasome recruitment and activation of CDC20 transcription in budding yeast. *Nature* 423:1009–1013.

Nakagawa, A., N. Miyazaki, J. Taka et al. 2003. The atomic structure of rice dwarf virus reveals the self-assembly mechanism of component proteins. *Structure* 11:1227–1238.

Nicaise, V., J. L. Gallois, F. Chafiai et al. 2007. Coordinated and selective recruitment of eIF4E and eIF4G factors for potyvirus infection in *Arabidopsis thaliana. FEBS Lett* 581:1041–1046.

Nilsen, H., I. Rosewell, P. Robins et al. 2000. Uracil-DNA glycosylase (UNG)-deficient mice reveal a primary role of the enzyme during DNA replication. *Mol Cell* 5:1059–1065.

Novak, B., A. Csikasz-Nagy, B. Gyorffy, K. Nasmyth and J. J. Tyson. 1998. Model scenarios for evolution of the eukaryotic cell cycle. *Phil Trans R Soc Lond B* 353:2063–2076.

Obaya, A. J. and J. M. Sedivy. 2002. Regulation of cyclin-CDK activity in mammalian cells. *Cell Mol Life Sci* 59:126–142.

Orchard, C. B., I. Siciliano, D. A. Sorrell et al. 2005. Tobacco BY-2 cells expressing fission yeast cdc25 bypass a G2/M block on the cell cycle. *Plant J* 44:290–299.

Ordiz, M. I., L. Magnenat, C. F. Barbas and R. N. Beachy. 2010. Negative regulation of the RTBV promoter by designed zinc finger proteins. *Plant Mol Biol* 72:621–630.

Pagano, M., R. Pepperkok, F. Verde, W. Ansorge and G. Draetta. 1992. Cyclin A is required at two points in the human cell cycle. *EMBO J* 11:961–971.

Pelletier, J. and N. Sonenberg. 1988. Internal initiation of translation of eukaryotic mRNA directed by a sequence derived from poliovirus RNA. *Nature* 334:320–325.

Peres, A., M. L. Churchman, S. Hariharan et al. 2007. Novel plant-specific cyclin-dependent kinase inhibitors induced by biotic and abiotic stresses. *J Biol Chem* 282: 5588–5596.

Peters, J. M. 1998. SCF and APC: The Yin and Yang of cell cycle regulated proteolysis. *Curr Opin Cell Biol* 10:759–768.

Petrie, B. L., H. B. Greenberg, D. Y. Graham and M. K. Estes. 1984. Ultrastructural localization of rotavirus antigens using colloidal gold. *Virus Res* 1:133–152.

Petruccelli, S., S. Dai, R. Carcamo, Y. Yin, S. Chen and R. N. Beachy. 2001. Transcription factor RF2a alters expression of the rice tungro bacilliform virus promoter in transgenic tobacco plants. *Proc Natl Acad Sci USA* 98:7635–7640.

Pirone, T. P. 1977. Accessory factors in nonpersistent virus transmission. In *Aphids as Virus Vectors*, ed. K. Maramorosch, pp. 221–235. New York, NY: Academic.

Piroux, N., K. Saunders, A. Page and J. Stanley. 2007. Geminivirus pathogenicity protein C4 interacts with *Arabidopsis thaliana* shaggy-related protein kinase AtSKeta, a component of the brassinosteroid signalling pathway. *Virology* 362:428–440.

Potuschak, T. and P. Doerner. 2001. Cell cycle controls: genome-wide analysis in *Arabidopsis*. *Curr Opin Plant Biol* 4:501–506.

Qu, R., M. Bhattacharyya, G. S. Laco et al. 1991. Characterization of the genome of rice tungro bacilliform virus: Comparison with *Commelina* yellow mottle virus and caulimoviruses. *Virology* 185:354–364.

Ramírez, B. C. and A. L. Haenni. 1994. Molecular biology of tenuiviruses, a remarkable group of plant viruses. *J Gen Virol* 75:467–475.

Reichheld, J. P., N. Chaubet, W. H. Shen, J. P. Renaudin and C. Gigot. 1996. Multiple A-type cyclins express sequentially during the cell cycle in *Nicotiana tabacum* BY2 cells. *Proc Natl Acad Sci USA* 93:13819–13824.

Rhim, J. S., L. E. Jordan and H. D. Mayor. 1962. Cytochemical, fluorescent-antibody and electron microscopic studies on the growth of reovirus (ECHO 10) in tissue culture. *Virology* 17:342–355.

Riou-Khamlichi, C., R. Huntley, A. Jacqmard and J. A. H. Murray. 1999. Cytokinin activation of *Arabidopsis* cell division through a D-type cyclin. *Science* 283:1541–1544.

Rossi, V. and S. Varotto. 2002. Insights into the G1/S transition in plants. *Planta* 215:345–356.

Rossignol, P., R. Stevens, C. Perennes et al. 2002. AtE2F-a and AtDP-a, members of the E2F family of transcription factors, induce *Arabidopsis* leaf cells to re-enter S phase. *Mol Genet Genomics* 266:995–1003.

Roudier, F., E. Fedorova, J. Gyorgyey et al. 2000. Cell cycle function of a Medicago sativa A2-type cyclin interacting with a PSTAIRE-type cyclin-dependent kinase and a retinoblastoma protein. *Plant J* 23:73–83.

Satoh, K., H. Kondoh, T. Sasaya et al. 2010. Selective modification of rice (*Oryza sativa*) gene expression by rice stripe virus infection. *J Gen Virol* 91:294–305.

Schnittger, A., U. Schöbinger, D. Bouyer et al. 2002. Ectopic D-type cyclin expression induces not only DNA replication but also cell division in *Arabidopsis* trichomes. *Proc Natl Acad Sci USA* 99:6410–6415.

Sebastian, L. S., R. Ikeda, N. Huang et al. 1996. Molecular mapping of resistance genes to rice tungro spherical virus and green leafhopper in rice. *Phytopathology* 86:25–30.

Sekiguchi, H., M. Isogai, C. Masuta and I. Uyeda. 2005. 3C-like protease encoded by rice tungro spherical virus is autocatalytically processed. *Arch Virol* 150:595–601.

Senthil, G., H. Liu, V. G. Puram, A. Clark, A. Stromberg and M. M. Goodin. 2005. Specific and common changes in *Nicotiana benthamiana* gene expression in response to infection by enveloped viruses. *J Gen Virol* 86:2615–2625.

Setiady, Y. Y., M. Sekine, N. Hariguchi et al. 1995. Tobacco mitotic cyclins: cloning, characterization, gene expression and functional assay. *Plant J* 8:949–957.

Shahjahan, M., A. H. Jalant, A. H. Zakri, T. Imbe and O. Othman. 1990. Inheritance of tolerance to rice tungro bacilliform virus (RTBV) in rice (*Oryza sativa* L.). *Theor Appl Genet* 80:513–517.

Shao, C.-G., H.-J. Lu, J.-H.Wu and Z.-X. Gong. 2004. Nucleic acid binding activity of Pns6 encoded by genome segment 6 of rice ragged stunt Oryzavirus *Acta Biochimica et Biophysica Sinica* 36:457–466.

Sheen, J. and S. Key. 2004. Cell signalling and gene regulation: Exploring new functions and actions of regulatory molecules. *Curr Opin Plant Biol* 7:487–490.

Shen, W. H. and L. Hanley-Bowdoin. 2006. Geminivirus infection up-regulates the expression of two *Arabidopsis* protein kinases related to yeast SNF1 and mammalian AMPK activating kinases. *Plant Physiol* 142:1642–1655.

Shen, P., M. Kaniewska, C. Smith and R. N. Beachy. 1993. Nucleotide sequence and genomic organization of rice tungro spherical virus. *Virology* 193:621–630.

Shibata, Y., R. C. Cabunagan, P. Q. Cabauatan and I.-R. Choi. 2007. Characterization of Oryza rufipogon-derived resistance to tungro disease in rice. *Plant Dis.* 91:1386–1391.

Shikata, E. 1972. Rice transitory yellowing virus. *CMI/AAB Descr. Plant Viruses No. 100.*

Shikata, E. 1974. Rice black-streaked dwarf virus. *CMI/AAB Descr. Plant Viruses No. 125.*

Shikata, E. and Y. Kitagawa. 1977. Rice black-streaked dwarf virus: Its properties, morphology and intracellular localization. *Virology* 77:826–884.

Shimizu, T., K. Satoh, S. Kikuchi and T. Omura. 2007. The repression of cell wall- and plastid-related genes and the induction of defense-related genes in rice plants infected with rice dwarf virus. *Mol. Plant-Microbe Interact.* 20:247–254.

Shimizu, T., M. Yoshii, T. Wei, H. Hirochika and T. Omura. 2009. Silencing by RNAi of the gene for Pns12, a viroplasm matrix protein of rice dwarf virus, results in strong resistance of transgenic rice plants to the virus. *Plant Biotechnol J.* 7:24–32.

Shimotohno, A., R. Ohno, K. Bisova et al. 2006. Diverse phosphoregulatory mechanisms controlling cyclin-dependent kinase-activating kinases in *Arabidopsis. Plant J* 47:701–710.

Silverstein, S. C. and P. H. Schur. 1970. Immunofluorescent localization of double-stranded RNA in reovirus-infected cells. *Virology* 41:564–566.

Sorrell, D. A., A. Marchbank, K. McMahon et al. 2002. A WEE1 homologue from *Arabidopsis thaliana. Planta* 215:518–522.

Spendlove, R. S., E. H. Lennette, J. N. Chin and C. O. Knight. 1964. Effect of antimitotic agents on intracellular reovirus antigen. *Cancer Res* 24:1826–1833.

Sta Cruz, F. C., H. Koganezawa and H. Hibino. 1993. Comparative cytology of rice tungro viruses in selected rice cultivars. *J Phytopathol* 138:274–282.

Stals, H. and D. Inzé. 2001. When plant cells decide to divide. *Trends Plant Sci* 6: 359–364.

Stevens, R., L. Mariconti, P. Rossignol et al. 2002. Two E2F sites in the *Arabidopsis* MCM3 promoter have different roles in cell cycle activation and meristematic expression. *J Biol Chem* 277:32978–32984.

Sun, Y., B. P. Dilkes, C. Zhang et al. 1999. Characterization of maize (*Zea mays* L.) Wee1 and its activity in developing endosperm. *Proc Natl Acad. Sci USA* 96:4180–4185.

Supyani, S., B. I. Hillman and N. Suzuki. 2007. Baculovirus expression of the 11 mycoreovirus-1 genome segments and identification of the guanylyltransferase-encoding segment. *J Gen Virol* 88:342–350.

Suzuki, N., M. Sugawara, D. L. Nuss and Y. Matsuura. 1996. Polycistronic (tri- or bicistronic) phytoreoviral segments translatable in both plant and insect cells. *J Virol* 70:8155–8159.

Swenson, K. I., K. M. Farrell and J. V. Ruderman. 1986. The clam embryo protein cyclin A induces entry into M phase and the resumption of meiosis in Xenopus oocytes. *Cell* 47:861–870.

Takahashi, M., S. Toriyama, C. Hamamatsu and A. Ishihama. 1993. Nucleotide sequence and possible ambisense coding strategy of rice stripe virus RNA segment 2 *J Gen. Virol* 74:769–773.

Thole, V. and R. Hull. 1996. Rice tungro spherical virus: Nucleotide sequence of the 3' genomic half and studies on the 3' two small open reading frames. *Virus Genes* 13:239–246.

Thole, V. and R. Hull. 1998. Rice tungro spherical virus polyprotein processing: Identification of a virus-encoded protease and mutational analysis of putative cleavage sites. *Virology* 247:106–114.

Thole, V. and R. Hull. 2002. Characterization of a protein from rice tungro spherical virus with serine proteinase-like activity. *J Gen Virol* 83:3179–3186.

Toriyama, S. 1983. Rice stripe virus. *CMI/AAB Description of Plant Viruses No. 269.*

Toriyama, S. 1986. An RNA-dependent RNA polymerase associated with the filamentous nucleoproteins of Rice Stripe Virus. *J Gen Virol* 67:1247–1255.

Toriyama, S. 1995. Viruses and molecular virology of tenuivirus. Rice tungro-associated viruses and their relationships to host plants and vector leafhoppers. In *Pathogenesis and Host Specificity in Plant Diseases, Histopathological, Biochemical, Genetic and Molecular Bases. Viruses and Viroids*, Vol. III, eds. P. R. Singh, U. S. Singh and K. Khomoto, 211–223. Oxford: Elsevier.

Toriyama, S., T. Kimishima and M. Takahashi. 1997. The proteins encoded by rice grassy stunt virus RNA5 and RNA6 are only distantly related to the corresponding proteins of other members of the genus Tenuivirus. *J Gen Virol* 78:2355–2363.

Toriyama, S., T. Kimishima, M. Takahashi et al. 1998. The complete nucleotide sequence of the rice grassy stunt virus genome and genomic comparisons with viruses of the genus Tenuivirus. *J Gen Virol* 79:2051–2058.

Toriyama, S., M. Takahashi, Y. Sano, T. Shimizu and A. Ishihama. 1994. Nucleotide sequence of RNA 1, the largest genomic segment of rice stripe virus, the prototype of the tenuiviruses. *J Gen Virol* 75:3569–3579.

Touris-Otero, F., J. Martínez-Costas, V. N. Vakharia and J. Benavente. 2004. Avian reovirus nonstructural protein μNS forms viroplasm-like inclusions and recruits protein NS to these structures. *Virology* 319:94–106.

Truniger, V. and M. A. Aranda. 2009. Recessive resistance to plant viruses. *Advances in Virus Research* 75:119–159.

Tyagi, H., S. Rajasubramaniam, M. V. Rajam and I. Dasgupta. 2008. RNA-interference in rice against rice tungro bacilliform virus results in its decreased accumulation in inoculated rice plants. *Transgenic Res* 17:897–904.

Ueda, S., C. Masuta and I. Uyeda. 1997. Hypothesis on particle structure and assembly of rice dwarf phytoreovirus: Interactions among multiple structural proteins. *J Gen Virol* 78:3135–3140.

Umeda, M., N. Iwamoto, C. Umeda-Hara et al. 1999. Molecular characterization of mitotic cyclins in rice plants. *Mol Gen Genet* 262:230–238.

Upadhyaya, N. M., M. Yang, W. Kositratana, A. Ghosh and P. M. Waterhouse. 1995. Molecular analysis of rice ragged stunt oryzavirus segment 9 and sequence conservation among isolates from Thailand and India. *Arch Virol* 140:1945–1956.

Upadhyaya, N. M., K. Ramm, J. A. Gellatly, Z. Li, W. Kositratana and P. M. Waterhouse. 1997. Rice ragged stunt oryzavirus genome segments S7 and S10 encode non-structural proteins of Mr 68,025 (Pns7) and Mr 32,364 (Pns10). *Arch Virol* 42:1719–1726.

Upadhyaya, N. M., K. Ramm, J. A. Gellatly, Z. Li, W. Kositratana and P. M. Waterhouse. 1998. Rice ragged stunt oryzavirus genome segment S4 could encode an RNA dependent RNA polymerase and a second protein of unknown function. *Arch Virol* 143:1815–1822.

Upadhyaya, N. M., E. Zinkowsky, W. Kositratana and P. M. Waterhouse. 1996. The M(r) 43K major capsid protein of rice ragged stunt oryzavirus is a posttranslationally processed product of a M(r) 67,348 polypeptide encoded by genome segment 8. *Arch Virol* 141:1689–1701.

Uyeda, I., H. Suga, S. Y. Lee et al. 1995. Rice ragged stunt oryzavirus genome segment 9 encodes a 38,600 Mr structural protein. *J Gen Virol* 76:975–978.

Uzarowska, A., G. Dionisio, B. Sarholz et al. 2009. Validation of candidate genes putatively associated with resistance to SCMV and MDMV in maize (*Zea mays* L.) by expression profiling. *BMC Plant Biol* 9:15.

Van Damme, D. and D. Geelen. 2008. Demarcation of the cortical division zone in dividing plant cells. *Cell Biol Int* 32:178–187.

Vandepoele, K., J. Raes, L. De Veylder et al. 2002. Genome-wide analysis of core cell cycle genes in *Arabidopsis. Plant Cell* 14:903–916

Ventelon-Debout, M., F. Delalande, J. P. Brizard et al. 2004. Proteome analysis of cultivar-specific deregulations of *Oryza sativa indica* and *O. sativa japonica* cellular suspensions undergoing Rice yellow mottle virus infection. *Proteomics* 4:216–225.

Verkest, A., C. L. Manes, S. Vercruysse et al. 2005. The cyclin-dependent kinase inhibitor KRP2 controls the onset of the endoreduplication cycle during *Arabidopsis* leaf development through inhibition of mitotic CDKA;1 kinase complexes. *Plant Cell* 17:1723–1736.

Wang, H., Q. Qi, P. Schorr et al. 1998. ICK1, a cyclin-dependent protein kinase inhibitor from *Arabidopsis thaliana* interacts with both Cdc2a and CycD3, and its expression is induced by abscisic acid. *Plant J* 15:501–510.

Wang, Z. H., S. G. Fang, J. L. Xu et al. 2003. Sequence analysis of the complete genome of rice black-streaked dwarf virus isolated from maize with rough dwarf disease. *Virus Genes* 27:163–168.

Wei, T., A., Kikuchi, Y. Moriyasu et al. 2006a. The spread of Rice dwarf virus among cells of its insect vector exploits virus-induced tubular structures. *J Virol* 80:8593–8602.

Wei, T., A. Kikuchi, N. Suzuki et al. 2006b. Pns4 of rice dwarf virus is a phosphoprotein, is localized around the viroplasm matrix, and forms minitubules. *Arch Virol* 151:1701–1712.

Wei, T., T. Shimizu, K. Hagiwara et al. 2006c. Pns12 protein of Rice dwarf virus is essential for formation of viroplasms and nucleation of viral-assembly complexes. *J Gen Virol* 87:429–438.

Weinl, C., S. Marquardt, S. J. Kuijt et al. 2005. Novel functions of plant cyclin-dependent kinase inhibitors, ICK1/KRP1, can act non-cell-autonomously and inhibit entry into mitosis. *Plant Cell* 17:1704–1722.

Whitham, S. A., S. Quan, H. S. Chang et al. 2003. Diverse RNA viruses elicit the expression of common sets of genes in susceptible *Arabidopsis thaliana* plants. *Plant J* 33:271–283.

Wu, S.-J., H. Zhong, Y. Zhou et al. 2009. Identification of QTLs for the resistance to rice stripe virus in indica rice variety Dular. *Euphytica* 165:557–565.

Xiong, R., J. Wu, Y. Zhou and X. J. Zhou. 2008. Identification of a movement protein of the tenuivirus rice stripe virus. *Virology* 82:12304–12311.

Xu, H., Y. Li, Z. J. Mao and Z. L. Chen. 1998. Rice dwarf phytoreovirus segment S11 encodes a nucleic acid binding protein. *Virology* 240:267–272.

Xu, J., W. ChunHong and L.Yi. 2009. Expression of rice dwarf phytoreovirus Pns6 and the specificity analysis of its monoclonal antibodies. *Sci China Ser C-Life Sci* 52:958–964.

Yan, J., H. Kudo, I. Uyeda, S. Y. Lee and E. Shikata. 1992. Conserved terminal sequences of rice ragged stunt virus genomic RNA. *J Gen Virol* 73:785–789.

Yan, J., H. Suga, I. Uyead, I. Kimura and E. Shikata. 1995. Molecular cloning and complete nucleotide sequence of rice ragged stunt oryzavirus genome segment 10. *Ann Phytopathol Soc Jpn* 61:189–193.

Yang, C., R. Guo, F. Jie et al. 2007. Spatial analysis of *Arabidopsis thaliana* gene expression in response to turnip mosaic virus infection. *Mol Plant Microbe Interact* 20:358–370.

Yin, Y. and R. N. Beachy. 1995. The regulatory regions of the rice tungro bacilliform virus promoter and interacting nuclear factors in rice (*Oryza sativa* L.). *Plant J* 7:969–980.

Yin, Y., L. Chen and R. Beachy. 1997a. Promoter elements required for phloem-specific gene expression from the RTBV promoter in rice. *Plant J* 12:1179–1188.

Yin, Y., Q. Zhu, S. Dai, C. Lamb and R. N. Beachy. 1997b. RF2a, a bZIP transcriptional activator of the phloem-specific rice tungro bacilliform virus promoter, functions in vascular development. *EMBO J* 16:5247–5259.

Yoshii, M., T. Shimizu, M. Yamazaki et al. 2009. Disruption of a novel gene for a NAC-domain protein in rice confers resistance to rice dwarf virus. *Plant J* 57:615–625.

Yoshii, M., M. Yamazaki, R. Rakwal et al. 2010. The NAC transcription factor RIM1 of rice is a new regulator of jasmonate signaling. *Plant J* 61:804–815.

Yu, Y., A. Steinmetz, D. Meyer, S. Brown and W. H. Shen. 2003. The tobacco A-Type cyclin, Nicta;CYCA3;2, at the nexus of cell division and differentiation. *Plant Cell* 15:2763–2777.

Zenna, N. S., P. Q. Cabauatan, M. Baraoidan, H. Leung and I.-R. Choi. 2008. Characterization of a putative rice mutant for reaction to rice tungro disease. *Crop Sci* 48:480–486.

Zenna, N. S., F. C. Sta. Cruz, E. L. Javier et al. 2006. Genetic analysis of tolerance to rice tungro bacilliform virus in rice (*Oryza sativa* L.) through agroinoculation. *J Phytopathol* 154:197–203.

Zhang, H. M., J. P. Chen and M. J. Adams. 2001. Molecular characterisation of segments 1 to 6 of rice black-streaked dwarf virus from China provides the complete genome. *Arch Virol* 146:2331–2339.

Zhang, L. D., Z. H. Wang, X. B. Wang et al. 2005. Two virus-encoded RNA silencing suppressors, P14 of Beet necrotic yellow vein virus and S6 of rice black streak dwarf virus. *Chin Sci Bull* 50:305–310.

Zhang, C., Y. Liu, L. Liu et al. 2008. Rice black streaked dwarf virus P9-1, an alpha-helical protein, self-interacts and forms viroplasms in vivo. *J Gen Virol* 89:1770–1776.

Zheng, H., L. Yu, C. Wei et al. 2000. Assembly of double-shelled, virus-like particles in transgenic rice plants expressing two major structural proteins of rice dwarf virus. *J Virol* 74:9808–9810.

Zhong, B., A. Kikuchi, Y. Moriyasu et al. 2003. A minor outer capsid protein, P9, of rice dwarf virus. *Arch Virol* 148:2275–2280.

Zhou, G. Y., X. B. Lu, H. J. Lu et al. 1999. Rice ragged stunt oryzavirus: role of the viral spike protein in transmission by the insect vector. *Ann Appl Biol* 135:573–578.

Zhou, F., Y. Pu, T. Wei et al. 2007. The P2 capsid protein of the nonenveloped rice dwarf phytoreovirus induces membrane fusion in insect host cells. *Proc Natl Acad Sci USA*, 104:19547–19552.

Zhou, Y., H. Wang, S. Gilmer et al. 2002. Control of petal and pollen development by the plant cyclin-dependent kinase inhibitor ICK1 in transgenic Brassica plants. *Planta* 215:248–257.

Zhu, H. T., X. Y. Chen, Z. L. Luo, R. X. Fang and D. M. Gao. 1997a. Nucleotide sequence of the rice yellow stunt rhabdovirus gene 2. *Chin J Virol* 13:369–375.

Zhu, S., F. Gao, X. Cao et al. 2005. The rice dwarf virus P2 protein interacts with *ent*-kaurene oxidases in vivo, leading to reduced biosynthesis of gibberellins and rice dwarf symptoms. *Plant Physiol* 139:1935–1945.

Zhu, Y., A. M. Hemmings, K. Iwasaki et al. 1997b. Details of the arrangement of the outer capsid of rice dwarf phytoreovirus, as visualized by two-dimensional crystallography. *J Virol* 71:8899–8901.

Zhu, Y., T. Hayakawa, S. Toriyama and M. Takahashi. 1991. Complete nucleotide sequence of RNA 3 of rice stripe virus: an ambisense coding strategy. *J Gen Virol* 72:763–767.

CHAPTER 18

Transcriptomics, Proteomics and Metabolomics

Integration of Latest Technologies for Improving Future Wheat Productivity

Michael G. Francki, Allison C. Crawford and Klaus Oldach

CONTENTS

18.1 INTRODUCTION

The current world population of approximately 6 billion people is predicted to increase to 9 billion by 2050 and will require high-yielding crop varieties to meet the future global demand for grain and food. Bread wheat (*Triticum aestivum* L.) is a major crop where flour produced from grain is used for a range of end-products suitable for human consumption. It has been reported that the annual production of wheat grain needs to increase from 600 million to 750 million tonnes to meet future demands (Lantican et al., 2002). Although the aim of breeding programmes is to develop commercial wheat varieties with significant yield improvements, they will be challenged with constant environmental changes that constrain grain production. Therefore, there is need to develop and apply modern technologies that will make a significant contribution towards wheat improvement to mitigate the effects of climate and seasonal variability and develop adaptable, high-yielding varieties to a range of different and changing environments for future grain production.

The challenge in developing high-yielding wheat varieties will be the exploitation of biological mechanisms responsible for adaptation to changing environments. There has been significant progress in understanding the genetic basis controlling phenotypic diversity for a range of important traits associated with adaptability. However, in many instances, trait variation is under polygenic control and phenotypic expression is often affected by environmental factors, complicating the analysis of interactions between biological mechanisms for trait improvement. For example, drought-prone environments will be predominant in major wheat-growing regions of the world and the physiological mechanism associated with tolerance to specific environments for both above-ground traits (leaf morphology, pigments, antioxidants, water use efficiency of leaf photosynthesis, spike photosynthesis) and below-ground traits (rapid ground cover, access to water by roots) are controlled by many genes that have altered responses and distinctive phenotypic expression in different environments (reviewed in Reynolds and Tuberosa, 2008). Identifying genetic diversity for trait variation, the genetic control and underlying biological mechanisms for genotype × environment interactions poses a significant challenge in understanding these complex mechanisms with the view to manipulate interactions for wheat improvement. Quantitative trait loci (QTL) mapping has been fundamental in identifying genomic regions controlling trait variation and developing molecular DNA markers that can be used to track alleles controlling trait variation (recently reviewed in Francki, 2009). However, research efforts are needed to extend our knowledge on discrete biological mechanisms including gene, protein and metabolite interactions, relationships between traits and the effects of environmental conditions in these interactions. Such research will complement our existing knowledge on the genetic control of trait variation and provide options for designing strategies to manipulate biological process for future trait improvement in wheat.

Genomic resources for wheat are continuing to evolve with the development of an extensive expressed sequence tags (EST) collection, location of expressed portions of the genome on the wheat cytogenetic map, extensive DNA marker resources and numerous genetic maps that have been applied for wheat improvement (reviewed in Francki, 2009). Wheat genomics will culminate with the release of a complete annotated genome sequence, currently under development with the co-operation of international research partner organizations (http://www.wheatgenome.org). Although the complete sequence for wheat will provide the fundamental basis for identifying gene content and those that contribute to trait variation, there will be a need to understand how environmental conditions influence gene function in controlling phenotypic expression. To this end, understanding how genes, proteins and metabolites interact in biological processes will be utilized to develop biological models defining molecular, biochemical and physiological pathways responsible for wheat adaptation in different and changing environments. This chapter reviews the current status of transcriptomics, proteomics and metabolomics in wheat and how the information will be used to construct systems biology networks to complement wheat genomics and provide targeted options for wheat improvement.

18.2 TRANSCRIPTOMICS

Transcriptomics captures spatial and temporal gene expression within plant tissues, and is commonly used to increase our knowledge on molecular events controlling biological processes. In many instances, tentative gene function is assigned through the association of differentially expressed genes or co-expression with plant phenotype, but often requires further experimentation (including mutational analysis, transgenics or expression in heterologous systems) to ascertain its biological role in controlling biochemical or physiological processes. Since the completion of genome sequencing for *Arabidopsis* and rice (The Arabidopsis Genome Initiative, 2000; Goff et al., 2002; Yu et al., 2002), the opportunity for analysis of gene function controlling biological processes and trait diversity is becoming more prevalent beyond 'guilt-by-association' principles (Saito et al., 2008). Although genomics of model plant species has provided the basis for identifying similar genes (orthologs) in other species, a comparative analysis of transcriptional regulation has limitations when identifying similar biological processes controlling trait diversity in major crops (reviewed in Francki and Appels, 2007). Ideally, the sequenced genome of hexaploid bread wheat (genome designation AABBDD) will make substantial contributions towards increasing our knowledge on gene expression controlling wheat phenotypes. Despite making considerable progress towards sequencing the wheat genome through an international consortium (http://www.wheatgenome.org), transcriptome analysis in the short to medium term will rely on the exploitation of 1 million wheat ESTs in the public domain and available through the National Centre for Biotechnology Information (NCBI) database (http://www.ncbi.nlm.nih.gov/).

A range of technologies is available to analyse gene expression and a clear trend towards the development of commercial platforms for genome-wide analysis. Transcript analysis is generally classified according to open (no prior sequence information needed) or closed (sequence-based) systems (reviewed by Leader, 2005). Briefly, open systems include RNA fingerprinting techniques such as differential display and cDNA-amplified fragment-length polymorphism, serial analysis of gene expression (SAGE) and massively parallel signature sequencing. Closed array-based systems include cDNAs, *in situ* synthesized oligo probes (e.g., 25 mers) and spotted long synthetic oligos (60–70 mers). Earlier transcriptome studies of wheat used cDNA arrays (Clarke and Rahman, 2005; Guo et al., 2007; Laudencia-Chingcuanco et al., 2006, 2007) but were limited in the number of transcripts that could be analysed for any given experiment, and the inability to discriminate transcripts that share in excess of 80% sequence identity (Poole et al., 2007). Microarray technology relies on hybridization of a fluorescently labelled cDNA or RNA to a target nucleic acid spotted as an array on a glass slide. Differences in transcript levels are monitored by the intensity of fluorescence and quantified either by the overall relative fluorescence or by the hybridization to specific reference probes on the array or both. The microarray-based systems have been more popular due to the higher density of transcripts that can be analysed for any given experiment and discrete hybridization intensities indicate differences in transcript abundance or discrimination between genes and alleles. The commercial oligo-based Wheat GeneChip® from Affymetrix comprises of 55,000 gene transcripts, each represented by 11 to 25-mer nucleotides and is estimated to cover about half of the expressed wheat genes (Wan et al., 2008). The use of the Wheat GeneChip® is a popular choice in transcript profiling studies (Aprile et al., 2009; Coram et al., 2008; Desmond et al., 2008; Houde and Diallo, 2008;

Hulbert et al., 2007; Jordan et al., 2007; Pandelova et al., 2009; Wan et al., 2008; Winfield et al., 2009); however, this platform is unable to detect low-abundance transcripts not represented on the array due to an incomplete knowledge of the gene content in wheat (Schreiber et al., 2009).

The complex organization of related genes in wheat (Akhunov et al., 2003) and the basis of technology development can have a significant effect on the reliability of transcript detection between closely related gene sequences. Gene homoeologs can best be distinguished by quantitative real-time reverse transcription polymerase chain reaction (PCR) (qPCR) with primers specifically designed for each member of a multi-gene family (Poole et al., 2007). This technique can also be used to determine the relative abundance of different gene homoeologs. A comparative analysis of qPCR and cDNA arrays indicated low-to-moderate correlation (up to 67%) of similar transcripts being detected despite the same target RNA being used in profiling (Poole et al., 2007). However, a medium-to-high correlation (67–99%) between similar transcript profiles (Aprile et al., 2009; Houde and Diallo, 2008; Poole et al., 2007; Winfield et al., 2009) allows reliable comparisons for integrating transcript profiles generated from oligo-based arrays and qPCR platform systems. Therefore, oligo arrays are more reliable in analysing the expression profile of individual sequences whereas a cDNA array is unable to discriminate the differential expression of similar genes.

SAGE is an open transcript analysis system that does not require prior gene sequence information for transcript quantification, although this detail is necessary for biological interpretation of transcripts. SAGE is based on generating short transcript tags (13–34 mers) by restriction of cDNA molecules and ligation to larger fragments prior to sequencing. After sequencing, the abundance of a specific tag is compared with the total number of sequenced fragments, reflecting the relative expression level of its corresponding gene transcript. SAGE is comparable to large-scale EST-sequencing projects (Mochida et al., 2006; Ogihara et al., 2003) and the complete transcriptome is estimated by sequencing as many gene transcripts as possible. The major difference is reduced costs associated with sequencing the smaller fragment sizes of SAGE transcripts. Tags of low-abundance transcripts can be detected by SAGE, but remain elusive to biological interpretation without a reference genome if the short tags cannot be linked to annotated genes (Leader, 2005; McIntosh et al., 2007).

18.2.1 Transcriptomics in Wheat Development

18.2.1.1 Transcriptomics in Transition Phase from Vegetative to Reproductive Growth

Yield improvement is a high priority in breeding programmes and a range of quantifiable traits including the time of maturity and the number of fertile florets in a wheat spike can have significant effects on yield *per se*. Increases in photoperiod and temperature regimes induce phase transition from vegetative to reproductive growth, and longer days affect the development of individual florets in a wheat spike. Therefore, the number of fertile florets is an important yield component that determines the final number of grains in a wheat spike (Fischer, 1984). In recent years, major progress in understanding the molecular basis of phase transition in wheat has been achieved by genetic analysis and map-based cloning of vernalization genes (*TaVRN1*, *TaVRN2* and *TaVRN3*) that contribute to the genetic control of reproductive growth and sensitivity to temperature and photoperiod regimes (Yan et al., 2003, 2004, 2006). In order to obtain further insights into the interactions of other genes in regulatory networks responding to environmental cues in the transition from vegetative to reproductive growth, Winfield et al. (2009) profiled the transcriptome in winter and spring wheat cultivars that have different vernalization temperature and light requirements for flowering. This study was a comprehensive analysis of transcripts regulating flowering time in wheat, and revealed similar molecular process to well characterized autonomous regulatory pathways controlling flowering time from model species, including the role of MADS-box genes. Opportunities to compare gene networks and translate interpretations from model species to wheat were also provided. Other studies have extended the gene regulatory network model to include transcripts encoding products involved in carbohydrate utilization during floret development (Ghiglione et al., 2008) that are responsible for regulating floret number and grain yield (Fischer, 1984). The comprehensive gene regulatory models controlling biochemical and physiological pathways will not only contribute towards building whole plant systems biological networks of genes, proteins and metabolites, but also provide an opportunity to target and manipulate specific transcripts controlling flowering time and improved yield performance in wheat.

18.2.1.2 Transcriptomics in Developing Grain

The transcripts identified during the transition from vegetative to reproductive phase in wheat provide a baseline for examining transcript abundance in developing grain, allowing an opportunity to identify interconnecting transcript networks between distinctive developmental phases and an in-depth understanding of how similar and different genes and transcripts influence grain development. To this end, transcriptomic analyses using a variety of technological platforms have been described that will link differing molecular processes responsible for grain yield and quality in wheat. Large-scale sequencing of approximately 3500 ESTs from developing grain (10–13 dpa) was first described by Ogihara et al. (2003) and extended to include an additional 16,000 ESTs from 22 different grain-specific cDNA libraries, and spotted on a cDNA array containing around 20,000 unique wheat genes (Wilson et al., 2004). Smaller cDNA arrays containing up to 7800 sequences are also available from six stages of seed development (Laudencia-Chingcuanco et al., 2006, 2007). These arrays however, are superseded by the Wheat GeneChip® (Jordan et al., 2007; Wan et al., 2008) which allows the simultaneous investigation of 55,000 gene transcripts. Nevertheless, the availability of partial and whole transcriptome arrays together with other platforms including SAGE (McIntosh et al., 2007) allow for both a reductionist approach to investigate specific transcript profiling at different stages and a holistic analysis of interconnecting gene systems networks to collectively provide a detailed tissue-specific transcriptome analysis of grain development.

Cell development during early grain development (4–10 dpa) involves rapid nuclear divisions in the endosperm tissue with the formation of cell walls and specialization of tissues to form the aleurone. Transcripts during early grain development encode proteins involved in cell-cycle regulation, aligning with physiological processes involved in condensation of maternal layers, expansion of the endosperm, embryo differentiation, biochemical processes to accumulate storage proteins (such as glutenins, gliadins and grain softness proteins), metabolic production and defence-related biochemical pathways to protect the immature grain vulnerable to biotic invasion (Drea et al., 2005; McIntosh et al., 2007). Measured transcript activity is high in the central endosperm and the aleurone where processes for cell wall biosynthesis and cell de-differentiation are located (Drea et al., 2005). The endosperm appears to be exclusively responsible for the expression of transcripts encoding storage proteins (Lamacchia et al., 2001) with the exception of the *puroindoline* (*Pin*) genes that are also expressed in the pericarp, the protective layer surrounding the embryo and endosperm, as shown when transcripts are used as probes during *in situ* analysis (Drea et al., 2005). The expression of *Pin* genes in the endosperm and pericarp relates to their dual function as storage proteins defining grain hardness (Beecher et al., 2002) and their role in plant defence (Faize et al., 2004). This intensive protein synthesis is likely to cause an increase in mis-folded proteins, which are removed by the accompanied expression of genes encoding ubiquitin-conjugated enzymes (McIntosh et al., 2007). The second most abundant gene transcripts encode defence-related proteins such as α-amylase inhibitors to prevent starch breakdown or non-specific lipid-transfer protein genes (Laudencia-Chingcuanco et al., 2007; McIntosh et al., 2007).

The second phase of grain development, the grain fill phase (10–28 dpa), is subdivided into three stages including medium milk (11–16 dpa), soft dough (17–21 dpa) and hard dough stage (21–28 dpa). At the medium milk stage, endosperm cells continue dividing, the large type-A starch granules are formed with lipid and protein bodies and the aleurone surrounds the embryo sac. At soft dough, the endosperm cells stop dividing and the smaller type-B starch granules appear. Finally, starch and storage proteins continue to be synthesized at the hard dough stage with antioxidant proteins synthesized for osmoprotection of grain components, and the embryo accumulates triacylglycerol lipid bodies as rich energy reserves. Profiling shows that in excess of 70% of the total gene activity during the grain fill stage is devoted to transcripts encoding products for seed storage protein synthesis and accumulation (McIntosh et al., 2007). Interestingly, a low abundance of transcripts encoding starch synthase genes *GbssIIa* and *GbssIII* was detected using SAGE analysis (McIntosh et al., 2007) during the second phase and over a longer period (3–35 dpa) during grain development (Laudencia-Chingcuanco et al., 2007), indicating that starch synthesis is governed by a stable biochemical process and slowly accumulates throughout seed development or that these genes have the ability to act as multi-functional enzymes for other biochemical processes during grain development. Not surprisingly, defence-related transcripts encoding protease inhibitors, pathogenesis-related (PR) protein genes and glycosyltransferases constitute approximately 25% of the total transcriptome during the second phase (McIntosh et al., 2007) to act as a protective mechanism to guard the rich nutrient source that is allocated for the next plant generation.

The final stage of grain development is the desiccation phase at 28–42 dpa, with dead endosperm cells and residual mRNA in embryo and aleurone tissues entering dormancy. In this phase, drastic changes occur in the transcript profile, marked by a decrease in transcripts for seed storage protein accumulation (5% of total transcripts) and a large increase (50%) in transcripts encoding defence-related proteins (Laudencia-Chingcuanco et al., 2007; McIntosh et al., 2007). Thus, the transcriptome underpins the molecular processes for the endosperm to load with nutrients for the germinating embryo and protection of the mature grain (Wan et al., 2008). The embryo itself has accumulated the storage protein globulin Gbl1 as a nitrogen and carbohydrate source required for germination, whereas transcripts encoding new PR (e.g., tritin, serpin) and heat-shock proteins (predominantly HSP70 and HSP80) that function in a manner similar to ubiquitin-conjugated enzymes in early grain development remove mis-folded or aggregated proteins in maturing seed. Transcripts encoding dehydrins and late embryogenesis-abundant storage proteins accumulate in the mature seed (Laudencia-Chingcuanco et al., 2007; McIntosh et al., 2007) and are considered to have an osmoprotective role.

The developments in profiling transcript abundance from early grain development through to seed maturity are now positioned to further explore the interconnecting networks and build whole transcriptome profiles that will decipher the molecular processes underpinning grain quality and yield. However, it is anticipated that microarray data will increase exponentially especially when further experiments are designed and implemented to understand environmental interactions on transcript abundance and effects on grain yield and quality. To this end, there will be a need to develop highly sophisticated computational methods to retrieve, integrate and interpret results, and construct gene regulatory networks that integrate with functional information, similar to Correlation Network (CORNET) being developed to analyse large microarray datasets from Arabidopsis (Bodt et al., 2010).

18.2.2 Transcriptomics of Wheat under Biotic Stress

The major threats to wheat production worldwide from *Fusarium* spp., rusts and powdery mildew requires a continual effort to develop genetic strategies for plant–host resistance to these pathogens. Transcriptomics provides an opportunity to explore the large-scale molecular processes governing disease resistance and susceptibility and to identify potential gene targets for manipulation in wheat improvement. The resistance locus *Lr34/Yr18* is an excellent example for multiple resistances, having broad resistance against leaf rust, stripe rust and powdery mildew (Krattinger et al., 2009; McIntosh, 1992), and is an ideal target to study transcript profiling under biotic stress.

In doing so, Hulbert et al. (2007) used the Wheat GeneChip® to analyse transcripts between near-isogenic lines (NILs) containing or devoid of the complex resistance locus. The rationale of using NILs was to target changes in gene expression specific to the *Lr34/Yr18*-resistance response rather than different transcripts expressed due to different genetic backgrounds and not necessarily associated with response to pathogen infection. Interestingly, no substantial transcriptional differences apart from moderate increases in abundance for defence-related genes (e.g., *PR-1*, β-*1,3-glucanase*, *chitinase*) were reported between susceptible and resistant NILs, which may reflect an under-representation of wheat genes on the Wheat GeneChip® (Hulbert et al., 2007). In fact, it is estimated that the genes represented on the Wheat GeneChip® accounts for only 50% of all wheat genes (Wan et al., 2008). Moreover, the recent successful map-based cloning of the resistance gene *Lr34* by Krattinger et al. (2009) has revealed that *Lr34* encodes an ATP-binding cassette (ABC) transporter protein constitutively expressed in susceptible and resistant cultivars, and could not have been identified as a differentially expressed gene between NILs. Resistance was due to only small sequence differences (single nucleotides or a 3-bp deletion) between the resistant and susceptible alleles of *Lr34* that did not affect transcript abundance (Krattinger et al., 2009).

Although the transcriptomics analysis by Hulbert et al. (2007) detected only small differences in expression between incompatible (resistance) and compatible (diseased) interactions, substantial differences in transcript abundance has been reported for other plant–pathogen systems using the Wheat GeneChip. Desmond et al. (2008) investigated transcript profiles in wheat infected with *Fusarium pseudograminearum*, a necrotrophic pathogen that destroys infected tissue and causes crown rot disease in wheat. The majority of differentially expressed transcripts between infected and uninfected plants represented genes involved in the production of reactive oxygen species (ROS), a biological mechanism which precedes cell death and is characteristic of host responses following infection by necrotrophic fungal pathogens such as *F. pseudograminearum* (Desmond et al., 2008) and *Pyrenophora tritici-repentis* (Pandelova et al., 2009). Studies to identify other host genes responding to infection with *Fusarium graminearum* (Bernardo et al., 2007; Golkari et al., 2007) or *Blumeria graminis* causing powdery mildew disease (Bruggmann et al., 2005) include transcripts encoding defence-related proteins (e.g., glucanase, chitinases, thaumatins), proteins involved in secondary metabolism (phenylpropanoid pathway components such as phenylalanine ammonia-lyase (PAL)), detoxification (e.g., cytochrome P450 or glutathione *S*-transferase) and components of signal transduction pathways (e.g., kinases and transcription factors). In summary, transcripts are generally up-regulated in response to pathogen infection in the incompatible rather than the compatible interaction irrespective of the pathogen being biotrophic or nectrotrophic. Interestingly, wheat transcripts differentially expressed in response to pathogen infection are in the same category as Arabidopsis genes involved in disease response, including genes involved in the production of ROS and encoding glucanases, chitinases, thionins, defensins, glutathione-*S*-transferases and PAL (Glazebrook et al., 1997). This indicates that similar biological mechanisms are likely to be involved in disease response across species. Although transcriptomics has provided substantial information about gene regulation in wheat, there are likely to be a proportion of genes constitutively expressed at a low level (Feuillet et al., 2003) that may not be readily identifiable as contributing to disease response by transcript analysis.

18.2.3 Transcriptomics of Wheat under Abiotic Stress

The need to improve grain production under increasing global temperatures and climate variability has led to greater attention of research on wheat yields under abiotic stress conditions. Among abiotic stresses, drought or dehydration responses have received most attention regarding transcriptomics analysis. More predictive but nevertheless important are soil constraints caused by aluminium toxicity in acidic soils and high saline levels that threaten crop production worldwide. In recent years, genome-scale expression studies have been reported on these three abiotic stress conditions.

18.2.3.1 Transcriptomics for Drought Stress Responses

Microarray platforms have been employed to study the transcriptome of drought-related responses in wheat. A long-oligo microarray representing 19,000 gene sequences was used by Mohammadi et al. (2007, 2008) to measure transcript changes in roots for deydration-tolerant genotypes upon withholding water. In comparison, Xue et al. (2006) utilised a cDNA array with 16,000 unique wheat genes to assess lines with contrasting transpiration efficiency, a trait associated with drought tolerance as it reflects the plant's ability to use water efficiently for biomass production. More recently, Aprile et al. (2009) used the Wheat GeneChip® for comprehensive transcript profiling during grain filling under terminal drought. Despite using a range of transcriptomic platforms and examining different traits associated with drought tolerance under various stressed conditions, certain genes known to be involved in dehydration stress have been reported in more than one of the aforementioned studies. These include the osmotic stress genes encoding dehydrins (Aprile et al., 2009; Mohammadi et al., 2007; Xue et al., 2006), and proteins of the proline, glycine-betaine and sorbitol pathways (Aprile et al., 2009; Mohammadi et al., 2007). Also of interest was the identification of differentially expressed genes reported to be involved in biotic stress including β-1,3-glucanase and chitinase (Aprile et al., 2009; Mohammadi et al., 2007; Xue et al., 2006), indicating that wheat under drought stress activates a set of non-specific stress responses in addition to specific drought-alleviating mechanisms. Therefore, transcripts related to general stress response mechanisms are prime targets for manipulation for the dual benefit of genetic improvement against biotic and abiotic stress.

18.2.3.2 Transcriptomics for Stress Responses to Aluminium and Salt

Similar to drought, tolerance to aluminium toxicity and salinity stress is complex involving more than one developmental, physiological or biochemical trait. A major achievement in tolerance to aluminium toxicity was the cloning of the wheat malate transporter *ALMT1* by a subtractive hybridization (SSH) approach (Sasaki et al., 2004) with its role in tolerance validated through functional analysis in transgenic barley (Delhaize et al., 2004). Although the location of *ALMT1* is on wheat chromosome 4DL (Ma et al., 2005; Raman et al., 2005), genetic analyses indicated that genes on other wheat chromosomes contribute to tolerance (Ma et al., 2006) and prompted further investigations in molecular processes involved with Al tolerance. In doing so, SSH was further used in combination with cDNA arrays (Guo et al., 2007) and the Wheat GeneChip® (Houde and Diallo, 2008) to identify differentially expressed transcripts encoding alternative oxidase, a lipid transfer protein and cytochrome P450 monooxygenase, suggesting a defence and detoxification mechanism being activated upon Al stress (Guo et al., 2007; Houde and Diallo, 2008). The future work using transgenics or further genetic analysis and mapping will indicate which of these genes has the greater impact in contributing to higher Al tolerance in wheat.

Although arable land is becoming increasingly affected by salinity and the importance of developing salt-tolerant wheat is of high importance, only few large-scale gene-expression studies have been described for wheat under salt stress. Kawaura et al. (2008) investigated transcript profiles of two-week-old wheat seedlings under salt stress using long-oligo arrays and identified 6000 differentially expressed genes. However, a study by Mott and Wang (2007) identified less than 60 differentially expressed transcripts using the Wheat GeneChip® when wheat seedlings were exposed to high salt stress under different experimental conditions. The remarkable difference in the number of expressed transcripts is difficult to interpret but may reflect differences in experimental approaches used to induce salt stress, transcriptomic platforms or both. The need to standardize protocols for experimental treatments and cross-platform compatibility would greatly assist integrating and interpreting results when developing accurate models of gene regulatory networks.

18.2.4 Future Directions in Wheat Transcriptomics

Commercial oligo arrays for wheat such as the Affymetrix Wheat GeneChip® are becoming more popular platforms for transcriptomic analysis. However, there is a need to develop compatible platforms and implement quality assurance to ensure that transcript profiling is of the highest standard for integrating and interpreting results. In this context, Brazma et al. (2001) developed guidelines on standards for microarray experiments (minimum information about microarray experiments, MIAME), whereby compliance has become common practice. However, data quality for integration, comparison and interpretation of results also relies on sound experimental practice for treatments and sample preparation. Winfield et al. (2009) provided evidence that gene-expression profiles greatly differ when cold treatment is either applied suddenly or gradually to wheat plants. Similarly, induction of salt stress is best applied by gradually increasing NaCl concentration with calcium ions to measure tolerance rather than shock response (Genc et al., 2010). Standardized practices for treatments and sample preparation under similar experimental conditions need to be agreed upon that would facilitate the integration of results and building of accurate gene regulatory models using transcript data generated from the international scientific community. Web-linked databases are developing where transcriptomics, genomics and phenotypic information is stored and publicly accessible. Typical examples include the plant expression database PLEXdb at http://www.plexdb.org/, massive parallel sequencing at http://mpss.udel.edu/ (no wheat data yet) and the Gene Expression Omnibus of the NCBI at http://www.ncbi.nlm.nih.gov/geo/.

The past decade has seen substantial developments in genetic resources for QTL analysis of agronomically important traits in wheat. An added benefit is that differentially expressed genes identified from transcriptomics studies can be treated as expression-level polymorphisms and integrated with existing genetic resources to identify expression QTL (eQTL). In doing so, gene transcripts can be assigned biological function based on known phenotypic variation and genetic mapping tools. In the first wheat example of eQTL, Jordan et al. (2007) identified 542 differentially expressed probe sets from the Wheat GeneChip® by hybridizing seed RNA from 39 selected progeny lines of a mapping population to the array. In excess of 40 QTL for grain quality and agronomic traits that had been previously mapped in this population (McCartney et al., 2005; McCartney et al., 2006) were assessed for their linkage to the eQTL. The genetic position of 17 eQTL coincided with grain protein content and yield and 28 eQTL matched loci linked to 1000 grain weight, maturity as well as flour and dough quality traits (Jordan et al., 2007). As transcriptomics becomes more popular, it is anticipated that amalgamation with genetic analysis studies will become a common approach to narrow down the individual genes responsible for phenotypic diversity from a larger pool.

18.3 PROTEOMICS

Critical to plant development are proteins controlling physiological, morphological and biochemical pathways for maintaining cellular homeostasis in a given environment. Unlike genes that are static within plant genomes, plant proteins are more difficult to characterize due to their dynamic expression and roles in controlling plant function and development responding to internal and external signals. However, increasing knowledge in protein profiling and quantification (absolute and relative), post-translational modification (PTM), protein–protein, protein–gene and protein–metabolite interactions using the science of 'proteomics' will provide strategies to unravel and manipulate biochemical processes to improve future plant performance and production. Current technologies provide capabilities to generate large volumes of information on the protein status of a given tissue at a given developmental time (the 'proteome map') but future transformational changes in technology platform development are crucial if we are to generate the complete protein complement of a species at any stage of plant development, and to accurately quantify protein regulation in stressed environments and proteins that have nutritional benefits or detrimental health effects. Furthermore, the integration of proteomics with accumulating but complementary genomics, transcriptomics and metabolomics data will allow for the development of comprehensive system networks and assist in the manipulation of biological processes for plant improvement.

Technology platforms are rapidly advancing for quantitative plant proteomic analysis and a number of recent reviews (Baginsky, 2009; Chen and Harmon, 2006; Jorrin-Novo et al., 2009; Oeljeklaus et al., 2009) provide details of sophisticated methods and their limitations for plant protein fractionation, separation, quantification, detection and identification using micro-analytical processes and database interrogations (see also Chapter 1). Moreover, methods in analysing plant PTM events such as protein phosphorylation, myristolation, glycosylation and ubiquination (reviewed in Kwon et al., 2006; Rossignol, 2006) and protein–protein interactions (reviewed in Morsy et al., 2008) will provide the relevant data needed to expand information about the abundance of protein isoforms in plant tissue and generate proteome maps to elucidate the functional role of proteins in biological processes.

The application of proteomic technologies has focused on rice (*Oryza sativa* L.) as the model cereal crop species for developing comprehensive proteome maps to integrate and decipher gene function using fully sequenced genomes. In this regard, the past decade has seen enormous advancements in developing proteome maps for rice plant development based on protein profiles from organelles, responses to biotic and abiotic stresses, disulphide proteins, lesion mimic mutants, hybrids and transgenic plants (Agrawal et al., 2006; Agrawal and Rakwal, 2006). There is no doubt that proteomic data will enhance annotation of the full rice genome sequence, and when combined with other resources (e.g., genetic mutants) will assist in the development of biological models aimed at elucidating protein function influencing plant phenotype. In comparison, proteomics analysis in wheat is not as advanced but considerable progress has been made in developing proteome maps for grain development and its relevance to wheat quality, and for biotic and abiotic stress in relation to yield potential in different grain production environments.

18.3.1 Proteomics for Wheat Grain Quality

Irrespective of whether destined for domestic or export market, flour produced from wheat grain must meet quality standards for pan, flat and steamed breads and various noodles. In this regard, quality traits such as protein content, grain hardness, dough strength, extensibility and starch characteristics are important for meeting market requirements. However, environmental interactions such as different fertilizer regimes, temperature and drought effects can have a profound effect on grain quality attributes such as starch accumulation, gluten composition (including the monomeric gliadens and polymeric high- and low-molecular-weight (HMW and LMW) glutenins), water-soluble albumins, salt-soluble globulins and puroindolines for grain hardness (Dupont and Altenbach, 2003). Therefore, the regulation of proteins during grain development coupled with the interaction of physiological effects represents a complex network of biochemical processes involved in synthesis and mobilization of starch and protein synthesis for grain composition in unstressed and stressed environments. The identification of proteins directly or indirectly involved in these processes will develop the ability to manipulate grain quality attributes for wheat improvement.

18.3.1.1 Proteome of the Developing Wheat Grain

A number of studies have focused on developing a proteome map of grain development. Skylas et al. (2000) generated one of the first proteome maps of developing wheat grain grown under controlled environmental conditions, whereby a total of 1300 polypeptides were identified at 17 days post anthesis (dpa) with 321 proteins submitted for detailed characterization. Many of the proteins identified were gliadins and glutenins, largely due to the strategy used to extract, fractionate and identify protein components. This study also identified 1125 expressed proteins in the mature grain, indicating a higher abundance of proteins in early grain development, providing the basis for a reference proteome map at early and late stages of wheat grain development. In further studies (using different separation and fractionation approaches), seed proteins of low abundance and those that are not well characterized were isolated and added to the reference proteome map of developing grain. A total of 68 thioredoxin targets were identified in immature and mature grain with functional roles in regulating metabolism, protein storage, oxidative stress, protein degradation and protein assembly (Wong et al., 2003; Wong et al., 2004). These studies collectively indicate that proteins are synthesized in the reduced state early in seed developed and oxidized during maturity. More recent studies of low-abundant proteins identified a further 250 proteins participating in 13 biochemical processes, with 10 of the biochemical pathways involved in proteins synthesized during early stages of seed development and the remaining processes involved in protein maintenance and storage (Vensel et al., 2005). Together, these studies provide the necessary information needed to build the complex proteome model controlling grain development and quality.

Earlier studies by Skylas et al. (2000), Wong et al. (2003) and Wong et al. (2004) have provided the basis for developing the wheat proteome map, and the increasing complexity of biochemical interactions revealed by Balmer et al. (2006) and Dupont (2008) demonstrates the advances made in proteome technology and strategies applied to decipher detailed biochemical interactions in developing grain. Proteins within the developing grain are located primarily in plastids (Lunn, 2007; Neuhaus and Emes, 2000) and amyloplasts (Bechtel and Wilson, 2003; Langeveld et al., 2000). Comprehensive proteomic analyses of amyloplasts in early grain development (10–12 dpa) have provided details of 132 enzymes representing at least 18 metabolic pathways (including the lipid, amino acid and nucleic acid pathways) and other biosynthetic processes for carbohydrate metabolism and starch production associated with grain-filling during the latter stages of grain development (Balmer et al., 2006; Dupont, 2008). However, thorough analysis of biochemical processes in amyloplasts during the final stages of grain filling is incomplete. Finally, preliminary studies have identified the feasibility of developing future proteomic reference maps of biochemical processes derived from peripheral and aleurone layers that are associated with flour constituents for bread-making, health benefits and protection from stressed conditions during wheat kernel formation (Laubin et al., 2008).

18.3.1.2 Proteomics for End-Products

The primary role of wheat grain production is to provide flour suitable for bread-making and other end-products, with different grain constituents essential for dough rheology, baking processes and end-product shelf life. Therefore, knowledge of processes leading to the synthesis of gluten proteins, starches, non-starch polysaccharides, non-storage proteins and enzymes is critical if significant gains are to be made to optimize constituents for wheat end-products (reviewed in Goesaert et al., 2005). Recent advances in proteomic technologies and strategies have allowed the generation of further knowledge regarding the variation and effects in biochemical processes associated with grain and flour quality. Physical and chemical reactions involving endogenous redox agents and enzymes have been identified that may be the causal effect for bread-making (Joye et al., 2009a). In this regard, the preferential application of endogenous compounds over chemical additives promises further gains in bread quality (Joye et al., 2009b). As proteomic technologies continue to improve resolution and sequence databases expand, identification of proteins with subtle or substantial changes will provide the much needed information to link causal relationships between biochemical processes with flour functionality for improved end-product use.

18.3.1.3 Protein Analysis of Grain under Abiotic Stress

Environmental factors including extreme temperature differences, drought and differing fertilizer regimes play a significant role in biochemical processes such as gluten and starch biosynthesis during grain development (Dupont and Altenbach, 2003). Proteomic investigations are now beginning to dissect the specific metabolic processes affected under different environmental conditions. The effects of high temperature (Majoul et al., 2003; Sancho et al., 2008; Skylas et al., 2002), drought (Hajheidari et al., 2007) and nutrient (Grove et al., 2009) stress on protein profiles during grain development are summarized in Table 18.1. Interestingly, the same proteins differentially expressed in grain within the same and between different environmental conditions have been consistently identified and include 70 KDa heat-shock protein (HSP), α-amylase inhibitors, Cys-peroxiredoxin, Serpin and Glyceraldehyde-3-phosphate dehydrogenase (GAPDH)

Table 18.1 Proteomic Analysis and Identification of Differentially Expressed Proteins in Wheat Grain under Stressed Conditions

Stress Conditions[a]	Protein Class	Reference
Heat-stressed grain 17 and 45 dpa (CE)	16.9 kDa HSP	Skylas et al. (2002)
	17.6 kDa HSP	
	18.1 kDa HSP	
Heat-stressed grain 313°, 488° and 763°Cd (F)	70 kDa HSP	Majoul et al. (2003)
	80 kDa HSP	
	Phosphoinositide-specific phopsholipase C	
	Glucose-6-phosphate isomerase cytosolic B	
	Granule-bound glycogen	
	Granule-bound starch synthase	
	Pectin methylesterase	
	Vacuolar ATP synthase Sub-unit B isoform 1	
	Catalase isozyme	
	DAHP synthetase 1	
	Lea protein	
	Dihyroflavonol-4-reductase	
	RPM1-like protein	
	Glucose-1-phosphate Adenyltransferase precursor	
Drought-stressed mature grain (F)	α-Amylase inhibitors	Hajheidari et al. (2007)
	Cold-regulated protein	
	1-Cys peroxiredoxin	
	Dehydroascorbate reductase	
	Glutathione-*S*-transferase	
	Serpin	
	Thioredoxin h	
	Xylanase inhibitor precursor	
	Elongation factor-1 alpha 3	
	Protein disulphide isomerase3 precursor	
	GAPDH	
	Methyl malonate semi-aldehyde dehydrogenase	
Heat/Cold-stressed mature grain (CE)	7S Vicilin	Sancho et al. (2008)

Table 18.1 (continued) Proteomic Analysis and Identification of Differentially Expressed Proteins in Wheat Grain under Stressed Conditions

Stress Conditions[a]	Protein Class	Reference
	α-Amylase inhibitor	
	Chitinase C	
	Serpin	
	Pathogen-related proteins	
	α-Globulin	
	Malate dehydrogenase	
	GAPDH	
	70 kDa HSP	
	Peroxidase	
Sulphur-fertilized mature grain (CE)	LMW glutenin subunits	
	g-Gliadins	Grove et al. (2009)
	Avenin-like b precursor	
	α-Amylase inhibitors	
	Manganese superoxide dismutase	
	Cys-peroxiredoxin	
	Isoflavone reductase	

Note: °Cd = accumulated averages of daily temperatures; CE = controlled environment conditions; F = field conditions.

[a] dpa = days post anthesis.

(Table 18.1). These proteins are therefore likely to be involved in general stress response in developing grain. Other proteins identified in any given study most likely represent those expressed in specific cultivars or in response to a particular environmental condition (Table 18.1), thereby providing information on biochemical mechanisms involved in genotype × environmental interactions. This information can be used to design strategies and manipulate biochemical processes to withstand the detrimental effects of stress response on wheat grain development and quality. It is expected that future studies in proteomics will link with functionality to investigate the role of specific proteins in overcoming the effects of stress conditions, maintaining good rheological dough-making and baking properties, and other quality characteristics for end-product use.

18.3.1.4 Proteomics of Grain for Health Benefits

A major health issue in society that is driven primarily by wheat products is the inability of genetically susceptible humans to tolerate seed storage proteins such as gluten (including gliadins and glutenins) in their diet. Ingestion of gluten triggers celiac disease (CD), a condition whereby gluten proteins damage the absorptive lining of the small intestine. The structural complexity of glutens and the unknown basis for toxicity makes it difficult to develop therapeutic approaches for patients with CD to coexist with gluten proteins in wheat products. In addition to identifying the underlying basis for CD immunopathogenesis, research in toxic wheat gluten provides strategies for prevention, diagnosis, prognosis and treatment of CD (reviewed in Ferranti et al., 2007). Proteomics and analysis of toxic epitopes of gluten proteins is one aspect that plays a major role in developing approaches for CD patients to coexist with food products containing wheat gluten. Mass spectrometry (MS) has been a vital tool in identifying toxic epitopes of gluten proteins that trigger an immune response in CD patients. A number of studies comprehensively reviewed by Ferranti et al. (2007)

have identified a range of toxic peptides associated with gluten protein families (particularly, A-gliadins and α2-gliadins) and their interactions with components of the human immune response. Further studies such as that described by Kasarda et al. (2008) identified traces of gluten associated with wheat starch granules used in food products that could trigger an immune response in CD patients. Therefore, MS and comprehensive proteomic analysis are indispensable tools not only to identify desirable gluten proteins in wheat cultivars suitable for CD patients but also to detect undesirable gluten proteins during food processing.

Although CD is a common disease in humans, allergens due to protein components of wheat flour can have anaphylactic responses. Atopic eczema/dermatitis syndrome and wheat-dependent exercise-induced anaphylaxis are typical examples. Therefore, proteomic approaches have been extended to characterize a range of water-soluble and insoluble proteins of wheat grain and flour thought to be responsible for IgE-mediated allergic reactions in an attempt to mitigate the effects of flour components causing detrimental human effects. IgE antibodies have been detected against glutenins and gliadins (Akagawa et al., 2007; Battais et al., 2003, 2005; Matsuo et al., 2005; Palosuo et al., 2001; Snégaroff et al., 2006; Varjonen et al., 2000), serine-protease inhibitors (serpins) (Akagawa et al., 2007; Sander et al., 2001), α-amylase and tripsin inhibitors (Akagawa et al., 2007; Kitta et al., 2006) and albumin/globulin (Battais et al., 2003). Proteomic analysis of wheat grain has provided a comprehensive analysis of flour allergens and provides opportunities to develop diagnostic assays to monitor protein contaminants in food products. Undoubtedly, further research is needed to characterize these proteins and the reactive protein domains in an attempt to manipulate them for their safe use by individuals susceptible to anaphylactic reactions of wheat flour products.

18.3.2 Proteomics for Agronomic Performance

18.3.2.1 Developmental Leaf Proteome

Proteomics of developing grain has provided knowledge on the biochemical properties leading to end-quality improvements. However, the application to agronomic plant performance is not as well developed and studies have commenced to decipher the proteome map of wheat tissue and protein profiles in response to environmental stress. A study by Donnelly et al. (2005) has investigated a detailed proteome map of tissue from wheat seedlings at the two-leaved stage. A total of 277 reproducibly visualized proteins were catalogued whereby 24% of the proteins were identified as related to energy production, 12% as protein metabolism, 12% as disease and defence related and 4% as storage proteins (Donnelly et al., 2005). The remaining could represent isoforms or products from PTM causing shifts in pI or molecular weight, rendering proteins not readily identifiable (Donnelly et al., 2005). Information on the proteome map of the leaf represents a building block by which to facilitate the identification of target proteins and shed further light on biochemical and physiological mechanisms that regulate plant development under biotic and abiotic stresses.

18.3.2.2 Proteomic Analysis under Abiotic Stress

Proteomic analysis of crop species under abiotic stress conditions have predominantly focused on protein differences in response to drought and salt, two of the major abiotic stresses affecting crop productivity (reviewed in Salekdeh and Komatsu, 2007). Interestingly, a comparison between durum and bread wheat cultivars under drought and salinity stress in controlled environmental conditions identified similar differentially expressed proteins related to glycolysis and gluconeogenesis (Caruso et al., 2008; Caruso et al., 2009; Peng et al., 2009). These studies provide experimental evidence that proteins involved in metabolic processes may be modulated for establishing homeostasis under general abiotic stress conditions in seedlings. However, the study by Peng et al. (2009) identified additional proteins associated with improved capacity for growth under abiotic stress, indicating that selected wheat genotypes have a superior capacity for ionic and osmotic homeostasis with more efficient removal of toxic by-products. Further proteomics analyses of abiotic stress response in wheat is required at the whole plant level, particularly for root and mature plant tissue that will complement the expanding proteomic map of seedlings and developing grain under stressed conditions.

18.3.2.3 Proteomics Analysis under Biotic Stress

The past two decades have seen the cloning of at least 40 genes (categorized into five classes) associated with disease resistance (*R*) from various plant species, with the role of *R* proteins subsequently evaluated in multi-protein recognition complexes, localization and downstream signal transduction events (reviewed in Martin et al., 2003). The advances made in developing models of biochemical events during disease response were largely due to the compilation of many independent studies across several plant species in response to different pathogens. Despite the accumulation of large volumes of information from different biological systems, models developed may not accurately reflect biochemical processes specific to a particular plant–pathogen interaction. However, proteomic technologies offer the potential to assess global protein changes during disease response, providing a more accurate proteome network for specific plant–pathogen interactions. Rice is the crop species where the majority of proteomic applications have provided further information on differential expression of the proteome response to fungal and viral infection. In many instances, rice proteins are related to defence/stress pathways in common with abiotic stress response (Agrawal and Rakwal, 2006). The challenge now will be relating this information to work done previously, building on models developed for specific disease-resistance mechanisms (reviewed in Martin et al., 2003) and extending proteomics applications for disease response in other cereal crops. The lack of detailed proteome studies of wheat in response to a particular pathogen infection is a reflection of recent concerted efforts to develop proteomic technology platforms, in order to develop the appropriate tools needed to understand the proteome during plant development prior to further application. Nevertheless, the technology is now having a wider application and it is expected that there will be an incremental release of information in the literature within the next 5 years on detailed protein profiles involved in disease response in wheat.

The initial studies on protein profiles focused on a major disease of wheat worldwide, *Fusarium* head blight (FHB). This disease is caused by the pathogen *F. graminearum* and is responsible for considerable yield reduction and mycotoxin contamination of infected grain which is unacceptable for feed and human consumption. Therefore, significant research efforts have focused on developing wheat varieties with suitable resistance against this grain pathogen. Although genetic analysis has identified numerous chromosomal regions controlling resistance to *F. graminearum* (Liu et al., 2009), genes responsible for resistance have not yet been cloned and functionally characterized. Proteomics offers an opportunity to bridge the gap between genome and phenotype by identifying proteins differentially expressed in response to pathogen infection and tracing the associated genes responsible for resistance. In a recent study by Wang et al. (2005), 117 proteins were differentially expressed in response to infection by *F. graminearum*, and 30 proteins identified by plant protein and wheat ESTs similarity were found to be related to carbon metabolism and photosynthesis, implying an adaptive change in physiology during infection. Although this study could not conclusively discriminate proteins involved in pathogen recognition or subsequent downstream biochemical events, it provides a representation of proteins involved in disease response and will assist in the development of a targeted approach to select genes responsible for resistance in forward and reverse genetic approaches. It is anticipated that further studies will be carried out as proteomic technologies become widely adopted for profiling protein responses to a range of wheat diseases.

18.3.3 Future Directions in Wheat Proteomics

Undoubtedly, the past 10 years have seen the development of proteomic technologies to improve sample throughput, protein resolution and identification for applications in wheat research. However, it is expected that proteomics will become more sophisticated and powerful particularly in analysing low-abundant proteins. In doing so, we will likely see an increase in the number of reported studies on specific effects on the wheat proteome in a range of tissues under stressed and non-stressed conditions and begin to unravel the complex protein networks controlling phenotypic variation. In doing so, there will be a demand among the science community to store, catalogue and retrieve information on wheat proteins and develop appropriate databases such as those for *Arabidopsis* and maize (Sun et al., 2009). Such a database would require a consortium of global researchers to prioritize protein profiling of wheat tissue under developmental and stressed conditions, design, articulate and standardize phenotyping methodologies for identifying protein profiles and assemble information in user-friendly graphical features with linkages to a range of genomic and protein databases of other crop species. As the wheat genome is being assembled (http://www.wheatgenome.org), linkages with a comprehensive wheat protein database will be a powerful resource for accurate sequence annotation. A wheat protein database will not be inconceivable in the near future considering recent technological advances, its increasing application to understanding the wheat proteome and strategies in developing analogous genomic databases.

18.4 METABOLOMICS

The field of metabolomics is increasingly viewed as a complementary approach to characterizing and improving the molecular features of important traits in cereal crop species such as wheat. It is estimated that plant systems are comprised of 5000–25,000 different metabolites, more than double that of humans and nearly an order of magnitude greater than that of yeast (Hegeman, 2010). Metabolomics tools can profile the manner in which metabolite levels change throughout plant development, or in response to biotic and abiotic stress, nutritional status and environmental perturbation (Dixon et al., 2006; Hall et al., 2008; Schauer and Fernie, 2006). In this regard, cereal crop research focuses primarily on improving the productivity/yield and nutritional profile of grains, and demonstrating ‘substantial equivalence’ of transgenic varieties. Although metabolomics studies for wheat and other cereals are only recently beginning to emerge in the literature (see Table 18.2), considerable gains in these areas have already been made (see also Chapter 8).

Several platforms are currently being utilized in metabolomics research, with profiling studies over the past decade dominated by nuclear magnetic resonance (NMR) and MS techniques (Fernie and Schauer, 2009; Hegeman, 2010; Ward et al., 2003). In this regard, no single technique is able to cover all metabolites present in plant tissues due to chemical differences in volatility, polarity, solubility and chromatographic behaviour (Ward et al., 2003). NMR is considered to be a relatively unbiased technique for identification and quantification of virtually any biomolecule, and is typically used for broad profiling of abundant metabolites and fingerprinting of large number of samples. MS technologies, however, are often preferred where greater sensitivity, dynamic range and quantification of metabolite concentrations are required. The disadvantage of MS techniques is that they are either limited by the types of analytes detected, or there are difficulties with database cross-comparisons for compound identification (Dixon et al., 2006; Hegeman, 2010).

Effective interpretation of metabolomic spectra is largely dependent on prior knowledge of metabolite identity, with the development of accessible databases key to streamlining this process. There are several web-based library resources that cater to identification of spectra from either NMR or MS platforms (Hegeman, 2010; Tohge and Fernie, 2009). The NIST 08 collection, for example, includes GC–MS retention indices for some 44,008 compounds, whereas the Madison

Table 18.2 Selected Metabolomic Studies from Wheat and Other Cereals

Cereals	Summary	Reference
Wheat	GC–MS reveals minimal effect of organic (versus conventional) agriculture on metabolite content	Zörb et al. (2009)
Wheat	GC–MS reveals variation in metabolites affecting flavour and aroma of cooked pasta for different cultivars	Beleggia et al. (2009)
Wheat	NMR of nitrogen and sulphur deficiency on nutrient remobilization in senescing canopy tissues	Howarth et al. (2008)
Wheat	NMR profiling in leaf and stem tissue correlating metabolite concentration with *F. graminearum* disease resistance	Browne and Brindle (2007)
Wheat	GC–MS of *F. graminearum* disease resistance	Hamzehzarghani et al. (2005) Hamzehzarghani et al. (2008a) Hamzehzarghani et al. (2008b) Paranidharan et al. (2008)
Wheat	NMR study of three genetically modified wheat lines with additional copies of glutenin HMW subunits grown in the field	Baker (2006)
Wheat	Dx5 glutenin overexpression in transgenic plants	Stamova et al. (2009)
Barley	GC–MS comparison of boron sensitive and tolerant varieties fails to explain tolerance mechanism	Roessner et al. (2006)
Barley	GC–MS and LC–MS of carbon and nitrogen metabolism in phosphate-deficient plants	Huang et al. (2008)
Rice	GC–MS of germinating rice seeds reveals dynamic changes in metabolites	Shu et al. (2008)
Rice	GC–MS study of changes during rice tillering using defined metabolite markers	Tarpley et al. (2005)
Rice	GC–MS reveals large metabolic differences between seed from different cultivars	Kusano et al. (2007)
Rice	LC–MS assessment of transgenic rice enhanced for tryptophan overproduction	Wakasa et al. (2006)
Maize	Metabolite variation due to genetic and environmental factors in maize grain	Harrigan et al. (2007a)
Maize	Impact of drought stress on kernel composition and stress metabolites	Harrigan et al. (2007b)
Maize	GC–MS reveals metabolite variation due to genetic and environmental factors in maize grain	Röhlig et al. (2009)

Metabolomics Consortium Database contains NMR spectral information for more than 19,700 metabolites in addition to standard compounds. Other studies (NMR and GC–MS) have further confirmed metabolite identity through spiking with pure compounds and acquiring further spectra (Browne and Brindle, 2007; Graham et al., 2009; Roessner et al., 2006). The continued development of metabolomics spectral libraries (for plants particularly) will allow faster identification of compounds involved in trait variation for agricultural crops, and enhanced integration with other genomic and proteomic platforms in a systems biology framework.

18.4.1 Metabolomics for Wheat Grain Quality

Metabolomic studies of crop species have generally focused on evaluating natural variance in metabolite composition (Fernie and Schauer, 2009; Hall et al., 2008). In this regard, metabolite content (and therefore grain quality) is influenced by both genotypic and environmental factors. This presents breeding opportunities for quality traits using conventional approaches. Establishing natural variation parameters for metabolites is also important for determining whether the biochemical composition of transgenic plants falls within appropriate ranges, otherwise known as 'substantial equivalence'.

18.4.1.1 Genotypic and Environmental Factors

Several metabolomic studies have specifically targeted metabolite composition in cereal grains. Howarth et al. (2008) examined metabolite content in developing wheat seed from a single variety (0, 7, 14, 21, 28 dpa) using NMR techniques. Strong clustering of samples according to stage was observed, indicating that the metabolite content of wheat grain is primarily developmentally controlled. Specifically, levels of glucose, maltose, sucrose and glutamine decreased during seed development. Changes in amino acid composition and aromatic compounds (tyrosine and tryptophan) were also observed. Although comparative studies are lacking from wheat, substantial differences in seed metabolite levels from different rice and maize cultivars or lines have been reported in several studies (Harrigan et al., 2007a; Kusano et al., 2007; Röhlig et al., 2009). This suggests that metabolomics strategies may be used to select crop breeding lines with enhanced nutritional profiles.

Environmental factors have also been implicated in seed metabolite variability in wheat and other cereal crops. Demands for greater productivity coupled with the need for sustainable fertilizer use has been identified as a key agronomic concern, with nitrogen and sulphur inputs in particular linked to wheat grain yield and quality. The impact of nitrogen (N) and sulphur (S) deficiency on nutrient remobilization from wheat canopy tissues to grain was examined by Howarth et al. (2008). Only a limited effect of fertilizer treatment on the grain metabolite content was observed, and it was suggested by the authors that further optimization of grain yield under reduced fertilizer inputs may depend on altering the timing of the process of senescence/remobilization. Similarly, organic (versus conventional) agricultural practices were found to have a negligible impact on metabolite concentration in mature wheat grain (Zörb et al., 2009). Together these studies suggest a high degree of homeostasis for wheat grain metabolites; however, metabolomics is clearly offering new directions and strategies for improving grain quality under different environmental constraints. Other studies have identified significant site/location, seasonal and drought treatment effects on metabolite composition in seed from different maize cultivars/lines (Harrigan et al., 2007a,b; Röhlig et al., 2009). In this regard, metabolite profiles from other cereal crop species will provide useful baseline information for future metabolomics studies of wheat.

18.4.1.2 End-Product Assessment and Human Health

Relatively few metabolomics studies have explored product-related issues in unmodified wheat grains or flour, due in part to the established role of proteomics in flour quality research. However, one recent GC–MS study of volatile components of semolina and pasta from four durum wheat cultivars found significant differences observed in the composition and level of metabolites (Beleggia et al., 2009). These findings suggest that the flavour and aroma of cooked pasta may vary significantly among cultivars, which presents some interesting prospects for future product improvement based on consumer preference. Metabolomics is likely to play an even more important role with regard to nutrient profile assessment of genetically modified crops, with the Organization for Economic Co-operation and Development recommending the measurement of specific plant components such as proximates, nutrients, anti-nutrients and other secondary metabolites (Harrigan et al., 2007a). In this regard, several studies have investigated transgenic wheat specifically modified for enhanced food functionality. Baker (2006) reported metabolite levels (determined by NMR) for wheat lines with added HMW glutenin genes to be generally within the range observed for parental material, with environmental differences (site and year) exerting a stronger influence on metabolite content than genotype. However, amino acid differences and increased levels of maltose and/or sucrose were associated with higher levels of transgene expression for some lines. Stamova et al. (2009) also observed significant differences for three metabolites (guanine, 4-hydroxycinnamic acid and an unidentified compound) in a similar study using GC–MS techniques. The significance of these differences was unclear, with the suggestion that metabolite levels may have normalized upon seed maturation. Other unintended effects on wheat metabolite levels were also observed by that study, which were attributed to the gene transformation process.

Other studies have sought to demonstrate substantial equivalence in cereal grains that have been genetically modified for agronomic traits unrelated to grain quality. For example, the GC–flame ionization detector (FID)/GC–MS examination of rice enhanced with *cryIAc* and *sck* insect-resistance genes demonstrated strong environmental and genetic effects for various metabolite levels, including glycerol-3-phosphate, citric acid, linoleic acid, oleic acid, hexadecanoic acid, 2,3-dihydroxypropyl ester, sucrose, 9-octadecenoic acid and (Z)-2,3-dihydroxypropyl ester (Zhou et al., 2009). Furthermore, sucrose, mannitol and glutamic acid all significantly increased in transgenic compared to wild-type grain. Similarly, Manetti et al. (2006) performed an NMR study of a maize line modified with the insect-resistance gene *CryIAb* and its wild-type homologue, grown under greenhouse conditions. A significant variation in metabolites was detected, including osmolytes and branched amino acids. The relative success of GM strategies in species such as rice and maize (in terms of metabolite composition) will play a strong role in deciding whether similar approaches are adopted for wheat.

Metabolomic assessment offers further opportunities to enhance the nutrient profile of cereal crops. In this regard, micronutrient malnutrition affects more than half the world's population, with many suffering the health effects of mineral and amino acid deficiency in particular (Hall et al., 2008). Increased biosynthesis of the essential amino acid tryptophan has been performed with transgenic rice, with only minor changes to the metabolome (Ishihara et al., 2007; Wakasa et al., 2006). There is also the potential to assess levels of anti-nutrients or toxicants in end-products. For wheat specifically, high levels of the amino acid asparagine have been detected in the endosperm of sulphur-deficient plants, which has been linked to the formation of the carcinogen acrylamide during the baking process (Howarth et al., 2008). The negative impact of phytic acid on mineral bioavailability in foods has also prompted the development of low phytic acid mutants for several crop species (including wheat). GC–MS assessment of low phytic acid rice mutants has revealed both expected and unexpected changes in metabolite levels (Frank et al., 2007).

18.4.2 Wheat Metabolome

The sensitivity of metabolic profiling to detect biochemical variability within crop species is becoming increasingly apparent, with differences observed between different plant tissues, over the course of plant development, and between different varieties. In this regard, distinct differences in metabolites have been observed between wheat leaf and grain tissues (Howarth et al., 2008). NMR-based methods have also been used to explore metabolite levels in coleoptile and two-week-old leaf tissue extracts in different European wheat cultivars (Graham et al., 2009). The dominant metabolites identified included carbohydrates, betaine and choline peaks, with clear biochemical differences between the two stages. Further studies were deemed feasible, based on the ability to separate metabolic profiles from different cultivars and the high repeatability of biological replicates.

Metabolomic analyses of tissues from related cereal species have also produced useful information. GC–MS analysis of barley has revealed greater differences in metabolite content between root tissues compared to leaf tissues for varieties with variable tolerance to boron toxicity (Roessner et al., 2006). Other GC–MS studies of rice have profiled biochemical changes over the tillering period (Tarpley et al., 2005) and significant differences in metabolite content were observed for rice seeds during the course of germination (Shu et al., 2008). Baseline studies such as these will help establish a framework from which to assess the impact of abiotic and biotic factors on the wheat metabolome.

18.4.2.1 Abiotic Interactions

The abiotic environment is critical for improved productivity in wheat and other cereal crops, with nutrient availability in particular linked to optimal grain size and weight. Howarth et al. (2008) found that suboptimal N and S supply had a significant effect on the metabolite content of wheat leaf tissue, with senescence occurring earlier in N-deficient plants. With only a limited effect of fertilizer treatment on the grain metabolite content, it was determined that the effective use of low fertilizer inputs may be dependent on synchronizing the timing of senescence with grain development. Carbon and nitrogen metabolism in phosphate-deficient barley plants has also been examined (Huang et al., 2008). In this study, polar metabolites were profiled using GC–MS and free amino acids were profiled using LC–MS. In treated plants, an increase in saccharides and ammonium metabolites was observed, and reduced levels of phosphorylated intermediates and organic acids. The authors proposed altering carbohydrate partitioning to enable more efficient carbon utilization in deficient plants.

Other abiotic factors have been identified as limitations to crop production in many agricultural regions, and in this regard metabolomic technologies have had variable success. GC–MS examination of barley cultivars varying in tolerance to boron toxicity did not provide a sufficient indication as to a putative biochemical mechanism for tolerant cultivars (Roessner et al., 2006). Variation in putrescine and six kestose levels were implicated in boron tolerance; however, this study failed to detect low levels of sugar alcohols and polyols that are known to be involved in boron complex formation. Despite potential shortcomings in metabolite detection, metabolomics technologies in general are successfully being used to provide insights into how rice and other plant species respond to salinity (Sanchez et al., 2008) and other abiotic stressors such as temperature, water deficit, oxidation and heavy-metal toxicity (Shulaev et al., 2008).

18.4.2.2 Biotic Interactions

Metabolomics approaches have identified compounds associated with the phenylpropanoid/shikimic acid pathway as playing a key role in FHB resistance in wheat (reviewed by Walter et al., 2010). Hamzehzarghani et al. (2005) examined metabolomic responses of resistant and susceptible wheat cultivars following *F. graminearum* inoculation using GC–MS. In particular, substantial differences in abundance were found for metahydroxycinnamic acid and myo-inositol, suggesting the involvement of cell wall strengthening and signaling pathways in disease resistance. The same cultivars were examined in a later GC–MS study, which examined metabolomic response following inoculation with either *F. graminearum* or virulence factor deoxynivalenol (Paranidharan et al., 2008). Similarly, this study identified the involvement of compounds linked to cell wall defence and signalling (cinnamic acid, ferulic acid and inositol) and differences in polyamine abundance. Examination of other wheat cultivars by Hamzehzarghani et al. (2008a) also identified a number of metabolites involved in signalling or cell wall strengthening, or with known anti-microbial properties. Further advances were made by Hamzehzarghani et al. (2008b) upon examination of two wheat NILs with alternate alleles for FHB-resistance QTL on chromosome 2DL, with *p*-coumaric acid and myo-inositol considered the most likely metabolites involved in this multi-gene locus. Unravelling the genetic constituents of these pathways and determining their physical genomic location in relation to known QTL represents the final stages of high-throughput crop improvement using molecular markers.

While these FHB-resistance studies focused solely on GC–MS techniques to obtain spectral information, one other study in this field has utilized NMR technology. Browne and Brindle (2007) identified a number of metabolites in FHB susceptible wheat plants such as choline, betaine, trans-aconitate, sucrose and the amino acids glutamine, glutamate and alanine. Metabolites related to increased disease resistance included glucose and other unassigned resonances in the carbohydrate and aromatic regions of the NMR spectra. Further work on the identification of metabolites via peak assignments will likely yield complementary information to that already published by

GC–MS studies; however, at this point NMR technologies appear to be lagging behind other spectrometry technologies due to reduced sensitivity and unidentified spectra.

Studies on fungal pathogens in other cereal species have revealed similar metabolomic responses. A comparative GC–MS metabolomic study on rice, barley and *Brachypodium distachyon* has revealed that the fungal pathogen *Magnaporthe grisea* (rice blast disease) induces common biochemical changes in plants that suppress specific defence responses and promote tissue colonization (Parker et al., 2009). Specifically, delayed activation of the phenylpropanoid pathway and other cell wall compounds were implicated in disease susceptibility. That study also reported altered amino acid biosynthesis, which has also been observed in LC–MS analyses of maize metabolomic response to smut disease (Horst et al., 2010). Together, these studies on disease resistance in wheat and other cereals are revealing some similarities with regard to plant defence responses, which may result in relatively quick delineation of gene targets for crop enhancement.

18.4.3 Future Directions in Wheat Metabolomics

Although only a few papers have been published in the field of metabolomics for improved wheat quality and productivity, this area of research is gaining increasing popularity as a tool for elucidating the causative mechanisms for particular traits. The progress made in FHB research best demonstrates this success. For other abiotic traits of interest, such as nutrient availability and other environmental limitations, metabolomics technologies may provide new insights. The usefulness of metabolomics technologies, however, is very much reliant on prior knowledge of the identity of acquired spectra. At present, GC–MS techniques are still preferred due to greater sensitivity and public availability of spectral information. In this regard, further identification and determination of plant metabolite peaks will facilitate even greater extrapolation of meaningful data within a systems biology framework. Whether GC–MS is a suitable platform for all plant-related biological questions is a different matter. GC–MS has been criticized in the studies on disease resistance in wheat as it lacks the ability to detect certain semi-polar defence compounds such as flavonoids, glucosinolates and saponins (Hamzehzarghani et al., 2008a). Furthermore, other compounds may only be present at very low levels and virtually undetectable.

18.5 TOWARDS BUILDING SYSTEMS BIOLOGY NETWORKS IN WHEAT

The '-omics' technologies in plants have made considerable progress towards developing a global analytical approach to integrate genetic, protein, metabolic, cellular and pathway events and decipher the interconnected relationships responsible for diversity in plant phenotypes. Although substantial biological information is available, the amount of data will continue to increase at a considerable rate as the '-omics' technologies become increasingly sophisticated and the need to discover a range of biological processes under different experimental conditions becomes more important. The ability to analyse and interpret data generated from the different '-omics' technologies and integration of datasets across platforms will need to be supported by high-level statistical and computational methodologies and further development of consensus theories for novel trends and biological processes based on prior knowledge of existing pathways.

The process of building system network models relies on moving away from a reductionist approach for identifying single entities that govern biological systems to integrating multi-dimensional, genome-scale datasets and modelling to investigate interactions involving multiple components (for recent reviews, see Fukushima et al., 2009; Long et al., 2008; Yuan et al., 2008). The general approach of building plant system networks involves statistical integration using multivariate regression methods (such as O2PLS described by Bylesjo et al., 2007) and incorporating discrete and kinetic models together with Bayesian and scale-free networks (Brady and Benfey,

2006). However, data modelling is a rate-limiting step that involves generalizations, simplification and assumptions and, in some instances, modelled networks need to be complemented by a reductionist approach (Yuan et al., 2008). Forward or reverse genetics, transgenic approaches (Yuan et al., 2008) or eQTL (Hansen et al., 2009) can be used to provide experimental complementation of theoretical models for transcript, protein and metabolite interactions. Platform-independent, user-friendly packages allowing robust statistical analysis and intuitive visualizations for plant systems biology will greatly facilitate generating new biological hypotheses for plant researchers. Plant MetGenMAP (Joung et al., 2009) is a typical example of an easy-to-use Web-based programme to construct systems biology networks of data generated from transcriptomics, proteomics and metabolomics, with demonstrated examples including interactions between gene expression and metabolite levels in the carotenoid biosynthesis pathway leading to changes in fruit colour in tomato.

To date, the model plant species have been the major beneficiaries from technologies associated with genome-scale transcriptomics, proteomics and metabolomics in determining the molecular and biochemical interactions controlling phenotypic diversity. The '-omics' technologies and know-how have been transferred to wheat but generation of datasets is not as extensive compared to other plant species. In this regard, models describing system networks for traits in wheat are in their infancy but as '-omics' technologies become more widely adopted and utilized for understanding the molecular and biochemical interactions controlling trait variation, there will be increasing development of systems biology networks for traits of high importance for wheat improvement. In the immediate future, the molecular, biochemical and physiological interactions for wheat grain development and tolerance mechanisms for FHB could be modelled using transcriptomics, proteomics and metabolomic data and preliminary systems biology networks established. Processes and knowledge to build systems biology networks in wheat will be invaluable but the ongoing challenge will be developing efficient strategies to use this knowledge for improving cultivar development for future adaptation and global wheat production under environmental constraints.

REFERENCES

Agrawal, G. K., N.-S. Jwa, Y. Iwahashi, M. Yonekura, H. Iwahashi and R. Rakwal. 2006. Rejuvenating rice proteomics: Facts, challenges and visions. *Proteomics* 6:5549–5576.

Agrawal, G. K. and R. Rakwal. 2006. Rice proteomics: A cornerstone for cereal food crop proteomes. *Mass Spec Rev* 25:1–53.

Akagawa, M., T. Handoyo, T. Ishii, S. Kumazawa, N. Morita and K. Suyama. 2007. Proteomic analysis of wheat flour allergens. *J Agric Food Chem* 55:6863–6870.

Akhunov, E. D., A. W. Goodyear, S. Geng et al. 2003. The organization and rate of evolution of wheat genomes are correlated with recombination rates along chromosome arms. *Genome Res* 13:753–763.

Aprile A., A. M. Mastrangelo, A. M. De Leonardis et al. 2009. Transcriptional profiling in response to terminal drought stress reveals differential responses along the wheat genome. *BMC Genom* 10:279.

Baginsky, S. 2009. Plant proteomics: Concepts, applications and novel strategies for data interpretation. *Mass Spec Rev* 28:93–120.

Baker, J. M. 2006. A metabolomic study of substantial equivalence of field-grown genetically modified wheat. *Plant Biotech J* 4:381–392.

Balmer, Y., W. H. Vensel, F. M. DuPont, B. B. Buchanan and W. J. Hurkman. 2006. Proteome of amyloplasts isolated from developing wheat endosperm presents evidence of broad metabolic capability. *J Exp Bot* 57:1591–1602.

Battais, F., F. Pineau, Y. Popineau et al. 2003. Food allergy to wheat: Identification of immunoglobulin E and immunoglobulin G-binding proteins with sequential extracts and purified proteins from wheat flour. *Clin Exp Allergy* 33:962–970.

Battais, F., T. Mothes, D. A. Moneret-Vautrin et al. 2005. Identification of IgE-binding epitopes on gliadins for patients with food allergy to wheat. *Allergy* 60:815–821.

Bechtel, D. B. and J. D. Wilson. 2003. Amyloplast formation and starch granule development in hard red winter wheat. *Cereal Chem* 80:175–183.

Beecher B., A. Bettge, E. Smidansky and M. J. Giroux. 2002. Expression of wild-type *pinB* sequence in transgenic wheat complements a hard phenotype. *Theor Appl Genet* 105:870–877.

Beleggia, R., C. Platani, G. Spano, M. Monteleone and L. Cattivelli. 2009. Metabolic profiling and analysis of volatile composition of durum wheat semolina and pasta. *J Cereal Sci* 49:301–309.

Bernardo A., G. H. Bai, P. G. Guo, K. Xiao, A. C. Guenzi and P. Ayoubi. 2007. *Fusarium graminearum*-induced changes in gene expression between *Fusarium* head blight-resistant and susceptible wheat cultivars. *Funct Integr Genom* 7:69–77.

Bodt, S., D. de Carvajal, J. Hollunder et al. 2010. CORNET: A user-friendly tool for data mining and integration. *Plant Physiol* 152:1167–1179.

Brady S. M. and P. N. Benfey. 2006. A systems approach to understanding root development. *Can J Bot* 84:695–701.

Brazma, A., P. Hingamp, J. Quackenbush et al. 2001. Minimum information about a microarray experiment (MIAME)-toward standards for microarray data. *Nat Genet* 29:365–371.

Browne, R. and K. Brindle. 2007. 1H NMR-based metabolite profiling as a potential selection tool for breeding passive resistance against *Fusarium* head blight (FHB) in wheat. *Mol Plant Path* 8:401–410.

Bruggmann, R., O. Abderhalden, P. Reymond and R. Dudler. 2005. Analysis of epidermis- and mesophyll-specific transcript accumulation in powdery mildew inoculated wheat leaves. *Plant Mol Biol* 58:247–267.

Bylesjo, M., D. Eriksson, M. Kusano, T. Mortiz and J. Trygg. 2007. Data integration in plant biology: The O2PLS method for combined modeling of transcript and metabolite data. *The Plant J* 52:1181–1191.

Caruso, G., C. Cavaliere, C. Guarino, R. Gubbiotti, P. Foglia and A. Laganà. 2008. Identification of changes in *Triticum durum* L. leaf proteome in response to salt stress by two-dimensional electrophoresis and MALDI-TOF mass spectrometry. *Anal Bioanal Chem* 391:381–390.

Caruso, G., C. Cavaliere, P. Foglia, R. Gubbiotti, R. Samperi and A. Laganà. 2009. Analysis of drought responsive proteins in wheat (*Triticum durum*) by 2D-PAGE and MALDI-TOF mass spectrometry. *Plant Sci* 177:570–576.

Chen, S. and A. C. Harmon. 2006. Advances in plant proteomics. *Proteomics* 6:5504–5516.

Clarke, B. and S. Rahman. 2005. A microarray analysis of wheat grain hardness. *Theor Appl Genet* 110: 1259–1267.

Coram, T. E., M. L. Settles, M. N. Wang and X. M. Chen. 2008. Surveying expression level polymorphism and single-feature polymorphism in near-isogenic wheat lines differing for the Yr5 stripe rust resistance locus. *Theor Appl Genet* 117:401–411.

Delhaize, E., P. R. Ryan, D. M. Hebb, Y. Yamamoto, T. Sasaki and H. Matsumoto. 2004. Engineering high-level aluminum tolerance in barley with the ALMT1 gene. *Proc Nat Acad Sci USA* 101:15249–15254.

Desmond, O. J., J. M. Manners, P. M. Schenk, D. J. Maclean and K. Kazan. 2008. Gene expression analysis of the wheat response to infection by *Fusarium pseudograminearum*. *Physiol Mol Plant Path* 73:40–47.

Dixon, R. A., D. R. Gang, A. J. Charlton et al. 2006. Applications of metabolomics in agriculture. *J Agric Food Chem* 54:8984–8994.

Donnelly, B. E., R. D. Madden, P. Ayoubi, D. R. Porter and J. W. Dillwith. 2005. The wheat (*Triticum aestivum* L.) leaf proteome. *Proteomics* 5:1624–1633.

Drea, S., D. J. Leader, B. C. Arnold, P. Shaw, L. Dolan and J. H. Doonan. 2005. Systematic spatial analysis of gene expression during wheat caryopsis development. *Plant Cell* 17:2172–2185.

Dupont, F. M. 2008. Metabolic pathways of the wheat (*Triticum aestivum*) endosperm amyloplast revealed by proteomics. *BMC Plant Biol* 8:39.

Dupont, F. M. and S. B. Altenbach. 2003. Molecular and biochemical impacts of environmental factors on wheat grain development and protein synthesis. *J Cereal Sci* 38:133–146.

Faize, M., S. Sourice, F. Dupuis, L. Parisi, M. F. Gautier and E. Chevreau. 2004. Expression of wheat puroindoline-b reduces scab susceptibility in transgenic apple (*Malus domestica* Borkh). *Plant Sci* 167:347–354.

Fernie, A. R. and N. Schauer. 2009. Metabolomics-assisted breeding: A viable option for crop improvement? *Trends Genet* 25:39–48.

Ferranti, P., G. Mamone, G. Picariello and F. Addeo. 2007. Mass spectrometry analysis of gliadins in celiac disease. *J Mass Spectrom* 42:1531–1548.

Feuillet, C., S. Travella, N. Stein, L. Albar, A. Nublat and B. Keller. 2003. Map-based isolation of the leaf rust disease resistance gene *Lr10* from the hexaploid wheat (*Triticum aestivum* L.) genome. *Proc Natl Acad Sci USA* 100:15253–15258.

Fischer, R. A. 1984. Wheat. In *Potential Productivity of Field Crops Under Different Environments*, eds. W. H. Smith and J. J. Banta, pp. 129–154. Los Banos, CA: IRRI.

Francki, M. G. 2009. Genomics for wheat improvement. In *Molecular Techniques in Crop Improvement* (2nd edition), eds. S. M. Jain and D. S. Brar, pp. 281–306. Berlin: Springer Verlag.

Francki, M. G. and R. Appels. 2007. Comparative genomics and crop improvement. In *Comparative Genomics: Fundamental and Applied Perspectives*, ed. J. R. Brown, pp. 323–342. Boca Raton, FL: Taylor & Francis.

Frank, T., B. S. Meuleye, A. Miller, Q. Y. Shu and K. H. Engel. 2007. Metabolite profiling of two low phytic acid (*lpa*) rice mutants. *J Agric Food Chem* 55:11011–11019.

Fukushima, A., M. Kusano, H. Redestig, M. Arita and K. Saito. 2009. Integrated omics approaches in plant systems biology. *Curr Op Chem Biol* 18:532–538.

Genc, Y., M. Tester and G. K. McDonald. 2010. Calcium requirement of wheat in saline and non-saline conditions. *Plant Soil* 327:331–345.

Ghiglione, H. O., F. G. Gonzalez, R. Serrago et al. 2008. Autophagy regulated by day length determines the number of fertile florets in wheat. *Plant J* 55:1010–1024.

Glazebrook, J., E. E. Rogers and F. M. Ausubel. 1997. Use of Arabidopsis for genetic dissection of plant defense responses. *Ann Rev Genet* 31:547–569.

Goesaert, H., K. Brijs, W. S. Veraverbeke, C. S. Courtin, K. Gebruers and J. A.Delcour. 2005. Wheat flour constituents: How they impact bread quality, and how to impact their functionality. *Trends Food Sci Tech* 16:12–30.

Goff, S. A., D. Ricke, T. H. Lan et al. 2002. A draft sequence of the rice genome (*Oryza sativa* L ssp *japonica*). *Science* 296:92–100.

Golkari, S., J. Gilbert, S. Prashar and J. D. Procunier. 2007. Microarray analysis of *Fusarium graminearum*-induced wheat genes: Identification of organ-specific and differentially expressed genes. *Plant Biotech J* 5:38–49.

Graham, S., E. Amigues, M. Migaud and R. Browne. 2009. Application of NMR based metabolomics for mapping metabolite variation in European wheat. *Metabolomics* 5:302–306.

Grove, H., K. Hollung, A. Moldestad, E. M. Faergestad and A. K. Uhlen. 2009. Proteome changes in wheat subjected to different nitrogen and sulfur fertilizations. *J Agric Food Chem* 57:4250–4258.

Guo, P. G., G. H. Bai, B. Carver, R. H. Li, A. Bernardo and M. Baum. 2007. Transcriptional analysis between two wheat near-isogenic lines contrasting in aluminum tolerance under aluminum stress. *Mol Genet Genom* 277:1–12.

Hajheidari, M., A. Eivazi, B. B. Buchanan, J. H. Wong, I. Majidi and G. H. Salekdeh. 2007. Proteomics uncovers a role for redox in drought tolerance in wheat. *J Prot Res* 6:1451–1460.

Hall, R. D., I. D. Brouwer and M. A. Fitzgerald. 2008. Plant metabolomics and its potential application for human nutrition. *Physiol Plant* 132:162–175.

Hamzehzarghani, H., A. C. Kushalappa, Y. Dion et al. 2005. Metabolic profiling and factor analysis to discriminate quantitative resistance in wheat cultivars against *Fusarium* head blight. *Physiol Mol Plant Path* 66:119–133.

Hamzehzarghani, H., V. Paranidharan, Y. Abu-Nada et al. 2008a. Metabolite profiling coupled with statistical analyses for potential high-throughput screening of quantitative resistance to *Fusarium* head blight in wheat. *Can J Plant Path* 30:24–36.

Hamzehzarghani, H., V. Paranidharan, Y. Abu-Nada, A. C. Kushalappa, O. Mamer and D. Somers. 2008b. Metabolic profiling to discriminate wheat near isogenic lines, with quantitative trait loci at chromosome 2DL, varying in resistance to *Fusarium* head blight. *Can J Plant Sci* 88:789–797.

Hansen B. G., B. A. Halkier and D. J. Kliebenstein. 2008. Identifying the molecular basis of QTLs: eQTLs add a new dimension. *Trends Plant Sci* 13:72–77.

Harrigan, G. G., L. G. Stork, S. G. Riordan et al. 2007a. Impact of genetics and environment on nutritional and metabolite components of maize grain. *J Agric Food Chem* 55:6177–6185.

Harrigan, G. G., L. G. Stork, S. G. Riordan et al. 2007b. Metabolite analyses of grain from maize hybrids grown in the United States under drought and watered conditions during the 2002 field season. *J Agric Food Chem* 55:6169–6176.

Hegeman, A. D. 2010. Plant metabolomics-meeting the analytical challenges of comprehensive metabolite analysis. *Brief Funct Genom* 9:139–148.

Horst, R. J., G. Doehlemann, R. Wahl et al. 2010. *Ustilago maydis* infection strongly alters organic nitrogen allocation in maize and stimulates productivity of systemic source leaves. *Plant Physiol* 152:293–308.

Houde, M. and A. O. Diallo. 2008. Identification of genes and pathways associated with aluminum stress and tolerance using transcriptome profiling of wheat near-isogenic lines. *BMC Genom* 9:400.

Howarth, J. R., S. Parmar and J. Jones. 2008. Co-ordinated expression of amino acid metabolism in response to N and S deficiency during wheat grain filling. *J Exp Bot* 59:3675–3689.

Huang, C. Y., U. Roessner, I. Eickmeier et al. 2008. Metabolite profiling reveals distinct changes in carbon and nitrogen metabolism in phosphate-deficient barley plants (*Hordeum vulgare* L.). *Plant Cell Physiol* 49:691–703.

Hulbert, S. H., J. Bai, J. P. Fellers, M. G. Pacheco and R. L. Bowden. 2007. Gene expression patterns in near isogenic lines for wheat rust resistance gene *Lr34/Yr18*. *Phytopath* 97:1083–1093.

Ishihara, A., F. Matsuda, H. Miyagawa and K. Wakasa. 2007. Metabolomics for metabolically manipulated plants: Effects of tryptophan overproduction. *Metabolomics* 3:319–334.

Jordan, M. C., D. J. Somers and T. W. Banks. 2007. Identifying regions of the wheat genome controlling seed development by mapping expression quantitative trait loci. *Plant Biotech J* 5:442–453.

Jorrin-Novo, J. V., A. M. Maldonado, S. Echevarria-Zomeno et al. 2009. Plant proteomics update (2007–2008): Second generation proteomic techniques, an appropriate experimental design, and data analysis to fulfil MIAPE standards, increase plant proteome coverage and expand biological knowledge. *J Proteom* 72:285–314.

Joung J.-G., A. M. Corbett, S. M. Fellman, D. M. Tieman, H. J. Klee, J. J. Giovannoni and Z. Fei. 2009. Plant MetGenMAP: An integrative analysis system for plant systems biology. *Plant Physiol* 151:1758–1768.

Joye, I. J., B. Lagrain and J. A. Delcour 2009a. Endogenous redox agents and enzymes that affect protein network formation during breadmaking—A review. *J Cereal Sci* 50:1–10.

Joye, I. J., B. Lagrain and J. A. Delcour 2009b. Use of chemical redox agents and exogenous enzymes to modify the protein network during breadmaking—A review. *J Cereal Sci* 50:11–21.

Kasarda, D. D., F. M. Dupont, W. H. Vensel et al. 2008. Surface-associated proteins of wheat starch granules: Suitability of wheat starch for celiac patients. *J Agric Food Chem* 56:10292–10302.

Kawaura, K., K. Mochida and Y. Ogihara. 2008. Genome-wide analysis for identification of salt-responsive genes in common wheat. *Funct Integr Genom* 8:277–286.

Kitta, K., M. Ohnishi-Kameyama, T. Moriyama, T. Ogawa and S. Kawamoto 2006. Detection of low-molecular weight allergens resolved on two-dimensional electrophoresis with acid–urea polyacrylamide gel. *Analyt Biochem* 351:290–297.

Krattinger, S. G., E. S. Lagudah, W. Spielmeyer et al. 2009. A putative ABC transporter confers durable resistance to multiple fungal pathogens in Wheat. *Science* 323:1360–1363.

Kusano, M., A. Fukushima, M. Kobayashi et al. 2007. Application of a metabolomic method combining one-dimensional and two-dimensional gas chromatography-Time-of-flight/mass spectrometry to metabolic phenotyping of natural variants in rice. *J Chromat B Analyt Tech Biomed Life Sci* 855:71–79.

Kwon, S. J., E. Y. Choi, Y. J. Choi, J. H. Ahn and O. K. Park. 2006. Proteomics studies of post-translational modifications in plants. *J Exp Bot* 57:1547–1551.

Lamacchia, C., P. R. Shewry, N. Di Fonzo et al. 2001. Endosperm-specific activity of a storage protein gene promoter in transgenic wheat seed. *J Exp Bot* 52:243–250.

Langeveld, S. M. J., R. van Wijk, N. Stuurman, J. W. Kijne and S. de Pater. 2000. B-type granule containing protrusions and interconnections between amyloplasts in developing wheat endosperm revealed by transmission electron microscopy and GFP expression. *J Exp Bot* 51:1357–1361.

Lantican, M. A., P. L. Pingali and S. Rajaram. 2002. Are marginal wheat environments catching up? In *World Wheat Overview and Outlook: Developing No-till Packages for Small-scale Farmers*, ed. J. Ekboir, CIMMYT 2000–2001. DF Mexico.

Laubin, B., V. Lullien-Pellerin, I. Nadaud, B. Gaillard-Martinie, C. Chambon and G. Branlard. 2008. Isolation of the wheat aleurone layer for 2D electrophoresis and proteomics analysis. *J Cereal Sci* 48:709–714.

Laudencia-Chingcuanco, D. L., B. S. Stamova, F. M. You, G. R. Lazo, D. M. Beckles and O. D. Anderson. 2007. Transcriptional profiling of wheat caryopsis development using cDNA microarrays. *Plant Mol Biol* 63:651–668.

Laudencia-Chingcuanco, D. L., B. S. Stamova, G. R. Lazo, X. Q. Cui and O. D. Anderson. 2006. Analysis of the wheat endosperm transcriptome. *J Appl Genet* 47:287–302.

Leader, D. J. 2005. Transcriptional analysis and functional genomics in wheat. *J Cereal Sci* 41:149–163.

Liu, S. Y, M. D. Hall, C.A. Griffey and A. L McKendry. 2009. Meta-analysis of QTL associated with *Fusarium* head blight in wheat. *Crop Sci* 49:1955–1968.

Long, T. A., S. M. Brady and P. N. Benfey. 2008. Systems approaches to identifying gene regulatory networks in plants. *Annu Rev Cell Dev Biol* 24:81–103.

Lunn, J. E. 2007. Compartmentation in plant metabolism. *J Exp Bot* 58:35–47.

Ma, H., G. Bai and W. Lu. 2006. QTLs for aluminum tolerance in wheat cultivar Chinese spring. *Plant Soil* 283:239–249.

Ma, H. X., G. H. Bai, B. F. Carver and L. L. Zhou. 2005. Molecular mapping of a quantitative trait locus for aluminum tolerance in wheat cultivar Atlas-66. *Theor Appl Genet* 112:51–57.

Majoul, T., E. Bancel, E. Triboi, J. B. Hamida and B. Branlard. 2003. Proteomic analysis of the effect of heat stress on hexaploid wheat grain: Characterization of heat-responsive proteins from total endosperm. *Proteomics* 3:175–183.

Manetti, C., C. Bianchetti, L. Casciani et al. 2006. A metabolomic study of transgenic maize (*Zea mays*) seeds revealed variations in osmolytes and branched amino acids. *J Exp Bot* 57:2613–2625.

Martin, G. B., A. J. Bogdanove and G. Sessa. 2003. Understanding the functions of plant disease resistance proteins. *Annu Rev Plant Biol* 54:23–61.

Matsuo, H., K. Kohno, H. Niihara and E. Morita. 2005. Specific IgE determination to epitope peptides of omega-5 gliadin and high molecular weight glutenin subunit is a useful tool for diagnosis of wheat-dependent exercise-induced anaphylaxis. *J Immunol* 175:8116–8122.

McCartney, C. A., D. J. Somers, D. G. Humphreys et al. 2005. Mapping quantitative trait loci controlling agronomic traits in the spring wheat cross RL4452 × 'AC Domain'. *Genome* 48:870–883.

McCartney, C. A., D. J. Somers, O. Lukow et al. 2006. QTL analysis of quality traits in the spring wheat cross RL4452 ×'AC Domain'. *Plant Breed* 125:565–575.

McIntosh, R. A. 1992. Close genetic linkage of genes conferring adult-plant resistance to leaf rust and stripe rust in wheat. *Plant Path* 41:523–527.

McIntosh, S., L. Watson, P. Bundock et al. 2007. SAGE of the developing wheat caryopsis. *Plant Biotech J* 5:69–83.

Mochida, K., K. Kawaura, E. Shimosaka et al. 2006. Tissue expression map of a large number of expressed sequence tags and its application to *in silico* screening of stress response genes in common wheat. *Mol Genet Genom* 276:304–312.

Mohammadi, M., H. N. V. Kav and M. K. Deyholos. 2007. Transcriptional profiling of hexaploid wheat (*Triticum aestivum* L.) roots identifies novel, dehydration-responsive genes. *Plant Cell Environ* 30:630–645.

Mohammadi, M., H. N. V. Kav and M. K. Deyholos. 2008. Transcript expression profile of water-limited roots of hexaploid wheat (*Triticum aestivum* 'Opata'). *Genome* 51:357–367.

Mott, I. W. and R. R. C. Wang. 2007. Comparative transcriptome analysis of salt-tolerant wheat germplasm lines using wheat genome arrays. *Plant Sci* 173:327–339.

Morsy, M., S. Gouthu, S. Orchard et al. 2008. Charting plant interactomes: Possibilities and challenges. *Trends Plant Sci.* 13:183–191.

Neuhaus, H. E. and M. J. Emes. 2000. Nonphotosynthetic metabolism in plastids. *Ann Rev Plant Physiol Plant Mol Biol* 51:111–140.

Oeljeklaus, S., H. Meyer and B. Warscheid. 2009. Advances in plant proteomics using quantitative mass spectrometry. *J Prot* 72:545–554.

Ogihara, Y., K. Mochida, Y. Nemoto et al. 2003. Correlated clustering and virtual display of gene expression patterns in the wheat life cycle by large-scale statistical analyses of expressed sequence tags. *Plant J* 33:1001–1011.

Palosuo, K., E. Varjonen, O. M. Kekki et al. 2001. Wheat omega-5 gliadin is a major allergen in children with immediate allergy to ingested wheat. *J Allergy Clin Immunol* 108:634–638.

Pandelova, I., M. F. Betts, V. A. Manning et al. 2009. Analysis of transcriptome changes induced by Ptr toxA in wheat provides insights into the mechanisms of plant susceptibility *Mol Plant* 2:1067–1083.

Paranidharan, V., Y. Abu-Nada, H. Hamzehzarghani et al. 2008. Resistance-related metabolites in wheat against *Fusarium graminearum* and the virulence factor deoxynivalenol (DON). *Botany* 86:1168–1179.

Parker, D., M. Beckmann and H. Zubair. 2009. Metabolomic analysis reveals a common pattern of metabolic re-programming during invasion of three host plant species by *Magnaporthe grisea. Plant J* 59:723–737.

Peng, Z., M. Wang, F. Li, H. Lv, C. Li and G. Xia. 2009. A proteomic study of the response to salinity and drought stress in an introgression strain of bread wheat. *Mol Cell Prot* 8:2676–2686.

Poole, R., G. Barker, I. D. Wilson, J. A. Coghill and K. J. Edwards. 2007. Measuring global gene expression in polyploidy: A cautionary note from allohexaploid wheat. *Funct Integr Genom* 7:207–219.

Raman, H., K. Zhang, M. Cakir et al. 2005. Molecular characterization and mapping of *ALMT1*, the aluminium tolerance gene of bread wheat (*Triticum aestivum* L.). *Genome* 48:781–791.

Reynolds, M. and R. Tuberosa. 2008. Translational research impacting on crop productivity in drought-prone environments. *Curr Op Plant Biol* 11:171–179.

Roessner, U., J. H. Patterson, M. G. Forbes, G. B. Fincher, P. Langridge and A. Bacic. 2006. An investigation of boron toxicity in barley using metabolomics. *Plant Physiol* 142:1087–1101.

Röhlig, R. M., J. Eder and K. H. Engel. 2009. Metabolite profiling of maize grain: Differentiation due to genetics and environment. *Metabolomics* 5:459–477.

Rossignal, M. 2006. Proteomic analysis of phosphorylated proteins. *Curr Opin Plant Biol* 9:538–543.

Saito, K., M. Y. Hirai and K. Yonekura-Sakakibara. 2008. Decoding genes by co-expression networks and metabolomics—'Majority report by precogs'. *Trends Plant Sci* 13:36–43.

Salekdeh, G. H. and S. Komatsu. 2007. Crop proteomics: Aim at sustainable agriculture of tomorrow. *Proteomics* 7:2976–2996.

Sanchez, D. H., M. R. Siahpoosh, U. Roessner, M. Udvardi and J. Kopka. 2008. Plant metabolomics reveals conserved and divergent metabolic responses to salinity. *Physiol Plant* 132:209–219.

Sancho, A. I., M. Gillabert, H. Tapp, P. R. Shewry, P. K. Skeggs and E. N. C. Mills. 2008. Effect of environmental stress during grain filling on the soluble protein of wheat (*Triticum aestivum*) dough liquor. *J Agric Food Chem* 56:5386–5393.

Sander, I., A. Flagg, R. Merget, T. M. Halder, H. E. Meyer and X. Baur. 2001. Identification of wheat flour allergens by means of 2-dimensional immunoblotting. *J Allergy Clin Immunol* 107: 907–913.

Sasaki, T., Y. Yamamoto, B. Ezaki et al. 2004. A wheat gene encoding an aluminium activated malate transporter. *Plant J* 37:645–653.

Schauer, N. and A. R. Fernie. 2006. Plant metabolomics: Towards biological function and mechanism. *Trends Plant Sci* 11:508–516.

Schreiber, A. W., T. Sutton, R. A. Caldo et al. 2009. Comparative transcriptomics in the *Triticeae. BMC Genom* 10:285.

Shu, X.-L., T. Frank, Q.-Y. Shu and K. H. Engel. 2008. Metabolite profiling of germinating rice seeds. *J Agric Food Chem* 56:11612–11620.

Shulaev, V., D. Cortes, G. Miller and R. Mittler. 2008. Metabolomics for plant stress response. *Physiol Plant* 132:199–208.

Skylas, D. J., J. A Mackintosh., S. J. Cordwell et al. 2000. Proteomic approach to the characterisation of protein composition in the developing and mature wheat-grain endosperm. *J Cereal Sci* 32:169–188.

Skylas, D. J., S. J. Cordwell, P. G. Hains et al. 2002. Heat shock of wheat during grain filling: Proteins associated with heat tolerance. *J Cereal Sci* 35:175–188.

Snégaroff, J., I. Bouchez-Mahiout., C. Pecquet, G. Branlard and M. Laurière. 2006. Study of IgE antigenic relationships in hyper sensitivity to hydrolyzed wheat proteins and wheat-dependent exercise induced anaphylaxis. *Int Arch Allergy Immunol* 139:201–208.

Stamova, B., U. Roessner, S. Suren, D. Laudencia-Chingcuanco, A. Bacic and D. Beckles. 2009. Metabolic profiling of transgenic wheat over-expressing the high-molecular-weight D × 5 glutenin subunit. *Metabolomics* 5:239–252.

Sun, Q., B. Zybailov, W. Majeran, G. Friso, P. D. B. Olinares and K. J. van Wijk. 2009. PPDB, the plant proteomics database at Cornell. *Nucl Acids Res* 37:D969–D974.

Tarpley, L., A. L. Duran, T. H. Kebrom and L. W. Sumner. 2005. Biomarker metabolites capturing the metabolite variance present in a rice plant developmental period. *BMC Plant Biology* 5:8.

The Arabidopsis Genome Initiative. 2000. Analysis of the genome sequence of the flowering plant *Arabidopsis thaliana. Nature* 498:796–815.

Tohge, T. and A. R. Fernie. 2009. Web-based resources for mass-spectrometry-based metabolomics: A user's guide. *Phytochemistry* 70:450–456.

Varjonen, E., E. Vainio and K. Kalimo. 2000. Antigliadin IgE—indicator of wheat allergy in atopic dermatitis. *Allergy* 55:386–391.

Vensel, W. H., C. K. Tanaka, N. Cai, J. H. Wong, B. B. Buchanan and W. J. Hurkman 2005. Developmental changes in the metabolic protein profiles of wheat endosperm. *Proteomics* 5:1594–1611.

Wakasa, K., H. Hasegawa and H. Nemoto. 2006. High-level tryptophan accumulation in seeds of transgenic rice and its limited effects on agronomic traits and seed metabolite profile. *J Exp Bot* 57:3069–3078.

Walter, S., P. Nicholson and F. M. Doohan. 2010. Action and reaction of host and pathogen during *Fusarium* head blight disease. *New Phytol* 185:654–663.

Wan, Y. F., R. L. Poole, A. K. Huttly et al. 2008. Transcriptome analysis of grain development in hexaploid wheat. *BMC Genom* 9:121.

Wang, Y., L. Yang, H. Xu, Q. Li, Z. Ma and C. Chu. 2005. Differential proteomic analysis of proteins in wheat spikes induced by *Fusarium graminearum. Proteomics* 5:4496–4503.

Ward, J. L., C. Harris, J. Lewis and M. H. Beale. 2003. Assessment of 1H NMR spectroscopy and multivariate analysis as a technique for metabolite fingerprinting of *Arabidopsis thaliana. Phytochemistry* 62:949–957.

Wilson, I. D., G. L. A. Barker, R. W. Beswick et al. 2004. A transcriptomics resource for wheat functional genomics. *Plant Biotech J* 2:495–506.

Winfield, M. O., C. G. Lu, I. D. Wilson, J. A. Coghill and K. J. Edwards. 2009. Cold- and light-induced changes in the transcriptome of wheat leading to phase transition from vegetative to reproductive growth. *BMC Plant Biol* 9:55.

Wong, J. H., N. Cai, Y. Balmer et al. 2004. Thioredoxin targets of developing wheat seeds identified by complementary proteomic approaches. *Phytochemistry* 65:1629–1640.

Wong, J. H., Y. Balmer, N. Cai et al. 2003. Unravelling thioredoxin-linked metabolic processes of cereal starch endosperm using proteomics. *FEBS Lett* 547:151–156.

Xue, G. P., C. L. McIntyre, S. Chapman et al. 2006. Differential gene expression of wheat progeny with contrasting levels of transpiration efficiency. *Plant Mol Biol* 61:863–881.

Yan, L., A. Loukoianov, G. Tranquilli, M. Helguera, T. Fahima and J. Dubcovsky. 2003. Positional cloning of the wheat vernalization gene VRN1. *Proc Natl Acad Sci USA* 100:6263–6268.

Yan, L., D. Fu, C. Li et al. 2006. The wheat and barley vernalization gene VRN3 is an orthologue of FT. *Proc Natl Acad Sci USA* 103:19581–19586.

Yan, L. L., A. Loukoianov, A. Blechl et al. 2004. The wheat VRN2 gene is a flowering repressor down-regulated by vernalization. *Science* 303:1640–1644.

Yu, J., S. Hu, J. Wang et al. 2002. A draft sequence of the rice genome (*Oryza sativa* L ssp indica). *Science* 296:79–91.

Yuan, J. S., D. W. Galbraith, S. Y. Dai, P. Griffith and C. N. Stewart Jr. 2008. Plant systems biology comes of age. *Trends Plant Sci* 13:165–171.

Zhou, J., C. Ma, H. Xu et al. 2009. Metabolic profiling of transgenic rice with *cryIAc* and *sck* genes: An evaluation of unintended effects at metabolic level by using GC–FID and GC–MS. *J Chrom B: Analyt Tech Biomed Life Sci* 877:725–732.

Zörb, C., K. Niehaus, A. Barsch, T. Betsche and G. Langenkämper. 2009. Levels of compounds and metabolites in wheat ears and grains in organic and conventional agriculture. *J Agric Food Chem* 57:9555–9562.

CHAPTER 19

Impact of Climatic Changes on Crop Agriculture
OMICS for Sustainability and Next-Generation Crops

Sajad Majeed Zargar, Muslima Nazir, Kyoungwon Cho, Dea-Wook Kim, Oliver Andrzej Hodgson Jones, Abhijit Sarkar, Shashi Bhushan Agrawal, Junko Shibato, Akihiro Kubo, Nam-Soo Jwa, Ganesh Kumar Agrawal and Randeep Rakwal

CONTENTS

19.1 INTRODUCTION

Agriculture has played a major role in socio-economic change worldwide, and was the key development that led to the rise of human civilization. Today, more than ever, agriculture remains the key factor in human growth and development. Without doubt, plants can be said to be our 'bread and butter', the fundamental basis of human and animal nutrition. Moreover, plants are one of the most important species on Earth, controlling our food production and human sustenance. With the predicted increases in human population to 9 billion by 2040 from 6.8 billion as of December 2009 (U.S. Census Bureau, International Data Base, www.un.org/esa/population/publications/longrange2/WorldPop2300final.pdf.), there is a looming uncertainty on the state of food production. The major agricultural products can be broadly grouped into foods, fibres, fuels and raw material (Figure 19.1). Plants are also being used as a source of biofuels, bioplastics (www.marketwatch.com/story/bioengineers-aim-to-cash-in-on-plants-that-make-green-plastics) and biopharmaceuticals. Hence, agriculture has a direct and indirect impact on the livelihood of all human beings. In the year 2007, about one-third of the world's workers were employed in agriculture. Introduction of various crops followed by their domestication and adoption in different regions of the world has played an important role in framing food habits of populations. In Figure 19.2, we highlight cultivation of major crops in different continents of the Earth. Looking at this figure, it can be said that many crops are common to continents and thus serve as the common source of food (and nutrition) to the people

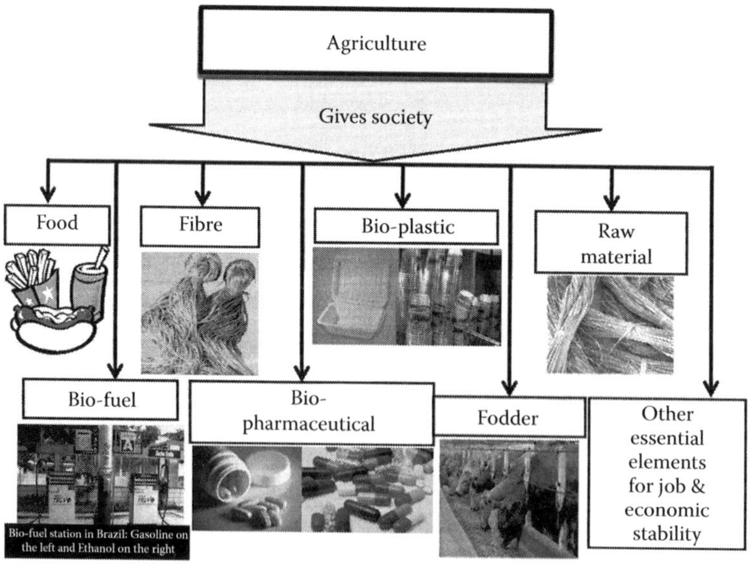

Figure 19.1 Utilization of agricultural products. Plants are mainly utilized as food, fodder and fibre. However, they are also the sources of bioplastic, biofuel, biofertilizer, biopharmaceuticals and so on.

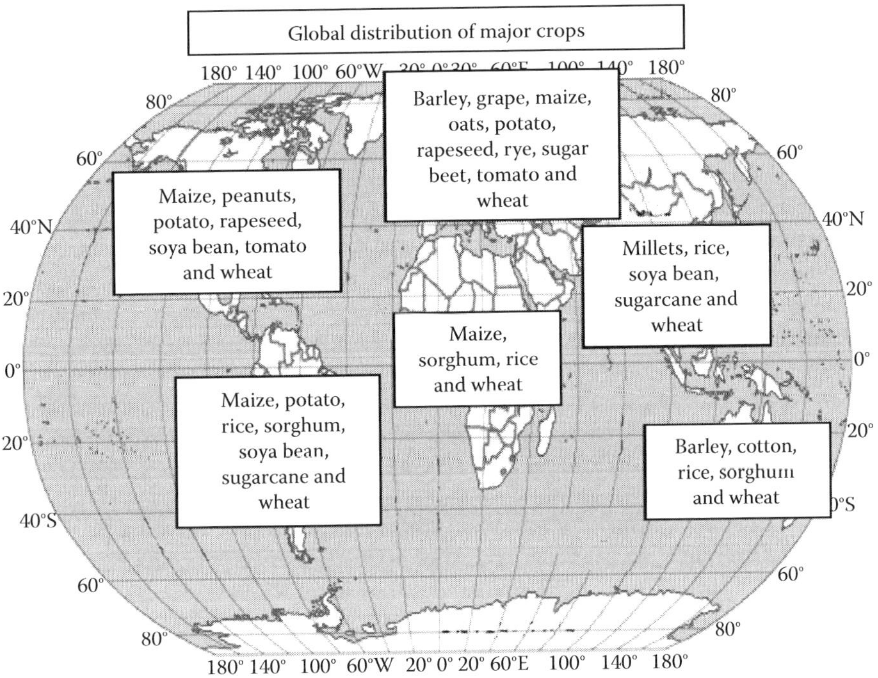

Figure 19.2 The major crops cultivated in the world regions. The figure shows the distribution of major crops in various continents of the world. It is evident that among different continents most of the crops are common.

inhabiting those regions. Thus, we have more in common than the differences in continents may seem to suggest. Nevertheless, it should be noted that there are unique or majorly cultivated crops to each continent, indicating that regional agriculture has its own importance. For example, rice is a crop of great socio-economic importance to South-East Asian countries, and can be rightly called the 'food for life' for billions who inhabit that part of the world.

Since crops sustain our growth and development, it is crucial that we devote our effort to study/investigate crops for their improved crop yield and seed quality. In the context of scientific duties/responsibilities in promoting and sustaining agriculture in the twenty-first century, we have to remember a key factor that may change the balance of food production, plant and animal life and thus human health and happiness, namely adverse changes in climate; called 'climate change'. This chapter deals with crop agriculture under the impact of climate change. Genomics-enabled technologies have the potential to accelerate crop improvement to overcome looming crisis in food production. It should be emphasized that 'environment' is another keyword in today's world, not just in name, but in practice, and therein, the question of sustainable agriculture is also of critical importance and thus worthy of debate.

19.2 SUSTAINABLE AGRICULTURE

The term 'sustainable agriculture' refers to the ability of a farm to produce food indefinitely without causing severe or irreversible damage to the ecosystem. The ideas about 'sustainability' can be seen in the oldest surviving writings from China, Greece and Rome (Cato, 1979; Pretty, 2002, 2005). Currently, the main focus of sustainability is to develop and adopt agriculture technologies and practices that (1) do not have adverse effects on environment, (2) are accessible to and effective for farmers and (3) lead to improvements in food productivity and have good impact on the environment. Looking at the trend of food production in the past few decades, a remarkable growth has been observed since 1960s. Although human population is also increasing at an incredible rate, there is still an additional 25% more food per person compared to the 1960s (Pretty, 2007). However, in totality, no reduction in the frequency of hunger has been observed. As stated by Richard Manning (2000, pp. 149–172), 'sustainable agriculture must be profitable and efficient enough to feed the world's population. It depends on the successful application of three main elements; plant nutrition, pest control and genetics. Hence the need is to make improvements in agriculture systems all over the world, partly by increasing the food production to answer the problems of hunger and partly by adjusting the increasing flow of environmental goods and services'.

The idea of sustainable agriculture does not rule out employment of any technology or practice, if a technology helps in improving agriculture production without having negative impact on the environment; then it is likely to have some sustainability benefits (Pretty, 2007). What will be the fate of various technologies with this prospective? Can genetically modified organisms (GMOs) be considered as a method/tool/approach for sustainable agriculture? About 40,000 people die of malnutrition everyday and half a million children become blind every year because of vitamin A deficiency (Sanderson, 2007a). Are we ready to take the challenge to solve this problem that is going to expand epidemically? Let us see the prospectus of GMOs and decide if this technology can fit in the picture of sustainable agriculture. Genetically modified (GM) plants are reducing costs, decreasing adverse environmental impact. It is believed that they will play an important part in achieving sustainable agriculture. Activists claim that the wide spread use of GM plants will lead to a decrease in biodiversity. But the question is: where is the biodiversity in Texas, East Anglia and Australian wheat belts? Huge areas have been reduced to very few plant species, and hence the need is to have a province of gene banks (Sanderson, 2007b). GM crops can contribute positively by reducing carbon dioxide (CO_2) emission and mitigating the impact of climate change on food security. Agriculture biotechnology can guide the farmers to go for sustainable production of food by reducing the emission of green house gas, crop protection and increased yield from less available arable land by means of using less fuel consumption on farms, by reducing the number of sprays, by carbon sequestration and by reducing use of fertilizers and nitrous oxide (N_2O) emission. As per the Barfoot and Brookes study (2009) (www.pgeconomics.co.uk), release of 1144 mkg of CO_2 in atmosphere was reduced due to the use of GM technology.

Owing to adaptation of GM technology, no-tilled area has been found doubled in the United States and a fivefold increase has been recorded in Argentina in 2007. This has an important impact in mitigating the increase in atmospheric CO_2 concentrations. GM rice and canola has been developed to efficiently utilize nitrogen from soil in order to reduce the applied amount of fertilizer. This 'nitrogen use efficiency' technology produces plants with yields equivalent to conventional varieties but will be requiring significantly less nitrogen fertilizer because they use it more efficiently (www.arcadiabio.com). By using GM crops, water can be used more sustainably as there will be a reduction in the amount of ploughing required before planting the crops. This also helps in reducing the use of fossil fuel resulting in less amount of CO_2 released. Agriculture accounts for 70% of all water use and it is expected that water stress area will be further enhanced in the coming years. Hence by using GM crops, it is again believed that the effects of drought tolerance can be alleviated.

Kwesi Atta-Krah (Deputy Director General of Italy-based research organization Biodiversity International) states, 'Plant breeders now need to use vast genetic resources present in gene banks and the wild varieties that can thrive well under changing climatic conditions (www.scidev.net)'. It is important to mention that new technologies, including, and most prominently, the 'genome revolution' will play a critical role in our quest towards crop improvement in the twenty-first century. Genome revolution, we believe, will constitute the 'second or next Green Revolution'.

19.3 CROP AGRICULTURE

19.3.1 Where We Were: A Brief History of the Green Revolution

The story begins in 1943, when Mexico was importing half its wheat to feed the population. However, by 1964, Mexico started exporting half a million ton of wheat owing to development of high-yielding disease-resistant wheat varieties. This turn-around in agriculture productivity for Mexico was owing to the lifetime work of Norman Ernest Borlaug who changed the scenario of agriculture and was named 'Green Revolution'. This term was first used in 1968 by the former USAID director William Gaud (speech on 8 March by William S. Gaud to the Society for International Development in Washington, DC, USA). Dr. Borlaug was an American agronomist, humanitarian and Nobel laureate and is known rightly as the father of the Green Revolution (Borlaug, 2007; Brown, 1970). The Green Revolution allowed food production to keep pace with worldwide population growth. The projects within Green Revolution spread technologies such as pesticides, irrigation, nitrogen fertilizers and development of improved crop varieties through conventional breeding methods (Borlaug and Dowswell, 2004). The Japanese wheat variety 'Norin 10', Indonesian rice variety 'Peta' and Chinese rice variety 'Dee-geo-woo-gen' were instrumental in developing Green Revolution wheat and rice varieties, respectively (http://en.wikipedia.org/wiki/Green_Revolution). With advances in molecular genetics, dwarf genes, namely wheat reduced height genes *Rht* (Appleford et al., 2007) and rice semi-dwarf genes *sd1* (Monna et al., 2002), were cloned. Increased crop yield owing to Green Revolution is widely credited with having helped to avoid widespread famine, and for feeding billions of people. In March 2005, Norman Borlaugh stated, 'we will have to double the world food supply by 2050'. He emphasized that it can be possible if we can increase crop immunity to large-scale diseases and pests. Dr. Borlaug considered that GMOs are the way to feed people on this planet.

19.3.2 Where We Are Today: The 'Second' Green Revolution

The question is what next? Population is increasing at an alarming rate and conditions for agriculture are becoming difficult due to changing climate. Hence, the need is to develop environmental-friendly technologies that can improve agriculture and feed the growing population. Keeping this in mind, it is now quiet evident that (1) we have to go for environmentally safe technologies; (2) we have to develop crops that can withstand adverse/harsh climatic conditions and at the same time (3) we have to conserve biodiversity which provides us useful genetic resources to withstand such potentially damaging (to crops) conditions. This is what we feel should be the next Green Revolution. However, there are two points to be kept in mind, namely the agriculture and the technology in developed and developing countries. In the case of developing countries like India and others in South-East Asia, in South America and Africa, one has to think in terms of immediate food output (production and yield) and importantly, with the threat of climate change hanging over their heads. Thus, developing countries need to take enhanced focus and efforts for crop improvement not only against climatic changes, but also against a wide variety of associated abiotic and biotic stresses. However, in the case of developing countries, the scenario is different; resources are there to withstand these changes and there are very less number of people to feed than in the developed world. Nevertheless, sooner or later, the developed world has to adapt to the changing climate and the problem of food insecurity is as much the same way as developing countries. Here, advanced science and technology resources will play a crucial role in the move towards global food security.

Genome revolution will be the next hope for us. By 2050, when human population is doubled of what it is at present, we have to be technologically sound to face the immense challenge of feeding humankind. During Green Revolution, various conventional breeding approaches were applied, and that by now are fully exploited for enhancing the crop productivity. Now we need to work to manipulate the genome (genes, proteins and metabolites) to further improve crop plants, and thus agriculture itself.

19.4 CLIMATE CHANGE AND ITS IMPACT ON CROPS

19.4.1 Climate Change

A lot is discussed about the climatic changes and the means to cope up with such changes. Sigmund Freud has very well stated that every crisis is potentially a stimulus to the positive side of the personality and is an opportunity to start afresh. Hence, there is a wide scope for improvement. As per the Declaration of the World Summit on Food Security (Document WSFS 2009/2; Rome, 16–18 November 2009) more than 1 billion people worldwide are hungry and poor, and this number, it further states, will increase if governments do not spend more on agriculture (Source-UN Food and Agriculture Organization; http://timesofindia.indiatimes.com/world/rest-of-world/Record-1-billion-go-hungry-says-UN/articleshow/5125942.cms). Looking at the gravity of the situation, it has been estimated that world agriculture output will have to increase by 70% between 2010 and 2050 to feed a world population exceeding 9 billion by 2050. Present climatic changes lead to decrease in timely precipitation, increase in temperature and have drastic effect on sustainability. Nicholas Stern, an economist, warned that temperature could increase by 2–3°C in the next 50 years and if nothing is done to prevent this, it will increase by as much as 5–8°C by the end of this century (Stern Review of the Economics of Climate Change, 2006). Water, food, health, land and environment, which are the essential needs of life, are all threatened by climate change. There will be a tremendous decline in the crop yield impacting all other aspects of life. It is believed that, 'new climate era' will have its winner and loser species and unfortunately many of the valuable crops are likely to fall in the category of losers.

The global warming could disrupt the timing of pollination in alpine environments, having serious negative impacts on both plants and pollinators (United Nations, 2004). It has been observed that flower timing has become earlier (Figure 19.3). Abundance of some flowers has changed and synchrony of plants and pollinators may be changing (www.sciencedaily.com/release/2006/08/060809234056.htm). Hence, one of the major tasks is to increase the crop production under such devastating conditions to feed the growing population.

Moreover, agricultural land area will decrease in the future due to rising temperature and desertification. It is again emphasized that there will be 2.5 billion more people in 2025 demanding 35% more food than today, and hence farmers will need to double their food production over next two decades to feed this population. As per the reports of European Association for bioindustries, climate change could cost the world at least 5% of GDP each year (Stern Review of the Economics of Climate Change, 2006).

People use different practices to maintain sustainability, such as recycling crop waste, growing legume crops for nitrogen fixation, adopting advanced irrigation systems and cultivating high-yielding varieties. However, for sustainability under the harsh changing climatic conditions, the potential of land races cannot be ignored. Among the most important factors for sustainable agriculture, the two, namely, water and soil quality and quantity, are most amenable to human intervention and will have to play a major role for higher crop production.

19.4.2 Impact of Climate Change

Devastating effects of climate change are most apparent on genetic resources and biodiversity, which represent one of the major limiting factors for crop improvement. The urgent need is to effectively conserve, manage and put in use such resources to answer climate change. As per the report

Figure 19.3 Impact of 'changing climate' on flowering time. Early blooming of flowers due to warmer temperatures during winter in Kashmir (India). (Adapted from Talib, A. H. 2007. *A Project on Climate Change and its Impact in Kashmir*, India: BfC. http://www.actionaidindia.org/download/On_the_brink.pdf)

published by science and development network, the diversity of Latin American's bean is shrinking where 12 out of 17 wild species are predicted to get extinct by 2055 due to climate change (www.scidev.net/en/features/can-crops-be-climateproofed.html). About 60% of land in India is already under crop cultivation and further increase in agriculture looks impossible (The Hindu Survey of the Indian Agriculture, 2005). The scenario of climate change will be depleting the crop production. Hence, the need is to look multidimensionally to address this issue. Variation in rainfall due to climate change is a threat for rice cultivation, as 50% of available irrigation water in India is used only for rice cultivation whereas other crops are cultivated with 50% of irrigation water (Kumar, 2006). Rice being the top most priority crop especially in South-East Asia, the need is to develop varieties that can grow in low moisture content.

Since 1800, the world's population has increased 20-fold, and their demand for the food has been fulfilled by the conversion of extensive land areas to agriculture and the introduction of numerous technological advances, including mechanization, synthetic fertilizers, chemical control of pests and diseases and improved high-yielding crop varieties. But the future looks very tough; the human population is predicted to increase 50% of the present number. It is believed that over the next 50 years, considerable stress will be placed on worldwide crop production (Rothstein, 2007). The food production is needed to be enhanced with much more pace and this will need to be accomplished in an environmentally and economically sustainable fashion.

As far as the impact of climate change on plant disease is concerned, various published reports suggest that the most likely impact of climate change will be felt in three areas: (1) in losses from plant diseases, (2) in the efficacy of disease management strategies and (3) in the geographical distribution of plant diseases (Chakraborty et al., 2000). Due to altered temperature and precipitation profiles, new combinations of pests and diseases may emerge. Any increase in the frequency or severity of extreme weather events, including droughts, heat waves, windstorms or floods, could also disrupt the predator–prey relationships that normally keep pest populations in check. The effect of climate on pests may add to the effect of other factors such as the overuse of pesticides and the loss of biodiversity due to pest and disease outbreaks (McMichael et al., 2003a). The negative effects of climate change on agriculture in poor countries could put an additional 40–300 million people at risk of hunger by 2060 (McMichael et al., 2003b).

19.5 CROP IMPROVEMENT: AN URGENT NEED

19.5.1 Journey from Classical Breeding to Molecular Breeding

As discussed in the earlier sections, it is a necessity rather than choice that we need to adopt and put in use new technologies to improve the present agriculture system. In the late 1960s, the agricultural scenario got changed by adapting Green Revolution technologies. After that, plant breeders got fully involved in breeding new varieties for overcoming the demand of food. The process of crop improvement started from selection which required more art than science followed by introduction of varieties from different locations and its domestication. Later, various hybridization methods were followed to incorporate the gene of interest in the desirable background through conventional breeding (Figure 19.4). Excellent progress has been made in crop improvement through conventional breeding, but it has been exploited almost to its highest potential. During the past two decades, and breeders are shifting from conventional to molecular approaches of breeding (called molecular breeding). Molecular breeding has its own advantages in comparison with conventional breeding. It helps in tracking the gene transferred to plant, and as a result, the population size is much lesser. Marker-based selection of plants can be done at seed, seed germination and seedling stages. Gene mapping and gene pyramiding have been easier tasks saving time and labour (Figures 19.5 and 19.6).

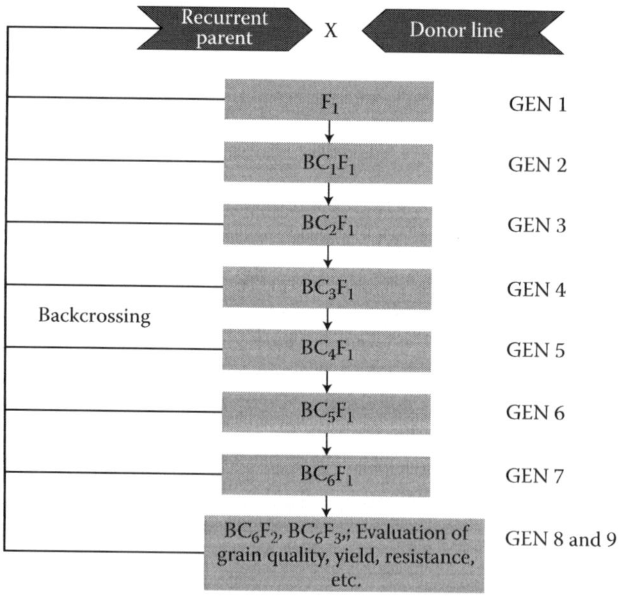

Figure 19.4 Conventional breeding approach for introgression of single gene. Introgression of a gene in an elite variety through conventional breeding needs eight to nine generations of backcrossing. In this breeding approach, large population size is required and phenotyping is must.

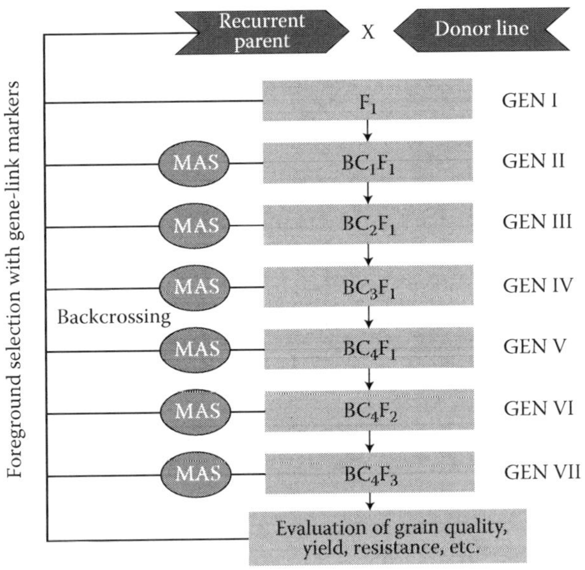

Figure 19.5 Molecular breeding for introgression of single gene. Molecular breeding required lesser number of generations as compared to conventional breeding for introgression of a gene. Selection is done using gene-linked markers that assures the transfer of trait of interest.

Plant breeding is based on the identification and utilization of genetic variation. The breeder first needs to decide and select appropriate parents to use for the initial cross or crosses and then the selection is done to identify the desirable individuals among the progeny of cross. A major cost and logistical issue in plant breeding is the population size that is used for evaluation and selection phases of a programme. Field trials can be expensive. Evaluation of some traits is very difficult and

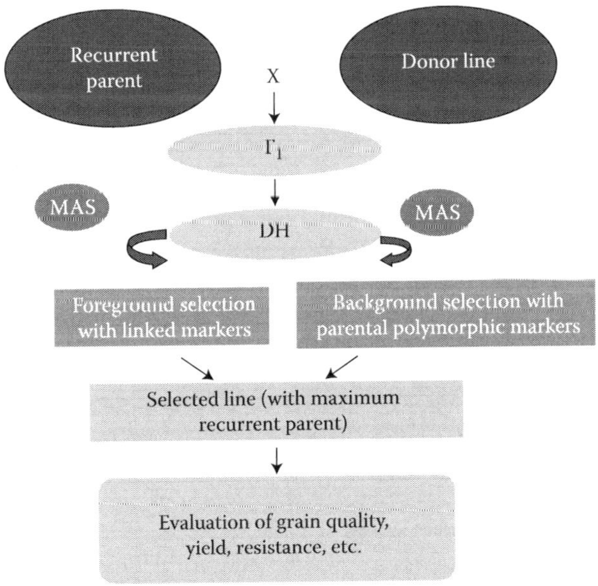

Figure 19.6 MAS with dihaploidy for introgression of single gene. Using DH population, the number of generations is tremendously reduced and utilizing the gene-linked markers will assure the transfer of trait.

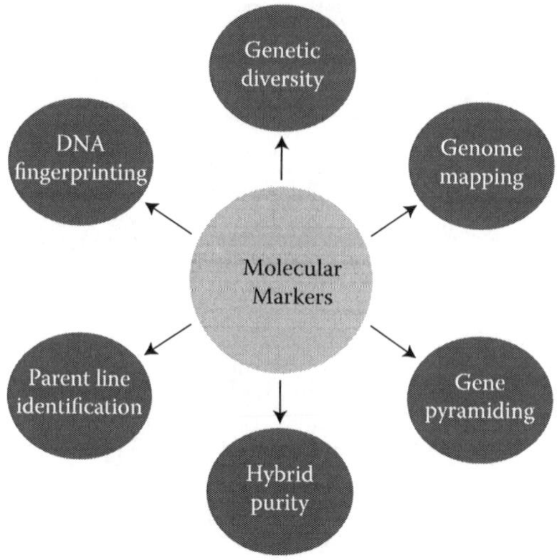

Figure 19.7 Application of molecular markers. Molecular markers are used in the determination of genetic diversity, hybrid purity, fingerprinting, parent line identification, gene mapping and pyramiding.

expensive to assess. Molecular markers have proved to be a powerful tool in replacing bioassays and to evaluate complex traits. Molecular markers once developed are used for parental line identification, genome mapping, gene pyramiding, determining the genetic diversity and hybrid purity and so on (Figure 19.7). For various important traits such as disease resistance, abiotic tolerance and quality, major loci are known. Only the markers that are closely linked to these traits showing polymorphism can be used in molecular breeding. Backcross breeding using markers have also proved very useful in transferring the desired traits in the genome of recurrent parent. So far, many improved varieties have been obtained through molecular breeding. Earlier, restriction fragment-length polymorphisms (RFLPs) were used as a molecular marker technique to discriminate the germplasm based on the location of restriction sites available in the genome. Later, different polymerase chain reaction (PCR)-based molecular marker techniques were developed. Among PCR-based techniques, simple sequence repeats (SSRs) were found to be very reliable. Currently, SNPs and DArT are being used in the characterization of germplasm, as they are much easier to use and more informative compared to other marker technologies. After sequencing of various plant genomes, it has become easier to understand the functionality of genes and gene products. Now in this era of OMICS, genes and gene products can be easily targeted and manipulated in a desirable manner. In 1920, Hans Winkler in his book *Verbreitung und Ursache der Parthenogenesis im Pflanzen- und Tierreiche* (Verlag Fischer, Jena): 'I propose the expression 'Genom' for the haploid chromosome set, which, together with the pertinent protoplasm, specifies the material foundations of the species ....' (p. 165). As a botanist, Winkler must have been familiar with a host of -ome words such as biome, rhizome, phyllome, thallome and tracheome—all of which predated 1920. They share in common the concept of -ome signifying the collectivity of the units as rhizome is the entire root system or modifications therein.

19.5.2 Genome Revolution and OMICS

This is the age (twenty-first century) of plant functional genomics. It can be stated that we are now in a position to use multiparallel approaches, and to apply a range of new technologies to the functional analysis of plant genomes in a high-throughput manner. This inevitably also leads to an

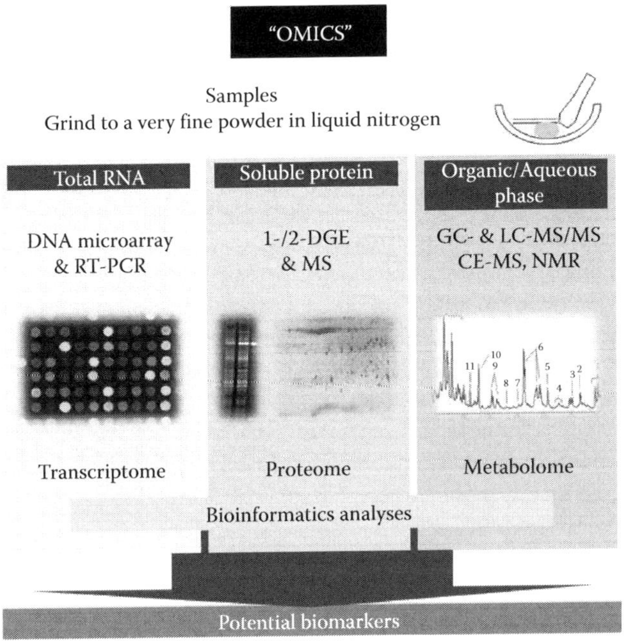

Figure 19.8 The OMICS approaches. Genomics, proteomics and metabolomics along with bioinformatics are ideal approaches for identification of biomarkers.

unprecedented pace (and efficiency) in analysis and deduction of gene (one or many) function in a short time, at the level of the transcript (transcriptomics), the protein (proteomics) and the metabolite (metabolomics). We refer these -omics to as the trinity of OMICS (Figure 19.8) that are discussed below in the genomics perspectives.

Looking at the plant genomics status, we can see a great progress in many sequenced plant genomes. The flowering plant *Arabidopsis thaliana* (L.) Heynh has been a model dicotyledonous plant and was the first plant for which the genome sequence was completed in December 2000 (The Arabidopsis Genome Initiative, 2000). Arabidopsis has now been adopted by plant biologists worldwide for conducting research into various aspects of plant biology, resulting in the development of various important tools, resources (DNA/clones, T-DNA collections, seed/germplasm and ecotypes) and experimental approaches (reverse genetics) for the scientific community (for further details, the readers are referred to The Arabidopsis Information Resource, TAIR: (www.arabidopsis.org/) and Somerville and Koornneef, 2002). The genome sequence of the monocotyledonous and crop plant model rice (*Oryza sativa* L.) was sequenced in 2002 (Goff et al., 2002; Yu et al., 2002), followed by the availability of more than 28,000 annotated cDNA clones (Kikuchi et al., 2003). It is worth mentioning here that it was agreed at the First International Symposium of Rice Molecular Biology in Kurashiki, Okayama Japan, in 1986 that the rice cultivar Nipponbare should be used as the standard japonica strain for the genome research. Like Arabidopsis, the resources for functional genomics of rice are also available and increasing, such as development of a large collection of rice-mutant resources for the functional analysis of rice genes (Krishnan et al., 2009; Miyao et al., 2007). It should be noted that with the completed Arabidopsis and rice genomes, researchers have been given the option to undertake high-throughput analysis of proteins by proteomics (Agrawal and Rakwal, 2008). With continuing progress on the genomics front, we now have the recently completed genomes of sorghum (Bedell et al., 2005), poplar (Tuskan et al., 2006), castor (Lu et al., 2007), grape (Jaillon et al., 2007) and soya bean (Schmutz et al., 2010; Umezawa et al., 2008). Furthermore, with the wide plant diversity available in the tropics, which forms the basis of the food security and sustenance for the human and animal population, it has been realized by the scientific community that there is a need to target more tropical plant/crop genomes, in the coming years. The plant genomics resources will be extremely helpful in the discovery of new genes, allele mining and large-scale SNP genotyping, which ultimately will help manipulating the plant genome to improve crop yield and seed quality.

To understand plant biology, not only under normal growth and development but also under the ever-changing environment, we need more than the current repertoire of classical genomics approaches. In the following text, we briefly discuss the technologies that are gaining prominence in the twenty-first century functional genomics era. These are genomics (transcriptomics), proteomics and metabolomics. Proteomics is one of the rapidly emerging and expanding fields of the functional genomics era, and as the term suggests, the approach involves a systematic and detailed analysis of protein population in a cell, subcellular compartment, tissue and whole organisms, and complements genomics and metabolomics. Proteomics is currently being used for a systematic identification, qualitative and quantitative profiling and functional characterization of all the expressed plant proteins (Agrawal and Rakwal, 2008). It should be emphasized again that the reason is partly due to its specific features compared to genomics or gene-expression profiling (Bradshaw, 2008). Historically, it was a graduate student at Macquarie University in Australia, Marc Wilkins who coined the term 'proteome'—PROTEin complement of the genOME (the complete set of proteins expressed and modified following their expression by the genome)—in the year 1994 (Wilkins et al., 1995). Since the year 2000, the term 'proteomics' as a whole has been widely applied to investigate the plant proteomes. In the past decade we have seen how plant proteomics has gained momentum, and has become a major area of research and development in the field of plant biology. Technically, and importantly, these studies have developed a number of specific techniques namely the gel-based and the gel-free approaches, to generate a large list of identified proteins. Among the gel-based proteomics approaches, one- and two-dimensional gel electrophoresis (1DGE and 2DGE) are the preferred techniques used in combination with mass spectrometry (MS) and are reviewed in Görg et al., 2009and Righetti et al., 2008. The gel-free proteomics approach, also known as MS-based proteomics approach (reviewed in Bergmuller et al., 2008; Leitner and Lindner, 2009), includes the widely used multidimensional protein identification technology (MudPIT) or semicontinuous MudPIT and isobaric tags for relative and absolute quantification. The readers are also referred to comprehensive reviews for further detailed reading on the proteomics technologies in the context of plant biology (for reviews, see Agrawal et al., 2005a,b,c, 2009, 2010; Agrawal and Rakwal, 2006; Canovas et al., 2004; Chen and Harmon, 2006; Jorrin et al., 2007; Kersten et al., 2006; Lilley and Dupree, 2006; Rossignol et al., 2006; Van Wijk, 2001; Weckwerth et al., 2008). We can here quote the elegant statement by Watson et al. (2003) 'as we seek to better understand the gene function and to study the holistic biology of systems, it is inevitable that we study the proteome'.

Finally, there is one more approach that completes the trinity of omics, namely metabolomics. In functional genomics studies, the study of the metabolites or 'metabolome' constitutes the field of metabolomics. Hence, metabolomics itself may be defined as the study of the complete set of small molecules, endogenous metabolites (such as sugars, organic acids, amino acids and nucleotides) present in cells at a particular instance in their life cycle (Fiehn, 2002; Griffin and Shore, 2007; Oliver et al., 1998; Saito et al., 2008). These compounds are the substrates and products of cell processes (such as enzymatic reactions) and as such have a direct effect on the phenotype. Metabolomics is therefore a powerful tool for the characterization of metabolic phenotypes in highly complicated cellular processes (Fiehn, 2002). Among the omics approaches, the subdiscipline of plant metabolomics is rapidly growing. Looking at the literature, it can be seen that some of the most significant developments in metabolic profiling and metabolomics-based work has been conducted on a diverse array of important food plant species (reviewed in Schauer and Fernie, 2006). Differing from genomics and proteomics approaches, where molecules (DNA, RNA and protein) can be analysed using established and standardized analytical platforms, metabolomics needs a variety of instrumentation based on different analytical principles. These are nuclear magnetic resonance (NMR) spectrometry, gas chromatography–mass spectrometry (GC–MS), liquid chromatography–mass spectrometry (LC–MS) and CE–MS (for further details, see Weckwerth, 2006). Currently, metabolomics is gaining importance in biotechnology applications, as exemplified by quantitative loci analysis, prediction of food quality and evaluation of GM crops (as reviewed in Saito and Matsuda, 2010). For example, among external stresses, water stress or drought is a great environmental stress, with implications to crop productivity and food security (Bartels and Sunkar, 2005; Umezawa et al., 2006). Here, the use of metabolomics (in combination with transcriptomics and proteomics) in understanding the global responses to dehydration stress could be critical in improving crop yields (see also Urano et al., 2009).

The integration of multiple omics techniques such as proteomics and transcriptomics with metabolomics would also help further elucidate the biochemical and common metabolic mechanisms used by plants not only under normal growth and development, but also in response to environmental stresses. Herein, bioinformatics and systems biology will play a crucial role in our quest for understanding the plant and its molecular components in response to the environment.

19.6 DEVELOPMENT OF MOLECULAR- AND BIOMARKERS IN CROPS AGAINST MAJOR COMPONENTS OF GLOBAL CLIMATE CHANGE

Understanding the crop response to climate variations has always been the focal theme for agriculturists, agro-meteorologists, plant biologists and environmentalists, as it directly affect the food security (Porter and Semenov, 2005). In search of higher yield and better grain quality, crop breeders had developed numerous cultivars through conventional breeding programmes. Diverse researches during the past two to three decades have revealed that these cultivars are not potential enough to cope up with the present climate change (Long et al., 2005). According to Singh et al. (2007), only wheat production must continue to increase at least 2% annually until 2020 to meet the future demand, but the major components of climate change, such as ozone (O_3), CO_2 and ultraviolet-B (UV-B) and so on, do not seem to let that happen. Keeping this realistic picture in mind, Dr. Norman Borlaugh had concluded that the next era of crop biotechnology would be governed by identifying genotypes that could maximally exploit the future environment for yield enhancement *vis-á-vis* improved stress-tolerant traits (Reynolds and Borlaug, 2006). Engineering crops for future world requires a basic understanding on the detailed network of induced biomolecular changes in different genotypes of present crops (Ainsworth et al., 2008). In this section, we provide examples relating to climate-induced changes in biomolecular network in crop plants, which can be exploited as bio- and molecular markers for crop improvement.

19.6.1 Study of Ozone Effect

Tropospheric O_3 is a notorious environmental gaseous pollutant that is produced by the photochemical reactions between volatile organic compounds and nitrogen oxides (NO_x) (Varotsos et al., 2004). It is believed that since the late 1970s, the annual increase in NO_x due to combustion of fossil fuels for industrialization has been a major factor in global O_3 increment (Fusco and Logan, 2003). The climate models have forecasted that the average level of ground O_3 will reach phytotoxic range in the future showing greater increase in its concentration in the major regions of the world, namely Asia, Africa and the United States, where crops such as rice, wheat, corn and soya bean are cultivated (Dentener et al., 2006; Emberson et al., 2007; Wang and Mauzerall, 2004). Therefore, O_3 is increasingly being recognized as the cause for considerable damages in both natural and cultivated plants including the major forest trees and crop plants (Kley et al., 1999). In recent years there have been many high-throughput omics approaches (transcriptomics, proteomics and metabolomics) to catalogue the O_3-responsive gene, protein or metabolite components in aspen (Bohler et al., 2007; Gupta et al., 2005), birch (Kontunen-Soppela et al., 2007; Ossipov et al., 2008), pepper (Lee and Yun, 2006), Arabidopsis (Mahalingam et al., 2006; Tamaoki et al., 2003), bean (Torres et al., 2007), maize (Torres et al., 2007) and rice (Agrawal et al., 2002; Cho et al., 2008). These studies continue to refine our understanding on the O_3-constitutive and -triggered signalling and metabolic responses. However, the global identification of genes tightly associated with the symptoms of foliar injury in natural and cultivated plants still remains largely unknown. In the above context, this 'black-box' of hidden gene information is critical to further our knowledge on the link between O_3-triggered morphological change (e.g., foliar injury) and gene expression (Kubo, 2002) in plants.

19.6.1.1 Rice

As an example, we briefly describe a study in rice, where a transcriptomics approach was applied for the investigation of molecular responses to O_3 (200 ppb) in two-week-old rice (*O. sativa* L. cv. *Nipponbare*; Japonica-type) seedling leaves (Cho et al., 2008). Using a 22K rice DNA microarray chip, expression profiles of genes were checked in leaves exposed to O_3 for 1, 12 and 24 h. A total of 1535 genes were differentially expressed more than fivefold over the control. Their functional categories suggested that genes involved in transcription, pentose phosphate pathway and signal transduction at 1 h, and genes related to antioxidant enzymes, ribosomal protein, post-translational modification (PTM), signal transduction, jasmonate, ethylene and secondary metabolism at 12 and 24 h play a crucial role in O_3 response (Cho et al., 2008).

Recently, our group has been using a combined approach of evaluating O_3-caused foliar injury symptom and global gene-expression profiling to identify potential genes whose expressions are tightly associated with severity of foliar injury on leaves of O_3 (200 ppb)-fumigated two-week-old rice seedlings. In addition, this study provides evidence for the presence of large number of genes tightly associated with the foliar injury symptom, which might serve as potential biomarkers to study mechanisms of visible injury development by O_3. Briefly, two-week-old rice seedlings were exposed to O_3 (200 ppb) and leaves collected at multitime point were used for evaluating the severity of foliar injury and large-scale transcript profiling. The results of this study (to be published elsewhere) report major achievements towards identification and dynamic expression profiles of 112 genes linked with increased foliar injury symptoms till 72 h and metabolic networks reconstruction (Kyoungwon Cho and co-workers, unpublished data). Thus, genomics in combination with physiological (phenotypical) studies can be useful to biomarker discovery in O_3 effects on rice, if approached in the right direction.

S. B. Agrawal and his co-worker at BHU (India) also found dependable phenotypical response, in the form of foliar injury, in mature rice plants (*O. sativa* L. cvs *Malviya dhan 36* and *Shivani*) followed by definite changes in leaf proteome against elevated O_3 exposures under near-natural condition using open top chambers (OTCs). It was concluded that foliar injury and protein response parameters can be utilized as *in vivo* biomarker in rice under normal agricultural condition against O_3 stress (Sarkar and Agrawal, 2010).

19.6.1.2 Wheat

Wheat (*Triticum aestivum* L.) is the third most important crop around the globe (Figure 19.2), and nearly two-third of the world population depends on this crop for their primary nutrition supplement. The group of S. B. Agrawal at BHU recently employed *field-based integrated-omics* approach to understand the background of O_3 response in two wheat cultivars (cvs Sonalika and HUW 510) against elevated O_3 concentrations (ambient + 10 and 20 ppb) under near natural conditions using OTCs. Results of that study (Sarkar et al., 2010) show significant down-regulation of some major photosynthetic proteins and up-regulation of defence/stress-related proteins and so on followed by decreased stomatal conductance and increased foliar injury in mature wheat plants after 50 days of O_3 exposure. However, the authors (S. B. Agrawal and his team) are planning for a more detailed and a large scale '-omics' study on wheat model; still, the initial report showed that up- and down-regulated proteins can be utilized as molecular markers and the foliar injury as phenotypic biomarker against O_3 in wheat model (for further details, see Chapter 21).

19.6.1.3 Maize

Maize (*Zea mays* L.) is another important crop at global context (Figure 19.2). Being a C_4 crop, its response to climate change has been always bit different from the others. Torres et al. (2007) have done detailed investigation of O_3 response in maize (cv. Guarare 8128) plants through gel based 'omics'. In that experiment, 16-day-old maize plants (grown in controlled environment at green house) were exposed to 200 ppb O_3 for 72 h, and then the response was compared with a controlled plant (grown under filtered pollutant-free air). Results showed that nearly 12 protein spots were differentially expressed under O_3 exposure, and can be exploited as marker proteins. Expression levels of catalase (increased), SOD (decreased) and APX (increased) were drastically changed by O_3 depending on the leaf stage, whereas cross-reacting heat-shock proteins (HSPs; 24 and 30 kD) and naringenin-7-*O*-methyltransferase (NOMT; 41 kD) proteins were strongly increased in O_3-stressed younger leaves. The study also enumerated leaf injury as biomarker under O_3 stress in maize leaves.

19.6.1.4 Bean

Torres et al. (2007) also extended their study on response of cultivated bean (*Phaseolus vulgaris* L. cv. IDIAP R-3) against O_3 stress. In that experiment, 16-day-old bean plants were exposed to 200 ppb O_3 for 72 h, and the effect was evaluated through integrated 'omics' approach using gel-based proteomics followed by MS and immunoblotting. Results showed that in bean leaves two SOD proteins (19 and 20 kD) were dramatically decreased, while APX (25 kD), small HSP (33 kD) and a NOMT (41 kD) were increased after O_3 fumigation.

19.6.2 Study of Carbon Dioxide Effect

CO_2 is another major component of global climate change. However, being a potent green house gas and main basis of global warming, CO_2 is also a key substrate for plant growth. Uncontrolled anthropogenic activities throughout the world have caused the atmospheric CO_2 to increase continuously from about 280 ppm at the beginning of the nineteenth century to 369 ppm at the beginning of the twenty-first century. Projections also range between about 450 and 600 ppm by the year 2050, but strongly depend on future scenarios of anthropogenic emissions (Woodward, 2002). Agriculture crops have been studied in detail against this atmospheric CO_2 enrichment for the past years (Wand et al., 1999), but most of them were based on physiological measurements and biochemical assays. Very few researchers have applied 'omics' approaches in crops to understand the impact of CO_2. There is a strong need to perform more research on this field to get a better and clear picture of CO_2 impact on crop plants.

19.6.2.1 Rice

Bokhari et al. (2007) had exposed 10-day-old rice (*O. sativa* L. spp *Indica* cv. 93–11) seedlings to 760, 1140 and 1520 ppm of CO_2 for 24 h, and assessed the response of test plants through 2D gel based proteomics followed by protein identification. Comparative analysis of leaf proteome revealed

57 differentially expressed proteins under elevated CO_2 in rice leaf proteome. Majority of these differentially expressed proteins were belonged to photosynthesis (34%), carbon metabolism (17%), protein processing (13%), energy pathway (11%) and antioxidants (4%). Several molecular chaperones and APX were found to be up-regulated under higher CO_2, whereas major photosynthetic proteins like RuBisCO and RuBisCO activase, different proteins of Calvin cycle were down-regulated.

19.6.2.2 Wheat

Hogy et al. (2009) have studied the effect of elevated CO_2 on the grain proteome of wheat (*T. aestivum* L. cv. Triso) in a completely free-air CO_2 enrichment set-up. The results of this experiment revealed that a total of 32 proteins were affected, out of them 16 proteins were up-regulated and 16 proteins were down-regulated. Among the up-regulated proteins, triticin precursor, putative avenin-like beta precursor, serpin, peroxidase 1, alpha-chain family 11 xylanase, starch synthase I and cytosolic glyceraldehyde-3-phosphate dehydrogenase were the major ones, whereas among down-regulated proteins, globulin (Glb 1) storage protein, low-molecular-weight glutenin, ATP synthase β subunit and alpha-tubulin were the major ones.

19.6.3 Study of UV-B Effect

UV-B is not an integral component of global climate change, but it can be described as a prime effect of climate change. Continuous depletion of stratospheric O_3 layer is resulting increase in solar UV-B (280–315 nm) at the Earth's surface. Although UV-B radiation has important regulatory and photomorphogenic roles (Ballare et al., 2001), excessive UV-B radiation is clearly harmful. In general, a high level of UV-B causes reduced photosynthesis and growth (Ruhland et al., 2005), oxidative damages (Yannarelli et al., 2006) and damage to DNA (Bray and West, 2005). Plants possess an array of adaptive responses to UV-B that allow them to prevent, mitigate or repair UV-B damage.

19.6.3.1 Maize

A flavonoid-deficient line (*b, pl W23*) of maize (*Z. mays* L.) was exposed to UV-B irradiance of 0.36 W m^{-2} normalized to 300 nm (9 kJ m^{-2} day^{-1}) for 21 days at normal field condition. Casati et al. (2005) have employed detailed proteomics and transcriptomics analysis to evaluate the UV-B regulation in the maize line. A total of 86 proteins were found to be differentially expressed under UV-B treatment, and majority of them belonged to translational and PTM group, photosynthesis and metabolism.

19.6.3.2 Soya Bean

Xu et al. (2008) have exposed soya bean (*Glycine max* L.) plants to UV-B for 12 days, and in that stage, first trifoliate were harvested for proteomics study. Statistical analysis showed that 67 protein spots were significantly ($P < 0.05$) affected by solar UV-B. The results also showed that some important proteins such as RuBisCO activase, triosephosphate isomerase, SOD, ATP synthase CF1, 60 kD chaperonin, PSI subunit D, RNA-binding protein, 50S ribosomal protein, 30S ribosomal protein, Cu-ZnSOD and so on were up-regulated, and alanine aminotransferase, glutamine synthetase, RuBisCO rbcS2, phosphoribulokinase, phosphoglycerate kinase, catalase and so on were down-regulated by the UV-B treatment.

However, it is true that sufficient reports were not found in this field till date. Still, the above studies on three different stresses and five different crops have showed how the internal protein network has been provoked by the external stresses. The most important thing is that all the stresses have their own 'signature' on the plant's response, and this can be well utilized as molecular markers.

These molecular markers have multiple important characters: they are stress specific, they respond rapidly and in some cases they provide protection to plants. Therefore, even these can be utilized for crop improvement (Figure 19.9).

19.6.4 Rice Blast and Host Interactions Using Genomics and Metabolomics Profiling

In plants and pathogen(s) interaction, critical differences between resistance and susceptible response are the timely recognition of the invading pathogen and the rapid and effective activation of host defences. Rice and its blast pathogen (*Magnaporthe grisea*) is a perfect model system for identifying diverse defence molecules such as genes, proteins and metabolic compounds. Nam-Soo Jwa and co-workers observed the infection process in detail according to a time-course profile with green fluorescent protein (GFP)-tagged blast pathogens and confirmed proper inoculation time using defence-related molecular markers including *OsPR10* and *OsPR10a* in compatible and incompatible interaction (Nam-Soo Jwa and co-workers, unpublished data). Transcription profiling of rice genes was performed in leaves inoculated with compatible and incompatible race of blast fungus for 24 and 48 h using rice oligo DNA microarray chip. In that experiment, a total of 238 non-redundant genes showed altered expression (≥/≤ 2.0/0.5-fold) representing 17 main functional categories in the incompatible over compatible interaction. Among the up-regulated genes, the defence-related gene categories (8%, 17%) were represented at 24 and 48 h, and transcription factors (0%, 10%) at 24 and 48 h, respectively. Down-regulated genes were categorized into defence-related genes (7%, 6%), transcription factors (14%, 6%) and signal transduction (0%, 25%), which were regulated at 24 and 48 h. Genes involved in the secondary metabolite biosynthesis (2%, 10%) were only represented in up-regulated gene categories at 24 and 48 h. The subcellular location of changed genes was mainly predicted to locate into the secretion pathway and others. These results reveal rapid changes in gene expressions in the incompatible interaction, where the incompatible interaction triggers a chain reaction of difference gene expression involved in defence-related processes in rice.

In a parallel study, Oliver Jones and co-workers utilized a metabolomics approach to study the infection of the Hwacheong rice cultivar with compatible (KJ201) and incompatible (KJ401) strains of the rice blast fungal pathogen *Magnaporthe grisea* (Nam-Soo Jwa and co-workers, unpublished data). The metabolic response of the rice plants to each strain was assessed 0, 6, 12, 24, 36 and 48 h post inoculation. NMR spectroscopy and gas and liquid chromatography tandem mass spectrometry (GC/LC–MS/MS) were used to study both aqueous and organic phase metabolites, collectively resulting in the identification of 93 compounds. Although clear metabolic profiles were observed at each time point there were no significant differences in the metabolic response elicited by each pathogen strain until 24 h post inoculation. The largest change was found to be in alanine, which was 30% higher in the leaves from the compatible compared to the resistant plants. Together with several other metabolites (malate, glutamine, proline, cinnamate and an unknown sugar) alanine revealed a good correlation between time of fungal penetration into the leaf and the divergence of metabolite profiles in each interaction. The present study demonstrates the strength of metabolomics as a tool to assess the overall effects of pathogen infection on plants. This work also emphasizes the necessity of multiple parallel metabolomics techniques for comprehensive identification of metabolites. Together with sample-based multivariate analysis (pattern recognition), this allows scientists to visualize the overall metabolic effects of stress by comparison with the untreated control, and forms a quantitative and functional basis for a discussion of the overall progression of the effects of infection that would have been difficult to establish by other means. Although the experimental design provides information about changes in relative concentrations of metabolites rather than the pathways involved, the results show that it can provide important information about the mechanisms of the infection response of plants as well as to help provide a more meaningful understanding of the plant–pathogen interaction of this important disease.

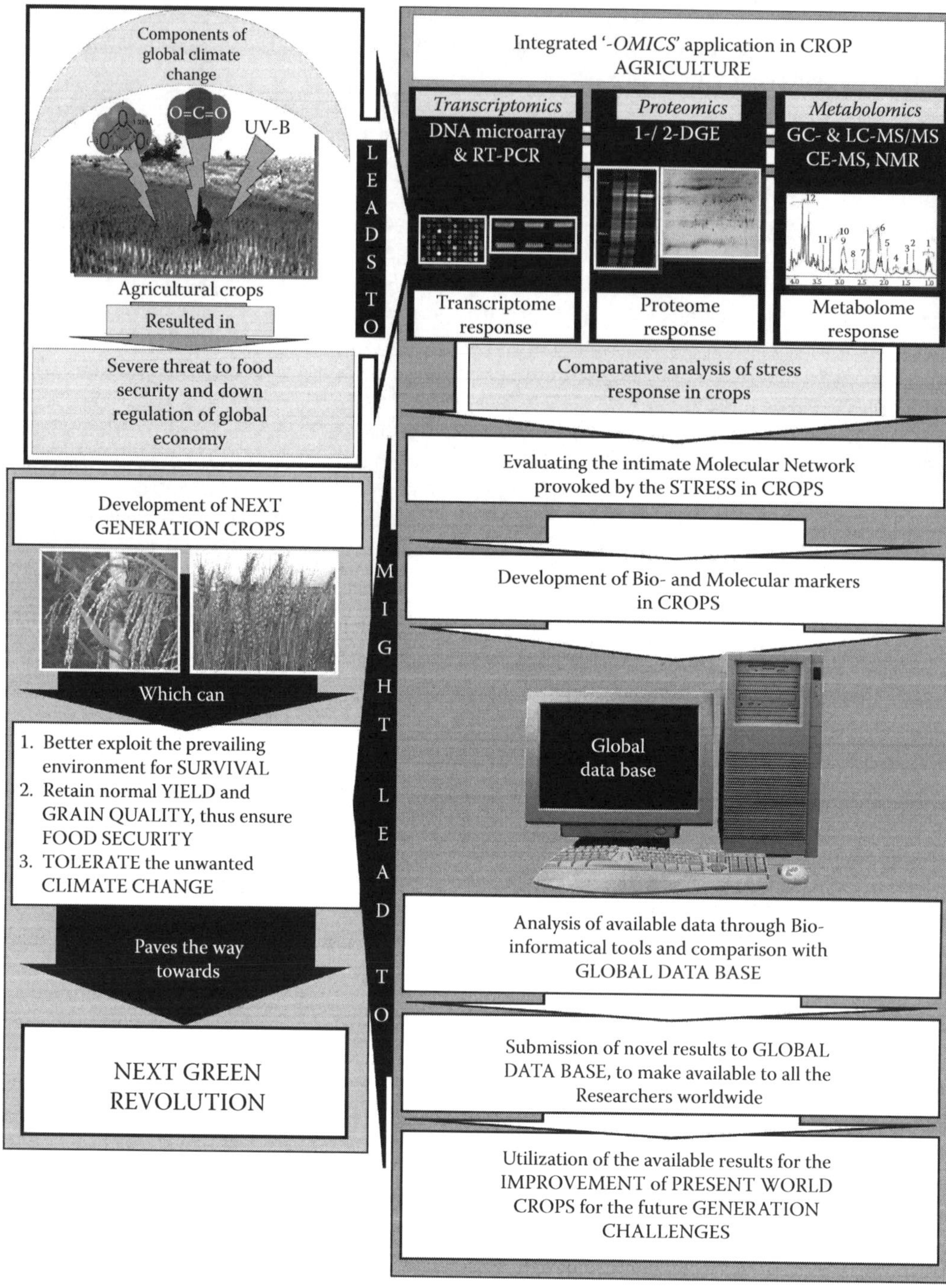

Figure 19.9 A schematic representation on the application of integrated 'OMICS' approach in agriculture for the improvement of crops. Agriculture is threatened by various stresses; so there is a need of second Green Revolution. Integrated OMICS approaches will help in tracking various biomolecules that can be used in the development of next-generation crops with improved traits.

19.7 EXPLOITATION OF NATURAL GENETIC RESOURCES USING ESTABLISHED MOLECULAR- AND BIOMARKERS

In the case of food crops, it takes more than 5–10 years to develop a new cultivar by using the conventional breeding method. For this reason, improving a breeding efficiency is an important issue for the crop breeders. Currently, the issue is becoming more serious because the crop cultivation environment is worsening due to the effects of a global climate change. With a rapid advance in biotechnology, more breeders are now applying the marker-assisted selection (MAS) method to their breeding projects. There are several molecular markers developed such as RFLP, SSR and amplified fragment-length polymorphism (AFLP). As each molecular marker technique has its advantage and disadvantage, they are chosen selectively by the scope of breeders (Jones et al., 2009). The DNA-based markers are a group of random DNA sequences and often not targeted to the core regions of the genome functioning (Van Tienderen et al., 2002). In this respect, proteomics can provide an alternative marker for breeders.

There have been good examples of using proteomics for crop breeding and genetic study. Kim et al. (2008) studied the variation in endosperm protein profiles of wheat lines lacking the null alleles for granule-bound starch synthase. In their study, protein spots were clustered into eight groups based on their expression that was detected or not detected from each wheat line, respectively. The same research group in 2009 used 2DGE to analyse a high-lysine barley-mutant M98 and a normal barley cultivar Chalssalbori for differences in seed protein profiles (Park et al., 2009). A total of 70 abundant protein spots deferentially expressed between M98 and Chalssalbori were analysed by nESI–LC–MS/MS, resulting in the identification of 46 non-redundant proteins. Among the identified proteins, they found 28 proteins, including serpin, B3-hordein and beta-amylase 1, which were known to be associated with the high-lysine trait of cereals. These results indicate that protein spot variation in unique breeding materials can be used as primary information to make a reliable biomarker.

In 2009, Finnie and co-workers analysed a total of 18 barley cultivars for differences in seed protein spot variation, SSR markers and malting quality parameters (Finnie et al., 2009). Their results indicate that malting quality is more closely correlated with seed proteomes than with SSR profiles. Proteome analysis of doubled haploid lines revealed genetic localization of 48 spot polymorphisms in relation to SSR and AFLP markers on the chromosome map.

Eivazi and co-workers evaluated the genetic diversity in 10 Iranian wheat genotypes using quality traits, DNA markers (AFLP and SSR) and proteome markers (Eivazi et al., 2008). Their results showed that the average genetic diversity based on quality traits was higher than the molecular markers. Compared to the other molecular markers, the cluster analysis based on AFLP data showed genetic relationships among the genotypes with better accuracy. Despite a few genotype-specific proteins found by MS analysis, their unique expression patterns were not considered as a result of gene expression.

Proteomics itself or coupling with other molecular markers will provide a highly efficient biomarker for crop breeders to design crosses with appropriate genotypes or to select a superior breeding line having favourable agronomic traits (Figure 19.10).

19.8 NEXT-GENERATION CROPS

Despite the great success of 'Green Revolution' in nineteenth century, more than 2 billion people worldwide still necessitate stable and reliable access to safe, nutritious and affordable food sources, and nearly 800 million of them are chronically malnourished (Fresco and Baudoin, 2002). In other words, 'Food Security' is still a prime issue today. However, in the present world,

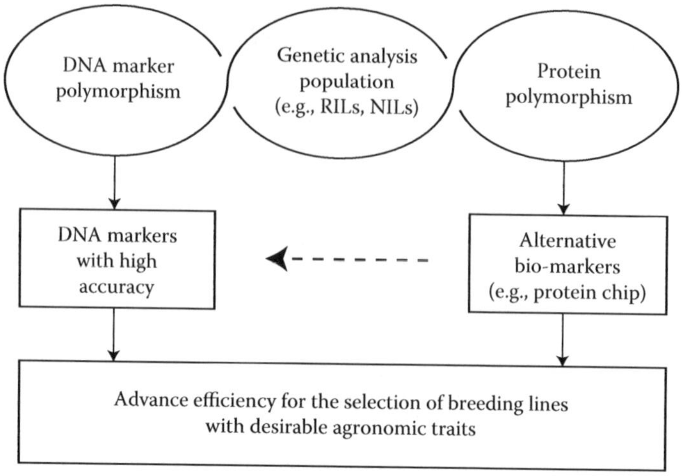

Figure 19.10 Enhancing selection efficiency of breeding lines by biomarkers. Primarily, a robust correlation analysis is needed between the defined traits (e.g., stress resistance, grain quality) of genetic analysis population and the polymorphism of DNA markers and protein spots. MS-based identification of the polymorphic protein spots can help to develop more accurate DNA markers targeting the functional region of genome. At the same time, the identified proteins can be alternative biomarkers in the form of protein chip.

'food security' is also being challenged by new multidimensional and upcoming risks, and 'global climate change' is the most hazardous among them (see also Figure 19.9). Global climate change will modify many elements of the future crop production environment. Atmospheric CO_2 concentration, average temperature, invasion of solar UV-B and tropospheric O_3 concentration will be higher, droughts and flooding will be more frequent and severe, soil fertility will degrade and climatic extremes will be more likely to occur (IPCC, 2007). Hence, there is an urgent need to invest in crop engineering to produce proper germplasms that sustain high yield under stressful and changing climatic conditions. Reynolds and Borlaug (2006) have well concluded that, in addition to the genetic challenges of crop improvement, agriculturists must also emphasis the problems associated with a highly heterogeneous and unpredictable environment. However, these crop-improvement programmes for superior stress tolerance have only been possible when we can better understand the crop–environment interaction at molecular level. The current integrated 'OMICS' approaches provide that platform for our understanding (see also Figure 19.9). Researches, analysis and reviews on the initial crop–environment interaction have pointed some important functional traits, which can play a crucial role in crop plants to combat with the emerging risks in future (Reynolds and Borlaug, 2006; Ainsworth et al., 2008).

- *Crops should have controlled but effective stomatal conductance:* Most of the gaseous pollutants, such as O_3, NOx, sulphur dioxide (SO_2) and so on, enter plants through stomata. Plants close stomata, as a first line of defence, to avoid the uptake of toxic air pollutants, but this stomatal closure also hinder the normal gaseous exchange of plant, and results in lower photosynthetic yield. Hence, according to Ainsworth et al. (2008), plant biotechnologists should try to optimize this unique property of plants, which on one side can avoid the uptake of toxic gaseous pollutant, as well as on the other side does not hamper the photosynthesis much.
- *Crops should have improved detoxification system and superior stress-tolerant molecular network within cell:* It was quite clear that the above-discussed stresses generate harmful compounds, such as reactive oxygen species, hydrogen peroxide and so on, which create toxic environment within plant cells. It is important for plant cells to have improved detoxification system and superior stress-tolerant molecular network, to fight against the future higher stressed world.

- *Crops should have stronger photosynthetic machinery with higher catalytic ability:* RuBisCO is the key carboxylating enzyme and frequently the rate-limiting factor of photosynthesis (Rogers and Humphries, 2000). Studies have showed that this becomes the primary victim under any kind of environmental stress. Even under higher CO_2 also, the activity of RuBisCO decreases. It is quite obvious that the plant biotechnologists should concentrate on this aspect to strengthen the photosynthetic machinery in crop plants.
- *Crops should possess a better and rapid translocation system, for transferring the major photosynthate to their grain part, with a shorter life span.*
- *Crops should have effective phenotypic traits.*

Keeping all these in mind, it will not be wrong to predict that if crop biotechnologists will be able to incorporate all these functional traits in crops, then surely the 'next-generation crops' will lead us to another 'Green Revolution' (Figure 19.9), which might help us create a new hunger free world.

19.9 CONCLUDING REMARKS

There is a need for *exploitation of natural genetic resources using established biomarkers* where the first Green Revolution was based on breeding. In the short term, a desired agronomic trait can be transferred to a next-generation crop, but in the long term, the next-generation crop basics needs to be understood and investigated using the biomarkers. Last but not the least, *creating the next-generation crops* is crucial, where integration of markers with breeding and recombinant technology will play a major role.

REFERENCES

Agrawal, G. K., N. S. Jwa, M.-H. Lebrun, D. Job and R. Rakwal. 2010. Plant secretome: Unlocking secrets of the secreted proteins. *Proteomics* 10:799–827.

Agrawal, G. K., N. S. Jwa and R. Rakwal. 2009. Rice proteomics: End of phase I and beginning of phase II. *Proteomics* 9:935–963.

Agrawal, G. K. and R. Rakwal. 2006. Rice proteomics: A cornerstone for cereal food crop proteomes. *Mass Spec Rev* 25:1–53.

Agrawal, G. K. and R. Rakwal. 2008. *Plant proteomics: Technologies, Strategies, and Applications*, 808pp. Hoboken, NJ: John Wiley & Sons, Inc.

Agrawal, G. K., R. Rakwal, M. Yonekura, A. Kubo and H. Saji. 2002. Proteome analysis of differentially displayed proteins as a tool for investigating ozone stress in rice (*Oryza sativa* L.) seedlings. *Proteomics* 2:947–959.

Agrawal, G. K., M. Yonekura, Y. Iwahashi, H. Iwahashi and R. Rakwal. 2005a. System, trends and perspectives of proteomics in dicot plants. Part I: Technologies in proteome establishment. *J Chromatogr B Analyt Technol Biomed Life Sci* 815:109–123.

Agrawal, G. K., M. Yonekura, Y. Iwahashi, H. Iwahashi and R. Rakwal. 2005b. System, trends and perspectives of proteomics in dicot plants. Part II: Proteomes of the complex developmental stages. *J Chromatogr B Analyt Technol Biomed Life Sci* 815:125–136.

Agrawal, G. K., M. Yonekura, Y. Iwahashi, H. Iwahashi and R. Rakwal. 2005c. System, trends and perspectives of proteomics in dicot plants. Part III: Unraveling the proteomes influenced by the environment, and at the levels of function & genetic relationships. *J Chromatogr B Analyt Technol Biomed Life Sci* 815:137–145.

Ainsworth, E. A., A. Rogers and A. D. B. Leakey. 2008. Targets for crop biotechnology in a future high-CO_2 and high-O_3 world. *Plant Physiol* 147:13–19.

Appleford, N. E., M. D. Wilkinson, Q. Ma et al. 2007. Decreased shoot stature and grain alpha-amylase activity following ectopic expression of a gibberellin 2-oxidase gene in transgenic wheat. *J Exp Bot* 58:3213–2326.

Ballaré, C. L., M. C. Rousseaux, P. S. Searles et al. 2001. Impacts of solar ultraviolet-B radiation on terrestrial ecosystems of Tierra del Fuego (southern Argentina)—An overview of recent progress. *J Photochem Photobiol B* 62: 67–77.

Barfoot, P. and G. Brookes. 2009. Global impact of biotech crops: Economic and environmental effects 1996–2007 (www.pgeconomics.co.uk).

Bartels, D. and R. Sunkar. 2005. Drought and salt tolerance in plants. *Crit Rev Plant Sci* 24:23–58.

Bedell, J. A., M. A. Budiman, A. Nunberg et al. 2005. Sorghum genome sequencing by methylation filtration. *PLoS Biology* 3:e13.

Bergmuller, E., S. Baginsky and W. Gruissem. 2008. Mass spectrometry-based proteomics: Identifying plant proteins. In *Plant Proteomics: Technologies, Strategies and Applications*, eds. G. K. Agrawal and R. Rakwal, pp. 33–44. Hoboken, NJ: John Wiley & Sons, Inc.

Bohler, S., M. Bagard, M. Oufir et al. 2007. A DIGE analysis of developing poplar leaves subjected to ozone reveals major changes in carbon metabolism. *Proteomics* 7:1584–1599.

Bokhari, S. A., X. Wan, Y. Yang et al. 2007. Proteomic response of rice seedling leaves to elevated CO_2 levels. *J Proteome Res* 6: 4624–4633.

Borlaug, N. E. 2007. Sixty-two years of fighting hunger: Personal recollections. *Euphytica* 157:287–297.

Borlaug, N. E. and C. R. Dowswell, 2004. Prospects for world agriculture in the twenty-first century. In *Sustainable Agriculture and the International Rice-Wheat System*, eds. R. La, P. R. Hobbs, N. Uphoff and D. O. Hansen, pp. 1–18. Madison, WI: Marcel Dekker Inc.

Bradshaw, R. A. 2008. An introduction to proteomics: Applications to plant biology. In *Plant Proteomics: Technologies, Strategies, and Applications*, eds. G. K. Agrawal and R. Rakwal, pp. 1–6. Hoboken, NJ: John Wiley & Sons, Inc.

Brown, L. R. 1970. Nobel Peace Prize: Developer of high-yield wheat receives award (Norman Ernest Borlaug). *Science* 170:518–519.

Bray, C. M. and C. E. West. 2005. DNA repair mechanisms in plants: Crucial sensors and effectors for the maintenance of genome integrity. *New Phytol* 168: 511–528.

Canovas, F. M., E. Dumas-Gaudot, G. Recorbet et al. 2004. Plant proteome analysis. *Proteomics* 4:285–298.

Casati, P., X. Zhang, A. L. Burlingam et al. 2005. Analysis of leaf proteome after UV-B irradiation in maize lines differing in sensitivity. *Mol Cell Proteomics* 4:1673–1685.

Cato, M. P. 1979. In *Di Agri Culture*, ed. W. D. Hooper. Cambridge: Harvard University Press.

Chakraborty, S., A. V. Tiedemann and P. S. Teng. 2000. Climate change: Potential impact on plant diseases. *Environ Pollut* 108:317–326.

Chen, S. and A. C. Harmon. 2006. Advances in plant proteomics. *Proteomics* 6:5504–5516.

Cho, K., J. Shibato, G. K. Agrawal et al. 2008. Integrated transcriptomics, proteomics, and metabolomics analyses to survey ozone responses in the leaves of rice seedling. *J Prot Res* 7:2980–2998.

Dentener F., D. Stevenson, K. Ellingsen et al. 2006. The global atmospheric environment for the next generation. *Environ Sci Tech* 40:3586–3594.

Eivazi, A. R., M. R. Naghavi, M. Hajheidari et al. 2008. Assessing wheat (*Triticum aestivum* L.) genetic diversity using quality traits, amplified fragment length polymorphisms, simple sequence repeats and proteome analysis. *Ann Appl Biol* 152:81–91.

Emberson L. D., M. R. Ashmore, H. Cambridge et al. 2007. Modelling stomatal flux across Europe. *Environ Pollut* 109:403–413.

Fiehn, O. 2002. Metabolomics—The link between genotypes and phenotypes. *Plant Mol Biol* 48:155–171.

Finnie, C., M. Bagge, T. Steenholdt et al. 2009. Integration of the barley genetic and seed proteome maps for chromosome 1H, 2H, 3H, 5H and 7H. *Funct Integr Genom* 9:135–143.

Fresco, L. O. and W. O. Baudoin. 2002. Food and nutrition security towards human security. In *ICV Souvenir Paper, International Conference on Vegetables: World Food Summit Five Years Later*. Rome, 11–13 June, 2002.

Fusco, A. C. and J. A. Logan. 2003. Analysis of 1970–1995 trends in tropospheric ozone at Northern Hemisphere midlatitudes with the GEOS-CHEM model. *J Geophys Res* 108:ACH4-1–ACH4-25.

Gupta, P., S. Duplessis, H. White et al. 2005. Gene expression patterns of trembling aspen trees following long-term exposure to interacting elevated CO_2 and tropospheric O_3. *New Phytol* 167:129–142.

Goff, S. A., D. Ricke, T. H. Lan et al. 2002. A draft sequence of the rice genome (*Oryza sativa* L. ssp. *japonica*). *Science* 296:92–100.

Görg, A., O. Drews, C. Lück et al. 2009. 2-DE with IPGs. *Electrophoresis* 30:S122–S132.

Griffin, J. L. and R. F. Shore. 2007. Applications of metabonomics within environmental toxicology. In *The Handbook of Metabonomics and Metabolomics*, eds. J. C. Lindon, J. K. Nicholson and E. Holmes, pp. 517–532. Kidlington: Elsevier Press.

Hogy P., C. Zorb, G. Langenkamper et al. 2009. Atmospheric CO_2 enrichment changes the wheat grain proteome. *J Cereal Sci* 50:248–254.

http://timesofindia.indiatimes.com/world/rest-of-world/Record-1-billion-go-hungry-says-UN/articleshow/5125942.cms. Accessed 15 October, 2009.

http://en.wikipedia.org/wiki/Green_Revolution. Accessed 28 April, 2010.

IPCC Climate change 2007: The physical science basis. In *Contribution of Working Group I to the Fourth Annual Assessment Report of the Intergovernmental Panel on Climate Change*, eds. S. Solomon, D. Qin, M. Manning, Z. Chen, M. Marquis, K. B. Averyt, M. Tignor and H. L. Miller, 996pp. Cambridge: University Press.

Jaillon, O., J. M. Aury, B. Noel et al. 2007. The grapevine genome sequence suggests ancestral hexaploidization in major angiosperm phyla. *Nature* 449:463–467.

Jones, N., H. Ougham, H. Thomas et al. 2009. Markers and mapping revisited: Finding your gene. *New Phytol* 183:935–966.

Jorrin, J. V., A. M. Maldonado and M. A. Castillejo. 2007. Plant proteome analysis: A 2006 update. *Proteomics* 7:2947–2962.

Kersten, B., G. K. Agrawal, H. Iwahashi et al. 2006. Plant phosphoproteomics: A long road ahead. *Proteomics* 6:5517–5528.

Kikuchi, S., K. Satoh, T. Nagata et al. 2003. Collection, mapping, and annotation of over 28,000 cDNA clones from japonica rice. *Science* 301:376–379.

Kim, D. W., H. Y. Heo, C. S. Park et al. 2008. Proteomics analysis in seed of waxy and non-waxy wheat using two-dimensional gel electrophoresis. Poster presented at the *Fifth International Crop Science Congress and Exhibition*, Jeju, Republic of Korea.

Kley, D., M. Kleinmann, H. Sanderman et al. 1999. Photochemical oxidants: State of the science. *Environ Pollut* 100:19–42.

Kontunen-Soppela, S., V. Ossipov, S. Ossipova et al. 2007. Shift in birch leaf metabolome and carbon allocation during long-term open-field ozone exposure. *Global Change Biol* 13:1053–1067.

Krishnan, A., E. Guiderdoni, G. An et al. 2009. Mutant resources in rice for the functional genomics of grasses. *Plant Physiol* 149:165–170.

Kubo, A. 2002. Effects of air pollutants on gene expression in plants. In *Air Pollution and Plant Biotechnology*, eds. K. Omasa, H. Saji, S. Youssefian and N. Kondo, pp. 121–139. Tokyo: Springer.

Kumar, S. 2006. Climate change and crop breeding objectives in the twenty first century. *Curr Sci* 8:90.

Lee, S. and S. C. Yun. 2006. The ozone stress transcriptome of pepper (*Capsicum annuum* L.). *Mol Cell* 21:197–205.

Leitner, A. and W. Lindner. 2009. Chemical tagging strategies for mass spectrometry—Based phosphoproteomics. *Methods Mol Biol* 527:229–243.

Lilley, K. S. and P. Dupree. 2006. Methods of quantitative proteomics and their application to plant organelle characterization. *J Exp Bot* 57:1493–1499.

Long, S. P., E. A. Ainsworth and A. D. B. Leakey. 2005. Global food insecurity: Treatment of major food crops with elevated carbon dioxide or ozone under large-scale fully open-air conditions suggests recent models may have overestimated future yields. *Philos Trans R Soc B* 360:2011–2020.

Lu, C., J. G. Wallis and J. Browse. 2007. An analysis of expressed sequence tags of developing castor endosperm using a full-length cDNA library. *BMC Plant Biol* 7:42.

Mahalingam R., N. Jambunathan, S. K. Gunjan et al. 2006. Analysis of oxidative signaling induced by ozone in *Arabidopsis thaliana*. *Plant Cell Environ* 29:1357–1371.

Manning, R. 2000. Food's frontier. In *The Next Green Revolution*, eds. Farrar, Straus and Giroux, pp. 149–172. CA: University of California Press.

McMichael, A. J., D. H. Campbell-Lendrum, C. F. Corvalan, K. L. Ebi, A. K. Githeko, J. D. Scheraga and A. Woodward. 2003a. *Climate Change and Human Health: Risks and Responses*. Chapter 6. Geneva: World Health Organization.

McMichael, A. J., D. H. Campbell-Lendrum, C. F. Corvalan, K. L. Ebi, A. K. Githeko, J. D. Scheraga and A. Woodward. 2003b. *Climate Change and Human Health: Risks and Responses*. Chapter 2. Geneva: World Health Organization.

Miyao, A., Y. Iwasaki, H. Kitano et al. 2007. A large-scale collection of phenotypic data describing an insertional mutant population to facilitate functional analysis of rice genes. *Plant Mol Biol* 63:625–635.

Monna, L., N. Kitazawa, R. Yoshino et al. 2002. Positional cloning of rice semidwarfing gene, *sd*-1: Rice "green revoltion gene" encodes a mutant enzyme involved in gibberellin synthesis. *DNA Res* 9:11–17.

Oliver, S. G., M. K. Winson, D. B. Kell et al. 1998. Systematic functional analysis of the yeast genome. *Trends Biotechnol* 16:373–378.

Ossipov V., S. Ossipova, V. Bykov et al. 2008. Application of metabolomics to genotype and phenotype discrimination of birch trees grown in a long-term open-field experiment. *Metabolomics* 4:39–51.

Park, H. H., D. W. Kim, H. S. Kim et al. 2009. Analyzing Seed protein property in a high-lysine barley mutant (*Hordeum vulgare*, cv.M98) using proteomics approach. Poster presented at *The 2008 Spring Symposium of the Korean Society of Crop Science*, Samcheok, Republic of Korea.

Porter, J. R. and M. A. Semenov. 2005. Crop responses to climatic variation. *Philos Trans R Soc B* 360:2021–2035.

Pretty, J. 2002. *Agri-culture: Reconnecting People, Land and Nature*. London: Earthscan.

Pretty, J. 2005. *The Earths can Reader in Sustainable Agriculture*. London: Earthscan.

Pretty, J. 2007. Agriculture sustainability: Concepts, principles and evidences. *Philos Trans R Soc B* 363:447–465.

Reynolds, M. P. and N. E. Borlaug. 2006. Applying innovations and new technologies for international collaborative wheat improvement. *J Agric Sci* 144:95–110.

Righetti, P. G., P. Antonioli and C. Simo et al. 2008. Gel-based proteomics. In *Plant Proteomics: Technologies, Strategies and Applications*, eds. G. K. Agrawal and R. Rakwal, pp. 11–30. Hoboken, NJ: John Wiley & Sons, Inc.

Rogers, A. and S. W. Humphries. 2000. A mechanistic evaluation of photosynthetic acclimation at elevated CO_2. *Global Change Biol* 6:1005–1011.

Rossignol, M., J. B. Peltier, H. P. Mock et al. 2006. Plant proteome analysis: A 2004–2006 update. *Proteomics* 6:5529–48.

Rothstein, S. J. 2007. Returning to our roots: Making plant biology research relevant to future challenges in agriculture. *Plant Cell* 19:2695–2699.

Ruhland, C. T., F. S. Xiong, W. D. Clark et al. 2005. The influence of ultraviolet-B radiation on hydroxycinnamic acids, flavonoids and growth of *Deschampsia antarctica* during the springtime ozone depletion season in Antarctica. *Photochem Photobiol* 81:1086–1093.

Saito, K. and F. Matsuda. 2010. Metabolomics for functional genomics, systems biology, and biotechnology. *Ann Rev Plant Biol* 61:463–489.

Saito, K., M. Y. Hirai and K. Yonekura-Sakakibara. 2008. Decoding genes with coexpression networks and metabolomics—'Majority report by precogs'. *Trends Plant Sci* 13:36–43.

Sanderson, C. J. 2007a. Biotech and sustainable agriculture. In *Understanding Genes and GMO's*, pp. 254–275. Singapore: World Scientific Publishing Co.

Sanderson, C. J. 2007b. GMO's in agriculture. In *Understanding Genes and GMO's*, pp. 276–314. Singapore: World Scientific Publishing Co.

Sarkar, A. and S. B. Agrawal. 2010. Identification of ozone stress in Indian rice through foliar injury and differential protein profile. *Environ Monit Assess* 161:205–215.

Sarkar, A., R. Rakwal, S. B. Agrawal et al. 2010. Investigating the impact of elevated levels of ozone on tropical wheat using integrated phenotypical, physiological, biochemical, and proteomics approaches. *J Proteome Res* 9:4565–4584.

Schauer, N. and A. R. Fernie. 2006. Plant metabolomics: Towards biological function and mechanism. *Trends Plant Sci* 11:508–516.

Schmutz, J., S. B. Cannon, J. Schlueter et al. 2010. Genome sequence of the palaeopolyploid soybean. *Nature* 463:178–183.

Singh, R. P., J. Huerta-Espino, R. Sharma et al. 2007. High yielding spring bread wheat germplasm for global irrigated and rainfed production systems. *Euphytica* 157:351–363.

Somerville, C. and M. Koornneef. 2002. A fortunate choice: The history of Arabidopsis as a model plant. *Nat Rev Genet* 3: 883–889.

Stern Review of the Economics of Climate Change. 2006. HM Treasury.

Talib, A. H. 2007. *A Project on Climate Change and its Impact in Kashmir*, Bangalore, India: BfC. http://www.actionaidindia.org/download/On_the_brink.pdf

Tamaoki M., N. Nakajima, A. Kubo et al. 2003. Transcriptome analysis of O_3-exposed Arabidopsis reveals that multiple signal pathways act mutually antagonistically to induce gene expression. *Plant Mol Biol* 53:443–456.

The Arabidopsis Genome Initiative. 2000. Analysis of the genome sequence of the flowering plant *Arabidopsis thaliana*. *Nature* 408:796–815.

The Hindu Survey of the Indian Agriculture. 2005. *The Hindu Survey of the Indian Agriculture*. Chennai: National Press.

Torres, N. L., K. Cho, J. Shibato et al. 2007. Gel-based proteomics reveals potential novel protein markers of ozone stress in leaves of cultivated bean and maize species of Panama. *Electrophoresis* 28:4369–4381.

Tuskan, G. A., S. DiFazio, S. Jansson et al. 2006. The genome of black cottonwood, *Populus trichocarpa* (Torr. & Gray). *Science* 313:1596–1604.

Umezawa, T., M. Fujita, Y. Fujita et al. 2006. Engineering drought tolerance in plants: Discovering and tailoring genes to unlock the future. *Curr Opin Biotechnol* 17:113–122.

Umezawa, T., T. Sakurai, Y. Totoki et al. 2008. Sequencing and analysis of approximately 40,000 soybean cDNA clones from a full-length-enriched cDNA library. *DNA Res* 15:333–346.

United Nations. 2004. UN World Population to 2300.

Urano, K., K. Maruyama, Y. Ogata et al. 2009. Characterization of the ABA-regulated global responses to dehydration in Arabidopsis by metabolomics. *Plant J* 57:1065–1078.

Van Tienderen, P. H., A. A. de Haan, C. G. van der Linden et al. 2002. Biodiversity assessment using markers for ecologically important traits. *Trends Ecol Evol* 17:577–582.

Van Wijk, K. J. 2001. Challenges and prospects of plant proteomics. *Plant Physiol* 126:501–508.

Varotsos, C., C. Cartalis, A. Vlamakis et al. 2004. The long term coupling between column ozone and tropopause properties. *J Climate* 17:3843–3854.

Wand, S. J. E., G. F. Midgley, M. H. Jones et al. 1999. Responses of wild C4 and C3 grass (Poaceae) species to elevated atmospheric CO_2 concentration: A meta-analytic test to current theories and perceptions. *Global Change Biol* 5:723–741.

Wang, X. and D. L. Mauzerall. 2004. Characterizing distribution of tropospheric ozone and its impacts on grain production in China, Japan and South Korea: 1990 and 2000. *Atmos Environ* 38:4348–4302.

Watson, B. S., V. S. Asirvatham, L. Wang and L. W. Sumner. 2003. Mapping the proteome of barrel medic (*Medicago truncatula*). *Plant Physiol* 131:1104–1123.

Weckwerth, W. 2006. *Metabolomics: Methods and Protocols*. Totowa, NJ: Humana Press Inc.

Weckwerth, W., S. Baginsky, K. van Wijk et al. 2008. The multinational Arabidopsis steering subcommittee for proteomics assembles the largest proteome database resource for plant systems biology. *J Proteome Res* 7:4209–4210.

Wilkins, M. R., J. C. Sanchez, A. A. Gooley et al. 1995. Progress with proteome projects: Why all proteins expressed by a genome should be identified and how to do it. *Biotechnol Genet Eng Rev* 13:19–50.

Woodward, F. I. 2002. Potential impacts of global elevated CO_2 concentrations on plants. *Curr Opin Plant Biol* 5:207–211.

www.arabidopsis.org. Accessed 16 July 2009.

www.arcadiabio.com. Accessed 16 July 2009.

www.marketwatch.com/story/bioengineers-aim-to-cash-in-on-plants-that-make-green-plastics. Accessed 16 July 2009.

www.pgeconomics.co.uk. Accessed 28 April 2010.

www.scidev.net. Accessed 16 July 2009.

www.scidev.net/en/features/can-crops-be-climateproofed.html. Accessed 16 July 2009.

www.sciencedaily.com/release/2006/08/060809234056.htm. Accessed 16 July 2009.

www.un.org/esa/population/publications/longrange2/WorldPop2300final.pdf. Accessed 16 July 2009.

Xu, C., J. H. Sullivan and W. M. Garrett. 2008. Impact of solar Ultraviolet-B on the proteome in soybean lines differing in flavonoid contents. *Phytochemistry* 69:38–48.

Yannarelli, G. G., G. O. Noriega, A. Batlle et al. 2006. Heme oxygenase up-regulation in ultraviolet-B irradiated soybean plants involves reactive oxygen species. *Planta* 224:1154–1162.

Yu, J., S. Hu, J. Wang et al. 2002. A draft sequence of the rice genome (*Oryza sativa* L. ssp. *indica*). *Science* 296:79–92.

CHAPTER 20

Designing Oilseeds for Biomaterial Production

Thomas A. McKeon

CONTENTS

20.1 INTRODUCTION

Fats and oils are an important part of the diet, providing caloric content and essential fatty acids to the consumer. However, fats and oils are also important feedstocks for the chemical industry. For untold centuries, they have served as a source of soaps, flammable fuel for illumination, lubricants, and drying oils for paints and coatings, to name the more prominent applications. The discovery of petroleum followed by expanded production, refining, and chemical conversion processes provided cheaper and more abundant substitutes for many of these applications. The relatively recent focus on increasing the use of renewable resources to manufacture products has led to a resurgence of nonpetroleum products to replace those derived from petroleum. This has resulted in increased interest in developing oils and fats with properties enhanced for chemical feedstock use. With the increased understanding of seed oil biochemistry and development of techniques to introduce genes, oilseed genetic engineering has expanded the possibilities of developing plants to meet industrial and chemical needs for producing biomaterials.

20.2 FATTY ACIDS PRESENT IN SEED OILS

Seed oils are composed of fatty acids esterified to each of the three hydroxy groups on glycerol, forming a triacylglycerol (TG). Five fatty acids are present in most commodity oils, although they vary in proportion depending on the plant and the environment. The five common fatty acids are palmitate (16 carbons) and stearate (18 carbons) which are saturated, oleate, linoleate, and linolenate which contain 18 carbons and 1, 2, and 3 methylene-interrupted *cis* double bonds, respectively. These fatty acids are depicted in Figure 20.1. They are important sources of energy and affect the quality and texture of foods. In addition to these roles, linolenate and linoleate are essential fatty acids in the human diet, with important functions in many physiological and developmental events. These five common fatty acids, as components of oil or fat, or as individual fatty acid derivatives, find numerous uses as chemical feedstocks, especially for the soap and detergent industry and in the manufacture of lubricants, biodiesel, paints and coatings, and polymer production. Oils with a high proportion of one fatty acid can be useful especially for specific types of applications. For example, linolenate is readily oxidized and therefore linseed oil, with linolenate content over 50%, is useful as a drying oil for coating porous surfaces such as wood. On the other hand, oil with high oleate content is much more stable to oxidation, and can be used in applications requiring stable liquid oil, such as hydraulic fluid.

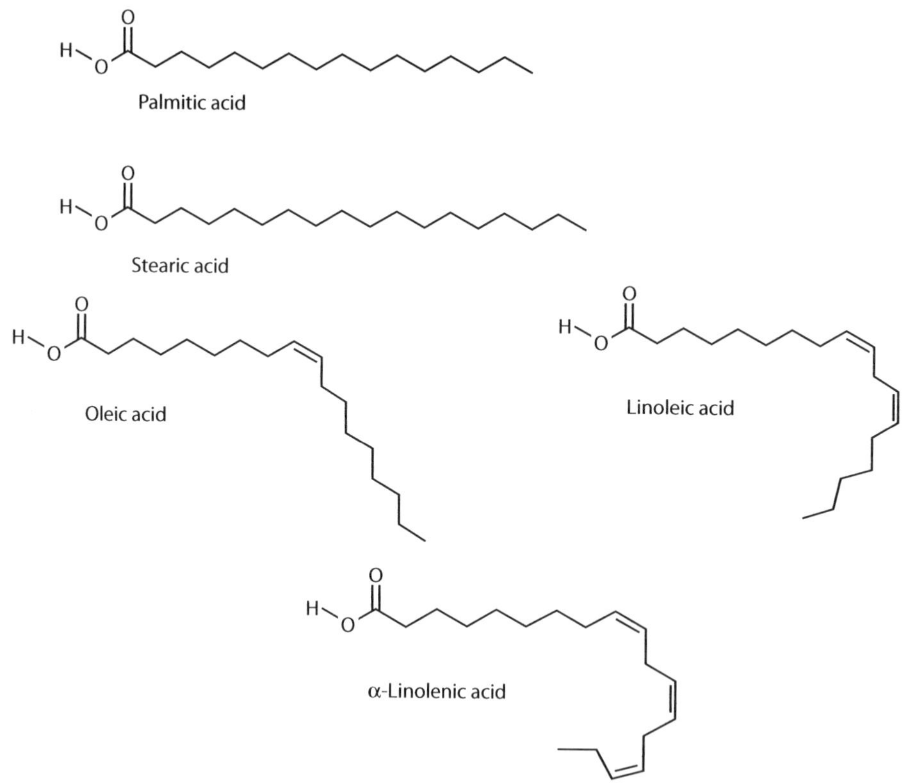

Figure 20.1 Common fatty acids in oil seeds.

Table 20.1 Industrially Useful Fatty Acids for Transgenic Plant Production

Fatty Acid	Functionality	Source	Use
Eleostearic	Conjugated double bonds	Tung, bitter melon	Drying oil
Octadeca-9c,11t,13t-trienoic			
Erucic	Very long chain (VLC)	Rapeseed, crambe	Lubricants, antislip agent
Docosa-13c-enoic			
Caproic to myristate	Medium-chain length	Cuphea, coconut, bay laurel	Detergents
6–14 carbons			
Oleic	Monounsaturate	Many	Hydraulic oil, oleochemicals
Octadeca-9c-enoic			
Ricinoleic	Hydroxylated	Castor	Lubricants, polymers
Octadeca-9c,12-OH-enoic			
Nervonyl Erucate	VLC wax ester	Jojoba	High-temperature lubricant

There are also seed oils that contain uncommon fatty acids. These fatty acids have either different chemical functionality or differ in chain length. As a result, the oils or uncommon fatty acids derived from the oil present physical or chemical differences making them highly suitable for specific applications that are not readily met by the common fatty acids. Table 20.1 lists a collection of seed oils containing uncommon fatty acids. Some of these are available as commodities and used as feedstocks for biomaterials, while others will have uses if the plant can be developed as a crop or the fatty acid can be produced in an alternate crop plant. Although these seed oils carry uncommon fatty acids, they are ultimately derived from the same biosynthetic pathway as the common fatty acids, with the addition of enzymes that direct the synthesis of the uncommon fatty acid (Cahoon and Schmid, 2008; McKeon, 2005).

20.3 ALTERING FATTY ACID COMPOSITION OF AN OIL

In some crops, traditional breeding and mutation have proven useful in generating oils with altered fatty acid composition, including high content of a single fatty acid (Damude and Kinney, 2007; Downey, 1964; Green, 1986). However, some crops are less amenable to these approaches due to the genetic limitations imposed by available germplasm. As a result, approaches that incorporate genetic engineering to develop oils containing high levels of desired fatty acids are being pursued, and this chapter focuses on that approach and the biomaterials that can be derived from successful engineering.

20.4 FATTY ACID AND OIL BIOSYNTHESIS

In order to modify oil composition through gene modification, it is necessary to understand the processes involved in oil biosynthesis. A combination of biochemical characterization of enzymes involved in fatty acid and oil biosynthesis plus generation of mutants in lipid biosynthesis provided the means to identify genes essential for directing the synthesis of the desired fatty acids and their incorporation in oil (Browse and Somerville, 1991).

Fatty acid biosynthesis in plants proceeds from acetyl-CoA which is converted into malonyl-CoA through the action of the acetyl-CoA carboxylase (ACC). The acetyl- and malonyl-CoA are

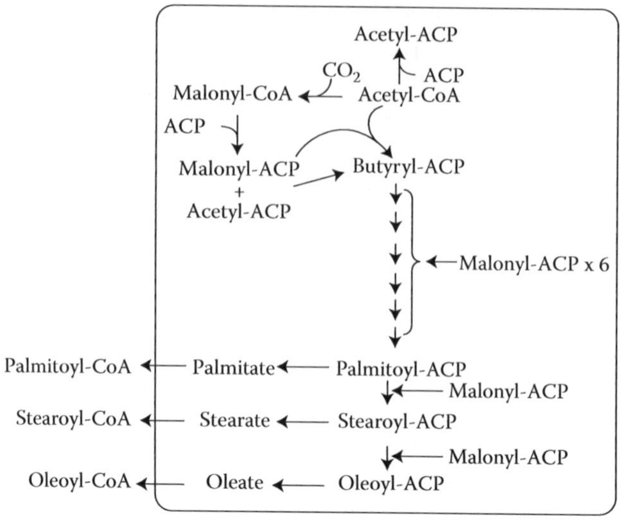

Figure 20.2 Fatty acids biosynthesis.

converted to the respective acyl-acyl carrier protein (ACP), and the condensing enzymes, including keto-acyl synthase III (KAS III) initiate a set of condensation reactions with malonyl-ACP, followed by 6 or 7 additional condensations with malonyl-ACP, to yield the saturated fatty acids palmitate or stearate, respectively, as depicted in Figure 20.2, which depicts the pathway of fatty acid biosynthesis to linoleic acid. Reactions leading to palmitate, stearate, and oleate occur in the plastid, separate from reactions leading to oil biosynthesis (Thelen and Ohlrogge, 2002). Given the dependence of fatty acid production on malonyl-CoA production (to provide malonyl-ACP), the acetyl-CoA carboxylase (ACCase) is generally thought to play a regulatory role in fatty acid production and oil biosynthesis. This hypothesis is supported by research in which ACCase from *Arabidopsis* was overexpressed in potato, leading to an increase in fatty acid production and a fivefold increase in triacylglycerol levels in the tuber (Klaus et al., 2004).

20.5 OIL BIOSYNTHESIS

Attempts to introduce genes that can modify the fatty acid composition of oil have been key in elucidating the steps involved in oil biosynthesis and the mechanisms by which fatty acid content is implemented by the seed. Much of this knowledge has been the result of gene introductions required to achieve a reasonable yield of the desired fatty acid, as seen in the case of Laurate Canola (Voelker et al., 1996). A simplified pathway for the incorporation of fatty acids into oil is presented in Figure 20.3. The original proposed pathway for oil biosynthesis is the Kennedy pathway to triacylglycerols, the principal component of seed oil. However, considerable research has demonstrated layers of complexity, including multiple enzymes carrying out the same basic reaction, for example, diacylglycerol acyltransferase (DGAT) types 1 and 2 (Shockey et al., 2006), as well as acyl exchange enzymes that switch fatty acids among acylglycerols and phospholipids (Banas et al., 2000; Dahlqvist et al., 2000). In terms of key enzymes in controlling the amount of oil produced and fatty acid composition, the DGATs appear to be leading contenders. The DGATs carry out the final step in oil biosynthesis, and the only step that is unique to oil biosynthesis as opposed to phospholipid synthesis. Several examples in the literature point to increased oil biosynthesis when DGAT type 1 is upregulated and decreased oil biosynthesis when it is downregulated (reviewed in He et al., 2005;

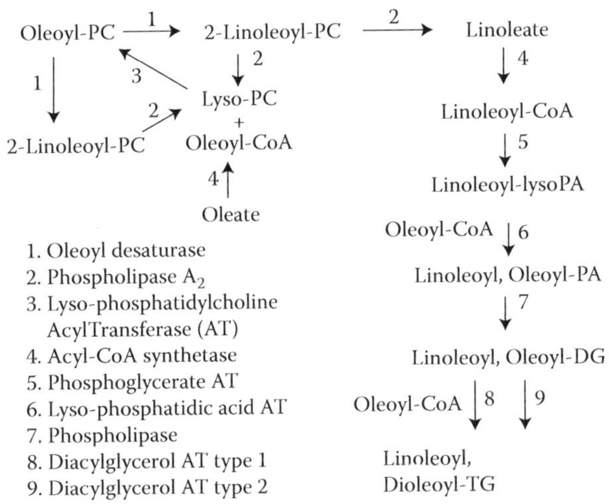

Figure 20.3 Oils biosynthesis.

Lung and Weselake, 2006). The DGAT type 2 is generally correlated with providing a higher degree of specificity for incorporating unusual fatty acids into the triacylglycerol fraction (Burgal et al., 2008; Shockey et al., 2006).

20.6 BIOMATERIALS DERIVED FROM COMMODITY SEED OILS

All the fatty acids produced in commonly available commodity seed oils have some value for use in industrial products. For example, palmitate, stearate, and oleate salts are common ingredients of soap and methyl esters provide biodiesel fuel that is stable to oxidation. Linoleate is the major component of soybean oil and is readily oxidized to the epoxy, and eventually used to make polyurethane or other polymeric materials. Oils with high linolenate content, such as flax, are useful as drying oils, providing a readily oxidized coating that serves to seal porous surfaces.

A considerable research effort has been expended on developing nonfood uses for soybean. The persistent carryover of 1–2 billion pounds annually in U.S. production of soybean oil necessitated alternatives to food use, and an array of uses including lubricants, biodiesel fuel, hydraulic fluids, inks, and drying oils use up to 15% of the soybean oil produced in the United States. Further developments included epoxidation followed by hydrolysis of the epoxy ring, producing polyhydroxy oils that are used in composites and to produce polyurethane foams for cushions and mattresses. Due to the availability of extensive germplasm resources for the soybean, the fatty acid composition of soybean oil has been modified through traditional breeding efforts to obtain oils high or moderately high in oleic acid, low in linolenate, and low in saturated fatty acids. Genetic engineering to silence or upregulate specific genes has also presented the means to alter the fatty acid composition of soybean oil. Both approaches have been used to develop soybean plants that produce oils with altered fatty acid composition (Clemente and Cahoon, 2009). However, those developed through traditional breeding and selection appear to be more favored by the market, although a high proportion of soybean cultivated in the United States, Brazil, Argentina, and China are transgenic varieties expressing genes for herbicide tolerance or insect resistance (James, 2009). Similar to soybean, other commodity seed oils have also been bred or engineered to alter the proportion of the five common fatty acids. Soybean oil, canola oil, corn oil, and animal tallow can all be used in production of certain fat-based nonfood products.

20.7 OTHER FATTY ACIDS USED FOR PRODUCING BIOMATERIALS

There are seed oils that contain fatty acids carrying additional functionality that enhance their use in generating biomaterials. In most temperate climate oilseeds, the oleate may be further desaturated to linoleate and α-linolenate. In rapeseed, crambe, and nasturtium, the oleate may be elongated to erucic acid by the action of an acylCoA-based elongation reaction. The products of elongation, usually 20:1 δ11 and 22:1 δ13 are incorporated into the triacylglycerol fraction (oil). In some plants, the oleate is oxidized to uncommon fatty acids. For example, in *Vernonia*, 18:1 Δ9, 12–13 epoxy (vernolate) is formed and then incorporated into TG (Voelker and Kinney, 2001). The possibilities resulting from oleate production provided the basis for the original concept of oleate as the central substrate in plant fatty acid biosynthesis (Stumpf et al., 1980). The set of modification reactions that can alter oleate is unusual in that it comprises a family of homologous enzymes that have evolved from the FAD2 genes, which encode the oleoyl desaturase in oilseeds (Broun et al., 1998a; Lee et al., 1998). Enzymes that have evolved from the FAD2 have been found to carry out an unusual array of conversions, using an oleoyl-phosphocholine (oleoylPC)-based substrate. These reactions include hydroxylation, epoxidation, desaturation–conjugation, and desaturation to a triple bond (Broun et al., 1998a; Cahoon et al., 1999). In the case of hydroxylation and desaturation, changes in as few as 4–6 amino acid residues result in an interchange of the two types of activities (Broadwater et al., 2002). In fact, the oleate 12-hydroxylase from *Lesquerella* has mixed functionality, and can introduce a hydroxyl group or a double bond (Broun et al., 1998b). Interestingly, introduction of genes to produce these uncommon fatty acids in the oil of a transgenic plant has also proven difficult.

20.8 SEED OILS AS SOURCES FOR PRODUCING BIOMATERIALS

20.8.1 Monounsaturated Fatty Acid Biosynthesis

In general, once saturated fatty acids are released from acylACP, they are incorporated into oil without any apparent modification except for, to a minor extent, elongation. In the plastid, though, the saturated fatty acyl-ACP can be desaturated by the δ9-desaturase, a class of soluble enzymes (as opposed to membrane-bound) formerly identified as the stearoyl-ACP desaturase, now identified as FAD1, which is the type present in most oilseeds. These enzymes share a considerable degree of amino acid sequence homology and the same type of active site in which the desaturation is carried out (Lindqvist et al., 1996). In most plants, the δ9-desaturase produces oleate which, for oil biosynthesis, is transported from the plastid to the endoplasmic reticulum and incorporated into CoA, phospholipid and acylglycerol. Oils high in oleic acid content have been considered desirable both for food and nonfood uses. A high-oleate soybean oil containing greater than 80% oleic acid was developed by suppressing expression of the desaturase enzyme that converts oleate to linoleate in soybean. This oilseed has been commercialized and for industrial purposes finds applications as a stable, biodegradable hydraulic oil and is likely useful for developing other bio-based lubricant applications (Kinney, 2002).

Some plants produce monounsaturated fatty acids of differing chain length or with the double bond in a different position on the carbon chain, or both. Many such desaturases have been cloned, the crystal structure of the soluble desaturase from castor (*Ricinus communis* L.) has been determined, and considerable insight into factors involved in chain length and positional specificity of the desaturase reaction have been revealed (Lindqvist et al., 1996). However, despite the apparent similarity of some products to oleate, for example, 18:1 δ6 (petroselenate), their production can differ from that of oleate, resulting in limited amounts of the product when introduced into a transgenic plant (Thelen and Ohlrogge, 2002). It has been shown that, in some cases, co-factors such as ferredoxin and ACP isoforms that interact specifically with the enzyme are required. Moreover, the FA may also require altered lipid metabolism to be suitably incorporated into TG. Thus, further understanding of lipid biochemistry leading to TG production underlies successful attempts to engineer oil composition.

20.8.1.1 Laurate Oils

In seeds of certain plants such as coconut, palm kernel, bay laurel, and cuphea, the flow of carbon through fatty acid synthesis is disrupted as the result of an acylACP thioesterase (product of the FAT B gene), which removes the ACP from the elongating fatty acid chain prior to achieving full length. This produces a medium-chain length fatty acid that is transported from the plastid and enters the oil biosynthetic pathway. This approach copied from nature led to the development of the first transgenic oilseed modified to produce an industrial oil product, namely Laurate Canola (Del Vecchio, 1996). Inserting the cDNA for a medium-chain-specific acyl-ACP thioesterase (Pollard et al., 1991) from California bay laurel, a plant which produces seeds containing >60% laurate in its oil, redirected fatty acid synthesis to laurate production, which was incorporated into the seed oil (Voelker et al., 1992). Although this achievement was a key early success in the contribution of genetic engineering to agriculture, the underlying science also pointed to a number of technical problems that have since been widely recognized. The production of a fatty acid not normally produced by the seed may trigger a "counter-reaction." In the case of laurate, considerable amounts of the laurate were beta-oxidized (Eccleston and Ohlrogge, 1998; Voelker et al., 1996). Despite increased fatty acid biosynthesis due to enhanced laurate production, overall, the canola cultivar had reduced oil yield, since some of the laurate produced was oxidized through the futile cycle of synthesis and oxidation.

The oil produced by the engineered canola also lacked laurate in the *sn*-2-position of the TG (Davies et al., 1995). This was the result of the canola seed lacking a suitable lyso-phosphatidic acid acyltransferase (LPAAT) that could use lauroyl-CoA as an acyl donor for the *sn*-2 position of glycerolipid. By crossing a canola plant containing an LPAAT gene from coconut (Davies et al., 1995), with a Laurate Canola plant (Knutzon et al., 1999), the resulting plant produced an oilseed in which laurate is distributed among all three positions of the TG. The resulting "high-laurate canola" had a laurate content of up to 70%. The successful design of a novel, temperate-climate industrial crop provided a great impetus to follow this approach for other industrially useful products, especially oils. It also provided a foreshadowing of the difficulties to be encountered in engineering production of uncommon fatty acids in oilseeds.

20.8.1.2 Castor Oil

The castor plant produces an oil high in ricinoleic acid (Figure 20.4), with up to 90% of the fatty acid content. This high content of hydroxy fatty acid imparts physical and chemical properties that make castor oil an important chemical feedstock (Figure 20.5) (Caupin, 1997). The mid-chain hydroxy group provides for interchain hydrogen bonding, resulting in considerably higher viscosity than "normal" vegetable oils. The hydroxy group and carboxy group can interact with metals, and this property underlies the value of the hydrogenated ricinoleate lithium soap (12-hydroxystearate) as a grease, known as lithium grease, and used worldwide as a lubricant for moving metal parts in vehicles and machines. Chemically, the hydroxy group can be modified to create an array of detergents. The first synthetic detergent, Turkey Red Oil, was produced by sulfation of castor oil, and this is still used in tanning leather. However, in terms of advanced biomaterials, castor oil is unmatched. The methyl ester can be converted into 10-undecenoate through pyrolysis at >500°C, and this product in turn brominated and substituted with an amino group to produce 11-undecanoic acid. This serves as the precursor for production of Nylon 11, also known as Rilsan™, an engineering Nylon that is used to coat metal parts for corrosion resistance, is resistant to organic solvents,

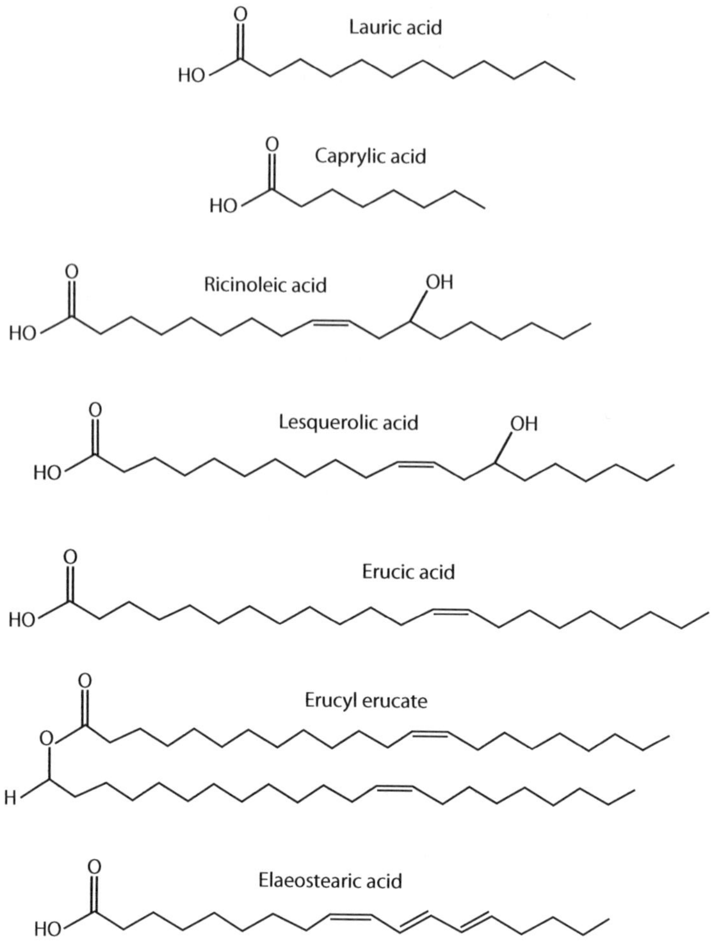

Figure 20.4 Industrially useful derivatives from oilseeds.

does not swell, maintaining dimensional stability in water, and has enough strength to allow its use in moving parts. Caustic fusion in sodium hydroxide at 225°C produces the 10-carbon diacid sebacic acid and 2-octanol as a coproduct. Sebacic acid can be converted into 1,10 diamino decane and together they are used to make Nylon 10,10, another engineering Nylon, similar in properties to Nylon 11, but with the added benefit of being moldable without the need for plasticizer. For many uses, especially when solvent resistance or dimensional stability is required, these Nylons are far superior to Nylon 6,6, the precursors of which are generally produced from petroleum. Since the triricinoleoyl triacylglycerol makes up 70% of the TG present in castor oil, and the rest is diricinoleoyl TG, castor oil is also an important commercial source of natural polyol for production of high-performance polyurethane foams and thermoplastic insulators.

The unique circumstances surrounding castor oil include 90% ricinoleate content, seeds with up to 60% oil, a seed containing a highly toxic protein, ricin, and the oil being used worldwide for a broad array of applications, with numerous potential applications unmet due to limited production of about 500,000 tons/year. As a result, the idea of producing castor oil in a transgenic oilseed has great attraction, and considerable research effort has been applied to understanding castor oil production in order to reproduce the high productivity of the castor plant in a temperate crop such as soy or canola.

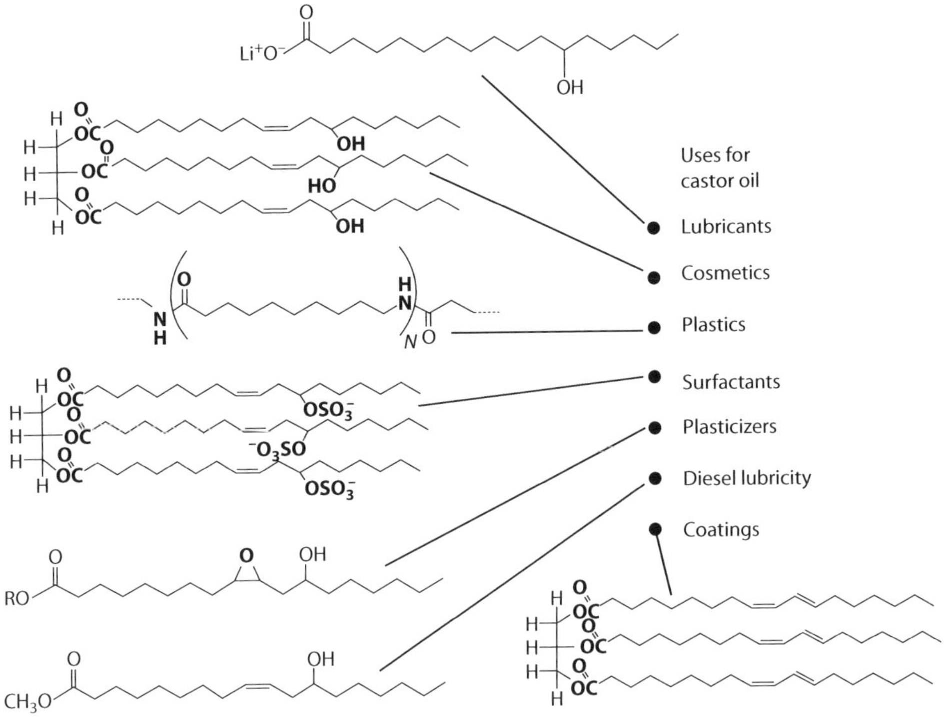

Figure 20.5 Uses for castor oil.

The key step in castor oil production is the hydroxylation of oleate to ricinoleate. A cDNA encoding this enzyme was cloned (Van deLoo et al., 1995) and, when inserted in transgenic plants, oils containing up to 20% hydroxy fatty acids were produced (Broun and Somerville, 1997). It has been hypothesized that the ricinoleate incorporated in lipid inhibits membrane function in most plants, and as in the case of laurate, is beta-oxidized (Moire et al., 2004). On the other hand, castor has evolved to produce and incorporate ricinoleate into oil. This led to the approach of identifying additional enzyme components in castor that enable it to produce an oil with 90% ricinoleate (McKeon and Lin, 2002). The cDNAs for several enzymes in the castor oil pathway have been cloned. While the enzymes expressed from the cDNAs appear to have some preference for using ricinoleate-based substrates (He et al., 2004, 2007; Kroon et al., 2006) to date, only the castor DGAT 2 has led to an increase in hydroxy fatty acid content when inserted in transgenic *Arabidopsis* expressing the oleoyl hydroxylase (Burgal et al., 2008). The hydroxy fatty acid content increased from 17% without the DGAT 2 to 30% hydroxy fatty acid content in lines that included the cDNA for DGAT 2. However, the goal of achieving 90% ricinoleate in a transgenic oilseed remains a daunting challenge. An alternate approach, transforming the castor plant to eliminate the toxic components has been proposed (McKeon et al., 2002).

20.8.1.3 Rapeseed Oil

Rapeseed oil is derived from *Brassica* species, including *Brassica napus*, *B. rapa*, and *B. juncea*. It generally contains 45% erucic acid (Figure 20.4) in addition to the common fatty acids. Erucic acid is an important industrial fatty acid, useful in lubricants, fabric softeners, hydraulic fluids with a special application as a slip agent for plastic films (Scarth and Tang, 2006). Erucic acid is produced biosynthetically from oleoyl-CoA by two sequential elongation steps, carried out by a specific fatty acid elongase (FAE).

Because erucic acid was found to cause myocardial damage, and the seed meal contains high levels of glucosinolates that are antifeedants, the rapeseed cultivar Canola was bred by traditional methods (Downey and Harvey, 1963) to develop a low erucic content oil (<2%) and low-glucosinolate seed meal. Canola is also known as low-erucic acid rape and is primarily used as food oil, as it has over 60% monounsaturated fatty acid content. The traditional rapeseed is known as high-erucic acid rape.

Because erucic acid is so useful for industrial purposes, cultivars containing higher levels are naturally desirable to increase available amounts and reduce the cost of purification. There has been some success using conventional breeding methods in developing rapeseed with up to 60% erucic acid (Scarth and Tang, 2006). Because the endogenous biosynthetic pathway limits erucoyl to the *sn*-1.3 positions of TG, there is a theoretical limit of 66% erucic acid content in rapeseed. Attempts to increase erucic acid content by introducing LPAATs from crops that do insert erucic acid into the *sn*-2 position of TG had little effect on increasing erucate content, but combining an LPAAT from *Limnanthes douglasii* with an FAE from *B. napus* (Nath et al., 2009) did increase erucate up to 72% content. The need for an FAE suggests that in addition to requiring an LPAAT that can insert erucate into the *sn*-2 position of TG, the limitation on increasing erucate is synthesis, with competing pathways such as oleate desaturation drawing away the substrate for erucic acid.

20.8.1.4 Cuphea Oil

Medium-chain-saturated fatty acids are used in soaps, surfactants, lubricants, and biodiesel. The current commercial sources of medium-chain fatty acids are coconut oil, which is rich in lauric acid (Figure 20.4), and palm kernel oil, which is rich in lauric and myristic acids. Laurate Canola was the first commercial oilseed with a transgenically altered output trait, with intended uses in the soaps and detergent industry as well as in foods as a cocoa butter substitute. Although Laurate Canola was not a commercial success, the continuing demand for temperate region crops that can produce medium-chain fatty acids remains. The *Cuphea* sp. produce such fatty acids, with *Cuphea llavea* producing up to 92% caprate in its oil, and many accessions of cuphea having >90% medium-chain content of varying composition (Phippen et al., 2006). While *Cuphea* sp. have been introduced as crops in the United States, agronomic considerations have limited its cultivation (Knothe et al., 2009). Therefore, genes from *Cuphea* sp. are being used to engineer oilseeds that can produce laurate as well as caprylate (C-8) and caprate (C-10).

As is the case with the Laurate Canola, fatty acid thioesterase cDNAs from *Cuphea* sp. have been isolated and introduced in *Arabidopsis* and canola plants. In addition, the ketoacyl synthase type 3 (KAS III), which is involved in elongation of medium-chain fatty acids, was blocked from expression in seed and replaced by a KAS from different *Cuphea* sp. The results in *Arabidopsis* (Filichkin et al., 2006) and in canola (Stoll et al., 2006) were a small increase in medium-chain fatty acid, but far from any commercially useful amounts.

20.8.1.5 Jojoba Wax

Unlike most plants, jojoba (*Simmondsia chinensis*) makes a wax ester instead of TG as a seed storage fat. The wax esters consist of very-long-chain monosaturated alcohols esterified to very-long-chain monounsaturated fatty acids. The waxes are mostly in the range of C38–C44 (Wisniak, 1994). As a result of its unique structure, it is very useful in lubricants and as an antifoaming agent, filling niches formerly occupied by sperm whale oil. The performance of jojoba factice (sulfurized wax) is equivalent to that of sperm whale oil factice, and so jojoba wax can be used to produce high-pressure lubricants for use in automotive transmission fluids, gear lubricants, and motor oils (Miwa et al., 1979). Because of its usefulness in these lubricant applications, there have been many proposals to develop jojoba as an industrial crop. However, its growth habit and agronomic features have prevented jojoba from taking hold as a crop to any great extent, and most of the current uses are in cosmetics where cost of ingredients is not always a major consideration. As a result, there is great interest in producing wax esters transgenically. Since rapeseed and crambe both produce very-long-chain monounsaturated fatty acids, they hold promise as potential platforms for wax ester production. To date, the model plant *Arabidopsis* has been transformed with cDNAs from jojoba that encode a fatty acyl-CoA reductase to produce the fatty alcohol, a wax synthase to carry out the esterification reaction, as well as a cDNA for a fatty acyl-CoA KAS from *Lunaria annua* to elongate fatty acyl-CoA (Lardizabal et al., 2000). Seeds containing as much as 70% wax by weight were detected, indicating the feasibility of producing wax esters equivalent to those in jojoba. Efforts are now underway to duplicate this success in the crop plants *Crambe abyssinica* and rapeseed.

20.8.1.6 Tung Oil

One of the principal industrial uses of seed oils is in paints and coatings. Oils containing multiple methylene-interrupted double bonds are much more susceptible to oxidation than oils composed of monounsaturated or saturated fatty acids. Thus, linseed oil from flax is a widely used coating for porous surfaces, as it contains approximately 55% α-linolenic acid. Oils containing fatty acids with conjugated double bonds are much more susceptible to oxidation, and provide a model for designing crops that produce drying oils. The tung tree produces an oil with up to 80% eleostearic acid, octadeca-9c, 11t, 13t-trienoic acid (Figure 20.4). Tung oil is higher priced than almost any other seed oil. It was formerly produced in the southeast United States until Hurricane Camille eliminated 50% of the tung trees in 1969 while reductions in crop supports and a series of damaging frosts further reduced economic incentive to produce tung oil in the United States.

Because eleostearic acid is a polyunsaturated fatty acid, structurally similar to oleic acid, it has been thought to be a good candidate for production in transgenic oilseed. In order to achieve this goal, research has been conducted on the enzymatic step required to produce eleostearic acid. The enzyme, a fatty acid conjugase, converts linoleate to α-eleostearate, although the enzyme appears to have additional activities, acting as a desaturase that introduces *trans* double bonds in the fatty acid backbone (Dyer et al., 2002). Studies of tung oil biosynthesis were among the first to demonstrate the importance of the DGAT type 2 in the incorporation of unusual fatty acids into the triacylglycerol fraction (Shockey et al., 2006). Attempts to produce transgenic seed oils containing α-eleostearate have met with limited success, with levels up to 13% accumulating in *Arabidopsis* seed (Cahoon et al., 2006). This low production is attributed to accumulation of α-eleostearate in phospholipid, where it may interfere with proper membrane function. This accumulation of unusual fatty acids in the membrane lipids appears to be a general phenomenon encountered in attempts to engineer transgenic oils containing unusual fatty acids. Research in this system will lead to understanding how plants that make unusual fatty acids efficiently remove the fatty acid from phospholipid, and should thereby improve the production of industrially useful fatty acids in common crops.

20.9 TECHNICAL AND SOCIAL ISSUES OF GENETIC ENGINEERING

There are currently 25 countries that plant genetically engineered food crops, with the United States leading the overall production, over 90% of soybean and cotton and 80% of corn engineered for insect resistance and/or herbicide tolerance (James, 2009). Therefore, at least for these 25 countries, the issue of whether or not to allow genetically engineered crops has already been answered in the affirmative. The history of genetic engineering is unique among technological advances. In 1975, years before it became possible to genetically transform organisms as complex as plants, a collection of guidelines were drawn up to govern the use of what was seen as an upcoming technological capability—the ability to introduce and express specific genes derived from an external source into organisms, namely, genetic engineering. As technology evolved, and actual generation of genetically engineered plants became possible, the regulations evolved as well, leading to the rule of substantial equivalence—any new gene introduced could not significantly reduce the safety or nutritional value of the nonengineered crop. In the United States, three regulatory agencies oversee genetically engineered crops, and each agency, U.S. Department of Agriculture Animal and Plant Health Inspection Service (USDA-APHIS), Environmental Protection Agency (EPA) and the Food and Drug Administration (FDA), must approve the application to introduce the modified crop.

There are numerous concerns raised about genetically engineered crops (Table 20.2) (reviewed in Lemaux, 2008; McKeon, 2003). The overall concern is that the engineered crop will be unsafe. Regulatory agencies require that introduced plants be, at a minimum, substantially equivalent in nutrient content to the unmodified crop. To address the issue of allergen introduction, proteins and products resulting from introduced genes must be demonstrated to be readily digestible or no different from food components generally regarded as safe. It had been common practice to include antibiotic resistance genes in the modification, in order to facilitate selection of transformants. The genes introduced usually conferred resistance to antibiotics not normally used in human therapy. It is thought extremely unlikely that genetic material surviving digestion would be transferred to gut bacteria. However, to address the concern, recombinase systems that can excise these genes and eliminate the selection markers during transformation obviate any possible gene transfer to confer antibiotic resistance on gut bacteria. Native germplasm has apparently been altered in some cases, due to pollen transfer from nearby genetically engineered plants. Generally, engineered crops are planted with a buffer zone of nonmodified crop to contain any outcrossing to native germplasm or related weeds. This is also true for nonengineered crop plants such as canola and rapeseed, where outcrossing can occur over some distance.

There have been problems resulting from misuse or poor planting protocols for transgenic crops. One famous case is the "Starlink" corn episode. The genetically modified variety was approved only for animal use, as there was some predicted potential that the introduced insect-resistance protein CrY9c would be allergenic. Some of this corn found its way into the human food supply and, as a result, approvals of genetically engineered crop introductions in the United States included the possibility of human consumption. A second episode resulted from a crop of corn engineered to produce an industrial protease. The following year, the field was planted with soy. Volunteer corn plants grew in the same field and some of the genetically engineered corn was coharvested with the soybeans. When the contamination was discovered, the soy affected was withdrawn from storage and destroyed, resulting in a multi-million dollar loss. These episodes demonstrated that there are problems associated even with crops engineered strictly for industrial purposes, with resulting increased oversight in granting of applications for introducing transgenic crops. In general, it is assumed that by-products of oilseeds that are strictly intended for industrial use of the oil may find

Table 20.2 Issues Related to GM Crops

Issues
• Introduction of antibiotic resistance
• "Random" insertion of gene
• Undesirably altered composition
• Outcrossing
• Introduction of allergens
• Crops for both food and industrial applications
• End uses of residues
• Identity preservation

their way into the food supply as food or feed from the residual seed meal after oil is extracted. Therefore, even potential by-products of transgenic oilseeds must pass the substantial equivalence test (Del Vecchio, 1996).

The economics of producing genetically engineered oilseeds is beyond the scope of this text. However, one key issue related to economics is identity preservation. The investment in research and development of a transgenic oilseed is considerable. It is generally recognized that only crops to be grown on a large scale will be profitable for the seed developer. Furthermore, the transgenic crop production derived must be kept separate from other related seeds in order to maintain its premium value. There are protocols developed for this process, identity preservation, currently applied to crops such as organic soybeans.

20.10 CONCLUSION

There is considerable potential for the development of oilseed crops that provide biomaterials and feedstocks to the chemical industry. Oleochemicals have already made an important contribution to biomaterial production in the form of soaps, surfactants, lubricants, paints, coatings, polymers, and biofuels. Successful development of approaches to genetically engineer crops to produce oils containing industrially useful fatty acids will aid the introduction of additional oil-derived biomaterials and provide renewable resources that can reduce the need for petrochemical production.

REFERENCES

Banas, A., A. Dahlqvist, U. Stahl, M. Lenman, and S. Stymne. 2000. The involvement of phospholipid:diacylglycerol acyltransferases in triacylglycerol production. *Biochem Soc Trans* 28:703–705.

Broadwater, J. A., E. Whittle, and J. Shanklin. 2002. Desaturation and hydroxylation. Residues 148 and 324 of Arabidopsis FAD2, in addition to substrate chain length, exert a major influence in partitioning of catalytic specificity. *Biol Chem* 277:15613–15620.

Broun, P. and C. Somerville. 1997. Accumulation of ricinoleic, lesquerolic, and densipolic acids in seeds of transgenic arabidopsis plants that express a fatty acyl hydroxylase cDNA from castor bean. *Plant Physiol* 113:933–942.

Broun, P., J. Shanklin, E. Whittle, and C. Somerville. 1998a. Catalytic plasticity of fatty acid modification enzymes underlying chemical diversity of plant lipids. *Science* 282:1315–1317.

Broun, P., S. Boddupalli, and C. Somerville. 1998b. A bifunctional oleate 12-hydroxylase: Desaturase from *Lesquerella fendleri*. *Plant J* 13:201–210.

Browse, J. and C. Somerville. 1991. Glycerolipid synthesis: Biochemistry and regulation. *Annu Rev Plant Physiol Plant Mol Biol* 42:467–506.

Burgal, J., J. Shockey, C. Lu, J. Dyer, T. Larson, I. Graham, and J. Browse. 2008. Metabolic engineering of hydroxy fatty acid production in plants: RcDGAT2 drives dramatic increases in ricinoleate levels in seed oil. *Plant Biotech J* 6:819–831.

Cahoon, E. B. and K. M. Schmid. 2008. Metabolic engineering of the content and fatty acid composition of vegetable oils. *Adv Plant Biochem Mol Biol* 1:161–200.

Cahoon, E. B., C. R. Dietrich, K. Meyer, H. G. Damude, J. M. Dyer, and A. J. Kinney. 2006. Conjugated fatty acids accumulate to high levels in phospholipids of metabolically engineered soybean and *Arabidopsis* seeds. *Phytochem* 67:1166–1176.

Cahoon, E. B., T. J. Carlson, K. G. Ripp et al. 1999. Biosynthetic origin of conjugated double bonds: Production of fatty acid components of high-value drying oils in transgenic soybean embryos, *Proc Natl Acad Sci USA* 96:12935–12940.

Caupin, H. J. 1997. Products from castor oil: Past, present, and future. In *Lipid Technologies and Applications*, eds. F. D. Gunstone, and F. B. Padley, 787–795. New York, NY: Marcel Dekker Inc.

Clemente, T. E. and E. B. Cahoon. 2009. Soybean oil: Genetic approaches for modification of functionality and total content. *Plant Physiol* 151:1030–1040.

Dahlqvist, A., U. Stahl, M. Lenman et al. 2000. Phospholipid:diacylglycerol acyltransferase: An enzyme that catalyzes the acyl-CoA-independent formation of triacylglycerol in yeast and plants. *Proc Natl Acad Sci USA* 97:6487–6492.

Damude, H. G. and A. J. Kinney. 2007. Engineering oilseed plants for a sustainable, land-based source of long chain polyunsaturated fatty acids. *Lipids* 42:179–185.

Davies, M. H., D. J. Hawkins, and J. S. Nelson 1995. Lysophosphatidic acid acyltransferase from immature coconut endosperm having medium chain length substrate specificity. *Phytochem* 39:989–996.

Del Vecchio, A. J. 1996. High-laurate canola. *INFORM* 7:230–243.

Downey, R. K. 1964. A selection of *Brassica campestris* L. containing no erucic acid in its seed oil. *Can J Plant Sci* 44:295.

Downey, R. K. and B. L. Harvey. 1963. Methods of breeding for oil quality in rape. *Can J Plant Sci* 43:271–275.

Dyer, J. M., D. C. Chapital, J.-C. W. Kuan et al. 2002. Molecular analysis of a bifunctional fatty acid conjugase/desaturase from tung. Implications for the evolution of plant fatty acid diversity. *Plant Physiol* 130:2027–2038.

Eccleston, V. and J. B. Ohlrogge. 1998. Expression of lauroyl-ACP thioesterase in *Brassica napus* seeds induces pathways for both fatty acid oxidation and biosynthesis and implies a set point for triacylglycerol accumulation, *Plant Cell* 10:613–621.

Filichkin, S. A., M. B Slabaugh, and S. J. Knapp. 2006. New FATB thioesterases from a high-laurate *Cuphea* species: Functional and complementation analyses. *Eur J Lipid Sci Tech* 108:979–990.

Green, A. G. 1986. A mutant genotype of flax (*Linum usitatissimum* L.) containing very low levels of linolenic acid in its seed oil. *Can J Plant Sci* 66:499–503.

He, X., G. Q. Chen, S. T. Kang, and T. A. McKeon. 2007. *Ricinus communis* contains an acyl-CoA synthetase that preferentially activates ricinoleate to its CoA thioester. *Lipids* 42:931–938.

He, X., G. Q. Chen, J. T. Lin, and T. A. McKeon. 2005. Molecular characterization of the acyl-CoA-dependent diacylglycerol acyltransferase in plants. *Rec Res Dev Appl Microbiol Biotechnol* 2:69–86.

He, X., C. Turner, G. Q. Chen, J. T. Lin, and T. A. McKeon. 2004. Cloning and characterization of a cDNA encoding diacylglycerol acyltransferase from castor bean. *Lipids* 39:311–318.

James, C. 2009. Global Status of Commercialized Biotech/GM Crops: 2009 The first fourteen years 1996–2009. ISAAA Brief 41–2009. www.isaaa.org. Accessed 10 September, 2010.

Kinney, A. J. 2002. Perspectives on the production of industrial oils genetically engineered oilseeds. In *Lipid Biotechnology*, eds. T. M. Kuo and H. W. Gardner, 85–93. New York, NY: Marcel Dekker Inc.

Klaus, D., J. B. Ohlrogge, H. E. Neuhaus, and P. Dormann. 2004. Increased fatty acid production in potato by engineering of acetyl-CoA carboxylase. *Planta* 219:389–396.

Knothe, G., S. C. Cermak, and R. L. Evangelista. 2009. *Cuphea* oil as source of biodiesel with improved fuel properties caused by high content of methyl decanoate. *En Fuels* 23:1743–1747.

Knutzon, D. S., T. R. Hayes, A. Wyrick, H. Xiong, H. M. Davies, and T. A. Voelker. 1999. Lysophosphatidic acid acyltransferase from coconut endosperm mediates the insertion of laurate at the sn-2 position of triacylglycerols in lauric rapeseed oil and can increase total laurate levels, *Plant Physiol* 120:739–746.

Kroon, J. T. M., W. Wei, W. J. Simon, and A. R. Slabas. 2006. Identification and functional expression of a type 2 acyl-CoA:diacylglycerol acyltransferase (DGAT2) in developing castor bean seeds which has high homology to the major triglyceride biosynthetic enzyme of fungi and animals. *Phytochem* 67:2541–2549.

Lardizabal, K. D., J. G. Metz, T. Sakamoto, W. C. Hutton, M. R. Pollard, and M. W. Lassner. 2000. Purification of a jojoba embryo wax synthase, cloning of its cDNA, and production of high levels of wax in seeds of transgenic *Arabidopsis*. *Plant Physiol* 122:645–655.

Lee, M., M. Lenman, A. Banas et al. 1998. Identification of non-heme diiron proteins that catalyze triple bond and epoxy group formation, *Science* 280:915–918.

Lemaux, P. G. 2008. Genetically engineered plants and foods: A scientist's analysis of the issues. *Ann Rev Plant Biol* 59:771–812.

Lindqvist, Y., W. Huang, G. Schneider, and J. Shanklin. 1996. Crystal structure of Δ^9 stearoyl-acyl carrier protein desaturase from castor seed and its relationship to other di-iron proteins, *EMBO* 15:4081–4092.

Lung, S.-C. and R. J. Weselake. 2006. Diacylglycerol acyltransferase: A key mediator of plant triacylglycerol synthesis. *Lipids* 41:1073–1088.

McKeon, T. A. 2005. Genetic modification of seed oils for industrial applications. In *Industrial Uses of Vegetable Oils*, ed. S. Z. Erhan, pp. 1–13, Champaign, IL: AOCS Press.

McKeon, T. A. 2003. Genetically modified crops for industrial products and processes and their effects on human health. *Trends Food Sci Technol* 14:229–241.

McKeon, T. A., J. T. Lin, and G. Chen. 2002. Developing a safe source of castor oil. *INFORM* 13:381–385.

McKeon, T. A. and J. T. Lin. 2002. Biosynthesis of ricinoleic acid for castor oil production. In *Lipid Biotechnology*, eds. T. M. Kuo and H. W. Gardner, pp. 129–139. New York, NY: Marcel Dekker, Inc.

Miwa, T. K., J. A. Rothfus, and E. Dimitroff. 1979. Extreme-pressure lubricant tests on jojoba and sperm whale oils. *JAOCS* 56:765–770.

Moire, L., E. Rezzonico, S. Goepfert, and Y. Poirier. 2004. Impact of unusual fatty acid synthesis on futile cycling through α-oxidation and on gene expression in transgenic plants. *Plant Physiol* 134:432–442.

Nath, U. K., J. A. Wilmer, E. J. Wallington, H. C. Becker, and C. Möllers. 2009. Increasing erucic acid content through combination of endogenous low polyunsaturated fatty acids alleles with Ld-LPAAT + Bn-fae1 transgenes in rapeseed (*Brassica napus* L.). *Theor Appl Gen* 118:765–773.

Phippen, W. B., T. A. Isbell, and M. E. Phippen. 2006. Total seed oil and fatty acid methyl ester contents of *Cuphea* accessions. *Ind Crops Prod* 24:52–59.

Pollard, M. R., L. Anderson, C. Fan, D. J. Hawkins, and H. M. Davies. 1991. A specific acyl-ACP thioesterase implicated in medium-chain fatty acid production in immature cotyledons of *Umbellularia californica*. *Arch Biochem Biophys* 284:306–312.

Scarth, R. and J. Tang. 2006. Modification of *Brassica* oil using conventional and transgenic approaches. *Crop Sci* 46:1225–1236.

Shockey, J. M., S .K. Gidda, D. C. Chapital et al. 2006. Tung tree DGAT1 and DGAT2 have nonredundant functions in triacylglycerol biosynthesis and are localized to different subdomains of the endoplasmic reticulum. *Plant Cell* 18:2294–2313.

Stoll, C., W. Lühs, M. K. Zarhloul, M. Brummel, F. Spener, and W. Friedt 2006. Knockout of KASIII regulation changes fatty acid composition in canola (*Brassica napus*). *Eur J Lipid Sci Technol* 108:277–286.

Stumpf, P. K., D. N. Kuhn, D. J. Murphy, M. R. Pollard, T. McKeon, and J. J. MacCarthy. 1980. Oleic acid—The central substrate. In *Biogenesis and Function of Plant Lipids*, eds. P. Mazliak, P. Benveniste, C. Costes, and R. Douce, R., pp. 3–10. North-Holland: Elsevier.

Thelen, J. J. and J. B. Ohlrogge. 2002. Metabolic engineering of fatty acid biosynthesis in plants. *Metab Eng* 4:12–21.

Van de Loo, F. J., P. Broun, S. Turner, and C. Somerville. 1995. An oleate 12-hydroxylase from *Ricinus communis* L. is a fatty acyl desaturase homolog. *Proc Natl Acad Sci USA* 92:6743–6747.

Voelker, T. and A. J. Kinney. 2001. Variations in the biosynthesis of seed-storage lipids. *Ann Rev Plant Physio Mol Biol* 52:335–336.

Voelker, T. A., T. R. Hayes, A. M. Cranmer, J. C. Turner, and H. M. Davies. 1996. Genetic engineering of a quantitative trait: Metabolic and genetic parameters influencing the accumulation of laurate in rapeseed. *Plant J* 9:229–241.

Voelker, T. A., A. C. Worrell, L. Anderson et al. 1992. Fatty acid biosynthesis redirected to medium chains in transgenic oilseed plants. *Science* 257:72–74.

Wisniak, J. 1994. Potential uses of jojoba oil and meal—A review. *Ind Crops Prod* 3:43–68.

CHAPTER 21

Bioenergy from Agricultural Biowaste
Key Technologies and Concepts

Suman Khowala and Swagata Pal

CONTENTS

21.1 INTRODUCTION

Energy is the prime need to sustain life on earth. It is obtained from two kinds of sources—non-renewable and renewable through nature. Non-renewable energy is mainly available in the form of fossil fuels like petroleum and gases stored deep below in earth's crust. Renewable form of energy in the form of sun, air, water and natural vegetation are at our disposal. At present, the importance of renewable energy has become even more necessary not only due to the continuous depletion of limited fossil fuel stock but also for a safe and better environment. Need for looking into alternative technologies using renewable energy sources such as wind, solar, hydrothermal, biomass and geothermal have increased. Biomass to bioethanol programme appears highly promising in this regard. Use of bioethanol as a source of energy would be complementing for solar, wind and other intermittent renewable energy sources in the long run (Dien et al., 2003; Lin and Tanaka, 2006).

Although renewable sources are considered as most promising to meet the ever-increasing demand of energy, at present most of them they are not technologically and economically ideally developed (Lin and Tanaka, 2006; Lynd and Wang, 2004). This chapter contains an overview of the solid biomass/biowaste—their nature and technologies associated with the conversion of lignocellulosic biomass into bioethanol.

21.2 RATIONALE OF CONVERTING BIOMASS INTO BIOENERGY (BIOETHANOL)

With increasing concentrations of carbon dioxide and other greenhouse gases in earth's atmosphere, there is widespread concern that this will ultimately lead to climate changes. The use of agricultural biowaste as a raw material also broadens the application range because it would (1) involve more biodegradable products and processes that create less pollution and have lesser harmful environmental impacts; (2) develop novel products not available from petroleum sources; (3) consume cheaper raw materials and (4) reduce the dependence on fossil fuels (Herrera, 2004; Wyman and Goodman, 1993). Biomass materials are solids and there are obvious difficulties associated with using these materials as direct replacements for liquid transportation fuels. Production of ethanol from surplus biomass seems to be a promising approach to address all these problems and to gain energy from waste biomass as well as to partly reduce dependence on fossil fuels while contributing to the greenhouse gas effect and improving urban air quality. Fuel ethanol from lignocelluloses may also open new employment opportunities in rural areas, and thus make a positive socio-economic impact (Kheshgi et al., 2000).

21.3 SOURCES OF BIOMASS/BIOWASTE

The term biomass/biowaste refers to all organic matter generated through photosynthesis and other biological processes. The ultimate source of this renewable biomass is the inexhaustible solar energy which is captured by plants through photosynthesis. It includes both terrestrial as well as aquatic matter such as wood, herbaceous plants, algae, aquatic plants and residues, such as straw, husks, corncobs, cow dung, saw-dust, wood shavings and other wastes such as disposable garbage, night soil, sewage solids, industrial refuse and so on (van Wyk, 2001). Biomass contributed over a third of primary energy in India (Table 21.1) (Asokan et al., 2007; NCAER, 1992). Some of the biowaste are categorized as below.

21.3.1 Crop Residue and Farm Wastes

The straw of cereals and pulses, stalks and seed coats of oil seeds, stalks and sticks of fibre crops, pulp and wastes of plantation crops, peelings, pulp and stalks of fruits and vegetables and other wastes such as sugarcane trash, rice husk, molasses, coconut shells and so on fall under this category. Most of the crop residues have higher ash content and mainly constitute carbon, oxygen and hydrogen.

Table 21.1 Major Biomass Resources in India

Biomass	Availability (tonnes/year)
Rice straw	9
Rice husk	19.9
Jute sticks	2.5
Wheat straw	50.5
Cattle dung	1335.00
Bagasse	28.1
Molasses	2.1
Oil seed cakes	6.7
Saw dust	2
Mahua flowers	1
Leaves, tops and so forth	3.3

21.3.2 Agricultural Industrial Wastes

Industrial waste is produced by agro-industrial activity, for example, from factories, mills and other agro-processing activities, and these wastes also include cotton seeds and fibres, bagasse, residue from legumes or starchy products, and similar residues such as beet pulp, soybean cake, other solid residues (including oil cake) resulting from the refining of soybean oil or peanut oil. Oil cake and solid residue from cotton seed, flax seed, sunflower seed, rape seed, coconut and copra, palm nuts and seeds and maize germ are also included.

21.3.3 Forest Wastes

Logs, chips bark and leaves together constitute forest wastes. Sawdust is the forest-based industry waste. Forest products are also used as a domestic fuel in many developing countries.

21.3.4 Logging Residues

Tree tops, small stems and roots removed from a standard logging operation and broken debris are generally considered as logging residues.

21.3.5 Animal Wastes

Available statistics indicates the production of more than 2000 million tonnes of dung annually from all types of animals. Of the total produce, 84% is of cow and buffalo dung and 13% goat and sheep droppings. Dung is used as a fuel in the form of cakes and biogas.

21.4 MANAGEMENT OF BIOMASS/BIOWASTE

The treatment of aqueous and solid wastes of agricultural origin, offers opportunities to apply a wide range of biotechnology techniques such as bioconversion, bioaugmentation, phytoremediation, biostimulation and biodestruction. Currently, solid biowastes are dealt with in different ways, including landfilling, composting, combustion, recycling and illegal dumping. Biowaste conversion into bioproducts needs to be used as non-renewable resources, as primarily fossil fuels are depleted. Waste causes health and environmental risks, when not managed properly. Recycling of waste proves to be an effective management option because it does not involve the emission of many greenhouse gases and water pollutants. This approach saves energy, supplies valuable raw materials to industry, stimulates the development of green technologies, conserves natural resources and reduces the need for new landfill sites and incinerators (van Wyk, 2001). The waste-to-energy conversion process has been proved to be safe, environment friendly and reduces the incoming volume by ~90%, the remaining ash can be used as a roadbed material or as a landfill material. By replacing fossil fuels, this technology has the additional advantage of reducing carbon dioxide emission and the combustion of biowaste would also reduce sulphur dioxide emissions over the level produced by a coal-driven energy-generating plant (Jefferson et al., 1991).

21.5 COMPOSITION OF BIOMASS

Lignocellulosic biomass represents the primary fraction of solid waste. Materials of organic origin are known as biomass (a term that describes energy materials that emanate from biological sources) and are of major importance to sustainable development because they are renewable as opposed to non-organic materials and fossil carbohydrates (Louwrier, 1998). The amount of organic agricultural waste, such as corn stalks, leaves and wheat straw from wheat-processing facilities, sawdust and other residues from wood mills, is considered as a component of solid waste and could be a principal resource for biodevelopment.

Biomass energy is mostly 'locked up' as inaccessible carbohydrates in the form of cellulose and hemicellulose. Biomass, or plant matter, is made up of three major components: cellulose, hemicellulose and lignin. *Cellulose* is a long chain of glucose molecules, linked to one another primarily by beta ($1 \rightarrow 4$) glycosidic bonds; the glucose units are linked at their 1- and 4-carbon positions by beta-glycosidic bonds into molecular chains varying in length from a few glucose units to as many as 3000 or more. A more technical name for cellulose can be 4-beta-polyanhydro glucopyranose. The linear molecules of cellulose are bound laterally by hydrogen bonds or Vander Waal forces into linear fibrils of 50–100 chains. The individual chains in these fibrils are associated in various degrees of parallelism. Regions containing highly oriented chains are called crystallites; those in which the chains are more randomly oriented are termed amorphous (Norkrans, 1950).

Hemicelluloses are branched polymers of xylose, arabinose, galactose, mannose and glucose. Hemicellulose differs from cellulose by composition of several sugar units, by the presence of shorter chains and by a branching of main xylose chain molecules, which is much easier to hydrolyse than cellulose. It binds with cellulose fibrils to form microfibrils, which enhance the stability of the cell wall, and also cross-link with lignin, creating a complex web of bonds that provide structural strength but also challenge microbial degradation (Lynch, 1992).

Lignin is a complex polymer of phenylpropane units, which are cross-linked to each other with a variety of different chemical bonds. Building units of lignin are *p*-coumaryl alcohol, coniferyl alcohol and sinapyl alcohol, which are polymerized creating amorphous lignin (Higuchi, 1990). This complexity makes it resistant to biochemical degradation. Because lignin is the most rigid component of the plant cell wall, its presence lowers the bioavailability of cellulose and hemicellulose for enzymatic penetration and activity.

Biosugars are the raw materials to make chemicals, consumer products, industrial products and energy from biomass. Per cent of monosugars present in lignocellulosic components are shown in Table 21.1. Cellulose comprises about 70% of the total solid biowaste, whereas rest 30% is constituted of plastic, rubber, leather, glass, ceramics, metals and miscellaneous items (van Wyk, 2001).

21.6 BIOMASS TO BIOETHANOL: ETHANOL PRODUCTION TECHNOLOGIES

During the last two decades, advances in technology for ethanol production from biomass have been developed to the extent of realization of large-scale production (Chandel et al., 2007). Ethanol production from biomass can be summarized briefly into the following steps: (1) pretreatment of the lignocellulosic biomass, (2) hydrolysis of cellulose and hemicellulose polymer into monomeric fermentable sugar, (3) fermentation of depolymerized substrates and (4) the distillation of the fermentation broth to obtain dehydrated ethanol (Mosier et al., 2005).

21.6.1 Pretreatment of Lignocellulosic Materials

The effect of pretreatment of lignocellulosic materials has been recognized for a long time (McMillan, 1994). The purpose of the pretreatment is to remove lignin and hemicellulose, reduce cellulose crystallinity and increase the porosity of the materials. Pretreatment must meet the following requirements: (1) improve the formation of sugars or the ability to subsequently form sugars by enzymatic hydrolysis; (2) avoid the degradation or loss of carbohydrate; (3) avoid the formation of by-products inhibitory to the subsequent hydrolysis and fermentation processes and (4) be cost effective. Physical, physico-chemical, chemical and biological processes have been used for pretreatment of lignocellulosic materials. Each method has its advantages and affects the lignocellulosic structures and are cost effective (Alvira et al., 2010).

21.6.1.1 Physical Pretreatment

21.6.1.1.1 Mechanical Comminution

This method reduces particle size, and the crystallinity of lignocellulosic is achieved in order to increase the specific surface and reduce the degree of polymerization. Waste materials can be comminuted by a combination of chipping, grinding and milling to reduce cellulose crystallinity. The size of the materials is usually 10–30 mm after chipping and 0.2–2 mm after milling or grinding. Vibratory ball milling has been found to be more effective in breaking down the cellulose crystallinity of spruce and aspen chips and improving the digestibility of the biomass than ordinary ball milling (Sun and Cheng, 2002). The power requirement of mechanical comminution of agricultural materials depends on the final particle size and the waste biomass characteristics. Because of high-power requirement this process is not economically feasible.

21.6.1.1.2 Extrusion

In the extrusion process the materials are subjected to heating, mixing and shearing, resulting in physical and chemical modifications during the passage through the extruder. Lignocellulose structure is disrupted causing defibrillation, fibrillation and shortening of the fibres—due to screw speed and barrel temperature, thereby increasing the accessibility of carbohydrates to enzymatic attack (Karunanithy et al., 2008). This is a novel and promising physical pretreatment method for biomass conversion to ethanol production including application of enzymes during the process.

21.6.1.2 Chemical Pretreatments

21.6.1.2.1 Alkali Pretreatments

Alkali pretreatments cause increased cellulose digestibility and lignin solubilization, as well as exhibit minor cellulose and hemicellulose solubilization than acid or hydrothermal processes

(Carvalheiro et al., 2008). The process is performed at room temperature and ranges from seconds to days and is reported to be more effective on agricultural residues than on wood materials (Kumar and Wyman, 2009). Sodium, potassium, calcium and ammonium hydroxides are mainly used for alkaline pretreatments, causing swelling, increasing the internal surface of cellulose and decreasing the degree of polymerization and crystallinity, by disruption of the lignin structure. Lignin removal increases enzyme effectiveness by reducing non-productive adsorption sites for enzymes and by increasing cellulose accessibility (Kim and Holtzapple, 2006).

21.6.1.2.2 Acid Pretreatment

Acid pretreatment enhances the solubilization of the hemicellulosic fraction of the biomass and renders the cellulose more accessible to enzymes. Acid pretreatments are performed with concentrated or diluted acid but the use of concentrated acid is not desired due to the formation of inhibiting compounds. The operational and maintenance costs are high and equipment corrosion problems and acid recovery are other drawbacks of the process using the concentrated acid pretreatment for commercial scale (Wyman, 1996).

21.6.1.2.3 Pyrolysis

When the lignocellulosic materials are treated at temperatures greater than 300°C, cellulose rapidly decomposes to produce gaseous products and residual char (Kilzer and Broido, 1965; Shafizadeh and Brad bury, 1979). The decomposition is much slower and less volatile products are formed at lower temperatures. Mild acid hydrolysis (1 N H_2SO_4, 97°C, 2.5 h) of the residues from pyrolysis pretreatment has resulted in 80–85% conversion of cellulose to reducing sugars with more than 50% glucose. The process can be enhanced with the presence of oxygen and by adding $ZnCl_2$ or Na_2CO_3 as a catalyst, the decomposition of pure cellulose can occur at lower temperatures (Shafizadeh and Brad bury, 1979).

21.6.1.2.4 Ionic Liquids Pretreatment

Ionic liquids (ILs) are known as 'green' solvents, since no toxic or explosive gases are formed. Carbohydrates and lignin dissolve simultaneously in ILs with anion activity, because ILs form hydrogen bonds between the non-hydrated chloride ions of the IL and the sugar hydroxyl protons in a 1:1 stoichiometry. As a result, the intricate network of non-covalent interactions among biomass polymers of cellulose, hemicellulose and lignin is effectively disrupted with minimum formation of degradation products. The ILs are salts, typically composed of large organic cations and small inorganic anions (e.g., 1-ethyl-3-methyl imidazolium diethyl phosphate), which exist as liquids at room temperature. Solvent properties of ILs like, chemical and thermal stability, non-flammability, low vapour pressures and a tendency to remain liquid in a wide range of temperatures can be varied by adjusting the anion and the alkyl constituents of the cation (Hayes, 2009). The use of ILs as solvents for pretreatment of cellulosic biomass has recently received much attention, but technology is expensive and commercial IL recovery methods need to be explored.

21.6.1.2.5 Ozonolysis

In this method, ozone is used as a powerful oxidant with high delignification efficiency and increases the yield in the following enzymatic hydrolysis (Sun and Cheng, 2002). The pretreatment is usually performed at room temperature and normal pressure, and does not lead to the formation of inhibitory compounds; however, due to requirement of the large amounts of ozone the process is not economically viable.

21.6.1.2.6 Organosolv Treatment

Numerous organic or aqueous solvent mixtures can be utilized, including methanol, ethanol, acetone, ethylene glycol and tetrahydrofurfuryl alcohol, in order to solubilize lignin and provide treated cellulose suitable for enzymatic hydrolysis (Zhao et al., 2009). Compared to other chemical pretreatments the main advantage of the organosolv process is the recovery of relatively pure lignin as a by-product (Zhao et al., 2009). But solvents need to be separated because they might be inhibitory to enzymatic hydrolysis and to fermentative microorganisms (Sun and Cheng, 2002). The high commercial price of solvents is another important factor to consider for industrial applications.

21.6.1.3 Physico-Chemical Pretreatment

21.6.1.3.1 Steam Explosion

Steam explosion is the most commonly used method for pretreatment of lignocellulosic materials. In this method, chipped biomass is treated with high-pressure saturated steam (160–260°C, corresponding pressure 0.69–4.83 MPa) for a short period and then the pressure is swiftly reduced, which makes the materials undergo an explosive decompression. This process may be described as a thermomechanochemical process, where the breakdown of structural components is aided by heat in the form of steam (thermo), shear forces due to the expansion of moisture (mechano) and hydrolysis of glycoside bonds (chemical) (Alvira et al., 2010). With decrease in pressure, condensed moisture evaporates and desegregation of lignocellulosic matrix occurs by breaking down inter- and intra-molecular linkages. The process causes hemicellulose degradation and lignin transformation due to high temperature, thus increasing the potential of cellulose hydrolysis. The factors that affect steam explosion pretreatment are residence time, temperature, chip size and moisture content. Optimal hemicellulose solubilization and hydrolysis was achieved by either high temperature and short residence time (270°C, 1 min) or lower temperature and longer residence time (190°C, 10 min). Variants of steam explosion are used when ammonia is also added in biomass in a variation of the steam explosion process (ammonia freeze explosion, AFEX) and produces only a pretreated solid material, whereas steam explosion produces slurry that can be separated in solid and liquid fractions. AFEX has been reported to decrease cellulose crystallinity and disrupt lignin–carbohydrates linkages (Laureano-Pérez et al., 2005). In another method, the addition of H_2SO_4 (or SO_2) or CO_2 in steam explosion effectively improved enzymatic hydrolysis, decreased the production of inhibitory compounds and led to a more complete removal of hemicellulose (Morjanoff and Gray, 1987; Schacht et al., 2008).

The advantages of steam explosion pre-treatment include the low-energy requirement compared to mechanical comminution and no recycling or environmental costs. Limitations of steam explosion include destruction of a portion of the xylan fraction, incomplete disruption of the lignin–carbohydrate matrix (leading to low lignin solubility) and generation of compounds that may be inhibitory to microorganisms used in downstream processes (Oliva et al., 2003).

21.6.1.3.2 Microwave Pretreatment

Microwave-based pretreatment is a physicochemical process involving both thermal and non-thermal effects. Slurry of biomass immersed in dilute chemical reagents were exposed to microwave radiation for time ranging from 5 to 20 min (Keshwani and Cheng, 2009). Sodium hydroxide was identified as the most effective alkali reagent for microwave-based pretreatment (Zhu et al., 2006).

21.6.1.4 *Biological Pretreatment*

In biological pretreatment processes, microorganisms such as brown-, white- and soft-rot fungi are used to degrade lignin and hemicellulose in waste materials. Brown rots mainly attack cellulose, whereas white and soft rots attack both cellulose and lignin. White-rot fungi are the most effective basidiomycetes for biological pretreatment of lignocellulosic materials. Several white-rot fungi such as *Phanerochaete chrysosporium, Ceriporia lacerata, Cyathus stercolerus, Ceriporiopsis subvermispora, Pycnoporus cinnarbarinus* and *Pleurotus ostreaus* have been examined on different lignocellulosic biomass showing high delignification efficiency (Kumar and Wyman, 2009; Shi et al., 2008). Conversion (35%) of wheat straw to reducing sugar was obtained in the pretreatment by *Phanerochaete sordida* 37 and *Pycnoporus cinnabarinus* 115 in four weeks. In order to prevent the loss of cellulose, a cellulase-less mutant of *Sporotrichum pulverulentum* was developed for the degradation of lignin in wood chips (Sun and Cheng, 2002). The white-rot fungus *P. chrysosporium* produces lignin-degrading enzymes, lignin peroxidases and manganese-dependent peroxidases, during secondary metabolism in response to carbon or nitrogen limitation (Van der Woude et al., 1993). Both enzymes have been found in the extracellular filtrates of many white-rot fungi for the degradation of wood cell walls (Kirk and Farrell, 1987). Other enzymes including polyphenol oxidases, laccases, H_2O_2 producing enzymes and quinone-reducing enzymes can also degrade lignin (Blanchette, 1991). The advantages of biological pretreatment include low-capital cost, no chemical requirement, low-energy requirement and mild environmental conditions (Sun and Cheng, 2002). However, the rate of hydrolysis in most biological pretreatment processes is very low.

21.6.2 Hydrolysis of Cellulose and Hemicellulose into Monomeric Fermentable Sugar

The sugar content of lignocellulose is contributed by cellulose (which also comprises 70% of total sold wastes) and hemicellulose (compositions given in Table 21.2). These polysaccharides can be converted into reducing sugars, glucose or xylose, respectively. Glucose can serve as the starting compound for biosynthesis of many bioproducts. The conversion of polysaccharide to sugars is known as the saccharification process could either be acid or enzyme catalysed, which decrystallizes the cellulose and hydrolyses it to glucose. Acid-catalysed reactions pollute the environment, are corrosive to the experimental set-up and lead to the production of furfural because of high temperatures used (Harris, 1949). Therefore, now the technology is oriented for developing economically viable enzyme-catalysed saccharification of cellulose and hemicellulose. Enzymatic reactions take place at mild conditions (pH 4.8 and temperature 45–50°C) and provide high yields with lesser side products apart from being biodegradable, environment friendly and can be reused (Sheeman and Himmal, 1999).

21.6.2.1 *Methods of Enzymes Production*

Generally, two basic methods are employed for the production of microbial enzymes through fermentation—solid-state fermentation (SSF) and submerged fermentation (SmF).

Table 21.2 Composition of Lignocellulosis Biomass

Component	Sugar Composition	Dry Weight (%)
Cellulose	Glucose	40–60
Hemicellulose	Xylose, mannose, galactose, glucose, arabinose	20–40
Lignin	Phenyl propane units	10–25

21.6.2.1.1 Solid-State Fermentation

In SSF the fermentation process is carried out on a non-soluble substrate that acts both as a physical support and source of nutrients, and microbial growth and product formation occurs at or near the surface of the solid substrate particle having low-moisture contents, which restricts the fermentation process viable for some limited microorganisms, mainly yeasts and fungi, although some bacteria have been used (Pandey et al., 2000). Most critical factors affecting the process are particle size and moisture level/water activity apart from pH, temperature and the bioreactor model chosen for the particular organism for the chosen product. SSF can be carried out in different bioreactors such as trays, drum, packed bed, fluidized bed and immersion bioreactors (Coutoa and Sanromána, 2006). Advantages of SSF are high productivities, extended stability of products and low production costs (Pandey et al., 2000), but there are some disadvantages such as difficulty in scale-up, control of process parameters (pH, heat, moisture, nutrient conditions, etc.) and maintenance of uniformity in the environment, higher impurity of the product, increased recovery product costs, need of more space and so on. It has been calculated and proved that in SSF processes the enzyme titres are higher than SmF, when comparing the same strain and fermentation broth (Holker and Lenz, 2005; Viniegra-González et al., 2003).

21.6.2.1.2 Submerged Fermentation

Enzyme production by SmF is commonly employed for enzyme production, due to the fact that many enzymes are subject to carbon-catabolite repression, carbon-limited chemostat and fed-batch cultivations are used to grow the microbes at low carbon concentrations. In case of no or little carbon-catabolite repression, carbon-excess batch cultivations are also used to attain the high specific growth rates and high enzyme productivities. SmF has the advantage of scaling up by using the soluble substrates with compact closed fermentor, needs minimum labour, low recovery costs and fine and easy control of parameters such as pH, temperature, nutrients and so on but the process entails considerable power, contamination problems, operational intricacies, require high-tech setup and so on (Raj and Karanth, 1999).

21.6.2.2 Enzyme Cocktails for Saccharification

Enzymatic hydrolysis of waste cellulose is performed using cellulase, a multi-component system produced by bacteria and fungi, which is highly specific with no by product formation and acts in concert to hydrolyse cellulose. Three general types of enzymes generally make up the cellulase enzyme complex: (1) endocellulase breaks internal beta (1–4) bonds to disrupt the crystalline structure of cellulose and expose individual cellulase polysaccharide chains, (2) exocellulase cleaves 2–4 units from the ends of the exposed chains produced by endocellulase, resulting in the tetrasaccharides or disaccharide such as cellobiose, (3) cellobiase or beta-glucosidase hydrolyses cellobiose the endocellulase product into individual monosaccharides. It has been found that enzyme preparations containing only endocellulases have little effect on native cellulose. On the other hand, those containing both endo and exocellulases will cause significant degradation of native cellulose. Thus, the endo and exocellulases appear to work in a synergistic or cooperative manner on native cellulose (Beguin and Aubert, 1994). Cellulases account for approximately 25% of the world enzyme market produced mainly from *Trichoderma* and *Aspergillus* strains (Kubicek and Penttilä, 1998) and other potential producers for cellulases are *Trichoderma viride*, *Penicillium*, *Fusarium*, *Humicola*, *Phanerochaete*, *Termitomyces clypeatus* and so on (Esterbauer et al., 1991; Mukherjee et al., 2006). These microorganisms can be aerobic or anaerobic, mesophilic or thermophilic. Bacteria belonging to *Clostridium*, *Cellulomonas*, *Bacillus*, *Thermomonospora*, *Ruminococcus*, *Bacteriodes*, *Erwinia*, *Acetovibrio*, *Microbispora* and *Streptomyces* can produce cellulases (Bisaria, 1991).

Hemicellulases include a combination of several enzymes such as endo-β-1, 4-xylanase, exo-β-1, 4-xylosidase, α-L arabinofuranosidase, endo-α-1, 5-arabinanase, α-glucuronidase, endo-β-1, 4-mannanase, exo-β-1,4-mannosidase, acetyl xylan esterase and cause a breakdown of complex heterogeneous polysaccharide hemicellulose to constituent sugars. Aerobic fungi *Trichoderma*, *Aspergillus* and *Termitomyces* secrete at high concentrations a large variety of hemicellulases that works synergistically to saccharify hemicellulose to xylan, glucose and other sugars (Ghorai et al., 2009; Khowala et al., 1992; Shallom and Shoham, 2003).

21.6.2.3 Different Hydrolysis Configurations and Fermentation for Ethanol Production

Cellulose and hemicellulase are degraded, respectively, by the cellulases and hemicellulases to reducing sugars that can be fermented by yeasts or bacteria to ethanol (McMillan, 1994). Biomass can be treated enzymatically in various ways such as sequential cellulase treatment, mixtures of cellulases from different origins (Kim et al., 1998) and the treatment of waste cellulose with different amounts of cellulase (Mandels et al., 1974). There are two main options for saccharification and fermentation—SHF (separate hydrolysis and fermentation) and SSF. During SHF, cellulase is added to pretreated biomass which results in the formation of glucose from the cellulose fraction with added yeast to ferment glucose into ethanol (Kim et al., 1998; Mandels et al., 1974). The other option that is more popular for the cellulose conversion process is direct microbial conversion by simultaneous saccharification and fermentation, when the microorganisms simultaneously produce cellulase, hydrolyse cellulose and ferment glucose into ethanol (Stenberg et al., 2000) while at the same time, fermentation by microorganisms converts the hemicellulose sugars into bioethanol (Lynd et al., 1991). If both hexose and pentose are present in SSF broth and pentose-fermenting yeast is used, the process is called simultaneous saccharification and co-fermentation. For glucose fermentation *Saccharomyces cerevisiae* is used whereas *Pichia stiptis*, *Candida shehatae* and *Pachysolen tannophilus* are natural pentose-fermenting yeasts (Pejo et al., 2008).

SSF has advantage over SHF in the reduction of end-product inhibition during hydrolysis step resulting in higher yields, shorter residence time and lower requirements of enzyme loading (Alfani et al., 2000). Ethanol production cost is lower in SSF than SHF since capital cost is reduced due to higher yield and employment of one single vessel. In an industrial process higher substrate concentration, lower enzyme and yeast loading need to be used to achieve higher final ethanol concentration (Pejo et al., 2008). There are some other bottlenecks, like enzyme hydrolysis optimally takes place at 50°C, whereas fermenting yeasts have an optimum temperature at 30–37°C. This can be sorted by using thermo-tolerant yeasts, for example, *Kluyveromyces marxianus* (Ballesteros et al., 1993). In SHF each step can be optimized whereas a compromise has to be maintained in SSF processes.

21.7 CONCLUDING REMARKS

It is widely accepted that there is a need to increase the potential of biomass as a renewable energy source all over the world. Although the production of bioethanol worldwide has increased from less than 20 billion litres in the year 2000 to above 75 billion litres in 2009, a lot is yet to be done to develop the processes with acceptable technological, economic and environmental sustainability. Currently, the importance of alternative energy sources has become even more necessary not only due to the continuous depletion of limited fossil fuel stock but also for the safe and better environment, with an inevitable depletion of the world's energy supply, there has been an increasing worldwide interest in alternative sources of energy (Chandel et al., 2007). Cost-effective and environmentally attractive means of generating fuels, chemicals, materials, foods and feeds from renewable plants and agricultural biomass need to be developed. Considering the main drawbacks, pretreatment conditions should be low in cost and avoid feedstock degradation (Alvira et al., 2010). In addition, several other parameters, for example, high substrate loadings, sugar recovery after pretreatment, tolerance to inhibitory compounds and xylose fermentation by yeast, must be optimized for successful industrial process for bioethanol (bioenergy) production from lignocellulose (biomass) (Pejo et al., 2008). Moreover, low enzyme and yeast loading are crucial for an economically viable process which will be attained by increasing enzyme efficiency and the choice of a suitable saccharification and fermentation method (Lee et al., 2008). Over the years, there have been substantial advances in enzyme-based technology for ethanol production and will increase more in the future as the pressure on waste management and biodevelopment increases.

REFERENCES

Alfani, A., F. Gallifuoco, A. Saporosi, A. Spera and M. Cantarella. 2000. Comparison of SHF and SSF process for the bioconversion of steam exploded wheat straw. *J Ind Microb Biotechnol* 25:184–192.

Alvira, P., E. Tomás-Pejó, M. Ballesteros and M. J. Negro. 2010. Pretreatment technologies for an efficient bioethanol production process based on enzymatic hydrolysis: A review. *Bioresour Technol* 101:4851–4861.

Asokan, P., M. Saxena and S. R. Asolekar. 2007. Solid wastes generation in India and their recycling potential in building materials. *Build Environ* 42:2311–2320.

Ballesteros, I., J. M. Oliva, M. Ballesteros and J. Carrasco. 1993. Optimization of the simultaneous saccharification and fermentation process using thermotolerant yeasts. *Appl Biochem Biotechnol* 39:201–211.

Beguin, P. and J. P. Aubert. 1994. The biological degradation of cellulose. *FEMS Microbiol Lett* 13:25–58.

Bisaria, V. S. 1991. Bioprocessing of agro-residues to glucose and chemicals. In *Bioconversion of Waste Materials to Industrial Products*, ed. A. M. Martin, pp. 210–233. London: Elsevier.

Blanchette, R. A. 1991. Delignification by wood-decay fungi. *Ann Rev Phytopathol* 29:381–398.

Carvalheiro, F., L. C. Duarte and F. M. Gírio. 2008. Hemicellulose biorefineries: A review on biomass pretreatments. *J Sci Ind Res* 67:849–864.

Chandel, A. K., E. S. Chan, R. Rudravaram et al. 2007. Economics and environmental impact of bioethanol production technologies: An appraisal, *Biotechnol Mol Biol Rev* 2:14–32.

Coutoa, S. R. and M. A. Sanromána. 2006. Application of solid-state fermentation to food industry—A review. *J Food Eng* 76:291–302.

Dien B. S., M. A. Cotta and T. W. JeVries. 2003. Bacteria engineered for fuel ethanol production: Current status. *Appl Microbiol Biotechnol* 63:258–266.

Esterbauer, H., W. Steiner, I. Labudova, A. Hermann and M. Hayn. 1991. Production of *Trichoderma* cellulase at laboratory and pilot scale. *Bioresour Technol* 36:51–65.

Ghorai, S., S. P. Banik, D. Verma, S. Chowdhury, S. Mukherjee and S. Khowala. 2009. Fungal biotechnology in food and feed processing. *Food Res Int* 42:577–587.

Harris, E. E. 1949. Wood saccharification. *Adv Carbohyd Chem* 4:153–158.

Hayes, D. J. 2009. An examination of biorefining processes, catalysts and challenges. *Catal Today* 145:138–151.

Herrera, S. 2004. Industrial biotechnology—A chance at redemption. *Nat Biotechnol* 22:671–675.

Higuchi, T. 1990. Lignin biochemistry: Biosynthesis and biodegradation. *Wood Sci Technol* 24:23–63.

Holker, U. and J. Lenz. 2005. Solid-state fermentation—Is there any biotechnological advantages? *Curr Opin Microbiol* 8:301–306.

Jefferson, W. T., O. W. David and T. W. Nancy 1991. *Energy and the Environment in the 21st Century*. Cambridge, MA: MIT Press.

Karunanithy, C., K. Muthukumarappan and J. L. Julson. 2008. Influence of high shear bioreactor parameters on carbohydrate release from different biomasses. In *Annual International Meeting, ASABE 084114*. St. Joseph, MI: Am Soc Agr., Biol Engineers.

Keshwani, D. R. and J. J. Cheng. 2009. Switchgrass for bioethanol and other value-added applications: A review. *Bioresour Technol* 100:1515–1523.

Kheshgi, H. S., R. C. Prince and G. Marland. 2000. The potential of biomass fuels in the context of global change: Focus on transportation fuels. *Ann Rev Energy Environ* 25:199–244.

Khowala, S., A. K. Ghosh and S. Sengupta. 1992. Saccharification of xylan by an amyloglucosidase of *Termitomyces clypeatus* and synergism in presence of xylanase. *Appl Microbiol Biotechnol* 37:287–292.

Kilzer, F. J. and A. Broido. 1965. The structure and properties of pyrolised cellulose carbon. *Pyrodynam* 2:151–163.

Kim, E., D. C. Irwin, L. P. Walker and D. B. Wilson. 1998. Factorial optimization of a six cellulase mixture. *Biotechnol Bioeng* 58:494–501.

Kim, S. and M. T. Holtzapple. 2006. Delignification kinetics of corn stover in lime pretreatment. *Bioresour Technol* 97:778–785.

Kirk, T. K. and R. L. Farrell. 1987. Enzymatic combustion: The microbial degradation of lignin. *Ann Rev Microbiol* 41:464–505.

Kubicek, C. P. and M. E. Penttilä. 1998. Regulation of production of plant polysaccharide degrading enzymes by *Trichoderma*. In *Trichoderma and Gliocladium, Enzymes, Biological Control and Commercial Applications*, eds. G. E. Harman and C. P. Kubicek, pp. 49–72. London: Taylor & Francis.

Kumar, R. and C. E., Wyman. 2009. Effects of cellulase and xylanase enzymes on the deconstruction of solids from pretreatment of poplar by leading technologies. *Biotechnol Prog* 25:302–314.

Laureano-Pérez, L., F. Teymouri, H. Alizadeh and B. E. Dale. 2005. Understanding factors that limit enzymatic hydrolysis of biomass. *Appl Biochem Biotechnol*. 121:1081–1099.

Lee, J. S., B. Parmeswaran, J. P. Lee and S. C. Park. 2008. Recent developments of key technologies on cellulose ethanol production. *J Sci Ind Res* 67:865–873.

Lin, Y. and S. Tanaka. 2006. Ethanol fermentation from biomass resources: Current state and prospects. *Appl Microbiol Biotechnol* 69:627–642.

Louwrier, A. 1998. Industrial products—Return to carbohydrate—Based industries. *Biotechnol Appl Biochem* 27:1–8.

Lynch, J. M. 1992. Substrate availability in the production of composts. In *Proceedings of the International Composting Research Symposium,* eds. H. A. J. Hoitink and H. Keener, pp. 24–35. Columbus, OH: The Ohio State University Press.

Lynd, L. R., H. J. Ahn, G. Anderson, P. Hill and D. Sean. 1991. Thermophilic ethanol production—Investigation of ethanol yield and tolerance in continuous culture. *Appl Biotech* 28:549–553.

Lynd, L. R. and M. Q. Wang. 2004. A product-nonspecific framework for evaluating the potential of biomass-based products to displace fossil fuels. *J Indust Ecol* 7:17–32.

Mandels, M. L., L. Hontz and J. Nistrom. 1974. Enzymatic hydrolysis of waste cellulose. *Biotechnol Bioeng* 16:1471–1493.

McMillan, J. D. 1994. Pretreatment of lignocellulosic biomass In *Enzymatic Conversion of Biomass for Fuels Production,* ACS Symposium Series 566, eds. M. E. Himmel, J. O. Baker and R. P. Overend, pp. 292–324. Washington, DC: ACS Publications.

Morjanoff, P. J. and P. P. Gray. 1987. Optimization of steam explosion as a method for increasing susceptibility of sugarcane bagasse to enzymatic saccharification. *Biotechnol Bioeng* 29:733–741.

Mosier, N., C. Wyman, B. Dale et al. 2005. Features of promising technologies for pretreatment of lignocellulosic biomass. *Bioresour Technol* 96:673–686.

Mukherjee, S., S. Chowdhury, S. Ghorai, Pal, S. and Khowala, S. 2006. Cellobiase from *Termitomyces clypeatus*: Activity and secretion in presence of glycosylation inhibitors. *Biotechnol Lett* 28:1773–1778.

NCAER. 1992. *Evaluation Survey of Household Biogas Plants Set Up During Seventh Five Year Plan*. New Delhi: National Council for Applied Economic Research.

Norkrans, B. 1950. Influence of cellulolytic enzymes from *Hymenocetes* on cellulose preparations of different crystallinity. *Physiol Physiol Plant* 3:75–81.

Oliva, J. M., F. Sáez, I. Ballesteros et al. 2003. Effect of lignocellulosic degradation compounds from steam explosion pretreatment on ethanol fermentation by thermotolerant yeast *Kluyveromyces marxianus*. *Appl Microbiol Biotechnol* 105:141–154.

Pandey, A., C. R. Soccol and D. A. Mitchell. 2000. New developments in solid-state fermentation: I Bioprocesses and products. *Proc Biochem* 35:1153–1169.

Pejo, E. T., J. M. Oliva and M. Ballesteros. 2008. Realistic approach for full scale bioethanol production from lignocellulose: A review. *J Sci Ind Res* 67:874–884.

Raj, A. E. and N. G. Karanth. 1999. Fermentation technology and bioreactor design, In *Food Biotechnology*, eds. K. Shetty, G. Paliyath, A. Pometto and R. E. Levin, pp. 33–86. Boca Raton, FL: CRC Press.

Schacht, C., C. Zetzl and G. Brunner. 2008. From plant materials to ethanol by means of supercritical fluid technology. *J Supercrit Fluids* 46:299–321.

Shafizadeh, F. and A. G. W. Brad bury. 1979. Smoldering combustion of cellulosic materials. *J Build Phys* 2:141–152.

Shallom, D. and Y. Shoham. 2003. Microbial hemicellulases. *Curr Opin Microbiol* 6:219–228.

Sheeman, J. and M. Himmal. 1999. Enzymes, energy and the environment: A strategic perspective on the US department of energy research and development activities for bioethanol. *Biotechnol Prog* 15:817–827.

Shi, J., M. S. Chinn and R. R. Sharma-Shivappa. 2008. Microbial pretreatment of cotton stalks by solid state cultivation of *Phanerochaete chrysosporium. Bioresour Technol* 99:6556–6564.

Stenberg, K., M. Bollók, K. Réczey, M. Galbe and G. Zacchi. 2000. Effect of substrate and cellulase concentration on simultaneous saccharification and fermentation of steam pretreated softwood for ethanol production. *Biotechnol Bioeng* 68:204–210.

Sun, Y. and J. Cheng. 2002. Hydrolysis of lignocellulosic materials for ethanol production: A review. *Bioresour Technol* 83:1–11.

Van der Woude, M. W., K. Boominathan and C. A. Reddy. 1993. Nitrogen regulation of lignin peroxidase and manganese-dependent peroxidase production is independent of carbon and manganese regulation in *Phanerochaete chrysosporium. Arch Microbiol* 160:1–4.

van Wyk, J. P. H. 2001. Biotechnology and the utilization of biowaste as a resource for bioproduct development: Review. *Trends Biotechnol.* 19:172–177.

Viniegra-González, G., E. Favela-Torres, C. N. Aguilar et al. 2003. Advantages of fungal enzyme production in solid state over liquid fermentation systems. *Biochem Eng J* 13:157–167.

Wyman, C. E. and B. J. Goodman. 1993. Biotechnology for production of fuels, chemicals, and materials from biomass. *Appl Biochem Biotechnol* 39:39–59.

Wyman, C. E., 1996. *Handbook on Bioethanol: Production and Utilization.* Washington, DC: Taylor & Francis.

Zhao, X., K. Cheng and D. Liu. 2009. Organosolv pretreatment of lignocellulosic biomass for enzymatic hydrolysis. *Appl Microbiol Biotechnol* 82:815–827.

Zhu, S., Y. Wu, Z. Yu et al. 2006. Comparison of three microwave/chemical pretreatment processes for enzymatic hydrolysis of rice straw. *Biosyst Eng* 93:279–283.

Index

A

C

D

E

F

G

J

K

L

N

O

P

Q

R

S

T

U

V

W

X

Y

Z